Statistical Sciences and Data Analysis

Editorial Board

STATISTICAL SCIENCES AND DATA ANALYSIS

PROCEEDINGS OF THE THIRD PACIFIC AREA STATISTICAL CONFERENCE

EDITORS
KAMEO MATUSITA
MADAN L. PURI
TAKESI HAYAKAWA

Utrecht, The Netherlands, 1993

VSP BV
P.O. Box 346
3700 AH Zeist
The Netherlands

First published in 1993

ISBN 90-6764-150-2

CIP-DATA KONINKLIJKE BIBLIOTHEEK, DEN HAAG

Statistical

Statistical sciences and data analysis : proceedings of
the third Pacific Area Statistical Conference / ed. by K.
Matushita ... [et al.]. - Utrecht [etc.] : VSP
Conference held in Makuhari on 11 - 13 December 1991.
ISBN 90-6764-150-2 bound
NUGI 815
Subject headings: statistical sciences / data analysis

Printed in The Netherlands by Koninklijke Wöhrmann BV, Zutphen.

Contents

Preface

The Third Pacific Area Statistical Conference was held in Makuhari in the outskirts of Tokyo, on 11-13 December 1991 under the auspices of the Pacific Statistical Institute and with the support and cooperation of the Foundation for Advancement of International Science, the Japan Statistical Society and the Institute of Statistical Mathematics. There were about 180 participants from the greater Pacific area as well as other continents. We are pleased to present herewith the Proceedings of this Conference.

The main theme of the Conference was "Statistical Sciences and Data Analysis." Its purpose was to bring together researchers in statistics and related fields from those areas to exchange results and problems in topics of mutual interest and concern. The papers comprising this volume were presented at the Conference. All papers were subsequently examined by referees before their inclusion here. These papers contain many recent developements in statistical sciences and data analysis and in application. Consequently, this book will be of interest both to statisticians and to researchers in other fields who apply statistical methods in their work.

The Conference has benefited greatly from the generous financial support of several industrial and commercial organizations, including those which are affiliated with the Japan Federation of Economic Organizations, the Japan World Exposition Commemorative Funds, the Kajima Foundation and Chiba Convention Bureau. To these organizations we express our deepest gratitute.

In the preparation of this volume we had the help and cooperation of many people in the refereeing of papers. We thank them sincerely.

K. Matusita

October 1992

Stat. Sci. & Data Anal., pp. 1-11
K. Matsusita *et al.* (Eds)

Kullback-Leibler Information for Ordering Genes Using Sperm Typing and Radiation Hybrid Mapping

HERMAN CHERNOFF
Harvard University, Cambridge, MA 02138, USA and Mathematical Sciences Research Institute, Berkeley, CA 94720, USA

Abstract. Two technologies applicable to gene mapping are those of sperm typing and radiation hybrid mapping. They are used to determine the ordering of the genes. For each of these methods, the analysis grows in complexity as the number of genes being considered increases. At the same time the accuracy of the probabilistic models used in the analysis becomes more questionable. On the other hand the ability to determine the order of three genes may be enhanced by the inclusion, in the analysis, of the data on nearby genes. For both of these methods, Kullback-Leibler information numbers are derived to test hypotheses involving the order of m genes. These information numbers are computed for testing hypotheses concerning the ordering of three genes with and without considering the presence of data involving other nearby genes. The results suggest when it pays to incorporate the additional data and how much radiation to use in radiation hybrid mapping.

AMS 1980 subject classifications. Primary 62B10; secondary 92D20.
Keywords and phrases. Kullback-Leibler Information; sperm typing, radiation hybrid mapping, gene ordering.

1. INTRODUCTION

Two technologies applicable to gene mapping are those of sperm typing and radiation hybrid mapping. Sperm typing makes use of the polymerase chain reaction, a biochemical technique which allows enormous amplification (production of multiple copies) of small, selected DNA fragments from a single chromosome. A sample of sperm from a single donor is analyzed to see which alleles (distinct forms of the various genes) are present in the individual sperms. The frequencies with which the various possibilities occur can be used to supply estimates of the ordering and of the recombination probabilities among the genes for which that donor is heterozygous (having different alleles of the same gene.) Radiation hybrid mapping employs a different technology where hybrid rodent cells containing a human chromosome are subjected to a dose of radiation, which leads to breaking the chromosome into segments, a fraction of which are retained in succeeding generations. The simultaneous presence or absence of various genes provides indirect information on how close together these genes are, and also on the ordering of these genes. The results suggest when it pays to incorporate the additional data and how much radiation to use in radiation hybrid mapping.

For each of these methods, the analysis grows in complexity as the number of genes being considered increases. At the same time the accuracy of the probabilistic models used in the analysis becomes more questionable. On the other hand the ability to determine the order of three genes may be enhanced by the inclusion, in the analysis, of the data on nearby genes. For both of these methods, we shall examine the relevant Kullback-Leibler information numbers for hypotheses concerning the ordering of three genes with and without considering the presence of data involving other nearby genes. The results suggest when it pays to incorporate the additional data, and how much radiation to use in radiation hybrid mapping.

In Section 2 we introduce the model for sperm typing and discuss the maximum likelihood estimates of the recombination probabilities. In Section 3 we derive expressions for the relevant Kullback-Leibler informations for sperm typing. In Section 4 we describe the model for radiation hybrids and derive the corresponding information numbers. The outcome of the calculations is described in Section 5. We terminate this introduction with a brief discussion of the Kullback-Leibler (KL) information.

Given two simple hypotheses concerning the (density) distribution $f(x)$ of the data X, $H_0 : f(x) = f_0(x)$ and $H_1 : f(x) = f_1(x)$ the KL information for discriminating between H_0 and H_1 is

$$K(f_0, f_1) = E_{f_0}\{\log[f_0(X)/f_1(X)]\}. \tag{1}$$

The subscript f_0 refers to the fact that the expectation is calculated for the case where the distribution of X is governed by f_0. The information K measures the exponential rate at which the posterior probability of H_1 approaches zero when H_0 is true, as independent observations on X are obtained. It is particularly relevant in the design of sequential experiments, such as were discussed by Goradia and Lange [1]. Suppose now that under our model the density of X can be described in terms of a parameter θ, i.e. $f(x) = f(x, \theta)$, and the underlying probability distribution is governed by $\theta = \theta_0$, and we are interested in a composite alternative $H_1 : \theta \in \Omega_1$ to the true hypothesis $H_0 : \theta = \theta_0$. Then the appropriate measure is

$$K(H_0, H_1) = \inf_{\theta_1 \in \Omega_1} E_{\theta_0}\{\log[f(X, \theta_0)/f(X, \theta_1)]\} \tag{2}$$

which can be decomposed into the following difference if either term is finite

$$K(H_0, H_1) = E_{\theta_0}\{\log f(X, \theta_0)\} - \sup_{\theta_1 \in \Omega_1} E_{\theta_0}\{\log f(X, \theta_1)\}.$$

We shall suppress the subscript θ_0 when there is no danger of ambiguity.

2. THE SPERM TYPING MODEL AND MAXIMUM LIKELIHOOD

Consider first the case of three genes for which the donor is heterozygous, and his two chromosomes have genes ABC and abc respectively. A sperm will have a chromosome providing one of the 8 following observations, $ABC, ABc, AbC, Abc, aBC, aBc, abC, abc$ with probabilities depending on the recombination probabilities and the ordering of the three genes on the chromosome. Suppose that the genes appeared in the order ABC rather than ACB or BAC. Suppose also that the recombination probabilities (indicating the probabilities that in the reproduction process, the chromosomes would separate and recombine) between A and B is ϕ_{ab} and between B and C is ϕ_{bc}. Finally suppose that the recombination events are independent. Then the probabilities associated with ABC, abc, and AbC, would be $(1 - \phi_{ab})(1 - \phi_{bc})/2$, $(1 - \phi_{ab})(1 - \phi_{bc})/2$, and $\phi_{ab}\phi_{bc}/2$ respectively. The probabilities associated with the other 5 events can be calculated similarly.

While the estimation of ϕ_{ab} and ϕ_{bc} are of interest and relevant, our main focus in the next section will be on deciding which is the correct one of the three possible orderings ABC, ACB, BAC. Note that without reference to other parts of the chromosome the orderings ABC and CBA are equivalent and we need consider only three, or half of the six possible permutations of ABC. It is also evident that the relevant information in the observed categories ABC and abc are equivalent, and thus we may combine these two observations into one equivalent one, $\bar{A}\bar{B}\bar{C}$ with probability $(1-\phi_{ab})(1-\phi_{bc})$ under the ordering ABC, and probability $(1-\phi^*_{ac})(1-\phi^*_{bc})$ under the ordering ACB, and probability $(1-\phi^{**}_{ab})(1-\phi^{**}_{ac})$ under the ordering BAC. Thus we need only consider 4 possible observations, e.g. $\bar{A}\bar{B}\bar{C}$, $\bar{A}\bar{B}\bar{c}$, $\bar{A}\bar{b}\bar{C}$, and $\bar{A}\bar{b}\bar{c}$, each representing a pair of the original 8 categories.

In our analysis it would seem important to bear in mind that the statistician does not know which alleles appear on the original chromosomes. Thus, even with the order ABC, it might be that the original chromosomes of the donor have AbC and Abc. For our problem involving relatively small recombination probabilities, the data would quickly and easily determine the form of the chromosome, for an original chromosome with ABC would lead to a great preponderance of the $\bar{A}\bar{B}\bar{C}$ observations independent of the order. Nevertheless it turns out that symmetry aspects of the analysis make it unimportant to hypothesize or estimate which alleles appear on each chromosome.

Goradia and Lange [1] analyze two sequential methods of selecting the correct order. They do not analyze the sequential probability ratio method, since the two approaches that they use are much easier for them to analyze. One may wonder whether there is a substantial loss of efficiency in using their methods. The related question that we address is whether there would be an increase in the efficiency of deciding the order of ABC if the analysis were extended to include 4 or 5 genes. Several complications arise in the use of KL numbers to address this question. One is that in ordering 4 (or 5) genes, there are $12 = 4!/2$ (or $60 = 5!/2$) possible orderings of concern. Another issue is that it is more difficult to find the donor who is heterozygous on four, rather than three, specified genes. Finally, technical problems in the technology may make the simple extension of the above probability model less reliable in the application to four or more genes.

In any case, when the KL numbers indicate that there is little to be gained by introducing 4 or 5 genes, then it makes sense to confine attention to three at a time. In case there is a potential gain of a great amount of information, then one ought to consider the relative merit of doing the possibly more complicated analysis required to deal with more than 3 genes.

Assuming the order ABC, the likelihood, based on n_{ABC}, n_{ABc}, n_{AbC}, and n_{Abc} observations $\bar{A}\bar{B}\bar{C}$, $\bar{A}\bar{B}\bar{c}$, $\bar{A}\bar{b}\bar{C}$, $\bar{A}\bar{b}\bar{c}$ respectively, is

$$\begin{aligned} L &= [(1-\phi_{ab})(1-\phi_{bc})]^{n_{ABC}}[(1-\phi_{ab})\phi_{bc}]^{n_{ABc}}[\phi_{ab}\phi_{bc}]^{n_{AbC}} \\ &\qquad \cdot[\phi_{ab}(1-\phi_{bc})]^{n_{Abc}} \\ &= \phi_{ab}^{n_{Ab}}(1-\phi_{ab})^{n_{AB}}\phi_{bc}^{n_{Bc}}(1-\phi_{bc})^{n_{BC}} \end{aligned}$$

where

$$n_{AB} = n_{ABC} + n_{ABc} = n - n_{Ab}$$

$$n_{BC} = n_{ABC} + n_{Abc} = n - n_{Bc}$$

and

$$n = n_{ABC} + n_{ABc} + n_{AbC} + n_{Abc}$$

is the total number of observations. The corresponding maximum likelihood estimates are

$$\hat{\phi}_{ab} = n_{Ab}/n$$

and

$$\hat{\phi}_{bc} = n_{Bc}/n$$

yielding the likelihood

$$L(ABC) = \{\hat{\phi}_{ab}^{\hat{\phi}_{ab}}(1-\hat{\phi}_{ab})^{(1-\hat{\phi}_{ab})}\hat{\phi}_{bc}^{\hat{\phi}_{bc}}(1-\hat{\phi}_{bc})^{(1-\hat{\phi}_{bc})}\}^n$$

with logarithm

$$\log L(ABC) = -n\{V(\hat{\phi}_{ab}) + V(\hat{\phi}_{bc})\} \tag{3}$$

where, for $0 < x < 1$,

$$V(x) = -\{x \log x + (1-x)\log(1-x)\} \tag{4}$$

is an entropy.

The likelihood corresponds to that calculated from observing the two sets of independent binomials corresponding to the recombinations from A to B and from B to C. Notice that if the original chromosomes had AbC and aBc, the estimates of $\hat{\phi}_{ab}$ and $\hat{\phi}_{bc}$ would be replaced by the complements $1-\hat{\phi}_{ab}$ and $1-\hat{\phi}_{bc}$ and $\log L$ would be unaltered.

These results help to understand the derivation of the KL numbers in the following section where we deal with expected log likelihoods for $n = 1$.

In generalizing to m genes, we could extend the alphabetic notation, but it seems more convenient to change the notation slightly. We label the genes 1 to m and consider those permutations, $\pi = (\pi_1, \pi_2, \ldots, \pi_m)$, for which 1 appears in the first half, or, in the case where m is odd, may appear in the center, but 2 appears in the first half. Thus, for $m = 3$, we have the permutations $(123, 132, 213)$ representing the 3 possible orderings.

A possible parametric point θ is described by a permutation π and a vector ϕ with components $\phi_{\pi_i,\pi_{i+1}}$ for $1 \le i \le m-1$, representing recombination probabilities. For the time being this notation seems mildly ambiguous since ϕ_{12} associated with $\pi^0 = (1,2,3,4,5)$ and ϕ_{12} associated with $\pi^1 = (1,2,5,4,3)$ should be designated separately, possibly with superscripts. Our observations will consist of n independent vectors of the form $X = (X_1, X_2, \ldots, X_m)$, where the i-th component of X is zero or one depending on which allele of the i-th gene is observed in the given sperm observation.

Supposing that the true ordering is π^0, and given the associated values of $\phi = (\phi_{12}, \phi_{23}, \ldots, \phi_{m-1,m})$, the likelihood is easily seen to be

$$L = \frac{1}{2}\prod_{i=1}^{m-1}[\phi_{i,i+1}^{n_{i,i+1}}(1-\phi_{i,i+1})^{n-n_{i,i+1}}] \tag{5}$$

where n_{ij} is the number of times that $X_i \ne X_j$ in the sample of n observations. Then the maximum likelihood estimates are

$$\hat{\phi}_{i,i+1} = n_{i,i+1}/n, \qquad 1 \le i \le m-1 \tag{6}$$

and the maximum likelihood under the ordering π^0 satisfies

$$\log 2L(\pi^0) = -\sum_{i=1}^{m-1} V(\hat{\phi}_{i,i+1}). \tag{7}$$

Given an alternate permutation π, it is clear that the corresponding MLE of the related recombination probabilities are given by

$$\hat{\phi}^*_{\pi_i\pi_{i+1}} = n_{\pi_i\pi_{i+1}}/n, \qquad 1 \leq i \leq m-1 \tag{8}$$

and the maximum likelihood satisfies

$$\log 2L(\pi) = -\sum_{i=1}^{m-1} V(\hat{\phi}^*_{\pi_i\pi_{i+1}}). \tag{9}$$

Note that if $\pi = (1,2,5,4,3)$, the MLE of ϕ^*_{12} under the ordering π is exactly the same as that of ϕ_{12} under π^0. Also, under the hypothesis $H_0 : \theta = \theta_0 = (\pi^0, \phi)$ where $\phi = (\phi_{12}, \phi_{23}, \ldots, \phi_{m-1,m})$ is specified, the variables n_{ij} are binomial random variables associated with probabilities

$$\phi_{ij} = P\{X_i \neq X_j\}, \qquad 1 \leq i,j \leq m. \tag{10}$$

Here, and later, we assume that H_0 applies and suppress the subscript θ_0 for P and E. Then ϕ_{ij} is the probability of an odd number of recombinations between the i-th and j-th genes. Thus $\phi_{ii} = 0$, $\phi_{i,i+1} = \phi_{i+1,i}$, and, for $1 \leq i < j \leq m-1$,

$$\phi_{i,j+1} = \phi_{j+1,i} = \phi_{ij}(1-\phi_{j,j+1}) + (1-\phi_{ij})\phi_{j,j+1}. \tag{11}$$

3. KULLBACK-LEIBLER INFORMATION FOR SPERM TYPING

To calculate the KL numbers, consider the case $n = 1$. Then $En_{ij} = \phi_{ij}$,

$$\begin{aligned} E2\log f(X,\theta_0) &= E\sum_{i=1}^{m-1}\{n_{i,i+1}\log\phi_{i,i+1} + (1-n_{i,i+1})\log(1-\phi_{i,i+1})\} \\ &= -\sum_{i=1}^{m-1} V(\phi_{i,i+1}). \end{aligned}$$

For specified $\theta = (\pi, \phi^*)$,

$$\begin{aligned} E2\log f(X,\theta) &= E\sum_{i=1}^{m-1}\{n_{\pi_i\pi_{i+1}}\log(\phi^*_{\pi_i\pi_{i+1}}) \\ &\qquad + (1-n_{\pi_i\pi_{i+1}})\log(1-\phi^*_{\pi_i\pi_{i+1}})\} \\ &= \sum_{i=1}^{m-1}\{\phi_{\pi_i\pi_{i+1}}\log\phi^*_{\pi_i\pi_{i+1}} \\ &\qquad + (1-\phi_{\pi_i\pi_{i+1}})\log(1-\phi^*_{\pi_i\pi_{i+1}})\} \end{aligned}$$

which is maximized with respect to ϕ^* by

$$\phi^*_{\pi_i\pi_{i+1}} = \phi_{\pi_i\pi_{i+1}}$$

Thus if H_1 corresponds to the composite hypothesis of the ordering π, we would have

$$K(H_0, H_1) = \sum_{i=1}^{m-1}[V(\phi_{\pi_i\pi_{i+1}}) - V(\phi_{i,i+1})]. \tag{12}$$

In particular, suppose that we are dealing with 3 genes and H_1 corresponds to the order $(1,3,2)$. Then

$$K(H_0, H_1) = V(\phi_{13}) + V(\phi_{23}) - V(\phi_{12}) - V(\phi_{23}) = V(\phi_{13}) - V(\phi_{12})$$

whereas for H_2 corresponding to $(2,1,3)$,

$$K(H_0, H_2) = V(\phi_{13}) - V(\phi_{23}).$$

Finally

$$K(H_0, H_1 \cup H_2) = V(\phi_{13}) - \max[V(\phi_{12}), V(\phi_{23})]. \tag{13}$$

More generally for m genes, let H_0 correspond to $\pi_0 = (1,2,\ldots,m)$ and $\phi = \{\phi_{i,i+1} : 1 \leq i \leq m-1\}$, and let $\mathcal{A}$ be a subset of the $m!/2 - 1$ other permutations corresponding to alternate orders. Then

$$K(H_0, H_{\mathcal{A}}) = \min_{\pi \in \mathcal{A}}\{\sum_{i=1}^{m-1} V(\phi_{\pi_i \pi_{i+1}})\} - \sum_{i=1}^{m-1} V(\phi_{i,i+1}). \tag{14}$$

We are mainly concerned with 3 cases. Given genes 1 to 5 with $\{\phi_{i,i+1} : 1 \leq i \leq 4\}$, we have

case 1: $m = 3$, $\phi = (\phi_{23}, \phi_{34})$, $\mathcal{A}$ is the set of 2 orderings of $(2,3,4)$ other than $(2,3,4)$

case 2: $m = 4$, $\phi = (\phi_{23}, \phi_{34}, \phi_{45})$, $\mathcal{A}$ is the set of 8 orderings of $(2,3,4,5)$ inconsistent with the ordering $(2,3,4)$ or its equivalent $(4,3,2)$

case 3: $m = 5$, $\phi = (\phi_{12}, \phi_{23}, \phi_{34}, \phi_{45})$, $\mathcal{A}$ is the set of 40 orderings inconsistent with the ordering $(2,3,4)$ or $(4,3,2)$.

These three cases give us the relevant KL numbers for the ordering of genes 2, 3, and 4 when considering data involving (1) the three genes $(2,3,4)$, (2) the four genes $(2,3,4,5)$, and (3) the five genes $(1,2,3,4,5)$.

4. RADIATION HYBRID MODEL

Another technology for estimating distances along the chromosome and for ordering genes is that of radiation hybrid mapping. Here again we introduce the model via cases involving few genes. This model was analyzed by Boehnke *et al.* [2] and Lange and Boehnke [3]. The technology consists of radiating a hybrid rodent cell which carries a human chromosome, thereby breaking the chromosome into several fragments, a proportion $r = 1 - \bar{r}$ of which are retained. The higher the rate of radiation, λ, the more fragments are made.

We will assume that r is known, and that the distance between two genes A and B is δ, unknown. Then the probability that the two genes will be on separate fragments is

$$\phi = 1 - \exp(-\lambda\delta) \tag{15}$$

assuming that breaks occur like a Poisson process with rate λ. The probabilities of observing, among the retained fragments, both A and B, A alone, B alone, and neither are

$$\begin{aligned} p_{11} &= r(1 - \bar{r}\phi) \\ p_{10} &= r\bar{r}\phi \\ p_{01} &= p_{10} \\ p_{00} &= \bar{r}(1 - r\phi) \end{aligned} \tag{16}$$

respectively. Here we have assumed that breaks and retention events are independent, and that $r = 1 - \bar{r}$ is constant.

It is relatively easy to calculate the Fisher Information for estimating ϕ. That is

$$\begin{aligned} J &= E\left\{\left[\frac{\partial(\log likelihood)}{\partial\phi}\right]^2\right\} \\ &= \frac{r\bar{r}(2-\phi)}{\phi(1-r\phi)(1-\bar{r}\phi)}. \end{aligned} \quad (17)$$

It follows that the Fisher information with respect to the distance δ is

$$J^* = J\left(\frac{d\phi}{d\delta}\right)^2 = \frac{r\bar{r}\lambda^2(1-\phi)^2(2-\phi)}{\phi(1-r\phi)(1-\bar{r}\phi)}. \quad (18)$$

For small $\lambda\delta$, $\phi \approx \lambda\delta$ and $J^* \approx 2r\bar{r}\lambda/\delta$. Insofar as $1/(nJ^*\delta^2)$ is the asymptotic *relative variance* of the large sample estimate of δ, it gives us a clue about what values of λ would be useful for ordering the genes. Uncertainty in the knowledge of r complicates matters somewhat. In that case the information matrix for δ and r should be evaluated and inverted.

To proceed with the ordering problem, suppose that the genes are arranged in order $\pi^0 = (1, 2, \ldots, m)$ and that the distances between successive genes $\delta_{i,i+1}$, give rise to separation probabilities $\phi_{i,i+1}$. Then let the observation be a vector $X = (X_1, X_2, \ldots, X_m)$ to indicate which genes are retained. That is $X_i = 1$ indicates retention of the i-th gene and otherwise $X_i = 0$. Then X is a Markov Process where

$$f_X(x) = r^{x_1}\bar{r}^{(1-x_1)}\prod_{i=1}^{m-1} g(x_i, x_{i+1}; \phi_{i,i+1}) \quad (19)$$

and

$$\begin{aligned} g(1,1;\tau) &= 1 - \bar{r}\tau \\ g(1,0;\tau) &= \bar{r}\tau \\ g(0,1;\tau) &= r\tau \\ g(0,0;\tau) &= 1 - r\tau. \end{aligned} \quad (20)$$

Further, for $i < j$,

$$P\{X_j = x_j | X_i = x_i\} = g(x_i, x_j; \phi_{ij}) \quad (21)$$

where

$$\phi_{ij} = 1 - \exp(-\lambda\delta_{ij}) = \phi_{ji} \quad (22)$$

and

$$\delta_{ij} = \sum_{k=i}^{j-1} \delta_{k,k+1} \quad (23)$$

is the distance between the i-th and j-th genes.

We shall be interested in maximizing

$$w_{ij}(\tau) = E\log g(X_i, X_j; \tau)$$

with respect to τ. Then

$$\begin{aligned} w_{ij}(\tau) = {} & r(1-\bar{r}\phi_{ij})\log(1-\bar{r}\tau) + r\bar{r}\phi_{ij}\log(\bar{r}\tau) + \bar{r}r\phi_{ij}\log(r\tau) \\ & + \bar{r}(1-r\phi_{ij})\log(1-r\tau) \end{aligned}$$

and

$$w'_{ij} = \frac{r\bar{r}(\phi_{ij} - \tau)(2 - \tau)}{\tau(1 - r\tau)(1 - \bar{r}\tau)}$$

vanishes only at $\tau = \phi_{ij}$ in the interval $(0, 1)$, and indeed, $w_{ij}(\tau)$ attains its maximum value

$$\begin{aligned} W(\phi_{ij}) &= r(1 - \bar{r}\phi_{ij}) \log(1 - \bar{r}\phi_{ij}) + r\bar{r}\phi_{ij} \log(r\bar{r}\phi_{ij}^2) \\ &\quad + \bar{r}(1 - r\phi_{ij}) \log(1 - r\phi_{ij}) \end{aligned} \tag{24}$$

at $\tau = \phi_{ij}$. Incidentally, this result could also be derived without calculating the derivative, by noting the relationship between w_{ij} and a Kullback-Leibler number and that a KL number is always nonnegative, and hence an expression of the form $E_{\theta_0}[\log f(x, \theta)]$ attains its maximum value when $\theta = \theta_0$.

We are now in position to calculate the KL information. Let
$H_0 : \theta = \theta_0$ correspond to the permutation π^0 and
$\phi = (\phi_{12}, \phi_{23}, \ldots, \phi_{m-1,m})$, and
$H_1 : \theta = \theta_1$ correspond to the permutation π^* and
$\phi^* = (\phi^*_{12}, \ldots, \phi^*_{m-1,m})$.
Then, with E_{θ_0} represented by E, we have

$$\begin{aligned} K(H_0, H_1) &= E \log f(X, \theta_0) - E \log f(X, \theta_1) \\ &= E \log \left[r^{X_1} \bar{r}^{(1-X_1)} \prod_{i=1}^{m-1} g(X_i, X_{i+1}; \phi_{i,i+1}) \right] \\ &\quad - E \log \left[r^{X_{\pi_1}} \bar{r}^{(1-X_{\pi_1})} \prod_{i=1}^{m-1} g(X_{\pi_i}, X_{\pi_{i+1}}; \phi^*_{\pi_i \pi_{i+1}}) \right] \\ &= -V(r) + \sum_{i=1}^{m-1} W(\phi_{i,i+1}) + V(r) - \sum_{i=1}^{m-1} w_{\pi_i \pi_{i+1}}(\phi^*_{\pi_i \pi_{i+1}}) \end{aligned}$$

which is minimized with respect to ϕ^* by $\phi^*_{\pi_i \pi_{i+1}} = \phi_{\pi_i \pi_{i+1}}$. Thus for H_1 corresponding to the ordering π,

$$K(H_0, H_1) = \sum_{i=1}^{m-1} W(\phi_{i,i+1}) - \sum_{i=1}^{m-1} W(\phi_{\pi_i \pi_{i+1}}).$$

Further when $\mathcal{A}$ is an arbitrary set of permutations,

$$K(H_0, H_{\mathcal{A}}) = \sum_{i=1}^{m-1} W(\phi_{i,i+1}) - \max_{\pi \in \mathcal{A}} \sum_{i=1}^{m-1} W(\phi_{\pi_i \pi_{i+1}}). \tag{25}$$

Thus we can evaluate the effect of considering neighboring genes for the case of radiation hybrids just as we did in the case of sperm typing with W playing the role of $-V$.

5. CALCULATIONS

The Kullback-Leibler numbers for ordering the three genes $(2, 3, 4)$ using sperm typing were calculated for various values of ϕ, yielding $S_3 = S_3(\phi_{23}, \phi_{34})$, $S_4 = S_4(\phi_{23}, \phi_{34}, \phi_{45})$ and $S_5 = S_5(\phi)$, when considering 3, 4, and 5 genes respectively. Table 1 presents the results for some cases and illustrates the following comments.

The values of $S_3(a, a)$ peak at $S_3(0.12, 0.12) = 0.14$, but this peak is rather broad, since $S_3(a, a)$ is 0.103 at $a = 0.04$ and 0.099 at $a = 0.25$. The function $S_3(a, b)$ drops rapidly as

a and b separate. It seems that $S_4(\phi_{23}, \phi_{34}, \phi_{45})$ is no improvement over $S_3(\phi_{23}, \phi_{34})$ when $\phi_{23} \leq \phi_{34}$. When $\phi_{23} > \phi_{34}$, there is room for substantial improvement by including gene 5. However, in those cases, including gene 1 also, rarely gives additional gain.

If $\phi_{23} = \phi_{34}$ is kept fixed, then S_5, regarded as a function of ϕ_{12} and ϕ_{45} is constant along squares for which the diagonal is along $\phi_{12} = \phi_{45}$. The function $S_5(a, a, a, a)$ attains a maximum value of 0.231 at $a = 0.10$. For fixed a, $S_5(b, a, a, b)$ peaks at $b = \tilde{b}(a)$ where $\tilde{b}(a) \approx 1.4a$. This value in turn has a peak of 0.258 at $a = 0.10$ and $b = 0.14$.

Table 1

KL information for ordering genes $(2, 3, 4)$ using sperm typing and considering 3, 4, and 5 genes

ϕ_{12}	ϕ_{23}	ϕ_{34}	ϕ_{45}	S_3	S_4	S_5
.01	.01	.01	.01	.041	.041	.077
.02	.02	.02	.02	.067	.067	.122
.04	.04	.04	.04	.103	.103	.180
.10	.10	.10	.10	.146	.146	.231
.14	.14	.14	.14	.147	.147	.217
.20	.20	.20	.20	.127	.127	.169
.25	.25	.25	.25	.099	.099	.123
.14	.10	.10	.14	.146	.146	.258
.20	.14	.14	.20	.147	.147	.239
.10	.01	.01	.01	.041	.041	.059
.10	.02	.02	.01	.067	.067	.096
.10	.02	.02	.10	.067	.067	.101
x	.02	.01	.01	.035	.067	.067
x	.04	.02	.02	.055	.101	.101
x	.04	.02	.04	.055	.109	.109
x	.04	.02	.10	.055	.088	.088
x	.10	.04	.02	.065	.092	.092
x	.10	.04	.04	.065	.117	.117
x	.10	.04	.10	.065	.130	.130
x	.10	.08	.04	.121	.162	.162
y	.10	.08	.08	.121	.168	.199
z	.10	.08	.10	.121	.168	.227
y	.10	.08	.15	.121	.168	.207
x	.10	.08	.25	.121	.161	.161

x represents any value in the interval $(0, \infty)$.
y represents any value in an interval containing $(0.04, 0.25)$.
z represents any value in an interval containing $(0.08, 0.20)$.

If ϕ_{23} is substantially larger than ϕ_{34}, then $S_5(\phi) = S_4(\phi_{23}, \phi_{34}, \phi_{45})$. If ϕ_{23} is not much larger than ϕ_{34}, the introduction of gene 1 begins to have some effect if ϕ_{45} is rather close to optimal for S_4 and ϕ_{12} is neither very small nor very large. There is another way to look at this phenomenon. If ϕ_{23} and $\phi_{34} < \phi_{23}$ are kept fixed, then S_5 is constant along rectangles in the (ϕ_{12}, ϕ_{45}) space. As ϕ_{34} decreases, these rectangles become elongated along the ϕ_{12} direction, and some of these rectangles degenerate to lines for small and large values of ϕ_{45}. When ϕ_{34} decreases enough, all the rectangles degenerate, and the level lines become parallel lines and S_5 is independent of ϕ_{12}.

In summary, if $\phi_{23} \approx \phi_{34}$, consideration of five genes is required to get improvement over that of three genes. If ϕ_{23} is considerably different than ϕ_{34}, one extra gene on the side of the two adjacent genes can give improvement, but the gene on the other side will not help.

Table 2

KL information for ordering genes $(2,3,4)$ using radiation hybrid mapping and considering 3, 4, and 5 genes

λ_{12}	λ_{23}	λ_{34}	λ_{45}	R_3	R_4	R_5
.01	.01	.01	.01	.023	.023	.044
.02	.02	.02	.02	.040	.040	.073
.10	.10	.10	.10	.109	.109	.188
.20	.20	.20	.20	.139	.139	.223
.30	.30	.30	.30	.144	.144	.217
.50	.50	.50	.50	.125	.125	.168
1.50	1.50	1.50	1.50	.024	.024	.025
.34	.23	.23	.34	.143	.143	.250
.20	.04	.04	.04	.064	.064	.099
.20	.10	.10	.04	.109	.109	.143
.20	.10	.10	.20	.109	.109	.188
x	.10	.04	.01	.048	.059	.059
x	.20	.10	.04	.079	.105	.105
x	.20	.10	.10	.079	.139	.139
x	.20	.10	.20	.079	.158	.158
x	.40	.20	.10	.084	.114	.114
x	.40	.20	.20	.084	.137	.137
x	.40	.20	.40	.084	.168	.168
x	.20	.15	.04	.111	.134	.134
y	.20	.15	.10	.111	.161	.164
z	.20	.15	.20	.111	.161	.206
w	.20	.15	.40	.111	.161	.178
x	.20	.15	.60	.111	.153	.153

x represents any value in the interval $(0, \infty)$.
y represents any value in an interval containing $(0.01, 1.50)$.
z represents any value in an interval containing $(0.10, 0.64)$.
w represents any value in an interval containing $(0.04, 1.10)$.

The qualitative results for the use of radiation hybrid mapping are similar to those for sperm typing. Table 2 presents some results. The KL information depends on $\lambda\delta$ and r. In our table we take $r = 0.4$, and letting $\lambda_{ij} = \lambda\delta_{ij}$, we present the KL numbers $R_3 = R_3(\lambda_{23}, \lambda_{34})$, $R_4 = R_4(\lambda_{23}, \lambda_{34}, \lambda_{45})$ and $R_5 = R_5(\lambda\delta)$, where $\lambda_{ij} = \lambda\delta_{ij}$. We note that the peak of $R_3(a,a)$ is 0.144 at $a = 0.28$ while values of a at 0.10 and 0.70 give 0.109 and 0.096. The peak value of $R_5(a,a,a,a)$ is 0.224 at $a = 0.22$, while the peak value of

$R_5(b, a, a, b)$ is 0.250 at $a = 0.23$, $b = 0.34$.

These results have obvious potential application in selecting appropriate doses of radiation to increase the information content. Of course, the broad peak of R_3 indicates that KL values are not very sensitive to the choice of λ. The tables indicate that there are circumstances where considering 4 or 5 genes may double the information content, but also suggest that often there is little to gain by considering five or more genes simultaneously. The tables can easily be supplemented, since the calculations of the KL numbers are easily implemented.

Acknowledgements
This research was supported in part by the U.S Office of Naval Research under contract N00014-91-J1005.

REFERENCES

1. T. M. Goradia and K. Lange, Multilocus ordering strategies based on sperm typing, *Annals of Human Genetics.* **54**, 49-77 (1990).

2. M. Boehnke, K. Lange and D. R. Cox, Statistical methods for multipoint radiation hybrid mapping. *American Journal of Human Genetics.* **49**, 1174-1188 (1991).

3. K. Lange and M. Boehnke, Bayesian methods and optimal experimental design for gene mapping by radiation hybrids. *Annals of Human Genetics*, **56** 119-144 (1992).

Stat. Sci. & Data Anal., pp. 13-24
K. Matsusita *et al.* (Eds)

Statistical Models For Forecasting Tornado Intensity

JOHN F. MONAHAN, KEVIN J. SCHRAB and CHARLES E. ANDERSON
Department of Statistics, North Carolina State University, Raleigh, North Carolina 27695
Department of Marine, Earth, and Atmospheric Sciences, North Carolina State University, Raleigh, North Carolina, USA

Abstract. Two measurements of thunderstorm cell circulation, both based on satellite imagery, show promise in forecasting the occurrence of tornadoes arising from the cell. Simple discriminant analysis clearly shows the ability of these two variables to predict the occurrence of a tornado. Censored regression models permit the analysis of tornado intensity and show variations from outbreak to outbreak. A third variable, vorticity, which characterizes the pre-storm environment, can account for most of that variation. Another satellite imagery variable, storm relative ambient wind, improves the model fit. Ordinal regression models are used to analyze the maximum intensity of tornadoes from a thunderstorm cell. The latent variable characterization motivates this model well by accounting for the arbitrariness of the intensity scale. Five methods for computing forecast probabilities appropriate for real-time field operation are also discussed.

Key Words: discriminant analysis, ordinal regression, Tobit regression, tornado forecasting

1. INTRODUCTION

Quite simply, tornadoes are difficult to forecast. They have a narrow path of destruction, are short-lived, move erratically, and out of tens of thousands of convective thunderstorms arising in the United States each year, only hundreds (~ 800) produce tourneyed. In this paper, we investigate the use of three measurements of thunderstorms from satellite images, along with regional prestorm conditions, to forecast the occurrence and intensity of tornadoes. The first methodology employed is standard discriminant analysis which demonstrates the promise of the two main predictors of tornadic thunderstorms. Next, censored regression models for average intensity are employed to investigate the performance of explanatory variables. Although good predictors could be clearly identified, these models proved inadequate, and an ordinal regression model was pursued. The methodology of forecasting with the ordinal regression model is illustrated with forecast probabilities for thunderstorm cells from an outbreak in South Carolina.

2. MEASUREMENTS

The occurrence and intensity of a tornado are determined from the pattern and severity of the ground destruction, and so these measurements are not available until the following day. Intensity is measured on the Fujita [1] scale, F0 to F5, relabeled here as levels from 1 to 6, with 0 denoting no tornado. While the Fujita scale orders the intensities of tornadoes, it does not correspond directly to a physical quantity, and so no linear ordering should be assumed. The problem of evaluator bias in the measurement is less important than others confronted here. Since the intensity can only be determined by the pattern of destruction, underreporting is likely, since a powerful tornado passing only over wheatfields may never show its real strength. Since a single thunderstorm cell may produce more than one tornado, both the average intensity (FBAR) and the maximum (FMAX) have been analyzed.

Any predictions on the occurrence and intensity of a tornado must be based on the characteristics of the parent thunderstorm cell. Three measurements can be made from satellite images in real-time as the outbreak is in progress, and take about one hour. UMAX measures the maximum anvil outflow from the thunderstorm and is found by fitting a simple two-dimensional outflow model to three successive images. MDA is the measured deviation angle of the anvil plume from its expected direction, that of the ambient wind. These two measurements have shown promise in predicting the occurrence and intensity of a tornado [2], [3], [4]. A third variable included in this analysis, the storm relative ambient wind (SRAW), measures the venting of the top of the cyclone. The square of UMAX, which brings a sense of kinetic energy to the outflow measurement, was included in the regression models in this paper.

Three measurements of the synoptic scale (300-500 km) environment are available from soundings: the potential buoyant energy (PBE), wind shear (SHEAR), and surface vorticity (SFCVOR). Although research on tornado models suggest that PBE and SHEAR are important determinants of tornadic behavior, our study does not show these two variables to have any predictive value, much to our surprise. SFCVOR appears to capture some of the synoptic characteristics, but will be the same for all tornadoes in a given outbreak.

For the sake of efficiency, this investigation consists of post hoc analysis of outbreaks of tornadoes with the hope that the methods can be applied successfully to single thunderstorm cells in real time. We have looked at eleven outbreaks over the years 1984-1990, consisting of 168 thunderstorm cells, of which 52 produced tornadoes. The smallest outbreak, from South Carolina in 1989 with 11 cells, was reserved to compare forecasting procedures. The other ten outbreaks (157 cells, 47 tornadic) was used as the estimation and training dataset.

3. DISCRIMINANT ANALYSIS

Simple discriminant analysis clearly shows that the main two measurements, UMAX and MDA, do a fine job of classifying thunderstorm as tornadic or not. Figure 1 illustrates this relationship well with UMAX and MDA plotted with the third variable FMAX denoting maximum intensity. The linear discriminant line on Figure 1 separates the non-tornadic zero cells from the tornadic cells. Clearly large values of UMAX and MDA are associated with nonzero, tornadic values of FMAX. Within sample, 95 of 110 non-tornadic thunderstorm cells are classified correctly, and 39 of 47 tornadic cells are classified as tornadic. For the South Carolina test outbreak, 6 of 6 non-tornadic and 5 of 5 tornadic cells are classified correctly. Moreover, classification probabilities can be used to incorporate the differing misclassification costs between false alarms and misses. Since misclassification costs should be different in this situation, with false alarms much less costly than missing a tornado, changing the cost ratio to 2 to 1 reduces the missed tornadoes from 8 to 3 within sample, and does not affect the perfect record in the South Carolina test cases. In one disturbing result, however, close examination shows that the break point between tornadic and non-tornadic cells appears to shift from one outbreak to another.

4. CENSORED REGRESSION MODELS

Following the success in predicting occurrence, a regression model appeared natural in order to assess tornado intensity. Fitting the average intensity FBAR appears plausible with the average suggesting a normal error distribution. But since FBAR was strictly positive, censoring at zero leads to the Tobit [5], [6] regression model:

$$Z_i = \beta_0 + \beta_1 x_{i1} + \dots + \beta_p x_{ip} + e_i = \beta^T x_{i.} + e_i \qquad e_i \sim N(0, \sigma^2) \tag{1}$$

$$FBAR_i = \begin{cases} Z_i & \text{if } Z_i > 0 \\ 0 & \text{if } Z_i \leq 0 \end{cases} \tag{2}$$

Two factors threaten the acceptability of the normality assumptions: first, typically only about three tornadoes intensities are averaged, and second, the smallest positive value is 1, a long way from the censoring zero value. However, while normality may be questionable, the most serious problem arises in using the Fujita scale as an interval scale. Nevertheless, this type of model is very convenient and allows for a closer examination of the forecasting power of the proposed variables since software is readily available for this kind of analysis. The following models are compared with results in

Model 0: $Z_i = \beta_0 + e_i$

Model 1: $Z_i = \beta_0 + \beta_1 UMAX_i + \beta_2 MDA_i + e_i$

Model 2: $Z_i = \beta_0 + \beta_1 UMAX_i + \beta_2 MDA_i + \beta_3 SRAW_i + \beta_4 UMAX2 + e_i$

Model 3: $Z_i = \beta_0 + \beta_1 UMAX_i + \beta_2 MDA_i + \beta_3 SRAW_i + \beta_4 UMAX2 + \beta_5 SFCVOR_{g(i)} + e_i$

Model 4: $Z_i = \alpha_{g(i)} + \beta_1 UMAX_i + \beta_2 MDA_i + \beta_3 SRAW_i + \beta_4 UMAX2 + e_i$

where g(i) denotes the outbreak for thunderstorm cell i, and $SFCVOR_g$ denotes the surface vorticity over the synoptic region for outbreak g. Model 0 is the "null" model with no predictive capability. Model 2 includes another satellite variable SRAW, as well as UMAX2, the square of UMAX. Model 4, with an intercept for each outbreak, can account for the apparent shift of the boundary between tornadic and nontornadic cells from outbreak to outbreak. Model 3 attempts to explain the changing boundary with the single synoptic variable of surface vorticity. While clearly Model 4 will fit the data better, it has no predictive capability, since forecasts must be made without any inference on the magnitude of the random outbreak effect. Even after measurements of characteristics of earlier cells in an outbreak, measurements on the intensities of any resultant tornadoes would not be available until the following day.

These five models were fit using SAS's PROC LIFEREG, where the Tobit regression is available using the DIST=NORMAL option. Table 1 gives the maximum log-likelihood for these five models, -2 log likelihood ratio test statistics for testing among these nested models, estimates of the scale parameter σ, and a heuristic R^2 following the analogue $R^2 = 1 - (\hat{\sigma}/\hat{\sigma}_0)^2$, where $\hat{\sigma}_0$ is the scale estimate under the null Model 0. We found UMAX, MDA, $UMAX^2$, SRAW and SFCVOR all to be useful explanatory variables. Quite surprisingly, the modeler choices PBE and SHEAR showed little or no predictive capability. Still disturbing is the significance of a random outbreak effect, which is impossible to accommodate in forecasting, although SFCVOR (constant for an outbreak) captures more than half of the effect. While this censored regression model can be used to compute forecast means and standard errors, the high forecast variance (over 1) leaves this approach unimpressive. Although this model fit the data reasonably well, answered some questions about explanatory variables, and could be computed with relative ease, a more appropriate model was sought to address these shortcomings.

Table 1.
Comparisons of Censored Regression Models

Model	Parameters	Scale est. $\hat{\sigma}$	log likelihood	heuristic R^2	χ^2 vs lower model
Model 0	2	3.432	−180.4589	N/A	
Model 1	4	1.815	−128.2725	0.72	104.37 vs $\chi^2(2)$
Model 2	6	1.591	−116.0292	0.79	12.24 vs $\chi^2(2)$
Model 3	7	1.254	−100.1176	0.87	31.82 vs $\chi^2(1)$
Model 4	15	1.028	− 87.9238	0.91	24.39 vs $\chi^2(8)$

Critical values at level 0.05 for the χ^2 distribution with 1, 2, and 8 degrees of freedom are 3.84, 5.99, and 15.5, respectively.

5. ORDINAL REGRESSION MODEL

McCullagh [7] motivated regression models for ordinal dependent variables along the lines of proportional odds and proportional hazards models. In a more direct approach [8], Anderson takes the viewpoint of a latent variable model similar to the censored regression model considered in Section 4. For the simpler case of a binary response leading to logistic regression, again pose Z_i as an unobserved variable with a regression mean function $\beta^T x_i$, but with error e_i which has the logistic distribution, $F(u) = 1/(1 + e^{-u})$. A cutoff point θ determines the border between success and failure, Pr(failure) = Pr($Z_i \le \theta$) = F($\theta - \beta^T x_i$), so that logit Pr(success) = $\beta^T x_i - \theta$, where logit p = ln (p/(1−p)). In this situation however, we wish to model the polytomous variable FMAX, the maximum intensity, which takes on values 0 to 6. To generalize, let there be K categories, where the variable I_i takes values 0 through K−1, and the model for the distribution of I_i is

$$\Pr(\, I_i = k \mid x_i \,) = \Pr(\, \theta_k < Z_i \le \theta_{k+1} \,) = F(\, \theta_{k+1} - \beta^T x_i) - F(\theta_k - \beta^T x_i) \qquad \textbf{(3)}$$

where naturally $\theta_0 = -\infty$ and $\theta_K = +\infty$. In this model, in addition to the regression parameters, there are K−1 unknown cutoff points θ_k to be estimated. These cutoff points subsume both intercept and scale parameters of the previous regression models, with the separation of the cutoff points dictating scale. Since the variance of the logistic noise variable e_i is constant, a low signal/high noise situation would be indicated by small separation of the cutoff points, low noise with large separation of the cutoff points and large variation in $\beta^T x_i$.

This generalized logistic model can incorporate the regression structure desired, as well as the arbitrariness of the Fujita scale. In contrast to both McCullagh [7] and Anderson [8], the explanatory variables here are continuous, and the latent variable is not an imaginary creature, but corresponds to the physical intensity on some more natural scale. However, like the usual logistic regression model, this model may still work well without corresponding to the truth. Since the dependent variable happens to be the maximum intensity, an alternative error distribution, the extreme value with

$F(u) = \exp(-\exp(-u))$, should also be considered; this corresponds to McCullagh's complementary log-log transformation. Since the only difference is the evaluation of the distribution function F, this extreme value model has also been used. Estimation and inference are all based on maximum likelihood, where the likelihood for the data ($I_i = k(i)$, x_i, $i = 1,...,N$) can be written as

$$L(\theta, \beta \mid \text{data}) = \prod_{i=1}^{N} [\, F(\theta_{k(i)+1} - \beta^T x_i) - F(\theta_{k(i)} - \beta^T x_i) \,] \tag{4}$$

In a fashion similar to Table 1, Table 2 provides maximum log-likelihoods for the same four models, as well as the -2 log likelihood ratio test statistics. Only the results from the logistic distribution are given; preliminary work using the extreme value error distribution differed little. As with the censored regression results, we can see the strength of the outbreak effect: testing H: Model 2 vs. A: Model 4 strongly rejects. However, while the synoptic surface vorticity variable accounts for about half of the difference in log likelihood, the test H: Model 3 vs. A: Model 4 rejects, indicating that substantial outbreak variation remains. Recall, of course, that Model 3 is still to be preferred for forecasting purposes. Figure 2 gives a sense of the strength of this model, with a plot of fitted values $\beta^T x_i$ against observed intensities, with the cutoff points θ_k given on the vertical scale with the fitted values. Another view can be obtained by constructing another heuristic R^2 by comparing the variance of $\hat{\beta}^T x_i$ and the variance of the logistic error distribution $\pi^2/3$. Taking the variance of $\hat{\beta}^T x_i$ over the sum with $\pi^2/3$ gives a heuristic $R^2 = .88$, similar to the censored regression results. Note that while the cutoff points θ_k must be in nondecreasing order, the maximum likelihood estimates were found without the need to impose such a restriction either explicitly, with a constrained search, or implicitly, with a reparameterization.

Table 2.
Comparisons of Logistic Ordinal Regression Models

Model	Parameters	log likelihood	χ^2 vs lower model
Model 0	6	-171.961	N/A
Model 1	8	-111.704	120.514 vs $\chi^2(2)$
Model 2	10	-102.772	17.864 vs $\chi^2(2)$
Model 3	11	-85.370	34.804 vs $\chi^2(1)$
Model 4	19	-72.430	25.880 vs $\chi^2(8)$

6. FORECASTING

The biggest advantage of the ordinal regression models is that they can be used to provide forecast probabilities for the maximum intensity of the tornadoes, if any, produced from a particular thunderstorm cell. These forecast probabilities can then be used by various decision makers to take whatever appropriate actions within their power. The construction of these forecast probabilities follows a Bayesian argument; other approaches may also lead to the same result. Using Model 3, and given the measurements x, in this case, UMAX, MDA, SFCVOR, UMAX2 and SRAW, the formula (3) above gives the probabilities for the intensity levels when the parameters θ and β are known. For the data currently available, while the sample is rather large, there is still substantial error in the estimates $\hat{\theta}$ and $\hat{\beta}$. But since the large sample overwhelms any prior information in this case, taking the posterior distribution $\pi(\theta,\beta \mid \text{data})$ equal to the likelihood $L(\theta,\beta \mid \text{data})$ expresses knowledge about θ and β. The forecast probability $\pi(k \mid \text{data, x})$ for intensity level k can then be computed by integrating the model probability given by (1) with respect to the posterior $\pi(\theta,\beta \mid \text{data})$,

$$\pi(k \mid \text{data}, x) = \int [\, F(\theta_{k+1} - \beta^T x) - F(\theta_k - \beta^T x)\,]\ \pi(\theta,\beta \mid \text{data})\, d\theta\, d\beta \,. \qquad \textbf{(5)}$$

For the case of the logistic distribution F, some shortcuts are available to make this a practical computation for real-time forecasting. The most straightforward route is to integrate (5) directly which would be generally impractical because of the high (11) dimension. The next most direct method would be to integrate by Monte Carlo. Since the posterior distribution $\pi(\theta,\beta \mid \text{data})$ should be asymptotically normal, this asymptotic normal approximation should serve as the importance distribution for an importance sampling scheme. This distribution is multivariate normal with a mean vector at the maximum likelihood point estimate vector, and a covariance matrix equal to the inverse of the sample information matrix, based on the Hessian matrix evaluated by numerical differences at the maximum. One shortcut here would be to ignore the approximate nature of this importance distribution and avoid the evaluation of the likelihood. Computationally, this would be equivalent to the same importance sampling scheme but with taking the weighting function identically equal to one. A second shortcut would be to use the results of Monahan and Stefanski (1992) for approximating the normal-logistic convolution. This method would permit calculation of forecast probabilities with minimal effort and would be idea for real-time use in the field.

To explain these methods in more mathematical detail, denote the parameter vector as $\gamma = (\theta,\beta)$ with dimension d. Then the asymptotic normal approximation $\hat{\pi}(\gamma \mid \text{data})$ to the posterior can be expressed as

$$\gamma \approx N_d(\, \hat{\gamma},\, [\nabla^2 L(\gamma \mid \text{data})\,]^{-1}\,)$$

where the Hessian matrix $\nabla^2 L(\gamma \mid \text{data})$ is computed by numerical differentiation at the maximum likelihood estimate $\hat{\gamma}$. The importance sampling method would be to generate $\{\gamma^{(j)}, j=1,...,M\}$ iid from the approximating normal distribution $\hat{\pi}(\gamma \mid \text{data})$, evaluating the weight function

$$w(\gamma) = \pi(\gamma \mid \text{data}) \,/\, \hat{\pi}(\gamma \mid \text{data})$$

and computing the mean of the Monte Carlo sample to estimate the forecast probability of intensity k:

$$\hat{\pi}_{IS}(\,k \mid \text{data}) = M^{-1} \sum_{j=1}^{M} w(\gamma^{(j)})\ [\, F(\theta_{k+1}^{(j)} - \beta^{(j)T}x)\ q - F(\theta_k^{(j)} - \beta^{(j)T}x)\,]\,.$$

This first estimate $\hat{\pi}_{IS}(\,k \mid \text{data})$ is exact in the sense that any error can be made arbitrarily small by increasing M without bound. The performance of a second estimate $\hat{\pi}_{NA}(\,k \mid \text{data})$ depends solely on the quality of the normal approximation to the posterior/likelihood. By taking the importance weight function identically equal to one would lead to forecast probabilities computed according the approximate normal posterior, and this estimate can be written as

$$\hat{\pi}_{NA}(\,k \mid \text{data}) = M^{-1} \sum_{j=1}^{M} [\, F(\theta_{k+1}^{(j)} - \beta^{(j)T}x) - F(\theta_k^{(j)} - \beta^{(j)T}x)\,]\,.$$

The saving with $\hat{\pi}_{NA}(k \mid \text{data})$ is due to avoiding the computation of the exact likelihood/posterior $\pi(\gamma|\text{data})$. The third method also follows the same normal approximation as the second method, but avoids the Monte Carlo integration. Note that assuming γ has a multivariate normal distribution leads to $(\theta_k - \beta^T x) = v(x)$ being univariate normal, with, say mean η and variance τ^2. Then computing the expectation of $F(v(x))$ is the distribution function of the convolution of a standard logistic and a normal with mean η and variance τ^2. Monahan and Stefanski [9] have computed accurate approximations of the logistic distribution function F of the form

$$F(t) = \sum_j p_j \, \Phi(t \, s_j)$$

which leads to the convolution approximation

$$\int_{-\infty}^{\infty} F(t) \, \phi((t-\eta)/\tau)/\tau \, dt \approx \sum_j p_j \, \Phi(\eta \, s_j \, / \, \sqrt{1 + \tau^2 s_j^2}).$$

For each cutoff point θ_k, we have $(\theta_k - \beta^T x) = v_k(x)$ approximately normal with mean η_k and variance τ_k^2, so that the forecast probability can be computed as

$$\hat{\pi}_{LNA}(k \mid \text{data}) = \sum_j p_j \, \Phi(\eta_{k+1} \, s_j \, / \, \sqrt{1 + \tau_{k+1}^2 s_j^2}) - \sum_j p_j \, \Phi(\eta_k \, s_j \, / \sqrt{1 + \tau_k^2 s_j^2}).$$

This method only requires numerical evaluation of the normal cdf, and can be done quite rapidly. Keep in mind that the accuracy of this method also depends on the quality of asymptotic normal approximation.

Table 3 gives the forecast probabilities for the different intensity levels for the eleven cells in the South Carolina outbreak, using the three methods described above. The Monte Carlo estimates $\hat{\pi}_{IS}(k \mid \text{data})$ and $\hat{\pi}_{NA}(k \mid \text{data})$ were computed using 10000 replicates. The third method used the Monahan-Stefanski approximation with eight pieces, which should have an error around 10^{-9}. One detail in the Monte Carlo integration left unattended is whether the normal approximation should impose the condition that $\theta_k < \theta_{k+1}$. If this constraint is not imposed in some form, negative probabilities could be calculated. However, the parameter space could be so constrained and the likelihood/posterior should be zero wherever the ordering condition is violated. In the calculations for Table 3, only 198 of the 10000 points violated the condition and were given zero weight, which speaks well for the normal approximation. Two versions of $\hat{\pi}_{NA}$ are included in Table 3 of forecast probabilities. One version (NAC) adheres to the ordering constraint on the cutoff points $\theta_j < \theta_{j+1}$; the other (NAU) does not. Finally, while the high dimension (11) of the integrations required here usually precludes any direct approach, the new adaptive-subregion method of Berntsen, Espelid, and Genz [10] appears promising for these kinds of Bayesian calculations. These results can be very accurate, and are referred in Table 3 by the notation EX. Upper bounds for the integration error are also available for this approach, and these values range only as high as .008, saying that the first two digits are certainly sound. As may be expected, this more accurate direct method required much longer computation time than the other methods. The fastest third method $\hat{\pi}_{LNA}(k \mid \text{data})$ which uses the convolution approximation appears to be the method of choice for real-time calculations since the computation time is about one second. However, the computed probabilities do not agree sufficiently, and so the importance sampling method must be used to refine the preliminary $\hat{\pi}_{LNA}(k \mid \text{data})$ in real-time forecasting. The most accurate $\hat{\pi}_{EX}$ requires too much time for practical field use in its present state. In spite of the relatively close agreement of these computations, one should bear in mind that decisions often depend on the ratios of probabilities which can be more sensitive to small changes.

Table 3.
Forecast Probabilities for the South Carolina Outbreak

Cell	True	Method	Intensity 0	1	2	3	4	5	6
SCA	2	LNA	0.2475	0.0882	0.3309	0.1691	0.1249	0.0380	0.0014
		NAU	0.2291	0.0908	0.3712	0.1817	0.1049	0.0219	0.0004
		NAC	0.2282	0.0929	0.3709	0.1814	0.1044	0.0218	0.0004
		IS	0.2043	0.1057	0.3893	0.1896	0.0955	0.0155	0.0001
		EX	0.2082	0.1009	0.3915	0.1859	0.0969	0.0165	0.0001
SCB	0	LNA	0.9163	0.0294	0.0426	0.0078	0.0033	0.0006	0.0000
		NAU	0.9158	0.0294	0.0431	0.0079	0.0031	0.0006	0.0000
		NAC	0.9158	0.0301	0.0426	0.0079	0.0031	0.0006	0.0000
		IS	0.9287	0.0299	0.0337	0.0055	0.0019	0.0003	0.0000
		EX	0.9266	0.0296	0.0356	0.0058	0.0021	0.0003	0.0000
SCC	0	LNA	0.9280	0.0256	0.0365	0.0066	0.0027	0.0005	0.0000
		NAU	0.9270	0.0257	0.0373	0.0068	0.0027	0.0005	0.0000
		NAC	0.9270	0.0262	0.0368	0.0068	0.0027	0.0005	0.0000
		IS	0.9391	0.0257	0.0287	0.0047	0.0016	0.0003	0.0000
		EX	0.9371	0.0255	0.0304	0.0049	0.0018	0.0003	0.0000
SCD	5	LNA	0.0006	0.0004	0.0056	0.0142	0.0764	0.5353	0.3675
		NAU	0.0002	0.0001	0.0013	0.0035	0.0281	0.6839	0.2828
		NAC	0.0002	0.0001	0.0013	0.0035	0.0280	0.6845	0.2823
		IS	0.0001	0.0001	0.0008	0.0022	0.0229	0.7355	0.2385
		EX	0.0001	0.0001	0.0008	0.0023	0.0217	0.7398	0.2352
SCE	0	LNA	0.8862	0.0390	0.0583	0.0109	0.0046	0.0009	0.0000
		NAU	0.8873	0.0388	0.0580	0.0108	0.0043	0.0008	0.0000
		NAC	0.8872	0.0397	0.0573	0.0107	0.0043	0.0008	0.0000
		IS	0.9015	0.0408	0.0469	0.0077	0.0027	0.0004	0.0000
		EX	0.8989	0.0401	0.0494	0.0082	0.0029	0.0005	0.0000
SCF	0	LNA	0.7997	0.0629	0.1047	0.0215	0.0093	0.0018	0.0000
		NAU	0.8049	0.0628	0.1022	0.0203	0.0082	0.0015	0.0000
		NAC	0.8048	0.0642	0.1012	0.0201	0.0081	0.0015	0.0000
		IS	0.8183	0.0707	0.0891	0.0156	0.0055	0.0008	0.0000
		EX	0.8159	0.0684	0.0925	0.0163	0.0060	0.0009	0.0000
SCG	0	LNA	0.7899	0.0647	0.1098	0.0232	0.0103	0.0020	0.0001
		NAU	0.7953	0.0648	0.1078	0.0217	0.0087	0.0016	0.0000
		NAC	0.7952	0.0662	0.1067	0.0215	0.0086	0.0016	0.0000
		IS	0.8075	0.0737	0.0951	0.0169	0.0059	0.0009	0.0000
		EX	0.8056	0.0711	0.0983	0.0176	0.0064	0.0010	0.0000

Table 3. (continued)
Forecast Probabilities for the South Carolina Outbreak

Cell	True	Method	Intensity 0	1	2	3	4	5	6
SCH	3	LNA	0.0113	0.0076	0.0725	0.1173	0.2980	0.4369	0.0564
		NAU	0.0068	0.0039	0.0392	0.0905	0.3709	0.4740	0.0147
		NAC	0.0067	0.0040	0.0392	0.0904	0.3709	0.4744	0.0144
		IS	0.0041	0.0031	0.0308	0.0808	0.3934	0.4807	0.0071
		EX	0.0044	0.0031	0.0321	0.0836	0.3891	0.4803	0.0074
SCI	0	LNA	0.9062	0.0328	0.0480	0.0087	0.0036	0.0007	0.0000
		NAU	0.9062	0.0323	0.0480	0.0091	0.0037	0.0007	0.0000
		NAC	0.9063	0.0331	0.0473	0.0090	0.0036	0.0007	0.0000
		IS	0.9195	0.0332	0.0382	0.0064	0.0023	0.0003	0.0000
		EX	0.9173	0.0330	0.0401	0.0068	0.0024	0.0004	0.0000
SCJ	5	LNA	0.4247	0.1045	0.2907	0.1045	0.0603	0.0148	0.0005
		NAU	0.4197	0.1124	0.3115	0.1000	0.0470	0.0091	0.0002
		NAC	0.4189	0.1150	0.3106	0.0996	0.0467	0.0090	0.0002
		IS	0.4017	0.1365	0.3218	0.0956	0.0386	0.0059	0.0000
		EX	0.4060	0.1297	0.3222	0.0956	0.0401	0.0063	0.0001
SCK	4	LNA	0.0276	0.0168	0.1321	0.1698	0.3214	0.3046	0.0277
		NAU	0.0185	0.0107	0.0981	0.1777	0.4197	0.2697	0.0057
		NAC	0.0183	0.0109	0.0980	0.1778	0.4200	0.2695	0.0055
		IS	0.0124	0.0097	0.0860	0.1743	0.4548	0.2604	0.0024
		EX	0.0130	0.0093	0.0877	0.1783	0.4499	0.2592	0.0026

Forecast probabilities are only given for the case of the logistic error distribution. Recall that the results for the extreme-value error distribution were very similar to those of the logistic. Moreover, no convolution approximations have been found for the extreme value-normal case, as yet. Nevertheless, the other forecast probability estimates could be computed for the extreme value case.

7. SUMMARY

Figure 1 and the discriminant analysis support the use of the two variables UMAX and MDA in predicting which thunderstorm cells may produce tornadoes. Analysis of the censored regression models for the average intensity of the tornadoes produced by a cell showed that the conditions for tornadicity of a cell changed from outbreak to outbreak, but also that a third synoptic variable, surface vorticity could account for much of these changes. The ordinal regression model is best suited for analyzing the maximum intensity of the tornadoes, and verified some of the results of the censored regression. Five methods are suggested for computing forecast probabilities, two which require Monte Carlo integration, and a third, faster method, which does not. While the normal approximation to the posterior works reasonably well, the results differ enough that this fastest method can only be used as preliminary values for real-time calculation of forecast probabilities in the field.

8. REFERENCES

1. T. F. Fujita "Tornadoes and Downbursts in the Context of Generalized Planetary Scales," *J. Atmospheric Science* **38**, pp. 1511-1534 (1981).

2. C. E. Anderson and K. J. Schrab "The Use of Satellite Imagery to Identify Tornadic Thunderstorms," in *Preprints 15th Conf. Severe Local Storms*, AMS, Boston, (1988) pp. 186-189 (1988).

3. D. R. Perry "Forecasting the Tornadic Intensities of Thunderstorms by Multivariate Techniques" M.S. Thesis, North Carolina State University (1989).

4. K. J. Schrab "A Study of the Use of Satellite Imagery to Identify Tornadic Intensity of Thunderstorms" M.S. Thesis, University of Wisconsin-Madison (1988).

5. T. Amemiya "Tobit Models: A Survey," *J. Econometrics* **24**, pp. 3-61 (1984).

6. J. Tobin "Estimation of Relationships for Limited Dependent Variables," *Econometrica* **26**, pp. 24-36 (1958).

7. P. McCullagh "Regression Models for Ordinal Data," *J. Royal Statist. Soc., Ser. B*, **42**, pp. 109-142 (1980).

8. J. A. Anderson "Regression and Ordered Categorical Variables," *J. Royal Statist. Soc., Ser. B,* **46**, pp.1-30 (with discussion) (1984).

9. J. F. Monahan and L. A. Stefanski "Normal Scale Mixture Approximations to $F^{*}(z)$ and Computation of the Logistic-Normal Integral," in *Handbook of the Logistic Distribution*, N. Balakrishnan (Ed.), Marcel-Dekker, New York (1992).

10. J. Berntsen, T. O. Espelid, and A. Genz "An Algorithm for the Approximate Calculation of Multiple Integrals," *ACM Trans. Math. Software* **17**, pp. 437-451 (1991).

11. K. J. Schrab, C. E. Anderson and J. F. Monahan "Techniques Used to Identify Tornado Producing Thunderstorms Using Geosynchronous Satellite Data," *Proc. Sixth Conf. Satellite Meteor. and Oceanog.*, AMS pp. 159-162 (1992).

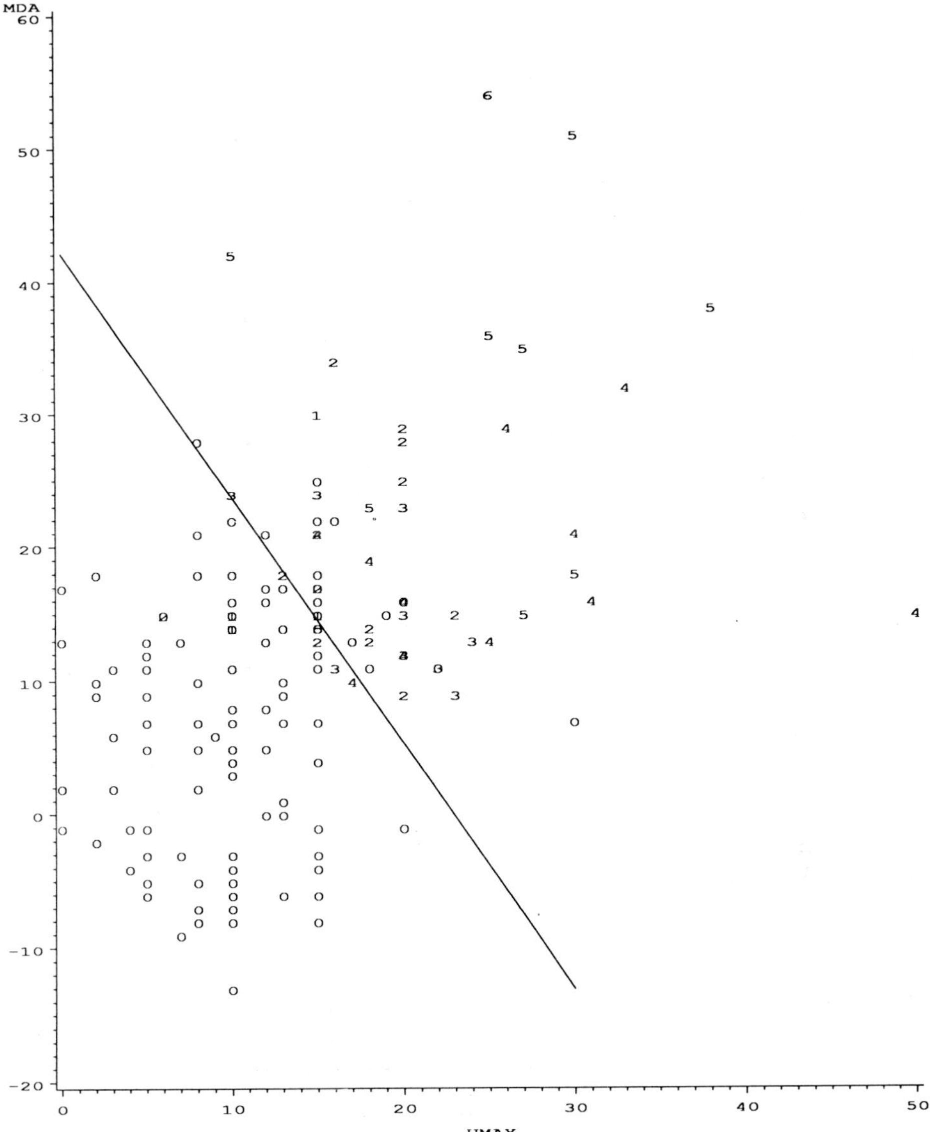

Figure 1. Tornado Intensity (FMAX) by UMAX and MDA. Line is linear discriminant line. Zeros denote non-tornadic cell.

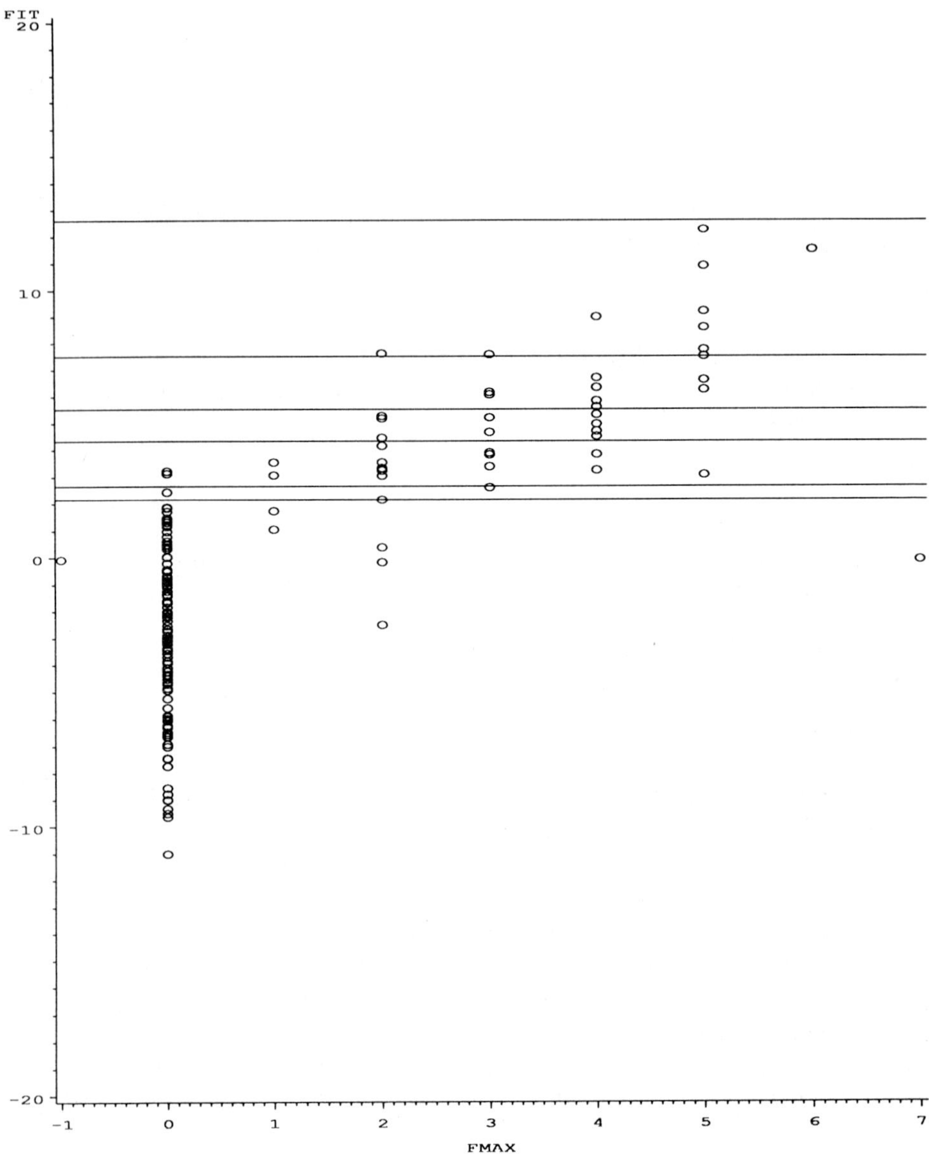

Figure 2. Fitted values from ordinal regression model by observed intensity FMAX. Horizontal lines are cutoff point estimates $\widehat{\theta}_k$.

Stat. Sci. & Data Anal., pp. 25-32
K. Matsusita *et al.* (Eds)

Incorporating Geographic Distribution into the Expected Number of Deaths in a Comparative Study

TAKASHI YANAGAWA* and DAVID G. HOEL**
* *Department of Mathematics, Kyushu University 33, Fukuoka 812, Japan*
** *Division of Biometry and Risk Assessment, National Institute of Environmental Health Sciences, Research Triangle Park, NC 27709, USA*

Abstract. A method is developed for incorporating geographic distribution into the expected number of deaths in the standardized mortality ratio (SMR) and comparative mortality figure (CMF). The impact of the geographic variation on these comparative measures is studied and shown to be substantial in an example.

Key words: standardized mortality ratio, comparative mortality figure, empirical Bayes

1. INTRODUCTION

In the analysis of data from a cohort study a comparison is undertaken of the observed number of deaths in a study group with the expected number of deaths which is calculated from the mortality rates in a reference population. This is a useful way of identifying diseases which occur at especially high or low frequency in the cohort, so that they may be studied further in relation to particular exposures (Breslow and Day [1]). In a comparison within an age group, simply the ratio of observed and expected is used. In a comparison across age groups, several age-time-specific measures have been introduced to facilitate the comparison. Among them the most frequently employed are the comparative mortality figure (CMF) and standardized mortality ratio (SMR). It is reported (Keiding [2]) that the SMR has been in service at least since 1786.

The mortality rates in the reference population are assumed constant in the calculation of the expected number of deaths. The assumption has been considered reasonable, since large geographic unit is typically employed for the reference population. However, the variation of mortality rates among geographic areas is well recognized, owing much to the contribution of geographic maps such as the atlas of cancer mortality among U.S. whites [3]. Clearly, we have uncertainty in the comparison which is caused by the selection of the reference population. The geographic variation of the mortality rates must be taken into account in the comparison of the observed and expected, if its impact is substantial.

The purpose of this paper is to develop a method for incorporating geographic distribution into the expected number of death, and to study the degree of impact of the geographic variation on the comparative measures. With practical data, it will be shown that this impact could be substantial.

2. MATHEMATICAL DEVELOPMENT

2.1 SMR AND CMF

Let d_j^* and n_j^* be the number of deaths and person-years at risk in the j-th age group in a study cohort, j=1,2,...,J. As Breslow and Day [1], we suppose that d_1^*, d_2^*,..., d_J^* are independent and that d_j^* follows a Poisson distribution with mean

$$E(d_j^*) = n_j^* \beta \lambda_{j1} \tag{2.1}$$

where β represents a common effect throughout all age groups and λ_{j1} represents an age specific mortality rate in a reference population. When λ_{j1}'s are known, the maximum likelihood estimator (m.l.e) of β throughout the age strata is given by

$$\hat{\beta}_o = \sum_{j=1}^{J} d_j^* / \sum_{j=1}^{J} n_j^* \lambda_{j1}.$$

Generally, λ_{j1} are unknown and are replaced by d_{j1} / n_{j1}, where d_{j1} and n_{j1} are the counter parts of d_j^* and n_j^* in a reference population. Then we have

$$SMR = \sum_{j=1}^{J} d_j^* / \sum_{j=1}^{J} n_j^* (d_{j1} / n_{j1}).$$

This estimator is called the standardized mortality ratio (SMR). The similar quantity

$$CMF = \sum_{j=1}^{J} n_{j1} (d_j^* / n_j^*) / \sum_{j=1}^{J} d_{j1}$$

is called the comparative mortality figure (CMF).

2.2 MATHEMATICAL MODEL

We consider variation of mortality rates among geographic units. Suppose that there are K+1 geographic units in the population, and denote by d_{jk} and n_{jk} the number of deaths and person-years at the j-th age group and in the k-th geographic unit. Without loss of generality we suppose that the first unit has been selected for the reference population. For simplicity, we concentrate ourself on the comparison of observed and expected number of deaths in one age group, and drop the suffix j in the sequel. We assume that within the k-th geographic unit the number of deaths d_k follows a Poisson distribution with mean

$$E(d_k \mid \lambda_k) = n_k \lambda_k , \quad \text{for } k=1,2,...,k+1,$$

and that $\{ d_k \}$ are conditionally independent conditioned on λ_k. For the adequacy of this Poisson assumption, see Manton and Stallard [4] and Brillinger [5]. Furthermore, we assume that the K+1 geographic units may be characterized by a realization of independent random variables λ_1, λ_2,...,λ_{K+1} with mean and variance given by

$$E(\lambda_k) = \theta \tag{2.2}$$

$$V(\lambda_k) = \alpha\,\theta / n_k\ ,\quad k=1,2,...,K+1. \tag{2.3}$$

Here the parameter α represents the variation of mortality rates among the geographic units; $\alpha=0$ shows homogeneity and larger the values of α, more heterogeneity is indicated among the mortality rates for fixed n_k. It is immediate to show that

$$E(d_k / n_k) = \theta,$$

$$V(d_k / n_k) = \theta(1+\alpha) / n_k\ .$$

It will be clear from these equations that the variance of λ_k is weighted by $1/n_k$ so that d_k / n_k converge to θ in probability as $n_k \rightarrow \infty$. Note that this was not assured in our previous paper (Yanagawa [6]).

2.3 EMPIRICAL BAYES ESTIMATE

Since that the population with suffix one, i.e., k=1, has been selected for the reference population, the expected mortality rate in the reference population is $E(\lambda_1)$. We consider the class of Bayes estimators

$$E(\lambda_1 \mid d_1) = A(d_1 / n_1) + b\ ,$$

where A and b are the quantities which are determined as follows:

$$\begin{aligned} \theta &= E(\lambda_1) = E\{ E(\lambda_1 \mid d_1)\} \\ &= E\{A(d_1 / n_1) + b\} = A\theta + b \end{aligned} \tag{2.4}$$

Furthermore

$$\begin{aligned} E\{E(\lambda_1 (d_1 / n_1))\} &= E\{E(\lambda_1 (d_1 / n_1) \mid \lambda_1)\} \\ &= E(\lambda_1^2) = (\alpha\theta / n_1) + \theta^2 \end{aligned} \tag{2.5}$$

On the other hand

$$\begin{aligned} E\{E(\lambda_1 (d_1 / n_1))\} &= E\{E(\lambda_1 (d_1 / n_1) \mid d_1)\} \\ &= E\{(d_1 / n_1)[A(d_1 / n_1) + b]\} \\ &= A\{[\theta(1+\alpha)/ n_1] + \theta^2\} + b\theta\ . \end{aligned} \tag{2.6}$$

From (2.4), (2.5) and (2.6) we have

$$A = \alpha/(1+\alpha)\ , \qquad b = (1 - A)\,\theta$$

Thus the Bayes estimate of the expected mortality rate in the reference population is

$$EB = A(d_1 / n_1) + (1 - A)\,\theta\ ,$$

which is a linear combination of the mortality rate in the reference population and the background rate. Alternatively, we may have the same estimator by assuming gamma distributions for λ's. Replacing λ_1 with this estimator, we have the following estimator of β:

$$\hat{\beta} = d^* / [n^* (EB)] \tag{2.7}$$

When the strata of age are taken into account, the following estimator of β is obtained throughout the strata:

$$\hat{\beta} = \sum_{j=1}^{J} d_j^* \left\{ \sum_{j=1}^{J} n_j^* \left[A_j \left(\frac{d_{j1}}{n_{j1}}\right) + (1 - A_j)\,\theta_j \right] \right\}^{-1},$$

When the mortality rates among the geographic units are homogeneous in all age groups, i.e., $\alpha_j = 0$ for j=1,2,..., J, then $A_j = 0$ and the estimator $\hat{\beta}$ shows no need of the data from the reference population. On the other hand, if the mortality rates among the geographic units are heterogeneous at any age so that α_j is large, A_j is close to one and the information from the reference population plays a dominant role. Similarly, the CMF is generalized to be

$$\hat{\beta} = \sum_{j=1}^{J} n_{j1}\left(\frac{d_j^*}{n_j^*}\right)\left\{\sum_{j=1}^{J} [A_j d_{j1} + n_{j1}(1 - A_j)\theta_j]\right\}^{-1}$$

The unknown parameters are involved in A and b. We estimate these parameters by a moment method using the data from the geographic units which are suffixed from k=2 to K+1. Putting

$$\bar{d}_j = d_{j+} / n_{j+}$$

$$S_j^2 = \frac{1}{K-1} \sum_{k=2}^{K+1} n_{jk}\left(\frac{d_{jk}}{n_{jk}} - \bar{d}_j\right)^2,$$

where + stands for the summation over the corresponding suffix, we have

$$E(\bar{d}_j) = \theta_j, \quad E(S_j^2) = (1 + \alpha_j)\theta_j$$

From this an estimator of A_j is given by $\hat{A}_j = 1 - (\bar{d}_j / S_j^2)$.

2.4 CHARACTERISTIC OF OF THE ESTIMATOR

Restricting to the case of a same age group, we suppose that A and θ are given constants. This could be justified since the population sizes of the geographic units are generally very large. Under this assumption the expectation of $\hat{\beta}$ in (2.7) is approximated by:

$$E(\hat{\beta}) \approx \beta\left[1 - \frac{\alpha A}{\theta n_1}\right] \tag{2.8}$$

Furthermore, assuming gamma distributions for λ and after some lengthy but simple computation an approximated variance of $\hat{\beta}$ is obtained as:.

$$V(\hat{\beta}) \approx \frac{\beta}{\theta n^*} + \frac{\beta^2}{\theta n_1} A \tag{2.9}$$

The bias and variance of this estimator in three important cases are separately discussed as follows:

(Case 1) We first note that the conventional estimator of β, which corresponds to the estimator $\hat{\beta}$ in (2.7) with A=1 and assumes the constant mortality rates in the reference population, has no bias and its variance is given by the first term of the right hand side of (2.9).

(Case 2) When we take into account the variation of the mortality rates in the reference population, but not the geographic distributions, the estimator is identical to the conventional estimator, but its approximate bias and variance are given by;

$$\text{bias} = (\alpha\beta)/(\theta n_1)$$

$$V(\hat{\beta}) = \frac{\beta}{\theta n^*} + \frac{\beta^2}{\theta n_1}$$

(Case 3) When both of the variation of the mortality rates in the reference population and its geographic distributions are taken into account, the estimator given in (2.7) has

$$\text{bias} = (\alpha\beta A) / (\theta n_1) \tag{2.10}$$

and the variance which is just give in (2.9).

Thus, denoting by MSE(1), MSE(2) and MSE(3) the mean squared error (MSE) of the estimator in Case 1, 2 and 3, respectively, the ratios MSE(2)/MSE(1) and MSE(3)/MSE (1) may be estimated as follows:

$$\frac{\text{MSE}(2)}{\text{MSE}(1)} = 1 + \frac{d^*}{d_1}\left(1 + \frac{\alpha^2}{d_1}\right) \tag{2.11}$$

$$\frac{\text{MSE}(3)}{\text{MSE}(1)} = 1 + \frac{d^*A}{d_1}\left(1 + \frac{\alpha^2 A}{d_1}\right). \tag{2.12}$$

The gain in incorporating the geographic distributions may be assessed by MSE(2) / MSE(3). It follows immediately from these equations that this gain is always larger than one, since $0<A<1$.

3. A NUMERICAL ILLUSTRATION

We use the Lee and Fraumeni [7] study of Montana copper smelter workers from the book of Breslow and Day [1], in which the latest follow-up data are given. In the Lee and Fraumeni study, 8047 male subjects were entered into study on 1 January 1938 if they had worked for at least one year and were still employed on that date, or at the end of their first year of employment for those hired later. The number of respiratory cancer deaths that occurred among the smelter workers at ages 40 - 79 in four calendar periods is listed in Table 2.8 in Breslow and Day [1]. These data are summarized and represented in Table 1.

Montana is selected as reference population, and the 48 states in the US, excluding Alaska and Hawaii, are taken as the geographic units. We did not collect the respiratory cancer mortality data in Montana from 1938 to 1977, but only collected the number of the white male respiratory cancer death in each state and four age groups in 1950 and 1970 from the Vital Statistics in the United States, and also collected the white male population of each state in four age groups in 1950 Census of population and 1970 Census of Population. For the purpose of illustration we ignore comparability and suppose that the 'expected number of deaths' is computed based respectively on 1950 and 1970 respiratory cancer mortality rates in Montana.

Table 2 summarizes the values of $\bar{d}_j$, S_j^2 and $1 - \hat{A}$ in each age stratum and

Table 1.

Respiratory Cancer Deaths (d*) and person-years at risk (n*) in the Montana Cohort, 1938 - 1977

age	person-years	deaths
40 - 49	49,363	21
50 - 59	41,811	80
60 - 69	24,394	117
70 - 79	9,423	58

Table 2.

Values of $\bar{d}_j$, S_j^2, α_j and $1 - \hat{A}_j$

age	year	$\bar{d}_j$ (x1000)	S_j^2 (x1000)	α_j	$1 - \hat{A}_j$
40-49	1950	0.190	0.402	1.114	0.473
	1970	0.348	1.099	2.155	0.317
50-59	1950	0.659	3.982	5.061	0.165
	1970	1.243	6.887	4.556	0.180
60-69	1950	1.145	10.402	8.091	0.110
	1970	2.866	48.755	15.949	0.059
70-79	1950	1.231	7.300	4.917	0.169
	1970	3.737	22.449	5.024	0.166

calender year which are computed from the geographic distributions of respiratory cancer deaths in 1950 and 1970 in the 48 states. The table shows that the variability of respiratory cancer mortality rates among 48 states increases monotonically till 69-70 age group and this variability is greater in 1970 than in 1950. Furthermore, the table shows substantially large values of 1 - A in younger age groups. Table 3 lists the age specific values of the estimated β and of their absolute bias.. Those crude estimates in the table are computed by means of the conventional method which assumes known mortality rates for the reference population, and adjusted estimates are computed by the method developed in this paper and the bias by using (2.10). Comparing these two estimates, it is shown that the impact of the geographic distributions could be substantial, in particular in the youngest age group, as is expected from the preceding table. Table 4 lists estimated ratios of the MSE in formulae (2.11) and (2.12), and the gain of incorporating the geographic distributions, i.e. MSE(2)/MSE(3).

5. DISCUSSION

The population at risk in the Montana smelter workers study was possible employees at the Montana smelter, and so ideally the proper reference group should have consisted

Table 3.

Crude[a] and Adjusted[b] estimates of β Adjusted for the geographic distributions

a: The conventional estimator of β, which assumes the constant mortality rates in the reference population

b: The estimator given in (2.7)

c: Obtained by (2.10)

age	year	crude[a]	adjusted[b]	|Bias|[c]
40-49	1950	3.13	2.63	.367
	1970	1.46	1.37	.195
50-59	1950	3.20	3.15	.752
	1970	1.77	1.73	.170
60-69	1950	5.15	5.02	1.425
	1970	2.29	2.24	.659
70-79	1950	6.35	6.07	.647
	1970	2.61	2.62	.342

Table 4.

The values of MSE(2)/MSE(1)[a], MSE(3)/MSE(1)[b] and gain in efficiency[c] in incorporating geographic distributions

a: computed by (2.11)
b: computed by (2.12)
c: c=b/a

age	year	MSE(2)/ MSE(1)[a]	MSE(3)/ MSE(1)[b]	gain in efficiency[c]
40-49	1950	6.24	3.50	1.78
	1970	3.72	2.68	1.39
50-59	1950	11.77	9.12	1.29
	1970	4.14	3.42	1.21
60-69	1950	16.83	13.98	1.20
	1970	14.26	12.86	1.11
70-79	1950	6.94	5.49	1.26
	1970	4.24	3.51	1.21

of those who were possibly employed but actually not at the smelter. However, in practice in epidemiologic study, it is frequently the case that the identification of such populations is not easy, or data are unavailable even if identified, and one is obliged to use the National Vital Statistics. If this is the case, the calculation of the expected number of death involves uncertainty in selecting a reference population. For example, in the original Montana study by Lee and Freumeni [7], the respiratory cancer mortality rates in Montana were used, but instead the three western states near the smelter, namely, Montana, Idaho and Wyoming, or the whole US could have been selected as a general population (Breslow and Day [1]). Naturally, the comparison of the observed and expected is influenced by the reference population which was selected in the study, because of the geographic variation of the mortality rates. The method developed in this paper resolves, although partially, this uncertainty, by incorporating the geographic distribution into the evaluation of the comparative measures. Note that we still have uncertainty in selecting the geographic units. For example, states were selected in our illustrative example because the data were relatively easy to access, but the respiratory cancer mortality data in counties might have been more reasonable.

We have shown that the geographic distribution can inflate the variance of comparative measures substantially, in other words the conventional method could be substantially anti-conservative. We note that, according to the atlas of U.S. cancer mortality among whites [8], the geographic distribution of the respiratory cancer which was used to show the impact in this paper is relatively uniform.

REFERENCES:

1. N.E. Breslow and N.E. Day, *Statistical Methods in Cancer Research: Vol II - The Design and Analysis of Cohort Studies*, IARC Scientific Publications No. 82, International Agency for Research on Cancer, Lyon (1987).
2. N. Keiding, *Int. Statist. Rev.*, **55**, 1 - 20 (1987).
3. L.W. Pickle, T.J. Mason, R. Hoover and J.F. Freumeni, Jr. *Atlas of U.S. cancer mortality among*

whites: 1950 - 1980, U.S. Department of Health and Human Services, Public Health Service, National Institute of Health, DHHS Publication No. (NIH) **87-2900** (1987).

4. K.G. Manton, N.A. Woodbury, E. Stalland, W. Riggan, J.P. Creason and A.C. Pellom, *J. of American Statistical Assoc.*, **84**, 637 - 650 (1989).
5. D.R. Brillinger, *Biometrics*, **42**, 693 -734 (1986).
6. T. Yanagawa, *Bull. Biometric Soc. of Japan*, **11**, 107 - 114, (1990).
7. A.M. Lee and J.F. Fraumeni, Jr. *J. Natl Cancer Inst.*, **42**, 1045 - 1052 (1969).
8. J.H. Lubin, L.M. Pottern, W.J. Blot, S. Tokudome, B.J. Stone and J.F. Fraumeni, Jr. *J. Occup. Med.*, **23**, 79 - 784 (1981).

Stat. Sci. & Data Anal., pp. 33-39
K. Matsusita *et al.* (Eds)

Fitting Random Coefficient Regression Models

RUDOLF BERAN
Department of Statistics, University of California, Berkeley, CA 94720, U.S.A.

Abstract. Linear regression models with random coefficients allow each individual sampled to have a different linear response function. Moreover, certain heteroscedastic linear models, random effects models, and deconvolution models can be viewed as instances of random coefficient regression. For both reasons, random coefficient models are becoming increasingly important in econometrics and statistics. This paper surveys recent work on statistical inference for such models, in a nonparametric or semiparametric setting. The focus is on estimating the joint distribution of the random coefficients, on testing the goodness-of-fit of the model, and on developing computational algorithms for these statistical procedures.

Key words: Minimum distance, empirical characteristic function, random effects, deconvolution, bootstrap.

1. INTRODUCTION

Recent research in both statistics and econometrics has called increasing attention to random coefficient regression models of the form

$$Y_i = A_i + X_i B_i \,. \tag{1.1}$$

Here Y_i and A_i are $p \times 1$ random vectors, B_i is a $q \times 1$ random vector, and X_i is a $p \times q$ random matrix. The triples $\{(A_i, B_i, X_i)\}$ are iid and (A_i, B_i) is independent of X_i for every value of i. The distribution of (A_i, B_i, X_i) is unknown, though it may be restricted in some applications. The sample S_n that we observe consists of the n pairs $\{(Y_i, X_i) : 1 \leq i \leq n\}$.

Model (1.1) may be interpreted as a multivariate linear regression model with random regressors and structured heteroscedastic errors. In the econometric literature, models like (1.1) have been used to investigate heteroscedasticity (cf. Hildreth and Houck [1], Goldfeld and Quandt [2], Chapter 3, and Amemiya [3]) and to analyze panel data (cf. Hsiao [4]). Recent surveys of work on random coefficient regression models, on their autoregressive analogs, and on models combining both features are given by Raj and Ullah [5], Chow [6], Nicholls and Pagan [7], and Newbold [8].

In the statistical literature, several special cases of model (1.1) are well established under several different labels. When the distribution of B_i is supported on one unknown point, then (1.1) is a multivariate linear model with random regressors and homoscedastic errors. When the distribution of X_i is supported on one known point, then (1.1) includes the random effects models of ANOVA (cf. Scheffé [9], Chapter 7) and the models studied in nonparametric deconvolution (cf. Fan [10], van Es [11]). When the $\{X_i : 1 \leq i \leq n\}$

are not observed but their distribution is known, then (1.1) becomes an affine mixture model.

This paper surveys some recent results on statistical inference in model (1.1). Section 2 discusses consistent nonparametric estimation of F_{AB}, the unknown distribution of (A_i, B_i), from the sample S_n. The emphasis here is on a minimum distance estimator $\hat{F}_{AB,n}$ recently developed in Beran and Millar [12]. The minimum distance approach is well-suited to testing the goodness-of-fit of a nonparametric random coefficient regression model, and has a tractable asymptotic theory. An immediate application of a consistent estimator for F_{AB} is to construct prediction regions for the future observable X_{n+1} in model (1.1), given the sample S_n and the condition that $X_{n+1} = x$. For details, see [12]. Earlier, Beran and Hall [13] proposed a method-of-moments technique for estimating consistently the marginal distributions of A_i and of B_i, when these are independent scalar random variables. However, it seems difficult to extend their results to multivariate cases or to establish rates of convergence for their estimators.

Section 3 gives new results on the asymptotic normality of a particular minimum distance estimator $\hat{F}_{AB,n}$. A key assumption is that the support of F_{AB} is finite, with known cardinality, and is restricted to a given compact, but is otherwise unknown. Neither the support points nor their probabilities are known. This assumption on support seems harmless from a practical point of view, because the number of support points can be taken large. Technically, the finite support assumption turns (1.1) into an interesting, yet tractable, semiparametric model under which $\hat{F}_{AB,n}$ is $n^{-1/2}$-consistent, as is shown in [12]. Section 3 also presents a goodness-of-fit test for the semiparametric form of model (1.1), based upon the minimum distance criterion that is used to define $\hat{F}_{AB,n}$ in the preceding section. The necessary critical values can be obtained by a suitable bootstrap algorithm.

The questions of how to calculate the estimator $\hat{F}_{AB,n}$ and how to bootstrap it efficiently are discussed briefly in Section 4. While highly computer-intensive, the methods proposed above are within the capability of modern scientific workstations.

2. CONSISTENT ESTIMATION OF F_{AB}

Let us introduce the following notations in model (1.1):

F_{AB} denotes the joint distribution of (A_i, B_i), which is restricted to a family of distributions $\mathcal{F}_{AB}$ on R^{p+q};

F_X denotes the distribution of X_i, which is restricted to a family of distributions $\mathcal{F}_X$ on R^{pq};

$P(F_{AB}, F_X)$ denotes the joint distribution of (Y_i, X_i) in model (1.1);

d denotes any metric for weak convergence of probabilities on R^{p+q}.

A sequence of distributions in $\mathcal{F}_{AB}$ will be indicated by $\{F_{AB,n}\}$, and similarly for a sequence in $\mathcal{F}_X$.

Let $\hat{P}_n$ be the empirical distribution which gives mass n^{-1} to each element of the sample S_n; and let $\hat{F}_{X,n}$ be the empirical distribution of the $\{X_i : 1 \leq i \leq n\}$. Suppose $\hat{F}_{AB,n}$ is any value of F_{AB} in $\mathcal{F}_{AB}$ that minimizes, up to an error of $o_p(n^{-1/2})$, the criterion

$$d[P(F_{AB}, \hat{F}_{X,n}), \hat{P}_n].$$

We will call any such $\hat{F}_{AB,n}$ a minimum distance estimator for F_{AB}. Let C' denote the transpose of any matrix C and let *supp*(G) denote the support of any distribution G.

Proposition 1. *Assume that $\mathcal{F}_{AB}$ consists of probabilities supported by a given compact and that $\{x't : x \in supp(F_X)\}$ contains an open set in R^q for every $t \neq 0$ in R^p. Suppose that the true distributions in model (1.1) at sample size n are given by $F_{AB,n}$ and $F_{X,n}$, where $d(F_{AB,n}, F_{AB,0}) \to 0$ and $d(F_{X,n}, F_{X,0}) \to 0$. Then*

$$d(\hat{F}_{AB,n}, F_{AB,0}) \to 0 \text{ in probability.}$$

This result is proved in [12]. The triangular array formulation entails that the convergence in probability of $\hat{F}_{AB,n}$ to F_{AB} is uniform over every d-compact subset of $\mathcal{F}_{AB}$. Almost sure convergence also holds, though only pointwise. The support assumptions in Proposition 1 ensure strong identifiability of the distribution F_{AB} from the distribution of (Y_i, X_i) under model (1.1). Such identifiability is, of course, a prerequisite to consistent estimation.

Proposition 1 allows great freedom in the selection of the metric d that determines the minimum distance method. How might we make a reasonable choice of d? Here are several points to consider:

a) To ensure consistency of $\hat{F}_{AB,n}$, the distance d should metrize weak convergence of distributions on R^{p+pq}.
b) For practicability, the distance d should be relatively easy to calculate.
c) To facilitate asymptotic analysis of $\hat{F}_{AB,n}$ beyond consistency, the distance d should be generated by a norm on a linear space.
d) A Hilbertian norm is especially convenient from the standpoint of (b) and (c).

These considerations motivate the following definition of d. Suppose P_1 and P_2 are any two distributions on R^{p+pq}, with characteristic functions $\phi_1(t,u)$ and $\phi_2(t,u)$ respectively, where $t \in R^p$ and $u \in R^{pq}$. Let $\|\cdot\|$ denote the $L_2(Q)$-norm on such characteristic functions, where Q is a probability on R^{p+pq} that has full support. Define

$$d(P_1, P_2) = \|\phi_1 - \phi_2\| . \tag{2.1}$$

Evidently, this d metrizes weak convergence.

The application to minimum distance estimation requires the characteristic functions of $\hat{P}_n$ and of $P(F_{AB}, \hat{F}_{X,n})$. The first of these is

$$\hat{\phi}_n(t,u) = n^{-1} \sum_{j=1}^{n} \exp[i\langle t, Y_j\rangle + i\langle u, X_j\rangle],$$

where $\langle\cdot,\cdot\rangle$ denotes the appropriate inner product. The other characteristic function is

$$\hat{\psi}_{n,AB}(t,u) = n^{-1} \sum_{j=1}^{n} \phi_{AB}(t, X_j't) \exp[i\langle u, X_j\rangle],$$

where ϕ_{AB} denotes the characteristic function of F_{AB}. With d as in (2.1), the minimum distance estimator $\hat{F}_{AB,n}$ is the value of F_{AB} in $\mathcal{F}_{AB}$ that minimizes $\|\hat{\phi}_n - \hat{\psi}_{n,AB}\|$, up to an error of $o_p(n^{-1/2})$. Section 4 discusses the computation of good approximations to this minimum distance estimator.

3. ASYMPTOTIC NORMALITY AND GOODNESS-OF FIT

This section treats two further questions: What is the asymptotic distribution of the minimum distance estimator $\hat{F}_{AB,n}$? What is the distribution under model (1.1) of the minimized distance $d(\hat{P}_n, \hat{F}_{AB,n})$? The first question is important for inferences about F_{AB}, such as confidence sets, while the second question concerns testing the fit of model (1.1) to the sample. Throughout this section, the metric d is that generated by the $L_2(Q)$-norm on characteristic functions; and the distribution F_{AB} is restricted as follows, for both technical and computational reasons (cf. Section 4):

> F_{AB} is a discrete distribution supported on r unknown sites $c_1, \ldots, c_r$, where $c_i = (a_i, b_i) \in R^{p+q}$. The number r is known. The probabilities $p_i = F_{AB}(\{c_i\}), 1 \leq i \leq r$, are unknown. The sites $\{c_i\}$ are distinct and lie within a known compact set K.

Thus, F_{AB} can be parametrized as $F_{AB}(\theta)$, where $\theta = (p_1, \ldots, p_r, c_1, \ldots, c_r)$, with $p_i \geq 0$ for every i and $\sum_{i=1}^r p_i = 1$. The distribution of (Y_i, X_i) under model (1.1) is then $P(F_{AB}(\theta), F_X)$, a semiparametric model in which the pair (θ, F_X) is unknown.

Let $\psi(\theta, F_X)$ denote the characteristic function of $P(F_{AB}(\theta), F_X)$:

$$\psi(t, u; \theta, F_X) = E\{\phi_{AB,\theta}(t, X_1' t) \exp[i\langle u, X_1\rangle]\},$$

where

$$\phi_{AB,\theta}(t, u) = \sum_{k=1}^{r} p_k \exp[i\langle t, a_k\rangle + i\langle u, b_k\rangle].$$

The minimum distance estimator of $F_{AB}(\theta)$ is now $F_{AB}(\hat{\theta}_n)$, where $\hat{\theta}_n$ minimizes $\|\hat{\phi}_n - \psi(\theta, \hat{F}_{X,n}\|$ over all possible θ, up to an error of order $o_p(n^{-1/2})$. Let T_n denote the minimized distance $\|\hat{\phi}_n - \psi(\hat{\theta}_n, \hat{F}_{X,n})\|$. Relatively large values of T_n are evidence that model (1.1) does not fit the sample.

Let $H_n(\theta, F_X)$ denote the distribution of $n^{1/2}(\hat{\theta}_n - \theta)$ and let $J_n(\theta, F_X)$ denote the distribution of $n^{1/2}T_n$, both calculated under the version of model (1.1) introduced in this section. Let η_0 be the derivative with respect to θ_0, in the $L_2(Q)$-norm, of the characteristic function $\psi(\theta_0, F_{X,0})$. This derivative is viewed as an $r \times 1$ vector-valued function on R^{p+pq}.

Proposition 2. *Suppose that the true distributions in model (1.1) at sample size n are given by $F_{AB}(\theta_n)$ and $F_{X,n}$, where $\{n^{1/2}(\theta_n - \theta_0)\}$ is a bounded sequence and $F_{X,n} \Rightarrow F_{X,0}$. Then*

$$H_n(\theta_n, F_{X,n}) \Rightarrow \mathcal{L}\{[\int \Re(\eta_0 \bar{\eta}_0')dQ]^{-1} \int \Re(\bar{\eta}_0 W_0)dQ\} = H(\theta_0, F_{X,0})$$

and

$$J_n(\theta_n, F_{X,n}) \Rightarrow \mathcal{L}\{\inf_s \|W_0 - \langle s, \eta_0\rangle\|\} = J(\theta_0, F_{X,0}),$$

where W_0 is a complex-valued process on R^{p+pq} whose real and imaginary parts form a bivariate gaussian process, with mean zero and covariance function depending upon $(\theta_0, F_{X,0})$.

A proof of this result and a complete description of the process W_0 appears in Beran [14]. The argument draws on ideas from Pollard [15]. The Proposition implies that the limiting distribution $H(\theta_0, F_{X,0})$ of $n^{1/2}(\hat{\theta}_n - \theta_n)$ is multivariate normal while the limiting distribution of nT_n^2 is the distribution of a weighted sum of independent chi-squared random variables. Both limit distributions depend on the unknown true $(\theta_0, F_{X,0})$ in a very complex way.

More usefully for statistical inference, Proposition 2 further implies that the natural bootstrap distributions $H_n(\hat{\theta}_n, \hat{F}_{X,n})$ and $J_n(\hat{\theta}_n, \hat{F}_{X,n})$ both converge in probability to the correct limit distributions. Moreover, convergence of a sequence $\{\theta_n\}$ to θ implies weak convergence of the distributions $\{F_{AB}(\theta_n)\}$ to $F_{AB}(\theta)$. Thus, confidence bands for F_{AB} can be derived from simultaneous bootstrap confidence intervals for the components of θ and have the intended asymptotic coverage probabilities. Additionally, the test which rejects model (1.1) whenever $n^{1/2}T_n$ exceeds the α-th quantile of $J_n(\hat{\theta}_n, \hat{F}_{X,n})$ has asymptotic probability α of rejection under the null hypothesis that the model (1.1) holds. See [14] for details of these applications.

4. COMPUTATIONAL METHODS

Let us consider first the calculation of the minimum distance estimator $\hat{F}_{AB,n}$ under the nonparametric form of model (1.1) that is used in Section 2. Minimizing the distance criterion over all distributions F_{AB} in $\mathcal{F}_{AB}$ is usually not feasible. Good approximate solutions may be found by restricting the minimization to distributions in a rich subclass $\mathcal{F}_{AB,0}$ of $\mathcal{F}_{AB}$. Two such choices for $\mathcal{F}_{AB,0}$ are:

a) Take $\mathcal{F}_{AB,0}$ to be the semiparametric model used in Section 3. The minimization is now over the r sites $\{c_i\}$ and the r probabilities $\{p_i\}$ that define the parameter θ.
b) Take $\mathcal{F}_{AB,0}$ to be the simpler semiparametric model in which each $p_i = 1/r$ and the corresponding sites are unknown. The minimization is now over the r sites $\{c_i\}$, which need not be distinct.

The minimization in either (a) or (b) may be accomplished by a general purpose routine for functions of several variables (cf. Press et al. [16], Chapter 10). Paper [12] gives an explicit algorithm and a numerical example for computational strategy (b). Proposition 2 and Proposition 3 below remain valid for semiparametric model (b) as well as for model (a).

Let us consider next approximations to the two bootstrap distributions $H_n(\hat{\theta}_n, \hat{F}_{X,n})$ and $J_n(\hat{\theta}_n, \hat{F}_{X,n})$ that were introduced at the end of Section 3. Suppose that the subclass $\mathcal{F}_{AB,0}$ is defined by (a) above. In principle, Monte Carlo approximations to the desired bootstrap distributions require repeatedly drawing a sample S_n^* of size n from the fitted model $P(F_{AB}(\hat{\theta}_n), \hat{F}_{X,n})$ and then calculating the minimum distance estimator θ_n^* for each such sample. In practice, repeated minimizations in high dimensions are extremely time-consuming. However, a good approximation to θ_n^*, which is accurate up to order $o_p(n^{-1/2})$, is found by using only one step of the Newton-Raphson minimization algorithm for θ_n^*, with the original minimum distance estimator $\hat{\theta}_n$ taken as starting point.

More precisely, let $\bar{\theta}_n^*$ denote this linearized approximation to θ_n^*. Correspondingly, let $\bar{H}_n(\hat{\theta}_n, \hat{F}_{X,n})$ denote the conditional distribution of $n^{1/2}(\bar{\theta}_n^* - \hat{\theta}_n)$, given the sample S_n. Let $\bar{J}_n(\hat{\theta}_n, \hat{F}_{X,n})$ denote the conditional distribution of $n^{1/2}\|\phi_n^* - \psi(\bar{\theta}_n^*, F_{X,n}^*\|$, given the sample S_n. Here $(\phi_n^*, F_{X,n}^*)$ are the values of $(\hat{\phi}_n, \hat{F}_{X,n})$ recalculated from the bootstrap sample S_n^*.

Proposition 3. *Suppose that the assumptions for Proposition 2 hold. Then*

$$\bar{H}_n(\hat{\theta}_n, \hat{F}_{X,n}) \Rightarrow H(\theta_0, F_{X,0}) \quad \textit{in probability}$$

and

$$\bar{J}_n(\hat{\theta}_n, \hat{F}_{X,n}) \Rightarrow J(\theta_0, F_{X,0}) \quad \textit{in probability,}$$

both convergences occurring under $P(F_{AB}(\theta_n), F_{X,n})$. *Moreover,*

$$n^{1/2}(\bar{\theta}_n^* - \theta_n^*) \to 0 \quad \textit{in probability}$$

under the joint distribution of (S_n, S_n^*).

A proof of this result appears in [14]. It follows from Propositions 2 and 3 that the estimated distributions $\bar{H}_n(\hat{\theta}_n, \hat{F}_{X,n})$ and $\bar{J}_n(\hat{\theta}_n, \hat{F}_{X,n})$ can be validly used as substitutes for the bootstrap distributions $H_n(\hat{\theta}_n, \hat{F}_{X,n})$ and $J_n(\hat{\theta}_n, \hat{F}_{X,n})$, when constructing confidence bands for F_{AB} and goodness-of-fit tests for model (1.1). The remarks in the second paragraph after Proposition 2 apply to these substitutes as well as to the bootstrap distributions. This fact is of practical importance because the distributions $\bar{H}_n(\hat{\theta}_n, \hat{F}_{X,n})$, $\bar{J}_n(\hat{\theta}_n, \hat{F}_{X,n})$ are much easier to approximate by Monte Carlo methods than are the respective bootstrap distributions. Schucany and Wang [17] introduced the idea of using one-step approximations in bootstrapping iterative estimators.

Acknowledgement
This research was supported in part by National Science Foundation Grant DMS 9001710.

REFERENCES

1. C. Hildreth and J.P. Houck, Some estimators for a linear model with random coefficients, *J. Amer. Statist. Assoc.* **63**, 584–595 (1968).
2. S.M. Goldfeld and R.E. Quandt, *Nonlinear Methods in Econometrics.* North-Holland, Amsterdam (1972).
3. T. Amemiya, A note on a heteroscedastic model, *J. Econometrics* **6**, 365–370 (1977).
4. C. Hsiao, *Analysis of Panel Data.* Econometric Society Monographs, Cambridge Univ. Press, Cambridge (1986).
5. B. Raj and A. Ullah, *Econometrics, A Varying Coefficients Approach.* Croom-Helm, London (1981).
6. G.C. Chow, Random and changing coeficient models, in: *Handbook of Econometrics*, Z. Griliches and M.D. Intrilgator (Eds.), vol. 2. North-Holland, Amsterdam (1983).
7. D.F. Nicholls and A.R. Pagan, Varying coefficient regression, in: *Handbook of Statistics*, E.J. Hannan, P.R. Krishnaiah, M.M. Rao (Eds.), vol. 5. North-Holland, Amsterdam (1985).
8. P. Newbold, Some recent developments in time series analysis—III, *Internat. Statist. Rev.* **56**, 17–29 (1988).
9. H. Scheffé, *The Analysis of Variance.* Wiley, New York (1959).
10. J. Fan, On the optimal rates for nonparametric deconvolution problems, *Ann. Statist.* **19**, 1257–1272 (1991).
11. A.J. van Es, Uniform deconvolution: nonparametric maximum likelihood and inverse estimation, in: *Nonparametric Functional Estimation and Related Topics*, G. Roussas (Ed.). Kluwer Academic Publishers, Dordrecht (1991).

12. R. Beran and P.W. Millar, Minimum distance estimation in random coefficient regression, submitted for publication (1991).

13. R. Beran and P. Hall, Estimating coefficient distributions in random coefficient regressions, *Ann. Statist.* **20**, in press (1992).

14. R. Beran, Semiparametric random coefficient regression models, submitted for publication (1992).

15. D. Pollard, The minimum distance method of testing, *Metrika* **27**, 43–70 (1980).

16. W.H. Press, B.P. Flannery, S.A. Teukolsky, and W.T. Vetterling, *Numerical Recipes: The Art of Scientific Computing.* Cambridge Univ. Press, Cambridge (1986).

17. W.R. Schucany and S. Wang, One-step bootstrapping for smooth iterative procedures, *J. Roy. Statist. Soc. B* **53**, 587–596 (1991).

Stat. Sci. & Data Anal., pp. 41-48
K. Matsusita *et al.* (Eds)

Design and Analysis in Repeated Measurement Experiment : on Use of Preliminary Knowledge about Covariance Structure of Observations

YASUNORI URAGARI and MASASHI GOTO

Shionogi Kaiseki Center, Shionogi & Co.,LTD.,22–41, 1–chome, Izumicho, Suita City, Osaka, Japan

Abstract. In this paper, we consider a method to determine required sample sizes in repeated measurement experiments, using preliminary knowledge on covariance structure of observations. Especially, we pay attention to two tests concerning to time effect and treatment–by–time interaction in repeated measures analysis of variance with degrees of freedom adjusted. We show that if the number of time points increases or the correlation between observations decreases then the required sample size becomes large, and that if the number of time points decreases or the correlation increases then the required sample size becomes small. We also show that the required sample size could be underestimated without the degrees of freedom adjusted. In the repeated measurement experiment for evaluating the degree and the durability of the effect of a hypotensive drug, we clarify the alternative hypothesis to be tested and calculate the required sample size.

Key words : Adjustment of degrees of freedom, ANOVA, covariance structure, design of experiment, preliminary knowledge, repeated measurement experiment, sample size.

1. INTRODUCTION

Repeated measurements are often taken in experimental studies, namely, repeated measurement experiments (RME, hereafter). It is widely said that success or failure of an experiment depends on its design, not excepting the RME(Nakahira[1], Uragari and Goto[2]). However, so far few literature has paid attention to the design of RME and the determination of sample size. One reason could be that the properties on the power of the mixed model univariate analysis of variance(ANOVA, hereafter), which is commonly used for analyzing repeated measurements, had not been clear until the research work by Muller and Barton [3], and any method to determine a required sample size had not been established. For the multivariate analysis of variance, the calculation of required sample size is investigated by Vonesh and Schork [4] for the case of single group and by Rochon [5] for the case of two groups, respectively. Another reason could be the difficulty to specify suitable time points for measuring response and hypotheses to be tested, without sufficient understanding of each phenomenon. In fact, each response index, such as blood pressure or bodily temperature, has its own desirable temporal change. In this paper, we discuss the design of the RME to evaluate the effect of a treatment or to compare the effect between two treatments. Here, it is assumed that we intend to apply the ANOVA, and use the test of time effect and one of treatment-by-time interaction in cases of single and two treatments, respectively. In fact, the univariate ANOVA has higher power than the multivariate analysis of variance. And, the results from the univariate ANOVA could be relatively easily interpreted. First, we explain a method to determine the required sample size in the RME, using preliminary knowledge on covariance structure of observations. Then, we clarify the influences of the number of time points and the correlation between repeated measurements observed on the required sample size. Finally, for an example, we present a design of the RME for evaluating the degree and the durability of the effect of a hypotensive drug.

2. ANOVA AND THE ADJUSTMENT OF DEGREES OF FREEDOM

Let Y_{ijk} be a response of j-th individual in i-th treatment group at k-th time point in an experiment with g treatments and τ time points($i=1,\cdots,g : j=1,\cdots,n_i : k=1,\cdots,\tau$), where n_i is the number of subjects in the i-th group, and $N=n_1+\cdots+n_g$. The model of the ANOVA is described as

$$Y_{ijk}=\mu+\alpha_i+\pi_{ij}+\beta_k+\gamma_{ik}+e_{ijk}, \quad i=1,\cdots,g : j=1,\cdots,n_i : k=1,\cdots,\tau, \tag{1}$$

where μ is the grand mean, α_i is the main effect of i-th treatment, β_k is the main effect of k-th time point, γ_{ik} is the treatment-by-time interaction effect, respectively. And, π_{ij} and e_{ijk} are random variables which denote variabilities due to subjects and experimental errors. In this model, it is assumed that $\boldsymbol{u}_{ij}=(u_{ij1},\cdots,u_{ij\tau})^{T}\sim \text{NID}(\mathbf{0},\Sigma)$, where $u_{ijk}=\pi_{ij}+e_{ijk}$, $\mathbf{0}$ is a τ-dimensional zero vector, and $\Sigma=\{\sigma_{kl}\}$is a $\tau\times\tau$ non-singular matrix. Let F_1, F_2 and F_3 be test statistics concerning to $H_1:\{\alpha_i=0,\ i=1,\cdots,g\}$, $H_2:\{\beta_k=0,\ k=1,\cdots,\tau\}$ and $H_3:\{\gamma_{ik}=0,\ i=1,\cdots,g: k=1,\cdots,\tau\}$, respectively. Geisser and Greenhouse [6] showed that

$$\begin{aligned} &F_1\sim F[g-1,\ N-g] \text{ under } H_1,\\ &F_2\sim_{ap} F[(\tau-1)\varepsilon,(N-g)(\tau-1)\varepsilon] \text{ under } H_2, \text{ and}\\ &F_3\sim_{ap} F[(g-1)(\tau-1)\varepsilon,(N-g)(\tau-1)\varepsilon] \text{ under } H_3, \end{aligned} \tag{2}$$

where $F[\nu_1,\nu_2]$ is the central F distribution with degrees of freedom (ν_1,ν_2) and "$\sim_{ap}$"denotes "being approximately distributed". The ε is called the adjusting factor of d.f. and defined as

$$\varepsilon=\{\mathrm{tr}(\Sigma-\boldsymbol{E}\Sigma)\}^2/\{(\tau-1)\mathrm{tr}(\Sigma-\boldsymbol{E}\Sigma)^2\}, \tag{3}$$

where $\boldsymbol{E}$ is a $\tau\times\tau$ matrix of all the elements $1/\tau$. We consider $H_1^{(A)}:\{\alpha_i\neq 0$ for some $i\}$, $H_2^{(A)}:\{\beta_k\neq 0$ for some $k\}$ and $H_3^{(A)}:\{\gamma_{ik}\neq 0$ for some$(i, k)\}$ as the alternative hypotheses against H_1, H_2 and H_3, respectively. Muller and Barton [3] showed that

$$F_1\sim F'[g-1, N-g, (g-1)f_{A1}] \text{ under } H_1^{(A)},$$
$$F_2\sim_{as} F'[(\tau-1)\varepsilon, (N-g)(\tau-1)\varepsilon, (\tau-1)f_{A2}\varepsilon] \text{ under } H_2^{(A)}, \text{ and} \tag{4}$$
$$F_3\sim_{as} F'[(g-1)(\tau-1)\varepsilon, (N-g)(\tau-1)\varepsilon, (g-1)(\tau-1)f_{A3}\varepsilon] \text{ under } H_3^{(A)},$$

where $F'[\cdot, \cdot, \delta]$ is the non-central F distribution with non-central parameter δ and "$\sim_{as}$" denotes "being asymptotically distributed". In addition,

$$f_{A1}=\tau\sum_{i=1}^{g}n_i\alpha_i^2/\{(g-1)\mathrm{tr}(\boldsymbol{E}\Sigma)\},$$
$$f_{A2}=N\sum_{k=1}^{\tau}\beta_k^2/\{\mathrm{tr}\Sigma-\mathrm{tr}(\boldsymbol{E}\Sigma)\},$$
$$f_{A3}=\sum_{i=1}^{g}n_i\sum_{k=1}^{\tau}\gamma_{ik}^2/[(g-1)\{\mathrm{tr}\Sigma-\mathrm{tr}(\boldsymbol{E}\Sigma)\}].$$

The adjusting factor ε can take the value between $1/(\tau-1)$ and 1. In case $\tau=2$, $\varepsilon=1$. In case $\tau>2$, $\varepsilon=1$ if and only if Σ is compound symmetric(i.e. , has equal variances and equal covariances). Being wrongly assumed to be $\varepsilon=1$ when Σ is not compound symmetric, the type I error rates of tests concerning to H_2 and H_3 will be inflated. To avoid the inflation, ε is estimated from observations and the degrees of freedom of the null distributions are reduced based on the estimate(adjustment of d.f.).

3. REQUIRED SAMPLE SIZE IN RME

3.1 Method to Determine Required Sample Size

From now on we explain a method to determine the required sample size in a RME. Concretely, we consider the experimental situation coming under the following (1)~(4).

(1) Numbers of treatments and time points are g and τ, respectively. Sample size of i-th treatment is $n_i=c_in$, where c_i is a constant $(i=1, \cdots, g)$. When the sample sizes of all g treatments are equal, we put $c_1=c_2=\cdots=c_g=1$. Let us denote the "forthcoming" observations be $\{Y_{ijk}, i=1, \cdots, g : j=1, \cdots, n_i : k=1, \cdots, \tau\}$.

(2) Power of the test concerning to the interaction(the test of H_3 toward $H_3^{(A)}$)at significance level α is P_w^* or more, where P_w^* is a required power.

(3) In the above test, estimator of ε

$$\hat{\varepsilon}=\{\mathrm{tr}(\boldsymbol{S}-\boldsymbol{ES})\}^2/\{(\tau-1)\mathrm{tr}(\boldsymbol{S}-\boldsymbol{ES})^2\} \tag{5}$$

is used for the adjustment of d.f., where

$$\boldsymbol{S}=\sum_{i=1}^{g}\sum_{j=1}^{n_i}(Y_{ij}-\overline{Y}_{i\cdot})(Y_{ij}-\overline{Y}_{i\cdot})^{\mathrm{T}}/(N-g), \quad \overline{Y}_{i\cdot}=\sum_{j=1}^{n_i}Y_{ij}/n_i,$$
$$Y_{ij}=(Y_{ij1}, Y_{ij2}, \cdots, Y_{ij\tau})^{\mathrm{T}}, \quad N=\sum_{i=1}^{g}n_i.$$

(4) Covariance matrix Σ of observations of each individual is unknown but one estimate Σ^* of Σ is given. The Σ^* can be used as preliminary knowledge.

In order to determine the required sample size, it is necessary to find a critical value $C_{\alpha 3}$ on the distribution of F_3 under H_3 and to calculate power using the distribution of

F_3 under $H_3^{(A)}$. The d.f. of the distribution of F_3 under H_3 depends on $\hat{\varepsilon}$ which will be obtained from the observations in the "forthcoming" experiment. According to Muller and Barton [3], we regard the null distribution of F_3 as

$$F[(g-1)(\tau-1)E(\hat{\varepsilon}), (N-g)(\tau-1)E(\hat{\varepsilon})].$$

The expected value $E(\hat{\varepsilon})$ is approximated by a function of $\tau-1$ eigenvalues $\lambda_1^*, \cdots, \lambda_{\tau-1}^*$ of $\Sigma^* - E\Sigma^*$:

$$E(\hat{\varepsilon}) \approx f(\lambda_1^*, \cdots, \lambda_{\tau-1}^*) + g(\lambda_1^*, \cdots, \lambda_{\tau-1}^*)/(N-\tau), \tag{6}$$

where

$$f(\lambda_1^*, \cdots, \lambda_{\tau-1}^*) = (\Sigma_{k=1}^{\tau-1} \lambda_k^*)^2 / \{(\tau-1)(\Sigma_{k=1}^{\tau-1} \lambda_k^{*2})\},$$

$$g(\lambda_1^*, \cdots, \lambda_{\tau-1}^*) = \Sigma_{k=1}^{\tau-1}(f_{kk}\lambda_k^{*2}) + \Sigma_{k=1}^{\tau-1}\Sigma_{l=1}^{\tau-1} h(\lambda_k^*, \lambda_l^*),$$

$$f_k = \partial f / \partial \lambda_k^*, \quad f_{kk} = \partial^2 f / \partial \lambda_k^{*2},$$

$$h(\lambda_k^*, \lambda_l^*) = \begin{cases} \lambda_k^* \lambda_l^* / (\lambda_k^* - \lambda_l^*), & \lambda_k^* \neq \lambda_l^*, \\ 0 & , \lambda_k^* = \lambda_l^*. \end{cases}$$

The critical value $C_{\alpha 3}$ of the test with significance level α is obtained as upper 100α per cent point of the null distribution. We use

$$F'[(g-1)(\tau-1)\varepsilon^*, (N-g)(\tau-1)\varepsilon^*, (g-1)(\tau-1)f_{A3}^*\varepsilon^*] \tag{7}$$

for the distribution of F_3 under $H_3^{(A)}$, where ε^* is the adjusting factor of d.f. calculated from Σ^*, and $f_{A3}^* = \{\Sigma_{i=1}^{g} n_i (\Sigma_{k=1}^{\tau} \gamma_{ik}^2)\} / [(g-1)\{\mathrm{tr}\Sigma^* - \mathrm{tr}(E\Sigma^*)\}]$. When the sample size of i-th treatment group is $n_i = c_i n (i=1, \cdots, g)$, the power $P_w(n)$ is obtained as the upper tail probability of $C_{\alpha 3}$ on the non-central F distribution (7). Steps of iteration for determination of the required sample size are as follows.

Step 0. Set $n=0$.

Step 1. Set $n=n+1$. Then, the sample size of i-th treatment is $n_i = c_i n (i=1, \cdots, g)$.

Step 2. Calculate critical value $C_{\alpha 3}$ of test. The $C_{\alpha 3}$ is upper 100α per cent point of null distribution (central F).

Step 3. Calculate power $P_w(n)$. The $P_w(n)$ is upper tail probability of $C_{\alpha 3}$ on the distribution of F_3 under $H_3^{(A)}$ (non-central F).

Step 4. If $P_w(n) \geqq P_w^*$, stop the iteration and put $n^* = n$. Then, the required sample size of i-th treatment is $n_i^* = c_i n^* (i=1, \cdots, g)$. Otherwise, return to Step 1.

Here we described the method to determine the required sample size for the test concerning to interaction effect, but similar manner can be taken for the test concerning to time effect. For the treatment effect, the method is more simple because the adjustment of d.f. is not necessary.

3.2 Some Properties of the Required Sample Size

In this subsection, we investigate the propriety of the adjustment of d.f. in sample size determination, and the influences of the number of time points and the correlation between repeated measurements observed on the required sample size. Let Σ^* be the estimate under such assumption that Σ has first order auto-regressive structure (AR

Table 1 Sample size per group required to detect $H_3^{(A)}$ under the AR covariance structure [significance level : 0.05, power : 0.80]

ρ^*/τ	3	4	5	6	7	8
0.9	6 (5)[a] 0.8244[b]	8 (6) 0.6817	10 (7) 0.5769	12 (8) 0.4998	15 (9) 0.4418	18 (10) 0.3970
0.7	14 (12) 0.8748	17 (14) 0.7689	21 (15) 0.6891	24 (17) 0.6293	28 (19) 0.5840	31 (21) 0.5488
0.5	20 (18) 0.9245	23 (20) 0.8588	26 (22) 0.8088	29 (24) 0.7714	32 (25) 0.7433	34 (27) 0.7218
0.3	25 (24) 0.9680	27 (25) 0.9397	29 (26) 0.9186	31 (28) 0.9034	33 (29) 0.8922	35 (31) 0.8838
0.1	29 (28) 0.9959	30 (29) 0.9922	31 (30) 0.9897	32 (31) 0.9880	34 (32) 0.9868	35 (33) 0.9859

a) required sample size without the adjustment of d.f., b) value of ε^*

structure, hereafter), and also the assumption be correct. Then, $\Sigma^*=\{\sigma_{kl}^*\}=\{(\sigma^*)^2(\rho^*)^{|k-l|}\}$. Here, we consider a situation where "the difference $\Delta(>0)$ between treatments exists only at one time point(say the first time point)" in a comparative experiment of two treatments. This is equivalent to the case where

$$H_1^{(A)}:\{\alpha_1=-\alpha_2=\Delta/(2\tau)\}$$

and

$$H_3^{(A)}:\{\gamma_{11}=-\gamma_{21}=(\Delta/2)(1-1/\tau),\quad \gamma_{12}=\cdots=\gamma_{1\tau}=-\gamma_{22}=\cdots=-\gamma_{2\tau}=-\Delta/(2\tau)\}.$$

Then, the noncentrality of the distribution of F_3 under $H_3^{(A)}$ is

$$\{n\Delta^2(\tau-1)^2/(2\tau)\}/\{\mathrm{tr}\,\Sigma^*-\mathrm{tr}\,(E\Sigma^*)\}.$$

We set $g=2$, $n_1=n_2=n$ and $\Delta=\sigma^*=1$. For each combination of ρ^* (5 levels : 0.1 (0.2) 0.9) and τ (6 levels : 3 (1) 8), we calculated the sample size required to detect $H_3^{(A)}$ by the test (significance level 0.05) concerning to H_3 at the power not less than 0.80. At the same time, we calculated the required sample size in the test without the adjustment of d.f. for the each combination (Table 1). As a result, it is seen that if ρ^* decreases of τ increases then n^* becomes large, and that if ρ^* increases or τ decreases then n^* becomes small. It is also seen that if ρ^* is high or τ is large then n^* could be underestimated without the adjustment of d.f.

Concerning to the covariance matrix in the mixed model for ANOVA, besides the AR structure assumed above, the compound symmetric (CS) structure is mainly investigated (Huynh and Feldt [7], Crowder and Hand [8]). The CS structured covariance matrix is denoted by

$$\Sigma^*=(\sigma^*)^2\{(1-\rho^*)\mathbf{I}_\tau+\rho^*\mathbf{1}_\tau\mathbf{1}_\tau^T\},$$

where $\mathbf{I}_\tau$ is a $\tau\times\tau$ unit matrix and $\mathbf{1}_\tau$ is a $\tau\times 1$ vector of all the elements 1. Under this covariance structure, $\varepsilon^*=1$, namely the adjustment of d.f. is not necessary. We assumed the same situation as considered above except for the CS structure of Σ^*, and calculated the sample size required to detect $H_3^{(A)}$. As a result, the relation between the required sample size and the correlation or the number of time points is almost same as the AR structure. Namely, if ρ^* decreases or τ increases then n^* becomes large.

In this section, we investigated the properties of required sample size for each of the two covariance structures. In practice, however, we must check the validity of such assumption for covariance structure. For example, we may apply the method by Hearne et al. [9] to the check of the AR assumption and Box [10] to the CS assumption, respectively. We should like to note the following fact. Namely, although we only considered the configulation of $H_3^{(A)}$ as a alternative hypothesis of treatment-by-time interaction, our results should also stand for any other configulation of alternative hypothesis. In fact, the degrees of all possible alternative hypotheses are indicated by the noncentrality of the distribution of F_3.

4. DESIGN OF RME OF A HYPOTENSIVE DRUG

In development of a hypotensive drug, not only the degree but also the durability of hypotensive effect of the drug are largely remarked. For an example, we present a design of the RME of one hypotensive drug administrated once a day in the morning. In this experiment, we intended to evaluate "inhibitive effect" of the drug against abrupt increasing of blood pressure before noon and "durability" of hypotensive effect till next administration in the next morning. The abrupt increasing before noon is a distinguishing feature of diurnal variation of blood pressure in essential hypertension, so insufficiency of the above "inhibitive effect" means low hypotensive effect. If the "durability" till next morning is not accepted, we must select twice or three times administlations per day.

(1) Preliminary information : In designing, we could use the results of the experiment of other hypotensive drug carried out before. In that experiment, for each of 39 patients of essential hypertension, blood pressure was observed at 9 time points (7 a.m., 9 a.m., 10 a.m., 12 a.m., 2 p.m., 4 p.m., 6 p.m., 8 p.m. and 7 a.m. in the next morning) in a day during the 2 nd week of medical treatment. The drug was administered at 8 a.m.

(2) Null hypothesis and alternative hypothesis : We adopted diastolic blood pressure (DBP) as main indicator. Concerning to the "inhibitive effect", we intended to evaluate ① "whether DBP rises more than some fixed value (say Δ) from 9 a.m. till 10 a.m. or 12 a.m". And, concerning to the "durability", we intended to evaluate ② "whether DBP at 7 a.m. in the next morning is higher than one at 9 a.m. by at least Δ". In order to investigate the variations of ① and ② between time points, we remarked 4 time points of 9 a.m., 10 a.m., 12 a.m. and 7 a.m. in the next morning. We adopted the test at significance level 0.05 of H_2. Putting μ_1, μ_2, μ_3 and μ_4 be population means of DBP at the 4 time points, we described the null hypothesis as "H_2 : $\mu_1=\mu_2=\mu_3=\mu_4$". The alternative hypothesis was set to be "$H_2^{(A)}$: the differ-

Table 2 Required sample size in the experiment of hypotensive drug

assumption	adjusting factor of d.f.	required sampl size	
		$\Delta = 5$ mmHg	$\Delta = 10$mmHg
none	0.797	56	16
CS structure	1.0	46	13
AR structure	0.750	72	20

ence more than Δ exists between at least one pair of μ_1, μ_2, μ_3 and μ_4". Under this alternative hypothesis, the noncentrality of the distribution of F_2 takes its minimum value in the case where there is a difference of Δ between some pair (say $\mu_{(1)}$ and $\mu_{(2)}$) of population means and both of the other two population means lie just in the center between $\mu_{(1)}$ and $\mu_{(2)}$. We used the minimum noncentrality for the following calculation of the required sample size. Of course, whenever significant time effect of ① or ② is recognized, the alternative hypothesis $H_2^{(A)}$ is necessarily accepted.

(3) Calculation of the required sample size : The unbiased estimate of covariance matrix obtained from observations at the 4 time points was

$$\begin{pmatrix} 194.19 & 157.95 & 140.55 & 85.54 \\ 157.95 & 198.54 & 143.45 & 86.53 \\ 140.55 & 143.45 & 149.42 & 77.55 \\ 85.54 & 86.53 & 77.55 & 120.46 \end{pmatrix}.$$

Both the assumptions of the CS structure and of the AR structure concerning to the covariance matrix were rejected (p values were 0.008 and 0.003, respectively). Therefore, we put the above estimate be Σ^*. Then, we have $\varepsilon^*=0.797$. For each of 2 levels (5 mmHg, 10mmHg) of Δ, we calculated the sample size required to detect it by the test concerning to H_2 at the power not less than 0.80 (Table 2). This table also shows the results under the assumptions of the CS and the AR structures. As a result, it was seen that in order to detect the difference more than 5 mmHg (10mm Hg), at least 56 patients (16 patients) was required.

5. CONCLUDING REMARKS

In the RME, if the number of time points increases or the correlation between repeated measurements decreases then the required sample size becomes large. Without the adjustment of d.f., the required sample size could be underestimated. In this paper, we supposed the situation where an estimete of covariance matrix has previously been obtained, and we did not refer to suitability of the estimate. However, if the number of time points is large or some outlier exists, accuracy of estimator of covariance matrix tends to be unstable or its precision does to be ruined. In order to avoid such situation, it may be useful to investigate whether some specific covariance structure such as the AR one could be assumed, or to adopt robust estimation method.

Acknowledgement

The authors would like to thank referees for constructive advices to the first draft of the paper, and editors for helpful suggestions.

REFERENCES

[1] Nakahira,M. (1988). Problems on repeated measurement analysis in clinical trials. Biometry : Clinical Trials and Related Topics, ed.by Okuno,T.,67–82, Excerpta Medica.

[2] Uragari,Y. and Goto, M.(1991). On comparing longitudinal data. Bulletin of the Biometric Soc. of Japan, 12, 35–45 (in Japanese).

[3] Muller, K.E. and Barton,C.N.(1989). Approximate power for repeated–measures ANOVA lacking sphericity. J.Amer.Statist. Assoc., 84, 549–555.

[4] Vonesh,E.F. and Schork,M.A.(1986). Sample sizes in the multivariate analysis of repeated measurements. Biometrics, 42, 601–610.

[5] Rochon,J. (1991). Sample size calculations for two–group repeated–measures experiments. Biometrics, 47, 1383–1398.

[6] Geisser,S. and Greenhouse,S.W.(1958). An extension of Box's results on the use of the F distribution in multivariate analysis. Ann.Math. Statist., 29, 885–891.

[7] Huynh,H. and Feldt,L.S.(1970). Conditions under which mean square ratios in repeated measurements designs have exact F–distributions. J.Amer.Statist. Assoc.,65, 1582–1589.

[8] Crowder,M.J. and Hand,D.J.(1990). Analysis of Repeated Measures. Chapman and Hall, London.

[9] Hearne,E.M. III., Clark, G.M. and Hatch, J.P.(1983). A test for serial correlation in univariate repeated–measures analysis. Biometrics, 39, 237–243.

[10] Box,G.E.P.(1949). A general distribution theory for a class of likelihood criteria. Biometrika, 36, 317–346.

Stat. Sci. & Data Anal., pp. 49-60
K. Matsusita *et al.* (Eds)

Least Squares Estimators of Regression Coefficients by using Misspecified Covariance Structure for Error Process

YOSHIHIRO USAMI and MITUAKI HUZII
Department of Information Sciences, Tokyo Institute of Technology, Tokyo 152, Japan

Abstract. We consider a problem of estimating coefficients of a linear regression by the method of the generalized least squares, when a covariance structure of error terms is unknown. In order to estimate the covariance matrix of the error terms, we fit a model to the error terms. Then we estimate the covariance matrix on the basis of a covariance structure of the fitted model. However, the fitted model, i.e., the fitted covariance structure might be incorrect. We are concerned with statistical properties of the generalized least squares estimators when we misspecify the covariance structure. Here we consider the case when the error term is a stochastic process. In particular, we deal with the case when we fit an autoregressive model to the estimated error sequence even when the error process is not necessarily autoregressive nor stationary. For this case, we show the consistency of the estimator and the numerical comparisons with the method of the ordinary least squares.

Key words: Regression, least squares estimators, generalized least squares estimators, model fitting, misspecified covariance structure.

1. INTRODUCTION

Consider a linear regression

$$y_t = \boldsymbol{x}_t'\boldsymbol{\beta} + \varepsilon_t \ , \ \ t = 1, 2, \dots \ ,$$

where y_t is an observable random variable, $\boldsymbol{x}_t = (x_{t,1}, x_{t,2}, \dots, x_{t,K})'$ is a vector of known explanatory variables, $\boldsymbol{\beta} = (\beta_1, \beta_2, \dots, \beta_K)'$ is a vector of unknown regression coefficients and ε_t is a random error term with mean zero and finite variance. We want to estimate $\boldsymbol{\beta}$ on the basis of T $(\geq K)$ observations $y_1, y_2, \dots, y_T$. Let $\boldsymbol{y} = (y_1, y_2, \dots, y_T)'$, $\boldsymbol{X} = (\boldsymbol{x}_1, \boldsymbol{x}_2, \dots, \boldsymbol{x}_T)'$ and $\boldsymbol{\varepsilon} = (\varepsilon_1, \varepsilon_2, \dots, \varepsilon_T)'$. Then we have

$$\boldsymbol{y} = \boldsymbol{X}\boldsymbol{\beta} + \boldsymbol{\varepsilon} \ . \tag{1.1}$$

We suppose that $\boldsymbol{X}$ has rank K . Let $\boldsymbol{\Sigma}_\varepsilon = E(\boldsymbol{\varepsilon}\boldsymbol{\varepsilon}')$ and $\boldsymbol{\Sigma}_\varepsilon$ be positive definite. If we know completely $\boldsymbol{\Sigma}_\varepsilon$, the method of the generalized least squares (GLS) is applicable for estimating $\boldsymbol{\beta}$ and we can obtain the best linear unbiased estimator of $\boldsymbol{\beta}$. The GLS estimator $\tilde{\boldsymbol{\beta}}_G$ is obtained by minimizing

$$S_G = (\boldsymbol{y} - \boldsymbol{X}\boldsymbol{\beta})'\boldsymbol{\Sigma}_\varepsilon^{-1}(\boldsymbol{y} - \boldsymbol{X}\boldsymbol{\beta})$$

with respect to $\boldsymbol{\beta}$. In most practical cases, however, $\boldsymbol{\Sigma}_\varepsilon$ is unknown. In such cases, we usually adopt the method of the ordinary least squares (OLS). Let $\boldsymbol{b}$ be the OLS estimator obtained by minimizing

$$S_O = (\boldsymbol{y} - \boldsymbol{X}\boldsymbol{\beta})'(\boldsymbol{y} - \boldsymbol{X}\boldsymbol{\beta})$$

with respect to $\boldsymbol{\beta}$.

We assume that we have no knowledge on $\boldsymbol{\Sigma}_\varepsilon$. In such a case, we sometimes fit a model to $\{\varepsilon_t\}$. Let the fitted model and the covariance structure of the model be described by a parameter $\boldsymbol{\phi} = (\phi_1, \phi_2, \ldots, \phi_p)'$. Since $\{\varepsilon_t\}$ is unobservable, we consider the case when, first, we estimate ε_t with $e_t = y_t - \boldsymbol{x}_t'\boldsymbol{b}$ for $t = 1, \ldots, T$ and, next, obtain an estimator $\hat{\boldsymbol{\phi}} = \left(\hat{\phi}_1, \hat{\phi}_2, \ldots, \hat{\phi}_p\right)'$ from $\{e_t\}$. Here we represent the covariance structure fitted to $\boldsymbol{\Sigma}_\varepsilon$ as a function $\boldsymbol{S}_\varepsilon\left(\hat{\boldsymbol{\phi}}\right)$ of $\hat{\boldsymbol{\phi}}$. Let $\tilde{\boldsymbol{\beta}}$ be the estimator obtained by minimizing

$$S = (\boldsymbol{y} - \boldsymbol{X}\boldsymbol{\beta})'\,\boldsymbol{S}_\varepsilon\left(\hat{\boldsymbol{\phi}}\right)^{-1}(\boldsymbol{y} - \boldsymbol{X}\boldsymbol{\beta})$$

with respect to $\boldsymbol{\beta}$ provided that $\boldsymbol{S}_\varepsilon\left(\hat{\boldsymbol{\phi}}\right)$ is nonsingular. Sometimes $\tilde{\boldsymbol{\beta}}$ is called an estimated generalized least squares estimator. The structure of the model fitted to $\{e_t\}$ might be different from the true one of $\{\varepsilon_t\}$. We are concerned with studying statistical properties of $\tilde{\boldsymbol{\beta}}$ when we use a misspecified covariance structure. In particular, we try to compare properties of $\tilde{\boldsymbol{\beta}}$ with those of $\boldsymbol{b}$.

In this paper, we regard $\{\varepsilon_t\}$ as a time series. When the error process $\{\varepsilon_t\}$ is stationary and satisfies some conditions, $\boldsymbol{b}$ is asymptotically efficient in a sense as $T \to \infty$ (See Grenander and Rosenblatt [1] and Anderson [2]). We mainly study a case when $\{\varepsilon_t\}$ is not necessarily stationary. Huzii and Toyooka [3] studied asymptotic properties of estimators of $\boldsymbol{\beta}$ when $\{\varepsilon_t\}$ is an uniformly modulated error process. Further Toyooka [4] gave an extension of the results in Huzii and Toyooka [3] and obtained an asymptotically efficient estimator. Here we consider the case when we fit a stationary autoregressive (AR) model to $\{\varepsilon_t\}$ and estimate $\boldsymbol{\beta}$ with an estimator of $\boldsymbol{\Sigma}_\varepsilon$, which is obtained on the basis of a covariance structure of the fitted AR model.

We derive the fitted covariance structure $\boldsymbol{S}_\varepsilon\left(\hat{\boldsymbol{\phi}}\right)$ and give a procedure for estimating $\boldsymbol{\beta}$ when we fit an autoregressive (AR(p)) model of order p to $\{\varepsilon_t\}$. Let $\boldsymbol{\phi}$ be the vector whose elements are the coefficients of the fitted AR(p) model. Since the error process $\{\varepsilon_t\}$ is unobservable, we use the estimated error sequence $\{e_t\}$ and obtain the estimator $\hat{\boldsymbol{\phi}}$ of $\boldsymbol{\phi}$, which minimizes

$$\sum_{t=p+1}^{T}\left(e_t - \sum_{i=1}^{p}\phi_i\, e_{t-i}\right)^2 = \boldsymbol{e}'\,(\boldsymbol{\Phi}^*)'\,\boldsymbol{\Phi}^*\boldsymbol{e}$$

with respect to $\boldsymbol{\phi}$, where $\boldsymbol{e} = \boldsymbol{y} - \boldsymbol{X}\boldsymbol{b}$ and $\boldsymbol{\Phi}^*$ is the $(T-p)\times T$ matrix such that

$$\boldsymbol{\Phi}^* = \begin{pmatrix} -\phi_p & -\phi_{p-1} & \cdots & -\phi_1 & 1 & 0 & \cdots & 0 & 0 \\ 0 & -\phi_p & \cdots & -\phi_2 & -\phi_1 & 1 & \cdots & 0 & 0 \\ \vdots & \vdots & & \vdots & \vdots & \vdots & & \vdots & \vdots \\ 0 & 0 & \cdots & 0 & 0 & 0 & \cdots & -\phi_1 & 1 \end{pmatrix}.$$

Then $\hat{\boldsymbol{\phi}}$ is the solution of the set of the normal equations

$$\sum_{t=p+1}^{T} e_t\, e_{t-m} = \sum_{i=1}^{p}\hat{\phi}_i \sum_{t=p+1}^{T} e_{t-i}\, e_{t-m} \tag{1.2}$$

for $m = 1, \ldots, p$. Further let $\hat{\boldsymbol{\Phi}}^*$ be the statistic obtained by replacing $\boldsymbol{\phi}$ in $\boldsymbol{\Phi}$ with $\hat{\boldsymbol{\phi}}$ Replacing $\boldsymbol{S}_\varepsilon \left(\hat{\boldsymbol{\phi}}\right)^{-1}$ in S with $\left(\hat{\boldsymbol{\Phi}}^*\right)' \hat{\boldsymbol{\Phi}}^*$, we estimate $\boldsymbol{\beta}$ with its estimator $\hat{\boldsymbol{\beta}}$ derived by minimizing

$$(\boldsymbol{y} - \boldsymbol{X\beta})' \left(\hat{\boldsymbol{\Phi}}^*\right)' \hat{\boldsymbol{\Phi}}^* (\boldsymbol{y} - \boldsymbol{X\beta})$$

with respect to $\boldsymbol{\beta}$. Then, provided that $\boldsymbol{X}' \left(\hat{\boldsymbol{\Phi}}^*\right)' \hat{\boldsymbol{\Phi}}^* \boldsymbol{X}$ is nonsingular, we have

$$\hat{\boldsymbol{\beta}} = \left[\boldsymbol{X}' \left(\hat{\boldsymbol{\Phi}}^*\right)' \hat{\boldsymbol{\Phi}}^* \boldsymbol{X}\right]^{-1} \boldsymbol{X}' \left(\hat{\boldsymbol{\Phi}}^*\right)' \hat{\boldsymbol{\Phi}}^* \boldsymbol{y} . \tag{1.3}$$

The asymptotic properties of an estimated generalized least squares estimator when $\{\varepsilon_t\}$ obeys an AR(p) process have been studied (See Fuller [5]). Here we consider the case when $\{\varepsilon_t\}$ might be different from an AR(p) process, but we fit an AR(p) model to $\{e_t\}$. In Section 2, we shall show that $\boldsymbol{b}$ is a consistent estimator of $\boldsymbol{\beta}$. Further, we shall show that $\hat{\beta}$ has the consistency under some assumptions even if we misspecify the covariance structure of $\{\varepsilon_t\}$. In Section 3, we consider relations between effectiveness of estimators and fitted covariance structures. Since theoretical consideration of the relations is difficult, we give some numerical examples in estimating a linear trend in order to compare $\hat{\boldsymbol{\beta}}$ with $\boldsymbol{b}$. In particular, we study the cases when the sample sizes are small and $\{\varepsilon_t\}$'s are nonstationary in a sense.

2. ASYMPTOTIC PROPERTIES OF ESTIMATORS

In this section, we shall discuss asymptotic properties of the estimators $\boldsymbol{b}$ and $\hat{\boldsymbol{\beta}}$. Let $[\boldsymbol{A}]_{i,j}$ denote the (i, j) element of a matrix $\boldsymbol{A}$.

First we consider the consistency of $\boldsymbol{b}$ under the following assumptions. Let $\boldsymbol{X}_k$ be the kth column of $\boldsymbol{X}$, i.e., $\boldsymbol{X} = (\boldsymbol{X}_1, \ldots, \boldsymbol{X}_K)$. Let $\boldsymbol{D}$ be the $K \times K$ diagonal matrix whose kth diagonal element is the norm $\|\boldsymbol{X}_k\|_T = \sqrt{\sum_{t=1}^T x_{t,k}^2}$.

Assumption 1. For $k = 1, \ldots, K$, $\lim_{T\to\infty} \|\boldsymbol{X}_k\|_T = \infty$.

Assumption 2. There exists $\lim_{T\to\infty} \boldsymbol{D}^{-1}\boldsymbol{X}'\boldsymbol{X}\boldsymbol{D}^{-1}$ which is nonsingular.

Assumption 3. For $t, s = 1, \ldots, T$,

$$\left| [\boldsymbol{\Sigma}_\varepsilon]_{t,s} \right| \leq C(|t - s|) ,$$

where $C(0)$,$C(1)$, ... are nonnegative constants which satisfy $\sum_{h=0}^\infty C(h) < \infty$.

Assumption 3 does not necessarily mean that the error process is stationary.

Let $\boldsymbol{0}_K$ be the K-dimensional vector whose elements are all zero and $\boldsymbol{0}_{K\times K}$ be the $K \times K$ matrix whose elements are all zero. Let " $\mathrm{p}\lim_{T\to\infty}$ " be an abbreviation of "convergence in probability" as T tends to infinity.

We have the following lemma.

Lemma 1. *Under Assumptions* 1-3,

$$\mathrm{p}\lim_{T\to\infty} \boldsymbol{b} = \boldsymbol{\beta} .$$

Proof. Let T tend to infinity on the both sides of $\boldsymbol{D}^{-1}\boldsymbol{X}'\boldsymbol{X}\boldsymbol{D}^{-1}\boldsymbol{D}(\boldsymbol{b}-\boldsymbol{\beta}) = \boldsymbol{D}^{-1}\boldsymbol{X}'\boldsymbol{\varepsilon}$. By Assumption 2, we have

$$\operatorname{p}\lim_{T\to\infty} \boldsymbol{D}(\boldsymbol{b}-\boldsymbol{\beta}) = \left(\lim_{T\to\infty} \boldsymbol{D}^{-1}\boldsymbol{X}'\boldsymbol{X}\boldsymbol{D}^{-1}\right)^{-1} \operatorname{p}\lim_{T\to\infty} \boldsymbol{D}^{-1}\boldsymbol{X}'\boldsymbol{\varepsilon} . \tag{2.1}$$

It holds

$$\begin{aligned} \left|[Cov(\boldsymbol{X}'\boldsymbol{\varepsilon})]_{i,j}\right| &\le \sum_{t=1}^{T} \left|x_{t,i}[\boldsymbol{\Sigma}_\varepsilon]_{t,t}x_{t,j}\right| + \sum_{t=2}^{T}\sum_{s=1}^{t-1} \left(|x_{t,i}x_{s,j}| + |x_{s,i}x_{t,j}|\right)\left|[\boldsymbol{\Sigma}_\varepsilon]_{t,s}\right| \\ &\le C(0)\,\|\boldsymbol{X}_i\|_T\,\|\boldsymbol{X}_j\|_T + 2\sum_{h=1}^{T-1} C(h)\,\|\boldsymbol{X}_i\|_T\,\|\boldsymbol{X}_j\|_T \end{aligned}$$

for $i,j = 1,\ldots,K$. Therefore, by Assumption 3, $[Cov(\boldsymbol{D}^{-1}\boldsymbol{X}'\boldsymbol{\varepsilon})]_{i,j}$ is bounded from above in its absolute value by $C(0)+2\sum_{h=1}^{\infty}C(h)$ for $i,j=1,\ldots,K$ as $T\to\infty$. Then, for $k = 1,\ldots,K$, $\operatorname{p}\lim_{T\to\infty} \|\boldsymbol{X}_k\|_T^{-1}\,\boldsymbol{D}^{-1}\boldsymbol{X}'\boldsymbol{\varepsilon} = \boldsymbol{0}_K$. Thus, noting (2.1), we obtain $\operatorname{p}\lim_{T\to\infty}(\boldsymbol{b}-\boldsymbol{\beta}) = \boldsymbol{0}_K$. Consequently, we derive the desired result.

We shall show an example of a set of explanatory variables which satisfies Assumptions 1 and 2. Further we shall give a few examples of $\{\varepsilon_t\}$ which satisfy Assumption 3.

Example 1. It is well-known that a set of explanatory variables satisfies Assumptions 1 and 2 in the case of a polynomial regression (See Anderson [2]). Let $x_{t,k} = t^{k-1}$ for $t=1,\ldots,T$ and $k=1,\ldots,K$. It has been shown that

$$\lim_{T\to\infty}\left[\boldsymbol{D}^{-1}\boldsymbol{X}'\boldsymbol{X}\boldsymbol{D}^{-1}\right]_{k,l} = \frac{\sqrt{(2k-1)(2l-1)}}{k+l-1} \tag{2.2}$$

for $k,l=1,\ldots,K$. In this case, it follows that $\lim_{T\to\infty}\boldsymbol{D}^{-1}\boldsymbol{X}'\boldsymbol{X}\boldsymbol{D}^{-1}$ is nonsingular.

Example 2. We shall give an example of a nonstationary error process which satisfies Assumption 3. The following model has a changing point T_0 and the value of the coefficient changes at the point T_0 :

$$\varepsilon_t = \begin{cases} \theta\varepsilon_{t-1} + u_t , & t < T_0 , \\ \theta^*\varepsilon_{t-1} + u_t , & t \ge T_0 , \end{cases} \tag{2.3}$$

where θ and θ^* are constants such that $|\theta|, |\theta^*| < 1$ and $\theta \ne \theta^*$ and $\{u_t\}$ is an independent sequence with $E(u_t)=0$ and $Var(u_t)=\sigma_u^2$. The covariance between ε_t and ε_s is given by

$$E(\varepsilon_t\varepsilon_s) = \begin{cases} \dfrac{1}{1-\theta^2}\theta^{|t-s|}\sigma_u^2 , & t,s < T_0 , \\ \dfrac{1}{1-\theta^2}\theta^{T_0-t-1}\theta^{*s-T_0+1}\sigma_u^2 , & t < T_0 ,\ s \ge T_0 , \\ \dfrac{1}{1-\theta^{*2}}\theta^{*|t-s|}\sigma_u^2\left\{1+\dfrac{\theta^2-\theta^{*2}}{1-\theta^2}\theta^{*2[\min(t,s)-T_0+1]}\right\} , & t,s \ge T_0 . \end{cases} \tag{2.4}$$

Let $\xi = (1-\theta^2)^{-1}\,|\theta^2-\theta^{*2}|$. Then we can show

$$\left|[\boldsymbol{\Sigma}_\varepsilon]_{t,s}\right| \le \frac{\theta_{max}^{|t-s|}}{1-\theta_{max}^2}\,\sigma_u^2\,(1+\xi) ,$$

where $\theta_{max} = \max(\ |\theta|\ ,\ |\theta^*|\ )$. Hence Assumption 3 is satisfied.

Example 3. Another example of a nonstationary error process which satisfies Assumption 3 is the following model which has a changing point T_0 :

$$\varepsilon_t = \begin{cases} u_t + \psi u_{t-1}\ , & t < T_0\ , \\ u_t + \psi^* u_{t-1}\ , & t \geq T_0\ , \end{cases} \tag{2.5}$$

where ψ and ψ^* are constants such that $|\psi|, |\psi^*| < 1$ and $\psi \neq \psi^*$ and $\{u_t\}$ is an independent process with $E(u_t) = 0$ and $Var(u_t) = \sigma_u^2$. In this case, it can be shown that

$$\left|[\Sigma_\varepsilon]_{t,s}\right| \begin{cases} \leq\ (\ 1 + \psi_{max}^2\ )\ \sigma_u^2\ , & t = s\ , \\ \leq\ \psi_{max}\ \sigma_u^2\ , & |t-s| = 1\ , \\ =\ 0\ , & |t-s| \geq 2\ , \end{cases}$$

where $\psi_{max} = \max(\ |\psi|\ ,\ |\psi^*|\ )$. Thus Assumption 3 is satisfied.

We have shown the consistency of $\boldsymbol{b}$. Since $\hat{\boldsymbol{\beta}}$ contains $\hat{\boldsymbol{\phi}}$ which is the solution of the set of the normal equations (1.2), we shall study asymptotic properties of

$$\frac{1}{T} \sum_{t=p+1}^{T} e_{t-l}\ e_{t-m}$$

for $l, m = 0, \ldots, p$ before discussing $\hat{\boldsymbol{\beta}}$. Let

$$\lambda_{l,m}(T)\ =\ \frac{1}{T} \sum_{t=p+1}^{T} \varepsilon_{t-l}\ \varepsilon_{t-m}$$

for $l, m = 0, \ldots, p$. In general, $\lambda_{l,m}(T)$ does not necessarily converge in probability to a constant as $T \to \infty$ under Assumption 3. Then, we make the following assumption.

Assumption 4. There exists $\mathrm{p}\lim_{T\to\infty} \lambda_{l,m}(T)$ for $l, m = 0, \ldots, p$.

Here we shall show the following lemma.

Lemma 2. *Under Assumptions* 1-4,

$$\mathrm{p}\lim_{T\to\infty} \frac{1}{T} \sum_{t=p+1}^{T} e_{t-l}\ e_{t-m}\ =\ \mathrm{p}\lim_{T\to\infty} \frac{1}{T} \sum_{t=p+1}^{T} \varepsilon_{t-l}\ \varepsilon_{t-m}$$

for $l, m = 0, \ldots, p$.

Proof. We can write

$$\begin{aligned} \sum_{t=p+1}^{T} e_{t-l}\ e_{t-m}\ =\ & \sum_{t=p+1}^{T} \varepsilon_{t-l}\ \varepsilon_{t-m}\ +\ (\boldsymbol{b}-\boldsymbol{\beta})'\boldsymbol{D}\boldsymbol{D}^{-1} \sum_{t=p+1}^{T} \boldsymbol{x}_{t-l}\ \boldsymbol{x}'_{t-m}\ \boldsymbol{D}^{-1}\boldsymbol{D}(\boldsymbol{b}-\boldsymbol{\beta}) \\ & - \sum_{t=p+1}^{T} \varepsilon_{t-l}\ \boldsymbol{x}'_{t-m}\ \boldsymbol{D}^{-1}\boldsymbol{D}(\boldsymbol{b}-\boldsymbol{\beta})\ -\ \sum_{t=p+1}^{T} \varepsilon_{t-m}\ \boldsymbol{x}'_{t-l}\ \boldsymbol{D}^{-1}\boldsymbol{D}(\boldsymbol{b}-\boldsymbol{\beta})\ . \end{aligned} \tag{2.6}$$

By using Schwarz' inequality, we have

$$\left| \left[\boldsymbol{D}^{-1} \sum_{t=p+1}^{T} \boldsymbol{x}_{t-l} \boldsymbol{x}'_{t-m} \, \boldsymbol{D}^{-1} \right]_{i,j} \right| \leq 1 \tag{2.7}$$

for i, j, l and m such that $1 \leq i, j \leq K$ and $0 \leq l, m \leq p$. Noting the way of the proof of Lemma 1, we can show

$$\operatorname{p}\lim_{T\to\infty} \frac{1}{T^{\gamma}} \boldsymbol{D}(\boldsymbol{b} - \boldsymbol{\beta}) = \boldsymbol{0}_K$$

for any positive constant γ . Hence, the second term on the right-hand side of (2.6) divided by T converges in probability to zero as $T \to \infty$. Moreover, we can show that, for $i, j = 1, \ldots, K$ and $l, m = 0, \ldots, p$, the covariance between $\|\boldsymbol{X}_i\|_T^{-1} \sum_{t=p+1}^{T} x_{t-l,i}\, \varepsilon_{t-m}$ and $\|\boldsymbol{X}_j\|_T^{-1} \sum_{t=p+1}^{T} x_{t-l,j}\, \varepsilon_{t-m}$ is bounded from above in its absolute value by $C(0) + 2\sum_{h=1}^{T-(p+1)} C(h)$. Thus,

$$\operatorname{p}\lim_{T\to\infty} \frac{1}{T^{\delta}\, \|\boldsymbol{X}_i\|_T} \sum_{t=p+1}^{T} x_{t-l,i}\, \varepsilon_{t-m} = 0$$

for any positive constant δ, $i = 1, \ldots, K$ and $l, m = 0, \ldots, p$. Therefore, the third and the forth terms on the right-hand side of (2.6) divided by T converge in probability to zero, respectively, as $T \to \infty$. Hence, Lemma 2 is proved.

By Lemma 2, we consider $\lambda_{l,m}(T) = T^{-1} \sum_{t=p+1}^{T} \varepsilon_{t-l}\, \varepsilon_{t-m}$ for $l, m = 0, \ldots, p$ in stead of $T^{-1} \sum_{t=p+1}^{T} e_{t-l}\, e_{t-m}$ for $l, m = 0, \ldots, p$ for the case when $T \to \infty$. In order to discuss an asymptotic property of $\hat{\boldsymbol{\beta}}$, we have to make further assumptions. Let

$$\boldsymbol{\Lambda}(T) = \begin{pmatrix} \lambda_{1,1}(T) & \cdots & \lambda_{p,1}(T) \\ \vdots & & \vdots \\ \lambda_{1,p}(T) & \cdots & \lambda_{p,p}(T) \end{pmatrix} \quad \text{and} \quad \boldsymbol{\lambda}(T) = \begin{pmatrix} \lambda_{0,1}(T) \\ \vdots \\ \lambda_{0,p}(T) \end{pmatrix} .$$

Assumption 5. $\operatorname{p}\lim_{T\to\infty} \boldsymbol{\Lambda}(T)$ is nonsingular.

In this paper, we have been dealing with the case when $\{\varepsilon_t\}$ does not necessarily obey an autoregressive model. Here we define $\boldsymbol{\phi}$ as the following:

$$\boldsymbol{\phi} = \left(\operatorname{p}\lim_{T\to\infty} \boldsymbol{\Lambda}(T) \right)^{-1} \operatorname{p}\lim_{T\to\infty} \boldsymbol{\lambda}(T) . \tag{2.8}$$

Further let

$$\boldsymbol{\Phi}^* = \operatorname{p}\lim_{T\to\infty} \hat{\boldsymbol{\Phi}}^* . \tag{2.9}$$

Moreover, we make the following assumption.

Assumption 6. There exists $\lim_{T\to\infty} \boldsymbol{D}^{-1}\boldsymbol{X}'(\boldsymbol{\Phi}^*)'\boldsymbol{\Phi}^*\boldsymbol{X}\boldsymbol{D}^{-1}$ which is nonsingular.

The following theorem demonstrates that $\hat{\boldsymbol{\beta}}$ is a consistent estimator of $\boldsymbol{\beta}$ under the above assumptions even if we misspecify the covariance structure of $\{\varepsilon_t\}$.

Theorem. *Under Assumptions* 1-6,

$$\operatorname{p}\lim_{T\to\infty} \hat{\boldsymbol{\beta}} = \boldsymbol{\beta} .$$

Proof. It follows from (1.1) and (1.3) that

$$D^{-1}X'\left(\hat{\Phi}^*\right)'\hat{\Phi}^*XD^{-1}D\left(\hat{\beta}-\beta\right) = D^{-1}X'\left(\hat{\Phi}^*\right)'\hat{\Phi}^*\varepsilon\,. \tag{2.10}$$

We have

$$\mathrm{p}\lim_{T\to\infty} D^{-1}X'\left[\left(\hat{\Phi}^*\right)'\hat{\Phi}^* - (\Phi^*)'\Phi^*\right]XD^{-1} = 0_{K\times K} \tag{2.11}$$

by noting Lemma 2, Assumption 5, (2.8), (2.9) and (2.7). We can also show

$$\mathrm{p}\lim_{T\to\infty} D^{-1}X'\left[\left(\hat{\Phi}^*\right)'\hat{\Phi}^* - (\Phi^*)'\Phi^*\right]\varepsilon = 0_K \tag{2.12}$$

by using the fact that the variance of $\|X_i\|_T^{-1}\sum_{t=p+1}^{T} x_{t-l,i}\,\varepsilon_{t-m}$ is bounded from above in its absolute value by $C(0)+2\sum_{h=1}^{\infty}C(h)$ for $i=1,\ldots,K$ and $l,m=0,\ldots,p$ as $T\to\infty$ and $\phi_l = \mathrm{p}\lim_{T\to\infty}\hat{\phi}_l$ for $l=1,\ldots,p$. It follows that

$$\mathrm{p}\lim_{T\to\infty} D\left(\hat{\beta}-\beta\right) = \left[\lim_{T\to\infty} D^{-1}X'(\Phi^*)'\Phi^*XD^{-1}\right]^{-1}\mathrm{p}\lim_{T\to\infty} D^{-1}X'(\Phi^*)'\Phi^*\varepsilon \tag{2.13}$$

from (2.10), (2.11), (2.12) and Assumption 6. Here we shall show that all the elements of the covariance matrix of $D^{-1}X'(\Phi^*)'\Phi^*\varepsilon$ are bounded from above in their absolute values as $T\to\infty$. For convenience sake, let $x_{t,k}=0$ for $t=1-p,\ldots,0,T+1,\ldots,T+p$ and $k=1,\ldots,K$. Then, it holds that

$$\left|\left[X'(\Phi^*)'\Phi^*\;\Sigma_\varepsilon\;(\Phi^*)'\Phi^*X\right]_{i,j}\right| \le (p+1)^2\;\phi_{max}^4\sum_{l=-p}^{p}\sum_{m=-p}^{p}\sum_{t=1}^{T}\sum_{s=1}^{T}\left|x_{t+l,i}\;[\Sigma_\varepsilon]_{t,s}\;x_{s+m,j}\right|$$

for $i,j=1,\ldots,K$, where $\phi_{max}=\max\left(|\phi_1|,\ldots,|\phi_p|,1\right)$. Moreover, we can show that, for $i,j=1,\ldots,K$ and $l,m=-p,\ldots,0,\ldots,p$, $\sum_{t=1}^{T}\sum_{s=1}^{T}\left|x_{t+l,i}\,[\Sigma_\varepsilon]_{t,s}\,x_{s+m,j}\right|$ divided by $\|X_i\|_T\;\|X_j\|_T$ is bounded from above in its absolute value by $C(0)+2\sum_{h=1}^{T-1}C(h)$. Hence, all the elements of the covariance matrix of $D^{-1}X'(\Phi^*)'\Phi^*\varepsilon$ are bounded from above in their absolute values by $C(0)+2\sum_{h=1}^{\infty}C(h)$ as $T\to\infty$. Then,

$$\mathrm{p}\lim_{T\to\infty}\frac{1}{\|X_k\|_T}D^{-1}X'(\Phi^*)'\Phi^*\varepsilon = 0_K$$

for $k=1,\ldots,K$. Thus, noting (2.13), we can obtain $\mathrm{p}\lim_{T\to\infty}\left(\hat{\beta}-\beta\right)=0_K$. This completes the proof.

It will be shown that Assumption 6 is satisfied in the case of the polynomial regression provided that $\{\varepsilon_t\}$ satisfies Assumptions 4 and 5. Moreover, we shall show that Assumptions 4 and 5 are satisfied when $\{\varepsilon_t\}$ obeys the nonstationary model (2.3) in Example 2 or (2.5) in Example 3, provided that $\{\varepsilon_t\}$ is a Gaussian process.

Example 4. In the case of the polynomial regression, we have

$$\sum_{t=p+1}^{T} x_{t-i,k}\,x_{t-j,l} = \sum_{t=p+1}^{T}(t-i)^{k-1}\,(t-j)^{l-1}$$

for $i,j=0,\ldots,p$ and $k,l=1,\ldots,K$. When we expand the right-hand side of this equation, divide the terms by $\|X_k\|_T\;\|X_l\|_T$ and let T tend to infinity, we can neglect all terms except $\sum_{t=p+1}^{T} t^{k+l-2}$. It follows that

$$\lim_{T\to\infty} \left[\boldsymbol{D}^{-1}\boldsymbol{X}'(\boldsymbol{\Phi}^*)'\boldsymbol{\Phi}^*\boldsymbol{X}\boldsymbol{D}^{-1}\right]_{k,l} = \frac{\sqrt{(2k-1)(2l-1)}}{k+l-1}\left[\left(\sum_{i=1}^{p}\phi_i\right)^2 - \sum_{i=1}^{p}\phi_i + 1\right]$$

for $k,l = 1,\ldots,K$ from (2.2), where $\boldsymbol{\phi}$ and $\boldsymbol{\Phi}^*$ are defined by (2.8) and (2.9), respectively, under Assumptions 4 and 5. The quantity in the brackets on the right-hand side is always positive for $-\infty < \sum_{i=1}^{p}\phi_i < \infty$. From Example 1, it follows that $\lim_{T\to\infty} \boldsymbol{D}^{-1}\boldsymbol{X}'(\boldsymbol{\Phi}^*)'\boldsymbol{\Phi}^*\boldsymbol{X}\boldsymbol{D}^{-1}$ is nonsingular. Consequently, Assumption 6 is satisfied.

Example 5. Here we assume that $\{\varepsilon_t\}$ obeys the model (2.3) with the changing point T_0, where $\{u_t\}$ is an independent and Gaussian process with $E(u_t) = 0$ and $Var(u_t) = \sigma_u^2$. Let τ or τ^* be positive integers such that $T = \tau + \tau^* + 1$ and the samples be obtained for $T_0 - \tau \le t \le T_0 + \tau^*$. We mean that τ and τ^* tend to infinity as the sample size T tends to infinity. Let $\alpha = \lim_{T\to\infty}(\tau/T) > 0$. In this case, noting (2.4), we can show

$$\lim_{T\to\infty} E\left(\lambda_{l,m}(T)\right) = \lim_{T\to\infty}\frac{1}{T}\sum_{t=T_0-\tau+p}^{T_0+\tau^*} E\left(\varepsilon_{t-l}\,\varepsilon_{t-m}\right) = \left[\alpha\frac{\theta^{|l-m|}}{1-\theta^2} + (1-\alpha)\frac{\theta^{*|l-m|}}{1-\theta^{*2}}\right]\sigma_u^2$$

and $\lim_{T\to\infty} Var\left(\lambda_{l,m}(T)\right) = 0$ for $l,m = 0,\ldots,p$. Here we do not show the way of the derivation of these equations because it is so much tedious. It follows that

$$\operatorname{p}\lim_{T\to\infty} \lambda_{l,m}(T) = \left[\alpha\frac{\theta^{|l-m|}}{1-\theta^2} + (1-\alpha)\frac{\theta^{*|l-m|}}{1-\theta^{*2}}\right]\sigma_u^2 \tag{2.14}$$

for $l,m = 0,\ldots,p$. The right-hand side of (2.14) is a weighted sum of two autocovariances of two stationary autoregressive processes. This type of autocovariance function is positive definite. Hence, in this case, $\operatorname{p}\lim_{T\to\infty}\boldsymbol{\Lambda}(T)$ is nonsingular. Then, Assumptions 4 and 5 are satisfied.

Example 6. Let $\{\varepsilon_t\}$ obey the model (2.5) with the changing point T_0 , where $\{u_t\}$ is an independent and Gaussian process. We suppose that the samples are derived in the same manner as in Example 5. We can obtain the similar result to that in the case of the process (2.3), i.e.,

$$\operatorname{p}\lim_{T\to\infty} \lambda_{l,m}(T) = \begin{cases} \left[\alpha(1+\psi^2) + (1-\alpha)(1+\psi^{*2})\right]\sigma_u^2 \,, & l = m \,, \\ \left[\alpha\psi + (1-\alpha)\psi^*\right]\sigma_u^2 \,, & |l-m| = 1 \,, \\ 0 \,, & 2 \le |l-m| \le p \,. \end{cases}$$

Thus, $\operatorname{p}\lim_{T\to\infty}\boldsymbol{\Lambda}(T)$ is written as a weighted sum of two autocovariances of two stationary moving-average processes and then is nonsingular. Hence, Assumptions 4 and 5 are satisfied.

3. RELATIONS BETWEEN ACCURACY OF THE ESTIMATORS AND COVARIANCE STRUCTURES

We shall consider relations between the accuracy of the estimators of the regression coefficients and the fitted covariance structures. We can write the difference $\tilde{\boldsymbol{\beta}} - \tilde{\boldsymbol{\beta}}_G$ as follows:

$$\tilde{\beta}-\tilde{\beta}_G=\left\{X'\Sigma_\varepsilon^{-1}X-X'\left[\Sigma_\varepsilon^{-1}-S\left(\hat{\phi}\right)^{-1}\right]X\right\}^{-1}X'\left[\Sigma_\varepsilon^{-1}-S\left(\hat{\phi}\right)^{-1}\right]\left(y-X\tilde{\beta}_G\right).$$

Thus $\tilde{\beta}-\tilde{\beta}_G$ is represented as a function of $\Sigma_\varepsilon^{-1}-S\left(\hat{\phi}\right)^{-1}$, which is considered to be a quantity measuring a kind of the closeness between the two covariance matrices.

Let us regard $\{\varepsilon_t\}$ as an AR(p) process expressed as $\varepsilon_t=\sum_{i=1}^{p}\phi_i\varepsilon_{t-i}+z_t$, where z_t's are considered to be mutually independent random variables with $E(z_t)=0$ and $Var(z_t)=\sigma_z^2$. In this case we use $\hat{\sigma}_z^{-2}\left(\hat{\Phi}^*\right)'\hat{\Phi}^*$ as $S\left(\hat{\phi}\right)^{-1}$, where $\hat{\sigma}_z^2$ is an estimator of σ_z^2 , and obtain the estimator $\hat{\beta}$ of β . Then the difference $\tilde{\beta}-\tilde{\beta}_G$ can be represented as a function of $\Sigma_\varepsilon^{-1}-\hat{\sigma}_z^{-2}\left(\hat{\Phi}^*\right)'\hat{\Phi}^*$. We can consider that this difference between two covariance structures is a measure of a kind of the closeness between the true error process and the fitted AR(p) model. On the other hand, if we consider the error process $\{\varepsilon_t\}$ to be a sequence of mutually independent random variables with $E(\varepsilon_t)=0$ and $Var(\varepsilon_t)=\sigma_\varepsilon^2$, we can apply the method of the OLS for estimating β . It follows that $S\left(\hat{\phi}\right)^{-1}$ is written as $\hat{\sigma}_\varepsilon^{-2}I_T$, where $\hat{\sigma}_\varepsilon^2$ is an estimator of σ_ε^2 . Then, $\tilde{\beta}-\tilde{\beta}_G$, i.e., $b-\tilde{\beta}_G$ is represented as a function of $\Sigma_\varepsilon^{-1}-\hat{\sigma}_\varepsilon^{-2}I_T$, which can be considered as a measure of a kind of the closeness between the true error process and the mutually independent sequence. It depends on the fact that which of the fitted covariance structures $\hat{\sigma}_z^{-2}\left(\hat{\Phi}^*\right)'\hat{\Phi}^*$ and $\hat{\sigma}_\varepsilon^{-2}I_T$ is closer to the true one Σ_ε^{-1} whether $\hat{\beta}$ is more effective than b or not.

Since, in our studies, it is difficult to investigate further properties of the estimators theoretically, we show some numerical results on comparison of $\hat{\beta}$ with b by simulation studies. If $\{\varepsilon_t\}$ is a stationary process satisfying some conditions, b is asymptotically efficient in a sense. However, in this paper, we treat cases when the sample sizes are small and $\{\varepsilon_t\}$'s obey one of the models (2.3) and (2.5), which are nonstationary.

Let $x_t=(1\ ,\ t)'$ and $\beta=(200\ ,\ 0.4)'$, i.e., $y_t=200+0.4\ t+\varepsilon_t$. We use the simulated data of $\{\varepsilon_t\}$'s which obey one of the models (2.3) and (2.5), where $\{u_t\}$ is a sequence of mutually independent and normal random variables with $E(u_t)=0$ and $Var(u_t)=5$. We set the changing point T_0 and the coefficients θ , θ^* , ψ and ψ^* at various values. We study the cases when the differences between θ and θ^* or ψ and ψ^* are large and small.

We predict the trend $E(y_{T+1})=x'_{T+1}\beta$ at time point $T+1$ by the two predictors $x'_{T+1}b$ and $x'_{T+1}\hat{\beta}$. We repeat this simulation study 2000 times by using 2000 different sets of realizations of $u_1,\cdots,u_T$. We compare $\hat{\beta}$ with b by their mean squared prediction errors (MSPE) . For example, the MSPE of b is given by

$$\mathrm{MSPE}\left(x'_{T+1}b\right)\ =\ \frac{1}{2000}\sum_{i=1}^{2000}\left[x'_{T+1}\left(b_{[i]}-\beta\right)\right]^2\ ,$$

where $b_{[i]}$ is the value of b obtained by the ith simulation. We put $T=50$.

Table 1 shows the results when $\{\varepsilon_t\}$'s obey the model (2.3) with several changing points and we set θ and θ^* to be 0.1 and 0.9 , respectively. We obtain the MSPE's of $\hat{\beta}$ by fitting AR models of orders 1 , 2 and 3 to $\{\varepsilon_t\}$'s. We also obtain the ratios $\mathrm{MSPE}\left(x'_{T+1}\hat{\beta}\right)$ / $\mathrm{MSPE}\left(x'_{T+1}b\right)$, which are shown in parentheses in the table. When the ratio is less than one, $\hat{\beta}$ is better than b . Furthermore, assuming that we know Σ_ε completely, we get $Var\left(x'_{T+1}b\right)$ and $Var\left(x'_{T+1}\tilde{\beta}_G\right)$, which are used for comparing with our simulation studies. Table 2 shows the simulation results when $\theta = 0.9$ and $\theta^* = 0.1$.

Table 1 .
Mean squared prediction errors of $\boldsymbol{b}$ and $\hat{\boldsymbol{\beta}}$.
$\{\varepsilon_t\}$: model (2.3) with $\theta = 0.1$, $\theta^* = 0.9$.

T_0	$\mathrm{MSPE}(\boldsymbol{x}'_{T+1}\boldsymbol{b})$	$\mathrm{MSPE}(\boldsymbol{x}'_{T+1}\hat{\boldsymbol{\beta}})$ AR(1)		AR(2)		AR(3)	
16	23.28	22.92	(0.98)	340.2	(14.6)	363.9	(15.6)
26	20.30	20.73	(1.02)	21.82	(1.07)	210.9	(10.4)
36	11.92	13.15	(1.10)	14.48	(1.22)	15.63	(1.31)

T_0	$Var\,(\boldsymbol{x}'_{T+1}\boldsymbol{b})$	$Var\,(\boldsymbol{x}'_{T+1}\tilde{\boldsymbol{\beta}}_G)$
16	22.61	13.67
26	19.72	5.362
36	11.53	1.899

Table 2 .
Mean squared prediction errors of $\boldsymbol{b}$ and $\hat{\boldsymbol{\beta}}$.
$\{\varepsilon_t\}$: model (2.3) with $\theta = 0.9$, $\theta^* = 0.1$.

T_0	$\mathrm{MSPE}(\boldsymbol{x}'_{T+1}\boldsymbol{b})$	$\mathrm{MSPE}(\boldsymbol{x}'_{T+1}\hat{\boldsymbol{\beta}})$ AR(1)		AR(2)		AR(3)	
16	2.697	2.375	(0.88)	2.215	(0.82)	2.111	(0.78)
26	2.569	2.311	(0.90)	2.335	(0.91)	2.389	(0.93)
36	4.826	2.977	(0.62)	2.854	(0.59)	2.851	(0.59)

T_0	$Var\,(\boldsymbol{x}'_{T+1}\boldsymbol{b})$	$Var\,(\boldsymbol{x}'_{T+1}\tilde{\boldsymbol{\beta}}_G)$
16	2.517	0.704
26	2.390	0.894
36	4.518	0.995

Table 1 shows that $\boldsymbol{b}$ is better than $\hat{\boldsymbol{\beta}}$ in all the cases except only one. When we fit AR(1) models to $\{\varepsilon_t\}$'s for $T_0 = 16$ and 26 , however, there are no great differences between the MSPE of $\hat{\boldsymbol{\beta}}$ and that of $\boldsymbol{b}$. On the other hand Table 2 shows that $\hat{\boldsymbol{\beta}}$ is much better than $\boldsymbol{b}$. In particular, it seems that, for $T_0 = 36$, $\{\varepsilon_t\}$'s are well fitted by AR models.

Table 3 shows the results when the differences of θ and θ^* are small and then $\{\varepsilon_t\}$'s might be regard as stationary AR(1) models. Then, we may expect that $\hat{\boldsymbol{\beta}}$ is better than $\boldsymbol{b}$. Here we set T_0 to be 26 . However, we have $\mathrm{MSPE}\left(\boldsymbol{x}'_{T+1}\boldsymbol{b}\right) < \mathrm{MSPE}\left(\boldsymbol{x}'_{T+1}\hat{\boldsymbol{\beta}}\right)$ in all the cases in Table 3. This means that $\hat{\boldsymbol{\beta}}$ is not better than $\boldsymbol{b}$ even if $\{\varepsilon_t\}$'s seem to obey nearly stationary AR(1) models.

Table 3 .
Mean squared prediction errors of $\boldsymbol{b}$ and $\hat{\boldsymbol{\beta}}$.
$\{\varepsilon_t\}$: model (2.3) with $T_0 = 26$.

θ	θ^*	$\mathrm{MSPE}(\boldsymbol{x}'_{T+1}\boldsymbol{b})$	$\mathrm{MSPE}(\boldsymbol{x}'_{T+1}\hat{\boldsymbol{\beta}})$ AR(1)		AR(2)		AR(3)	
0.10	0.15	0.571	0.582	(1.02)	0.602	(1.05)	0.622	(1.09)
0.50	0.55	1.882	1.937	(1.03)	2.000	(1.06)	2.054	(1.09)
0.85	0.90	23.37	23.43	(1.00)	28.05	(1.20)	7×10^8	(3×10^7)

θ	θ^*	$Var\,(\boldsymbol{x}'_{T+1}\boldsymbol{b})$	$Var\,(\boldsymbol{x}'_{T+1}\tilde{\boldsymbol{\beta}}_G)$
0.10	0.15	0.555	0.553
0.50	0.55	1.835	1.769
0.85	0.90	22.58	18.75

In Tables 1 and 3, MSPE's of $\hat{\boldsymbol{\beta}}$ have extremely large values for the several cases when we fit AR(2) and AR(3) models to $\{\varepsilon_t\}$. Since we deal with the cases when the sample sizes are small, the prediction errors occasionally could be large in their absolute values. We newly obtain MSPE's of $\hat{\boldsymbol{\beta}}$ after removing a few sequences of normal random numbers such that the absolute values of the prediction errors are extremely large, i.e., $\left|\boldsymbol{x}'_{T+1}\left(\hat{\boldsymbol{\beta}} - \boldsymbol{\beta}\right)\right| > 50.0$ out of 2000 repetitions. The result shows that we always have $\text{MSPE}\left(\boldsymbol{x}'_{T+1}\boldsymbol{b}\right) < \text{MSPE}\left(\boldsymbol{x}'_{T+1}\hat{\boldsymbol{\beta}}\right)$ but there are no great differences between $\boldsymbol{b}$ and $\hat{\boldsymbol{\beta}}$.

The following Tables 4-6 are the results when $\{\varepsilon_t\}$'s obey the model (2.5) with several changing points. We study the cases when the differences between ψ and ψ^* are large and small.

Table 4 .
Mean squared prediction errors of $\boldsymbol{b}$ and $\hat{\boldsymbol{\beta}}$.
$\{\varepsilon_t\}$: model (2.5) with $\psi = 0.1$, $\psi^* = 0.9$.

		$\text{MSPE}\left(\boldsymbol{x}'_{T+1}\hat{\boldsymbol{\beta}}\right)$					
T_0	$\text{MSPE}(\boldsymbol{x}'_{T+1}\boldsymbol{b})$	AR(1)		AR(2)		AR(3)	
16	1.348	1.432	(1.06)	1.417	(1.05)	1.490	(1.11)
26	1.333	1.402	(1.05)	1.410	(1.06)	1.468	(1.10)
36	1.221	1.281	(1.05)	1.304	(1.07)	1.360	(1.11)

T_0	$Var\left(\boldsymbol{x}'_{T+1}\boldsymbol{b}\right)$	$Var\left(\boldsymbol{x}'_{T+1}\tilde{\boldsymbol{\beta}}_G\right)$
16	1.344	1.289
26	1.332	1.208
36	1.208	0.952

Table 5 .
Mean squared prediction errors of $\boldsymbol{b}$ and $\hat{\boldsymbol{\beta}}$.
$\{\varepsilon_t\}$: model (2.5) with $\psi = 0.9$, $\psi^* = 0.1$.

		$\text{MSPE}\left(\boldsymbol{x}'_{T+1}\hat{\boldsymbol{\beta}}\right)$					
T_0	$\text{MSPE}(\boldsymbol{x}'_{T+1}\boldsymbol{b})$	AR(1)		AR(2)		AR(3)	
16	0.613	0.625	(1.02)	0.634	(1.04)	0.652	(1.06)
26	0.627	0.642	(1.02)	0.645	(1.03)	0.666	(1.06)
36	0.745	0.752	(1.01)	0.759	(1.02)	0.770	(1.03)

T_0	$Var\left(\boldsymbol{x}'_{T+1}\boldsymbol{b}\right)$	$Var\left(\boldsymbol{x}'_{T+1}\tilde{\boldsymbol{\beta}}_G\right)$
16	0.602	0.563
26	0.614	0.568
36	0.738	0.604

Table 6 .
Mean squared prediction errors of $\boldsymbol{b}$ and $\hat{\boldsymbol{\beta}}$.
$\{\varepsilon_t\}$: model (2.5) with $T_0 = 26$.

			$\text{MSPE}\left(\boldsymbol{x}'_{T+1}\hat{\boldsymbol{\beta}}\right)$					
ψ	ψ^*	$\text{MSPE}(\boldsymbol{x}'_{T+1}\boldsymbol{b})$	AR(1)		AR(2)		AR(3)	
0.10	0.15	0.540	0.552	(1.02)	0.565	(1.05)	0.582	(1.08)
0.50	0.55	0.968	1.014	(1.05)	1.012	(1.05)	1.052	(1.09)
0.85	0.90	1.452	1.546	(1.06)	1.512	(1.04)	1.600	(1.10)

ψ	ψ^*	$Var\left(\boldsymbol{x}'_{T+1}\boldsymbol{b}\right)$	$Var\left(\boldsymbol{x}'_{T+1}\tilde{\boldsymbol{\beta}}_G\right)$
0.10	0.15	0.534	0.533
0.50	0.55	0.960	0.949
0.85	0.90	1.442	1.411

As shown in the above tables, $\boldsymbol{b}$ is always better than $\hat{\boldsymbol{\beta}}$ in the cases of the model (2.5) with the changing points. Since $[\Sigma_\varepsilon]_{i,j} = 0$ for i and j such that $|i-j| \geq 2$, the covariance structure of $\{\varepsilon_t\}$ can be regarded to be nearer to that of an independent process than that of an AR process.

It seems that the accuracy of $\hat{\boldsymbol{\beta}}$ depends on the closeness between the covariance structure of $\{\varepsilon_t\}$ and that of a fitted AR model and the accuracy of $\boldsymbol{b}$ depends on the closeness between the covariance structure of $\{\varepsilon_t\}$ and that of a sequence of independent errors with the same variances . Even if we misspecify the covariance structure of $\{\varepsilon_t\}$ and fit an AR(1) model to $\{\varepsilon_t\}$, $\hat{\boldsymbol{\beta}}$ is as accurate as $\boldsymbol{b}$ for almost all the cases which we studied.

REFERENCES

1. U. Grenander and M. Rosenblatt, *Statistical Analysis of Stationary Time Series*, John Wiley, New York (1957).
2. T. W. Anderson, *The Statistical Analysis of Time Series*, John Wiley, New York (1971).
3. M. Huzii and Y. Toyooka, *Rep. Statist. Appl. Res. JUSE*, **24**, 191-199 (1977).
4. Y. Toyooka, *Math. Japon.*, **25**, 525-532 (1980).
5. W. A. Fuller, *Introduction to Statistical Time Series*, John Wiley, New York (1976).

Stat. Sci. & Data Anal., pp. 61-75
K. Matsusita *et al.* (Eds)

Estimation of Regression Parameters When the Errors are Autocorrelated

EDWARD J. CHEN and A. K. Md. E. SALEH
Statistics Canada
Carleton University, Ottawa, Canada

Abstract. For the linear multiple regression model $y_j = \beta_1 x_{1j} + \ldots + \beta_p x_{pj} + e_j$, where $e_j = \rho e_{j-1} + v_j$, $|\rho| < 1$, $j = 1, 2, \ldots, T$ and v_j is $N(0, \sigma^2)$, we consider the risk analysis of five estimators, namely: (i) unrestricted estimator (UE), (ii) restricted estimator (RE), (iii) Judge and Bock [1] preliminary test estimator (PTE), (iv) a proposed shrinkage estimator (SE) and (v) a proposed shrinkage preliminary test estimator (PT2). By simulation study, we show that the shrinkage estimator performs uniformly better than the unrestricted estimator as well as the two PTE's under the risk analysis.

Key words: Preliminary test; Shrinkage estimation; Risk analysis; Simulation

1. INTRODUCTION

Consider the linear multiple regression model

$$\mathbf{Y} = X\beta + \mathbf{e}, \tag{1}$$

where $\beta = (\beta_1, \beta_2, \ldots, \beta_p)'$ is a vector of unknown parameters, X is an $T \times p$ design matrix of known constants, $T > p \geq 1$. The error terms $\mathbf{e} = (e_1, e_2, \ldots, e_T)'$ are autocorrelated which normally distributed with mean $\mathbf{0}$ and variance $\sigma^2\Phi$, i.e., $\mathbf{e} \sim N(\mathbf{0}, \sigma^2\Phi)$ and

$$\phi_{T\times T} = \begin{bmatrix} 1 & \rho & \rho^2 & \ldots & \rho^{T-1} \\ \rho & 1 & \rho & \ldots & \rho^{T-2} \\ \vdots & \vdots & \vdots & & \vdots \\ \rho^{T-1} & \rho^{T-2} & \rho^{T-3} & \ldots & 1 \end{bmatrix} \tag{2}$$

The jth element of $\mathbf{e}$ is given by a first-order stationary autoregressive process $e_j = \rho e_{j-1} + v_j$, where $|\rho| < 1$ and $\mathbf{v} = (v_1, v_2, \ldots, v_T)'$ has the properties $E(\mathbf{v}) = 0$ and $E(\mathbf{vv}') = \sigma_v^2 I$.

We are primarily interested in the estimation of the unknown parameters β when the variance σ^2 and correlation coefficient ρ are both unknown. It is also suspected that $\rho = 0$ and $\beta = 0$ may hold in some cases. Let $\tilde{\beta}$ be the generalized least squares estimates (GLS) when ρ is estimated consistently by $\hat{\rho}$ and $\hat{\beta}$ be the ordinary least square (OLS) estimator of β when $\rho = 0$. Then, in general $\tilde{\beta}$ has smaller dispersion than $\hat{\beta}$ and the dispersion increases with the increasing value of ρ. In this context, it is reasonable to consider a preliminary test for the null hypothesis $H_0 : \rho = 0$ and then use $\hat{\beta}$ or $\tilde{\beta}$ according to whether null hypothesis is accepted or rejected. Such a compromised estimator is in term described as a preliminary test estimator (PTE). We denote the estimation of β under

PTE as $\hat{\beta}^{PT} \cdot \hat{\beta}^{PT}$ has been studied by Judge and Bock [1]. The theory of PTE was first advocated by Bancroft [2] and the asymptotic theory have been thoroughly studied by Saleh and Sen [3]. The finite sample simulation of Judge and Bock [1] show that the risks of PTE lies in between the risks of $\tilde{\beta}$ and $\hat{\beta}$. Thus there is not much gain in using PTE of β as opposed to $\tilde{\beta}$.

The objective of this paper is to suggest a new estimator of β which will outperform $\tilde{\beta}$ uniformly as well as outperform $\hat{\beta}^{PT}$. In this study, we follow Saleh and Sen [4] and Sen and Saleh [5] to derive new estimators of β. Let $\hat{\beta}^S$ (when $\rho = 0$) and $\tilde{\beta}^S$ (when $\rho \neq 0$) be the Saleh and Sen estimator of β,, we then show by simulation that $\hat{\beta}^S$ and $\tilde{\beta}^S$ have uniformly smaller risks than $\hat{\beta}$ and $\tilde{\beta}$ respectively. Also, we consider a new PTE, $\hat{\beta}^{PT2}$ which is taken as the value $\hat{\beta}^S$ or $\tilde{\beta}^S$ according whether H_0 is accepted or not. In this situation $\hat{\beta}^{PT2}$ has uniform smaller risks than $\hat{\beta}^{PT}$, and $\tilde{\beta}^S$ dominates all the estimators uniformly.

In section 2, we discuss various two-stage estimators of the correlation coefficient ρ, i.e. the widely used Cochran-Orcutt, Durbin and Prais-Winsten estimators [6,7]. The two different test statistics, namely, Durbin and Watson's statistics and Berenblutt and Webb g-statistic for testing $H_0 : \rho = 0$ against the alternative $H_1 : \rho > 0$ are also presented in this section. The new estimators of β, namely, $\hat{\beta}^S$ or $\tilde{\beta}^S$ and the proposed preliminary estimator, $\hat{\beta}^{PT2}$, which is based on the outcome of the testing (i.e., either $\hat{\beta}^S$ or $\tilde{\beta}^S$) are presented in section 3. The simulation study, which is designed to examine the performance of various estimators, is discussed in section 4. Finally, the concluding remarks and the supporting tables and graphs are contained in section 5 and appendix respectively.

2. ESTIMATION AND TESTS FOR AUTOCORRELATION

The unknown parameters β in (1.1) are usually estimated by Generalized Lease Squares (GLS) or Ordinary Least Squares (OLS) i.e. for $\rho \neq 0$ or $\rho = 0$ when ρ is known. However, the assumption of a known value for ρ is unrealistic. We will discuss various consistent estimation procedures to estimate ρ along with the estimate of the unknown parameters β. The statistical tests for identifying the presence of autocorrelation are also discussed in this section.

2.1 Estimation of β and Various Estimators of ρ

We consider the estimation of the unknown parameters β in (1.1). If the autocorrelation coefficient ρ is known, then (1.1) can be rewritten as

$$PY = PX\beta + Pe, \tag{3}$$

where

$$P'P = (1-\rho^2)\Phi^{-1} \tag{4}$$

and Φ^{-1} is the inverse of Φ and the transformation matrix P is defined as

$$\mathbf{P} = \begin{bmatrix} \sqrt{1-\rho^2} & 0 & 0 & \dots & 0 & 0 \\ -\rho & 1 & 0 & \dots & 0 & 0 \\ \vdots & \vdots & \vdots & & \vdots & \vdots \\ 0 & 0 & 0 & \dots & -\rho & 1 \end{bmatrix} \tag{5}$$

then the GLS of β from (2.1) is

$$\tilde{\beta} = (X'\Phi^{-1}X)^{-1}X'\Phi^{-1}\mathbf{Y} \tag{6}$$

and has mean β and covariance matrix $\sigma^2(X'\Phi^{-1}X)^{-1}$. When $\rho = 0$, (2.4) reduces to the ordinary least square estimator, i.e.

$$\hat{\beta} = (X'X)^{-1}X'\mathbf{Y} \tag{7}$$

which is also the best linear unbiased estimator of β.

However, in most practical situations, the autocorrelation coefficient is usually unknown. If a consistent estimator of $\hat{\rho}$ can be found, then a form of GLS estimation (2.4) is employed by replacing ρ by $\hat{\rho}$, i.e., the elements ρ's in Φ are replace by $\hat{\rho}$'s. There are numerous methods available to estimate ρ. Judge and Bock [1] consider the following three two-stage estimators when ρ is unknown.

The three two-stage estimator of ρ are: (1) the Cochrane-Orcutt Estimator (CO), (2) the Durbin Estimator (Durbin) and (3) the Prais-Winsten Estimator (PW). These three estimators basically use different forms of the transformation matrix of P in (2.3).

Two test-statistics are used to test the existence of autocorrelation, namely, the Durbin and Watson $\hat{\mathrm{d}}$ statistic and the Berenblutt and Webb g_1 statistic. The decision rules for the test statistics are based on the value of the lower limit d_L and upper limit d_U in the usual context. They are discussed in more details in Judge and Bock [1]. These two test statistics will be employed in the study to examine their relative powers in the performance of the estimators.

3. PROPOSED ESTIMATORS AND RISK ANALYSIS

3.1 The Estimators

In this section, we consider six different estimators of β and evalute them using the unweighted and the weighted *risk functions*. The six different estimators of which the first three are as follows:

- (i) The *Unrestricted Estimators* (UE): $\tilde{\beta}$, where $\tilde{\beta}$ is given by (2.4) and $\hat{\rho}$ is an estimate obtained by one of the two-stage consistent estimators discussed in section 2.1.
- (ii) The *Restricted Estimator* (RE): $\hat{\beta}$, where $\hat{\beta}$ is given by (2.5) and ρ is assumed to be zero.
- (iii) The *Preliminary Test Estimator* (PTE): $\hat{\beta}^{PT}$, where $\hat{\beta}^{PT}$ is of the form [

$$\hat{\beta}^{PT} = I(L^* < C_\alpha)\hat{\beta} + I(L^* \geq C_\alpha)\tilde{\beta} \tag{8}$$

  where L^* is the test-statistic for hypothesis $H_0 : \rho = 0$ against the one-sided alternative hypothesis $H_1 : \rho > 0$. The indicator function I has the values 1 or 0 depending on the outcome of testing at significant level of α with critical value C_α. The test statistic is modified later in such a way that the acceptance of H_0 is made only when the computed statistic exceeds the upper limit d_U in the tests. Finally, we propose the two shrinkage estimators, as follows.

We follow Saleh and Sen [4] and Sen and Saleh [5] to derive the following estimators of β when $\rho \neq 0$ and $\rho = 0$. The estimators of β^S are of the form

- (iv) when $\rho \neq 0$, then the shrinkage estimator $\tilde{\beta}^S$ is defined by

$$\tilde{\beta}^S = \{1 - (p-2)(T-p+1)(T-p-1)^{-1}\tilde{\sigma}^2(\tilde{\beta}'(X'\hat{\Phi}_{\hat{\rho}}^{-1}X)\tilde{\beta})^{-1}\}\tilde{\beta} \tag{9}$$

where

$$\tilde{\sigma}^2 = \frac{(\mathbf{Y} - X\tilde{\beta})'\hat{\Phi}_{\hat{\rho}}^{-1}(\mathbf{Y} - X\tilde{\beta})}{T-p+1}$$

- (v) when $\rho = 0$, then the second shrinkage estimator $\hat{\beta}^S$ is given by

$$\hat{\beta}^S = \{1 - (p-2)(T-p+2)(T-p)^{-1}\hat{\sigma}^2(\hat{\beta}'(X'X)\hat{\beta})^{-1}\}\hat{\beta} \tag{10}$$

where

$$\hat{\sigma}^2 = \frac{(\mathbf{Y} - X\hat{\beta})'(\mathbf{Y} - X\hat{\beta})}{T-p+2}$$

The value p is the number of parameters in the model, ($p > 2$) with the number of observations T. Next, we consider the new preliminary test estimator which is defined as follows using $\tilde{\beta}^S$ and $\hat{\beta}^S$. Thus,

$$\hat{\beta}^{PT2} = \hat{\beta}^S I(L^* < C_\alpha) + \tilde{\beta}^S I(L^* \geq C_\alpha) \tag{11}$$

3.2 Risk Analysis

Let β^* denote any estimator of β from section 3.1 and then the *unweighed loss function* is defined as

$$L = (\beta^* - \beta)'(\beta^* - \beta) \tag{12}$$

then the risk function is the average of the loss function

$$R(\beta^*) = E[(\beta^* - \beta)'(\beta^* - \beta)] \tag{13}$$

Similarly, we have the weighted loss function is defined to be

$$L = (\beta^* - \beta)'W(\beta^* - \beta) \tag{14}$$

where we take the weight matrix as $W = (X'\Phi^{-1}X)\sigma^{-2}$ and the weighted risk function is given by

$$R(\beta^*) = E[(\beta^* - \beta)'W(\beta^* - \beta)] \tag{15}$$

when ρ is known, the weighted risk function for estimator based on generalized least square is equal to the number of the parameters in the model, i.e. p.

We are going to evaluate the performance of different estimators of β by their relative magnitude of the risk values.

4. DESIGN OF MONTE CARLO EXPERIMENT

We use a Monte Carlo experiment to evaluate and compare the sampling performance of different estimators for the unknown parameters β.

The criterion used to evaluate the estimator's performance is based on the expected unweighed square error loss as well as the weighted square error loss, namely, we compare them on the basis of estimators' unweighed risk values and quadratic risk values. For example, the unweighed risk function for $\hat{\beta}^{PT}$ based on the Durbin-Watson test statistic is

$$E[(\hat{\beta}^{PT}-\beta)'(\hat{\beta}^{PT}-\beta)] = E[(\; (\; I_{(-2,0]}(\hat{d}-d_U)\tilde{\beta} + I_{(0,2)}(\hat{d}-d_U)\hat{\beta}) - \beta)' \\ (\; I_{(-2,0]}(\hat{d}-d_U)\tilde{\beta} + I_{(0,2)}(\hat{d}-d_U)\hat{\beta}) - \beta)]$$

and the weighted risk function is

$$E[(\hat{\beta}^{PT}-\beta)'W(\hat{\beta}^{PT}-\beta)] = E[(\; (\; I_{(-2,0]}(\hat{d}-d_U)\tilde{\beta} + I_{(0,2)}(\hat{d}-d_U)\hat{\beta}) - \beta)'W \\ (\; I_{(-2,0]}(\hat{d}-d_U)\tilde{\beta} + I_{(0,2)}(\hat{d}-d_U)\hat{\beta}) - \beta)]$$

where $W = (X'\Phi^{-1}X)\sigma^{-2}$.

In the Monte Carlo study, data of two hundred samples of size 25 were generated using the orthornormal statistical model

$$\mathbf{Y}_t = X\beta_0 = \mathbf{e}_t \tag{16}$$

where

$$\beta_0' = (13.9, \;\; 10.79, \;\; 6.13, \;\; 3.01, \;\; 10.81)$$

and

$$X'X = I_T.$$

The $\mathbf{e}_t$ in (4.3) are generated by a first order auto-regressive process

$$\mathbf{e}_t = \rho\mathbf{e}_{t-1} + \mathbf{v}_t, \tag{17}$$

where $\rho \geq 0$ is the autocorrelation coefficient. The $\mathbf{v}$'s are generated from $N(0,\ 60.8)$ and assumed to be i.i.d.. The generated $\mathbf{v}$'s are then used to create $\mathbf{e}_t$ of (4.4) by varying the autocorrelation in steps of tenth from 0.9 to 0.9. We obtain the maximum likelihood estimators of the parameters for the 200 samples in order to compute the test statistics. Note that we only consider the case that ρ is non-negative, similar results can be obtained for $\rho < 0$.

The test statistics discussed in section 2 have been modified such that the acceptance of H_0 is made only when the computed statistic exceeds the d_U for a specified significant level. This practice is recommended especially when the model does not include the intercept term. The test is based on the difference between the d_U and the test-statistics for significance levels of 0.01 and 0.05. $\tilde{\beta}$ or $\hat{\beta}$ as well as $\tilde{\beta}^S$ or $\hat{\beta}^S$ are chosen depending upon whether the null hypothesis of zero autocorrelation is rejected or not in the two preliminary test estimators.

5. EMPIRICAL RESULTS AND CONCLUDING REMARKS

Most of the empirical results in the areas of using the usual estimators (i.e., before using the proposed shrinkage estimators) from our simulation study have been reached by several previous studies. In particular, the study by Griliches and Rao [8] in comparison of different two-stage autocorrelation estimators, the study by Berenblutt and Webb [9] on the relative powers of the test for autocorrelation. More importantly, the simulation study results by Judge and Bock [1] in the performances of the different pre-test estimators. In addition, we have followed Judge and Bock [1] approach as much as possible in order to have close comparisons between the studies.

Therefore, we are going to discuss the sampling performance results of the proposed shrinkage estimators in comparison to the usual regression estimators and Judge and Bock pre-test estimator. We will only discuss the empirical results based on the unweighed risk functions. Similar observations can be made to the ones based on weighted risk values. The results of the weighted risks are presented in Table 6 to Table 12.

The following results are based on one sampling experiment and the model chosen in our simulation study, one should therefore be careful in generalizing the inferences to other linear regression models, especially on the magnitudes of the results reported here.

The major findings include the following:

5.1 Empirical Risks Between the Proposed Shrinkage Estimators and the Usual Estimators Prior to Testing

In Table 1 and Table 2, we present the empirical risk for different estimators and the proposed shrinkage estimators. The shrinkage estimators have uniformly lower risks than their counter parts in the usual estimators across the entire ρ parameter space. The GLS estimator has the smallest risks values than the other two-stage estimators, the Durbin two-stage estimator of ρ has the lowest risk. This is true in both tables. The same results have been reported by Griliches and Rao [8] and Judge and Bock [1]. The risk values of Table 1 and Table 2 are plotted against ρ in Fig. 1 and Fig. 2.

5.2 Empirical Risks Between the Two Test Statistics $\hat{d}$ and g_1 Under Shrinkage Estimation

In Table 3 to Table 6, we present the empirical results for the two pre-test estimators $\hat{\beta}^{PT}$ and $\hat{\beta}^{PT2}$ under the two test statistics, namely, Durbin-Watson $\hat{d}$ statistic and Berenblutt and Webb g_1 statistic. We show the results with the significant levels of 0.01 and 0.05 respectively. Again, the pre-test estimator with shrinkage estimates $\hat{\beta}^{PT2}$ have uniformly lower risks than their counter parts $\hat{\beta}^{PT}$ across the entire ρ parameter space. In comparison of the two different test statistics $\hat{d}$ and g_1, the results tend to support that g_1 is superior to $\hat{d}$ for higher values of ρ, as claimed by Berenblutt and Webb [9]. However, the risks gains are relatively small when the tests are based on $\alpha = 0.05$. (See Fig. 3). Our results suggest that the use of $\hat{d}$ and g_1 yield the same risks when ρ is above 0.6 and under $\alpha = 0.05$.

5.3 Empirical Results Between the Two Preliminary Test Estimators, $\hat{\beta}^{PT}$ and $\hat{\beta}^{PT2}$ Under the Durbin Estimation of ρ

As reported by other studies, the Durbin estimator has the higher risk gains among the other two-stage estimators of ρ. In Fig. 4, we present different empirical functions for the Durbin estimator of ρ under the two pre-test estimators when $\alpha = 0.05$. The shrinkage estimator $\hat{\beta}^{PT2}$ has lower risks than the Judge and Bock pre-test estimator $\hat{\beta}^{PT}$ under the g_1 statistic. The results also show that GLS shrinkage estimator has the uniformly lowest risks across the entire ρ parameter space.

Acknowledgement

This research has been carried out under the NSERC grant No. A3088.

REFERENCES

1. G.G. Judge and M.E. Bock, *The Statistical Implications of Pretest ad Stein-Rule Estimators in Economics*. North-Holland, Amsterdam (1978).

2. T.A. Bancroft, On biases in estimation due to use of preliminary tests of significance. *Ann. Math. Statist.*, **15**, 190-204 (1944).

3. A.K.Md.E. Saleh and P.K. Sen, Nonparametric estimation of location parameter after a preliminary test on regression. *Ann. Statist.*, **6**, 154-168 (1978).

4. A.K.Md.E. Saleh and P.K. Sen, On shrinkage M-estimators of location parameters. *Commun. Statist. A-theory Methods*, **14**, 2313-2329 (1985).

5. P.K. Sen and A.K.Md.E. Salhe, On some shrinkage estimators of multivariate location. *Ann. Statist.*, **13**, 272-281 (1985).

6. J. Durbin, Estimation of parameters in time series regression models. *J. Roy. Statist. Soc.*, Series B., **22**, 139-153 (1960).

7. S.J. Prais and Winsten, Trend estimators and series correlation. *Cowles Commission Discussion Paper*, No. 383, Chicaco (1954).

8. Z. Griliches and P. Rao, Small sample properties of several two-stage regression methods in the context of autocorrelated errors. *J. Amer. Statist. Assoc.*, **64**, 253-272 (1969).

9. I.I. Berenblutt and G.I. Webb, A new test for autocorrelated errors in the linear regression model. *J. Roy. Statist. Soc.*, Series B., **35**, 33-50 (1973).

Fig. 1. UNWEIGHTED RISKS PRIOR TO TESTS

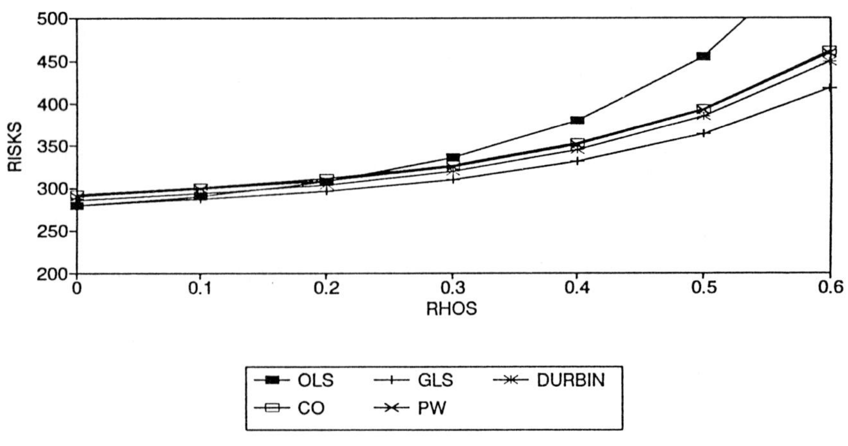

Fig. 2. UNWEIGHTED RISKS PRIOR TO TESTS
SHRINKAGE ESTIMATES

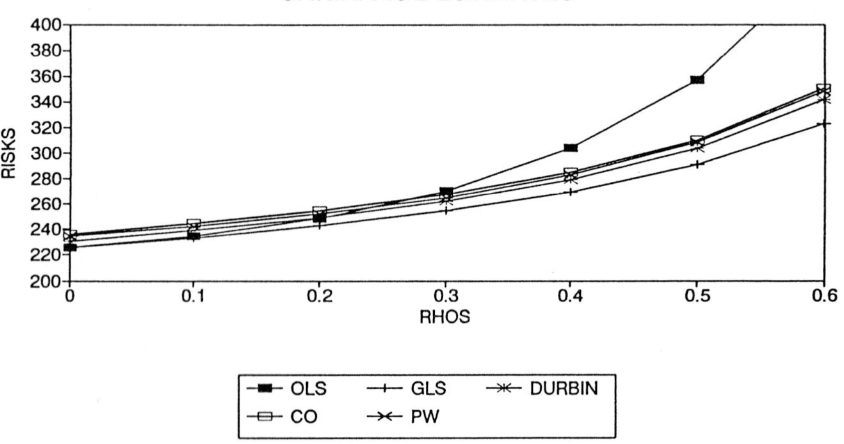

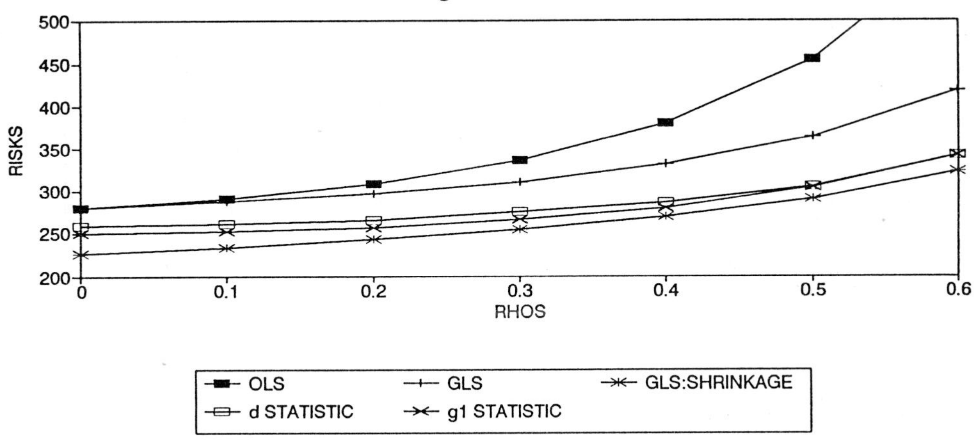
Fig 3. COMP. OF DIFF. TEST STATISTICS
BETWEEN d AND g1 UNDER SHRINKAGE EST.
500
450
400
350
300
250
200
RISKS
0
0.1
0.2
0.3
0.4
0.5
0.6
RHOS
OLS
GLS
GLS:SHRINKAGE
d STATISTIC
g1 STATISTIC

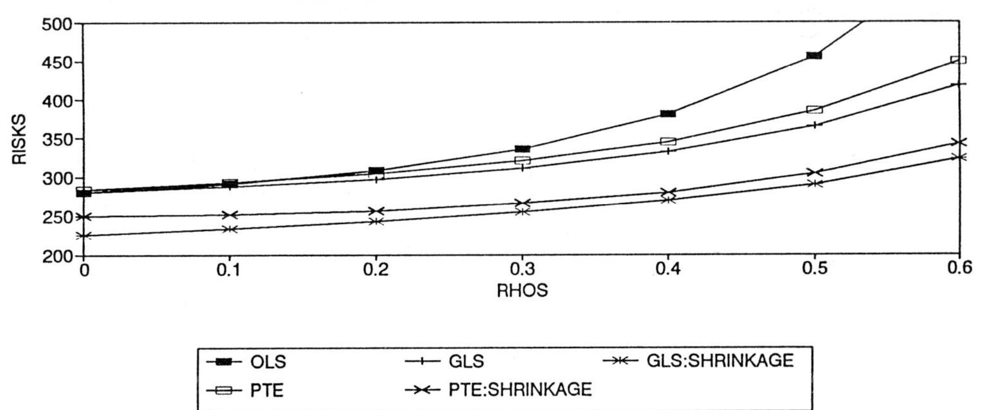
Fig. 4 COMP. OF DIFF. PTE ESTIMATORS
BETWEEN PTE AND SHRINKAGE PTE
500
450
400
350
300
250
200
RISKS
0
0.1
0.2
0.3
0.4
0.5
0.6
RHOS
OLS
GLS
GLS:SHRINKAGE
PTE
PTE:SHRINKAGE

Table 1. Empirical Risks for Different Estimators
Prior to Testing

RHO	OLS	GLS	Durbin	CO	PW
0.0	280.56	280.56	285.98	292.91	290.82
0.1	290.94	287.68	294.15	301.25	298.98
0.2	308.07	296.94	304.65	311.81	309.37
0.3	335.57	310.44	320.08	327.67	325.17
0.4	380.18	331.31	344.73	353.14	350.75
0.5	454.95	364.49	385.04	394.43	392.15
0.6	587.68	418.03	448.87	461.60	458.88
0.7	846.43	504.92	545.12	573.35	568.80
0.8	1436.42	643.91	716.11	777.24	768.53
0.9	3208.83	848.30	958.54	1244.43	1225.25

Table 2. Empirical Risks for Different Estimators
Prior to Testing - Shrinkage Estimates

RHO	OLS	GLS	Durbin	CO	PW
0.0	225.91	225.80	230.67	236.30	234.57
0.1	234.84	233.23	238.86	244.57	242.52
0.2	248.73	242.82	249.18	254.66	252.29
0.3	270.24	254.83	262.27	267.65	265.24
0.4	303.86	269.61	278.98	284.93	282.83
0.5	357.15	290.40	303.12	310.22	308.28
0.6	449.66	322.93	341.52	350.36	348.02
0.7	639.65	375.14	403.59	418.13	414.54
0.8	1100.84	457.53	507.77	548.22	542.14
0.9	2609.99	571.12	652.46	865.44	852.51

Table 3. Empirical Risk Values for PTE
Based on D-W and G1 statistic, alpha = 1%

RHO	Durbin		CO		PW	
	DW	G1	DW	G1	DW	G1
0.0	283.43	283.51	286.48	287.51	285.91	286.43
0.1	294.56	292.75	298.16	297.16	297.28	295.67
0.2	307.24	304.49	310.75	308.97	309.58	307.34
0.3	324.10	320.68	328.14	326.45	326.84	324.42
0.4	350.87	346.35	355.61	352.91	353.96	351.03
0.5	394.16	385.45	398.91	394.63	397.25	392.38
0.6	458.03	449.06	466.23	461.67	464.02	458.98
0.7	545.95	545.61	573.78	573.59	569.22	569.04
0.8	717.06	716.11	777.57	777.24	768.86	768.53
0.9	1009.81	958.54	1252.77	1244.43	1232.51	1225.25

Table 4. Empirical Risk Values for Shrinkage PTE
Based on D-W and G1 statistic, alpha = 1%

RHO	Durbin		CO		PW	
	DW	G1	DW	G1	DW	G1
0.0	271.15	260.54	273.59	263.43	273.17	262.68
0.1	274.03	262.93	276.62	266.14	276.10	265.07
0.2	280.15	268.05	282.67	271.06	281.78	269.84
0.3	289.07	277.00	291.59	280.29	290.63	278.41
0.4	298.38	286.41	301.16	290.28	299.77	288.74
0.5	326.04	305.72	329.05	312.55	327.66	310.69
0.6	357.54	343.72	362.50	352.42	360.52	350.11
0.7	407.99	407.07	422.07	421.34	418.51	417.78
0.8	508.92	507.77	548.93	548.22	542.85	542.14
0.9	707.72	652.46	881.06	865.44	866.75	852.51

Table 5. Empirical Risk Values for PTE
Based on D-W and G1 statistic, alpha = 5%

RHO	Durbin		CO		PW	
	DW	G1	DW	G1	DW	G1
0.0	283.90	283.96	288.19	288.69	286.90	287.11
0.1	292.65	292.25	296.77	297.11	295.38	295.32
0.2	304.14	303.76	308.71	309.59	306.95	307.42
0.3	321.25	320.21	326.36	326.94	324.42	324.66
0.4	346.52	344.84	353.05	353.23	351.05	350.85
0.5	385.30	385.04	394.48	394.42	392.21	392.15
0.6	448.87	448.87	461.59	461.59	458.88	458.88
0.7	545.12	545.12	573.35	573.35	568.80	568.80
0.8	716.11	716.11	777.24	777.24	768.53	768.53
0.9	958.54	958.54	1244.43	1244.43	1225.25	1225.25

Table 6. Empirical Risk Values for Shrinkage PTE
Based on D-W and G1 statistic, alpha = 5%

RHO	Durbin		CO		PW	
	DW	G1	DW	G1	DW	G1
0.0	258.32	249.92	261.46	253.44	260.66	252.33
0.1	260.79	252.11	263.78	255.61	262.82	254.07
0.2	265.52	255.71	268.70	259.65	267.01	257.65
0.3	275.13	266.14	278.29	270.51	276.41	268.34
0.4	286.18	279.71	290.04	285.61	288.32	283.52
0.5	305.37	303.93	312.30	311.01	310.37	309.08
0.6	341.81	341.81	350.64	350.64	348.29	348.29
0.7	403.59	403.59	418.13	418.13	414.54	414.54
0.8	507.77	507.77	548.22	548.22	542.14	542.14
0.9	652.46	652.46	865.44	865.44	852.51	852.51

Table 7. Empirical Weighted Risks for Different Estimators Prior to Testing

RHO	OLS	GLS	Durbin	CO	PW
0.0	5.15	5.15	5.76	6.02	5.98
0.1	5.32	5.19	5.77	6.03	5.99
0.2	5.51	5.23	5.79	6.03	5.99
0.3	5.75	5.28	5.82	6.02	5.99
0.4	6.11	5.33	5.84	6.02	5.99
0.5	6.69	5.39	5.86	6.03	6.00
0.6	7.68	5.43	5.87	6.07	6.04
0.7	9.48	5.47	5.92	6.17	6.15
0.8	13.10	5.50	6.02	6.47	6.46
0.9	21.31	5.56	6.04	7.38	7.39

Table 8. Empirical Weighted Risks for Different Estimators Prior to Testing -- Shrinkage Estimates

RHO	OLS	GLS	Durbin	CO	PW
0.0	3.84	3.81	4.32	4.52	4.50
0.1	3.98	3.82	4.30	4.51	4.48
0.2	4.12	3.85	4.32	4.50	4.48
0.3	4.28	3.90	4.34	4.49	4.47
0.4	4.51	3.95	4.36	4.48	4.46
0.5	4.88	4.02	4.38	4.48	4.46
0.6	5.52	4.09	4.41	4.50	4.48
0.7	6.80	4.18	4.48	4.57	4.56
0.8	9.61	4.26	4.59	4.80	4.79
0.9	16.60	4.33	4.61	5.54	5.55

Table 9. Empirical Weighted Risks for PTE
Based on D-W and G1 Statistic, alpha = 1%.

RHO	Durbin		CO		PW	
	DW	G1	DW	G1	DW	G1
0.0	5.40	5.39	5.54	5.56	5.54	5.54
0.1	5.46	5.44	5.62	5.62	5.61	5.60
0.2	5.52	5.49	5.68	5.66	5.67	5.64
0.3	5.56	5.54	5.71	5.69	5.69	5.67
0.4	5.61	5.57	5.74	5.71	5.72	5.69
0.5	5.64	5.57	5.78	5.73	5.76	5.70
0.6	5.66	5.58	5.82	5.76	5.80	5.74
0.7	5.64	5.63	5.87	5.87	5.85	5.85
0.8	5.74	5.72	6.15	6.14	6.14	6.13
0.9	5.84	5.73	7.06	7.01	7.07	7.02

Table 10. Empirical Weighted Risks for Shrinkage PTE
Based on D-W and G1 Statistic, alpha = 1%

RHO	Durbin		CO		PW	
	DW	G1	DW	G1	DW	G1
0.0	4.23	4.21	4.34	4.33	4.34	4.32
0.1	4.27	4.24	4.40	4.38	4.40	4.37
0.2	4.32	4.29	4.44	4.42	4.44	4.40
0.3	4.37	4.36	4.47	4.47	4.47	4.45
0.4	4.42	4.38	4.50	4.47	4.50	4.46
0.5	4.47	4.38	4.54	4.48	4.53	4.46
0.6	4.48	4.41	4.56	4.50	4.54	4.48
0.7	4.50	4.49	4.58	4.57	4.57	4.56
0.8	4.60	4.59	4.80	4.80	4.80	4.79
0.9	4.68	4.61	5.61	5.54	5.63	5.55

Table 11. Empirical Weighted Risks for PTE
Based on D-W and Gl Statistic, alpha = 5%.

RHO	Durbin		CO		PW	
	DW	Gl	DW	Gl	DW	Gl
0.0	5.40	5.40	5.57	5.58	5.55	5.55
0.1	5.43	5.43	5.61	5.63	5.59	5.60
0.2	5.48	5.48	5.66	5.68	5.63	5.65
0.3	5.53	5.53	5.69	5.70	5.67	5.68
0.4	5.58	5.55	5.71	5.72	5.69	5.69
0.5	5.57	5.56	5.73	5.73	5.70	5.70
0.6	5.58	5.58	5.76	5.76	5.74	5.74
0.7	5.63	5.63	5.86	5.86	5.84	5.84
0.8	5.72	5.72	6.14	6.14	6.13	6.13
0.9	5.73	5.73	7.01	7.01	7.02	7.02

Table 12. Empirical Weighted Risks for Shrinkage PTE
Based on D-W and Gl Statistic, alpha = 5%

RHO	Durbin		CO		PW	
	DW	Gl	DW	Gl	DW	Gl
0.0	4.22	4.21	4.34	4.35	4.33	4.33
0.1	4.24	4.23	4.38	4.38	4.37	4.36
0.2	4.29	4.29	4.42	4.44	4.40	4.42
0.3	4.35	4.34	4.46	4.47	4.45	4.46
0.4	4.40	4.36	4.48	4.48	4.47	4.46
0.5	4.38	4.38	4.48	4.48	4.46	4.46
0.6	4.41	4.41	4.50	4.50	4.48	4.48
0.7	4.48	4.48	4.57	4.57	4.56	4.56
0.8	4.59	4.59	4.80	4.80	4.79	4.79
0.9	4.61	4.61	5.54	5.54	5.55	5.55

Stat. Sci. & Data Anal., pp. 77-86
K. Matsusita *et al.* (Eds)

Centering and Scaling in Ridge Regression

MASAYUKI JIMICHI and NOBUO INAGAKI
Osaka University, Department of Applied Mathematics, Faculty of Engineering Science, Toyonaka, Osaka 560, JAPAN.

Abstract. We consider ridge regression in the usual multiple linear regression model (say M_Z) or its normalized (centered and scaled) model (say M_X). We propose several ridge type estimators for the model M_Z by using ridge estimators for the model M_X and compare their total mean squared errors (TMSE). We present some new ridge rules and test them with a Monte Carlo experiment on some data sets.

Key words: Ridge regression, Centered and scaled regression model

1. INTRODUCTION

The usual multiple linear regression model may be defined as

$$(\mathrm{M}_Z) \qquad \mathbf{y} = \mathbf{Z}\boldsymbol{\theta} + \boldsymbol{\varepsilon},$$

where $\mathbf{y}$ is an $n \times 1$ vector of response variables, $\mathbf{Z} = (\mathbf{1}, \mathbf{Z}_p) = (\mathbf{1}, \mathbf{z}_1, ..., \mathbf{z}_p)$ is an $n \times (p+1)$ matrix of linearly independent explanatory variables, $\mathbf{1} = (\overbrace{1, ..., 1}^{n \text{ times}})'$ and $\mathbf{z}_j = (z_{1j}, ..., z_{nj})'$, $(j = 1, ..., p)$, $\boldsymbol{\theta} = (\theta_0, \boldsymbol{\theta}_p')' = (\theta_0, \theta_1, ..., \theta_p)'$ is a $(p+1) \times 1$ vector of regression coefficient parameters, and $\boldsymbol{\varepsilon}$ is an $n \times 1$ vector of errors with $\mathrm{E}(\boldsymbol{\varepsilon}) = \mathbf{0}$ and $\mathrm{Var}(\boldsymbol{\varepsilon}) = \sigma^2 \mathbf{I}_n$ denoting the $n \times 1$ zero vector by $\mathbf{0}$, the $n \times n$ identity matrix by $\mathbf{I}_n$ and a variance parameter of error by σ^2. In this paper, we consider ridge type estimation of $\boldsymbol{\theta}$.

In ridge regression(cf. Hoerl & Kennard[1],[2].), it is used to treat the model

$$(\mathrm{M}_X) \qquad \mathbf{y} = \mathbf{X}\boldsymbol{\beta} + \boldsymbol{\varepsilon},$$

where $\mathbf{X} = (\mathbf{1}, \mathbf{X}_p)$, $\boldsymbol{\beta} = (\beta_0, \boldsymbol{\beta}_p')'$ and $\mathbf{X}_p$ is the normalization (centering and scaling) of $\mathbf{Z}_p$. That is,

$$\mathbf{X} = \mathbf{Z}\mathbf{T}^{-1}, \tag{1.1}$$

$$\boldsymbol{\beta} = \mathbf{T}\boldsymbol{\theta}, \tag{1.2}$$

where $\mathbf{T} = \begin{pmatrix} 1 & \mathbf{m}_p' \\ \mathbf{0} & \mathbf{S}_p \end{pmatrix}$, and thus, $\mathbf{T}^{-1} = \begin{pmatrix} 1 & -\mathbf{m}_p'\mathbf{S}_p^{-1} \\ \mathbf{0} & \mathbf{S}_p^{-1} \end{pmatrix}$, with

$$\mathbf{m}_p = (m_1, ..., m_p)', \quad \mathbf{S}_p = \mathrm{diag}(s_1, ..., s_p),$$

letting

$$m_j = \frac{1}{n}\sum_{i=1}^{n} z_{ij}, \quad s_j^2 = \sum_{i=1}^{n}(z_{ij} - m_j)^2, \ j = 1, ..., p.$$

Therefore, it is noted that $\mathbf{X}_p'\mathbf{X}_p$ is correlation form and that $\mathbf{1}'\mathbf{X}_p = \mathbf{0}$.

Ridge regression users have been recommended to use the model M_X in order to estimate $\boldsymbol{\beta}_p$. However, there were few papers which consider the estimation of β_0 at the same time.

Brown[3] discussed the simultaneous estimation of $\boldsymbol{\beta}$ by the following (usual) estimator:

$$\hat{\boldsymbol{\beta}}(0) = \begin{pmatrix} \hat{\beta}_0(0) \\ \hat{\boldsymbol{\beta}}_p(k) \end{pmatrix} \equiv \begin{pmatrix} m_0 \\ (\mathbf{X}_p'\mathbf{X}_p + k\mathbf{I}_p)^{-1}\mathbf{X}_p'\mathbf{y} \end{pmatrix}, \tag{1.3}$$

where $m_0 = \sum_{i=1}^n y_i/n$ and $k \geq 0$, and pointed out that this estimator is invariant for location. The first coordinate, $\hat{\beta}_0(0) = m_0$, is the ordinary least squares estimator for intercept β_0 and so, is not shrunk in the same way as the residual part, $\hat{\boldsymbol{\beta}}_p(k)$, which is the ordinary ridge regression estimator of $\boldsymbol{\beta}_p$. The estimator may be written by using the notation of general ridge regression as

$$\hat{\boldsymbol{\beta}}(0) = (\mathbf{X}'\mathbf{X} + k\mathrm{diag}(0,1,..,1))^{-1}\mathbf{X}'\mathbf{y}, \tag{1.4}$$

where diag(0,1,..,1) is a $(p+1) \times (p+1)$ diagonal matrix with ones everywhere on the diagonal except in the first location.

We consider the slightly general ridge estimation of $\boldsymbol{\beta}$. We propose the following estimator :

$$\hat{\boldsymbol{\beta}}(\nu) = \begin{pmatrix} \hat{\beta}_0(\nu) \\ \hat{\boldsymbol{\beta}}_p(k) \end{pmatrix} \equiv (\mathbf{X}'\mathbf{X} + \mathrm{diag}(\nu, k, .., k))^{-1}\mathbf{X}'\mathbf{y} \tag{1.5}$$

$$= \begin{pmatrix} \dfrac{n}{n+\nu} m_0 \\ (\mathbf{X}_p'\mathbf{X}_p + k\mathbf{I}_p)^{-1}\mathbf{X}_p'\mathbf{y} \end{pmatrix}, \qquad \nu \geq 0.$$

Note that the estimator, $\hat{\boldsymbol{\beta}}(k)$, is the ordinary ridge regression estimator for $\boldsymbol{\beta}$:

$$\hat{\boldsymbol{\beta}}(k) = \begin{pmatrix} \hat{\beta}_0(k) \\ \hat{\boldsymbol{\beta}}_p(k) \end{pmatrix} \equiv (\mathbf{X}'\mathbf{X} + k\mathbf{I}_{p+1})^{-1}\mathbf{X}'\mathbf{y} \tag{1.6}$$

$$= \begin{pmatrix} \dfrac{n}{n+k} m_0 \\ (\mathbf{X}_p'\mathbf{X}_p + k\mathbf{I}_p)^{-1}\mathbf{X}_p'\mathbf{y} \end{pmatrix}.$$

Now, let us consider ridge estimation of $\boldsymbol{\theta}$ in the model M_Z. We see the following equivalent equation to (1.2) :

$$\boldsymbol{\theta} = \mathbf{T}^{-1}\boldsymbol{\beta} \iff \begin{cases} \theta_0 = \beta_0 - \dfrac{m_1}{s_1}\beta_1 - \cdots - \dfrac{m_p}{s_p}\beta_p, \\ \theta_j = \dfrac{\beta_j}{s_j}. \qquad (j = 1, ..., p) \end{cases} \tag{1.7}$$

We call the following estimators for $\boldsymbol{\theta}$ obtained by substituting the previous ridge estimators of $\boldsymbol{\beta}$ into (1.7) *ridge type estimators of* $\boldsymbol{\theta}$:

$$\hat{\boldsymbol{\theta}}(0) \equiv \mathbf{T}^{-1}\hat{\boldsymbol{\beta}}(0) = \begin{pmatrix} \hat{\beta}_0(0) - \mathbf{m}_p'\mathbf{S}_p{}^{-1}\hat{\boldsymbol{\beta}}_p(k) \\ \mathbf{S}_p{}^{-1}\hat{\boldsymbol{\beta}}_p(k) \end{pmatrix},$$

$$\hat{\boldsymbol{\theta}}(\nu) \equiv \mathbf{T}^{-1}\hat{\boldsymbol{\beta}}(\nu) = \begin{pmatrix} \hat{\beta}_0(\nu) - \mathbf{m}_p'\mathbf{S}_p{}^{-1}\hat{\boldsymbol{\beta}}_p(k) \\ \mathbf{S}_p{}^{-1}\hat{\boldsymbol{\beta}}_p(k) \end{pmatrix}, \tag{1.8}$$

$$\hat{\boldsymbol{\theta}}(k) \equiv \mathbf{T}^{-1}\hat{\boldsymbol{\beta}}(k) = \begin{pmatrix} \hat{\beta}_0(k) - \mathbf{m}_p'\mathbf{S}_p{}^{-1}\hat{\boldsymbol{\beta}}_p(k) \\ \mathbf{S}_p{}^{-1}\hat{\boldsymbol{\beta}}_p(k) \end{pmatrix}.$$

The first ridge type estimator in (1.8), $\hat{\boldsymbol{\theta}}(0)$, may be often used as a usual estimator of $\boldsymbol{\theta}$. (For example, see Draper & Smith[4], and Goto[5]).

Our main aim is to determine the ridge coefficient ν which minimizes the total mean squared errors(TMSE) of the ridge type estimator $\hat{\boldsymbol{\theta}}(\nu)$ proposed by us.

Remark. *The estimator, $\hat{\boldsymbol{\theta}}$, obtained by substituting the ordinary least squares estimator, $\hat{\boldsymbol{\beta}}$, into (1.7) is the ordinary least squares estimator of $\boldsymbol{\theta}$:*

$$\hat{\boldsymbol{\theta}} = (\mathbf{Z}'\mathbf{Z})^{-1}\mathbf{Z}'\mathbf{y} = \mathbf{T}^{-1}\hat{\beta}.$$

However, any ridge type estimator $\hat{\boldsymbol{\theta}}(\nu)$ for $\nu \geq 0$ and $k > 0$ is not equal to the ordinary ridge estimator of $\boldsymbol{\theta}$:

$$\tilde{\boldsymbol{\theta}}(k) \equiv (\mathbf{Z}'\mathbf{Z} + k\mathbf{I}_{p+1})^{-1}\mathbf{Z}'\mathbf{y}.$$

2. TOTAL MEAN SQUARED ERRORS OF RIDGE TYPE ESTIMATORS

2.1 Calculation of Total Mean Squared Errors of Ridge Type Estimators

The total mean squared errors(TMSE) of the ridge type estimator $\hat{\boldsymbol{\theta}}(\nu)$ is divided into two parts consisting of its variance and bias:

$$\begin{aligned}\rho(\nu) \equiv \mathrm{TMSE}(\hat{\boldsymbol{\theta}}(\nu)) &= \mathrm{E}[(\hat{\boldsymbol{\theta}}(\nu) - \boldsymbol{\theta})'(\hat{\boldsymbol{\theta}}(\nu) - \boldsymbol{\theta})] \\ &= \mathrm{tr}[\mathrm{Var}(\hat{\boldsymbol{\theta}}(\nu))] + [\mathrm{bias}(\hat{\boldsymbol{\theta}}(\nu))]'[\mathrm{bias}(\hat{\boldsymbol{\theta}}(\nu))]\end{aligned} \tag{2.1}$$

where $\mathrm{Var}(\hat{\boldsymbol{\theta}}(\nu))$ is the covariance matrix of $\hat{\boldsymbol{\theta}}(\nu)$ and

$$\mathrm{bias}(\hat{\boldsymbol{\theta}}(\nu)) = \mathrm{E}(\hat{\boldsymbol{\theta}}(\nu)) - \boldsymbol{\theta}.$$

In this section, we calculate the TMSE's of the ridge type estimators described in Section 1 and make a comparison among them. At first, we prepare the following notations. For nonnegative numbers $k \geq 0$ and $\nu \geq 0$, put

$$\mathbf{W} = \mathbf{X}'\mathbf{X}, \quad \mathbf{W}(\nu) = \mathbf{W} + \mathrm{diag}(\nu, k, ..., k).$$

Let $\boldsymbol{\Gamma}$ be the $(p+1)$-dimensional orthogonal matrix such that

$$\boldsymbol{\Gamma}'\mathbf{W}\boldsymbol{\Gamma} = \boldsymbol{\Lambda} = \mathrm{diag}(n, \lambda_1, ..., \lambda_p)$$

where $(n >)\lambda_1 \geq \cdots \geq \lambda_p(> 0)$ which are the eigen values of the cross-product (information) matrix $\mathbf{W}$. Then, we see

$$\boldsymbol{\Gamma}'\mathbf{W}(\nu)\boldsymbol{\Gamma} = \boldsymbol{\Lambda} + \mathrm{diag}(\nu, k, ..., k) \equiv \boldsymbol{\Lambda}(\nu).$$

The matrices $\mathbf{W}$, $\boldsymbol{\Gamma}$, $\boldsymbol{\Lambda}$ have the following partitioned forms, respectively:

$$\mathbf{W} = \begin{pmatrix} n & \mathbf{0} \\ \mathbf{0} & \mathbf{W}_p \end{pmatrix}, \quad \boldsymbol{\Gamma} = \begin{pmatrix} 1 & \mathbf{0} \\ \mathbf{0} & \boldsymbol{\Gamma}_p \end{pmatrix}, \quad \boldsymbol{\Lambda} = \begin{pmatrix} n & \mathbf{0} \\ \mathbf{0} & \boldsymbol{\Lambda}_p \end{pmatrix} \tag{2.2}$$

where $\mathbf{W}_p = \mathbf{X}_p'\mathbf{X}_p$ and $\boldsymbol{\Gamma}_p$ is the p -dimensional orthognal matrix such that

$$\boldsymbol{\Gamma}_p{}'\mathbf{W}_p\boldsymbol{\Gamma}_p = \boldsymbol{\Lambda}_p = \mathrm{diag}(\lambda_1, ..., \lambda_p).$$

We also represent the partitioned forms of the matrices $\mathbf{W}(\nu)$, $\boldsymbol{\Lambda}(\nu)$ in the following manner:

$$\boldsymbol{\Gamma}'\mathbf{W}(\nu)\boldsymbol{\Gamma} = \begin{pmatrix} 1 & \mathbf{0} \\ \mathbf{0} & \boldsymbol{\Gamma}_p{}' \end{pmatrix}\begin{pmatrix} n+\nu & \mathbf{0} \\ \mathbf{0} & \mathbf{W}_p(k) \end{pmatrix}\begin{pmatrix} 1 & \mathbf{0} \\ \mathbf{0} & \boldsymbol{\Gamma}_p \end{pmatrix} = \begin{pmatrix} n+\nu & \mathbf{0} \\ \mathbf{0} & \boldsymbol{\Lambda}_p(k) \end{pmatrix} \tag{2.3}$$

where

$$\mathbf{W}_p(k) = \mathbf{W}_p + k\mathbf{I}_p \quad \boldsymbol{\Lambda}_p(k) = \boldsymbol{\Lambda}_p + k\mathbf{I}_p.$$

By using the partitioned forms, (2.2) and (2.3), the variance and bias terms of the TMSE (2.1) are represented as follows :

$$\begin{aligned}
&\mathrm{trVar}(\hat{\boldsymbol{\theta}}(\nu)) \\
&= \sigma^2 \mathrm{tr}[\mathbf{T}^{-1}\mathbf{W}(\nu)^{-1}\mathbf{W}\mathbf{W}(\nu)^{-1}(\mathbf{T}^{-1})'] \\
&= \sigma^2 \mathrm{tr}[\boldsymbol{\Lambda}(\nu)^{-1}\boldsymbol{\Lambda}\boldsymbol{\Lambda}(\nu)^{-1}\boldsymbol{\Gamma}'(\mathbf{T}^{-1})'\mathbf{T}^{-1}\boldsymbol{\Gamma}] \\
&= \sigma^2 \mathrm{tr}[\begin{pmatrix} \frac{1}{n+\nu} & \mathbf{0} \\ \mathbf{0} & \boldsymbol{\Lambda}_p(k)^{-1} \end{pmatrix} \begin{pmatrix} n & \mathbf{0} \\ \mathbf{0} & \boldsymbol{\Lambda}_p \end{pmatrix} \begin{pmatrix} \frac{1}{n+\nu} & \mathbf{0} \\ \mathbf{0} & \boldsymbol{\Lambda}_p(k)^{-1} \end{pmatrix} \\
&\quad \times \begin{pmatrix} 1 & -\mathbf{m}_p'\mathbf{S}_p{}^{-1}\boldsymbol{\Gamma}_p \\ -\boldsymbol{\Gamma}_p'\mathbf{S}_p{}^{-1}\mathbf{m}_p & \boldsymbol{\Gamma}_p'(\mathbf{S}_p{}^{-1}\mathbf{m}_p\mathbf{m}_p'\mathbf{S}_p{}^{-1} + \mathbf{S}_p{}^{-2})\boldsymbol{\Gamma}_p \end{pmatrix}] \\
&= \sigma^2 \left\{ \frac{n}{(n+\nu)^2} + \mathrm{tr}[\boldsymbol{\Lambda}_p(k)^{-1}\boldsymbol{\Lambda}_p\boldsymbol{\Lambda}_p(k)^{-1}\boldsymbol{\Gamma}_p'(\mathbf{S}_p{}^{-1}\mathbf{m}_p\mathbf{m}_p'\mathbf{S}_p{}^{-1} + \mathbf{S}_p{}^{-2})\boldsymbol{\Gamma}_p] \right\},
\end{aligned}$$

and, letting $\boldsymbol{\alpha} \equiv (\alpha_0, \boldsymbol{\alpha}_p')' = \boldsymbol{\Gamma}'\boldsymbol{\beta}$,

$$\begin{aligned}
&[\mathrm{bias}(\hat{\boldsymbol{\theta}}(\nu))]'[\mathrm{bias}(\hat{\boldsymbol{\theta}}(\nu))] \\
&= [\boldsymbol{\beta}'(\mathbf{W}\mathbf{W}(\nu)^{-1} - \mathbf{I}_{p+1})(\mathbf{T}')^{-1}][\mathbf{T}^{-1}(\mathbf{W}(\nu)^{-1}\mathbf{W} - \mathbf{I}_{p+1})\boldsymbol{\beta}] \\
&= [\boldsymbol{\beta}'\boldsymbol{\Gamma}(\boldsymbol{\Lambda}\boldsymbol{\Lambda}(\nu)^{-1} - \mathbf{I}_{p+1})\boldsymbol{\Gamma}'(\mathbf{T}')^{-1}][\mathbf{T}^{-1}\boldsymbol{\Gamma}(\boldsymbol{\Lambda}(\nu)^{-1}\boldsymbol{\Lambda} - \mathbf{I}_{p+1})\boldsymbol{\Gamma}'\boldsymbol{\beta}]. \\
&= \boldsymbol{\alpha}'(\boldsymbol{\Lambda}(\nu)^{-1}\boldsymbol{\Lambda} - \mathbf{I}_{p+1})'\boldsymbol{\Gamma}'(\mathbf{T}^{-1})'\mathbf{T}^{-1}\boldsymbol{\Gamma}(\boldsymbol{\Lambda}(\nu)^{-1}\boldsymbol{\Lambda} - \mathbf{I}_{p+1})\boldsymbol{\alpha} \\
&= (\alpha_0,\ \boldsymbol{\alpha}_p') \begin{pmatrix} -\frac{\nu}{n+\nu} & \mathbf{0} \\ \mathbf{0} & \boldsymbol{\Lambda}_p\boldsymbol{\Lambda}_p(k)^{-1} - \mathbf{I}_p \end{pmatrix} \\
&\quad \times \begin{pmatrix} 1 & -\mathbf{m}_p'\mathbf{S}_p{}^{-1}\boldsymbol{\Gamma}_p \\ -\boldsymbol{\Gamma}_p'\mathbf{S}_p{}^{-1}\mathbf{m}_p & \boldsymbol{\Gamma}_p'(\mathbf{S}_p{}^{-1}\mathbf{m}_p\mathbf{m}_p'\mathbf{S}_p{}^{-1} + \mathbf{S}_p{}^{-2})\boldsymbol{\Gamma}_p \end{pmatrix} \\
&\quad \times \begin{pmatrix} -\frac{\nu}{n+\nu} & \mathbf{0} \\ \mathbf{0} & \boldsymbol{\Lambda}_p(k)^{-1}\boldsymbol{\Lambda}_p - \mathbf{I}_p \end{pmatrix} \begin{pmatrix} \alpha_0 \\ \boldsymbol{\alpha}_p \end{pmatrix} \\
&= \frac{\nu^2}{(n+\nu)^2}\alpha_0^2 + 2\frac{\nu}{n+\nu}\alpha_0\mathbf{m}_p'\mathbf{S}_p{}^{-1}\boldsymbol{\Gamma}_p(\boldsymbol{\Lambda}_p(k)^{-1}\boldsymbol{\Lambda}_p - \mathbf{I}_p)\boldsymbol{\alpha}_p \\
&+ \boldsymbol{\alpha}_p'(\boldsymbol{\Lambda}_p\boldsymbol{\Lambda}_p(k)^{-1} - \mathbf{I}_p)\boldsymbol{\Gamma}_p'(\mathbf{S}_p{}^{-1}\mathbf{m}_p\mathbf{m}_p'\mathbf{S}_p{}^{-1} + \mathbf{S}_p{}^{-2})\boldsymbol{\Gamma}_p(\boldsymbol{\Lambda}_p(k)^{-1}\boldsymbol{\Lambda}_p - \mathbf{I}_p)\boldsymbol{\alpha}_p\ .
\end{aligned}$$

Hence, the TMSE (2.1) are represented by using the partitioned form as follows :

$$\begin{aligned}
&\rho(\nu) \equiv \mathrm{TMSE}(\hat{\boldsymbol{\theta}}(\nu)) \qquad (2.4) \\
&= \sigma^2 \left\{ \frac{n}{(n+\nu)^2} + \mathrm{tr}\boldsymbol{\Lambda}_p(k)^{-1}\boldsymbol{\Lambda}_p\boldsymbol{\Lambda}_p(k)^{-1}\boldsymbol{\Gamma}_p'(\mathbf{S}_p{}^{-1}\mathbf{m}_p\mathbf{m}_p'\mathbf{S}_p{}^{-1} + \mathbf{S}_p{}^{-2})\boldsymbol{\Gamma}_p \right\} \\
&+ \frac{\nu^2}{(n+\nu)^2}\alpha_0^2 + 2\frac{\nu}{n+\nu}\alpha_0\mathbf{m}_p'\mathbf{S}_p{}^{-1}\boldsymbol{\Gamma}_p(\boldsymbol{\Lambda}_p(k)^{-1}\boldsymbol{\Lambda}_p - \mathbf{I}_p)\boldsymbol{\alpha}_p \\
&+ \boldsymbol{\alpha}_p'(\boldsymbol{\Lambda}_p\boldsymbol{\Lambda}_p(k)^{-1} - \mathbf{I}_p)\boldsymbol{\Gamma}_p'(\mathbf{S}_p{}^{-1}\mathbf{m}_p\mathbf{m}_p'\mathbf{S}_p{}^{-1} + \mathbf{S}_p{}^{-2})\boldsymbol{\Gamma}_p(\boldsymbol{\Lambda}_p(k)^{-1}\boldsymbol{\Lambda}_p - \mathbf{I}_p)\boldsymbol{\alpha}_p.
\end{aligned}$$

Especially, by substituting $\nu = k$ and $\nu = 0$ into (2.4), respectively, the TMSE's of $\hat{\boldsymbol{\theta}}(k)$ and $\hat{\boldsymbol{\theta}}(0)$ are obtained, as follows:

$$\begin{aligned} \rho(0) &\equiv \mathrm{TMSE}(\hat{\boldsymbol{\theta}}(0)) \qquad (2.5)\\ &= \sigma^2\left\{\frac{1}{n} + \mathrm{tr}\boldsymbol{\Lambda}_p(k)^{-1}\boldsymbol{\Lambda}_p\boldsymbol{\Lambda}_p(k)^{-1}\boldsymbol{\Gamma}_p{}'(\mathbf{S}_p{}^{-1}\mathbf{m}_p\mathbf{m}_p{}'\mathbf{S}_p{}^{-1} + \mathbf{S}_p{}^{-2})\boldsymbol{\Gamma}_p\right\}\\ &+ \boldsymbol{\alpha}_p{}'(\boldsymbol{\Lambda}_p\boldsymbol{\Lambda}_p(k)^{-1} - \mathbf{I}_p)\boldsymbol{\Gamma}_p{}'(\mathbf{S}_p{}^{-1}\mathbf{m}_p\mathbf{m}_p{}'\mathbf{S}_p{}^{-1} + \mathbf{S}_p{}^{-2})\boldsymbol{\Gamma}_p(\boldsymbol{\Lambda}_p(k)^{-1}\boldsymbol{\Lambda}_p - \mathbf{I}_p)\boldsymbol{\alpha}_p, \end{aligned}$$

and

$$\begin{aligned} \rho(k) &\equiv \mathrm{TMSE}(\hat{\boldsymbol{\theta}}(k)) \qquad (2.6)\\ &= \sigma^2\left\{\frac{n}{(n+k)^2} + \mathrm{tr}\boldsymbol{\Lambda}_p(k)^{-1}\boldsymbol{\Lambda}_p\boldsymbol{\Lambda}_p(k)^{-1}\boldsymbol{\Gamma}_p{}'(\mathbf{S}_p{}^{-1}\mathbf{m}_p\mathbf{m}_p{}'\mathbf{S}_p{}^{-1} + \mathbf{S}_p{}^{-2})\boldsymbol{\Gamma}_p\right\}\\ &+ \frac{k^2}{(n+k)^2}\alpha_0^2 + 2\frac{k}{n+k}\alpha_0\mathbf{m}_p{}'\mathbf{S}_p{}^{-1}\boldsymbol{\Gamma}_p(\boldsymbol{\Lambda}_p(k)^{-1}\boldsymbol{\Lambda}_p - \mathbf{I}_p)\boldsymbol{\alpha}_p\\ &+ \boldsymbol{\alpha}_p{}'(\boldsymbol{\Lambda}_p\boldsymbol{\Lambda}_p(k)^{-1} - \mathbf{I}_p)\boldsymbol{\Gamma}_p{}'(\mathbf{S}_p{}^{-1}\mathbf{m}_p\mathbf{m}_p{}'\mathbf{S}_p{}^{-1} + \mathbf{S}_p{}^{-2})\boldsymbol{\Gamma}_p(\boldsymbol{\Lambda}_p(k)^{-1}\boldsymbol{\Lambda}_p - \mathbf{I}_p)\boldsymbol{\alpha}_p. \end{aligned}$$

2.2 Comparison of Total Mean Squared Errors of Ridge Type Estimators

We regard $\rho(\nu) \equiv \mathrm{TMSE}(\hat{\boldsymbol{\theta}}(\nu))$ as a function of ν for a fixed k and investigate its variation by differentiation. Letting

$$g(k) = \alpha_0\mathbf{m}_p{}'\mathbf{S}_p{}^{-1}\boldsymbol{\Gamma}_p(\boldsymbol{\Lambda}_p(k)^{-1}\boldsymbol{\Lambda}_p - \mathbf{I}_p)\boldsymbol{\alpha}_p,$$

we have

$$\begin{aligned} \rho'(\nu) &= \frac{d}{d\nu}\rho(\nu)\\ &= -\frac{2n\sigma^2}{(n+\nu)^3} + \frac{2n\alpha_0^2\nu}{(n+\nu)^3} + \frac{2ng(k)}{(n+\nu)^2}\\ &= \frac{2n}{(n+\nu)^3}(\alpha_0^2 + g(k))\left\{\nu - \frac{n}{\alpha_0^2 + g(k)}\left(\frac{\sigma^2}{n} - g(k)\right)\right\}\\ &= \frac{2n}{(n+\nu)^3}(\alpha_0^2 + g(k))(\nu - \nu(k)), \end{aligned}$$

where

$$\nu(k) = \frac{n}{\alpha_0^2 + g(k)}\left(\frac{\sigma^2}{n} - g(k)\right).$$

If $g(k) \le -\alpha_0^2$, we see $\rho'(\nu) < 0$ for $\nu > 0$, and thus, $\rho(\nu)$ is monotonically decreasing.

If $-\alpha_0^2 < g(k) < \frac{\sigma^2}{n}$, we see

$$\rho'(\nu)\begin{cases} < 0, & \text{if } 0 \le \nu < \nu(k),\\ = 0, & \text{if } \nu = \nu(k),\\ > 0, & \text{if } \nu(k) < \nu, \end{cases}$$

and thus, $\rho(\nu)$ is minimized at $\nu(k)$.

If $\frac{\sigma^2}{n} < g(k)$, we see $\rho'(\nu) > 0$ for $\nu > 0$, and thus, $\rho(\nu)$ is monotonically increasing. Hence, we have the following theorem that enable us to determine the ridge coefficient ν which minimizes the TMSE of the ridge type estimator $\hat{\boldsymbol{\theta}}(\nu)$.

Theorem. *For a fixed value of* $k \geq 0$,

(i) $g(k) \leq -\alpha_0^2 \Rightarrow TMSE(\hat{\boldsymbol{\theta}}(0)) > TMSE(\hat{\boldsymbol{\theta}}(k)) > TMSE(\hat{\boldsymbol{\theta}}(\nu)),$ *for* $k < \nu$

(ii)

$$-\alpha_0^2 < g(k) < \frac{\sigma^2}{n} \Rightarrow \begin{cases} TMSE(\hat{\boldsymbol{\theta}}(0)) & > TMSE(\hat{\boldsymbol{\theta}}(\nu(k))) \\ TMSE(\hat{\boldsymbol{\theta}}(k)) & \geq TMSE(\hat{\boldsymbol{\theta}}(\nu(k))) \end{cases}$$

(The equality holds when $k = \nu(k)$ *.)*

(iii) $$g(k) = \frac{\sigma^2}{n} \Rightarrow \begin{cases} TMSE(\hat{\boldsymbol{\theta}}(0)) & < TMSE(\hat{\boldsymbol{\theta}}(\nu)), \text{ for } \nu > 0 \\ TMSE(\hat{\boldsymbol{\theta}}(0)) & < TMSE(\hat{\boldsymbol{\theta}}(k)) \end{cases}$$

(iv) $\frac{\sigma^2}{n} < g(k) \Rightarrow TMSE(\hat{\boldsymbol{\theta}}(0)) < TMSE(\hat{\boldsymbol{\theta}}(\nu)) < TMSE(\hat{\boldsymbol{\theta}}(k)),$ *for* $0 < \nu < k$.

From this theorem, we propose the following rule for determining ν of the ridge type estimator $\hat{\boldsymbol{\theta}}(\nu)$:

$$\nu = \begin{cases} \text{any value greater than } k, & \text{if } g(k) \leq -\alpha_0^2 \\ \nu(k), & \text{if } -\alpha_0^2 < g(k) < \frac{\sigma^2}{n} \\ 0, & \text{if } \frac{\sigma^2}{n} \leq g(k). \end{cases} \tag{2.7}$$

Equivalently, the ridge type estimator $\hat{\boldsymbol{\theta}}(\nu)$ is as follows :

$$\hat{\boldsymbol{\theta}}(\nu) = \begin{cases} \hat{\boldsymbol{\theta}}(\nu), \text{ (any } \nu > k) & \text{if } g(k) \leq -\alpha_0^2, \\ \hat{\boldsymbol{\theta}}(\nu(k)), & \text{if } -\alpha_0^2 < g(k) < \frac{\sigma^2}{n}, \\ \hat{\boldsymbol{\theta}}(0), & \text{if } \frac{\sigma^2}{n} \leq g(k). \end{cases} \tag{2.8}$$

Now, there are two problems if we practically use this rule. First problem is that the rule(2.7) depends on unknown parameters, $\boldsymbol{\alpha} = \boldsymbol{\Gamma}'\boldsymbol{\beta}$ and σ^2. Second problem is that how do we determine the ridge coefficient k of the ridge type estimator $\hat{\boldsymbol{\theta}}(\nu)$.

In the next section, we consider these problems and compare estimated TMSE's of the ridge type estimators by simulation for Hoerl[6]'s data.

3. RULES OF CHOOSING THE RIDGE COEFFICIENT AND SIMULATION

First of all, we discuss the first problem in Section 2. Since the rule(2.7) is depends on unknown parameters $\boldsymbol{\alpha}$ and σ^2, we substitute least square estimators for them, we denote the corresponding values of by putting hat '^' on them. That is, if

$$\hat{\boldsymbol{\alpha}} = (\hat{\alpha}_0, \hat{\boldsymbol{\alpha}}_p')' \equiv \boldsymbol{\Gamma}'\hat{\boldsymbol{\beta}}, \quad \hat{\sigma}^2 \equiv \frac{\|\mathbf{y} - \mathbf{Z}\hat{\boldsymbol{\theta}}\|^2}{n-p-1},$$

then

$$\hat{g}(k) \equiv \hat{\alpha}_0 \mathbf{m}_p{}' \mathbf{S}_p{}^{-1} \mathbf{\Gamma}_p (\mathbf{\Lambda}_p(k)^{-1} \mathbf{\Lambda}_p - \mathbf{I}_p) \hat{\boldsymbol{\alpha}}_p, \tag{3.1}$$

$$\hat{\nu}(k) \equiv \frac{n}{\hat{\alpha}_0^2 + \hat{g}(k)} \left(\frac{\hat{\sigma}^2}{n} - \hat{g}(k) \right), \tag{3.2}$$

where $\hat{\boldsymbol{\theta}}$ and $\hat{\boldsymbol{\beta}}$ are the ordinary least squares estimators for $\boldsymbol{\theta}$ and $\boldsymbol{\beta}$, respectively.

Next, we consider rules for determining the ridge coefficient k of the ridge type estimator $\hat{\boldsymbol{\theta}}(\nu)$. The following classical rules have been used to determine the ridge coefficient k :

Type I rules (Hoerl, Kennard & Baldwin[7])

$$\begin{cases} \hat{k}_{\mathrm{I}1} & = \dfrac{(p+1)\hat{\sigma}^2}{\|\hat{\boldsymbol{\beta}}\|^2}, \\ \hat{k}_{\mathrm{I}2} & = \dfrac{p\hat{\sigma}^2}{\|\hat{\boldsymbol{\beta}}_p\|^2}, \end{cases}$$

Type II rules (Lee & Campbell[8], Lee[9])

$$\begin{cases} \hat{k}_{\mathrm{II}1} & ; \; k \text{ s.t. } \hat{M}_1'(k) = 0, \\ \hat{k}_{\mathrm{II}2} & ; \; k \text{ s.t. } \hat{M}_2'(k) = 0, \end{cases}$$

where

$$\hat{M}_1(k) \equiv \frac{n\hat{\sigma}^2 + k^2\hat{\alpha}_0^2}{(n+k)^2} + \sum_{j=1}^{p} \frac{\lambda_j \hat{\sigma}^2 + k^2 \hat{\alpha}_j^2}{(\lambda_j + k)^2},$$

$$\hat{M}_2(k) \equiv \frac{\hat{\sigma}^2}{n} + \sum_{j=1}^{p} \frac{\lambda_j \hat{\sigma}^2 + k^2 \hat{\alpha}_j^2}{(\lambda_j + k)^2},$$

and each of them is an estimator for the TMSE of $\hat{\boldsymbol{\beta}}(k)$(, say $M_1(k)$) , $\hat{\boldsymbol{\beta}}(0)$(, say $M_2(k)$), respectively. $M_1(k)$ and $M_2(k)$ are

$$M_1(k) = \mathrm{TMSE}(\hat{\boldsymbol{\beta}}(k)) = \frac{n\sigma^2 + k^2\alpha_0^2}{(n+k)^2} + \sum_{j=1}^{p} \frac{\lambda_j \sigma^2 + k^2 \alpha_j^2}{(\lambda_j + k)^2}, \tag{3.3}$$

$$M_2(k) = \mathrm{TMSE}(\hat{\boldsymbol{\beta}}(0)) = \frac{\sigma^2}{n} + \sum_{j=1}^{p} \frac{\lambda_j \sigma^2 + k^2 \alpha_j^2}{(\lambda_j + k)^2}. \tag{3.4}$$

Note that the rules $\{\hat{k}_{\mathrm{I}1}, \hat{k}_{\mathrm{II}1}\}$ and $\{\hat{k}_{\mathrm{I}2}, \hat{k}_{\mathrm{II}2}\}$ are used to determine ridge coefficient for $\hat{\boldsymbol{\theta}}(k)$ and $\hat{\boldsymbol{\theta}}(0)$, respectively.

Now, let us improve the Type II (Lee & Campbell's) rules. We propose two kind of rules for determining the ridge coefficient k. First rules(,say Type II-U rules,) are obtained by using unbiased estimators of TMSE. Second rules(,say Type II-S rules,) are obtained by using estimators of TMSE for which a ridge estimator is substituted. That is,

Type II-U (Unbiased Lee & Campbell's) rules

$$\begin{cases} \hat{k}_{\mathrm{IIU}1} & ; \; k \text{ s.t. } \hat{M}_{1U}'(k) = 0, \\ \hat{k}_{\mathrm{IIU}2} & ; \; k \text{ s.t. } \hat{M}_{2U}'(k) = 0, \end{cases}$$

Type II-S (Shrinkaged Lee & Campbell's) rules

$$\begin{cases} \hat{k}_{\mathrm{IIS1}} & ; k \text{ s.t. } \hat{M}'_{1\mathrm{S}}(k) = 0, \\ \hat{k}_{\mathrm{IIS2}} & ; k \text{ s.t. } \hat{M}'_{2\mathrm{S}}(k) = 0, \end{cases}$$

where

$$\begin{cases} \hat{M}_{1\mathrm{U}}(k) & \equiv \sum_{j=0}^{p} \frac{1}{(\lambda_j + k)^2} \left\{ \lambda_j \sigma^2 + k^2 \left(\hat{\alpha}_j^2 - \frac{\hat{\sigma}^2}{n} \right) \right\}, \\ \hat{M}_{1\mathrm{S}}(k) & \equiv \frac{n\hat{\sigma}^2 + k^2 \hat{\alpha}_0^2(\hat{k}_{\mathrm{III1}})}{(n+k)^2} + \sum_{j=1}^{p} \frac{\lambda_j \hat{\sigma}^2 + k^2 \hat{\alpha}_j^2(\hat{k}_{\mathrm{III1}})}{(\lambda_j + k)^2}, \end{cases}$$

$$\begin{cases} \hat{M}_{2U}(k) & \equiv \frac{\hat{\sigma}^2}{n} + \sum_{j=1}^{p} \frac{1}{(\lambda_j + k)^2} \left\{ \lambda_j \sigma^2 + k^2 \left(\hat{\alpha}_j^2 - \frac{\hat{\sigma}^2}{n} \right) \right\}, \\ \hat{M}_{2S}(k) & \equiv \frac{\hat{\sigma}^2}{n} + \sum_{j=1}^{p} \frac{\lambda_j \hat{\sigma}^2 + k^2 \hat{\alpha}_j^2(\hat{k}_{\mathrm{II2}})}{(\lambda_j + k)^2}, \end{cases}$$

and $\hat{M}_{1U}(k)$, $\hat{M}_{1S}(k)$ are estimators for $M_1(k)$ (TMSE of $\hat{\boldsymbol{\beta}}(k)$), $\hat{M}_{2U}(k)$, $\hat{M}_{2S}(k)$ for $M_2(k)$ (TMSE of $\hat{\boldsymbol{\beta}}(0)$). Note that $\hat{M}_{1U}(k)$, $\hat{M}_{2U}(k)$ are unbiased estimators for $M_1(k)$, $M_2(k)$, respectively, and $\{\hat{k}_{\mathrm{IIU1}}, \hat{k}_{\mathrm{IIS1}}\}$ are used to determine the ridge coefficient for $\hat{\boldsymbol{\theta}}(k)$, $\{\hat{k}_{\mathrm{IIU2}}, \hat{k}_{\mathrm{IIS2}}\}$ for $\hat{\boldsymbol{\theta}}(0)$.

Then, we propose the following rule for determining ν of $\hat{\boldsymbol{\theta}}(\nu)$:

$$\hat{\nu} = \begin{cases} \text{any value greater than } k, & \text{if } \hat{g}(k) \leq -\hat{\alpha}_0^2 \\ \hat{\nu}(k), & \text{if } -\hat{\alpha}_0^2 < \hat{g}(k) < \frac{\hat{\sigma}^2}{n} \\ 0, & \text{if } \frac{\hat{\sigma}^2}{n} \leq \hat{g}(k). \end{cases} \tag{3.5}$$

Equivalently, the ridge type estimator $\hat{\boldsymbol{\theta}}(\nu)$ is as follows:

$$\hat{\boldsymbol{\theta}}(\hat{\nu}) = \begin{cases} \hat{\boldsymbol{\theta}}(\nu), \ (\text{any } \nu > k) & \text{if } \hat{g}(k) \leq -\hat{\alpha}_0^2, \\ \hat{\boldsymbol{\theta}}(\hat{\nu}(k)), & \text{if } -\hat{\alpha}_0^2 < \hat{g}(k) < \frac{\hat{\sigma}^2}{n}, \\ \hat{\boldsymbol{\theta}}(0), & \text{if } \frac{\hat{\sigma}^2}{n} \leq \hat{g}(k). \end{cases} \tag{3.6}$$

Now, let us simulate by Hoerl[6]'s data. Suppose that $\boldsymbol{\theta} = (10, 2, 3, 5)'$, $\sigma^2 = 1$, $n = 10$ and $p = 3$, and the following data are used :

$$\text{Data 1} \begin{cases} \boldsymbol{\alpha} = (27.790000, 8.038114, 1.634923, -2.030949)' \\ (\lambda_1, \lambda_2, \lambda_3) = (2.86683730, 0.08973179, 0.04343091) \end{cases}$$

$$\text{Data 2} \begin{cases} \boldsymbol{\alpha} = (28.620000, 7.373473, 2.970877, -1.251313)' \\ (\lambda_1, \lambda_2, \lambda_3) = (2.88781535, 0.09852193, 0.01366272) \end{cases}$$

Our simulation procedures is that first of all, we generate one thousand normal pseudo random numbers of size ten with the expectation of ten dimensional zero vector and variance-covariance matrix of ten dimensional identity matrix. Next, the following criterion is computed for each the ridge type estimators :

$$\mathrm{TASE}(\hat{\boldsymbol{\theta}}(\hat{\nu})) \equiv \frac{1}{N}\sum_{t=1}^{N} \|\hat{\boldsymbol{\theta}}(\hat{\nu})_t - \boldsymbol{\theta}\|^2$$

; Total Average Squared Error of a ridge type estimator $\hat{\boldsymbol{\theta}}(\hat{\nu})$,

where $\hat{\boldsymbol{\theta}}(\hat{\nu})_t$ is the ridge type estiamtor for t th generated pseudo random numbers and $N(=1000)$ is the number of simulations. Note that $\mathrm{TASE}(\hat{\boldsymbol{\theta}}(\hat{\nu}))$ is an estimated $\mathrm{TMSE}(\hat{\boldsymbol{\theta}}(\nu))$.

Simulation result for the Data 1 is given by the Table 1. All of TASE of the ridge type estimators are less than the TASE of the ordinary least squares estimator($\mathrm{TASE}(\hat{\boldsymbol{\theta}}) = 19.9838$). The smallest TASE is given by determining the ridge coefficients $k = \hat{k}_{\mathrm{IIS1}}$ and $\nu = \hat{\nu}$. This TASE is about half of the TASE of the ordinary least squares estimator.

The result for the Data 2 is similar to the Data 1. (See Table 2.) The smallest TASE is, also, given by detemining the ridge coefficients $k = \hat{k}_{\mathrm{IIS1}}$ and $\nu = \hat{\nu}$. This is less than half of the TASE of the ordinary least squares estimator ($\mathrm{TASE}(\hat{\boldsymbol{\theta}}) = 59.19531$). So our simulation concludes again that the rule for the ridge coefficient ν is improved on the othe rules, even if the ridge coefficient ν is estimated.

4. CONCULUSION

We proposed the ridge type estimators (1.8) for the vector of regression coefficients parameters $\boldsymbol{\theta}$ in the model M_Z and denoted that the ridge type estimator $\hat{\boldsymbol{\theta}}(\nu)$ has the smaller TMSE than $\hat{\boldsymbol{\theta}}(0)$ and $\hat{\boldsymbol{\theta}}(k)$ if the ridge coefficient $k \geq 0$ is fixed and ν is determined by the rule (2.7).

When we practically use the rule (2.7), the estimators $\hat{\boldsymbol{\alpha}}$, $\hat{\sigma}^2$ are substituted for the unknown parameters $\boldsymbol{\alpha}$, σ^2, respectively and k is determined by the Type I, II, II-U, II-S rules. That is, we recommend to use the rule (3.5).

These new ridge rules were tested with a Monte Carlo experiment on some data sets which showed that the smallest TASE is given by detemining the ridge coefficients $k = \hat{k}_{\mathrm{IIS1}}$ and $\nu = \hat{\nu}$.

Table 1

TASE of Ridge Type Estimators for Data 1

k	$\mathrm{TASE}(\hat{\boldsymbol{\theta}}(k))$	$\mathrm{TASE}(\hat{\boldsymbol{\theta}}(0))$	$\mathrm{TASE}(\hat{\boldsymbol{\theta}}(\hat{\nu}))$
$\hat{k}_{\mathrm{I1}}$	17.53737		17.59499
$\hat{k}_{\mathrm{II1}}$	12.10160		12.08178
$\hat{k}_{\mathrm{IIU1}}$	11.07079		11.01131
$\hat{k}_{\mathrm{IIS1}}$	10.50987		10.28206
$\hat{k}_{\mathrm{I2}}$		12.64316	12.64513
$\hat{k}_{\mathrm{II2}}$		12.80375	12.09650
$\hat{k}_{\mathrm{IIU2}}$		12.01220	11.03110
$\hat{k}_{\mathrm{IIS2}}$		12.45649	10.35746

Table 2

TASE of Ridge Type Estimators for Data 2

k	TASE($\hat{\boldsymbol{\theta}}(k)$)	TASE($\hat{\boldsymbol{\theta}}(0)$)	TASE($\hat{\boldsymbol{\theta}}(\hat{\nu})$)
$\hat{k}_{\mathrm{I1}}$	41.38838		41.39733
$\hat{k}_{\mathrm{II1}}$	32.77508		32.81626
$\hat{k}_{\mathrm{IIU1}}$	27.76340		27.81167
$\hat{k}_{\mathrm{IIS1}}$	25.76138		25.75913
$\hat{k}_{\mathrm{I2}}$		25.87620	25.85381
$\hat{k}_{\mathrm{II2}}$		33.08111	32.82483
$\hat{k}_{\mathrm{IIU2}}$		28.17628	27.82893
$\hat{k}_{\mathrm{IIS2}}$		26.91279	25.83154

Acknowledgements
The authors wish to thank the referee for his helpful comments.

REFERENCES

1. A.E. Hoerl and R.W. Kennard, *Technometrics* **12**, 55–67 (1970).
2. A.E. Hoerl and R.W. Kennard, *Technometrics* **12**, 69–82 (1970).
3. P.J. Brown, *Technometrics* **19**, 35–36 (1977).
4. N.R. Draper and H. Smith, *Applied regression analysis 2nd ed.* John Wiley, New York (1981).
5. M. Goto, *Evaluation of ordinary and generalized ridge regressions.* Doctor thesis(1981).
6. A.E. Hoerl, *Chemical Engineering Progress* **58**, 54–59 (1962).
7. A.E. Hoerl, R.W. Kennard and K.F. Baldwin, *Communications in Statistics* **4**, 105–123 (1975).
8. T.S. Lee and D.B. Campbell, *Communications in Statistics***A14**, 1589–1604 (1985).
9. T.S. Lee, *Applied Statistics* **36**, 112–118(1987).

Stat. Sci. & Data Anal., pp. 87-95
K. Matsusita *et al.* (Eds)

Reparametrization Methods in Linear Minimax Estimation

HILMAR DRYGAS

FB Mathematik/Informatik, University of Kassel
Heinrich-Plett-Str. 40
D-W-3500 Kassel, Germany

Abstract. We consider the linear model $Y=X\beta+\varepsilon$, $E\varepsilon=0$, Cov $\varepsilon=\sigma^2V$ under the ellipsoidal constraints $(\beta-\beta_0)'T(\beta-\beta_0)\leq 1$ and the affine constraints $R\beta=r$. The reparametrization $\beta=R^-R\beta+(I-R^-R)\beta$, R^- a g-inverse of R, is substituted into the regression function as well as into the ellipsoidal constraints. Minimax estimaton after this reparametrization yields appropriate results according to the latter reparametrization.

Key words: Linear models, restrictions, affine and ellipsoidal, reparametrization, 62J05.

1. INTRODUCTION

We consider the linear model

$$\underset{n\times 1}{y} = \underset{n\times k}{X}\ \underset{k\times 1}{\beta} + \underset{n\times 1}{\sigma\varepsilon}, \qquad E\varepsilon=0, E\varepsilon\varepsilon'=V \tag{1.1}$$

under the constraints of ellipsoidal type

$$\beta\in\mathcal{B}=\{\beta:(\beta-\beta_0)'T(\beta-\beta_0)\leq 1\} \tag{1.2}$$

and the affine constraints

$$\beta\in\mathcal{A}=\{\beta:R\beta=r\}, \tag{1.3}$$

where T is a positive definite (p.d.) $k\times k$-matrix and R is an $m\times k$-matrix of rank m. We want to discuss Linear Minimax Estimators in such models. Linear Minimax-Estimators were introduced by Kuks [1] and Kuks and Olman [2] (see also Pilz [3]) in the case $\mathcal{A}=\mathbb{R}^k(m=0)$ as an estimator $\widehat{\beta}=Cy+d$ such that

$$\sup_{\beta\in\mathcal{B}} E((a'(Cy+d-\beta))^2)$$

is minimized simultaneously for all $a\in\mathbb{R}^k$, thus yielding the ridge-type estimator

$$\widehat{\beta}=(\sigma^2I+X'V^{-1}X)^{-1}X'V^{-1}(y-X\beta_0)+\beta_0\,. \tag{1.4}$$

There are many ways to incorporate the affine restrictions (1.3) into a minimax-approach. It is the purpose of this paper to discuss these different approaches. One way of applying the minimax-principle is to consider r as an additional observation, thus arriving at the model

$$\widehat{y}=\begin{pmatrix} y \\ \cdots \\ r \end{pmatrix}, E\widehat{y}=\widehat{X}\beta, \widehat{X}=\begin{pmatrix} X \\ \cdots \\ R \end{pmatrix}, \operatorname{Cov}\widehat{y}=\widehat{V}, \widehat{V}=\begin{pmatrix} \sigma^2 V & \vdots & 0 \\ \cdots & \cdots & \cdots \\ 0 & \vdots & 0 \end{pmatrix}. \tag{1.5}$$

The Minimax-Estimator in this model under the ellipsoidal constraints is given by (see Drygas [4]):

$$\widehat{\beta} = \beta_0 + T^{-1}\widehat{X}'(\widehat{X}T^{-1}\widehat{X}' + \widehat{V})^{-}(y - \widehat{X}\beta_0). \tag{1.6}$$

This estimator has been called the naive Minimax-Estimator. Its shortcomings are that it does not react on an empty intersection $\mathcal{A} \cap \mathcal{B}$ or on the situation that $\mathcal{A} \cap \mathcal{B}$ only consists of a single point. However, it was shown in Drygas [4] that if $R\beta_0 = r$ there is a version of the naive Minimax-Estimator which coincides with the Minimax-Estimator to be defined below.

The minimax-estimator is obtained by minimizing

$$\sup_{\beta \in \mathcal{A} \cap \mathcal{B}} E((a'(Cy + d - \beta))^2)$$

subject to C, d. The solution was found by Stahlecker and Trenkler [5]. The estimator is defined in such a way that if $\mathcal{A} \cap \mathcal{B}$ is a single point t_* then $Cy = t_*$, and if $\mathcal{A} \cap \mathcal{B} = \emptyset$ some constant α becomes less than zero and the algorithm stops.

2. ELIMINATION AND REPARAMETRIZATON METHODS

A natural approach to the Minimax-approach under affine and ellipsoidal restrictions is the following: Write the restrictions in the form $R\beta = R_1\beta_1 + R_2\beta_2 = r$, where $R_1 \in \mathbb{R}_{m \times m}$ is regular. This implies that $\beta_1 = R_1^{-1}(r - R_2\beta_2)$ and leads to the model

$$\begin{aligned} E(y - X_1 R_1^{-1} r) &= (X_2 - X_1 R_1^{-1} R_2)\beta_2\,, \\ \operatorname{Cov}(y - X_1 R_1^{-1} r) &= \sigma^2 V\,. \end{aligned} \tag{2.1}$$

Applying the Minimax-approach to this model leads to the correct estimators in the sense of the Minimax-Estimators. The elimination method is a special case of a more general elimination method, see Drygas [4].

Finally we are considering two reparametrization methods. Let R^- be any g-inverse of R (see Rao and Mitra [6]), i.e., any matrix such that $RR^-R = R$. Then $Ey = X\beta = XR^-r + X(I - R^-R)\beta$. Let $\widetilde{\beta}_0$ be the Minimax-Estimator of β in this model under the ellipsoidal constraints. (It is defined analogously to formula (1.6).) Consider

$$\begin{aligned} \widetilde{\beta} &= R^-r + (I - R^-R)\widetilde{\beta}_0 \qquad (2.2) \\ &= R^-r + (I - R^-R)\beta_0 + (I - R^-R)T^{-1}(I - R^-R)'X' \\ &\qquad \left[X(I - R^-R)T^{-1}(I - R^-R)X' + \sigma^2 V\right]^-(y - XR^-r - X(I - R^-R)\beta_0)\,. \end{aligned}$$

Evidently $R\widehat{\beta} = RR^{-}r = r$ if the equation $R\beta = r$ is consistent. $\widehat{\beta}$ can also be written in the form

$$\widetilde{\beta} = C_{11}(y - X\beta_0) + C_{21}(r - R\beta_0) + \beta_0 \,, \tag{2.3}$$

where

$$C_{11} = (I - R^{-}R)T^{-1}(I - R^{-}R)X' \\ \left[X(I - R^{-}R)T^{-1}(I - R^{-}R)'X' + \sigma^2 V\right]^{-} \tag{2.4}$$

$$C_{21} = (I - C_{11}X)R^{-} \,. \tag{2.5}$$

The reparametrization estimator

$$\widetilde{\beta} = R^{-}r + (I - R^{-}R)\beta_0 + (I - R^{-}R)T^{-1} \\ (I - R^{-}R)X' \left[X(I - R^{-}R)T^{-1}(I - R^{-}R)X' + \sigma^2 V\right]^{-} \\ \left[y - XR^{-}r - X(I - R^{-}R)\beta_0\right] \tag{2.6}$$

is equal to the naive minimax-estimator $\widehat{\beta}$ if $R^{-} = T^{-1/2}(RT^{-1/2})^{-} = T^{-1}R'\times$ $\times(RT^{-1}R')^{-}$. To see this let $S = RT^{-1/2}$ and note that $(I - R^{-}R) = (I - T^{-1/2}S^{+}ST^{1/2}) = T^{-1/2}(I - S^{+}S)T^{1/2}$ and $(I - R^{-}R)' = T^{1/2}(I - S^{+}S)T^{-1/2}$, $(I - R^{-}R)T^{-1}(I - R^{-}R)' = T^{-1/2}(I - S^{+}S)T^{-1/2}$. Thus

$$\widetilde{\beta} = (I - G_1X)R^{-}(r - R\beta_0) + G_1(y - X\beta_0) + \beta_0 \,, \tag{2.7}$$

where $G_1 = T^{-1/2}(I - S^{+}S)T^{-1/2}X'E^{-}$, $E = \sigma^2 V + XT^{-1/2}(I - S^{+}S)T^{-1/2}X'$. This formula coincides with the naive minimax-estimator as given by formula (1.6) (see also Drygas [4]).

The reparametrization estimator (2.6) depends on the choice of the generalized inverse R^{-} of R. We will give an example in (2.4) in order to illustrate this fact. Because of the dependence of $\widetilde{\beta}$ on R^{-} one might think of optimizing with respect to R^{-}. If we consider the model

$$E(y - XR^{-}r) = X(I - R^{-}R)\beta \,, \operatorname{Cov} y = \sigma^2 V \tag{2.8}$$

$$(\beta - \beta_0)'T(\beta - \beta_0) \le 1 \,, \tag{2.9}$$

then the approximate minimax-risk of the reparametrization estimator $\widetilde{\beta}$ of β is given by

$$\begin{aligned} \sup_{\beta \in B} E((\beta - \widetilde{\beta})'A(\beta - \widetilde{\beta})) &\le \sigma^2 \operatorname{tr}(AG_1VG_1') + \\ &+ \operatorname{tr}(A(G_1X(I - R^{-}R) - I)T^{-1}(G_1X(I - R^{-}R) - I)') \\ &= \sigma^2 \operatorname{tr}(AG_1(\sigma^2 V + X(I - R^{-}R)T^{-1}(I - R^{-}R)'X')G_1') \\ &- 2\operatorname{tr}(AG_1X(I - R^{-}R)T^{-1}) + \operatorname{tr}(AT^{-1}) \,. \end{aligned} \tag{2.10}$$

Now $(\sigma^2 V + X(I - R^{-}R)T^{-1}(I - R^{-}R)'X')G_1' = X(I - R^{-}R)T^{-1}$. Thus it follows that

$$\sup_{\beta\in B} E((\widetilde{\beta}-\beta)'A(\widetilde{\beta}-\beta)) \le \mathrm{tr}(AT^{-1}) - \mathrm{tr}(AT^{-1}(I-R^-R)'X' \\ (\sigma^2V + X(I-R^-R)T^{-1}(I-R^-R)'X')^-X(I-R^-R)T^{-1}). \tag{2.11}$$

In the expression on the right hand side of (2.11) we choose $A = T$ and show that this expression is maximized if $R^- = T^{-1/2}(RT^{-1/2})^- = T^{-1}R'(RT^{-1}R')^-$. Moreover, we assume that V and a fortiori $\sigma^2V + X(I-R^-R)T^{-1}(I-R^-R)'X'$ is non-singular because singularity of V implies linear restrictions on β. We can assume that all linear constraints are contained in the restriction $R\beta = r$. Thus the right hand side of (2.11) becomes

$$\mathrm{tr}(I_k) - \mathrm{tr}((\sigma^2V + X(I-R^-R)T^{-1}(I-R^-R)'X')^{-1} \\ X(I-R^-R)T^{-1}(I-R^-R)X') \\ = k - n + \mathrm{tr}(\sigma^2V(\sigma^2V + X(I-R^-R)T^{-1}(I-R^-R)'X')^{-1}). \tag{2.12}$$

Now $A \le B$ implies $A^{-1} \ge B^{-1}$ (there exists a regular matrix C and a diagonal matrix Λ such that $A = C\Lambda C'$ and $B = CC'$. $A \le B$ is equivalent to $\Lambda \le I$. But then $\Lambda^{-1} \ge I$ and $A^{-1} = C'^{-1}\Lambda^{-1}C^{-1} \ge C'^{-1}C^{-1} = (CC')^{-1} = B^{-1}$). Thus (2.12) is maximized if

$$X(I-R^-R)T^{-1}(I-R^-R)'X' \tag{2.13}$$

is minimized. Let

$$V(A,B) = X(I-AR)T^{-1}(I-BR)'X'. \tag{2.14}$$

Then $V(A,B) - V(A,0) = X(I-AR)T^{-1}R'B'X'$. $V(A,A)$ is optimal in the sense of Loewner ordering subject to $RAR = R$ iff $\mathrm{tr}(V(A,A)C)$ is minimized simultaneously for all n.n.d. C. This is obtained from the thory of quasi-inner products (see Rao and Kleffe [7], Drygas [8]). By (2.14) this is the case iff $X(I-AR)T^{-1}R'B'X' = 0$ for all B such that $RBR = 0$. A solution is independent of X iff $(I-AR)T^{-1}R' = 0$ or $A(RT^{-1}R') = T^{-1}R'$. Thus $A = T^{-1}R(RT^{-1}R')^- = T^{-1/2}(RT^{-1/2})^+$ maximizes (2.12) and $\widetilde{\beta}$ becomes the naive minimax-estimator $\widehat{\beta}$.

2.1. Theorem: *The approximate minimax-risk for the loss function* $(\beta - \widetilde{\beta})'T(\beta-\widetilde{\beta}) = l(\beta,\widetilde{\beta})$ *is maximized for* $R^- = T^{-1}R'(TR^{-1}R')^- = T^{-1/2}\times(RT^{-1/2})^+$, *thus resulting in the naive minimax-estimator.*

2.2. Remark: Theorem 2.1 is an interesting result from the mathematical point of view. It shows that for the approximate loss function (2.10) we have an upper bound for the class of estimators under consideration. Moreover, this upper bound is attained by an estimator already known to us, namely the naive Minimax-Estimator. However, the theorem is no statistical recommandation because it is hard to imagine that a user of linear models would like to maximize some risk. On the other hand this is quite natural because this kind of reparametrization is not a convenient one. These estimators should not not be used at all!

In the model $Ey = X\beta$, $R\beta = r$, $\mathrm{Cov}\, y = \sigma^2V$, $(\beta-\beta_0)'T(\beta-\beta_0) \le 1$ we could also try to reparametrize the ellipsoidal restrictions by $\beta = R^-r + (I-R^-R)\beta$, R^- again being an arbitrary g-inverse of R. Will this lead to different estimators? It will not! Let us see: With — again — $s = r - R\beta_0$ we get $\beta - \beta_0 = R^- s + (I - R^- R)(\beta - \beta_0)$ and

$$(R^- s + (I - R^- R)(\beta - \beta_0))'T(R^- s + (I - R^- R)(\beta - \beta_0)) \leq 1 \tag{2.15}$$

iff $s'(R^-)'TR^- s + 2(\beta - \beta_0)'(I - R^- R)'TR^- s + (\beta - \beta_0)'(I - R^- R)'T(I - R^- R)(\beta - \beta_0) \leq 1$. Let β_1 be determined in such a way that

$$(I - R^- R)'TR^- s = (I - R^- R)'T(I - R^- R)\beta_1 \,. \tag{2.16}$$

Then $(\beta - \beta_0)'T(\beta - \beta_0) \leq 1$ and $R\beta = r$ imply

$$\begin{aligned}(\beta - (\beta_0 - \beta_1))'(I - R^- R)'T(I - R^- R)(\beta - (\beta_0 - \beta_1))& \\ \leq 1 - s'(R^-)'TR^- s + \beta_1'(I - R^- R)'T(I - R^- R)\beta_1 = d_- \,.&\end{aligned} \tag{2.17}$$

We show that $d_- = 1 - s'(RT^{-1}R')^- s$. Indeed,

$$\begin{aligned} d_- &= 1 - s'(R^-)'TR^- s + s'(R^-)'T(I - R^- R) \\ &\qquad ((I - R^- R)'T(I - R^- R))^-(I - R^- R)'TR^- s \\ &= 1 - s'(R^-)'TR^- s + s'(R^-)'T^{1/2}P_{\text{im}(T^{1/2}(I - R^- R))}T^{1/2}R^- s \,. \end{aligned} \tag{2.18}$$

But $\text{im}(I - R^- R) = R^{-1}(\{0\})$, and $x \in \text{im}(T^{1/2}R^{-1}(\{0\}))$ iff $T^{-1/2}x \in R^{-1}(\{0\})$, i.e., $RT^{-1/2}x = Sx = 0$. Thus $P_{\text{im}(T^{1/2}(I - R^- R))} = I - S^+ S$ and

$$\begin{aligned} d_- &= 1 - s'(R^-)'T^{1/2}(I - (I - S^+ S))T^{1/2}R^- s \\ &= 1 - s'(R^-)'T^{1/2}S^+ ST^{1/2}R^- s \\ &= 1 - s'(R^-)'T^{1/2}T^{-1/2}R'(RT^{-1}R')^- RT^{-1/2}T^{1/2}R^- s \\ &= 1 - s'(R^-)'R'(RT^{-1}R')^- RR^- s \,. \end{aligned} \tag{2.19}$$

Using $s = R\beta - R\beta_0$ for some β if the equation $R\beta = r$ is consistent, we get the desired result.

Let $\beta_- = \beta_0 - \beta_1$, $T_- = (I - R^- R)'T(I - R^- R)$. In order to find the minimax estimator of β in the model $Ey = X\beta$, $R\beta = r$, $\text{Cov}\, y = \sigma^2 V$, $(\beta - \beta_-)'T_-(\beta - \beta_-) \leq d_-$, let us consider the artificial model

$$E = \begin{pmatrix} y \\ \cdots \\ r \\ \cdots \\ D\beta_- \end{pmatrix} = \begin{pmatrix} X \\ \cdots \\ R \\ \cdots \\ D \end{pmatrix} \beta, \text{Cov} \begin{pmatrix} y \\ \cdots \\ r \\ \cdots \\ D\beta_- \end{pmatrix} = \begin{pmatrix} \sigma^2 V & \vdots & 0 & \vdots & 0 \\ \cdots & \cdots & \cdots & \cdots & \cdots \\ 0 & \vdots & 0 & \vdots & 0 \\ \cdots & \cdots & \cdots & \cdots & \cdots \\ 0 & \vdots & 0 & \vdots & d_- P \end{pmatrix}, \tag{2.20}$$

where $D = T^{1/2}(I - R^- R)$, $D'D = T_-$, $D(T_-)'D' = P_{\text{im}(S)} = I - S^+ S = P$. β is estimable in the model (2.20) since $T^{-1/2}D + R^- R = I$. The model (2.20) is a model with singular covariance matrix. Its consideration was suggested in Drygas [9] in order to cope with the problem of estimating regression parameters in a linear model with singular ellipsoidal constraints. It was shown that the BLUE of any estimable function in the artificial regression model just yields the minimax-estimator model with a singular matrix defining the ellipsoidal constraints. We now consider a BLUE $G_1y+G_2r+G_3D\beta_-$ in the artificial model (2.20). Then we have the following conditions

$$\text{(i)} \qquad G_1X + G_2R + G_3D = I \quad \text{(unbiasedness)} \tag{2.21}$$

$$\text{(ii)} \qquad G_1(\sigma^2V)z_1 + G_2 0 \cdot z_2 + d_- G_3\, Pz_3 = 0 \quad \text{(optimality)} \tag{2.22}$$

whenever $X'z_1 + R'z_2 + D'z_3 = 0$ (orthogonality). This implies that $G_1(\sigma^2V)z_1 + d_-G_3D(T_-)^-(-X'z_1 - R'z_2)) = G_1(\sigma^2V)z_1 + d_-(I - G_1X - G_2R)(T_-)^- \times$ $\times(-X'z_1 - R'z_2) = 0$ whenever $X'z_1 + R'z_2 \in \operatorname{im}(D')$. For given $z_1 \in \mathbb{R}^n$ we set $z_2 = -(R^-)'X'z_1$, $z_3 = -T^{-1/2}X'z_1$. Then $X'z_1 + R'z_2 + D'z_3 = 0$ and it follows that

$$0 = G_1(\sigma^2V) - d_-(I - G_1X - G_2R)(T_-)^-(I - R^-R)'X' \tag{2.23}$$

or $G_1(\sigma^2V + d_-X(T_-)^-(I - R^-R)'X') = d_-(I - G_2R)(T_-)^-(I - R^-R)'X'$. From $G_1X + G_2R + G_3D = I$ or $(I - G_1X) = G_2R + G_3D = G_2R + G_3T^{1/2}(I - R^-R)$ we get by postmultiplying with R^-RR^- that

$$(I - G_1X)R^-RR^- = G_2RR^- . \tag{2.24}$$

A solution of this equation is given by $G_2 = (I - G_1X)R^-$ implying $I - G_2R = I - (I - G_1X)R^-R = (I - R^-R) + G_1XR^-R$ and (2.23) becomes thus the equation

$$\begin{aligned} G_1(\sigma^2V + d_-X(I - R^-R)(T_-)^-(I - R^-R)'X') \\ = d_-(I - R^-R)(T_-)^-(I - R^-R)'X' \end{aligned} \tag{2.25}$$

or

$$G_1(\sigma^2V + XT^{-1/2}(I - S^+S)T^{-1/2}X') = d_- \cdot T^{-1/2}(I - S^+S)X' . \tag{2.26}$$

From $G_1X + G_2R + G_3D = I$ we get by postmultiplying with $I - R^-R$ that

$$G_1X(I - R^-R) + G_3D = I - R^-R \tag{2.27}$$

or $(I - G_1X)(I - R^-R) = G_3T^{1/2}(I - R^-R)$. A solution of this equation is $G_3 = (I - G_2X)(I - R^-R)T^{-1/2}$. We now prove that with these values of G_1, G_2 and G_3 the two conditions (2.21) and (2.22) are met. Indeed

$$\begin{aligned} G_1X + (I - G_1X)R^-R + (I - G_1X)(I - R^-R)T^{-1/2}T^{1/2}(I - R^-R) \\ = G_1X + (I - G_1X)R^-R + (IG_1X)(I - R^-R) = I . \end{aligned} \tag{2.28}$$

Now let $X'z_1 + R'z_2 + D'z_3 = 0$. Premultiplying with $(R^-)'R'(R^-)'$ yields

$$(R^-)'R'(R^-)'X'z_1 + (R^-)'R'z_2 = 0 , \tag{2.29}$$

$$\begin{aligned} (I - R^-R)'(X'z_1 + R'z_2) = (I - R^-R)'X'z_1 \\ = (I - R'(R^-)'R'(R^-)')(X'z_1 + R'z_2) = X'z_1 + R'z_2 . \end{aligned} \tag{2.30}$$

Thus

$$\begin{aligned}
G_1(\sigma^2 V)z_1 &+ d_-(I-G_1X)(I-R^-R)(T_-)^-(I-R^-R)'T^{1/2}z_3 \\
&= G_1(\sigma^2 V)z_1 + d_-(I-G_1X)(I-R^-R)(T_-)^-(-X'z_1-R'z_2) \\
&= G_1(\sigma^2 V)z_1 + d_-(I-G_1X)(I-R^-R)(T_-)^-(I-R^-R)'(-X'z_1) \\
&= G_1(\sigma^2 V)z_1 + d_-(I-G_1X)T^{-1/2}(I-S^+S)T^{-1/2}(-X'z_1) \\
&= G_1(\sigma^2 V + d_-XT^{-1/2}(I-S^+S)T^{-1/2}X')z_1 \\
&\qquad - d_-T^{-1/2}(I-S^+X)T^{-1/2}X'z_1 = 0\,.
\end{aligned}$$

Thus optimality of the estimator $G_1y + G_2r + G_3D\beta_-$ is proved. This estimator is equal to

$$\begin{aligned}
G_1y+&(I-G_1X)R^-r + (I-G_1X)(I-R^-R)T^{-1/2}(\beta_0-\beta_1) \\
&= G_1(y-X\beta_0) + (I-G_1X)(R^-(r-R\beta_0) - (I-R^-R) \\
&\qquad ((I-R^-R)'T(I-R^-R))^-(I-R^-R)'TR^-(r-R\beta_0)) + \beta_0\,.
\end{aligned} \tag{2.31}$$

Since $(I-R^-R)((I-R^-R)'T(I-R^-R))^-(I-R^-R)' = T^{-1/2}(I-S^+S)T^{-1/2}$, the middle term on the right hand side of (2.31) is also equal to

$$\begin{aligned}
(I-G_1X)&T^{-1/2}S^+ST^{1/2}R^-(r-R\beta_0) \\
&= (I-G_1X)T^{-1/2}R'(RT^{-1}R')^-RR^-(r-R\beta_0) \\
&= (I-G_1X)T^{-1}R'(RT^{-1}R')^-(r-R\beta_0)\,,
\end{aligned}$$

if $r \in \mathrm{im}(R)$ (consistency of the model). Therefore (2.31) becomes

$$\beta_0 + G_1(y-X\beta_0) + (I-G_1X)T^{-1}R'(RT^{-1}R')^-(r-R\beta_0) \tag{2.32}$$

and this is just the Minimax-Estimator derived by Stahlecker and Trenkler [5].

2.3. Theorem: *Consider the linear model $Ey = X\beta$, $R\beta = r$, $(\beta-\beta_0)'T(\beta-\beta_0) \le 1$, $\operatorname{Cov} y = \sigma^2 V$. There exists a Minimax-Estimator of β in the model*

$$E\begin{pmatrix} y \\ \cdots \\ r \end{pmatrix} = \begin{pmatrix} X \\ \cdots \\ R \end{pmatrix}\beta\,, \operatorname{Cov}\begin{pmatrix} y \\ \cdots \\ r \end{pmatrix} = \begin{pmatrix} \sigma^2 V & \vdots & 0 \\ \cdots & \cdots & \cdots \\ 0 & \vdots & 0 \end{pmatrix}$$

under the reparametrized ellipsoidal restrictions (2.17) which coincides with the Minimax-Estimator in the original model.

2.4. Example: We consider as a simple example the model

$$E\begin{pmatrix} y_1 \\ y_2 \end{pmatrix} = \begin{pmatrix} x_1 & 1 \\ x_2 & 1 \end{pmatrix}\begin{pmatrix} \beta_1 \\ \beta_2 \end{pmatrix}, \operatorname{Cov}\begin{pmatrix} y_1 \\ y_2 \end{pmatrix} = \begin{pmatrix} \sigma^2 & 0 \\ 0 & 1 \end{pmatrix} \tag{2.33}$$

i.e., $\beta = (\beta_1, \beta_2)'$ under the restrictions $\beta_1 + \beta_2 = 1$, $\beta_1^2 + \beta_2^2 \le 1$. We assume that $\rho^2 = (1-x_1)^2 + (1-x_2)^2 > 0$. Computational details can be found in Drygas [10]. The

Best Linear Unbiased Estimator (BLUE) or Gauss-Markov Estimator (GME) is then given by

$$\begin{aligned} \widehat{\beta}_1 &= \frac{(x_1-1)(y_1-1)+(x_2-1)(y_2-1)}{\rho^2} \\ \widehat{\beta}_2 &= 1-\widehat{\beta}_1 \,. \end{aligned} \tag{2.34}$$

The naive minimax-estimator is computed as

$$\begin{aligned} \widetilde{\beta}_1 &= \frac{1}{2}+\frac{(x_1-1)(y_1-\frac{x_1+1}{2})+(x_2-1)(y_2-\frac{x_2+1}{2})}{2\sigma^2+\rho^2}\,, \\ \widetilde{\beta}_2 &= 1-\widetilde{\beta}_1 \,. \end{aligned} \tag{2.35}$$

Since $\beta_0 = 0$, $R\beta_0 = 0 \neq 1 = r$, the Minimax-Estimator will therefore not coincide with the Naive Minimax-Estimator. The Minimax-Estimator is found as

$$\begin{aligned} \widehat{\beta}_M^{(1)} &= \frac{(x_1-1)(y_1-1)+(x_2-1)(y_2-1)+2\sigma^2}{\rho^2+4\sigma^2}\,, \\ \widehat{\beta}_M^{(2)} &= 1-\widehat{\beta}_M^{(1)} \,. \end{aligned} \tag{2.36}$$

In order to find the reparametrization parameter we change the restrictions $\beta'\beta \leq 1$ to $\beta'\beta \leq \delta$. Then if $\delta = \frac{1}{2}$, $\mathcal{B}\cap\mathcal{A} = \{\beta'\beta \leq \delta\}\cap\{(1,1)\beta = 1\} = \{(\frac{1}{2},\frac{1}{2})'\}$, and $\mathcal{B}\cap\mathcal{A} = \emptyset$ iff $\delta < \frac{1}{2}$. Since $R = (1,1)$, $R^- = (\alpha,\beta)'$, $\alpha+\beta = 1$. Consider the linearly sufficient statistic $w_0 = \sum_{i=1}^{2}(x_i-1)(y_i-1)$, then the reparametrization estimator is found to be equal

$$\widehat{\beta}_{\text{rep}}^{(1)} = (\alpha\sigma^2+\sigma_0^2 w_0)/(\rho\sigma_0^2+\sigma^2)\,, \tag{2.37}$$

where $\sigma_0^2 = \delta(\alpha^2+(1-\alpha)^2)$. $\alpha = \frac{1}{2}$ yields $R^- = R^+$ and thus the naive Minimax-Estimator. The latter formula can be given a Bayesian interpretation as well. Since $E(w_0) = \beta_1\rho^2$, $\text{Var}(w_0) = \sigma^2\rho^2$, the linear Bayes estimator of β_1 is equal to

$$\widehat{\beta}_1^B = \left(\sigma^2(E\beta_1)+\text{Var}(\beta_1)w_0\right) / \left(\rho^2\,\text{Var}(\beta_1)+\sigma^2\right)\,. \tag{2.38}$$

Thus $\widehat{\beta}_{\text{rep}}^{(1)}$ has a Bayesian interpretation with $\alpha = E\beta_1$ and

$$\sigma_0^2 = \delta(\alpha^2+(1-\alpha)^2) = \text{Var}(\beta_1)\,.$$

Acknowledgement. The author thanks the anonymous referee for his very stimulating suggestions.

REFERENCES

1. J. Kuks, (Russian), *Izv. Akad. Nauk Est. SSSR* **21**, 73–78 (1972).

2. J. Kuks and W. Olman, (Russian), *Izv. Akad. Nauk Est. SSSR* **21**, 66–72 (1972).

3. J. Pilz, *J. Statist. Planning and Inference* **13**, 297–318 (1986).

4. H. Drygas, *Computational Statistics and Data Analysis* **12**, 101–113 (1991).

5. P. Stahlecker and G. Trenkler, *Linear Algebra and Its Applications* **111**, 279–292 (1988).

6. C.R. Rao and S.K. Mitra, *Generalized Inverse of Matrices and Its Applications.* Wiley, New York (1971).

7. C.R. Rao and J. Kleffe, *Estimation of Variance Components and Applications.* North Holland, Elsevier, Amsterdam (1988).

8. H. Drygas, in: *Advances in Multivariate Statistical Analysis (Pillai Memorial Volume)*, A.K. Gupta (Ed.), pp 13–30, D. Reidel Publishing Company, Dordrecht/NL (1987).

9. H. Drygas, in: *Linear Statistical Inference, Proceedings, Poznań 1984*, T. Caliński, W. Klonecki (Eds.), Lecture Notes in Statistics No. **35**, Springer Verlag Berlin, pp. 48–60 (1986).

10. H. Drygas,*A note on Minimax-Estimation in Regression Models with affine Restrictions.* Kasseler Mathematische Schriften (1988).

11. C.R. Rao, *Linear Statistical Inference and its Applications.* Wiley, New York, Second Edition (1973).

Stat. Sci. & Data Anal., pp. 97-107
K. Matsusita *et al.* (Eds)

Robust Tests for Linear Models

STEPHAN MORGENTHALER

Swiss Federal Institute of Technology
Mathematics Department
1015 Lausanne, Switzerland

Abstract. Morgenthaler and Tukey [1] describe point estimation and interval estimation for parameters of a linear model based on robust modelling. The shape of the error distribution is modeled by a few extreme choices rather than a continuous family of possibilities, thus gaining in simplicity and flexiblity without loosing much generality. They derive optimal inferential procedures and exhibit their small sample behavior. The resulting optimal confidence intervals are, however, computationally very complex. The present paper describes simpler, approximate procedures and points out some open problems.

Key Words: conditional inference, extreme types, asymptotic expansions.

1. INTRODUCTION

Consider observations $(y_1, \ldots, y_n) = (y_i)$ of the form

$$y_i = x_i^T \beta + e_i \sigma \,, \tag{1.1}$$

where $e_1,\ldots,e_n$ are i.i.d. from a fixed continuous distribution F with density f. The vectors $x_1,\ldots,x_n$ in $\mathbf{R}^p$ are known, whereas $(\beta_1, \ldots, \beta_p) = (\beta_i)$ and $\sigma > 0$ are unknown parameters. In this paper we discuss tests of the hypothesis $H_0 : \beta_p = 0$ against alternatives $H_A : \beta_p > 0$. Invariant tests are based on the distribution of the pivot

$$piv = piv(y_i) = (\hat{\beta}_p(y_i) - \beta_p)/\hat{\sigma}(y_i) \,, \tag{1.2}$$

where $\hat{\beta}_p(y_i)$ is an estimate of β_p based on the observations (y_i) and $\hat{\sigma}(y_i)$ is an estimate of the standard deviation. Such pivots have a distribution that is independent of the underlying parameters β and σ, whenever $\hat{\beta}_p(y_i)$ and $\hat{\sigma}(y_i)$ satisfy the equivariance conditions

$$\hat{\beta}_p(sy_i + x_i^T r) = s\hat{\beta}_p(y_i) + r_p$$
$$\hat{\sigma}(sy_i + x_i^T r) = s\hat{\sigma}(y_i)$$

for all $s > 0$ and for all $r \in \mathbf{R}^p$. This special property follows immediately from

$$piv(y_i) = (\sigma\hat{\beta}_p(e_i) + \beta_p - \beta_p)/(\sigma(\hat{\sigma}(e_i))) = piv(e_i) \,.$$

Our model has another basic property in the fact that

$$(a_i) = ([y_i - x_i^T \hat{\beta}_o(y_i)]/\hat{\sigma}_o(y_i)) \tag{1.3}$$

is an ancillary statistic for any fixed choice of equivariant statistics $\hat{\beta}_o \in \mathbf{R}^p$, $\hat{\sigma}_o$. This entreats us to apply conditional inference arguments, which means that we must base our conclusions on the conditional distribution of the observations (y_i) given the observed ancillary (a_i), rather than the simpler unconditional law of (y_i). The data vectors (y_i) that remain feasible for a given ancillary (a_i) are of the form

$$y_i = s(a_i + x_i^T t)\,, \tag{1.4}$$

for $s > 0$ and $t = (t_1, \ldots, t_p)^T \in \mathbf{R}^p$. Since this transformation from (a_i) to (y_i) has Jacobian proportional to s^{n-1}, the conditional density of the data (y_i) given (a_i) for $\beta = 0$ and $\sigma = 1$ is

$$k(s, t_1, \ldots, t_p) \propto s^{n-1} \prod_{i=1}^{n} f(s(a_i + x_i^T t))\,. \tag{1.5}$$

The conditional density of the pivot $piv(y_i)$ is – up to a translation and a rescaling – equivalent to the conditional density $k(t_p)$, since, for $\beta = 0$, (1.4) implies that

$$\begin{aligned} piv(y_i) &= (s(\hat{\beta}_p(a_i) + t_p) - 0)/(s(\hat{\sigma}(a_i))) \\ &= \hat{\beta}_p(a_i)/\hat{\sigma}(a_i) + t_p/\hat{\sigma}(a_i) \\ &= Const(a_i) + Const^*(a_i) t_p\,. \end{aligned} \tag{1.6}$$

The classical conditional test rejects the hypothesis, if $piv(y_i) > cv^* = Const(a_i) + Const^*(a_i) cv$, where the critical value cv is such that

$$\int_{cv}^{\infty} k(t_p)\, dt_p = \alpha\,,$$

the predetermined size. This rejection rule is equivalent to $t_p > cv$, i.e., the choice of $\hat{\beta}$ and $\hat{\sigma}$ in (1.2) plays no rôle whatsoever.

The relation between the ancillary (a_i) and the observed data (y_i) is given by

$$y_i = \hat{\sigma}_o(y_i) a_i + x_i^T \hat{\beta}_o(y_i) = \hat{\sigma}_o(y_i) \left(a_i + x_i^T (\hat{\beta}_o(y_i)/\hat{\sigma}_o(y_i))\right)$$

(see (1.3)). The value of t_p is, therefore, equal to $\hat{\beta}_{o,p}(y_i)/\hat{\sigma}_o(y_i)$, whereas $s = \hat{\sigma}_o(y_i)$, and the rejection rule $t_p > cv$ is the same as

$$\hat{\beta}_{o,p}(y_i)/\hat{\sigma}_o(y_i) > cv\,. \tag{1.7}$$

<u>Example</u>: Let $F = \Phi$, the standard normal distribution and suppose $\hat{\beta}_o$ and $\hat{\sigma}_o$ are the usual least squares estimates. Eq. (1.5) leads to

$$\begin{aligned} k(s, t_1, \ldots, t_p) &\propto s^{n-1} \exp\left(-\frac{s^2}{2}\sum_{i=1}^{n}(a_i + x_i^T t)^2\right) \\ &\propto s^{n-1} \exp\left(-\frac{s^2}{2}(t^T X^T X t + 2t^T X^T a + a^T a)\right) \\ &\propto s^{n-1} \exp\left(-\frac{s^2}{2}(t^T X^T X t + (n-p))\right), \end{aligned}$$

where X denotes the design matrix with ith row equal to x_i^T and where we made use of our special choices of $\hat{\beta}_o$ and $\hat{\sigma}_o$. For fixed s, this density is evidently a multivariate normal, which implies that

$$k(s, t_p) \propto s^{n-1} s^{-(p-1)} \exp\left(-\frac{s^2}{2}\left(t_p^2/(X^T X)^{-1}_{pp} + (n-p)\right)\right).$$

The integral with respect to s leads to

$$k(t_p) \propto \left((n-p) + t_p^2/(X^T X)^{-1}_{pp}\right)^{-(n-p+1)/2},$$

i.e., a Student's t_{n-p} density, scaled by $((X^T X)^{-1}_{pp})^{1/2}$. The critical value in this example does not depend on the ancillary, $cv = t_{n-p}(1-\alpha)((X^T X)^{-1}_{pp})^{1/2}$, where $t_{n-p}(1-\alpha)$ denotes the $1-\alpha$ quantile of the corresponding Student's t density. By (1.7), the hypothesis $H_0 : \beta_p = 0$ is rejected, if

$$\hat{\beta}_{o,p}(y_i)/\left(\hat{\sigma}_o(y_i)((X^T X)^{-1}_{pp})^{1/2}\right) > t_{n-p}(1-\alpha).$$

In this example, we find the classical unconditional t-test for β_p. ∎

The theory we sketched above can of course be extended to cover two-sided tests as well as tests for linear combinations rather than single components of β. And, we can construct from these tests conditional confidence intervals. The symmetric conditional confidence interval based on the pivot (1.2) derives from

$$P_{\sigma,\beta}\{cv_l^* \le piv \le cv_u^*\} = 1 - \alpha,$$

simultaneously for all β and all σ. In this equation, cv_l^* and cv_u^* are the lower and upper $\alpha/2$ quantiles of the conditional pivot distribution, which is linked to the conditional distribution of t_p via (1.6). One obtains from this the interval

$$\left[\hat{\beta}_p(y_i) - \hat{\sigma}(y_i)cv_u^*, \hat{\beta}_p(y_i) - \hat{\sigma}(y_i)cv_l^*\right]. \qquad \textbf{(1.8)}$$

From $cv_u^* = \hat{\beta}_p(a_i)/\hat{\sigma}(a_i) + cv_u/\hat{\sigma}(a_i)$, with cv_u such that

$$\int_{cv_u}^{\infty} k(t_p)\, dt_p = \alpha/2,$$

one gets $\hat{\beta}_p(y_i) - \hat{\sigma}(y_i)cv_u^* = s(t_p - cv_u)$. The interval (1.8) is, therefore, the same as

$$\left[\hat{\beta}_{o,p}(y_i) - \hat{\sigma}_o(y_i)cv_u, \hat{\beta}_{o,p}(y_i) - \hat{\sigma}_o(y_i)cv_l\right] . \qquad \mathbf{(1.9)}$$

In the remainder of the paper we extend the above conditional inferences to robust tests based on more flexible, robust, models and study easily computable approximations.

2. HUBER'S ROBUST TEST

Huber [2] and [3] studies robust analogues of the classical Neyman-Pearson tests. Replacing the simple alternatives F_0 and F_1 by sets $\mathcal{F}_0$ and $\mathcal{F}_1$, he considers the optimality problem

$$\begin{aligned} &\max_{\text{Tests}} \min_{F\in\mathcal{F}_1} \quad \text{Power}_F(\text{Test}) \\ &\text{subject to } \max_{F\in\mathcal{F}_0} \quad \text{Size}_F(\text{Test}) \le \alpha\,. \end{aligned} \qquad \mathbf{(2.1)}$$

In some special cases, this problem has a simple solution independent of the sample size.

<u>Example</u>: The most powerful test of $F_0 = \Phi(x+\mu)$ against $F_1 = \Phi(x-\mu)$ $(\mu > 0)$ based on a sample $y_1, \ldots, y_n$, rejects F_0 for large values of $y_1 + \cdots + y_n$. If we consider the neighborhoods $\mathcal{F}_i = \{F : F_i - \varepsilon \le F \le F_i + \varepsilon\}$ $(i = 0, 1)$, the maximin test (2.1) rejects for large values of $\psi_k(y_1) + \cdots + \psi_k(y_n)$, where $\psi_k(u) = \max(-k, \min(k, u))$ is the well-known Huber function. The value of k depends on μ and on ε and is tabulated on p. 272 of [3]. Note that this solution corresponds simply to a truncation of the Neyman-Pearson probability ratio. The optimal values of k are, however, surprisingly small, which is best illustrated by an example. Consider a particular alternative shape, namely $G_0(x) = \text{ArcTan}((x+\mu)/b) + 0.5$, i.e., a scaled Cauchy distribution centered at $-\mu$. The smallest value of ε such that $G_0 \in \mathcal{F}_0$ is $\varepsilon \approx 0.056$, achieved at $b = 0.49$. For that value, one obtains a constant $k \approx 0$ for a separation of $2\mu = 0.3$. A neighborhood large enough to include the Cauchy shape contains already too much diversity for this theory of robust tests to be practical. The theory essentially tells us to use the sign test $(k = 0)$ for local alternatives. ∎

For arbitrary sets $\mathcal{F}_0, \mathcal{F}_1$ the maximin test will generally be much more complicated than the truncated probability ratio test.

3. ROBUST EQUIVARIANT TESTS

We now return to the linear model introduced previously and to the hypothesis H_0 : $\beta_p = 0$. Optimal robust tests in the spirit of (2.1) can be derived when the model class $\mathcal{F}$ of distributional shapes is simple. Morgenthaler and Tukey [1] discuss such simple models, in particular the case of two extreme types $\mathcal{F} = \{F, G\}$. Any equivariant test is completely determined by a critical value function $cv(a_i)$, and rejects for $t_p = \hat{\beta}_{o,p}(y_i)/\hat{\sigma}_o(y_i) > cv(a_i)$. The conditional sizes of the test are given by

$$s_F(a_i) = \int_{cv(a_i)}^{\infty} k_F(t_p)\, dt_p \text{ and } s_G(a_i) = \int_{cv(a_i)}^{\infty} k_G(t_p)\, dt_p\,,$$

where $k_F(t_p)$ and $k_G(t_p)$ denote the conditional densities of t_p when sampling from F or G, respectively. The conditional power functions are

$$m_F(\beta_p, a_i) = \int_{cv(a_i)}^{\infty} \int_0^{\infty} k_F(s, t_p - \beta_p/s)\, ds\, dt_p \,,$$

and analogously for $m_G(\beta_p, a_i)$. This follows from

$$P_\beta\{t_p > cv(a_i)\} = \int_{cv(a_i)}^{\infty} \int_0^{\infty} s^{n-1} \prod_{i=1}^{n} f(s(a_i + x_i^T t) - x_i^T \beta)\, ds\, dt_p \,.$$

The unconditional size and power are given by an average of the conditional sizes and powers with respect to the distribution of the ancillary. Note that this distribution depends only on the distributional shape, but not on the parameters β and σ.

Tests that are admissible with respect to a quality criterion such as the power at a certain alternative can now be determined.

Example: The slope of the conditional power function at $\beta_p = 0$ is equal to

$$\begin{aligned} \gamma_F(a_i) = dm_F(0, a_i)/d\beta_p &= \int_{cv(a_i)}^{\infty} \int_0^{\infty} k_{F,2}(s, t_p)(-s^{-1})\, ds\, dt_p \\ &= \int_0^{\infty} k_F(s, cv)/s\, ds \\ &= \int_{-\infty}^{\infty} k_F(\exp(u), cv)\, du \,, \end{aligned}$$

where $k_{F,2}(s, t_p) = dk_F(s, t_p)/dt_p$. An analogous expression holds for $\gamma_G(a_i)$. The unconditional local powers γ_F, γ_G are computed as the average of $\gamma_F(a_i), \gamma_G(a_i)$ with respect to their corresponding distributions of the ancillary.

Tests that are admissible with respect to local power within the robust shape model $\{F, G\}$ maximize $\gamma_F + \pi\gamma_G$ with $\pi \geq 0$ fixed. The solution satisfies

$$\begin{aligned} w_F(a_i) \int_{-\infty}^{\infty} k_{F,2}(\exp(u), cv)\, du + \pi w_G(a_i) \int_{-\infty}^{\infty} k_{G,2}(\exp(u), cv)\, du \\ = \lambda_F w_F k_F(cv) + \lambda_G w_G k_G(cv) \,, \end{aligned} \quad (\mathbf{3.1})$$

where $w_F(a_i), w_G(a_i)$ denote the densities at the ancillary (a_i), and λ_F, λ_G are Lagrange multipliers that must be adjusted to yield a test that achieves unconditional size α for both shapes. ∎

4. ANALYTIC APPROXIMATIONS FOR CONDITIONAL PROBABILITIES

The tests described in Section 3 are computationally too demanding since difficult and numerous numerical quadratures are in general needed. Several simple approximations to the conditional density (1.5) are, however, possible. Note that with

$$L(\beta, \sigma) = \log\left(\prod_{i=1}^{n} [f((a_i - x_i^T \beta)/\sigma)/\sigma]\right) \quad (\mathbf{4.1})$$

denoting the usual log-likelihood function at the ancillary (a_i) for the model (1.1), we have

$$k(s, t_1, \ldots, t_p) \propto \exp\left[L(-t, 1/s)\right] s^{-1}$$
$$\propto \exp\left[L(-t, 1/s) - \log(s)\right] .$$

In transforming the variable s to $u = \log(s)$, we find the transformed density

$$k(u, t_1, \ldots, t_p) \propto \exp\left[L(-t, \exp(-u))\right] . \tag{4.2}$$

Expansions can be based on the fact that $L(\beta, \sigma)$ becomes increasingly peaked with increasing sample size. A simple normal approximation is obtained in expanding

$$M(\beta, \tau) = L(\beta, \exp(\tau))$$

in a neighborhood of is maximum $\hat{\beta}_m = \hat{\beta}_m(a_i)$ and $\hat{\tau}_m = \hat{\tau}_m(a_i) = \log(\hat{\sigma}_m)$. One has the following expression up to quadratic terms

$$M(\beta, \tau) \approx M(\hat{\beta}_m, \hat{\tau}_m) + \frac{1}{2}\begin{pmatrix}\beta - \hat{\beta}_m & \tau - \hat{\tau}_m\end{pmatrix}\begin{pmatrix}\hat{M}_{11} & \hat{M}_{12} \\ \hat{M}_{12}^T & \hat{M}_{22}\end{pmatrix}\begin{pmatrix}\beta - \hat{\beta}_m \\ \tau - \hat{\tau}_m\end{pmatrix},$$

where $\hat{M}_{11}$ is the Hessian of M with respect to $\beta_1, \ldots, \beta_p$ evaluated at $\hat{\beta}_m, \hat{\tau}_m$, etc.

Substitution in (4.2) gives the approximate formula

$$\begin{aligned}\tilde{k}(u, t_1, \ldots, t_p) &\propto \exp\left[\frac{1}{2}\begin{pmatrix}-t - \hat{\beta}_m & -u - \hat{\tau}_m\end{pmatrix}\begin{pmatrix}\hat{M}_{11} & \hat{M}_{12} \\ \hat{M}_{12}^T & \hat{M}_{22}\end{pmatrix}\begin{pmatrix}-t - \hat{\beta}_m \\ -u - \hat{\tau}_m\end{pmatrix}\right] \\ &\propto \exp\left[-\frac{1}{2}\begin{pmatrix}t + \hat{\beta}_m & u + \hat{\tau}_m\end{pmatrix}\begin{pmatrix}-\hat{M}_{11} & -\hat{M}_{12} \\ -\hat{M}_{12}^T & -\hat{M}_{22}\end{pmatrix}\begin{pmatrix}t + \hat{\beta}_m \\ u + \hat{\tau}_m\end{pmatrix}\right] .\end{aligned} \tag{4.3}$$

The conditional distribution of (t_i) and $u = \log(s)$ is approximated by a multivariate normal with mean $-\hat{\beta}_m, -\hat{\tau}_m$ and covariance

$$\hat{\Sigma} = \begin{pmatrix}-\hat{M}_{11} & -\hat{M}_{12} \\ -\hat{M}_{12}^T & -\hat{M}_{22}\end{pmatrix}^{-1} .$$

It follows that the conditional density of t_p is approximately normal with mean $-\hat{\beta}_{m,p}$ and variance equal to the corresponding diagonal element $\hat{\Sigma}_{pp}$ of $\hat{\Sigma}$. The weight $w(a_i)$ can be approximated by the normalizing constant in (4.3) multiplied by $\exp\left[M(\hat{\beta}_m, \hat{\tau}_m)\right]$, the maximal value of the likelihood.

Note that from (4.1)

$$\hat{M}_{11} = \frac{1}{\hat{\sigma}_m^2} \cdot X^T \mathrm{diag}\left(\left(\frac{f'}{f}\right)'(r_1), \ldots, \left(\frac{f'}{f}\right)'(r_n)\right) X ,$$

where f is the underlying density and where $r_i = (a_i - x_i^T \hat{\beta}_m)/\hat{\sigma}_m$. Similarly,

$$\hat{M}_{12} = \frac{1}{\hat{\sigma}_m} \cdot X^T \left(\left(\frac{f'}{f}\right)'(r_i) r_i + \left(\frac{f'}{f}\right)(r_i)\right) ,$$

and

$$\hat{M}_{22} = \sum_{i=1}^{n} \left(\left(\frac{f'}{f}\right)' (r_i) r_i^2 + \left(\frac{f'}{f}\right)(r_i) r_i \right).$$

Example: In the Gaussian case $f = \varphi$, we have $(\varphi'/\varphi)(u) = -u$ and $(\varphi'/\varphi)'(u) = -1$. This implies

$$\hat{M}_{11} = -\frac{1}{\hat{\sigma}_m^2} \cdot X^T X, \; \hat{M}_{12} = -\frac{1}{\hat{\sigma}_m} \cdot X^T r, \; \hat{M}_{22} = -2r^T r,$$

where $r = (r_1, \ldots, r_n)^T$. The maximum likelihood estimates are $\hat{\beta} = (X^T X)^{-1} X^T a$ and $\hat{\sigma}_m^2 = \sum_{i=1}^n (a_i - x_i^T \hat{\beta}_m)^2 / n$. Therefore, we have

$$\hat{M}_{12} = 0 \text{ and } \hat{M}_{22} = -2n.$$

Thus, the conditional density of t_p is approximated by a normal with mean $-\hat{\beta}_{m,p}$ and variance $(X^T X)_{pp}^{-1} \cdot \hat{\sigma}_m^2$.

The condition (1.7) for rejection, $\hat{\beta}_{o,p}(y_i)/\hat{\sigma}_o(y_i) = t_p(y_i) > cv = -\hat{\beta}_{m,p}(a_i) + z(1-\alpha)\left((X^T X)_{pp}^{-1}\right)^{1/2} \cdot \hat{\sigma}_m(a_i)$ is equivalent to $\hat{\beta}_{o,p}(y_i) > \hat{\beta}_{o,p}(y_i) - \hat{\beta}_{m,p}(y_i) + z(1-\alpha)\left((X^T X)_{pp}^{-1}\right)^{1/2} \cdot \hat{\sigma}_m(y_i)$ because of the equivariance of the maximum likelihood estimator and (1.4). We therefore reject if $\hat{\beta}_{m,p}(y_i) > z(1-\alpha)\left((X^T X)_{pp}^{-1}\right)^{1/2} \cdot \hat{\sigma}_m(y_i)$. ∎

A more sophisticated expansion can be based on

$$L(\beta, \sigma) = L(\tilde{\beta}, \sigma) + \frac{1}{2}(\beta - \tilde{\beta})^T \tilde{L}_{11}(\beta - \tilde{\beta}) + \text{ terms of higher order}, \tag{4.4}$$

where $\tilde{\beta}(\sigma)$ is the maximizer of the likelihood function for fixed σ. This would give an approximation of the conditional density $k(s,t)$ whose behavior with regard to s can be complex. We could also expand around the global maxima $\hat{\beta}_m$, which would lead to

$$L(\beta, \sigma) = L(\hat{\beta}_m, \sigma) + (L_1^*)^T(\beta - \hat{\beta}_m) + \frac{1}{2}(\beta - \hat{\beta}_m)^T L_{11}^*(\beta - \hat{\beta}_m) + \text{ terms of higher order}, \tag{4.5}$$

where L_1^* is the gradient of $L(\beta, \sigma)$ with respect to β, evaluated at $(\hat{\beta}_m, \sigma)$, etc. This too will generally lead to an approximation of $k(s,t)$ that is complicated in s such that integration with respect to s cannot be done analytically. In general, (4.5) leads to

$$\begin{aligned} k(s, t_1, \ldots, t_p) &\propto m(s) s^{-1} \exp\left[\frac{1}{2}(-t - \hat{\beta}_m)^T L_{11}^*(-t - \hat{\beta}_m) + (-t - \hat{\beta}_m)^T L_1^*\right] \\ &\propto m(s) s^{-1} \exp\left[-\frac{1}{2}\left((t + \hat{\beta}_m)^T (s^2 X^T D X)(t + \hat{\beta}_m) + 2(t + \hat{\beta}_m)^T L_1^*\right)\right], \end{aligned}$$

where $D = \text{diag}\left(-(f'/f)'(r_i)\right)$ and $m(s) = \exp\left[L(\hat{\beta}_m, 1/s)\right] = \left(\prod_{i=1}^n f(s(a_i - x_i^T \hat{\beta}_m))\right) s^n$.

Example: In the Gaussian case $f = \varphi$ both of the above expansions lead to the same result, since

$$L_1(\beta, \sigma) = -\frac{1}{\sigma} X^T \left(\left(\frac{f'}{f} \right) (r_i) \right)$$

evaluates to zero at $\hat{\beta}_m$. One obtains the approximation

$$k(s, t_p) \propto \exp\left(-\frac{s^2}{2} \sum_{i=1}^{n} (a_i - x_i^T \hat{\beta}_m)^2 \right) s^{n-p} \exp\left(-\frac{s^2}{2} (t_p + \hat{\beta}_{m,p})^2 / (X^T X)_{pp}^{-1} \right).$$

In this case, one can do the integral with respect to s rather easily and obtains

$$k(t_p) \propto \left(\sum_{i=1}^{n} (a_i - x_i^T \hat{\beta}_m)^2 + (t_p + \hat{\beta}_{m,p})^2 / (X^T X)_{pp}^{-1} \right)^{-(n-p+1)/2}$$

$$\propto \left(1 + \left((t_p + \hat{\beta}_{m,p}) \Big/ \left(\sum_{i=1}^{n} (a_i - x_i^T \hat{\beta}_m)^2 (X^T X)_{pp}^{-1} \right)^{1/2} \right)^2 \right)^{-(n-p+1)/2}.$$

This expansion is exact, one obtains a Student's t distribution with $(n-p)$ degrees of freedom for the ratio

$$\frac{t_p + \hat{\beta}_{m,p}(a_i)}{\left(\frac{1}{n-p} \sum_{i=1}^{n} (a_i - x_i^T \hat{\beta}_m)^2 (X^T X)_{pp}^{-1} \right)^{1/2}}.$$

In the next section, we will explore the use of the approximation (4.3) in deriving approximate versions of the robust tests described in Section 3.

5. APPROXIMATE ROBUST TESTS: AN EXAMPLE

If we substitute the simple normal approximation for $(u = \log(s), t_1, \ldots, t_p)$ into (3.1) we obtain a messy expression, namely

$$w_F(a_i) \int_{-\infty}^{\infty} \tilde{k}_{F,2}(u, cv) e^{-u} \, du + \pi w_G(a_i) \int_{-\infty}^{\infty} \tilde{k}_{G,2}(u, cv) e^{-u} \, du$$

$$= \lambda_F w_F k_F(cv) + \lambda_G w_G k_G(cv).$$

Procedures that are based on such asymptotic approximations have been named likelihood compromises by Easton [4].

A further approximation that is valid for symmetric error densities, when the asymptotic covariance between s and t is zero, leads to

$$w_F(a_i) \frac{d}{dt_p} \tilde{k}_F(t_p) \big|_{t_p = cv} \int_{-\infty}^{\infty} \tilde{k}_F(u) e^{-u} \, du + \pi w_G(a_i) \frac{d}{dt_p} k_F(t_p) \big|_{t_p = cv} \int_{-\infty}^{\infty} k_G(u) e^{-u} \, du$$

$$= \lambda_F w_F k_F(t_p) \big|_{t_p = cv} + \lambda_G w_G k_G(t_p) \big|_{t_p = cv}.$$

This in turn is approximately equal to

$$w_F(a_i)\frac{d}{dt_p}k_F(t_p)\big|_{t_p=cv}\hat{\sigma}_{m,F}(a_i)+\pi w_G(a_i)\frac{d}{dt_p}k_F(t_p)\big|_{t_p=cv}\hat{\sigma}_{m,G}(a_i)$$

$$=\lambda_F w_F k_F(cv)+\lambda_G w_G k_G(cv)\,.$$

Using the approximation (4.3) for $k_F(t_p), k_G(t_p)$, this is equivalent to

$$w_F(a_i)\left(\frac{cv+\hat{\beta}_{m,F,p}(a_i)}{\hat{\Sigma}_{F,pp}}\right)k_F(cv)\hat{\sigma}_F(a_i)+\pi w_G(a_i)\left(\frac{cv+\hat{\beta}_{m,G,p}(a_i)}{\hat{\Sigma}_{G,pp}}\right)k_G(cv)\hat{\sigma}_{m,F}(a_i)$$

$$=\lambda_F w_F k_F(t_p)\big|_{t_p=cv}+\lambda_G w_G k_G(t_p)\big|_{t_p=cv}\,, \qquad (5.1)$$

where $-\hat{\beta}_{m,F,p}$, $\hat{\Sigma}_{F,pp}$ denote the conditional mean and the conditional variance of t_p when sampling from F.

This equation can still not be solved explicitly for the critical value cv, but it could be used as a basis for a simple numerical approximation. We can go further, however. The ratio of the weights $w_F(a_i)/w_G(a_i)$ converges exponentially fast to 0 or 1 depending on the true underlying sampling distribution. In the case where $w_F(a_i)$ dominates, the approximate solution to (5.1) is

$$cv(a_i)=-\hat{\beta}_{m,F,p}(a_i)+\lambda_F\frac{\hat{\Sigma}_{F,pp}}{\hat{\sigma}_{m,F}(a_i)}\,.$$

With $\lambda_F\approx n^{1/2}z(1-\alpha)$ this defines a critical value that is asymptotically of size α. When the weights are similar and the ratio $w_F(a_i)/w_G(a_i)$ is not close to 0 or 1, the two conditional densities $k_F(t_p), k_G(t_p)$ are not to far from each other. We, therefore, might neglect the factors $k_F(cv), k_G(cv)$, which leads to

$$\begin{aligned} cv = &-\frac{w_F(a_i)\hat{\sigma}_{m,F}(a_i)\beta_{F,p}(a_i)/\hat{\Sigma}_{F,pp}+\pi w_G(a_i)\hat{\sigma}_{m,G}(a_i)\beta_{G,p}(a_i)/\hat{\Sigma}_{G,pp}}{w_F(a_i)\hat{\sigma}_{m,F}(a_i)/\hat{\Sigma}_{F,pp}+\pi w_G(a_i)\hat{\sigma}_{m,G}(a_i)/\hat{\Sigma}_{G,pp}} \\ &+\frac{\lambda_F w_F(a_i)+\lambda_G w_G(a_i)}{w_F(a_i)\hat{\sigma}_{m,F}(a_i)/\hat{\Sigma}_{F,pp}+\pi w_G(a_i)\hat{\sigma}_{m,G}(a_i)/\hat{\Sigma}_{G,pp}}\,. \end{aligned} \qquad (5.2)$$

The asymptotic theory of the tests (3.1) is determined by the behavior of the weights $w_F(a_i)$ and $w_G(a_i)$. The tests are asymptotically equivalent to the likelihood test obtained from the better of the extreme models (see Morgenthaler [5] for related work). The same can be said of the approximate test (5.2). The real check of these methods is provided by their finite sample behavior. In the last section, we will discuss such a study.

6. THE GAUSSIAN AND THE CAUCHY AS EXTREME TYPES: A LIMITED SIMULATION

Chapter 8 of [1] discusses optimal confidence intervals for a location parameter using the extreme types Gaussian and slash, where the slash distribution is defined as the law of X/U for X being standard normal and U being uniform on the interval $[0, 1]$ independent of X. To check the likelihood compromise tests discussed in the last section, we will use this same framework of interval estimation. Equation (1.9) shows how to compute a confidence interval based on the conditional critical values $cv_l(a_i)$ and $cv_u(a_i)$. The expected length of this confidence interval can be used as a criterion of its quality.

In the location case the model (1.1) simplifies to $y_i = \beta + e_i\sigma$, where β is a real, and the conditional density (1.5) of the data for a fixed ancillary (a_i) becomes

$$k(s,t) \propto s^{n-1} \prod_{i=1}^{n} f(s(a_i + t)),$$

where t is a real variable.

The extreme types we use in our study are the Gaussian together with the Cauchy. The Gaussian is a good choice for an overly nice case, whereas the Cauchy is an overly heavy-tailed shape. It is not a smart choice for robust modelling, mainly because of the peakedness of its density in the center. However, the density has a simple form that renders the closed-form computation of $k(t)$ possible up to a normalizing constant. The values of the two Lagrange multipliers if we are interested in 95% confidence should be roughly equal to $2n^{1/2}$. Calibration at the Gaussian and the slash, i.e., choosing these multipliers such that the unconditional confidence is approximately 95% in these two cases leads to the choices shown in Table 1.

Table 1.
Values of the Lagrange multipliers to be used in (5.2) when $\pi = 1$.

$n = 5$		$n = 10$		$n = 20$	
Gau	Sla	Gau	Sla	Gau	Sla
17.0	2.6	8.6	6.0	9.8	7.6

The expected lenght of these intervals when sampling from the Gaussian and the slash are given in Table 2. That table contains also the expected lengths of some other interval procedures.

Table 2.
Expected Lengths of a Variety of Interval Estimators.

	$n = 5$		$n = 10$		$n = 20$	
	Gau	Sla	Gau	Sla	Gau	Sla
Student's t	2.33	∞	1.39	∞	0.93	∞
Likelihood Compr.	3.81	15.3	1.54	4.80	0.94	2.65
biweight interval	3.75	18.6	1.69	5.47	0.97	3.09
bioptimal interval	2.44	16.2	1.42	4.72	0.935	2.47
bioptimal interval	2.62	13.7	1.49	4.44	0.944	2.38

The biweight interval is centered at the one-step biweight M-estimate with scaling equal to nine times the median absolute deviation. Its width is proportional to an estimate of the asymptotic standard error. The bioptimal intervals show a range of the best that can be achieved for the {Gaussian, slash} model. All procedures except the Student's t are calibrated to yield 95% (unconditional) coverage probability at the Gaussian and at the Slash.

It is clear that the likelihood compromise is not doing sufficiently well for small sample sizes. A simpler interval such as the biweight is substantially shorter. It is not clear, however, whether this is due to our use of the Cauchy as the other shape, or whether it is caused by the poor quality of the approximation. As the sample size grows, the likelihood compromise interval starts to be competitive.

7. OPEN PROBLEMS

A personal list of the most interesting open problems is as follows.

(1) The approximations to bioptimal tests and interval procedures based on asymptotic expansions of conditional densities have to be further developed. A similar approach in the point estimation case is very competitive as shown in Easton (1991). We derived a very simple such procedure in Section 5, but there are clearly other possibilities. First, with respect to approximate solutions of (5.1) and also with respect to optimality criteria for tests.

(2) We need a refined analysis that allows us to compute asymptotic approximations to the Lagrange multipliers λ_F and λ_G as a function of π and n.

(3) Comparison to existing testing and interval procedures in the regression setting are to be done in sufficiently great detail to be able to draw conclusions for practitioners. This means a spectrum of error distributions and designs.

Acknowledgements

The research for this article was supported in part by Swiss National Science Foundation Grant 20–31218.91.

REFERENCES

1. Morgenthaler, S. and Tukey, J.W., *Configural Polysampling*, Wiley, New York (1991).

2. Huber, P.J., A Robust Version of the Probability Ratio Test, *Annals of Math. Statist.*, **36**, 1753–1758 (1965).

3. Huber, P.J., *Robust Statistics*, Wiley, New York (1981).

4. Easton, G.S., Compromise Maximum Likelihood Estimators for Location, *J. Amer. Statist. Assoc.*, **86**, 1051–1064 (1991).

5. Morgenthaler, S., Asymptotics for Configural Location Estimators, *Annals of Statist.*, **14**, 174-187 (1986).

Stat. Sci. & Data Anal., pp. 109-128
K. Matsusita *et al.* (Eds)

CIRCULAR REGRESSION

Y. R. SARMA and S.RAO JAMMALAMADAKA

Indian Statistical Institute, Calcutta, India
Department of Statistics and Applied Probability, University of California, Santa Barbara, California, USA.

Abstract. The classical methods of regression analysis cannot be applied when observations are directions, measured as angles in a plane with reference to a fixed sense of rotation and a fixed zero direction. In this paper, a concept of circular regression is introduced and estimation of the regression parameters is discussed. Some empirical methods and asymptotic tests for the determination of the degree of the regression equation, are discussed and some recursive algorithms developed. Tests and estimation of regression parameters in some parametric models are also presented.

Keywords: Directions, circular data, regression, trigonometric polynomials, tests of hypotheses, circular models.

1. INTRODUCTION

Observations in the form of directions on two or more variables, are made in several experiments. Sometimes one may consider fixing the direction of one (independent) variable and observe the direction of the other (dependent) variable. In such situations, one is naturally interested in the problem of regression or predicting one variable using the values of the other. It is clear that the classical regression analysis developed for linear variables cannot be applied in this situation because of the natural restrictions one has for directional variables. In what follows, a method of predicting one circular variable based on another circular variable is presented. The concept of circular regression has not received much attention and only recently attention is drawn to this problem (see Jupp and Mardia [1], Batschelet [2], Rivest [3]). In Section 2, we present a concept of circular regression and discuss how the parameters of the model may be estimated. In Section 3, some criteria for the determination of the degree of the trigonometric polynomials introduced in Section 2 are given along with some empirical approaches and large sample tests for the same. Some recursive algorithms for the computation of the estimates are also presented. In Section 4, regression parameters in some specific parametric models as well as some tests of regression parameters in such models are presented. For a discussion of a "correlation coefficient" which can be used as a measure of association between two circular variables, the reader is referred to Jammalamadaka and Sarma [4].

2. CIRCULAR REGRESSION

Let (α, β) be a pair of random variables which are directions, both measured with reference to the same zero direction and the same sense of rotation. The angles can be treated as unit vectors in the plane in terms of their sine and cosine components. Let $f(\alpha, \beta)$ be the joint density of (α, β) on the torus $0 \leq \alpha, \beta < 2\pi$.

To predict β for a given value of α, the vector corresponding to β is predicted by the conditional expectation (or regression) of $e^{i\beta}$ given α, namely

$$\text{(i.e.)} \qquad E(e^{i\beta} | \alpha) = g(\alpha) = g_1(\alpha) + ig_2(\alpha) = \rho(\alpha)\, e^{i\mu(\alpha)}, \text{ say.} \tag{2.1}$$

Or equivalently,

$$E(\cos\beta | \alpha) = g_1(\alpha)$$

and

$$E(\sin\beta | \alpha) = g_2(\alpha) \tag{2.2}$$

from which β is predicted as

$$\mu(\alpha) = \hat{\beta} = \begin{cases} \tan^{-1} \dfrac{g_2(\alpha)}{g_1(\alpha)} & \text{if} \quad g_1(\alpha) \geq 0 \\ \pi + \tan^{-1} \dfrac{g_2(\alpha)}{g_1(\alpha)} & \text{if} \quad g_1(\alpha) \leq 0 \\ \text{undefined} & \text{if } g_1(\alpha) = g_2(\alpha) = 0\,. \end{cases} \tag{2.3}$$

Here $\mu(\alpha)$ represents the conditional mean direction of β given α and $0 \leq \rho \leq 1$ the conditional concentration towards this direction. Predicting β as in (2.1) is optimal in the sense that it minimizes the distance $E\|e^{i\beta} - g(\alpha)\|^2$ and is very similar to the "least squares" idea.

Note that in this approach, we do not need the knowledge of the joint distribution of (α, β), but rather the conditional distribution of β given α. This allows us to take the conditional distribution for instance, to be a wrapped stable law, as discussed in Section 4.

In the absence of further specifications on the structure of $(g_1(\alpha), g_2(\alpha))$ it is, in general, difficult to estimate them from the data although one might adopt nonparametric curve estimation methods. Even with further specifications, the determination can be quite difficult, a situation often encountered even in linear analysis. So, as in linear analysis, $g_1(\alpha)$ and $g_2(\alpha)$ are approximated by suitable functions. Since the functions $g_1(\alpha)$ and $g_2(\alpha)$ are both periodic functions with period 2π, they can be expressed in terms of their Fourier series expansions. This leads to the method of approximating $g_1(\alpha)$ and $g_2(\alpha)$ by trigonometric polynomials of "degree m", viz.

$$g_1(\alpha) \approx \sum_{k=0}^{m} (A_k \cos k\alpha + B_k \sin k\alpha)$$

and (2.4)

$$g_2(\alpha) \approx \sum_{k=0}^{m} (C_k \cos k\alpha + D_k \sin k\alpha)$$

for a suitable choice of m. From this, one has the following model:

$$\cos \beta = \sum_{k=0}^{m} (A_k \cos k\alpha + B_k \sin k\alpha) + \epsilon_1$$

and (2.5)

$$\sin \beta = \sum_{k=0}^{m} (C_k \cos k\alpha + D_k \sin k\alpha) + \epsilon_2$$

where $\underset{\sim}{\epsilon}' = (\epsilon_1\ \epsilon_2)$ is the vector of errors which is assumed to be random with mean vector zero and covariance matrix Σ, which in general is unknown. The problem of estimating the various parameters $\{A_k, B_k, C_k, D_k, k = 0, 1, ..., m\}$, their standard errors as well as the matrix Σ, is considered here using the generalized least squares. Large sample tests are proposed for the regression coefficients and for determining the degree m. When specific parametric models are assumed for the conditional distribution of β given α, like the von Mises, wrapped Cauchy etc, $g_1(\alpha)$ and

$g_2(\alpha)$ have a known parametric form. Problems of estimation and tests of hypotheses about the regression parameters, in these cases are explored.

Let $(\alpha_1, \beta_1), \dots, (\alpha_n, \beta_n)$ be a random sample of size n . The observational equations can be written as

$$
\begin{aligned}
Y_{1i} &= \cos \beta_i = \sum_{k=0}^{m} (A_k \cos k\alpha_i + B_k \sin k\alpha_i) + \epsilon_{1i} \\
Y_{2i} &= \sin \beta_i = \sum_{k=0}^{m} (C_k \cos k\alpha_i + D_k \sin k\alpha_i) + \epsilon_{2i}
\end{aligned}
\tag{2.6}
$$

for $i = 1, \dots, n$. We assume that $B_0 = D_0 = 0$, to ensure identifiability.
Writing

$$
\begin{aligned}
\underset{\sim}{Y}^{(1)} &= (Y_{11}, \cdots, Y_{1n})' \\
\underset{\sim}{Y}^{(2)} &= (Y_{21}, \cdots, Y_{2n})' \\
\underset{\sim}{\epsilon}^{(1)} &= (\epsilon_{11}, \cdots, \epsilon_{1n})' \\
\underset{\sim}{\epsilon}^{(2)} &= (\epsilon_{21}, \cdots, \epsilon_{2n})',
\end{aligned}
\tag{2.7}
$$

$$
\underset{n x(2m+1)}{X} = \begin{bmatrix} 1 & \cos \alpha_1 & \cdots & \cos m\alpha_1 & \sin \alpha_1 & \cdots & \sin m\, \alpha_1 \\ 1 & \cos \alpha_2 & \cdots & \cos m\alpha_2 & \sin \alpha_2 & \cdots & \sin m\, \alpha_2 \\ \vdots & \vdots & & \vdots & \vdots & & \vdots \\ 1 & \cos \alpha_n & \cdots & \cos m\alpha_n & \sin \alpha_n & \cdots & \sin m\, \alpha_n \end{bmatrix}
$$

and

$$
\begin{aligned}
\underset{\sim}{\lambda}^{(1)} &= (A_0, A_1, \dots, A_m, B_1, \dots, B_m)' \\
\underset{\sim}{\lambda}^{(2)} &= (C_0, C_1, \dots, C_m, D_1, \dots, D_m)'
\end{aligned}
$$

the observational equations (2.6) can be written in matrix notation as

$$
\begin{aligned}
\underset{\sim}{Y}^{(1)} &= X\, \underset{\sim}{\lambda}^{(1)} + \underset{\sim}{\epsilon}^{(1)} \\
\underset{\sim}{Y}^{(2)} &= X\, \underset{\sim}{\lambda}^{(2)} + \underset{\sim}{\epsilon}^{(2)}
\end{aligned}
\tag{2.8}
$$

or

$$\underset{\sim}{Y}^* = X^* \underset{\sim}{\lambda}^* + \underset{\sim}{\epsilon}^* \tag{2.9}$$

where $\underset{\sim}{Y}^{*\prime} = (\underset{\sim}{Y}^{(1)\prime} : \underset{\sim}{Y}^{(2)\prime})$, $\underset{\sim}{\epsilon}^{*\prime} = (\underset{\sim}{\epsilon}^{(1)\prime} : \underset{\sim}{\epsilon}^{(2)\prime})$

$$X^* = \begin{bmatrix} X & 0 \\ 0 & X \end{bmatrix}, \lambda^{*\prime} = (\underset{\sim}{\lambda}^{(1)\prime} : \underset{\sim}{\lambda}^{(2)\prime}).$$

The covariance matrix of $\underset{\sim}{\epsilon}^*$ can be written as $\Sigma \otimes I_n$ where $\otimes$ denotes the usual Kronecker product and I_n is the identity matrix of order n.

One can now obtain the generalized least squares estimates (possibly multistage) of $\underset{\sim}{\lambda}^*$ from the observational equations (2.9).

However, from the structure of equations (2.8), it can be seen (C.R. Rao [5], Section 8C.2, Schmidt [6], Section 2.6), that the generalized least squares estimates obtained from (2.9) coincide with the ordinary least squares estimates of $\underset{\sim}{\lambda}^{(1)}$ and $\underset{\sim}{\lambda}^{(2)}$ obtained separately from (2.8). The least squares estimates are given, on the assumption that X is of full rank $(2m + 1 < n)$, by

$$\hat{\underset{\sim}{\lambda}}^{(1)} = (X' X)^{-1} X' \underset{\sim}{Y}^{(1)}$$

$$\hat{\underset{\sim}{\lambda}}^{(2)} = (X' X)^{-1} X' \underset{\sim}{Y}^{(2)} \tag{2.10}$$

Remark 2.1: If the values of α's are taken to be equally spaced over the circle, the computations become much simpler (see eg. Guest [7]). For instance, if $\alpha_i = \frac{2\pi(i-1)}{n}$, $1 \le i \le n$, we have the explicit estimates (for $1 \le j \le m$):

$$\hat{A}_0 = \frac{1}{n}\sum_{i=1}^{n} \cos\beta_i, \quad \hat{A}_j = \frac{1}{n}\sum_{i=1}^{n} \cos\beta_i \cos j\alpha_i, \quad \hat{B}_j = \frac{1}{n}\sum_{i=1}^{n} \cos\beta_i \sin j\alpha_i$$

$$\hat{C}_0 = \frac{1}{n}\sum_{i=1}^{n} \sin\beta_i, \quad \hat{C}_j = \frac{1}{n}\sum_{i=1}^{n} \sin\beta_i \cos j\alpha_i, \quad \hat{D}_j = \frac{1}{n}\sum_{i=1}^{n} \sin\beta_i \sin j\alpha_i .$$

Remark 2.2: If $(\alpha, \beta_1), \ldots, (\alpha, \beta_n)$ is a sample for the same given value of α, then the regression coefficients (i.e. the coefficients in the Fourier series expansions of $g_1(\alpha)$ and $g_2(\alpha)$) will all be zero except A_0 and C_0 and these are given by $\hat{A}_0 = \frac{1}{n} \Sigma \cos\beta_i$ and $\hat{C}_0 = \frac{1}{n} \Sigma \sin \beta_i$ by taking the unique Moore–Penrose inverse of $X'X$ since X is not of full rank. Thus, when several β's are observed corresponding to the same α, then their resultant mean direction is obtained as the best predictor of β, as one would expect. Suppose for each α_i, $i = 1, \dots, n$ there are n_i observations $(\alpha_i, \beta_{i1}), \dots, (\alpha_i, \beta_{in_i})$. In view of Remark 2.2 we can first obtain the resultant mean direction β_i^* for each α_i and consider (α_i, β_i^*), $i = 1, \dots, n$ to obtain the regression equation.

Remark 2.3: The parameters can also be estimated by minimizing the quantity

$$\sum_{i=1}^{n} \left[\cos \beta_i - \sum_{k=0}^{m} (A_k \cos k\alpha_i + B_k \sin k\alpha_i)\right]^2 + \sum_{i=1}^{n} \left[\sin \beta_i - \sum_{k=0}^{m} (C_k \cos k\alpha_i + D_k \sin k\alpha_i)\right]^2 . \tag{2.11}$$

This method was called the method of "minimum distance" (J.S. Rao [8]) and it was shown there that the estimators obtained by this method are consistent and asymptotically normal under mild conditions. Since the estimators (2.10) coincide with the minimum distance estimators, one can conclude that these estimators are consistent and asymptotically normal. In fact, it is explicitly shown in the next section that $\hat{\underset{\sim}{\lambda}}^{(1)}$ and $\hat{\underset{\sim}{\lambda}}^{(2)}$ are jointly asymptotically normal and asymptotic tests regarding the regression coefficients are carried out for the determination of the degree m .

The covariance matrix Σ can be estimated from the least squares theory as follows writing:

$$\begin{aligned} R_0(i, j) &= \underset{\sim}{Y}^{(i)\prime} \underset{\sim}{Y}^{(j)} - \underset{\sim}{Y}^{(i)\prime} X(X'X)^{-1} X' \underset{\sim}{Y}^{(j)} \\ &= \underset{\sim}{Y}^{(i)\prime} (I - M) \underset{\sim}{Y}^{(j)} \end{aligned} \tag{2.12}$$

where $M = X(X'X)^{-1}X'$ and $R_o = ((R_0(i, j))_{i,j=1,2}$, we have $\hat{\Sigma} =$

$(n - 2(2m + 1))^{-1}R_0$ is an unbiased and consistent estimator of Σ. From this, the standard errors of the estimators can also be obtained.

Remark 2.4: Since the number of parameters to be estimated is quite large even for small values of m, the conventional method of using the residual sum of squares and cross–product matrix to estimate Σ and the standard errors of the estimates may not provide sufficiently accurate estimators. It is shown by several authors that the bootstrap method provides much better estimates of the standard errors of the estimates if resampling is done a large number of times (see eg. Wu [9]). For the bootstrap, we start with the estimates $\hat{\lambda}^*$ obtained above in (2.10) and using these estimates, the error vectors $\epsilon' = (\epsilon_{1i}, \epsilon_{2i})$ are computed from the model (2.6). From the emprical distribution of $\epsilon_1, \ldots, \epsilon_n$ a random sample of size n, say $\epsilon_1^*, \ldots, \epsilon_n^*$ is drawn. Using these, pseudo data $(\tilde{Y}^{(1)'}, \tilde{Y}^{(2)'})$ can be generated from

$$\tilde{Y}^{(1)} = X\hat{\lambda}^{(1)} + \epsilon^{*(1)}$$
$$\tilde{Y}^{(2)} = X\hat{\lambda}^{(2)} + \epsilon^{*(2)}$$

where $\epsilon^{*(i)} = (\epsilon_{i1}^*, \ldots, \epsilon_{in}^*)'$ for $i = 1, 2$.

Assuming now that the data $(\tilde{Y}^{(1)'}, \tilde{Y}^{(2)'})$ are generated from the model (2.6), the regression coefficients λ^* are estimated as before. This procedure is repeated a large number of times. This simulation preserves the assumptions that the errors are independent and identically distributed. In each repetition, a new set of sample of error vectors from the empirical distribution is generated and pseudo data are generated from model (2.6). Using these, a new set of estimators of λ^* are obtained. From these sets thus obtained, the sample covariance matrix can be obtained, providing an estimate of the dispersion matrix Σ.

Remark 2.5: It is shown in Bhimasankaram and Jammalamadaka [10] how recursive estimation and testing in linear models can be carried out when the sample size is increased. The methods developed there can be adapted to the present case also and the details are omitted as they can be directly worked out from those in the above cited paper.

Remark 2.6: The hypothesis that β does not depend on α or $g_1(\alpha) = A_0$ and $g_2(\alpha) = C_0$ can be tested by an asymptotically χ^2 test, following the discussion in Section 3.2.

3. DETERMINATION OF m IN THE REGRESSION EQUATION

A general problem in fitting any polynomial regression is the determination of the degree m of the polynomial. In what follows, some criteria for choosing m are suggested. Large sample tests for the validity of the criteria are also given. A further problem would be to update the estimates every time the degree is changed. In the usual polynomial regression for linear variables this is easily done with the use of orthogonal polynomials. Such orthogonality holds good in the present case also if α_i's are taken to be equally spaced (see Remark 2.1). In the more general case, the recursive method suggested below can be used to determine the degree m as well as to update the estimates without having to compute all the coefficients afresh when the degree is increased. This updating involves minimal additional computations. It is of interest to note that the different criteria discussed below leading to the determination of the degree, all lead to the same computations.

3.11. One of the considerations in determining the degree m could be to see the reduction in error sum of squares by increasing m. The effect of augmenting the design matrix X on the residual sum of squares involves standard calculations (see Seber [11] and is done in our special case as follows (see also Bhimasankaram and Jammalamadaka [10] when X is not of full rank).

Deciding to take $(m+1)^{th}$ degree trigonometric polynomial amounts to adding columns $(\cos(m+1)\alpha_1, \ldots, \cos(m+1)\alpha_n)'$ and $(\sin(m+1)\alpha_1, \ldots, \sin(m+1)\alpha_n)'$ as the $(2m+2)^{th}$ and the last columns to the design matrix X given in (2.7). The augmented matrix can be written as

$$X_{(1)} = (X : W)\,P \quad \text{where} \quad W = \begin{bmatrix} \cos(m+1)\alpha_1, & \sin(m+1)\alpha_1 \\ \vdots & \vdots \\ \cos(m+1)\alpha_n, & \sin(m+1)\alpha_n \end{bmatrix}$$

and P is a suitable permutation matrix which keeps the columns of W in correct places. The corresponding model is now

$$\underset{\sim}{Y}^{(1)} = X_{(1)} \underset{\sim}{\lambda}^{(1)}_{(1)} + \underset{\sim}{\epsilon}^{(1)}$$

$$\underset{\sim}{Y}^{(2)} = X_{(1)} \underset{\sim}{\lambda}^{(2)}_{(1)} + \underset{\sim}{\epsilon}^{(2)} \tag{3.1}$$

where $\underset{\sim}{\lambda}^{(1)}_{(1)}$ and $\underset{\sim}{\lambda}^{(2)}_{(1)}$ are $(2m + 3) \times 1$ vectors of coefficients which can be estimated by the method of least squares. The least squares estimates are given for $i = 1,2$, by

$$\hat{\underset{\sim}{\lambda}}^{(i)}_{(1)} = [X'_{(1)} X_{(1)}]^{-1} X'_{(1)} \underset{\sim}{Y}^{(i)} .$$

Now,

$$X'_{(1)} X_{(1)} = P' \begin{bmatrix} X'X & X'W \\ W'X & W'W \end{bmatrix} P .$$

In order to invert $X'_{(1)} X_{(1)}$ the following result is used.

<u>Theorem</u> (Rohde [12]). For any partitioned symmetric matrix $\begin{bmatrix} A & B \\ C & D \end{bmatrix}$ we have

$$\begin{bmatrix} A & B \\ C & D \end{bmatrix}^{-1} = \begin{bmatrix} A^{-1} + A^{-1}BF^{-1}CA^{-1} & -A^{-1}BF^{-1} \\ -F^{-1}CA^{-1} & F^{-1} \end{bmatrix}$$

where $F = D - C A^{-1}B$. Using this result, we can write

$$\begin{bmatrix} X'X & X'W \\ W'X & W'W \end{bmatrix}^{-1} = \begin{bmatrix} (X'X)^{-1} & 0 \\ 0 & 0 \end{bmatrix} + \begin{bmatrix} (X'X)^{-1} & X'W \\ & I \end{bmatrix} H^{-1} \begin{bmatrix} (X'X)^{-1} & X'W \\ & I \end{bmatrix}^{-1} \tag{3.2}$$

where

$$\begin{aligned} H &= W'W - W'X(X'X)^{-1}X'W \\ &= W'(I - M)W \quad \text{writing } M = X(X'X)^{-1}X' . \end{aligned}$$

The least squares estimates can then be written as

$$\underset{\sim}{\lambda}\{^{i}_{1}\} = P'\{X'_{(1)}X_{(1)}\}^{-1} P P' \begin{bmatrix} X' \\ \cdots \\ W' \end{bmatrix} \underset{\sim}{Y}^{(i)}$$

$$= P' \begin{bmatrix} (X'X)^{-1} X' \underset{\sim}{Y}^{(i)} - (X'X)^{-1} X'N(I-M) Y^{(i)} \\ H^{-1} W' (1-M) \underset{\sim}{Y}^{(i)} \end{bmatrix} \quad (3.3)$$

where $N = W H^{-1} W'$. The error sum of squares will now be

$$\underset{\sim}{Y}^{(i)'}\underset{\sim}{Y}^{(i)} - \underset{\sim}{Y}^{(i)'}(X \vdots W)\, \hat{\underset{\sim}{\lambda}}\{^{i}_{1}\} = Y^{(i)'}(I-M)\underset{\sim}{Y}^{(i)} - Y^{(i)'}(I-M)N(I-M)\underset{\sim}{Y}^{(i)} \quad (3.4)$$

on substitution from (3.3) and simplifying. From (3.4) it can be seen that the reduction in $R_0(i,i)$, the error sum of squares corresponding to $\hat{\underset{\sim}{\lambda}}^{(i)}$, by taking the additional terms is

$$\underset{\sim}{Y}^{(i)'}(I-M)\, N(I-M)\, \underset{\sim}{Y}^{(i)} \geq 0 . \quad (3.5)$$

Thus before deciding to take the $(m+1)^{th}$ degree terms, we may first compute (3.5) and if this is sufficiently large from a practical point of view, we may decide to include the $(m+1)^{th}$ degree terms. When n is sufficiently large, a test for testing whether the proposed additional parameters are significantly different from zero, is available and is discussed in section 3.2. Notice that the computations (3.5) use $(I-M)\ \underset{\sim}{Y}^{(i)}$ which are already available and we only need to find N for which only a second order matrix H has to be inverted. Further, if the additional terms are deemed necessary, the computations we have already made are enough to estimate $\hat{\underset{\sim}{\lambda}}\{^{i}_{1}\}$ using (3.3).

To decide whether to include yet another term, we start with $X_{(1)}$ and add two columns $(\cos (m+2)\, \alpha_1, \cdots, \cos (m+2)\, \alpha_n)'$ and $(\sin (m+2)\, \alpha_1, \cdots,$ $\sin (m+2)\, \alpha_n)'$ as $(2m+4)^{th}$ and the last columns of $X_{(1)}$ and proceed as before. This procedure can be repeated successively.

Remark 3.1.1: In the special case when the column vector(s) W that are adjoined to the original design matrix X , are orthogonal to the columns of the design matrix X , the coefficients that we have already obtained are unaffected as the correction term

$-(X'X)^{-1}X'WH^{-1}\ W'(I-M)\underset{\sim}{Y}^{(i)} = 0$ and the additional coefficient(s) is (are) given by $H^{-1}W'(I-M)\underset{\sim}{Y}^{(i)} = (W'W)^{-1}W'Y^{(i)}$ since $W'M = 0$ and $H = W'(I-M)W = W'W$.

3.1.2: An alternative approach which is algebraically equivalent to the above in the determination of m is to assess the increase in the estimate of the concentration parameter of the distribution of β given α.

Having computed the mth degree trigonometric polynomial approximations, say $g_{im}(\alpha)$ of $g_i(\alpha)$ for $i = 1,2$ the concentration parameter is given by

$$\rho_m(\alpha) = \{g_{1m}^2(\alpha) + g_{2m}^2(\alpha)\}^{\frac{1}{2}}. \tag{3.6}$$

Notice that $\rho_m(\alpha)$ is an increasing function of m and $\rho_m(\alpha) \uparrow \rho(\alpha)$ as $m \to \infty$ for any given α. Since $0 \le \rho(\alpha) \le 1$, we have $0 \le \rho_m(\alpha) \le 1$ for any m and any given α. Computing $\rho_m(\alpha_i)$ for each α_i, $i = 1, \cdots, n$ after an mth degree trigonometric polynomial is fitted, we can decide whether an additional term is to be included depending on whether there is a significant change in the estimate of the concentration parameter or not. Note

$$\begin{aligned} \hat{\rho}_m^2 &= \frac{1}{n}\sum_{i=1}^{n} \rho_m^2(\alpha_i) = \frac{1}{n}\sum_{i=1}^{n} [g_{1m}^2(\alpha_i) + g_{2m}^2(\alpha_i)] \\ &= \frac{1}{n}\sum_{j=1}^{2} \underset{\sim}{Y}^{(j)\prime} M\, \underset{\sim}{Y}^{(j)} = \frac{1}{n}\sum_{j=1}^{2} [\underset{\sim}{Y}^{(j)\prime}\underset{\sim}{Y}^{(j)} - R_0(j,j)]. \end{aligned} \tag{3.7}$$

Working with the augmented matrix as above, it can be seen that the increase in the estimate of the square of the concentration parameter is given by

$$\frac{1}{n}\sum_{j=1}^{2} [\underset{\sim}{Y}^{(j)\prime}(I-M)\ N\ (I-M)\ \underset{\sim}{Y}^{(j)}].$$

which is equal to the reduction in error sum of squares given in (3.5). This shows that this criterion is equivalent to the one discussed in Section 3.1.1.

3.1.3. The extent to which the accuracy of prediction of β would be improved by bringing in additional terms in the Fourier series approximation, can also be ascertained in a way similar to the linear analysis where partial correlation ratio is considered (C.R. Rao, [5] p. 268). It may be recalled that in linear analysis, if we are considering only linear predictors, the proportional reduction in the mean square error when variables $X_1, \cdots, X_p$ are used to predict Y rather than $X_1, \cdots, X_k$ where $p > k$, is given by

$$\rho^2_{0(k+1,\ldots,p)(1,\ldots,k)} = \frac{\rho^2_{0(1,\ldots,p)} - \rho^2_{0(1,\ldots,k)}}{1-\rho^2_{0(1,\ldots,k)}} \tag{3.8}$$

where $\rho^2_{0(1,\ldots,m)}$ is the square of the multiple correlation coefficient between Y and $(X_1, \ldots, X_m)$. If $\underset{\sim}{\sigma}_0$ the vector of covariances of Y and $X_1,\ldots,X_m$ and C is the dispersion matrix of $(X_1,\ldots,X_m)$ assumed positive definite, then we have $\rho^2_{0(1,\ldots,m)} = \underset{\sim}{\sigma}_0' C^{-1} \underset{\sim}{\sigma}_0 / \sigma^2_Y$. An estimate of this is obtained by taking the sample variances and covariances. In the present situation, we consider

$$\rho^{*2}_{iom} = \frac{\underset{\sim}{Y}^{(i)'} M \underset{\sim}{Y}^{(i)}}{\underset{\sim}{Y}^{(i)'} \underset{\sim}{Y}^{(i)}} = \frac{\hat{\underset{\sim}{\lambda}}^{(i)'} (X'X) \hat{\underset{\sim}{\lambda}}^{(i)}}{\underset{\sim}{Y}^{(i)'} \underset{\sim}{Y}^{(i)}} \tag{3.9}$$

for $i = 1, 2$ which is analogous to the sample multiple correlation coefficient in the linear case. Notice that $0 \leq \rho^{*2}_{iom} \leq 1$ since $\underset{\sim}{Y}^{(i)'} M \underset{\sim}{Y}^{(i)} \leq \underset{\sim}{Y}^{(i)'} \underset{\sim}{Y}^{(i)}$. This is easily seen since M is idempotent and for any positive definite matrix A, $\sup \underset{\sim}{Z}' A \underset{\sim}{Z} / \underset{\sim}{Z}' \underset{\sim}{Z} \leq \lambda_{max}$ where λ_{max} is the largest eigen value of A. The stated result follows since the eigen values of M are all $+1$. Consider

$$\rho^{*2}_{iom(m+1)} = \frac{\rho^{*2}_{io(m+1)} - \rho^{*2}_{iom}}{1 - \rho^{*2}_{iom}} \qquad \text{for } i = 1, 2 \tag{3.10}$$

which is the proportional reduction in the regression sum of squares to ascertain the accuracy of the prediction of β. If the quantities are significantly large, we decide to include the higher order term also. A large sample test to see whether this is significantly large is give in Section 3.2. From (3.9) it can be deduced that

$$\rho^{*2}_{iom(m+1)} = \frac{\underset{\sim}{Y}^{(i)'}(I-M)\,N(I-M)\,\underset{\sim}{Y}^{(i)}}{\underset{\sim}{Y}^{(i)'}(I-M)\underset{\sim}{Y}(i)} \tag{3.11}$$

using the computations made earlier. It may be noticed that here again we are led to the same computations as in the above two criteria.

3.2 Asymptotic tests for the determination of the degree m:

From the discussion in Section 3.1 it is seen that the degree m can be determined on the basis of $S_i = \underset{\sim}{Y}^{(i)'}(I-M)N(I-M)\underset{\sim}{Y}^{(i)}$ or some suitable function thereof. If the sample size is sufficiently large the following asymptotic tests can be used for the determination of m.

Proposition: Let $\Sigma = \begin{bmatrix} \sigma_1^2 & \rho\sigma_1\sigma_2 \\ \rho\,\sigma_1\sigma_2 & \sigma_2^2 \end{bmatrix}$ and let $\lim_{n\to\infty} \left[\frac{X'X}{n}\right] = Q$ which is finite and nonsingular. Then for any fixed m, (1) $n^{-1/2} X' \underset{\sim}{\epsilon}^{(i)}$ converges in distribution to $N(0, \sigma_i^2 Q)$ as $n \to \infty$ for $i = 1,2$ (2) $\sqrt{n}(\hat{\underset{\sim}{\lambda}}^{(i)} - \underset{\sim}{\lambda}^{(i)})$ converges in distribution to $N(0, \sigma_i^2 Q^{-1})$ as $n \to \infty$ for $i = 1, 2$.

This proposition can be proved as in Schmidt [6] section 2.4 and the details are omitted.

Since $R_0(i, i)/[n-(2m+1)]$ is a consistent estimator of σ_i^2, for sufficiently large n, we can approximate the distribution of

$$\frac{n-(2m+1)}{R_0(i,i)}(\hat{\underset{\sim}{\lambda}}_n^{(i)} - \underset{\sim}{\lambda}^{(i)})'(X'X)(\hat{\underset{\sim}{\lambda}}_n^{(i)} - \underset{\sim}{\lambda}^{(i)})$$

as $\chi^2(2m+1)$.

Remark 3.2.1: If one uses a bootstrap estimator σ_i^{*2} for σ_i^2, then the distribution of $\frac{1}{\sigma_i^{*2}}(\hat{\underset{\sim}{\lambda}}_n^{(i)} - \underset{\sim}{\lambda}^{(i)})'(X'X)(\hat{\underset{\sim}{\lambda}}_n^{(i)} - \underset{\sim}{\lambda}^{(i)})$ can be approximated as $\chi^2(2m+1)$ and similar modification can be made in all the computations discussed below.

Further, the marginal distribution of any subset of $\hat{\underset{\sim}{\lambda}}^{(i)}$ is also asymptotically normal, and tests for determining the degree m can be based on this result.

Let ${}_m\underset{\sim}{\lambda}_{(1)}^{(i)} = (\lambda_{(1),2m+2}^{(i)}, \lambda_{(1),2m+3}^{(i)})$ the last two components of $\underset{\sim}{\lambda}_{(1)}^{(i)}$. Let their least squares estimates be denoted by ${}_m\hat{\underset{\sim}{\lambda}}_{(1)}^{(i)}$.

From (3.3), we have ${}_m\hat{\underset{\sim}{\lambda}}_{(1)}^{(i)} = H^{-1}W'(I-M)\underset{\sim}{Y}^{(i)}$ which is unbiased for ${}_m\underset{\sim}{\lambda}_{(1)}^{(i)}$. By direct computation we have the dispersion matrix of ${}_m\hat{\underset{\sim}{\lambda}}_{(1)}^{(i)}$ as $V({}_m\hat{\underset{\sim}{\lambda}}_{(1)}^{(i)}) = \sigma_i^2 H^{-1}$. To examine whether an $(m+1)^{th}$ degree polynomial improves the prediction significantly, we test,

$$H_0^{(i)} : {}_m\underset{\sim}{\lambda}_{(1)}^{(i)} = 0 \text{ against } H_1^{(i)} : {}_m\underset{\sim}{\lambda}_{(1)}^{(i)} \neq 0 .$$

From the above discussion, we have

$$T^{(i)} = \frac{n - (2m+1)}{R_0(i,i)} \, {}_m\hat{\underset{\sim}{\lambda}}_{(1)}^{(i)\prime} \, H \, {}_m\hat{\underset{\sim}{\lambda}}_{(1)}^{(i)} \tag{3.12}$$

is asymptotically distributed as a $\chi^2(2)$ under $H_0^{(i)}$. Substituting the estimates, the test statistic has the form

$$T^{(i)} = (n-(2m+1))\frac{\underset{\sim}{Y}^{(i)\prime}(I-M)N(I-M)\underset{\sim}{Y}^{(i)}}{\underset{\sim}{Y}^{(i)\prime}(I-M)\underset{\sim}{Y}^{(i)}} . \tag{3.13}$$

Thus for all the three criteria proposed in Section (3.1) to determine the degree m , asymptotic tests can be obtained using $T^{(i)}$'s as defined in (3.13). We settle for an mth degree polynomial, only if both $H_0^{(i)}$ $i = 1, 2$ are accepted.

Remark 3.2.2: In the above discussion we have used a χ^2 test assuming the consistency of of $R_0(i, i)$. For moderately large n , one can also justify the use of an F test for $H_0^{(i)}$ as follows:

Lemma: Under $H_0^{(i)}$, $\underset{\sim}{Y}^{(i)\prime}(I - M)\underset{\sim}{Y}^{(i)}$ and $\underset{\sim}{Y}^{(i)\prime}(I - M)N(I - M)\underset{\sim}{Y}^{(i)}$ are asymptotically independent.

Proof. Writing $R_{0k}(i, i)$ as the residual sum of squares when a kth degree polynomial is fitted, we have

$$\frac{R_{0m}(i,i)}{\sigma_i^2} = \frac{\underset{\sim}{Y}^{(i)\prime}(I - M)\,\underset{\sim}{Y}^{(i)}}{\sigma_i^2} \quad \text{asymptotically } \chi^2(n - (2m + 1)) \text{ as } n \to \infty$$

for $i = 1, 2$. Further, by direct computation using (3.3)

$$\frac{R_0(m + 1)^{(i,i)}}{\sigma_i^2} = \frac{\underset{\sim}{Y}^{(i)\prime}(I - M)\underset{\sim}{Y}^{(i)}}{\sigma_i^2} - \frac{\underset{\sim}{Y}^{(i)\prime}(I - M)N(I - M)\underset{\sim}{Y}^{(i)}}{\sigma_i^2}$$

which is asymptotically $\chi^2(n - (2m + 3))$. Moreover, under $H_0^{(i)}$, $\dfrac{\underset{\sim}{Y}^{(i)\prime}(I - M)N(I - M)\underset{\sim}{Y}^{(i)}}{\sigma_i^2}$ is asymptotically $\chi^2(2)$. From these we conclude the stated result.

Using the above lemma

$$F^{(i)} = \frac{\underset{\sim}{Y}^{(i)\prime}(I - M)N(I - M)\underset{\sim}{Y}^{(i)}}{\underset{\sim}{Y}^{(i)\prime}(I - M)\underset{\sim}{Y}^{(i)}} \, \frac{n - (2m + 1)}{2} \tag{3.14}$$

has an $F(2, n - (2m + 1))$ distribution under $H_0^{(i)}$. Thus $H_0^{(i)}$ can be tested using $F^{(i)}$ and the degree m can be determined.

Remark 3.2.3: To decide the degree m , it is suggested above to consider $H_0^{(i)}$ separately for $i = 1, 2$. It may be pointed out that we can derive the joint asymptotic distribution of $(\hat{\underset{\sim}{\lambda}}^{(1)}, \hat{\underset{\sim}{\lambda}}^{(2)})$, and a test can be obtained for $H_0 : {}_m\underset{\sim}{\lambda}^{(1)}_{(1)} = 0$ and

$$m\underset{\sim}{\lambda}_{(1)}^{(2)} = 0 .$$

4. REGRESSION IN SOME PARAMETRIC MODELS

In this section exact forms of $g_1(\alpha)$ and $g_2(\alpha)$ in some special cases when the conditional distribution of β given α is of known form are obtained and the problems of estimation and testing hypotheses about regression parameters in these cases are considered.

4.1: Suppose the conditional distribution of β given α is von Mises with density function

$$f(\beta | \alpha) = \frac{1}{2\pi I_0(\kappa)} \exp \kappa \cos\{\beta - \mu(\alpha)\} \quad \textbf{(4.1)}$$

where $\mu(\alpha)$ is the conditional mean and κ is the concentration parameter which is also possibly dependent on α. In this case

$$\begin{aligned} g_1(\alpha) &= A(\kappa) \cos \mu(\alpha) \\ g_2(\alpha) &= A(\kappa) \sin \mu(\alpha) \end{aligned} \quad \textbf{(4.2)}$$

where $A(\kappa) = I_1(\kappa)/I_0(\kappa)$. Here $\mu(\alpha)$ is of interest and without further structure, one may have to resort to the nonparametric regression methods or estimate $g_i(\alpha)$'s by the method described in Section 2.

(a) The hypothesis that β does not depend on α is equivalent to

$$H_0 : g_1(\alpha) = A_0 \text{ and } g_2(\alpha) = C_0 . \quad \textbf{(4.3)}$$

Estimating A_0 and C_0 as described in Section 2,

$$\hat{A}_0 = \frac{1}{n}\sum_1^n \cos \beta_i \text{ and } \hat{C}_0 = \frac{1}{n}\sum_1^n \sin \beta_i .$$

Writing $\underset{\sim}{\epsilon}_i' = (\cos \beta_i - \hat{A}_0 , \sin \beta_i - \hat{C}_0)$ and $\bar{\underset{\sim}{\epsilon}} = \frac{1}{n} \Sigma \underset{\sim}{\epsilon}_i$, one has that $\sqrt{n}\, \hat{\Sigma}^{-1/2} \bar{\underset{\sim}{\epsilon}}$ is asymptotically distributed as $N_2(0, I)$ under H_0 and consequently

$$S = n\, \bar{\epsilon}'\, \hat{\Sigma}^{-1} \bar{\epsilon} \tag{4.4}$$

is asymptotically distributed as a χ^2 variable on 2 degrees of freedom under H_0, where $\hat{\Sigma}$ is an unbiased, consistent estimate of Σ obtained from (2.12).

If data are available in the form that for each α_i there are $(\beta_{i1},\ldots,\beta_{in_i})$, $i = 1,\ldots,p$ then $\hat{E}(\beta|\alpha_i) = \hat{\mu}(\alpha_i) = \bar{\beta}_i$ where $\bar{\beta}_i$ is the circular mean direction of $(\beta_{i1},\ldots,\beta_{in_i})$. Here one can test the hypothesis that $\mu(\alpha)$ does not depend on α (i.e., the regression is constant) by the appropriate Anova procedure described in Mardia [13] p. 163).

(b) Suppose $\mu(\alpha) = \mu + a\alpha \pmod{2\pi}$ where μ and a are both unknown. A test for $a = 0$ corresponds to the regression being independent of α.

For the more general hypothesis

$$H_0 : a = a_0 \text{ (not an integer)}, \quad \mu = \mu_0, \text{ and } \kappa = \kappa_0 \tag{4.5}$$

one can proceed as follows. Under H_0,

$$\begin{aligned} g_1(\alpha) &= A(\kappa_0) \cos(\mu_0 + a_0\alpha) \\ &= A(\kappa_0) \cos \mu_0 \cos a_0\alpha - A(\kappa_0) \sin \mu_0 \sin a_0\alpha \end{aligned} \tag{4.6}$$

$$\begin{aligned} g_2(\alpha) &= A(\kappa_0) \sin(\mu_0 + a_0\alpha) \\ &= A(\kappa) \cos \mu_0\alpha + A(\kappa_0) \sin \mu_0 \cos a_0\alpha . \end{aligned}$$

But

$$\cos a_0\alpha = \frac{2a_0}{\pi} \sin a_0\pi \left[\frac{1}{2a_0^2} + \frac{\cos \alpha}{1^2 - a_0^2} - \frac{\cos 2\alpha}{2^2 + a_0^2} + \ldots + (-1)^{m-1} \frac{\cos m\alpha}{m^2 + a_0^2} + \ldots\right]$$

and

$$\sin a_0\alpha = \frac{2}{\pi}\sin a_0\, a_0\pi\left[\frac{\sin\alpha}{1^2-a_0^2} - \frac{2\sin 2\alpha}{2^2-a_0^2} + \dots + (-1)^{m-1}\frac{m\ \sin m\alpha}{m^2-a_0^2} + \right. \tag{4.7}$$

(cf. Carslaw [14] p. 265, problem 6). From (4.7) one can obtain the Fourier series expansions for $g_1(\alpha)$ and $g_2(\alpha)$. From this one can test the given hypothesis by comparing the estimates of the Fourier coefficients as obtained from the least squares theory with the values obtained from the above expansions (after fixing a suitable value of m) by using the proposition in Section 3.2.

(ii) If a_0 is an integer, say r, then we have

$$g_1(\alpha) = A(\kappa_0)\cos\mu_0\cos r\alpha - A(\kappa_0)\sin\mu_0\sin r\alpha$$
$$g_2(\alpha) = A(\kappa_0)\sin\mu_0\cos r\alpha + A(\kappa_0)\cos\mu_0\sin r\alpha\ .$$

In this case the hypothesis becomes

$$H_0:\quad A_r = A(\kappa_0)\cos\mu_0,\ B_r = -A(\kappa_0)\sin\mu_0$$
$$C_r = A(\kappa_0)\sin\mu_0,\ D_r = A(\kappa_0)\cos\mu_0 \tag{4.8}$$

and all the other coefficients are zero. This again can be tested using the results of the proposition in Section 3.2.

4.2: Suppose the conditional distribution of β given α is wrapped Cauchy with frequency function

$$f(\beta|\alpha) = \frac{1}{2\pi}\,\frac{1-\rho^2}{1+\rho^2 - 2\rho\cos(\beta-\mu(\alpha))} \tag{4.9}$$

where $0 \le \rho \le 1$, which may depend on α. From the characteristic function

$$\phi_p(\alpha) = e^{i\mu(\alpha)p-ap}$$

where $\rho = e^{-a}$, for all positive integer values of p, one has

$$g_1(\alpha) = \rho\cos\mu(\alpha)$$
$$g_2(\alpha) = \rho\sin\mu(\alpha)\ . \tag{4.10}$$

Here too one can consider the problems of testing as in the above case.

4.3: Suppose the conditional distribution of β given α is wrapped normal with the characteristic function

$$\phi_p(\alpha) = e^{i\mu(\alpha)p-(\sigma^2/2)p^2} \tag{4.11}$$

for all positive integer values of p. In this case

$$g_1(\alpha) = e^{-\sigma^2/2} \cos \mu(\alpha) , \quad g_2(\alpha) = e^{-\sigma^2/2} \sin \mu(\alpha) . \tag{4.12}$$

4.4: As a final example, suppose the conditional distribution of β given α is, more generally, a wrapped version of a stable law on the line with characteristic function given by

$$\phi_p(\alpha) = \exp\{i\mu(\alpha)p - c\,|p|^r\,(1 - \delta\,\mathrm{sgn}(p)\,w(p,r)\} \tag{4.13}$$

for some $\mu(\alpha)$ real, $c > 0$, $|\delta| < 1$ and $0 < r \leq 2$ which may also depend on α, and

$$w(p,r) = \begin{cases} \tan \frac{r\pi}{2} & \text{if } r \neq 1 \\ -\frac{2}{\pi} \log |p| & \text{if } r = 1 \end{cases}$$

for any integral value of p. Special cases of this include the wrapped Cauchy distribution when $r = 1$, and the wrapped Normal distribution if $r = 2$. When $r = 0$ one has $c = 0$ and one gets the characteristic function of the degenerate distribution with mass concentrated at $\mu(\alpha)$. In the general case,

$$\begin{aligned} g_1(\alpha) &= e^{-c} \cos\{\mu(\alpha) + \delta \tan \tfrac{r\pi}{2}] \\ g_2(\alpha) &= e^{-c} \sin[\mu(\alpha) + \delta \tan \tfrac{r\pi}{2}] . \end{aligned} \tag{4.14}$$

Again several estimation and testing problems concerning various parameters can be handled by taking suitable trigonometric polynomial expansions of $g_1(\alpha)$ and $g_2(\alpha)$.

Remark: Throughout the discussion, we have considered the regression of one circular variable on another. The problem of regressing one circular variable on more than one circular variable, a case analogous to the multiple regression in the linear context, can also be dealt with along the lines of this paper. The results are more involved and will be considered elsewhere.

REFERENCES

[1] P.E. Jupp and K.V. Mardia, A General Correlation Coefficient for Directional Data and Related Regression Problems, *Biometrika*, **67**, 163–173, (1980).

[2] E. Batschelet, *Circular Statistics in Biology*, Acad. Press, New York. (1981).

[3] P. Rivest, Some statistical methods for bivariate circular data. *J. Roy. Statist. Soc.,* Ser. B, 37, 81–90, (1982).

[4] S.R. Jammalamadaka and Y.R. Sarma, A Correlation Coefficient for Angular Variables, *Statistical Theory and Data Analysis II.* Ed. K. Matusita, North Holland, New York, 349–364, (1988).

[5] C.R. Rao, *Linear Statistical Inference and Its Applications*, 2nd edition, John Wiley, New York (1973).

[6] P. Schmidt, *Econometrics.* Marcel Dekker, New York, (1976).

[7] P.H. Guest, *Numerical Methods of Curve Fitting*, Cambridge University Press, Cambridge (1961).

[8] J.S. Rao, *Some contributions to the analysis of circular data.* Unpublished Ph.D. thesis, Indian Statistical Institute, Calcutta, (1969).

[9] C.F.J. Wu, Jacknife bootstrap and other resmpling methods in regression analysis (with discussion). *Ann. Statist.* **14**, 1261–1350, (1986).

[10] P. Bhimasankaram and S.R. Jammalamadaka, *Recursive Estimation and Testing in General Linear Models with Applications to Regression Diagnostics*, Technical Report No. 29, Department of Statistics, University of California, Santa Barbara, (1987).

[11] G.A.F. Seber, *Applied regression analysis.* John Wiley, New York, (1977).

[12] C.A. Rohde, Generalized inverses of partitioned matrices. *SIAM J. Appl. Math,* **13**, 1033–1035, (1965).

[13] K.V. Mardia, *Statistics of Directional Data*, Acad. Press, New York (1972).

[14] H.S. Carslaw, *Introduction to the Theory of Fourier Series and Integrals*, 3rd revised edition, Dover, New York (1930).

Stat. Sci. & Data Anal., pp. 129-141
K. Matsusita *et al.* (Eds)

Properties of Least Squares Methods for Choosing the Parameter of the Simple Exponential Smoothing Predictor

CHUNHANG CHEN and MITUAKI HUZII
Department of Information Sciences, Tokyo Institute of Technology, Tokyo 152, Japan

Abstract. It has not been clearly known whether or not the simple exponential smoothing predictor is a reasonable predictor for a wide class of time series. In order to make this problem clear, we need to clarify the statistical properties of the chosen values of the unknown smoothing parameter. This paper discusses some asymptotic properties of the chosen values $\tilde{\beta}_n$ and $\tilde{\beta}_{n,k}$, which are obtained by minimizing the average squared error of the one-step-ahead forecasts and that of the truncated one-step-ahead forecasts up to k terms, respectively. As the first approach to showing the asymptotic properties of $\tilde{\beta}_n$ and $\tilde{\beta}_{n,k}$ for a wide class of time series, we show these for two cases. The first case is when the time series is stationary and the second case is when the time series deviates from a stationary process and is an autoregressive-integrated moving average process with difference order 1. We show that, if $\{k\}$ is taken suitably, then $\tilde{\beta}_{n,k}$ has the same consistent properties as $\tilde{\beta}_n$, besides, $\tilde{\beta}_{n,k}$ is asymptotically normally distributed.

Key words: Simple exponential smoothing predictor; smoothing parameter; prediction.

1. INTRODUCTION

For a time series $\{X_t\}$, we consider the case when we have observed $X_1, X_2, \ldots, X_n$ and are required to predict the value X_{n+1} of the next period. In this case, we have many forecasting methods (Makridakis, etc.[1]). However, it is not easy to decide which method we should use, because each method may have different properties with others, and the accuracy of a forecasting method is often affected by the properties of the data.

In Makridakis, etc.[1], they carried out a forecasting competition to compare the forecasting behavior of some major time series methods, including the Box-Jenkins approach and exponential smoothing methods, etc., by applying them to real time series data. Concerning exponential smoothing methods, they pointed out that the simple and Holt's exponential smoothing methods did extremely well and, in particular, that the (deseasonalized) simple exponential smoothing method was the best method overall for a one-period forecast. And it was remarked that the best advantage of exponential smoothing methods is their adaptability to any and all types of data and forecasting situations (see Chapters 1 and 8 of Makridakis, etc.[1] and Gardner [2]).

In spite of the usefulness of exponential smoothing methods in forecasting, their statistical properties have not been clarified theoretically. The exponential smoothing predictors involve some unknown smoothing parameters, which are often chosen to minimize the average squared error of the one-step-ahead forecasts in practice (Makridakis, etc.[1]). In this paper, we restrict our discussion to the simple exponential smoothing method. We consider it would be essential and necessary to clarify the statistical properties of the chosen value of the smoothing parameter.

Let $\tilde{X}_{n+1}$ denote the simple exponential smoothing predictor of X_{n+1}, based on X_1, X_2,

$\ldots, X_n$. $\tilde{X}_{n+1}$ is defined recursively by the following formula:

$$\begin{aligned} \tilde{X}_1 &= X_0, \\ \tilde{X}_t &= (1-\beta)X_{t-1} + \beta\tilde{X}_{t-1}, \quad 2 \le t \le (n+1), \end{aligned}$$

where β, $|\beta| \le 1$, is an unknown smoothing parameter, which should be chosen adequately. And X_0 is an initial value, which is assumed to be a random variable such that $EX_0^2 < \infty$.

To choose the value for β, we can adopt the least squares method by minimizing the average squared error of the one-step-ahead forecasts

$$S_n(\beta) = n^{-1}\sum_{t=1}^{n}(X_t - \tilde{X}_t)^2 \tag{1.1}$$

(Makridakis, etc.[1]). Here we put $\tilde{\beta}_n$ as a value of β, which minimizes $S_n(\beta)$ on $[-1, 1]$. In order to show the statistical properties of $\tilde{\beta}_n$, we need to evaluate the convergence in probability of $S_n(\beta)$ as n tends to infinity. But this is not easy, since $\tilde{X}_t$ takes the form

$$\tilde{X}_t = (1-\beta)\sum_{j=1}^{t-1}\beta^{j-1}X_{t-j} + \beta^{t-1}X_0$$

and the number of the terms of the right-hand-side increases with t. In addition, it is not easy to show the asymptotic distribution of $\tilde{\beta}_n$. In this paper we propose another method as follows. For any positive integer n, we truncate $\tilde{X}_t$ up to k terms. Put

$$\tilde{X}_{t,k} = (1-\beta)\sum_{j=1}^{k}\beta^{j-1}X_{t-j}, \quad (k+1) \le t \le n$$

and choose the value of β which minimizes

$$Q_{n,k}(\beta) = (n-k)^{-1}\sum_{t=k+1}^{n}(X_t - \tilde{X}_{t,k})^2. \tag{1.2}$$

Let $\tilde{\beta}_{n,k}$ be a value of β, which minimizes $Q_{n,k}(\beta)$.

In comparison with $S_n(\beta)$, $Q_{n,k}(\beta)$ is easier to handle. We can more easily show the statistical properties of $\tilde{\beta}_{n,k}$ than $\tilde{\beta}_n$. In practice, it is more convenient to choose the value of β by minimizing $Q_{n,k}(\beta)$. We shall show that $Q_{n,k}(\beta)$ has the same asymptotic properties as $S_n(\beta)$ when n tends to infinity, if we take $\{k\}$ to be a subsequence of $\{n\}$ such that $1 \le k < n$, $k = O(n^\delta)$ for some δ, $0 < \delta < 1$, and $k \to \infty$ as $n \to \infty$. Hereafter, we let $\{k\}$ be such a subsequence unless it is other specified.

We are interested in considering whether or not the simple exponential smoothing predictor is robust. When $\{X_t\}$ is a stationary process with a known model, one can often find the best linear predictor. We try to consider whether or not $\tilde{X}_{n+1}$ is a reasonable predictor of X_{n+1} for a wide class of process $\{X_t\}$. This means that we consider whether or not $\tilde{X}_{n+1}$ is still a reasonable predictor when $\{X_t\}$ deviates from a stationary process and we have no knowledge about the true model. This paper concerns itself with the primary stage of this investigation. We shall show some statistical properties of the chosen values of β under the assumption that $\{X_t\}$ is a stationary process or an autoregressive-integrated moving average (ARIMA) process as a first step for considering the case when $\{X_t\}$ deviates from a stationary process.

In the next section, we first consider a stationary case. We show that $Q_{n,k}(\beta)$ converges in probability to a function $S(\beta)$ for $\beta \in (-1, 1]$, diverges at $\beta = -1$ and $S_n(\beta)$ has the same asymptotic properties as $Q_{n,k}(\beta)$ if we put $X_0 = 0$, when n tends to infinity. And we show that, for any minimal point β_0 of $S(\beta)$, there exist local minimal points $\hat{\beta}_{n,k}$ and $\hat{\beta}_n$ of $Q_{n,k}(\beta)$ and $S_n(\beta)$, respectively, such that $\hat{\beta}_{n,k}$ and $\hat{\beta}_n$ converge in probability to β_0 when n tends to infinity. We also show that $\sqrt{n}(\hat{\beta}_{n,k} - \beta_0)$ is asymptotically normally distributed. Finally, we consider the case of an ARIMA process.

2. ASYMPTOTIC PROPERTIES OF $\hat{\beta}_{n,k}$ AND $\hat{\beta}_n$

In this section, we consider the asymptotic properties of $\hat{\beta}_{n,k}$ and $\hat{\beta}_n$ when n tends to infinity.

2.1. CASE WHEN $\{X_t\}$ IS STATIONARY

Let $\{X_t\}$ be a stationary process such that

$$X_t = \sum_{j=0}^{\infty} \psi_j e_{t-j}, \tag{2.1}$$

where $\sum_{j=0}^{\infty} |\psi_j| < \infty$ and $\{e_t\}$ is a sequence of independently identically distributed (iid) random variables such that $Ee_t = 0$, $Ee_t^2 = \sigma^2$ and $Ee_t^4 = \eta\sigma^4$ for some $\eta > 0$.

Here we introduce the notation used throughout this paper. "$p\lim$" or "$\xrightarrow{P}$" denotes convergence in probability and "$\Rightarrow$" denotes convergence in distribution. Now we show some asymptotic properties of $Q_{n,k}(\beta)$. We have the following lemma.

Lemma 1 *It holds that*

$$p\lim_{n\to\infty} Q_{n,k}(\beta) = \gamma(0) + \frac{1-\beta}{1+\beta}[\gamma(0) - 2\sum_{j=1}^{\infty} \beta^{j-1}\gamma(j)] \quad \text{for } \beta \in (-1, 1], \tag{2.2}$$

where $\gamma(h) = E(X_t X_{t+h})$.

Proof. We can write

$$\begin{aligned} Q_{n,k}(\beta) &= \frac{1}{n-k}\sum_{t=k+1}^{n} X_t^2 + (1-\beta)^2 \frac{1}{n-k}\sum_{t=k+1}^{n}\sum_{i=1}^{k}\sum_{j=1}^{k} \beta^{i+j-2} X_{t-i}X_{t-j} \\ &\quad -2(1-\beta)\frac{1}{n-k}\sum_{t=k+1}^{n}\sum_{j=1}^{k} \beta^{j-1} X_t X_{t-j} \\ &= A_n + B_n - C_n, \end{aligned}$$

say. It is easy to show that $A_n \xrightarrow{P} \gamma(0)$ as $n \to \infty$. Thus (2.2) holds when $\beta = 1$. In what follows, we assume $|\beta| < 1$. Now we show that

$$B_n \xrightarrow{P} \frac{1-\beta}{1+\beta}[\gamma(0) + 2\sum_{j=1}^{\infty} \beta^j \gamma(j)] \quad \text{as } n \to \infty. \tag{2.3}$$

We have

$$EB_n = (1-\beta)^2 \sum_{i=1}^{k}\sum_{j=1}^{k} \beta^{i+j-2}\gamma(i-j) \longrightarrow \frac{1-\beta}{1+\beta}[\gamma(0) + 2\sum_{j=1}^{\infty} \beta^j\gamma(j)]$$

as $n \to \infty$ and

$$\begin{aligned} Var(B_n) &= (1-\beta)^4(n-k)^{-2} \sum_{t=k+1}^{n} \sum_{s=k+1}^{n} \sum_{i=1}^{k}\sum_{j=1}^{k}\sum_{u=1}^{k}\sum_{v=1}^{k} \beta^{i-1}\beta^{j-1}\beta^{u-1}\beta^{v-1} \\ &\qquad \times[2\gamma(s-t+i-u)\gamma(s-t+j-v) + \\ &\qquad (\eta-3)\sigma^4 \sum_{f=0}^{\infty} \psi_f\psi_{f+i-j}\psi_{f+s-t+i-u}\psi_{f+s-t+i-v}] \\ &= K_{n1} + K_{n2}, \end{aligned}$$

say. Now

$$\begin{aligned} |K_{n1}| &\le \frac{8(1-\beta)^2\gamma(0)}{(1+\beta)^2(1-|\beta|)}(n-k)^{-1}\sum_{i=0}^{k}|\beta|^i[\sum_{m=0}^{n-1}|\gamma(m)| + \sum_{m=-k}^{n-k-1}|\gamma(m)|] \\ &\longrightarrow 0 \quad \text{as} \quad n \to \infty \end{aligned}$$

and

$$\begin{aligned} |K_{n2}| &\le (\eta+3)\sigma^4\Big(\frac{1-\beta}{1-|\beta|}\Big)^4(n-k)^{-1}(\sum_{j=0}^{\infty}|\psi_j|)^4 \\ &\longrightarrow 0 \quad \text{as} \quad n \to \infty. \end{aligned}$$

Thus $Var(B_n) \to 0$ as $n \to \infty$ and (2.3) is obtained.

Similarly, we can show that $C_n \xrightarrow{P} 2(1-\beta)\sum_{j=1}^{\infty}\beta^{j-1}\gamma(j)$ as $n \to \infty$. Thus we establish (2.2).

Now we consider the case when $\beta = -1$. We make the following assumptions:

Assumption 1. $\sum_{j=0}^{\infty} j|\psi_j| < \infty$;
Assumption 2. $\sum_{j=0}^{\infty}(-1)^j\psi_j \neq 0$.

We obtain the following lemma.

Lemma 2 *Under Assumptions* 1 *and* 2,

$$\lim_{n\to\infty} Pr\{Q_{n,k}(-1) \le A\} = 0 \quad \textit{for any } A,\ A > 0. \tag{2.4}$$

Proof. Let A_n, B_n and C_n be as given in the proof of Lemma 1. When $\beta = -1$, we can show that $C_n \xrightarrow{P} 4\sum_{j=1}^{\infty}(-1)^j\gamma(j)$ as $n \to \infty$. Thus

$$\frac{1}{k}Q_{n,k}(-1) - \frac{4}{(n-k)k}\sum_{t=k+1}^{n}\sum_{i=1}^{k}\sum_{j=1}^{k}(-1)^{i+j-2}X_{t-i}X_{t-j} = o_p(1). \tag{2.5}$$

Define

$$\xi_{n,k} = \frac{1}{(n-k)k}\sum_{t=k+1}^{n}\sum_{i=1}^{k}\sum_{j=1}^{k}(-1)^{i+j-2}X_{t-i}X_{t-j}.$$

We show that $\xi_{n,k} \xrightarrow{P} \sum_{j=-\infty}^{\infty}(-1)^j\gamma(j)$ as $n \to \infty$. In fact, under Assumption 1,

$$E\xi_{n,k} = \sum_{|j|<k}(-1)^j\gamma(j) + \frac{1}{k}\sum_{|j|<k}(-1)^j|j|\gamma(j) \longrightarrow \sum_{|j|<\infty}(-1)^j\gamma(j)$$

as $n \to \infty$ and

$$\begin{aligned} Var(\xi_{n,k}) &= \frac{1}{(n-k)^2k^2} \sum_{t=k+1}^{n} \sum_{s=k+1}^{n} \sum_{i=1}^{k} \sum_{j=1}^{k} \sum_{u=1}^{k} \sum_{v=1}^{k} (-1)^{i+j+u+v} \\ &\qquad \times [2\gamma(s-t+i-u)\gamma(s-t+j-v) + \\ &\qquad \sum_{w=0}^{\infty} \psi_w \psi_{w+i-j} \psi_{w+s-t+i-u} \psi_{w+s-t+i-v} (\eta - 3)\sigma^4] \\ &= 2K_{n1} + K_{n2}, \end{aligned}$$

say. We can show that

$$|K_{n1}| \leq \frac{2k}{n-k} [\sum_{j=-\infty}^{\infty} |\gamma(j)|]^2 \longrightarrow 0 \quad \text{as} \quad n \to \infty,$$

since $k = O(n^{\delta})$. Also we have

$$|K_{n2}| \leq 2(\eta + 3)\sigma^4 (n-k)^{-1} (\sum_{j=0}^{\infty} |\psi_j|)^4 \quad \text{as} \quad \longrightarrow 0.$$

It follows that $Var(\xi_{n,k}) \to 0$ and $\xi_{n,k} \xrightarrow{P} \sum_{|j|<\infty} (-1)^j \gamma(j)$ as $n \to \infty$. Now put $\xi = \sum_{|j|<\infty} (-1)^j \gamma(j)$. Under Assumption 2, we have $\xi > 0$. From the formulation (2.5), we obtain

$$\frac{1}{k} Q_{n,k}(-1) \xrightarrow{P} 4\xi > 0.$$

Thus, for any positive number A, we have

$$\Pr\{Q_{n,k}(-1) \leq A\} = \Pr\{\frac{1}{k} Q_{n,k}(-1) \leq \frac{A}{k}\} \longrightarrow 0$$

as $n \to \infty$.

In the next, we consider the asymptotic properties of $S_n(\beta)$. We obtain the following lemma.

Lemma 3 *It holds that*

$$p \lim_{n\to\infty} S_n(\beta) = \gamma(0) + \frac{1-\beta}{1+\beta} [\gamma(0) - 2 \sum_{j=1}^{\infty} \beta^{j-1} \gamma(j)] \quad \textit{for } \beta \in (-1,\ 1) \tag{2.6}$$

and

$$p \lim_{n\to\infty} S_n(1) = \gamma(0) + X_0^2. \tag{2.7}$$

We omit the proof of Lemma 3, since it can be proved in the same way as Lemma 1. Now we consider the asymptotic properties of $S_n(\beta)$ when $\beta = -1$. We make the following assumption.

Assumption 3. e_t is symmetrically distributed.

We prepare some lemmas. With $\{X_t\}$ defined in (2.1), we define ξ_t as

$$\xi_t = \sum_{j=1}^{t} X_j, \quad t = 1, 2, \ldots \tag{2.8}$$

and put $\hat{\gamma}_\xi(0) = n^{-1} \sum_{t=1}^{n} \xi_t^2$. We have the following lemma.

Lemma 4 *Under Assumption 1,*

$$n^{-1}\hat{\gamma}_\xi(0) = c_0^2 n^{-2}\sum_{t=1}^{n} Z_t^2 + O_p(n^{-\frac{1}{2}}), \tag{2.9}$$

where $c_0 = \sum_{i=0}^{\infty}\psi_i$ *and* $Z_t = \sum_{j=1}^{t} e_j$.

Proof. We can show that

$$\xi_t = c_0 Z_t - W_t + \zeta_0,$$

where $Z_t = \sum_{j=1}^{t} e_j$, $W_t = \sum_{j=0}^{\infty} c_{j+1}e_{t-j}$, $\zeta_0 = \sum_{j=0}^{\infty} c_{j+1}e_{-j}$ and $c_j = \sum_{i=j}^{\infty}\psi_i$ for $j = 0, 1, \ldots$.

Under Assumption 1, we have $\sum_{j=0}^{\infty}|c_{j+1}| \le \sum_{i=1}^{\infty} i|\psi_i| < \infty$. Thus $\{W_t\}$ and ζ_0 are well-defined and $\{W_t\}$ is a linear stationary process. Now

$$\begin{aligned}
\hat{\gamma}_\xi(0) &= c_0^2 n^{-1}\sum_{t=1}^{n} Z_t^2 + n^{-1}\sum_{t=1}^{n} W_t^2 + \zeta_0^2 \\
&\quad -2c_0 n^{-1}\sum_{t=1}^{n} Z_t W_t + 2c_0\zeta_0 n^{-1}\sum_{t=1}^{n} Z_t - 2\zeta_0 n^{-1}\sum_{t=1}^{n} W_t \\
&= A_n + B_n + C_n - D_n + E_n - F_n,
\end{aligned}$$

say. It is easy to show that B_n, C_n and F_n are of $O_p(1)$ and D_n and E_n of $O_p(n^{\frac{1}{2}})$. This completes the proof.

By Lemma 4, $n^{-1}\hat{\gamma}_\xi(0)$ and $c_0^2 n^{-2}\sum_{t=1}^{n} Z_t^2$ have the same asymptotic distributions. On the other hand, the asymptotic distribution of $n^{-2}\sum_{t=1}^{n} Z_t^2$, when $\sigma^2 = 1$, is given in the following lemma, which was originally shown by Hasza and Fuller [3].

Lemma 5 *Let* $U = \sum_{i=1}^{\infty}\delta_i^2 V_i^2$ *and* $Z_t = \sum_{j=1}^{t} e_j$, *where* $\delta_i = (-1)^{i+1}[(2i-1)\pi]^{-1}$ *and* $\{V_i\}$ *is a sequence of mutually independent normal* (0, 1) *random variables and* $\{e_j\}$ *a sequence of iid*(0, 1) *random variables. Then*

$$n^{-2}\sum_{t=1}^{n} Z_t^2 \Rightarrow U \quad as \quad n \to \infty.$$

Now we obtain the following lemma.

Lemma 6 *Under Assumptions* 1, 2 *and* 3,

$$\lim_{n\to\infty} Pr\{S_n(-1) \le A\} = 0 \quad for\ any\ A > 0. \tag{2.10}$$

Proof. We can show that

$$n^{-1}S_n(-1) - 4n^{-2}\sum_{t=1}^{n}\sum_{i=1}^{t-1}\sum_{j=1}^{t-1}(-1)^{i+j}X_{t-i}X_{t-j} = o_p(1). \tag{2.11}$$

Define

$$\eta_n = n^{-2}\sum_{t=1}^{n}\sum_{i=1}^{t-1}\sum_{j=1}^{t-1}(-1)^{i+j}X_{t-i}X_{t-j} = n^{-2}\sum_{t=1}^{n}[\sum_{j=1}^{t-1}(-1)^j X_j]^2.$$

Let $Y_t = (-1)^t X_t$ and $\xi_t = \sum_{j=1}^{t} Y_j$, $t \ge 1$. Then

$$\eta_n = n^{-2}\sum_{t=1}^{n-1}\xi_t^2 = n^{-1}\hat{\gamma}_\xi(0) + o_p(1),$$

where $\hat{\gamma}_\xi(0) = n^{-1}\sum_{t=1}^n \xi_t^2$. Under the assumptions of the model (2.1), we have

$$Y_t = \sum_{j=0}^{\infty}(-1)^j\psi_j(-1)^{t-j}e_{t-j}.$$

By Lemmas 4 and 5, it is easy to show

$$n^{-1}\hat{\gamma}_\xi(0) \Rightarrow \sigma^2[\sum_{j=0}^{\infty}(-1)^j\psi_j]^2 U$$

under Assumptions 1, 2 and 3. Here U is given in Lemma 5. Consequently, from (2.11), we have

$$\frac{1}{n}S_n(-1) \Rightarrow \eta,$$

where $\eta = 4\sigma^2[\sum_{j=0}^{\infty}(-1)^j\psi_j]^2 U$. It can be seen that U is a random variable with a continuous distribution function. Under Assumption 2, η is a random variable with a continuous distribution function $F_\eta(x)$, say. Now for any $A > 0$ and $\varepsilon > 0$, we have

$$\begin{aligned}\lim_{n\to\infty}\Pr\{S_n(-1) \le A\} &= \lim_{n\to\infty}\Pr\{\frac{1}{n}S_n(-1) \le \frac{A}{n}\} \\ &\le \lim_{n\to\infty}\Pr\{\frac{1}{n}S_n(-1) \le \varepsilon\} \le F_\eta(\varepsilon).\end{aligned}$$

As ε, $\varepsilon > 0$, can be taken arbitrarily small, we obtain

$$\lim_{n\to\infty}\Pr\{S_n(-1) \le A\} = 0.$$

By Lemmas 1, 2, 3 and 6, we see that $S_n(\beta)$ has the same asymptotic properties as $Q_{n,k}(\beta)$ for $\beta \in [-1,\ 1]$, except that $p\lim_{n\to\infty} S_n(1) = \gamma(0) + X_0^2$ while $p\lim_{n\to\infty} Q_{n,k}(1) = \gamma(0)$. In the following, we take $X_0 = 0$. We now consider the asymptotic properties of $\hat{\beta}_{n,k}$ and $\hat{\beta}_n$. Put

$$S(\beta) = \begin{cases} p\lim_{n\to\infty} Q_{n,k}(\beta) & \text{for } |\beta| \le 1 \\ +\infty & \text{for } \beta = -1. \end{cases}$$

We can show that $S(\beta)$ is continuous on $(-1,\ 1]$. It thus has at least one minimal point β_0, say, on $[-1,\ 1]$ and $\beta_0 \ne -1$. We have the following result.

Theorem 1 *Suppose* $|\beta_0| < 1$. *Then there exists a local minimal point* $\hat{\beta}_{n,k}$ *of* $Q_{n,k}(\beta)$, *such that*

$$\hat{\beta}_{n,k} \xrightarrow{p} \beta_0 \quad as \quad n \to \infty$$

under Assumptions 1 *and* 2.

Proof. For shorthand notations, we shall use $Q_{n,k}^{(i)}(\beta_0)$, $i = 1, 2, 3$, to denote the value of the ith partial derivative of $Q_{n,k}(\beta)$ with respect to β at $\beta = \beta_0$. For any β, $|\beta| < 1$, Taylor series expansion of $Q_{n,k}^{(1)}(\beta)$ about β_0 gives

$$Q_{n,k}^{(1)}(\beta) = Q_{n,k}^{(1)}(\beta_0) + (\beta - \beta_0)Q_{n,k}^{(2)}(\beta_0) + \frac{1}{2}(\beta - \beta_0)^2 Q_{n,k}^{(3)}(\beta_n^*),$$

where $\beta_n^* = \theta_n\beta + (1-\theta_n)\beta_0$, for some θ_n, $0 < \theta_n < 1$. In Lemma 3, we showed that $Q_{n,k}(\beta) \xrightarrow{P} S(\beta)$ for $\beta \in (-1,\ 1]$ as $n \to \infty$. We can also show that

$$Q_{n,k}^{(1)}(\beta) \xrightarrow{P} S^{(1)}(\beta), \tag{2.12}$$

$$Q_{n,k}^{(2)}(\beta) \xrightarrow{P} S^{(2)}(\beta) \tag{2.13}$$

and

$$Q_{n,k}^{(3)}(\beta) \xrightarrow{P} S^{(3)}(\beta) \tag{2.14}$$

for $\beta \in (-1,\ 1)$. For any ϵ, $0 < \epsilon < 1$, we can show that $S^{(3)}(\beta)$ is bounded on $[-1+\epsilon,\ 1-\epsilon]$. Thus there exists a constant M, $M > 0$, such that $|S^{(3)}(\beta)| \le M$ for $\beta \in [-1+\epsilon,\ 1-\epsilon]$. Consequently, from (2.14), we obtain

$$\Pr\{|Q_{n,k}^{(3)}(\beta_n^*)| \ge 2M\} \longrightarrow 0 \quad \text{as} \quad n \to \infty.$$

We have $S^{(1)}(\beta_0) = 0$ and $S^{(2)}(\beta_0) > 0$, since β_0, $|\beta_0| < 1$, is a minimal point of $S(\beta)$ on $[-1,\ 1]$. Thus from (2.12) and (2.13), $Q_{n,k}^{(1)}(\beta_0) \xrightarrow{P} 0$ and $\Pr\{Q_{n,k}^{(2)}(\beta_0) \le c\} \longrightarrow 0$ as $n \to \infty$, where c is any positive number such that $c < S^{(2)}(\beta_0)$. We conclude that, for any positive numbers δ and ε, there exists a positive integer $n_0 = n_0(\delta,\ \varepsilon)$ such that

$$\begin{aligned} P_1 &= \Pr\{|Q_{n,k}^{(1)}(\beta_0)| \ge \delta^2\} < \frac{1}{3}\varepsilon, \\ P_2 &= \Pr\{Q_{n,k}^{(2)}(\beta_0) \le c\} < \frac{1}{3}\varepsilon \end{aligned}$$

and

$$P_3 = \Pr\{|Q_{n,k}^{(3)}(\beta_n^*)| \ge 2M\} < \frac{1}{3}\varepsilon$$

for $n > n_0$. Now we define

$$\begin{aligned} S_n = & \{\mathbf{X}_n : \mathbf{X}_n = (x_1,\ x_2,\ \dots,\ x_n) \in \mathbf{R}^n \text{ such that} \\ & |Q_{n,k}^{(1)}(\beta_0)| < \delta^2,\ Q_{n,k}^{(2)}(\beta_0) > c \text{ and } |Q_{n,k}^{(3)}(\beta_n^*)| < 2M\} \end{aligned}$$

and let S_n^* be the complementary space of S_n. Then

$$\Pr\{\mathbf{X}_n \in S_n^*\} \le P_1 + P_2 + P_3 < \varepsilon$$

and consequently

$$\Pr\{\mathbf{X}_n \in S_n\} > 1 - \varepsilon$$

for any n, $n > n_0$. Following the similar argument of Cramér ([4], pp.503), we can show that, for any n, $n > n_0$ and any $\mathbf{X}_n$, $\mathbf{X}_n \in S_n$, there exists a point $\hat{\beta}_{n,k}$, $\beta_0 - \delta < \hat{\beta}_{n,k} < \beta_0 + \delta$, such that $Q_{n,k}^{(1)}(\beta_0 - \delta) < 0$, $Q_{n,k}^{(1)}(\hat{\beta}_{n,k}) = 0$ and $Q_{n,k}^{(1)}(\beta_0 + \delta) > 0$ if $\delta < (M+1)^{-1}c$. This means that $\hat{\beta}_{n,k}$ is a local minimal point of $Q_{n,k}(\beta)$. Now as δ and ε can be made arbitrarily small, we obtain $\hat{\beta}_{n,k} \xrightarrow{P} \beta_0$ as $n \to \infty$.

We proceed to consider the limiting distribution of $\hat{\beta}_{n,k}$. We have the following theorem.

Theorem 2 *Suppose $\hat{\beta}_{n,k}$ is a local minimal point of $Q_{n,k}(\beta)$ such that $\hat{\beta}_{n,k} \xrightarrow{P} \beta_0$ as $n \to \infty$, where β_0, $|\beta_0| < 1$, is a minimal point of $S(\beta)$. Then*

$$\sqrt{n}(\hat{\beta}_{n,k} - \beta_0) \Rightarrow N(0,\ 4V_1/V_2) \quad \textit{as } n \to \infty,$$

where $V_2 = [S^{(2)}(\beta_0)]^2$
and

$$V_1 = (\eta - 3)\sigma^4\Big(\sum_{i=0}^{\infty} \theta_i \phi_i\Big)^2 + \sigma^4 \sum_{h=-\infty}^{\infty} \sum_{i=0}^{\infty} \sum_{j=0}^{\infty} \theta_i \phi_i (\theta_{i+h}\phi_{j+h} + \theta_{j+h}\phi_{i+h}).$$

Here $\theta_0 = 1$, $\phi_0 = 0$, $\theta_i = \phi_i = 0$ *for* $i < 0$ *and*

$$\theta_i = \psi_i - (1-\beta_0)\sum_{u=0}^{i-1} \beta_0^{i-u-1}\psi_u,$$
$$\phi_i = \sum_{u=1}^{i}(1-u(1-\beta_0))\beta_0^{u-2}\psi_{i-u}$$

for $i \geq 1$.

Proof. We have

$$n^{\frac{1}{2}}Q_{n,k}^{(1)}(\beta_0) = n^{\frac{1}{2}}Q_{n,k}^{(1)}(\hat{\beta}_{n,k}) + n^{\frac{1}{2}}(\beta_0 - \hat{\beta}_{n,k})Q_{n,k}^{(2)}(\beta_n^*) = n^{\frac{1}{2}}(\beta_0 - \hat{\beta}_{n,k})Q_{n,k}^{(2)}(\beta_n^*),$$

where $\beta_n^* = \theta_n\hat{\beta}_{n,k} + (1-\theta_n)\beta_0$ and $0 < \theta_n < 1$. Write

$$n^{\frac{1}{2}}(\beta_0 - \hat{\beta}_{n,k}) = n^{\frac{1}{2}}[Q_{n,k}^{(1)}(\beta_0) - EQ_{n,k}^{(1)}(\beta_0)]/Q_{n,k}^{(2)}(\beta_n^*) + n^{\frac{1}{2}}EQ_{n,k}^{(1)}(\beta_0)/Q_{n,k}^{(2)}(\beta_n^*).$$

The second term on the right-hand side of the above equation converges in probability to 0 as $n \to \infty$, since $EQ_{n,k}^{(1)}(\beta_0) = O(k|\beta_0|^k)$ and

$$Q_{n,k}^{(2)}(\beta_n^*) \xrightarrow{P} S^{(2)}(\beta_0) \quad \text{as } n \to \infty.$$

Consequently, it suffices to show that

$$n^{\frac{1}{2}}[Q_{n,k}^{(1)}(\beta_0) - EQ_{n,k}^{(1)}(\beta_0)] \Rightarrow N(0,\ 4V_1) \quad \text{as } n \to \infty.$$

We have

$$Q_{n,k}^{(1)}(\beta_0) = 2(n-k)^{-1}\sum_{t=k+1}^{n} Y_{t,k}Z_{t,k},$$

where

$$Y_{t,k} = X_t - (1-\beta_0)\sum_{j=1}^{k}\beta_0^{j-1}X_{t-j},$$
$$Z_{t,k} = \sum_{j=1}^{k}[1-j(1-\beta_0)]\beta_0^{j-2}X_{t-j}.$$

Under the assumptions of the model (2.1), we can write

$$Y_{t,k} = \sum_{i=0}^{\infty}\theta_i^{(k)}e_{t-i} \quad \text{and} \quad Z_{t,k} = \sum_{j=0}^{\infty}\phi_j^{(k)}e_{t-j},$$

where

$$\theta_0^{(k)} = \psi_0 = 1,$$
$$\theta_i^{(k)} = \psi_i - (1-\beta_0)\sum_{u=0}^{i-1}\beta_0^{i-u-1}\psi_u, \quad 1 \leq i \leq k,$$
$$\theta_i^{(k)} = \psi_i - (1-\beta_0)\sum_{u=i-k}^{i-1}\beta_0^{i-u-1}\psi_u, \quad k+1 \leq i \leq \infty,$$
$$\phi_0^{(k)} = 0,$$
$$\phi_j^{(k)} = \sum_{v=1}^{j}[1-v(1-\beta_0)]\beta_0^{v-2}\psi_{j-v}, \quad 1 \leq j \leq k,$$
$$\phi_j^{(k)} = \sum_{v=1}^{k}[1-v(1-\beta_0)]\beta_0^{v-2}\psi_{j-v}, \quad k+1 \leq j \leq \infty.$$

Define

$$W_n = 2(n-k)^{-1} \sum_{t=k+1}^{n} \sum_{i=0}^{\infty} \sum_{j=0}^{\infty} \theta_i \phi_j e_{t-i} e_{t-j}.$$

Now we show that

$$K_n = (n-k)^{\frac{1}{2}} [(W_n - EW_n) - \{Q_{n,k}^{(1)}(\beta_0) - EQ_{n,k}^{(1)}(\beta_0)\}] \xrightarrow{P} 0 \qquad (2.15)$$

as $n \to \infty$. We have

$$\begin{aligned} W_n - Q_{n,k}^{(1)}(\beta_0) &= 2(n-k)^{-1} \sum_{t=k+1}^{n} [\sum_{i=0}^{k} \sum_{j=k+1}^{\infty} \theta_i(\phi_j - \phi_j^{(k)}) + \sum_{i=k+1}^{\infty} \sum_{j=0}^{k} (\theta_i - \theta_i^{(k)}) \phi_j \\ &\quad + \sum_{i=k+1}^{\infty} \sum_{j=k+1}^{\infty} (\theta_i \phi_j - \theta_i^{(k)} \phi_j^{(k)})] e_{t-i} e_{t-j} \\ &= 2(K_{n1} + K_{n2} + K_{n3}), \end{aligned}$$

say. Thus

$$EK_n^2 \leq 12(n-k)[Var(K_{n1}) + Var(K_{n2}) + Var(K_{n3})].$$

Put $\eta_j = \phi_j - \phi_j^{(k)}$. We have

$$\begin{aligned} (n-k)Var(K_{n1}) &\leq 2\sigma^4 \sum_{j=k+1}^{\infty} |\eta_j| (\sum_{i=0}^{\infty} |\theta_i|)^2 (\sum_{j=0}^{\infty} |\phi_j| + \sum_{j=0}^{\infty} |\phi_j^{(k)}|) \\ &\longrightarrow 0 \text{ as } n \to \infty. \end{aligned}$$

Following the similar argument, we can show that $\lim_{n\to\infty}(n-k)Var(K_{n2}) = 0$ and $\lim_{n\to\infty}(n-k)Var(K_{n3}) = 0$. It follows that $EK_n^2 \to 0$ as $n \to \infty$ and we obtain (2.15).

In the next, we show that

$$(n-k)^{\frac{1}{2}}[W_n - EW_n] \Rightarrow N(0,\ 4V_1)$$

as $n \to \infty$. For a fixed positive integer m, define $\xi_{tm} = \sum_{i=0}^{m} \sum_{j=0}^{m} \theta_i \phi_j e_{t-i} e_{t-j}$ and $W_{n,m} = 2(n-k)^{-1} \sum_{t=k+1}^{n} \xi_{tm}$. Now consider the limiting distribution of $(n-k)^{\frac{1}{2}}[W_{n,m} - EW_{n,m}]$ when $n \to \infty$. It can be easily shown that $\{\xi_{tm}\}$ is a strictly stationary m-dependent sequence of random variables. By applying the central limit theorem for strictly stationary m-dependent sequences (cf. Brockwell and Davis [5], pp.206), we obtain

$$(n-k)^{\frac{1}{2}}[W_{n,m} - EW_{n,m}] \Rightarrow N(0,\ 4v_m),$$

where

$$\begin{aligned} v_m = \sum_{h=-m}^{m} \gamma_{\xi_{tm}}(h) &= (\eta - 3)\sigma^4 \sum_{h=-m}^{m} \sum_{i=0}^{m} \theta_i \theta_{i+h} \phi_i \phi_{i+h} \\ &\quad + \sigma^4 \sum_{h=-m}^{m} \sum_{i=0}^{m} \sum_{j=0}^{m} \theta_i \phi_j (\theta_{i+h} \phi_{j+h} + \theta_{j+h} \phi_{i+h}). \end{aligned}$$

We have $v_m \to V_1$ as $m \to \infty$. Hence

$$N(0,\ 4v_m) \Rightarrow N(0,\ 4V_1) \quad \text{as } n \to \infty.$$

Define

$$R_{n,m} = (n-k)^{\frac{1}{2}}[(W_n - EW_n) - (W_{n,m} - EW_{n,m})].$$

We can show that

$$ER_{n,m}^2 \leq K_1 \sum_{j=m+1}^{\infty} |\phi_j| + K_2 \sum_{j=m+1}^{\infty} |\theta_j|,$$

where K_1 and K_2 are positive numbers. The terms on the right-hand side of the above inequality are independent of n and tend to 0 as $m \to \infty$. Applying Corollary 7.7.1 of Anderson [6], we complete the proof of Theorem 2.

Corollary 1 *Suppose* $|\beta_0| < 1$. *Then there exists a local minimal point* $\hat{\beta}_n$ *of* $S_n(\beta)$ *such that*

$$\hat{\beta}_n \xrightarrow{P} \beta_0 \quad as \quad n \to \infty$$

under Assumptions 1, 2 *and* 3.

However, it is difficult to show the limiting distribution of $\hat{\beta}_n$.

2.2. CASE OF AN ARIMA(p, 1, q) PROCESS

In this section, we consider the asymptotic properties of $\hat{\beta}_{n,k}$ and $\hat{\beta}_n$ in the case of an ARIMA(p,1,q) process. We shall see that the results of Section 2.1 can be easily extended to such a case.

Suppose that the underlying process $\{Y_t\}$ satisfies

$$Y_t = \xi_0 + \sum_{j=1}^{t} X_j, \quad t \geq 1, \tag{2.16}$$

where $\{X_t\}$ is an ARMA(p, q) process, which satisfies the assumptions of the model (2.1), and ξ_0 is a random variable such that $E\xi_0^2 < \infty$, which is independent of $\{X_t,\ t \geq 1\}$. For simplicity, we assume $\xi_0 = 0$. We further make the following assumption:
Assumption 4. $\sum_{j=0}^{\infty} \psi_j \neq 0$.

We obtain the following lemmas:

Lemma 7 *It holds that*

$$p \lim_{n\to\infty} Q_{n,k}(\beta) = \frac{1}{1-\beta^2}[\gamma(0) + 2\sum_{j=1}^{\infty} \beta^j \gamma(j)] \quad for\ \beta,\ |\beta| < 1 \tag{2.17}$$

and, under Assumption 4,

$$\lim_{n\to\infty} Pr\{Q_{n,k}(1) \leq A\} = 0 \tag{2.18}$$

and, under Assumptions 1 *and* 2,

$$\lim_{n\to\infty} Pr\{Q_{n,k}(-1) \leq A\} = 0 \tag{2.19}$$

for any positive number A.

Proof. Note that

$$\begin{aligned} Q_{n,k}(\beta) &= \frac{1}{n-k} \sum_{t=k+1}^{n} [Y_t - (1-\beta)\sum_{j=1}^{k} \beta^{j-1} Y_{t-j}]^2 \\ &= \frac{1}{n-k} \sum_{t=k+1}^{n} (\beta^k \sum_{i=1}^{t-k} X_i + \sum_{i=t-k+1}^{t} \beta^{t-i} X_i)^2 \end{aligned}$$

$$= \frac{1}{n-k}\sum_{t=k+1}^{n}[\beta^{2k}(\sum_{i=1}^{t-k}X_i)^2 + (\sum_{i=t-k+1}^{t}\beta^{t-i}X_i)^2 + 2\beta^k\sum_{i=1}^{t-k}\sum_{j=t-k+1}^{t}\beta^{t-j}X_iX_j]$$

$$= A_n + B_n + C_n, \text{ say,}$$

for β, $|\beta| < 1$. It can be shown that the terms A_n and C_n converge in probability to 0 as $n \to \infty$. Following the similar argument as in Lemma 1, we can show that

$$B_n \xrightarrow{P} \frac{1}{1-\beta^2}[\gamma(0) + 2\sum_{j=1}^{\infty}\beta^j\gamma(j)]$$

as $n \to \infty$. Thus we obtain (2.17).

Note that $Q_{n,k}(1) = (n-k)^{-1}\sum_{t=k+1}^{n} Y_t^2$ and

$$\frac{1}{n-k}Q_{n,k}(-1) = \frac{1}{(n-k)^2}\sum_{t=k+1}^{n}[(-1)^k Y_{t-k} + \sum_{j=0}^{k-1}(-1)^j X_{t-j}]^2$$

$$= \frac{1}{(n-k)^2}\sum_{t=k+1}^{n} Y_{t-k}^2 + A_n,$$

say. Following the similar argument of (2.4) as in Lemma 2, we can show that A_n converges in probability to 0 as $n \to \infty$. On the other hand, by Lemmas 4 and 5, the distribution function of the first term on the right-hand-side of the above equation converges to a continuous distribution function. Then following the similar argument as (2.10), we can show (2.18) and (2.19).

Lemma 8 *It holds that*

$$p\lim_{n\to\infty} S_n(\beta) = \frac{1}{1-\beta^2}[\gamma(0) + 2\sum_{j=1}^{\infty}\beta^j\gamma(j)] \quad for\ \beta,\ |\beta| < 1$$

and, under Assumption 4,

$$\lim_{n\to\infty} Pr\{S_n(1) \le A\} = 0$$

and, under Assumptions 1, 2 and 3,

$$\lim_{n\to\infty} Pr\{S_n(-1) \le A\} = 0$$

for any positive number A.

Proof. Noting that

$$S_n(\beta) = n^{-1}\sum_{t=1}^{n}[(\sum_{j=1}^{t}\beta^j X_{t-j})^2 - 2Y_0\beta^{t-1}\sum_{j=1}^{t}\beta^j X_{t-j} + Y_0^2\beta^{2(t-1)}],$$

$$n^{-1}S_n(1) = n^{-2}\sum_{t=1}^{n} Y_t^2 + o_p(1)$$

and

$$n^{-1}S_n(-1) = n^{-2}\sum_{t=1}^{n}[\sum_{j=1}^{t}(-1)^j X_j]^2 + o_p(1).$$

Lemma 8 can be proved similarly as Lemmas 3 and 6.

By Lemmas 7 and 8, it is seen that $S_n(\beta)$ has the same asymptotic properties as $Q_{n,k}(\beta)$ for $\beta \in [-1,\ 1]$. Now we show the asymptotic properties of $\hat{\beta}_{n,k}$ and $\hat{\beta}_n$. Put

$$S(\beta) = \begin{cases} p\lim_{n\to\infty} Q_{n,k}(\beta) & \text{for } |\beta| < 1 \\ +\infty & \text{for } \beta = \pm 1. \end{cases}$$

Clearly, $S(\beta)$ has at least one minimal point β_0 on $[-1,1]$ and $\beta_0 \neq \pm 1$. We obtain the following theorems. These theorems can be shown by tracing the similar ways as we took in the proof of Theorems 1 and 2. Here we omit the proof.

Theorem 3 *There exist local minimal points $\hat{\beta}_{n,k}$ and $\hat{\beta}_n$ of $Q_{n,k}(\beta)$ and $S_n(\beta)$, respectively, such that*

$$\hat{\beta}_{n,k} \xrightarrow{p} \beta_0 \quad as \quad n \to \infty$$

under Assumptions 1 and 2, and

$$\hat{\beta}_n \xrightarrow{p} \beta_0 \quad as \quad n \to \infty$$

under Assumptions 1, 2, 3 and 4.

Theorem 4 *Suppose $\hat{\beta}_{n,k}$ is a local minimal point of $Q_{n,k}(\beta)$ such that $\hat{\beta}_{n,k} \xrightarrow{P} \beta_0$ as $n \to \infty$. Then*

$$\sqrt{n}(\hat{\beta}_{n,k} - \beta_0) \Rightarrow N(0,\ 4V_1/V_2) \quad as\ n \to \infty,$$

where $V_2 = [S^{(2)}(\beta_0)]^2$ and

$$V_1 = (\eta - 3)\sigma^4 (\sum_{i=0}^{\infty} \theta_i \phi_i)^2 + \sigma^4 \sum_{h=-\infty}^{\infty} \sum_{i=0}^{\infty} \sum_{j=0}^{\infty} \theta_i \phi_i (\theta_{i+h}\phi_{j+h} + \theta_{j+h}\phi_{i+h}).$$

Here $\theta_i = \phi_i = 0$ for $i < 0$ and $\theta_i = \sum_{j=0}^{i} \beta_0^j \psi_{i-j}$, $\phi_i = \sum_{j=0}^{i} j\beta_0^{j-1}\psi_{i-j}$ for $i \geq 0$.

REFERENCES

1. S. Makridakis, A. Andersen, R. Carbone, R. Fildes, M. Hibon, R. Lewandowski, J. Newton, E. Parzen and R. Winkler, *The Forecasting Accuracy of Major Time Series Methods.* John Wiley, New York (1984).
2. E.S. Gardner, *J. Forecasting*, **1**, 1-28 (1985).
3. D.P. Hasza and W.A. Fuller, *Ann. Statist.* **5**, 1106-1120 (1979).
4. H. Cramér, *Mathematical Methods of Statistics.* Princeton University Press, Princeton (1973).
5. P.J. Brockwell and R.A. Davis, *Time Series: Theory and Methods.* Springer-Verlag, New York (1987).
6. T.W. Anderson, *The Statistical Analysis of Time Series.* John Wiley, New York (1971).

Stat. Sci. & Data Anal., pp. 143-150
K. Matsusita *et al.* (Eds)

Characterization of MTV Model and Its Diagnostic Checking

TAKEAKI KARIYA

Institute of Economic Research, Hitotsubashi University, Kunitachi, Tokyo 186, *Japan*

Abstract. The MTV model Kariya [1] proposed is characterized on which a diagnostic checking procedure for the model is proposed.

Key words: Characterization of MTV model, VARMA model, diagnostic checking

1. INTRODUCTION

Financial and economic time series have the following features and aspects;

(1) they are multi-dimensional phenomena,

(2) economic structure evolves gradually but constantly, and

(3) profitable variational structure and opportunities will be exploited, which will change variational structure of financial time series.

Hence these points must be taken into account in efficiently analyzing such volatile time series as stock prices, interest rates, exchange rates, etc. In particular, a multivariate approach is more appropriate and a proper length of sample period is required to be chosen. A typical multivariate time series model sometimes used in practice is VARMA (vector-valued autoregressive moving average) model. However, as Tiao and Tsay [2] pointed out, the model has the two major difficulties in applications;

(a) overflow of parameters the estimates of which are highly correlated,

(b) lack of identifiable models.

In fact, for example, a p-dimensional AR(k) model contains $p^2 \times k$ parameters as the coefficients and $p(p+1)/2$ parameters in covariance matrix. Further a VARMA model is not uniquely identified in general. Tiao and Tsay [2] made a great deal of contributions to settle these problems in VARMA model and developed a statistical procedure to find a simpler structure for a given VARMA model. However, it seems to me that the procedure still require a very large sample size when p is of moderate size, say $p=20$.

Kariya [1] proposed the MTV (multivariate time series variance component) model as an alternative model for analysis and prediction of multi-dimensional volatile time series phenomena. It is supposed in the MTV model that there are a comparatively small number of major common time series variance components through which the cross-sectional correlation structure and the serial and cross-serial correlation structure are formed. Peña and Box [3] also proposed a similar multivariate factor model by combining a factor analysis model with a VARMA model. The model lacks the identifiability associated with a factor analysis model and with a VARMA model. On the other hand, the MTV model can be regarded as a dynamic generalization of the PC (principal component) model, but in the context of multivariate stationary time series theory it is the simplest time-reversible model so that the spectral density matrix is real and symmetric, and diagonalized by a certain orthogonal matrix. But the model itself allows a heteroscedastic nonstationarity. Another feature of the MTV model is that the variance components can be any nonlinear models such as GARCH (generalized auto-regressive conditional heteroscedastic) model, TARCH (threshold autoregressive conditional heteroscedastic) model, etc.

In Section 2, we characterize a stationary MTV model as a multivariate stationary process with symmetric autocovariance matrices which commute each other, and review some theoretical features of the model. Based on this characterization, in Section 3 a diagnostic checking procedure is proposed about whether a given process $\{x_t\}$ follows an MTV model.

2. MTV MODEL AND ITS PROPERTIES

In this section, we characterize the MTV model as the simplest multivariate time-reversible stationary process that has a spectral density matrix (Fourier transform) and is generated by p uncorrelated stationary processes.

First, recall that a p-dimensional (weak) stationary process $\{\boldsymbol{x}_t\}$ with constant mean vector $\boldsymbol{\mu}$ is described by its autocovariance matrices $\{\boldsymbol{\Sigma}(k):k=0,\pm 1,\pm 2,\cdots\}$ where

$$\boldsymbol{\Sigma}(k) = \mathrm{Cov}(\boldsymbol{x}_t,\ \boldsymbol{x}_{t-k}) = E(\boldsymbol{x}_t-\boldsymbol{\mu})(\boldsymbol{x}_{t-k}-\boldsymbol{\mu})'. \tag{2.1}$$

It is noted that when $p>1$, $\boldsymbol{\Sigma}(k)$ is not symmetric but

$$\boldsymbol{\Sigma}(-k) = \boldsymbol{\Sigma}(k)', \tag{2.2}$$

implying that $\{\boldsymbol{x}_t\}$ is not time-reversible in general. Assume that the spectral density matrix $H(\omega)$ of $\{\boldsymbol{x}_t\}$ exists;

$$H(\omega) = \frac{1}{2\pi}\sum_{k=-\infty}^{\infty} \exp(i\omega k)\boldsymbol{\Sigma}(k). \tag{2.3}$$

Here when $p>1$, $H(\omega)$ is not symmetric but Hermitian, i.e., $H(\omega)'=\overline{H(\omega)}$ because of (2.2).

Now first assume that $\{x_t\}$ is time-reversible. Then $\Sigma(k)'=\Sigma(k)$ and hence $H(\omega)$ is symmetric and real. Hence there exists a $p\times p$ orthogonal matrix $A(\omega)$ which diagonalizes $H(\omega)$ as

$$A(\omega)'H(\omega)A(\omega) = \operatorname{diag}\{h_1(\omega),\cdots,h_p(\omega)\} \equiv D_k(\omega) \qquad (2.4a)$$

where

$$\gamma_1(0) \geqq \gamma_2(0) \geqq \cdots \geqq \gamma_p(0) \qquad \text{with} \qquad (2.4b)$$

$$\gamma_j(k) = \int_{-\pi}^{\pi} \exp(-ki\omega)h_j(\omega)\,d\omega. \qquad (2.4c)$$

For the identifiability of $h_j(\omega)$'s, we need the strict inequalities in (2.4b)

$$\gamma_1(0) > \cdots > \gamma_p(0). \qquad (2.5a)$$

In fact, if (2.5) holds, $A(\omega)$ is unique up to transformation $A(\omega)\rightarrow A(\omega)E$ with $E\in\varepsilon$ where

$$\varepsilon = \{E=\operatorname{diag}\{e_1,\cdots,e_p\} : e_i=1 \text{ or } -1\}. \qquad (2.5b)$$

Further assume that $A(\omega)\equiv A$ does not depend on ω. This assumption holds if and only if $\Sigma(k)$'s are simultaneous diagonalized. Then by (2.4a)

$$H(\omega) = AD_k(\omega)A' = h_1(\omega)\boldsymbol{\alpha}_1\boldsymbol{\alpha}_1'+\cdots+h_p(\omega)\boldsymbol{\alpha}_p\boldsymbol{\alpha}_p'. \qquad (2.6a)$$

with

$$A = [\boldsymbol{\alpha}_1,\cdots,\boldsymbol{\alpha}_p] \qquad (2.6b)$$

Under (2.6a), let $\{f_{jt}\}$ represent a process with spectral density $h_j(\omega)$. Then we obtain the MTV model as a model representing the spectral density matrix $H(\omega)$;

$$\boldsymbol{x}_t = \boldsymbol{\mu}+\boldsymbol{\alpha}_1 f_{1t}+\cdots+\boldsymbol{\alpha}_p f_{pt} \qquad (2.7)$$

with the conditions

(A) $F_j=\{f_{jt}\}$ is stationary $(j=1,\cdots,p)$

(B) F_j and F_k are uncorrelated $(j\neq k)$

(C) $\gamma_1(0)>\cdots>\gamma_p(0)$ when (2.5a) is assumed rather than (2.4b)

(D) $A'A=I$ where A is unique up to $\{AE\}$ with $E\in\varepsilon$.

We summarize this as

Theorem 2.1. *A regualr stationary process $\{x_t\}$ follows an MTV model if and only if*

(a) $\{x_t\}$ is time-reversible, i.e., $\Sigma(-k)=\Sigma(k)=\Sigma(k)'$

(b) $\Sigma(k)$'s are simultaneously diagonalized, i.e., $\Sigma(k)\Sigma(\ell)=\Sigma(\ell)\Sigma(k)$ for all k and ℓ.

Clearly x_t in (2.7) is linearly generated by p uncorrelated processes, which can be nonlinear processes such as ARCH, TARCH, etc. In (2.7) the components f_{jt}'s are identified by the condition (C), and the variational sizes of the components f_{jt}'s are standardized by (D) so that

$$\gamma_j(0)\to 1 \quad (j=1,\cdots,p) \quad \text{if and only if} \quad \Sigma(0)=\mathrm{Var}(x_t)\to I. \tag{2.8}$$

Conversely supppose x_t is linearly generated by p uncorrelated processes as

$$x_t = \mu+\beta_1 g_{1t}+\cdots+\beta_p g_{pt} = \mu+Bg_t$$

For this model to be well-defined, we need a condition on B or g_t. With almost no loss of generality we can assume (2.5a) for the variances of g_{jt}'s $(j=1,\cdots,p)$. Then if we require the normalization (2.8) between $\gamma_j(0)$'s and $\Sigma(0)$, then $BB'=I$, which yield the MTV model. In this sense the MTV model is the simplest model generated by p uncorrelated stationary processes which has the property (2.8).

Some other properties of the model including an optimality are discussed in Kariya [1].

3. DIAGNOSTIC CHECKING FOR MTV MODEL

As has been discussed in Section 2, a p-dimensional stationary process $\{x_t\}$ follows an MTV model if and only if the following two hypotheses hold;

$$H_1: \quad \Sigma(h)=\Sigma(h) \quad \text{for all} \quad h=0,\pm1,\pm2,\cdots$$
$$H_2: \quad \Sigma(h)\Sigma(\ell)=\Sigma(\ell)\Sigma(h) \quad \text{for all} \quad h \text{ and } \ell$$

where $\Sigma(h)$ is the h lag autocovariance

$$\Sigma(h) = E(x_t-\mu)(x_{t+h}-\mu)' \equiv (\sigma_{ij}(h)). \tag{3.1}$$

Here μ is assumed to be constant. In this section, a diagnostic checking procedure is considered about whether a given stationary process $\{x_t\}$ really follows an MTV model. Our procedure tests the hypotheses H_1 and H_2 stepwise.

Test for $H_1: \Sigma(h)'=\Sigma(h)$ for all h

For notation, let

$$\nu_{ijkm}(h, r, s) = E(\tilde{x}_{it}\tilde{x}_{jt+h}\tilde{x}_{kt+r}\tilde{x}_{mt+s}) \tag{3.2}$$

where $x_t = (x_{1t}, \cdots, x_{pt})'$ and $\tilde{x}_{it} = x_{it} - \mu_i$. If $\{x_t\}$ is a normal process, $\nu_{ijkm}(h, r, s)$ equals

$$\nu^*_{ijkm}(h, r, s) = \sigma_{ij}(h)\sigma_{km}(s-r) + \sigma_{ik}(r)\sigma_{jm}(s-h) + \sigma_{im}(s)\sigma_{jk}(h-r) \tag{3.3}$$

Let

$$\kappa_{ijkm}(h, r, s) = \nu_{ijkm}(h, r, s) - \nu^*_{ijkm}(h, r, s) \tag{3.4}$$

so that $\kappa_{ijkm}(h, r, s) = 0$ if $\{x_t\}$ is normal. To construct a test for H_1, we estimate $\sigma_{ij}(h)$ by

$$c_{ij}(h) = \frac{1}{T}\Sigma_{t=1}^{T-h}(x_{it} - \bar{x}_i)(x_{jt+h} - \bar{x}_j). \tag{3.5}$$

Then it follows from Hannan ([4] p209) that

$$\begin{aligned} &\lim_{T\to\infty} T\cdot\mathrm{Cov}(c_{ij}(h), c_{km}(g)) \\ &= \Sigma_{r=-\infty}^{\infty}[\sigma_{ik}(r)\sigma_{jm}(r+h-g) \\ &\quad + \sigma_{im}(r-g)\sigma_{jk}(r+h) + \kappa_{ijkm}(h, -r, g-r)] \\ &\equiv \delta_{ijkm}(h, g). \end{aligned} \tag{3.6}$$

Hence, assuming that a given process $\{x_t\}$ is strongly linear with 8th order moments, it follows from a central limit theorem that under H_1

$$\sqrt{T}[c_{ij}(h) - c_{ji}(h)] \to N(0, \Delta_{ijji}(h, h)) \tag{3.7}$$

as $T\to\infty$, where

$$\begin{aligned} \Delta_{ijkm}(h, g) &= \lim T\cdot\mathrm{Cov}(c_{ij}(h) - c_{ji}(h),\ c_{km}(g) - c_{mk}(g)) \\ &= \delta_{ijkm}(h, g) - \delta_{ijmk}(h, g) - \delta_{jikm}(h, g) \\ &\quad + \delta_{ijmk}(h, g) \end{aligned} \tag{3.8}$$

with $\delta_{ijkm}(h, g)$ in (4.6). To consider all the elements simultaneously, let

$$\begin{aligned} d(h) = (&c_{12}(h) - c_{21}(h),\ c_{13}(h) - c_{31}(h), \cdots, \\ &c_{1p}(h) - c_{p1}(h); \end{aligned} \tag{3.9}$$

$$c_{23}(h)-c_{32}(h),\ c_{24}(h)-c_{42}(h),\cdots,\ c_{2p}(h)-c_{p2}(h);\ \cdots:\ c_{p-1,p}(h)-c_{p,p-1}(h))'$$

which is of order $p(p-1)/2\times 1$. Then under H_1

$$\sqrt{T}d(h)\rightarrow N(0,\ \Delta(h))\quad \text{as}\quad T\rightarrow\infty, \tag{3.10}$$

where $\Delta(h)=(\Delta_{ijkm}(h,h))$. Finally we consider the total elements involved in lags up to n by

$$d=(d(1)',d(2)',\cdots,d(n)')':\ np(p-1)/2\times 1. \tag{3.11}$$

Then under H_1

$$\sqrt{T}d\rightarrow N(0,\ \Delta)\quad \text{as}\quad T\rightarrow\infty \tag{3.12}$$

where $\Delta=(\Delta_{ijkm}(h,g))$ with $\Delta_{ijkm}(h,g)=\mathrm{Cov}(d_{ij}(h),\ d_{km}(g))$ as in (3.8). Here we estimate $\Delta_{ijkm}(h,g)$ by replacing $\delta_{ijkm}(h,g)$ in (3.8) by

$$\hat{\delta}_{ijkm}(h,g)=\Sigma_{r=-N}^{N}[c_{ik}(r)c_{jm}(r+h-g)+c_{im}(r-g)c_{jk}(r+h)+\hat{\kappa}_{ijkm}(h,-r,g-r)] \tag{3.13}$$

where

$$\hat{\kappa}_{ijkm}(h,r,g)=\hat{\nu}_{ijkm}(h,r,g)-c_{ij}(h)c_{km}(g-r)-c_{ik}(r)c_{jm}(g-h)-c_{im}(g)c_{jk}(r-h) \tag{3.14}$$

with

$$\hat{\nu}_{ijkm}(h,r,g)=\frac{1}{T}\Sigma_{t=1}^{T_0}(x_{it}-\bar{x}_i)(x_{jt+h}-\bar{x}_j)\times(x_{kt+r}-\bar{x}_k)(x_{mt+g}-\bar{x}_m) \tag{3.15}$$

($T_0=T-\max(h,r,g)$). Therefore a test statistic for testing the hypothesis H_1 is constructed as

$$U=Td'\hat{\Delta}^{-1}d \tag{3.16}$$

and H_2 is rejected when U is large. The null distribution of U will be approximated by a χ^2 distribution with degrees of freedom $np(p-1)/2$, i.e., $\chi^2_{np(p-1)/2}$. This hypothesis can be also tested by the approach due to Taniguchi and Kondo [5].

Test for H_2 : $\Sigma(h)\Sigma(\ell)=\Sigma(\ell)\Sigma(h)$ for all h and ℓ

We assume that H_1 is true. Let

$$\boldsymbol{y}_t = (\boldsymbol{x}_t', \boldsymbol{x}_{t+1}', \cdots, \boldsymbol{x}_{t+n-1}')' : np\times 1$$

$$\Omega = \mathrm{Cov}(\boldsymbol{y}_t) = \begin{pmatrix} \Sigma(0), & \Sigma(1), & \cdots, & \Sigma(n-1) \\ \Sigma(-1), & \Sigma(0), & \cdots, & \Sigma(n-2) \\ \cdot & & & \\ \cdot & & & \\ \cdot & & & \\ \cdot & & & \\ \Sigma(-n+1), & \cdots\cdots & & , \Sigma(0) \end{pmatrix} \tag{3.17}$$

Then when H_2 is not true, the unconstrained covariance matrix Ω is estimated by

$$\hat{\Omega} = (\hat{\Sigma}_{ij}) \quad \text{with} \quad \hat{\Sigma}_{ij} = \hat{\Sigma}(i-j) = (c_{km}(i-j)) \tag{3.18}$$

where $\Sigma_{ij}: p\times p$ is the (i, j) block matrix of Ω. Next, we estimate Ω in the case where H_2 is true. When H_2 is true with the assumed H_1, there exists a $p\times p$ common orthogonal matrix Γ such that

$$\Gamma' \Sigma(h)\Gamma = \Lambda(h) = \mathrm{diag}\{\lambda_1(h), \cdots, \lambda_p(h)\} \tag{3.19}$$

for $h=0,1,2,\cdots,n-1$, where $\lambda_j(h)$'s are the characteristic roots of $\Sigma(h)$. Hence the covariance matrix of $\boldsymbol{y}_t$ is expressed under H_1 and H_2 as

$$\Omega_H = (I_n \otimes \Gamma)\Lambda(I\otimes \Gamma)' \tag{3.20}$$

where $\Lambda = (\Lambda_{ij})$ with $\Lambda_{ij} = \Lambda(i-j): p\times p \quad (i, j=1,\cdots, n)$. Thus we estimate Ω_H by

$$\hat{\Omega}_H = (I\otimes \hat{\Gamma}_0)\hat{\Lambda}(I\otimes \hat{\Gamma}_0)' \tag{3.21}$$

where $\hat{\Lambda} = (\hat{\Lambda}(i-j))$ with $\hat{\Lambda}(h)$ being the diagonal matrix of the characteristic roots of

$$\frac{1}{2}[\hat{\Sigma}(h) + \hat{\Sigma}(-h)]$$

and $\hat{\Gamma}_0$ is a $p\times p$ orthogonal matrix such that

$$\hat{\Gamma}_0' \hat{\Sigma}(0)\hat{\Gamma}_0 = \hat{\Lambda}(0).$$

Hence as a test for H_2 when H_1 is true or accepted, we can propose

$$Q = \mathrm{tr}\, \hat{\Omega}_H \hat{\Omega}^{-1} \tag{3.22}$$

which is regarded as a generalized F ratio between the constrained $\hat{\Omega}_H$ and unconstrained $\hat{\Omega}$. Of course, H_2 is rejected for large values of Q. The null distribution of Q will be approximated by χ^2_f with $f \equiv p(p-1)(n-1)$, though the proof is difficult, where f is the difference of the numbers of parameters between Ω and Ω_H. In asset allocation, p is usually large and so f is larger. Hence the null distribution of

$$z = (Q - f)/\sqrt{2f}$$

will be approximated by normal distrinbution N(0,1), and when z is large, H_2 is rejected.

Consequently, the above two testing procedures for H_1 and H_2 can be applied to a diagonostic checking about whether data is consistent with the MTV time series structure. But it involves a heavy computational task. Another rather direct checking is to apply the MTV model directly to data and examine its predictive power of the model.

Acknowledgements

This research is financially supported by the New Japan Securities Science Promotion Foundation.

REFERENCES

1. T.Kariya, MTV model and its application to prediction of stock prices, *Proc. Second International Tampere Conference in Statistics* (ed. by Pukkila, T. and Puntanen, S.), 161-176 (1987).
2. G.C.Tiao and R.S.Tsay, Model specification in multivariate time series, *Jour. Roy. Statist. Soc.* B51, 157-213 (1989).
3. D. Peña and G.E.P.Box, Identifying a simple structure in time series, *Jour. Amer. Statist. Assoc.* 82, 836-843 (1987).
4. E.J.Hannan, *Multiple Time Series*, John Wiley (1970).
5. M.Taniguchi and M.Kondo, Nonparametric approach in time series analysis, To appear from *Jour. Time Series Analysis*, (1991).

Stat. Sci. & Data Anal., pp. 151-163
K. Matsusita *et al.* (Eds)

Limit theorems for statistical inference on stationary processes with strong dependence

Yuzo Hosoya
Faculty of Economics, Tohoku University, Kawauchi, Aoba-ku, Sendai 980, Japan

Abstract This paper gives some central limit theorems for multivariate stationary processes whose spectral density may possibly have unbounded peaks at the origin or at some other frequencies. The results are generalization of the auther's previous results for stationary processes with weak dependence. The central limit theorems are proved on the assumption that the innovations of the stationary processes satisfy certain mixing conditions for their conditional moments. Thus the usual assumptions of exact Martingale difference or the (transformed) Gaussianity for the innovation process are dispensed with. For the asymptotic distribution of the quasi-likelihood estimate, the bracketing condition approach is proposed in order to deal with the central limit theorem for the estimate even in such situations where peaks of spectral density appear at unknown frequencies.

Key words: Central limit theorems, mixing conditions, Martingale differences, serial covariances, non-standard conditions, strong dependence.

0. Introduction.

The paper gives central limit theorems for vector-valued linear stationary processes with unbounded spectral peaks; namely processes with strong dependence. It deals with the processes $z(t) = \sum_{j=0}^{\infty} G(j)e(t-j), t \in J$, where $\{e(t)\}$ is a mean zero white noise vector-valued process satisfying some mixing conditions for its conditional moments, but it is not assumed Gaussian. The literatures abound recently in the field of limit theorems for stationary processes with strong dependence, each with its own different point of emphasis. This paper presents an approach to deal with that problem, assuming weaker conditions on $\{e(t)\}$.

Section 1 deals with the asymptotic normality of the standarized sample average. Yajima [1] gave the limit theorems for the finite Fourier transform for non-Gaussian processes with strong dependence under the assumption of summability of cumulant functions for all order. Theorem 1.1 gives the central limit theorem by an alternative approach which has more similarity to the one for weakly dependent processes.

Section 2 shows the central limit theorem for sample serial covariances from the vector-valued process $\{z(t)\}$. Theorem 2.1 gives it in a general form and Section 3 gives an example. That theorem shows that those statistics have different asymptotic performance according as peaks of spectral density occur at the origin or at some other frequencies.

For the purpose of illustration of Theorem 2.1, Section 3 considers the standardized sample autocovariance for the scalar process $z(t) = \sum_{j=1}^{\infty} j^{-3/4} e(t-j)$ for which the autocovariance $\gamma(r) = E(z(t)z(t+r))$ is of the form $\gamma(r) = r^{-1/2}L(r)$ for a slowly varing function $L(r)$. For such a case where the spectral density is not square-integrable Breuer and Major [2] gave the central limit theorem for sample autocovariances of Gaussian processes, as a part of their results. Fox and Taqqu [3] and Giraitis and Surgailis [4] gave the central limit theorem for general quadratic forms for Gaussian and non-Gaussian processes, respectively, but as far as autocovariance is concerned, their assumption reduces to that a spectral density of the process $\{z(t)\}$ is square-integrable.

Since Yajima [5] initially showed the central limit theorem for the Whittle estimate of the parameters of spectral density of a scalar-valued Gaussian process with strong dependence, there have been produced many related results aiming at extension [see Fox and Taqqu [6], Dahlhaus [7] and Giraitis and Surgailis [4] for instance]. For the statistical estimation of parameters in spectral density and for its asymptotic theory, Section 4 proposes an approach given in Hosoya [8] which is based on the methods explored respectively by Daniels [9] , Huber [10] and Pollard [11] to deal with non-regular statistical models. The necessity for use of their version of asymptotic theory is due to the fact the the spectral density which has unbounded peaks at unknown frequencies constitutes a non-regular statisitcal model. The section disuesses a possible asymptotic theory which contains such models.

Except for Section 3 which deals with a specific case in some detail in order to justify the assumptions of Section 2, this paper is mainly a sketch of an approach to the limit theorems for stationary processes with strong dependence. For rigorous proofs, see Hosoya [12] and a forthcoming paper by the auther. Throughout the paper, J denotes the set of all integers; $\delta(\cdot,\cdot)$ indicates a function such that $\delta(x,y) = 1$ if $x = y$ and $\delta(x,y) = 0$ otherwise. The conjugate transpose of a matrix A is denoted A^*, and I is the indicator function.

1. Central limit theorems for stationary processes.

Let $\{z(t); t \in J\}$ be a vecter-valued linear process generated by

$$z(t) = \sum_{j=0}^{\infty} G(j)e(t-j), \quad t \in J \tag{1.1}$$

where the $z(t)$'s have q components and the $e(t)$'s are p-vectors such that $E\{e(m), e(n)^*\}$ $= \delta(m,n)K$ for K a nonsingular p by p matrix; the matrices $G(j)$ are q by p and the components of z, e and G are all real. Assume thoughout that

$$\sum_{j=0}^{\infty} trG(j)KG(j)^* < \infty$$

so that the process $\{z(t)\}$ is a second order stationary process and has a spectral density matrix $f(\omega)$ representable as

$$f(\omega) = \frac{1}{2\pi} k(\omega) K k(\omega)^*, \quad -\pi \leq \omega \leq \pi,$$

where $k(\omega) = \sum_{j=0}^{\infty} G(j) e^{i\omega j}$. Denote by the (α, β) component of $G(j)$ by $G_{\alpha\beta}(j)$ and α-th component of $z(t)$ and $e(t)$ by $z_\alpha(t)$ and $e_\alpha(t)$.
Let $K_n(\omega)$ be the Fejér kenel. For an integrable possibly matrix-valued function g define $L_n(g)$ by

$$L_n(g) = n \int_{-\pi}^{\pi} K_n(\omega) g(\omega) d\omega$$

and define $\phi_\alpha(n)$ by $L_n(f_{\alpha\alpha}) = n\phi_\alpha(n)$ where $f_{\alpha\alpha}$ is the (α, α) component of the spectral density of f. Let D_n be the s by s diagonal matrix with $L_n(f_{\alpha\alpha})^{1/2}$ as the (α, α)th element. Let $\{m(n)\}$ be a sequence of positive integers. Set $k^{(2)}(\omega) = \sum_{j=m_2(n)+1}^{\infty} G(j) e^{i\omega j}$.
Assumption 1.1. There exists a choice of $m_1(n)$ and $m_2(n)$ such that

$$(i) \quad \lim_{n\to\infty} m_1(n) = \lim_{n\to\infty} m_2(n) = \lim_{n\to\infty} \frac{n}{m_1(n) + m_2(n)} = \infty,$$

$$\lim_{n\to\infty} m_1(n)/m_2(n) = \infty,$$

$$(ii) \quad \lim_{n\to\infty} D_n^{-1} L_{n'}(k^{(2)} k^{(2)*}) D_n^{-1} = 0,$$

$$(iii) \quad \lim_{n\to\infty} \phi_\alpha(m_1(n))/\phi_\alpha(n) = 1.$$

Assumption 1.2. $D_n^{-1} L_n(kKk^*) D_n^{-1}$ tends to a positive-definite matrix Ω as n tends to infinity.

Let $\mathcal{F}(t)$ be the ω-field generated by $\{e_\alpha(s);\ 1 \leq \alpha \leq p, s \leq t\}$; then,
Assumption 1.3. For any $t_1, t_2 > t$, there is $\delta > 0$ such that uniformly in t and α, β.

$$(i) \quad E \mid E\{e_\alpha(t_1) \mid \mathcal{F}(t)\} E\{e_\beta(t_2) \mid \mathcal{F}(t)\} \mid = O(\mid (t_1 - t)(t_2 - t) \mid^{-(1+\delta)}) \text{ and}$$

$$(ii) \quad E \mid E\{e_\alpha(t_1) e_\beta(t_2) \mid \mathcal{F}(t)\} - E(e_\alpha(t_1) e_\beta(t_2)) \mid = O(\mid (t_1 - t)(t_2 - t) \mid^{-(1+\delta)}).$$

Set $S_{n\alpha}(r) = \sum_{t=1}^{n} z_\alpha(t + r)/L_n(f_{\alpha\alpha})^{1/2}$ and let $S_n(r)$ be the q-vector with components $S_{n\alpha}(r)$.
Theorem 1.1. *Suppose that Assumptions 1.1-1.3 hold and suppose that for any $\varepsilon > 0$, there exists $M_\varepsilon > 0$ such that for all sufficiently large n,*

$$E\{S_n(r)^* S_n(r) I(S_n(r)^* S_n(r) > M_\varepsilon)\} < \varepsilon \tag{1.2}$$

uniformly in r. Then $S_n(0)$ has the asymptotically normal distribution with mean vector 0 and covariance matrix Ω.

Example 1.1. Let $\{z(t);\ t \in J\}$ be a scaler-valued linear process which has the representation

$$z(t) = \sum_{j=1}^{\infty} \frac{1}{j}\, e\,(t - j),$$

where J is the set of all integers, $\{e(t); t \in J\}$ is a white noise process such that $E\{e(t)\} = 0$ and $Var\{e(t)\} = 1$. Set $S_n(r) = \sum_{t=1+r}^{n+r} z(t)$ and $z''(t) = \sum_{j=m(n)+1}^{\infty} \frac{1}{j} e(t-j)$ where $\{m(n)\}$ is a sequence of positive integers such that $m(n) \leq n$. Set $L_n = Var\{S_n(0)\}$.

Then it holds that

$$L_n = (\log n)^2 n(1 + o(n))$$

and

$$Var\{\sum_{t=1}^{n} z''(t)\}/L_n = O(\{\log(n/m(n))/\log n\}^2). \tag{1.3}$$

The proof is gives as this. The variance $Var\{S_n(0)\}$ has the expression:

$$\begin{aligned} L_n &= \sum_{j=1}^{\infty} \frac{1}{j^2} + 2 \sum_{l=1}^{n-1} \sum_{j=1}^{\infty} (n-l) \frac{1}{j(j+l)} \\ &= 2n\{\sum_{l=1}^{n-1} \sum_{j=1}^{\infty} \frac{1}{j(j+l)}\}(1 + o(1)), \end{aligned}$$

where

$$\sum_{j=1}^{\infty} \frac{1}{j(j+l)} = \frac{1}{l}(1 + \frac{1}{2} + \cdots + \frac{1}{l}) = \frac{\log l}{l}(1 + o(1)).$$

Therefore the first result follows since

$$\sum_{l=1}^{n-1} \frac{\log l}{l} = \frac{1}{2}(\log n)^2 (1 + o(1)).$$

On the other hand,

$$Var\{\sum_{t=1}^{n} z''(t)\} = n \sum_{j=m(n)+1}^{\infty} \frac{1}{j^2} + 2 \sum_{l=1}^{n-1} \sum_{j=m(n)+1}^{\infty} (n-l) \frac{1}{j(j+l)}$$

where

$$\begin{aligned} \sum_{j=m(n)+1}^{\infty} \frac{1}{j(j+l)} &= \frac{1}{l}\{\log(m(n)+l) - \log(m(n)+1)\}(1 + o(1)) \\ &= \frac{1}{l} \log\{1 + \frac{l}{m(n)}\}(1 + o(1)). \end{aligned}$$

But

$$\begin{aligned} \sum_{l=1}^{n-1} \frac{1}{l} \log\{1 + \frac{l}{m(n)}\} &= c_1 \int_0^{n/m(n)} \frac{1}{x} \log(1+x) dx \\ &\leq c_2 \{\log(\frac{n}{m(n)})\}^2 (1 + o(1)), \end{aligned}$$

whence (1.3) follows. The left-hand member of (1.3) tends to 0 as $n \to \infty$ if $m_2(n) = n(\log n)^{-1}$ for example and then Assumption 1.1 (ii) is satisfied. Also if $m_1(n) = n(\log n)^{-1+\varepsilon} (0 < \varepsilon < 1)$, then Assumption 1.1 (i) and (iii) are satisfied.

Example 1.2. Let $\{z_1(t), z_2(t)\}$ be a bivariate stationary process such that

$$z_1(t) = \sum G_{11}(j) e_1(t-j) + \sum G_{12}(j) e_2(t-j)$$

$$z_2(t) = \sum G_{21}(j)e_1(t-j) + \sum G_{22}(j)e_2(t-j)$$

where $\{e_1(t)\}$ and $\{e_2(t)\}$ are white noise processes with variance $(e_1(t)) = Var(e_2(t)) = 1$, $Cov(e_1(t), e_2(t)) = 0$.
Let $h(\omega) = \sum_{j=1}^{\infty} \frac{1}{j} e^{i\omega j}$ and suppose that

$$G_{ij}(\omega) = h(\omega) g_{ij}(\omega) \quad (i, j = 1, 2)$$

where the $g_{ij}(\omega)$ are bounded on $(-\pi, \pi]$, contiruous at $\omega = 0$ and $g_{11}(0) = g_{22}(0)$ are non-zero. Then

$$\lim_{n\to\infty} \frac{L_n(f_{12})}{(L_n(f_{11})L_n(f_{22}))^{12}} = \frac{Re[\{g_{11}(0) + g_{12}(0)\}\{g_{21}(0) + g_{22}(0)\}]}{\mid g_{11}(0) + g_{12}(0) \mid\mid g_{21}(0) + g_{22}(0) \mid}.$$

Corollary 1.1. If Assumption 1.3 and (1.2) hold and if the density $f(\omega)$ is continuous at $\omega = 0$ and $f(0)$ is a nondegenerate matrix, then $S(0)$ has the asymptotically normal distribution with mean 0 and covariance matrix Ω , where $\Omega = D^{-1} f(0) D^{-1}$ for the diagonal matrix $D = diag\{f_{\alpha\alpha}(0)^{1/2}; \alpha = 1, \cdots, q\}$.

2. Limit theorems for serial covariances.

For the process $\{z(t)\}$ given in (1.1), assume moreover that $\{e(t)\}$ is fourth-order stationary and that

$$\sum_{t_1,t_2,t3=-\infty}^{\infty} \mid Q^e_{\alpha_1,\cdots,\alpha_4}(t_1,t_2,t_3) \mid < \infty, \quad (1 \leq \alpha_1, \cdots, \alpha_4 \leq p) \tag{2.1}$$

for $Q^e_{\alpha_1,\cdots,\alpha_4}$ the joint fourth cumulant of $e_{\alpha_1}(t), e_{\alpha_2}(t+t_1), e_{\alpha_3}(t+t_2), e_{\alpha_4}(t+t_3)$, so that the procss $\{e(t)\}$ has a fourth-order spectral density $\tilde{Q}^e_{\alpha_1,\cdots,\alpha 4}(\omega_1,\omega_2,\omega_3)$ such that

$$\tilde{Q}^e_{\alpha_1,\cdots,\alpha 4}(\omega_1,\omega_2,\omega_3) = \frac{1}{(2\pi)^3} \sum_{t_1,t_2,t_3=-\infty}^{\infty} exp\{-i(\omega_1 t_1 + \omega_2 t_2 + \omega_3 t_3)\} Q^e_{\alpha_1,\cdots,\alpha_4}(t_1,t_2,t_3).$$

Assumption 2.1. For any $t_1, t_2, t_3, t_4 > t$ and for some $\delta_2 > 0$,

$$(i) E|E\{e_\alpha(t_1)e_\beta(t_2)|\mathcal{F}(t)\}E\{e_\alpha(_3)e_\beta(t_4)|\mathcal{F}(t)\} - E\{e_\alpha(t_1)e_\beta(t_2)\}E\{e_\alpha(_3)e_\beta(t_4)\}|$$
$$= O(\Pi_{j=1}^4 (t_j - t)^{-(1+\delta_2)})$$

uniformly in t.

$(ii) E|E\{e_\alpha(t_1)e_\beta(t_2)e_\gamma(_3)e_\delta(t_4)|\mathcal{F}(t)\} - E\{e_\alpha(t_1)e_\beta(t_2)e_\gamma(_3)e_\delta(t_4)\}| = O(\Pi_{j=1}^4 (t_j - t)^{-(1+\delta_2)})$

uniformly in t.

Set

$$\varphi^{r_1 r_2}_{\alpha_1\alpha_2\alpha_3\alpha_4}(\omega) = 2\pi \int_{-\pi}^{\pi} [f_{\alpha_1\alpha_3}(\omega - \omega_2)\overline{f_{\alpha_2\alpha_4}(\omega_2)} exp\{-ir_2\omega_2 + ir_1(\omega - \omega_2)\}$$

$$+f_{\alpha_1\alpha_4}(\omega-\omega_2)\overline{f_{\alpha_2\alpha_3}(\omega_2)}exp\{ir_1(\omega-\omega_2)+ir_2\omega_2\}]d\omega_2$$

$$+2\pi\sum_{\beta_1\cdots\beta_4=1}^{s}\int_{-\pi}^{\pi}\int_{-\pi}^{\pi}exp(ir_1\omega_1+ir_2(\omega-\omega_2))k_{\alpha_1\beta_1}(\omega_1+\omega)k_{\alpha_3\beta_3}(-\omega_2)$$

$$\cdot k_{\alpha_2\beta_2}(-\omega_1)k_{\alpha_3\beta_3}(\omega-\omega_2)(\tilde{Q}^e_{\beta_1\cdots\beta_4}(\omega_1+\omega,\omega_2,\omega-\omega_2)d\omega_1 d\omega_2 \qquad (2.2)$$

and set $T^m_{\alpha_1\alpha_2}(n,r)=\sum_{t=1+r}^{n+r-|m|}\{z_{\alpha_1}(t)z_{\alpha_2}(t+m)-E(z_{\alpha_1}(t)z_{\alpha_2}(t+m))\}$. Then in view of Hosoya and Taniguchi [13],

$$Cov\{T^{r_1}_{\alpha_1\alpha_2}(n,r),T^{r_2}_{\alpha_3\alpha_4}(n,r)\}=L_n(\varphi^{r_1r_2}_{\alpha_1\alpha_2\alpha_3\alpha_4}).$$

Set $N_\alpha(n)=L_n(\varphi^{00}_{\alpha\alpha\alpha\alpha})$, and let $z''_\alpha(t)=\sum_{j=m_2(n)+1}^{\infty}\sum_\beta G_{\alpha\beta}(j)e_\beta(t-j)$.

Assumption 2.2. There exists a choice of $m_1(n)$ and $m_2(n)$ for which

(i) $\lim_{n\to\infty}E\{(\sum_{t=1}^{n}z''_\alpha(t)^2)^2\}/N_\alpha(n)=0$

(ii) $\lim_{n\to\infty}(n_1m_1)^{1/2}/m_2^{1+2\delta_2}=\lim_{n\to\infty}m_1/m_2^{1+2\delta}=0$

(iii) $g_\alpha(n)=N_\alpha(n)/n$ is bounded away from 0 and $\lim_{n\to\infty}g_\alpha(m_1(n))/g_\alpha(n)=1$

where $n_1=[n/(m_1(n)+m_2(n))]$.

Assumption 2.3. $L_n(\varphi^{r_1r_2}_{\alpha_1\alpha_2\alpha_3\alpha_4})/\{\Pi_{j=1}^4N_{\alpha_j}(n)\}^{1/4}$ tends to $\Phi^{r_1r_2}_{\alpha_1\alpha_2\alpha_3\alpha_4}$ as $n\to\infty$ for each fixed r_1,r_2 and $\alpha_1,\alpha_2,\alpha_3,\alpha_4$ such that $1\le\alpha_j\le q$.

For each fixed $r\ge 0$ and $L\ge 0$, let $T_n(r)$ be the vector whose components are ordered $T^m_{\alpha_1\alpha_2}(n,r)/\{N_\alpha(n)N_\beta(n)\}^{1/4}$ where α_1,α_2 and m vary on $1\le\alpha_1\le\alpha_2\le q$ and $0\le m\le L$.

Theorem 2.1. *Suppose that Assumptions 2.1 to 2.3 are satisfied and suppose that the Lindeberg type condition (1.2) holds for $T_n(r)$. Then $T_n(0)$ has the limit normal distribution with mean 0 and the covariances $\Phi^{r_1r_2}_{\alpha_1\alpha_2\alpha_3\alpha_4}$ $(1\le\alpha_i\le q,$ and $0\le r_1\le r_2\le L)$.*

Condition A. A cross spectral density $f_{\alpha\beta}(\omega)$ has the property

$$\int_{(-\varepsilon,\varepsilon)}|f_{\alpha\beta}(\omega)|^2d\omega=\infty \quad\text{and}\quad \int_{(-\pi,-\varepsilon)\cup(\varepsilon,\pi)}|f_{\alpha\beta}(\omega)|^2d\omega<\infty$$

for any $\varepsilon>0$.

Condition B. A spectral density $f_{\alpha\alpha}(\omega)$ has the unbounded peak at the origin and is bounded a.e. for $\omega\in[-\pi,-\varepsilon)\cup(\varepsilon,\pi)$ for each ε and $\int_{-\pi}^{\pi}|f_{\alpha\alpha}(\omega)|^2d\omega=\infty$.

Corollary 2.1. (i) If $f_{\alpha\alpha}(\omega)$ satisfies Condition A and if $f_{\beta\beta}(\omega)$ satisfies Condition B, or (ii) if $f_{\alpha\alpha}(\omega)$ is square-integrable and if $f_{\beta\beta}(\omega)$ satisfies Condition A, then the joint distribution of $T^0_{\alpha\beta}(0)$ and $T^r_{\alpha\beta}(0)$ asymptotically degenerates.

The next corollary is a comequence of Theorem 2.2 of Hosoya-Taniguchi [13].

Corollary 2.2. If $f_{\alpha_i\alpha_i}(\omega), i=1,\cdots,4$, are square-integrable,

$$\Phi^{r_1r_2}_{\alpha_1\alpha_2\alpha_3\alpha_4} = \varphi^{r_1r_2}_{\alpha_1\alpha_2\alpha_3\alpha_4}/\{\Pi^4_{i=1}\varphi_{\alpha_j}(0)\}^{1/4}$$

where $\varphi_{\alpha_j}(0) = \varphi^{00}_{\alpha_j\alpha_j\alpha_j\alpha_j}(0)$.

Remark 2.1. It should be noted that Corollary 2.1 does not necessarily hold if Condition A is violated. For instance, if $\int_{-\pi}^{\pi} f_{11}(\omega)^2 d\omega = \infty$ but if the unbounded peaks appear away from the origin, then under the assumptions of the theorem, $T^0_{11}(n,0)$ and $T^m_{11}(n,0)(m \neq 0)$ have not the same limit distribution. Let $\{x_t, t \in J\}$ be a scalar-valued fourth-order stationary process $x_t = \sum_{j=0}^{\infty} \beta_j e(t-j)$ with spectral density such that for nonnegative ω

$$\begin{aligned} f(\omega) &= c|\omega-\omega_0|^{-\alpha} && \omega_0-\varepsilon \le \omega \le \omega_0+\varepsilon \\ &= m && \text{otherwise} \end{aligned}$$

where $1/2 < \alpha < 1, \omega_0 \geq 0$ and c is sufficiently large and m sufficiently small so that the mass of $g(\omega_1) = f(\omega-\omega_1)f(\omega_1)$ for small $|\omega|$ is mostly concentrated on $\pm\omega_0 - \varepsilon \le \omega_1 \le \pm\omega_0 + \varepsilon$. The process $\{e_t\}$ is assumed to satisfy the fourth-order condition (2.1). Set $y_t = x_t - x_{t+r}$; then since the joint spectral density of the process $\{x_t, y_t\}$ is given by

$$\begin{bmatrix} f(\omega) & (1-e^{-ir\omega})f(\omega) \\ (1-e^{ir\omega})f(\omega) & |1-e^{ir\omega}|^2 f(\omega) \end{bmatrix}$$

it follows that

$$E\{\sum_{t=1}^{n}(x_t y_t - E(x_t y_t))\}^2 = n\int_{-\pi}^{\pi} K_n(\omega)\varphi(\omega)d\omega + Q_n$$

where

$$\begin{aligned} \varphi(\omega) = 2\pi\int_{-\pi}^{\pi} &[f(\omega-\omega_1)|1-e^{ir\omega}|^2 f(\omega_1) \\ &+(1-e^{-ir(\omega-\omega_1)})f(\omega-\omega_1)(1-e^{ir\omega_1}f(\omega_1)]d\omega_1 \end{aligned}$$

and

$$\begin{aligned} Q_n = 2\pi n\int_{-\pi}^{\pi}\cdots\int_{-\pi}^{\pi} &K_n(\omega_2+\omega_3)k_1(\omega_1+\omega_2+\omega_3)\cdot k_2(-\omega_1) \\ &\cdot k_1(-\omega_2)k_2(-\omega_3)\tilde{Q}^e(\omega_1+\omega_2+\omega_3,\omega_2,\omega_3)d\omega_1 d\omega_2 d\omega_3, \end{aligned}$$

for $k_1(\omega) = \sum\beta_j e^{-ij\omega}$ and $k_2 = (1-e^{ir\omega})\sum\beta_j e^{-ij\omega}$.
Then if $|\omega|$ is sufficiently small,

$$\varphi(\omega) \approx 2\pi(8sin^2(\omega_0 r/2))\int_{-\pi}^{\pi} f(\omega-\omega_1)f(\omega_1)d\omega_1,$$

and $Q_n = O(n)$ in view of the condition (2.1). Similarly,

$$E\{\sum(x_t^2 - E(x_t^2))\}^2 = n\int_{-\pi}^{\pi} K_n(\omega)\varphi_1(\omega)d\omega + Q'_n$$

where $\varphi_1(\omega) = 2\pi \int_{-\pi}^{\pi} 2f(\omega - \omega_1)f(\omega_1)d\omega_1$ and $Q_n' = O(n)$.
Consequently,

$$\lim_{n\to\infty} E\{\sum\{x_t^2 - E(x_t^2)\} - \sum\{x_{t+r}x_t - E(x_{t+r}x_t)\}]^2 / E[\{x_t^2 - E(x_t^2)\}^2]$$

$$\approx 4sin^2(\omega_0 r/2). \tag{2.3}$$

The relationship (A.1) shows that the cases $\omega_0 = 0$ and $\omega_0 > 0$ respectively give examples where Corollary 2.1 does and does not hold.

Remark 2.2. If the process $\{z(t)\}$ in Theorem 1.1 is fourth-order stationary process whose innovation process $\{e(t)\}$ satisfies the condition (2.1), then the Lindeberg type condition (1.2) is not needed for that theorem to hold, since then for each α there exists $M > 0$ such that

$$E\{S_{n\alpha}(0)\}^4 = E\{\sum_{t=1}^{n} z_\alpha(t)\}^4 / L_n(f_{\alpha\alpha})^2 < M$$

and consequently in view of fourth-order stationarily the Liapounov type condition

$$E\{S_n(r)^* S_n(r)\}^2 < \infty \tag{2.4}$$

holds for any r and (1.2) follows from it.

Remark 2.3. The question how to check the validity of the assumptions 1.3 and 2.1 for given finite observation of $\{z(t)\}$ seems to raise two important problems which are yet to be explored. If the process $\{e(t)\}$ itself in those assumptions is observable, the conditional moments involved are able to be estimated either by the kernel method or by the near-neighborhood method (see Robinson [14] for example) and the moments in those assumptions are estimated, and pertinently standardized estimated moments would be used for test statistics. Since in general no direct observation of $\{e(t)\}$ is available, some estimation of it from observed $z(t)$ is needed. But there is yet not known a suitable general estimation method of the innovation process $\{e(t)\}$ for a strongly dependent process, in contract to weakly dependent processes such as ARMA processes. Even if an estimate of the sequence $\{e(t)\}$ is available, the problem of how to modify the asymptotics of the above nonparametric estimation of conditional moments and the involved moments in Assumption 1.3 and 2.1. is not so straightforward as in the case of weakly dependent processes. This is the one problem. In stead of the direct inference in the moment conditions $\{e(t)\}$, the validity might be checked indirectly by the induced moment conditions of the obsevable process $\{z(t)\}$. Then there arises the second problem of asymptotics of involved estimated conditional moments under strong dependency. Nonparametric estimation of time-series with strong dependence is the problem which needs to be explored.

3. An example.

This section considers a scalar-valued prpocess whose spectral density is not square-integrable and gives an outline of the proof for the central limit theorem 2.1 for the autocovariances. Suppose that the stationary process $\{z(t), t \in J\}$ is generated by $z(t) = \sum_{j=1}^{\infty} \frac{1}{j^{3/4}} e(t-j)$ where the process $\{e(t)\}$ is fourth-order stationary and satisfies that $E(e(t)) = 0, Cov(e(t), e(s)) = \delta(t,s)$. Assume that $\{e(t)\}$ satisfies Assumption 2.1. Let $Q^e(t_1, t_2, t_3)$ be the fourth-order cumulant of $e(t), e(t+t_1), e(t+t_2), e(t+t_3)$ and assume as in the preceding section

$$\sum_{t_1,t_2,t_3=-\infty}^{\infty} |Q^e(t_1,t_2,t_3)| < \infty,$$

and so the process $\{e(t)\}$ has a fourth-order spectral density $\tilde{Q}^e(t_1, t_2, t_3)$ such that

$$\tilde{Q}^e(\omega_1,\omega_2,\omega_3) = \frac{1}{(2\pi)^3} \sum_{t_1,t_2,t_3=-\infty}^{\infty} exp\{-i(\omega_1 t_1 + \omega_2 t_2 + \omega_3 t_3)\} Q^e(t_1,t_2,t_3).$$

Set

$$T^r(n,s) = \sum_{t=1+s}^{n+s-|r|} \{z(t)z(t+r) - E(z(t)z(z+r))\},$$

and set $k(\omega) = \sum \frac{1}{j^{3/4}} e^{-i\omega j}$ and denote the spectral density of the process $\{z(t)\}$ by $f(\omega)$; namely $f(\omega) = \frac{1}{2\pi}|k(\omega)|^2$. Set $M_n = Var\{T_n(0,0)\}$; then it is expressed as

$$M_n = 2\sum_{t=-n+1}^{n-1} (n-|t|)\gamma(t)^2 + 2\pi n \int_{-\pi}^{\pi} \cdots \int_{-\pi}^{\pi} K_n(\omega_2+\omega_3) k(\omega_1+\omega_2+\omega_3)$$

$$\cdot k(-\omega_1)k(-\omega_2)k(-\omega_3)\tilde{Q}^e(\omega_1+\omega_2+\omega_3,\omega_2,\omega_3) d\omega_1 d\omega_2 d\omega_3$$

where $\gamma(t) = E\{z(0)z(t)\}$ and $K_n(\cdot)$ is the Féjer kernel (see Hosoya and Taniguchi [13]).
Lemma 3.1. *If $k_j(\omega), j = 1,2,3,4,$ are square integrable,*

$$2\pi \int_{-\pi}^{\pi} \cdots \int_{-\pi}^{\pi} K_n(\omega_2+\omega_3) k_1(\omega_1+\omega_2+\omega_3) k_2(-\omega_1) k_3(-\omega_2) k_4(-\omega_3)$$

$$\cdot \tilde{Q}^e(\omega_1+\omega_2+\omega_3,\omega_2,\omega_3) d\omega_1 d\omega_2 d\omega_3 = O(1).$$

Proof. In view of Hosoya-Taniguchi [13] p.145, the left-hand side integral above converges to

$$2\pi \int_{-\pi}^{\pi} \int_{-\pi}^{\pi} k_1(\omega_1) k_2(-\omega_1) k_3(\omega_2) k_4(-\omega_2) \tilde{Q}^e(\omega_1,-\omega_2,\omega_2) d\omega_1 d\omega_2$$

whereras the latter integral is finite since $\tilde{Q}^e$ is bounded. □

Hence $M_n = 2\sum_{t=-n+1}^{n-1}(n-|t|)\gamma(t)^2 + O(n)$. It also holds that:
Lemma 3.2. *There exists $c_1, c_2 > 0$ such that $c_1 n \log n \leq M_n \leq c_2 n \log n$. Also it holds that*

$$n \sum_{r=1}^{n} [\sum_{j=m}^{\infty} \frac{1}{\{j(j+r)\}^{3/4}}]^2 / M_n = O(\log(n/m)/\log n). \quad \textbf{(3.1)}$$

Proof. It follows from the theory of elliptic integration that

$$\frac{2}{\sqrt{r}}F\{\arcsin\sqrt{1-\sqrt{m/(m+r)}},1/\sqrt{2}\} \leq \sum_{j=m}^{\infty}\frac{1}{\{j(j+r)\}^{3/4}}$$

$$\leq \frac{c_5}{\sqrt{r}}F\{\arcsin\sqrt{1-\sqrt{m/(m+r)}},1/\sqrt{2}\} \quad (3.2)$$

where

$$F(\varphi,k)=\int_0^{\varphi}\frac{1}{\sqrt{1-k^2\sin^2\theta}}d\theta = c_3(k)\varphi - \frac{1}{2}c_4(k)\sin(2\pi); \quad (3.3)$$

whence the first proposition of the lemma follows by substitution $m=1$. On the other hand, it follows from (2.2) and the last inequality of (2.3) that

$$\sum_{r=1}^{n}\{\sum_{j=m}^{\infty}\frac{1}{\{j(j+r)\}^{3/4}}\}^2 \leq c_6^2\sum_{r=1}^{n}\frac{1}{r}\{\sqrt{1-\sqrt{\frac{m}{m+r}}}\}^2$$

$$\leq c_7\sum_{r=1}^{n}\frac{1}{m+r} = O\{\log\frac{n}{m}\}.$$

Since $M_n \geq c_1 n\log n$, the relationship (3.1) follows. □

Set $z'(t)=\sum_{j=1}^{m}\frac{1}{j^{3/4}}e(t-j)$ and set $z''(t)=z(t)-z'(t)$ for m $(m<n)$. Then

Lemma 3.3. $Var\{\sum_{t=1}^{n}(z(t)^2 - z'(t)^2)\}/M_n = O([\log(n/m)/\log n]^{1/2})$.

Choose in what follows $m_1(n)$ and $m_2(n)$ in such a way that $\lim_{n\to\infty} m_2(n)/m_1(n)=0$, and each of $\log(n/m_2(n))/\log n$, $m_1(n)n_1/(m_2(n)^{1+2\delta_2}\log m_1(n))$, and $(n_1 m_1(n))^{1/2}/\{m_2(n)^{1+2\delta}(\log m_1(n))^{1/2}\}$ tends to 0. For example $m_1(n)=n(\log\log n)^{-1}$,

$m_2(n) = n(\log n)^{-1}$ will do. Let I_k and J_k be the sets of integers and let $\mathcal{B}_k$ and $\mathcal{B}'_k$ be the σ-fields given as,

$$I_k=\{i;(k-1)(m_1(n)+m_2(n))+1\leq i\leq km_1(n)+(k-1)m_2(n)\}$$

$$J_k=\{j;km_1(n)+(k-1)m_2(n)+1\leq j\leq k(m_1(n)+m_2(n))\} \quad k=1,\cdots,n_1,$$

and set

$$\xi_k=\sum_{t\in I_k}z'(t)^2/M_{m_1(n)}^{1/2} \quad\text{and}\quad \xi'_k=\sum_{t\in J_k}z'(t)^2/M_{m_1(n)}^{1/2}.$$

By the choice of $m_1(n)$ and $m_2(n)$ above, the following lemma holds.

Lemma 3.4.

(*i*) $E|\sum_{k=1}^{n_1}(E(\xi_k|\mathcal{B}'_{k-1})-E(\xi_k))|/n_1^{1/2}\to 0,$

(*ii*) $\sum_{k=1}^{n_1}E\{E(\xi_k|\mathcal{B}_{k-1})-E(\xi_k)\}^2/n_1$ *tends to 0 as* $n\to\infty,$

$$(iii) \quad \lim_{n\to\infty} E\{\sum_{k=1}^{n_1}(\xi_k - E(\xi_k|\mathcal{B}_{k-1}))^2/n_1\} = 1,$$

$$(iv) \quad \lim_{n\to\infty} E\sum_{k=1}^{n_1} |[E\{(\xi_k - E(\xi_k|\mathcal{B}_{k-1}))^2|\mathcal{B}_{k-1}\} - E\{\xi_k - E(\xi_k|\mathcal{B}_{k-1})\}^2]|/n_1 = 0,$$

$$(v) \quad \lim_{n\to\infty} E\sum_{k=1}^{n_1} |E\{(\xi_k - E(\xi_k))^2|\mathcal{B}_{k-1}\} - E\{\xi_k - E(\xi_k)\}^2|/n_1 = 0.$$

The next lemma is a version of the Martingale central limit theorem given by Brown [15].

Lemma 3.5. *Let $\{\mathcal{F}_n(k), k = 1, 2, \cdots\}$ be a filtration of σ-fields for each $n = 1, 2, \cdots$ and suppose that $\{u_n(k), \mathcal{F}_n(k), k = 1, \cdots, m(n)\}$ is a sequence Martingale differences. Suppose that*

$$\lim_{i\to\infty} \frac{1}{m(n)} \sum_{k=1}^{m(n)} E[u_n(k)^2 I\{|u_n(k)| \geq \varepsilon m(n)^{1/2}\}] = 0 \tag{3.4}$$

for any $\varepsilon > 0$ and $\{m(n)\}$ tends to infinity as $n \to \infty$. Also suppose that

$$\sum_{k=1}^{m(n)} E\{u_n(k)^2|\mathcal{F}_n(k-1)\} / \sum_{k=1}^{m(n)} E\{u_n(k)^2\} \to 1 \tag{3.5}$$

in probability as $n \to \infty$. Then $\sum_{k=1}^{m(n)} u_n(k)/[\sum_{k=1}^{m(n)} E\{u_n(k)^2\}]^{1/2}$ is asymptotically normally distributed with mean 0 and variance 1.

Theorem 3.1. *If for any $\varepsilon > 0$ there exsits N_ε such that*

$$E[\frac{T^0(n,s)^2}{M_n} I\{\frac{T^0(n,s)}{{M_n}^{1/2}} > N_\varepsilon\}] < \varepsilon \tag{3.6}$$

uniformly in s and n and if Assumptions 2.1 and 2.2 hold, $T_n(0,0)/{M_n}^{1/2}$ has the asymptotically nornal distribution with mean 0 and variance 1.

Now Lemmas 3.3 and 3.4 (i) imply that $T^0(n,0)/M_n^{1/2}$ has the same limit distribution as $\sum_{k=1}^{n_1}\{\xi_k - E(\xi_k|\mathcal{B}_{k-1})\}/{n_1}^{1/2}$, since $\lim_{n\to\infty} n_1 M_{m_1(n)}/M_n = 1$. So what is necessary to prove is that the Martingale difference process $\{\xi_k - E(\xi_k|\mathcal{B}_{k-1}), \mathcal{B}_k\}$ satisfies the conditions of Lemma 3.5. Lemma 3.4 (iv) and (v) imply that (3.5) is satisfied for $u_n(k) = \xi_k - E(\xi_k|\mathcal{B}_{k-1})$. Also Lemma 3.4 and (3.6) below imply that the Lindeberg type condition (3.4) is satisfied.

Theorem 3.2. *For any $r > 0$,*

$$\lim_{n\to\infty} E[\sum_{t=1}^{n} z(t)(z(t) - z(t+r)) - E\{\sum_{t=1}^{n} z(t)(z(t) - z(t+r))\}]^2/M_n = 0;$$

as a consequence, the join distribution of $T^0(n,0)/M_n^{1/2}$ and $T^r(n,0)/M_n^{1/2}$ asymptotically degenerates and $T^r(n,0)/M_n^{1/2}$ has the same limit distribution as $T^0(n,0)/M_n^{1/2}$.

4. Statistical inference.

Let $\{z(t), t \in J\}$ be a stationary process given by (1.1) and with a spectral density matrix $f(\omega)$ and suppose that statistical inference such as estimation or testing is conducted on a parametric spectral model $f_\theta(\omega), \theta \in \Theta \subset R^s$; that is, the parameter space is an open subset of a Euclidean s-space. In order to fit a strong dependence model, $f_\theta(\omega)$ would possibly have unbounded peaks for some ω.

As a general estimation procedure of the parameter θ, Hosoya and Taniguchi [13] proposed the use of the criterion

$$D(f_\theta, I_n) = \int_{-\pi}^{\pi} [\log det f_\theta(\omega) + tr\{f_\theta(\omega)^{-1} I_n(\omega)\}] d\omega$$

where $I_n(\omega)$ is the periodogram matrix defined by

$$I_n(\omega) = \frac{1}{2\pi n}\{\sum_{t=1}^{n} z(t)e^{i\omega t}\}\{\sum_{t=1}^{n} z(t)e^{i\omega t}\}^*,$$

and proposed, as the estimate of θ, the use of call a minimizing value of $D(f_\theta, I_n)$ which is termed a quasi-maximum likelihood (ML) estimate $\tilde{\theta}$ of θ. Furthermore Hosoya [8] proposed the use of that estimate for testing purpose. In any case, the performance of the quasi-M.L. estimate is to be determined. In such a situation as $f(\omega)$ is square-integrable, Hosoya and Taniguchi gave the central limit theorem for $\tilde{\theta}$ of regular f_θ and Hosoya [8] of non-regular f_θ, respectively, assuming some mixing conditions for conditional moments of the innovation process $\{e(t)\}$, without the assumption of Gaussianity. Therefore, even when $\{z(t)\}$ has strong dependence, as long as $f(\omega)$ is square-integrable, there arises no new problem in view of Theorem 2.1 except that some mild modification of Lemma A3.3 of Hosoya and Taniguchi [13] and the proof of Thorem 1.2 in Hosoya [8] is needed.

It should be noted that if the model spectral density $f_\theta(\omega)$ has unbounded peaks at some unknown frequencies, then it is non-regular and needs a pertinent approach to deal with it. For be specific, suppose for example that the process $\{z(t)\}$ is scaler-valued and a model for the spectral density is specified as

$$f_\theta(\omega) = |\omega - \theta_1|^{-\alpha} g(\omega, \theta_2)$$

where $g(\omega, \theta_2)$ is a regular positive function with θ_2 possibly vector-valued $0 < \alpha < 1$ and $\theta = (\theta_1, \theta_2, \alpha)$. If θ_1 is known, $f_\theta(\omega)$ is then regular and for such a case there are established a variety of central limit theorems for the Wittle estimate (see Yajima [16] for the references), although results so far known seem to be concentrated on the case $\theta_1 = 0$, and the true density $f(\omega)$ has also peaks at $\omega = 0$. If $f(\omega)$ has peaks other than at the origin and satisfies the conditions of Theorem 2.1, then Theorem 2.1 implies that a slightly different treatment is necessary.

In general, for vector-valued processes, the situation where $f(\omega)$ is square integrable could be dealt with by Hosoya's approach even if θ_1 is unknown. The remaining problem is the central limit theorem of the quasi ML estimate $\tilde{\theta}$ in the case where some elements of $f(\omega)$ are not square integrable and still satisfies the conditions of Theorem 3.1. As was seen in that theorem, the central limit theorem for serial covariances still holds for such a case, but for the limit theorem for $\tilde{\theta}$, some more results are necessary in order to fill the gap so that the central limit theorem for serial covariances is transferred to that of $\tilde{\theta}$. The author's forthcoming paper will deal with this problem.

Acknowledgments.
This research was partially supported by the Ministry of Education of Japan Grant No. 04630012. The author would like to thank a referee for his careful reading and many helpful comments.

References

1. Y. YAJIMA, A central limit theorem of Fourier transforms of strongly dependent stationary processes. *J. Time Series Anal.*, 1989, **10**, 375-83.
2. P. BREUER and P. MAJOR, Central limit theorems for non-linear functionals of Gaussian fields. *J. Multivariate Anal.*, 1983, **13**, 425-441.
3. R. FOX and M. S. TAQQU, Central limit theorems for quadratic forms in random variables having long-range dependence. *Probab. Th. Rel. Fields*, 1982, **74**, 213-40.
4. L. GIRAITIS and D. SURGAILIS, A Central limit theorem for quadratic forms in strongly dependent linear variables and its application to asymptotical normality of Whittle's estimate. *Probab. Th. Rel. Fields*, 1990, **86**, 87-104.
5. Y. YAJIMA, On estimation of long-memory time series models. *Austral. J. Statist.*, 1985, **27**, 303-320.
6. R. FOX and M. S. TAQQU, Large sample properties of parameter estimates for strongly dependent stationary Gaussian time series. *Ann. Statist.*, 1986, **14**, 517-532.
7. R. DAHLHAUS, Efficient parameter estimation for self-similar processes. *Ann. Stat.*, 1989, **17**, 1749-1766.
8. Y. HOSOYA, The bracketing condition for limit theorems on stationary linear processes. *Ann. Statist.*, 1989, **17**, 401-18.
9. H. E. DANIELS, The asymptotic efficiency of a maximum likelihood estimator. *Proc. Fourth Berkeley Symp. Math. Statist. Probab.*, 1961, **1**, 151-163, Univ. California press.
10. P. HUBER, The behavior of maximum-likelihood estimates under nonstandard conditions. *Proc. Fifth Berkeley Symp. Math. Statist. Probab.*, 1967, **1**, 221-231, Univ. California Press.
11. D. POLLARD, New ways to prove central limit theorems. *Econometric Theory*, 1985, **1**, 295-314.
12. Y. HOSOYA, Some limit theorem for stationary processes with strong dependence. *Tohoku. Univ. Faculty of Econo. Disccusion Paper, No.100*, 1992.
13. Y. HOSOYA and M. TANIGUCHI, A central limit theorem for stationary processes and the parameter estimation of linear processes. *Ann. Statist.*, 1982, **10**, 193-53.
14. P. M. ROBINSON, Nonparametric estimators for time series. *J. Time Series Anal.*, 1983, **4**, 185 -207.
15. B. M. BROWN, Martingale central limit theorems. *Ann. Math. Statist.*, 1971, **42**, 59-66.
16. Y. YAJIMA, Asymptotic properties of estimates in incorrect ARMA models for long-memory time series, to appear, 1990.

Stat. Sci. & Data Anal., pp. 165-174
K. Matsusita *et al.* (Eds)

Two Sample Problem in Time Series Analysis

MASAO KONDO and MASANOBU TANIGUCHI
Department of Statistics, College of Liberal Arts, Kagoshima University, Kagoshima 890, Japan
Department of Mathematical Science, Faculty of Engineering Science, Osaka University, Toyonaka 560, Japan

Abstract. Suppose that $\{X_t\}$ and $\{Y_t\}$ are Gaussian stationary processes with the spectral densities $f(\lambda)$ and $g(\lambda)$, respectively. Here we consider the testing problem

$$H : \int_{-\pi}^{\pi} K\{f(\lambda)\}d\lambda = \int_{-\pi}^{\pi} K\{g(\lambda)\}d\lambda,$$

against

$$A : \int_{-\pi}^{\pi} K\{f(\lambda)\}d\lambda \neq \int_{-\pi}^{\pi} K\{g(\lambda)\}d\lambda,$$

where $K(\cdot)$ is an appropriate function. This setting of test is unexpectedly wide, and can be applied to many problems in time series. For this problem we propose a test based on $\int_{-\pi}^{\pi} K\{\hat{f}_n(\lambda)\}d\lambda$ and $\int_{-\pi}^{\pi} K\{\hat{g}_n(\lambda)\}d\lambda$ where $\hat{f}_n(\lambda)$ and $\hat{g}_n(\lambda)$ are nonparametric spectral estimators of $f(\lambda)$ and $g(\lambda)$,respectively, and evaluate the asymptotic power under a sequence of nonparametric contiguous alternatives. We compare the asymptotic power of our test with the other one, and show some good properties of ours.

Key words : Nonparametric hypothesis testing, two sample problem, Gaussian stationary process, spectral density, nonparametric spectral estimator, contiguous alternative, efficacy, asymptotic relative efficiency, Burg's entropy, spectral moment, interpolation error, exponential spectral model.

1. INTRODUCTION

The ordinary nonparametric approach for *i.i.d.* random variables has developed in various directions. For example Hallin, Ingenbleek and Puri[1] introduced a class of linear serial rank statistics for the problem of testing white noise against alternatives of ARMA serial dependence. They derived the asymptotic distributions of the proposed test statistics under the null as well as alternative hypotheses, and gave an explicit formulation of the asymptotically most powerful test under a sequence of contiguous ARMA alternatives. Dzhaparidze[2] considered a class $\mathcal{F}$ of goodness-of-fit tests for testing a simple hypothesis about the form of the spectral density. He investigated the asymptotic properties of test $T \in \mathcal{F}$ under a sequence of nonparametric contiguous alternatives.

Suppose that $\{X_t\}$ and $\{Y_t\}$ are mutually independent Gaussian stationary processes with the spectral densities $f(\lambda)$ and $g(\lambda)$, respectively. Here we consider the testing problem

$$H : \int_{-\pi}^{\pi} K\{f(\lambda)\}d\lambda = \int_{-\pi}^{\pi} K\{g(\lambda)\}d\lambda,$$

against

$$A : \int_{-\pi}^{\pi} K\{f(\lambda)\}d\lambda \neq \int_{-\pi}^{\pi} K\{g(\lambda)\}d\lambda,$$

where $K(\cdot)$ is an appropriate function. This setting of test is unexpectedly wide, and can be applied to many problems in time series. For this problem we propose a test based on $\int_{-\pi}^{\pi} K\{\hat{f}_n(\lambda)\}d\lambda$ and $\int_{-\pi}^{\pi} K\{\hat{g}_n(\lambda)\}d\lambda$ where $\hat{f}_n(\lambda)$ and $\hat{g}_n(\lambda)$ are nonparametric spectral estimators of $f(\lambda)$ and $g(\lambda)$, respectively, and evaluate the asymptotic power under a sequence of nonparametric contiguous alternatives.

In Section 3.1, we consider the case where $K(x) = \log x$. Then our problem is that Burg's entropy of $\{X_t\}$ is equal to that of $\{Y_t\}$ or not (*i.e.* $H : \int_{-\pi}^{\pi} \log f(\lambda)d\lambda = \int_{-\pi}^{\pi} \log g(\lambda)d\lambda$). It is shown that our test behaves well. In Section 3.2, we consider the case where $K(x) = x^{\beta}$. Then our problem is that the spectral moment of $\{X_t\}$ is equal to that of $\{Y_t\}$ or not (*i.e.* $H : \int_{-\pi}^{\pi} f(\lambda)^{\beta}d\lambda = \int_{-\pi}^{\pi} g(\lambda)^{\beta}d\lambda$). In Section 3.3, we consider the case where $K(x) = x^{-1}$. Then our problem is that the interpolation error of $\{X_t\}$ is equal to that of $\{Y_t\}$ or not (*i.e.* $H : \int_{-\pi}^{\pi} f(\lambda)^{-1}d\lambda = \int_{-\pi}^{\pi} g(\lambda)^{-1}d\lambda$). In Section 4, some numerical studies will be given for a sequence of exponential spectral alternatives. They confirm the theoretical results and show the goodness of our test.

2. BASIC THEOREMS

In this section we formulate some basic theorems concerning a nonparametric testing problem. Let $\{X_t; t = 0, \pm 1, \ldots\}$ and $\{Y_t; t = 0, \pm 1, \ldots\}$ be mutually independent Gaussian stationary processes with $E(X_t) = 0$ and $E(Y_t) = 0$ and spectral densities $f(\lambda)$ and $g(\lambda)$, respectively. Initially, we make the following assumptions.

Assumption 1. (i) *There exists a positive number δ such that $f(\lambda) \geq \delta$ and $g(\lambda) \geq \delta$, $-\pi \leq \lambda \leq \pi$.*
(ii)

$$\sum_{t=1}^{\infty} t|R_X(t)| < \infty, \qquad \sum_{t=1}^{\infty} t|R_Y(t)| < \infty,$$

where $R_X(t) = E\{X(s)X(s+t)\}$ and $R_Y(t) = E\{Y(s)Y(s+t)\}$.

Assumption 2. *$K(x)$ is continuously 3 times differentiable and $K'(x) \neq 0$ a.e. on $(0, \infty)$.*

Consider the testing problem

$$H : \int_{-\pi}^{\pi} K\{f(\lambda)\}d\lambda = \int_{-\pi}^{\pi} K\{g(\lambda)\}d\lambda,$$

against

$$A : \int_{-\pi}^{\pi} K\{f(\lambda)\}d\lambda \neq \int_{-\pi}^{\pi} K\{g(\lambda)\}d\lambda. \tag{2.1}$$

As we shall see in Section 3, this setting of test can be applied to many problems in time series analysis. Since the testing problem is written nonparametrically, we will construct a nonparametric test statistic.

Denote by $I_n^X(\lambda)$ and $I_n^Y(\lambda)$ the periodograms constructed from partial realizations $\{X_1, \ldots, X_n\}$ and $\{Y_1, \ldots, Y_n\}$; namely

$$I_n^X(\lambda) = \frac{1}{2\pi n}|\sum_{t=1}^{n} X_t e^{it\lambda}|^2, -\pi \leq \lambda \leq \pi,$$

$$I_n^Y(\lambda) = \frac{1}{2\pi n}|\sum_{t=1}^{n} Y_t e^{it\lambda}|^2, -\pi \le \lambda \le \pi.$$

To estimate $f(\lambda)$ and $g(\lambda)$ we use

$$\hat{f}_n(\lambda) = \int_{-\pi}^{\pi} W_n(\lambda - \mu) I_n^X(\mu) d\mu,$$

$$\hat{g}_n(\lambda) = \int_{-\pi}^{\pi} W_n(\lambda - \mu) I_n^Y(\mu) d\mu. \tag{2.2}$$

Here $W_n(\cdot)$ satisfies the following assumption.

Assumption 3. (i) *$W(x)$ is bounded, even, non-negative and such that*

$$\int_{-\infty}^{\infty} W(x) dx = 1.$$

(ii) *For $M = O(n^\alpha)$, $(1/4 < \alpha < 1/2)$, the function $W_n(\lambda) = MW(M\lambda)$ can be expanded as*

$$W_n(\lambda) = \frac{1}{2\pi}\sum_l w(\frac{l}{M}) e^{-il\lambda},$$

where $w(x)$ is a continuous, even function with $w(0) = 1$, $|w(x)| \le 1$ and $\int_{-\infty}^{\infty} w(x)^2 dx < \infty$, and satisfies

$$\lim_{|x|\to 0} \frac{1 - w(x)}{|x|^2} = k_2 < \infty.$$

By using $\hat{f}_n(\lambda)$ and $\hat{g}_n(\lambda)$ we make a test statistic for (2.1)

Theorem 1. *Suppose that Assumptions 1, 2 and 3 hold. Then*
(a) *Under the null hypothesis H,*

$$S = \sqrt{n}[\int_{-\pi}^{\pi} K\{\hat{f}_n(\lambda)\} d\lambda - \int_{-\pi}^{\pi} K\{\hat{g}_n(\lambda)\} d\lambda],$$

has, asymptotically, a normal distribution with mean zero and variance

$$v(f, g) = 4\pi \int_{-\pi}^{\pi} [K'\{f(\lambda)\}]^2 f(\lambda)^2 d\lambda + 4\pi \int_{-\pi}^{\pi} [K'\{g(\lambda)\}]^2 g(\lambda)^2 d\lambda.$$

(b) *Under the null hypothesis H,*

$$T = \frac{\sqrt{n}[\int_{-\pi}^{\pi} K\{\hat{f}_n(\lambda)\} d\lambda - \int_{-\pi}^{\pi} K\{\hat{g}_n(\lambda)\} d\lambda]}{\sqrt{4\pi \int_{-\pi}^{\pi} [K'\{\hat{f}_n(\lambda)\}]^2 \hat{f}_n(\lambda)^2 d\lambda + 4\pi \int_{-\pi}^{\pi} [K'\{\hat{g}_n(\lambda)\}]^2 \hat{g}_n(\lambda)^2 d\lambda}}$$

has, asymptotically, the standard normal distribution $N(0,1)$.

[proof] (a) Under H, S is written as

$$S = \sqrt{n}[\int_{-\pi}^{\pi} [K\{\hat{f}_n(\lambda)\} - K\{f(\lambda)\}] d\lambda - \sqrt{n}[\int_{-\pi}^{\pi} [K\{\hat{g}_n(\lambda)\} - K\{g(\lambda)\}] d\lambda.$$

It is known that

$$\hat{f}_n(\lambda) - f(\lambda) = O_p\{(M/n)^{1/2}\}, \text{and } \hat{g}_n(\lambda) - g(\lambda) = O_p\{(M/n)^{1/2}\}, \tag{2.3}$$

(*e.g.*, Hannan[3], Brillinger[4], Taniguchi[5]).
From Corollary 5.1.5 of Fuller[6] and continuously twice differentiability of $K(\cdot)$, we have

$$K\{\hat{f}_n(\lambda)\} - K\{f(\lambda)\} = K'\{f(\lambda)\}\{\hat{f}_n(\lambda) - f(\lambda)\} + O_p(M/n),$$

and

$$K\{\hat{g}_n(\lambda)\} - K\{g(\lambda)\} = K'\{g(\lambda)\}\{\hat{g}_n(\lambda) - g(\lambda)\} + O_p(M/n). \tag{2.4}$$

Recalling Assumption 3 (ii), we see that

$$S = \sqrt{n}[\int_{-\pi}^{\pi} K'\{f(\lambda)\}\{\hat{f}_n(\lambda) - f(\lambda)\}d\lambda] - \sqrt{n}[\int_{-\pi}^{\pi} K'\{g(\lambda)\}\{\hat{g}_n(\lambda) - g(\lambda)\}d\lambda] + o_p(1). \tag{2.5}$$

By virtue of Theorem 3 of Taniguchi[5] the proof for (a) follows from (2.5).
(b)It is easy to show that $v(\hat{f}_n, \hat{g}_n)$ converges to $v(f, g)$ in probability. Therefore the assertion follows from Slutsky's theorem.

Here it should be noted that $\sqrt{n}$-consistency holds in Theorem 1. This is due to the fact that integration of $\hat{f}_n$ recovers $\sqrt{n}$-consistency. In view of Theorem 1 we can propose the test of H, given by critical region

$$[|T| > t_\alpha], \tag{2.6}$$

where t_α is defined by

$$\int_{t_\alpha}^{\infty} (2\pi)^{-1/2} \exp(-x^2/2)dx = \alpha/2.$$

Next we shall evaluate the asymptotic power of the test (2.6) under a sequence of spectral densities. Henceforth we denote the probability density function of $(\mathbf{X}_n, \mathbf{Y}_n) = ((X_1, \ldots, X_n), (Y_1, \ldots, Y_n))$ by $p^n_{(h_1,h_2)}(\cdot)$ if the processes $\{X_t\}$ and $\{Y_t\}$ are assumed to have spectral densities $h_1(\lambda)$ and $h_2(\lambda)$, respectively. Let $a(\lambda)$ and $b(\lambda)$ be square integrable functions on $[-\pi, \pi]$.
Putting

$$f_n(\lambda) = f(\lambda)\{1 + a(\lambda)/\sqrt{n}\}$$

and

$$g_n(\lambda) = g(\lambda)\{1 + b(\lambda)/\sqrt{n}\},$$

we define

$$\Lambda((f, g), (f_n, g_n)) = \log\{p^n_{(f_n,g_n)}(\mathbf{X}_n, \mathbf{Y}_n)/p^n_{(f,g)}(\mathbf{X}_n, \mathbf{Y}_n)\}.$$

The following lemma is essentially due to Dzhaparidze[2].
Lemma 1. *Suppose that Assumptions 1, 2 and 3 hold and that*

$$\sum_{t=1}^{\infty} t|\rho_f(t)|^2 < \infty, \qquad \sum_{t=1}^{\infty} t|\rho_g(t)|^2 < \infty, \tag{2.7}$$

where

$$\rho_f(t) = \frac{1}{2\pi}\int_{-\pi}^{\pi} \frac{a(\lambda)}{f(\lambda)} e^{i\lambda t} d\lambda, \qquad \rho_g(t) = \frac{1}{2\pi}\int_{-\pi}^{\pi} \frac{b(\lambda)}{g(\lambda)} e^{i\lambda t} d\lambda.$$

Then
(a) $p^n_{(f,g)}$ *and* $p^n_{(f_n,g_n)}$ *are contiguous.*

(b)

$$\Lambda((f,g),(f_n,g_n)) - \frac{\sqrt{n}}{4\pi}\int_{-\pi}^{\pi}\frac{I_n^X(\lambda)-f(\lambda)}{f(\lambda)}a(\lambda)d\lambda + \frac{1}{8\pi}\int_{-\pi}^{\pi}a^2(\lambda)d\lambda$$

$$-\frac{\sqrt{n}}{4\pi}\int_{-\pi}^{\pi}\frac{I_n^Y(\lambda)-g(\lambda)}{g(\lambda)}b(\lambda)d\lambda + \frac{1}{8\pi}\int_{-\pi}^{\pi}b^2(\lambda)d\lambda \longrightarrow 0 \text{ in } p^n_{(f,g)}. \tag{2.8}$$

Using this lemma we can prove the following theorem.

Theorem 2. *Suppose that Assumptions 1, 2 and 3 and the condition (2.7) hold. Then* T *defined in Theorem 1 is under* $p^n_{(f_n,g_n)}$ *asymptotically normal* $N\{\mu(f,g,a,b),1\}$ *where*

$$\mu(f,g,a,b) = \int_{-\pi}^{\pi} v(f,g)^{-1/2}K'\{f(\lambda)\}a(\lambda)f(\lambda)d\lambda$$

$$-\int_{-\pi}^{\pi} v(f,g)^{-1/2}K'\{g(\lambda)\}b(\lambda)g(\lambda)d\lambda.$$

Therefore the asymptotic power is given by

$$\lim_{n\to\infty} P_{(f_n,g_n)}[|T|\geq t] = \int_{|x|\geq t}\frac{1}{\sqrt{2\pi}}\exp -\frac{1}{2}\{x-\mu(f,g,a,b)\}^2dx. \tag{2.9}$$

[Proof] In view of Theorem 1 we see that

$$T = \sqrt{n}\int_{-\pi}^{\pi} v(f,g)^{-1/2}K'\{f(\lambda)\}\{\hat{f}_n(\lambda)-f(\lambda)\}d\lambda$$

$$-\sqrt{n}\int_{-\pi}^{\pi} v(f,g)^{-1/2}K'\{g(\lambda)\}\{\hat{g}_n(\lambda)-g(\lambda)\}d\lambda + o_p(1),$$

under $p^n_{(f,g)}$ (*i.e.*, under H). It follows from Theorem 3 of Taniguchi[5] that

$$T = \sqrt{n}\int_{-\pi}^{\pi} v(f,g)^{-1/2}K'\{f(\lambda)\}\{I_n^X(\lambda)-f(\lambda)\}d\lambda$$

$$-\sqrt{n}\int_{-\pi}^{\pi} v(f,g)^{-1/2}K'\{g(\lambda)\}\{I_n^Y(\lambda)-g(\lambda)\}d\lambda + o_p(1). \tag{2.10}$$

By using Lemma 1 and recalling the asymptotic theory of integrals of $I_n(\lambda)$ (see Brillinger[4], Taniguchi[5]), we can show that $(T,\Lambda((f,g),(f_n,g_n)))$ is under $p^n_{(f,g)}$ asymptotically normal with mean vector $(0,-\frac{1}{2}B)'$ and covariance matrix

$$\begin{pmatrix} 1, & \mu(f,g,a,b) \\ \mu(f,g,a,b), & B \end{pmatrix},$$

where $B = \frac{1}{4\pi}\int_{-\pi}^{\pi}a^2(\lambda)d\lambda + \frac{1}{4\pi}\int_{-\pi}^{\pi}b^2(\lambda)d\lambda$. This, together with LeCam's third lemma (see Hájek and Šidák[7], p208), implies that T is under $p^n_{(f_n,g_n)}$ asymptotically normal $N\{\mu(f,g,a,b),1\}$.

This theorem enables us to evaluate the asymptotic power of T for a sequence of 'nonparametric alternatives'. Let $E_{(f_n,g_n)}(\cdot)$ and $V_{(f,g)}(\cdot)$ denote the expectation under

$p^n_{(f_n,g_n)}$ and the variance under $p^n_{(f,g)}$, respectively . It is natural to define an efficacy of T by

$$eff(T) = \lim_{n\to\infty} \frac{\{E_{(f_n,g_n)}(T)\}^2}{V_{(f,g)}(T)},$$

in line with the usual definition for a sequence of 'parametric alternatives '(see Randles and Wolfe[8], p.147-149, or Gibbons[9], p.281). Then we see that

$$eff(T) = \mu(f, g, a, b)^2. \tag{2.11}$$

For another test T^* we may define an asymptotic relative efficiency(ARE) of T relative to T^* by

$$ARE(T, T^*) = \frac{eff(T)}{eff(T^*)}. \tag{2.12}$$

3. EXAMPLES

3.1 Testing for Burg's entropy

Our testing problem can be applied to many situations in time series analysis. If we put $K(x) = \log x$, the testing problem becomes

$$H : \int_{-\pi}^{\pi} \log f(\lambda)d\lambda = \int_{-\pi}^{\pi} \log g(\lambda)d\lambda,$$

against

$$A : \int_{-\pi}^{\pi} \log f(\lambda)d\lambda \neq \int_{-\pi}^{\pi} \log g(\lambda)d\lambda. \tag{3.1}$$

It is known that the quantity $\int_{-\pi}^{\pi} \log f(\lambda)d\lambda$ is important and is called Burg's entropy. This is a key parameter of $\{X_t\}$ which is specified by the variance, σ^2 of the prediction error for the one step ahead best linear predictor of X_t, *i.e.*,

$$\sigma^2 = \exp\{\frac{1}{2\pi}\int_{-\pi}^{\pi} \log 2\pi f(\lambda)d\lambda\}.$$

(*e.g.* Hannan[3], p.137). Then the test statistic for (3.1) becomes

$$T = \frac{\sqrt{n}[\int_{-\pi}^{\pi} \log \hat{f}_n(\lambda)d\lambda - \int_{-\pi}^{\pi} \log \hat{g}_n(\lambda)d\lambda]}{4\pi}. \tag{3.2}$$

In fact, from Theorem 2 we have

$$\lim_{n\to\infty} P_{(f_n,g_n)}[|T| \geq t] = \int_{|x|\geq t} dN[\frac{1}{4\pi}\{\int_{-\pi}^{\pi} a(\lambda)d\lambda - \int_{-\pi}^{\pi} b(\lambda)d\lambda\}, 1].$$

The efficacy of T is given by

$$eff(T) = \frac{1}{16\pi^2}\{\int_{-\pi}^{\pi} a(\lambda)d\lambda - \int_{-\pi}^{\pi} b(\lambda)d\lambda\}^2. \tag{3.3}$$

For the testing problem (3.1) we can construct another test . Let

$$T^* = \sqrt{n}[\int_{-\pi}^{\pi} \log \delta I_n^X(\lambda)d\lambda - \int_{-\pi}^{\pi} \log \delta I_n^Y(\lambda)d\lambda]\sqrt{3}/2\sqrt{2}\pi^2, \tag{3.4}$$

where $\delta = \exp\gamma$ ($\gamma \cong 0.57721$, Euler's constant). In view of estimation for $\int_{-\pi}^{\pi}\log f(\lambda)d\lambda - \int_{-\pi}^{\pi}\log g(\lambda)d\lambda$, it is shown that T^* is, under H, asymptotically normal with mean 0 and variance 1 (see Taniguchi[10], p.82, Hannan and Nicholls[11], Bloomfield[12]). Then the efficacy of T^* is

$$eff(T^*) = \frac{3}{8\pi^4}\{\int_{-\pi}^{\pi} a(\lambda)d\lambda - \int_{-\pi}^{\pi} b(\lambda)d\lambda\}^2.$$

Therefore,

$$ARE(T, T^*) = eff(T)/eff(T^*) = \frac{\pi^2}{6} > 1,$$

which implies that T is better than T^*.

3.2 Testing for spectral moment

If we put $K(x) = x^\beta \quad (\beta > 0)$, the testing problem becomes

$$H : \int_{-\pi}^{\pi} f(\lambda)^\beta d\lambda = \int_{-\pi}^{\pi} g(\lambda)^\beta d\lambda,$$

against

$$A : \int_{-\pi}^{\pi} f(\lambda)^\beta d\lambda \neq \int_{-\pi}^{\pi} g(\lambda)^\beta d\lambda. \tag{3.5}$$

Then the test statistic for (3.5) becomes

$$T = \frac{\sqrt{n}[\int_{-\pi}^{\pi} \hat{f}_n(\lambda)^\beta d\lambda - \int_{-\pi}^{\pi} \hat{g}_n(\lambda)^\beta d\lambda]}{\sqrt{4\pi\int_{-\pi}^{\pi}\beta^2 \hat{f}_n(\lambda)^{2\beta}d\lambda + 4\pi\int_{-\pi}^{\pi}\beta^2\hat{g}_n(\lambda)^{2\beta}d\lambda}}. \tag{3.6}$$

The efficacy of T is given by

$$eff(T) = \frac{\{\int_{-\pi}^{\pi} f(\lambda)^\beta a(\lambda)d\lambda - \int_{-\pi}^{\pi} g(\lambda)^\beta b(\lambda)d\lambda\}^2}{4\pi\{\int_{-\pi}^{\pi} f(\lambda)^{2\beta}d\lambda + \int_{-\pi}^{\pi} g(\lambda)^{2\beta}d\lambda\}}. \tag{3.7}$$

For the testing problem (3.5) we can construct another test. Let

$$T^* = \frac{\sqrt{n}[\int_{-\pi}^{\pi}\frac{I_n^X(\lambda)^\beta}{\Gamma(\beta+1)}d\lambda - \int_{-\pi}^{\pi}\frac{I_n^Y(\lambda)^\beta}{\Gamma(\beta+1)}d\lambda]}{\sqrt{4\pi\Psi(\beta)\int_{-\pi}^{\pi}\frac{I_n^X(\lambda)^{2\beta}}{\Gamma(2\beta+1)}d\lambda + 4\pi\Psi(\beta)\int_{-\pi}^{\pi}\frac{I_n^Y(\lambda)^{2\beta}}{\Gamma(2\beta+1)}d\lambda}}, \tag{3.8}$$

where $\Psi(\beta) = (\frac{\Gamma(2\beta+1)}{\Gamma(\beta+1)^2} - 1)$. In view of estimation for $\int_{-\pi}^{\pi} f(\lambda)^\beta d\lambda - \int_{-\pi}^{\pi} g(\lambda)^\beta d\lambda$, it is shown that T^* is, under H, asymptotically normal with mean 0 and variance 1 (see Taniguchi[10]). Then the efficacy of T^* is

$$eff(T^*) = \frac{\beta^2\{\int_{-\pi}^{\pi} f(\lambda)^\beta a(\lambda)d\lambda - \int_{-\pi}^{\pi} g(\lambda)^\beta b(\lambda)d\lambda\}^2}{4\pi\Psi(\beta)\{\int_{-\pi}^{\pi} f(\lambda)^{2\beta}d\lambda + \int_{-\pi}^{\pi} g(\lambda)^{2\beta}d\lambda\}}.$$

Therefore,

$$ARE(T, T^*) = \frac{\Psi(\beta)}{\beta^2} > 1, \quad (\beta > 1).$$

3.3 Testing for interpolation error

If we put $K(x) = x^{-1}$, the testing problem becomes

$$H : \int_{-\pi}^{\pi} f(\lambda)^{-1} d\lambda = \int_{-\pi}^{\pi} g(\lambda)^{-1} d\lambda,$$

against

$$A : \int_{-\pi}^{\pi} f(\lambda)^{-1} d\lambda \neq \int_{-\pi}^{\pi} g(\lambda)^{-1} d\lambda. \tag{3.9}$$

The quantity $[\frac{1}{2\pi}\int_{-\pi}^{\pi}\{2\pi f(\lambda)\}^{-1}d\lambda]^{-1}$ is known to be the interpolation error of X_0 by the best linear interpolation based on X_t, $t = \pm 1, \pm 2, \ldots$(see Hannan[3], p.165).
Then the test statistic for (3.9) becomes

$$T = \frac{\sqrt{n}[\int_{-\pi}^{\pi} \hat{f}_n(\lambda)^{-1} d\lambda - \int_{-\pi}^{\pi} \hat{g}_n(\lambda)^{-1} d\lambda]}{\sqrt{4\pi \int_{-\pi}^{\pi} \hat{f}_n(\lambda)^{-2} d\lambda + 4\pi \int_{-\pi}^{\pi} \hat{g}_n(\lambda)^{-2} d\lambda}}. \tag{3.10}$$

The efficacy of T is given by

$$eff(T) = \frac{\{\int_{-\pi}^{\pi} f(\lambda)^{-1} a(\lambda) d\lambda - \int_{-\pi}^{\pi} g(\lambda)^{-1} b(\lambda) d\lambda\}^2}{4\pi\{\int_{-\pi}^{\pi} f(\lambda)^{-2} d\lambda + \int_{-\pi}^{\pi} g(\lambda)^{-2} d\lambda\}}. \tag{3.11}$$

If we assume the parametric spectral densities $f(\lambda) = f_\theta(\lambda)$ and $g(\lambda) = g_\mu(\lambda)$, we can construct another test. Let $\hat{\theta}$ and $\hat{\mu}$ be the quasi-maximum likelihood estimators of θ and μ , respectively(see Hosoya and Taniguchi[13]). Then we can propose

$$T^* = \frac{\sqrt{n}[\int_{-\pi}^{\pi} f_{\hat{\theta}}(\lambda)^{-1} d\lambda - \int_{-\pi}^{\pi} g_{\hat{\mu}}(\lambda)^{-1} d\lambda]}{\sqrt{v(f_{\hat{\theta}}, g_{\hat{\mu}})}}, \tag{3.12}$$

where

$$v(f_\theta, g_\mu) = 4\pi \mathbf{f}'\mathbf{M}_f^{-1}\mathbf{f} + 4\pi \mathbf{g}'\mathbf{M}_g^{-1}\mathbf{g},$$

$$\mathbf{f}' = \int_{-\pi}^{\pi} \frac{\partial f_\theta(\lambda)^{-1}}{\partial \theta'} d\lambda, \qquad \mathbf{g}' = \int_{-\pi}^{\pi} \frac{\partial g_\mu(\lambda)^{-1}}{\partial \mu'} d\lambda,$$

$$\mathbf{M}_f = \int_{-\pi}^{\pi} \frac{\partial \log f_\theta(\lambda)}{\partial \theta} \frac{\partial \log f_\theta(\lambda)}{\partial \theta'} d\lambda,$$

$$\mathbf{M}_g = \int_{-\pi}^{\pi} \frac{\partial log g_\mu(\lambda)}{\partial \mu} \frac{\partial log g_\mu(\lambda)}{\partial \mu'} d\lambda.$$

It is shown that T^* is under H asymptotically normal with mean 0 and variance 1, and that the efficacy of T^* is

$$eff(T^*) = \frac{\{4\pi \mathbf{f}'\mathbf{M}_f^{-1} \int_{-\pi}^{\pi} \frac{\partial}{\partial \theta} f_\theta(\lambda)^{-1} f_\theta(\lambda) a(\lambda) d\lambda - 4\pi \mathbf{g}'\mathbf{M}_g^{-1} \int_{-\pi}^{\pi} \frac{\partial}{\partial \mu} g_\mu(\lambda)^{-1} g_\mu(\lambda) b(\lambda) d\lambda\}^2}{v(f_\theta, g_\mu)}.$$

4. NUMERICAL STUDIES

In this section we give three numerical studies of the results in Sections 3.1, 3.2 and 3.3.

The first example is the one for testing for Burg's entropy. The testing problem is

$$H : \int_{-\pi}^{\pi} \log f(\lambda) d\lambda = \int_{-\pi}^{\pi} \log g(\lambda) d\lambda,$$

against

$$A : \int_{-\pi}^{\pi} \log f(\lambda) d\lambda \neq \int_{-\pi}^{\pi} \log g(\lambda) d\lambda.$$

We consider the following sequence of alternatives

$$f_n(\lambda) = \frac{1}{2\pi} \exp(0.5 \cos \lambda), \tag{4.1}$$

and

$$g_n(\lambda) = \frac{1}{2\pi} \exp(0.5 \cos \lambda)\{1 + \frac{1}{\sqrt{n}} \frac{0.5}{2\pi} \exp(\theta \cos \lambda)\}. \tag{4.2}$$

Under the sequences of (4.1) and (4.2), for n=1024 and M=32, we calculated the tests T and T^* defined by (3.2) and (3.4), respectively, and repeated this procedure 1000 times. Then the following table gives the powers of T and T^* with level $\alpha = 0.05$

Table 4.1

θ	*power of* T	*power of* T^*
2.5	0.55	0.41
3.0	0.82	0.62
3.5	0.96	0.85

This shows that the simulation agrees with the theoretical results.

We next consider the example of testing for spectral moment. The problem is

$$H : \int_{-\pi}^{\pi} f(\lambda)^{\beta} d\lambda = \int_{-\pi}^{\pi} g(\lambda)^{\beta} d\lambda,$$

against

$$A : \int_{-\pi}^{\pi} f(\lambda)^{\beta} d\lambda \neq \int_{-\pi}^{\pi} g(\lambda)^{\beta} d\lambda.$$

We also evaluate the power of tests under the sequences of (4.1) and (4.2). For β =2, n=1024 and M=32, we calculated the tests T and T^* defined by (3.6) and (3.8), respectively, and repeated this procedure 1000 times. Then the following table gives the powers of T and T^* with level $\alpha = 0.05$.

Table 4.2

θ	*power of* T	*power of* T^*
1.5	0.35	0.31
2.0	0.54	0.48
2.5	0.84	0.78

This simulation result confirms the theoretical results.

Finally we consider the following testing problem ;

$$H : \int_{-\pi}^{\pi} f(\lambda)^{-1} d\lambda = \int_{-\pi}^{\pi} g(\lambda)^{-1} d\lambda,$$

against

$$A : \int_{-\pi}^{\pi} f(\lambda)^{-1} d\lambda \neq \int_{-\pi}^{\pi} g(\lambda)^{-1} d\lambda.$$

Here we use the parametric spectral densities

$$f(\lambda) = f_\theta(\lambda) = \frac{\sigma_1^2}{2\pi|1 - \beta_1 \exp(i\lambda)|^2},$$

and

$$g(\lambda) = g_\mu(\lambda) = \frac{\sigma_2^2}{2\pi|1 - \beta_2 \exp(i\lambda)|^2}.$$

The following sequences of alternatives are considered.

$$f_n(\lambda) = \frac{1}{2\pi|1 - 0.8\exp(i\lambda)|^2}. \tag{4.3}$$

$$g_n(\lambda) = \frac{1}{2\pi|1 - 0.8\exp(i\lambda)|^2}\{1 + \frac{1}{\sqrt{n}} \frac{4}{2\pi|1 - \beta\exp(i\lambda)|^2}\}. \tag{4.4}$$

Under the sequences of (4.3) and (4.4), for n=1024 and M=32, we calculated the tests T and T^* defined by (3.10) and (3.12), respectively, and repeated this procedure 1000 times. Then the table below gives the powers of T and T^* with level $\alpha = 0.05$.

Table 4.3

β	*power of* T	*power of* T^*
-0.4	0.51	0.46
-0.5	0.58	0.52
-0.6	0.69	0.62

This simulation result also shows the goodness of our test.

REFERENCES

1. M. Hallin, J-F. Ingenbleek, and M.L. Puri, Linear serial rank tests for randomness against ARMA alternatives. *Ann. Statist.*, **13**, 1156-1181 (1985).
2. K. Dzhaparidze, *Parameter Estimation and Hypothesis Testing in Spectral Analysis of Stationary Time Series.* Springer-Verlag (1986).
3. E.J. Hannan, *Multiple Time Series.* Wiley, New York (1970).
4. D.R. Brillinger, *Time Series Data Analysis and Theory.* Holt, Rinehart and Winston (1975).
5. M. Taniguchi, Minimum contrast estimation for spectral densities of stationary processes. *J. Roy. Stat. Soc.*, Ser. **B**, **49**, 315-325 (1987).
6. W.A. Fuller, *Introduction to Statistical Time Series.* Wiley, New York (1976).
7. J. Hájek, and Z. Šidák, *Theory of Rank Tests.* Academic Press, New York (1976).
8. R.H. Randles, and D.A. Wolfe, *Introduction to the Theory of Nonparametric Statistics.* Wiley, New York (1979).
9. J.D. Gibbons, *Nonparametric Statistical Inference.* McGraw-Hill (1971).
10. M. Taniguchi, On estimation of the integrals of certain functions of spectral density. *J. Appl. Prob.*, **17**, 73-83 (1980).
11. E.J. Hannan, and D.F. Nicholls, The estimation of the prediction error variance. *J. Amer. Stat. Assoc.*, **72**, 834-840 (1977).
12. P. Bloomfield, An exponential model for the spectrum of a scalar time series. *Biometrika*, **60**, 217-226 (1973).
13. Y. Hosoya, and M. Taniguchi, A central limit theorem for stationary processes and the parameter estimation of linear processes. *Ann. Statist.*,**10**, 132-153 (1982).
14. M. Taniguchi, and M. Kondo, Nonparametric approach in time series analysis. *J. Time Ser. Anal.* in press

Stat. Sci. & Data Anal., pp. 175-180
K. Matsusita *et al.* (Eds)

Malliavin Calculus and Higher Order Statistical Inference

NAKAHIRO YOSHIDA
The Institute of Statistical Mathematics, 4-6-7 Minami-Azabu, Minato-ku, Tokyo 106, Japan

Abstract. An application of the Malliavin calculus to statistics is discussed with illustrative examples of diffusion processes driven by small noise.

Key words: Malliavin calculus, asymptotic expansion, small diffusion, second order efficiency.

1. INTRODUCTION

The purpose of this paper is to discuss an application of the Malliavin calculus (analysis of Wiener functionals) to certain problems in statistics. Assume that we have an asymptotic expansion of a family of random variables in some sense:

$$F_\epsilon(w) \sim f_0(w) + \epsilon f_1(w) + \epsilon^2 f_2(w) + \cdots \tag{1.1}$$

as $\epsilon \downarrow 0$. In a classical way one can obtain the asymptotic expansion for $F_\epsilon(w)$ such as

$$P[F_\epsilon(w) \in A] \sim c_0(A) + \epsilon c_1(A) + \epsilon^2 c_2(A) + \cdots,$$

where A is any Borel set, using the characteristic function $E[e^{itF_\epsilon(w)}]$ of $F_\epsilon(w)$ and inverting the expansion

$$E[e^{itF_\epsilon(w)}] \sim \phi_0(t) + \epsilon\phi_1(t) + \epsilon^2\phi_2(t) + \cdots.$$

This method has been elaborated and succeeded mainly for independent observations. See for example Bhattacharya and Rao [1]. Here we adopt a somewhat different approach with so-called Malliavin's calculus in probability theory. Again assume that $F_\epsilon(w)$ has the asymptotic expansion (1.1) as $\epsilon \downarrow 0$ in some sense. For smooth function $T(x)$ we have the Taylor expansion

$$\begin{aligned} T(F_\epsilon(w)) &= T(f_0(w) + \epsilon f_1(w) + \cdots) \\ &\sim T(f_0(w)) + \epsilon f_1(w)\tfrac{\partial}{\partial x}T(f_0(w)) + \cdots \end{aligned} \tag{1.2}$$

as $\epsilon \downarrow 0$. Then the expectation yields the expansion

$$E[T(F_\epsilon(w))] \sim E[T(f_0(w))] + \epsilon E[f_1(w)\frac{\partial}{\partial x}T(f_0(w))] + \cdots.$$

If we can substitute the indicator function I_A of Borel set A in T, we obtain automatically the asymptotic expansion of the distribution of $F_\epsilon(w)$

$$P(F_\epsilon(w) \in A) \sim E[I_A(f_0(w))] + \epsilon E[f_1(w)\frac{\partial}{\partial x}I_A(f_0(w))] + \cdots$$

as $\epsilon \downarrow 0$. For this purpose, the asymptotic expansion (1.2) for $I_A(F_\epsilon(w))$ has to make sense. Then there are problems how to interpret and define the composite functionals such as $\frac{\partial}{\partial x} I_A(f_0(w))$ and their asymptotic expansions. S. Watanabe ([2],[3]) justified the composite functionals of the Schwartz distributions and random variables using Sobolev spaces on a probability space, and developed their asymptotic expansions in these Sobolev spaces. By this method various asymptotic expansions for heat kernels can be derived. Also in statistics we can use his program to derive asymptotic expansions for statistical estimators.

2. PROCESSES WITH SMALL DIFFUSIONS

Now let us discuss concrete examples. One of them is a family of diffusion processes defined by the stochastic differential equation

$$\begin{aligned} dX_t^\epsilon &= V_0(X_t^\epsilon, \theta)dt + \epsilon V(X_t^\epsilon)dw_t, t \in [0,T], \\ X_0^\epsilon &= x_0, \end{aligned} \tag{2.1}$$

where V_0 and V are given smooth functions, θ is a real unknown parameter and w is a Wiener process. The process X_t^ϵ is observed up to a fixed time T, and θ requires to be estimated from this observation. We consider asymptotics when the parameter ϵ tends to zero. It is known that the maximum likelihood estimator and the Bayes estimator are consistent, asymptotically normal and efficient in the first order. See, e.g., Kutoyants [4]. Furthermore, it can be proved that they are optimal in the second order in a sense from the asymptotic expansions of their distributions.

When θ_0 is the true value of θ, the log likelihood function is given by the formula

$$l_\epsilon(\theta;\theta_0) = \int_0^T \epsilon^{-2} V_0 V^{-2}(X_t^\epsilon, \theta) dX_t^\epsilon - \frac{1}{2}\int_0^T \epsilon^{-2} V_0^2 V^{-2}(X_t^\epsilon, \theta)dt,$$

where X_t^ϵ is the solution of (2.1) for θ_0 and ϵ. Roughly speaking, the maximum likelihood estimator $\hat{\theta}_\epsilon(w;\theta_0)$ is a root of the estimating equation

$$\frac{\partial}{\partial \theta} l_\epsilon(\theta;\theta_0) = 0.$$

From this, by the implicit function theorem, we obtain the stochastic expansion of the maximum likelihood estimator:

$$\psi_\epsilon(w)\epsilon^{-1}(\hat{\theta}_\epsilon(w;\theta_0) - \theta_0) \sim f_0(w) + \epsilon f_1(w) + \cdots$$

as $\epsilon \downarrow 0$, where $\psi_\epsilon(w)$ is a functional nearly equal to one and it does not affect the expansion essentially but it ensures the existence of the functional $\psi_\epsilon(w)\epsilon^{-1}\hat{\theta}_\epsilon(w;\theta_0)$ defined on the whole sample space, while $\hat{\theta}_\epsilon(w;\theta_0)$ may not exist. This expansion is valid in certain Sobolev spaces. Then we have the asymptotic expansion (if necessary, with truncation)

$$I_A(\psi_\epsilon(w)\epsilon^{-1}(\hat{\theta}_\epsilon(w;\theta_0) - \theta_0)) \sim I_A(f_0(w)) + \epsilon f_1(w)\frac{\partial}{\partial x} I_A(f_0(w)) + \cdots$$

as $\epsilon \downarrow 0$ and by generalized expectation

$$P(\epsilon^{-1}(\hat{\theta}_\epsilon(w;\theta_0) - \theta_0) \in A) \sim E[I_A(f_0(w)] + \epsilon E[f_1(w)\frac{\partial}{\partial x} I_A(f_0(w))] + \cdots$$

as $\epsilon \downarrow 0$ for any Borel set A. More precisely, we have the following theorem under some regularity conditions. The Fisher information measure at θ_0 is denoted by I.

Theorem.

$$P(\epsilon^{-1}(\hat{\theta}_\epsilon(w;\theta_0) - \theta_0) \in A) \sim \int_A p_0(x)dx + \epsilon \int_A p_1(x)dx + \cdots$$

as $\epsilon \downarrow 0$ for any Borel set A, where $p_0(x) = \phi(x; 0, I^{-1})$ (the normal density with mean zero and variance I^{-1}), and a smooth function $p_1(x) = P(x)\phi(x; 0, I^{-1})$ with $P(x)$ a polynomial of degree 3 whose coefficients are determined by V_0, V, x_0, T and θ_0.

Let $\hat{\theta}^*_\epsilon(w;\theta_0)$ be a bias corrected maximum likelihood estimator, namely,

$$\hat{\theta}^*_\epsilon(w;\theta_0) = \hat{\theta}_\epsilon(w;\theta_0) - \epsilon^2 b(\hat{\theta}_\epsilon(w;\theta_0)),$$

where $b(\theta)$ is a given function. We can also show the asymptotic expansion for this bias corrected maximum likelihood estimator. For details see Yoshida [5],[6]. In like manner, we have asymptotic expansions for Bayes and the bias corrected Bayes estimators (Yoshida [7]). Furthermore, the asymptotic expansions can be derived under contiguous alternatives, that is, when $\theta_0 + \epsilon h$, $h \in \mathbf{R}$, is the true value.

3. SIMULATION

This section presents a result of a numerical study. Consider the diffusion process defined by the stochastic differential equation

$$dX^\epsilon_t = \theta_0 X^\epsilon_t dt + \epsilon dw_t, \ t \in [0,1],$$
$$X_0 = x_0.$$

The maximum likelihood estimator $\hat{\theta}_\epsilon(w;\theta_0)$ is given by

$$\hat{\theta}_\epsilon(w;\theta_0) = \frac{\int_0^1 X^\epsilon_t dX^\epsilon_t}{\int_0^1 X^{\epsilon 2}_t dt}.$$

We know that the distribution of $\epsilon^{-1}(\hat{\theta}_\epsilon(w;\theta_0) - \theta_0)$ converges to the normal distribution $N(0, I^{-1})$ as $\epsilon \downarrow 0$. We also have the Edgeworth expansion up to ϵ-order from Section 2. In the case where $x_0 = 1$, $\epsilon = 0.3$ and $\theta_0 = 1$, Table 1 compares the distribution functions of $\epsilon^{-1}(\hat{\theta}_\epsilon(w;\theta_0) - \theta_0)$ obtained from simulation (100,000 repeated samples), the Edgeworth expansion up to ϵ-order and the normal approximation. Here the interval $[0,1]$ is divided into 100,000 subintervals and we use an approximation to this diffusion by an AR(1) model. Table 2 gives the result in the case where $x_0 = 1$, $\epsilon = 0.3$ and $\theta_0 = -1$. We also obtained similar results for $\epsilon = 0.1$ and $\theta_0 = \pm 1$.

4. SECOND ORDER EFFICIENCY OF ESTIMATORS

Let a bounded open interval Θ denote the space of the unknown parameter θ. An estimator T_ϵ of θ is second order asymptotically median unbiased (second order AMU) if for any θ_0 and any positive c

$$\lim_{\epsilon\downarrow 0} \sup_{\substack{\theta\in\Theta \\ |\theta-\theta_0|<\epsilon c}} \epsilon^{-1}|P_{\epsilon,\theta}(T_\epsilon - \theta \leq 0) - \frac{1}{2}| = 0$$

and

$$\lim_{\epsilon\downarrow 0} \sup_{\substack{\theta\in\Theta \\ |\theta-\theta_0|<\epsilon c}} \epsilon^{-1}|P_{\epsilon,\theta}(T_\epsilon - \theta \geq 0) - \frac{1}{2}| = 0,$$

Akahira-Takeuchi [8]. It can be proved that there exist functions $\Phi_+^\epsilon(h,\theta_0)$ and $\Phi_-^\epsilon(h,\theta_0)$ of h such that for any second order AMU estimator T_ϵ

$$\liminf_{\epsilon\downarrow 0} \epsilon^{-1}\{\Phi_+^\epsilon(h,\theta_0) - P_{\epsilon,\theta_0}(\epsilon^{-1}(T_\epsilon - \theta_0) \leq h)\} \geq 0$$

for $h > 0$, and

$$\limsup_{\epsilon\downarrow 0} \epsilon^{-1}\{\Phi_-^\epsilon(h,\theta_0) - P_{\epsilon,\theta_0}(\epsilon^{-1}(T_\epsilon - \theta_0) \leq h)\} \leq 0$$

for $h < 0$. These inequalities give the bounds $\Phi_+^\epsilon(h,\theta_0)$ and $\Phi_-^\epsilon(h,\theta_0)$ of distributions for any second order AMU estimator. According to Takeuchi and Akahira it is said that a second order AMU estimator attaining these bounds is second order efficient. By the asymptotic expansions for the bias corrected maximum likelihood estimator and the bias corrected Bayes estimator, we can verify that these estimators are second order AMU and second order efficient if bias corrections $b(\theta)$ are chosen appropriately.

5. FURTHER TOPICS

Consider the diffusion process X_t^ϵ defined by (2.1). Let

$$F_T^\epsilon = F_0^\epsilon + \int_0^T f_0^\epsilon(X_t^\epsilon)dt + \epsilon\int_0^T f^\epsilon(X_t^\epsilon)dw_t,$$

where $F_0^\epsilon \in \mathbf{R}$, $f_0^\epsilon(x)$ and $f^\epsilon(x)$ are smooth functions. It is possible to show the asymptotic expansion of the mean value

$$E[\varphi^\epsilon(\tilde{F}^\epsilon)I_A(\tilde{F}^\epsilon)], \tag{5.1}$$

where I_A is the indicator function of Borel set A, $\varphi^\epsilon(x)$ is a smooth function and

$$\tilde{F}^\epsilon = \epsilon^{-1}(F_T^\epsilon - f_{-1})$$

with $f_{-1} = \lim_{\epsilon\downarrow 0} F_T^\epsilon$. For example, the distribution of the likelihood ratio statistic takes the form of (5.1).

In economics the price of the underlying security is supposed to satisfy the stochastic differential equation

$$dX_t^\epsilon = cX_t^\epsilon dt + \epsilon X_t^\epsilon dw_t,$$
$$X_0^\epsilon = x_0,$$

where c and x_0 are constants. The mean value

$$E[Max\{Z_T^\epsilon - K, 0\}] \tag{5.2}$$

is necessary in the problem of how to price average options at time $t = 0$, where

$$Z_T^\epsilon = \frac{1}{T}\int_0^T X_t^\epsilon dt$$

and K is a constant. See Kunitomo-Takahashi [9]. We can obtain the asymptotic expansion of (5.2) from that of (5.1), Yoshida [7].

The method used here is also applicable to models perturbed by small noises realized on a Wiener space.

Acknowledgments
The author thanks Miss N.Noguchi of ISM for her computer programming.

References

[1] R. N. Bhattacharya and R. Ranga Rao, *Normal approximation and asymptotic expansions,* Wiley, New York (1976).

[2] S. Watanabe, in: Lecture Notes in Control and Inf. Sci., **49**, Springer, New York (1983).

[3] S. Watanabe, *Ann. Probab.,* **15**, 1-39 (1987).

[4] Yu. A. Kutoyants, *Parameter estimation for stochastic processes,* Heldermann, Berlin (1984).

[5] N. Yoshida, Research Memorandum, **383**, Institute of Statistical Mathematics, Tokyo (1990).

[6] N. Yoshida, Research Memorandum, **392**, Institute of Statistical Mathematics, Tokyo (1990), To appear in *Probab. Theory Related Fields.*

[7] N. Yoshida, Research Memorandum, **415**, Institute of Statistical Mathematics, Tokyo (1991).

[8] M. Akahira and K. Takeuchi, *Asymptotic efficiency of statistical estimators: concepts and higher order asymptotic efficiency,* Lect. Notes Stat., **7**, Springer, Berlin Heidelberg New York (1981).

[9] N. Kunitomo and A. Takahashi, *Pricing average options, discussion paper,* Faculty of Economics, University of Tokyo.(in Japanese) (1990).

Table 1: Edgeworth and normal approximations. $\epsilon = 0.3$, $\theta_0 = 1$.

x	Simulation	Edgeworth	Normal
-1.300	0.058	0.034	0.012
-1.200	0.071	0.047	0.018
-1.100	0.087	0.063	0.028
-1.000	0.106	0.083	0.041
-0.900	0.129	0.107	0.059
-0.800	0.155	0.135	0.082
-0.700	0.187	0.168	0.112
-0.600	0.224	0.206	0.148
-0.500	0.267	0.249	0.192
-0.400	0.315	0.299	0.243
-0.300	0.370	0.354	0.301
-0.200	0.430	0.414	0.364
-0.100	0.495	0.479	0.431
0.000	0.562	0.548	0.500
0.100	0.629	0.617	0.569
0.200	0.695	0.686	0.636
0.300	0.756	0.752	0.699
0.400	0.813	0.812	0.757
0.500	0.863	0.865	0.808
0.600	0.904	0.910	0.852
0.700	0.936	0.945	0.888
0.800	0.959	0.971	0.918

Table 2: Edgeworth and normal approximations. $\epsilon = 0.3$, $\theta_0 = -1$

x	Simulation	Edgeworth	Normal
-2.400	0.132	0.122	0.058
-2.200	0.153	0.144	0.075
-2.000	0.175	0.168	0.095
-1.800	0.202	0.195	0.119
-1.600	0.231	0.224	0.147
-1.400	0.264	0.256	0.180
-1.200	0.300	0.291	0.216
-1.000	0.340	0.329	0.256
-0.800	0.381	0.371	0.300
-0.600	0.427	0.415	0.347
-0.400	0.474	0.462	0.397
-0.200	0.525	0.512	0.448
-0.000	0.576	0.563	0.500
0.200	0.629	0.616	0.552
0.400	0.683	0.669	0.603
0.600	0.734	0.720	0.653
0.800	0.782	0.770	0.700
1.000	0.828	0.817	0.744
1.200	0.869	0.859	0.784
1.400	0.904	0.897	0.820
1.600	0.932	0.929	0.853
1.800	0.954	0.956	0.881

Stat. Sci. & Data Anal., pp. 181-185
K. Matsusita *et al.* (Eds)

Estimation of Levels of Intensity in a Simple Self-Correcting Point Process

T. HAYASHI [1] and N. INAGAKI [2]

[1] *University of Osaka Prefecture, College of Integrated Arts and Sciences, Sakai, Osaka 593, JAPAN.*
[2] *Osaka University, Faculty of Engineering Science, Toyonaka, Osaka 560, JAPAN.*

Abstract. We consider a self-correcting point process with the conditional intensity which takes only r values (levels). We give explicit expressions of the likelihood, Fisher information and the maximum likelihood estimator (MLE) of the levels of intensity and show asymptotic normality of the MLE.

Key words: Intensity, Maximum likelihood estimator, Self-correcting point process

1. INTRODUCTION

We consider a point process $N(\cdot)$ on the time axis $[0, \infty)$ and assume that the process $N(\cdot)$ is orderly and has the conditional intensity

$$\begin{aligned} \lambda(t) &= \lim_{\Delta t \downarrow 0} \frac{1}{\Delta t} P(N(t+\Delta t) - N(t-) \geq 1 \,|\, \mathcal{F}_{t-}) \\ &= \lim_{\Delta t \downarrow 0} \frac{1}{\Delta t} P(N(t+\Delta t) - N(t-) = 1 \,|\, \mathcal{F}_{t-}), \end{aligned} \tag{1.1}$$

where $\mathcal{F}_t$ is the σ-field generated by $\{N(s)\,;\, 0 \leq s \leq t\}$. We also assume that the process $N(t-)$ has the Markov property, which is equivalent to $\lambda(t) = E[\lambda(t) \,|\, N(t-)\,]$.

Let ρ be a positive constant and $\psi(\cdot)$ be a function satisfying that

(C1) $0 \leq \psi(x) < \infty$ for any $x \in \mathbf{R}$ (the real line),
(C2) there exists a constant $c_0 > 0$ such that $\psi(x) \geq c_0$ for all $x > 0$,
(C3) $\liminf_{x \to \infty} \psi(x) > 1$ and $\limsup_{x \to -\infty} \psi(x) < 1$.

Isham and Westcott [1] considered a point process $N(\cdot)$ having the conditional intensity

$$\lambda(t) = \rho\, \psi(\rho t - N(t-)\,). \tag{1.2}$$

From (C3) the process $N(\cdot)$ corrects the conditional intensity so that the absolute value of $\rho t - N(t-)$ may tend to become small when it is large. They called such a process a self-correcting point process and obtained that

$$\limsup_{t \to \infty} \left| \rho t - E[N(t)] \right| < \infty,$$

$$\limsup_{t \to \infty} E \left| N(t) - E[N(t)] \right|^p < \infty \qquad (p > 0).$$

The law of large numbers for the process $\rho t - N(t)$ was given by Vere-Jones and Ogata [2] when $\psi(x) = e^x$. Hayashi [3] showed it for all locally bounded function $\psi(\cdot)$ satisfying (C1)–(C3). Ogata and Vere-Jones [4] and Hayashi [5] investigated asymptotic properties of the maximum likelihood estimator (MLE) in the self-correcting point processes having parametrized intensities. However, they studied only when $\psi(\cdot)$ is a known function.

In the present paper, we suppose that $\psi(\cdot)$ is a piecewise constant function:

$$\psi(x) = \begin{cases} \theta_1 & \text{if} \quad x \in K_1 = (-\infty, c_1], \\ \vdots & \\ \theta_i & \text{if} \quad x \in K_i = (c_{i-1}, c_i], \\ \vdots & \\ \theta_r & \text{if} \quad x \in K_r = (c_{r-1}, \infty), \end{cases} \tag{1.3}$$

where $r(\geq 2)$ is a given integer, c_i's are given constants satisfying that $-\infty = c_0 < c_1 < \cdots < c_{r-1} < c_r = \infty$ and θ_i's are unknown positive parameters. To hold (C3), we also assume that $0 < \theta_1 < 1 < \theta_r < \infty$. In the present model, the intensity takes only r values (levels). To concentrate our interest on the estimation of levels, we treat ρ in (1.2) as a known constant. The authors will discuss the estimation of ρ elsewhere. Since we can choose the scale of the time axis so that $\rho = 1$, the intensity of the process $N(\cdot)$ considered here is written as

$$\lambda(t, \boldsymbol{\theta}) = \psi(X(t-)), \tag{1.4}$$

where $\boldsymbol{\theta} = (\theta_1, \cdots, \theta_r)'$ and $X(t) = t - N(t)$.

In section 2 we show that the information matrix converges to a positive definite matrix. In Section 3, we investigate the asymptotic properties of the MLE of $\boldsymbol{\theta}$ and obtain its asymptotic normality of the MLE. Furthermore, the family of measures induced by $N(\cdot)$ is locally asymptotically normal.

2. LIKELIHOOD AND INFORMATION MATRIX

It is well known that the log likelihood based on the observation $(N(t)\,;\, 0 < t \leq T)$ up to the time T is given by

$$\ell(T, \boldsymbol{\theta}) = \int_0^T \log \lambda(t, \boldsymbol{\theta})\, dN(t) - \int_0^T \lambda(t, \boldsymbol{\theta})\, dt$$

(see e.g. Kutoyants [6]). For $i = 1, 2, \cdots, r$, let

$$D_i = \{t \geq 0\,;\, X(t-) \in K_i\}, \quad D_i^{(T)} = D_i \cap [0, T],$$

where K_i is the interval given in (1.3). Then we have that

$$N(D_i^{(T)}) = \int_{D_i^{(T)}} dN(t) = \int_0^T \chi_{K_i}(X(t-))dN(t),$$

$$|D_i^{(T)}| = \int_{D_i^{(T)}} dt = \int_0^T \chi_{K_i}(X(t-))dt,$$

where $\chi_{K_i}(\cdot)$ is the indicator of K_i and $|\cdot|$ denotes the Lebesgue measure. We easily see that

$$\ell(T,\boldsymbol{\theta}) = \sum_{i=1}^{r}\left\{N(D_i^{(T)})\log\theta_i - \theta_i|D_i^{(T)}|\right\}, \tag{2.1}$$

$$\begin{aligned}\frac{\partial}{\partial\theta_i}\ell(T,\boldsymbol{\theta}) &= \frac{N(D_i^{(T)})}{\theta_i} - |D_i^{(T)}| \\ &= \frac{1}{\theta_i}\int_0^T \chi_{K_i}(X(t-))\,dN(t) - \frac{1}{\theta_i}\int_0^T \chi_{K_i}(X(t-))\,\lambda(t,\boldsymbol{\theta})\,dt \\ &= \frac{1}{\theta_i}\int_0^T \chi_{K_i}(X(t-))\,dM(t),\end{aligned} \tag{2.2}$$

where $M(t) = N(t) - \int_0^t \lambda(s,\boldsymbol{\theta})\,ds$. Therefore, the MLE of θ_i is given by

$$\widehat{\theta}_i^{(T)} = \frac{N(D_i^{(T)})}{|D_i^{(T)}|} \tag{2.3}$$

and the MLE of $\boldsymbol{\theta}$ is

$$\widehat{\boldsymbol{\theta}}_T = (\widehat{\theta}_1^{(T)},\cdots,\widehat{\theta}_r^{(T)})'. \tag{2.4}$$

Let $I_T(\boldsymbol{\theta})$ be the information matrix, that is,

$$I_T(\boldsymbol{\theta}) = E\left[\left(\frac{\partial}{\partial\boldsymbol{\theta}}\ell(T,\boldsymbol{\theta})\right)\left(\frac{\partial}{\partial\boldsymbol{\theta}}\ell(T,\boldsymbol{\theta})\right)'\right]. \tag{2.5}$$

Then we have that

$$\begin{aligned}&(i,j)\text{- element of } I_T(\boldsymbol{\theta}) \\ &= E\left[\frac{1}{\theta_i}\int_0^T \chi_{K_i}(X(t-))\,dM(t)\,\frac{1}{\theta_j}\int_0^T \chi_{K_j}(X(t-))\,dM(t)\right] \\ &= \frac{1}{\theta_i\theta_j}E\left[\int_0^T \chi_{K_i}(X(t-))\chi_{K_j}(X(t-))\,\lambda(t,\boldsymbol{\theta})\,dt\right] \\ &= \begin{cases}\dfrac{1}{\theta_i}E|D_i^{(T)}| & i=j, \\ 0 & i\neq j.\end{cases}\end{aligned}$$

Hence we get

$$I_T(\boldsymbol{\theta}) = \operatorname{diag}\left(\frac{E|D_1^{(T)}|}{\theta_1},\cdots,\frac{E|D_r^{(T)}|}{\theta_r}\right). \tag{2.6}$$

By using the law of large numbers (see Hayashi [3]), we get that

$$\frac{|D_i^{(T)}|}{T} = \frac{1}{T}\int_0^T \chi_{K_i}(X(t))\,dt$$

$$\to \sum_{j=-\infty}^{\infty} \pi_j E\left[\int_0^1 \chi_{K_i}(X(t))\,dt \,\Big|\, X(0)=j\right] \;(= R_i,\text{ say}) \tag{2.7}$$

in probability as $T \to \infty$, where $\{\pi_j\}$ is the invariant distribution of the discrete time Markov chain $\{X(n)\}_{n=0}^{\infty}$. From (2.6) and (2.7), we obtain

PROPOSITION 2.1.

$$\frac{1}{T} I_T(\boldsymbol{\theta}) \to I(\boldsymbol{\theta}) \quad \text{as } T \to \infty, \tag{2.8}$$

where $I_T(\boldsymbol{\theta})$ is given in (2.5) and

$$I(\boldsymbol{\theta}) = \operatorname{diag}\left(\frac{R_1}{\theta_1}, \cdots, \frac{R_r}{\theta_r}\right). \tag{2.9}$$

Remark 2.2. *In $r=2$, the following expression is given by Inagaki and Hayashi [7]:*

$$I(\boldsymbol{\theta}) = \begin{pmatrix} \dfrac{\theta_2 - 1}{\theta_1(\theta_2-\theta_1)} & 0 \\ 0 & \dfrac{1-\theta_1}{\theta_2(\theta_2-\theta_1)} \end{pmatrix}.$$

3. ASYMPTOTIC NORMALITY OF THE MLE

From (2.2) and (2.3), we easily see that

$$\sqrt{T}\left(\widehat{\theta}_i^{(T)} - \theta_i\right) = T\frac{\theta_i}{|D_i^{(T)}|}\frac{1}{\sqrt{T}}\frac{\partial}{\partial\theta_i}\ell(T,\boldsymbol{\theta}).$$

and that

$$\sqrt{T}(\widehat{\boldsymbol{\theta}}_T - \boldsymbol{\theta}) = T\operatorname{diag}\left(\frac{\theta_1}{|D_1^{(T)}|}, \cdots, \frac{\theta_r}{|D_r^{(T)}|}\right)\Delta_T(\boldsymbol{\theta}), \tag{3.1}$$

where

$$\Delta_T(\boldsymbol{\theta}) = \frac{1}{\sqrt{T}}\frac{\partial}{\partial\boldsymbol{\theta}}\ell(T,\boldsymbol{\theta}). \tag{3.2}$$

Let

$$\boldsymbol{f}_T(t) = \frac{1}{\sqrt{T}}\left(\frac{\chi_{K_1}(X(t-))}{\theta_1}, \cdots, \frac{\chi_{K_r}(X(t-))}{\theta_r}\right)'.$$

From (2.7), we see that

$$\int_0^T \boldsymbol{f}_T(t)\boldsymbol{f}_T(t)'\lambda(t,\boldsymbol{\theta})dt \to I(\boldsymbol{\theta})$$

in probability as $T \to \infty$. Since $\|\boldsymbol{f}_T(t)\| \le \text{const.}/\sqrt{T}$, for every $\varepsilon > 0$,

$$\lim_{T\to\infty} \int_0^T E\left[\, \|\boldsymbol{f}_T(t)\|^2 \, \chi_{(\varepsilon,\infty)}(\|\boldsymbol{f}_T(t)\|) \, \lambda(t,\boldsymbol{\theta}) \,\right] dt = 0.$$

Using Corollary 4.5.1 in Kutoyants [6], we have that

$$\mathcal{L}[\Delta_T(\boldsymbol{\theta})] \Longrightarrow N(0, I(\boldsymbol{\theta})) \tag{3.3}$$

in distribution as $T \to \infty$. From (2.7), we get that

$$T \operatorname{diag}\left(\frac{\theta_1}{|D_1^{(T)}|}, \cdots, \frac{\theta_r}{|D_r^{(T)}|} \right) \to I(\boldsymbol{\theta})^{-1} \tag{3.4}$$

in probability as $T \to \infty$. Hence we obtain the following theorem.

THEOREM 3.1. *The MLE $\widehat{\boldsymbol{\theta}}_T$ is asymptotically normal, that is,*

$$\mathcal{L}[\sqrt{T}(\widehat{\boldsymbol{\theta}}_T - \boldsymbol{\theta})] \Longrightarrow N(0, I(\boldsymbol{\theta})^{-1}) \tag{3.5}$$

in distribution as $T \to \infty$, where $\widehat{\boldsymbol{\theta}}_T$ and $I(\boldsymbol{\theta})$ are given in (2.4) and (2.9), respectively.

Remark 3.2. *Let $P_{\boldsymbol{\theta}}^{(T)}$ denote the measure induced by $\{N(t)\,;\, 0 \le t \le T\}$. Then the family $\{P_{\boldsymbol{\theta}}^{(T)}\}$ is locally asymptotically normal (LAN).*

In fact, we can easily see that the conditions 1–6 of Theorem 4.5.3 in Kutoyants [6], which gives LAN for general point processes of Poisson type. In our model, we directly obtain LAN from (2.1), (2.7), (3.3) and the expansion of the function $\log(1 + x)$ into the series.

Acknowledgments
The authors wish to thank the referee for his helpful comments.

REFERENCES

1. V. Isham and M. Westcott, A self-correcting point process, *Stoch. Proc. Appl.* (1979) **8**, 335–347.
2. D. Vere-Jones and Y. Ogata, On the moments of a self-correcting process, *J. Appl. Prob.* (1984) **21**, 335–342.
3. T. Hayashi, Laws of large numbers in self-correcting point processes, *Stoch. Proc. Appl.* (1986) **23**, 319–326.
4. Y. Ogata and D. Vere-Jones, Inference for earthquake models: a self-correcting model, *Stoch. Proc. Appl.* (1984) **17**, 337–347.
5. T. Hayashi, Local asymptotic normality in self-correcting point processes, in: *Statistical Theory and Data Analysis II*, K. Matusita (Ed.), North Holland, Amsterdam, New York (1988).
6. Yu.A. Kutoyants, *Parameter estimation for stochastic processes*, translated by B.S.L.Prakasa Rao, Heldermann Verlag, Berlin (1984).
7. N. Inagaki and T. Hayashi, Parameter estimation for the simple self-correcting point process, *Annl. Inst. Statist. Math.* (1990) **42**, 89–98.

Stat. Sci. & Data Anal., pp. 187-196
K. Matsusita *et al.* (Eds)

Additional Information and Precision of Estimators in Multivariate Structural Models

YUTAKA KANO, PETER M. BENTLER, and AB MOOIJAART

Department of Mathematical Sciences, College of Engineering, University of Osaka Prefecture, Sakai, Osaka 593, JAPAN,
Department of Psychology, University of California, Los Angeles, CA 90024-1563, U.S.A.,
and
Department of Psychometrics and Research, Faculty of Social Sciences, University of Leiden, Leiden, THE NETHERLANDS

Abstract. This paper investigates the effect of additional information upon parameter estimation in multivariate structural models. It is shown that the asymptotic covariances of estimators based on a model with additional variables are smaller than those based on a model with no additional variables, where the estimation methods employed are the methods of maximum likelihood and minimum chi-square. Some applications to moment structure models are provided.

Key words: Additional observation, asymptotic covariance matrix, Fisher information matrix, method of maximum likelihood, method of minimum chi-square, nuisance parameters.

1. INTRODUCTION

In multivariate analysis it is important to investigate effects of additional variables as well as the latent structure of given observed variables. In principal component analysis, discriminant analysis, and canonical correlation analysis, this issue has been studied in some detail and the results can be applied to variable selection (see e.g., Fujikoshi [1]; Fujikoshi, Krishnaiah, and Schmidhammer [2]; Wijsman [3]). In principal component analysis, for example, the largest eigenvalues of the sample covariance matrix of a set of variables are usually smaller than the corresponding ones based on the covariance matrix of original variables augmented by additional variables. By statistically observing the increase in the magnitude of the eigenvalues, one can investigate the effect of the additional variables, e.g., whether the added variables significantly contribute to the principal components. Rao [4] investigated the effect of additional variables upon the power of tests in multivariate regression.

In this article, we study a similar but alternative type of effects of additional variables in multivariate structural analysis. We define a small model and a large model in which the variables in the small model are part of those in the large model, and then estimators for a common parameter θ based on the two models are compared. A more specific setup is the following.

Let $\mathbf{y}_1$ and $\mathbf{y}_2$ be random p_1- and p_2-vectors. Assume that the distribution of $\mathbf{y}_1$ involves an unknown parameter vector θ; the distribution of $\mathbf{y}_2$ relates not only to the parameter θ but to a nuisance parameter ϕ as well. The joint distribution of $[\mathbf{y}_1', \mathbf{y}_2']'$ then depends on $[\theta', \phi']'$. We will call $\mathbf{y}_1$ and $[\mathbf{y}_1', \mathbf{y}_2']'$ a small model and a large model, respectively. The subject of this note is to investigate whether inference regarding θ is better based on the small model or based on the large model. The (asymptotic) covariance matrices of estimators for θ based on the two models are compared. The large model may have more information about θ than the small one, but the nuisance parameter ϕ involved in the large model may disturb estimation of θ.

The above situation appears in many models in multivariate structural analysis as described in Section 3. Here let us take a multiple regression model as an example. Let

$$\mathrm{E}[\mathbf{y}_1] = X_{11}\beta_1, \tag{1.1}$$

where X_{11} is a design matrix of order $p_1 \times k_1$ and β_1 is a k_1-vector of regression coefficients. Assume that the covariance matrix of $\mathbf{y}_1$ is $\sigma^2 I_{p_1}$. This is a small model. Then the best linear unbiased estimator (BLUE) for β_1 is given as $\widehat{\beta}_1 = (X_{11}'X_{11})^{-1}X_{11}'\mathbf{y}_1$, and $\mathrm{Var}(\widehat{\beta}_1) = \sigma^2(X_{11}'X_{11})^{-1}$.

A large model is constructed by adding extra observations $\mathbf{y}_2$ to $\mathbf{y}_1$. The covariance matrix of $[\mathbf{y}_1', \mathbf{y}_2']'$ is assumed to be $\sigma^2 I_{p_1+p_2}$. When the independent variables are the same as in the small model, the large model can be expressed in the form:

$$\mathrm{E}\begin{bmatrix}\mathbf{y}_1\\ \mathbf{y}_2\end{bmatrix} = \begin{bmatrix}X_{11}\\ X_{21}\end{bmatrix}\beta_1, \tag{1.2}$$

where X_{21} is of $p_2 \times k_1$. The covariance matrix of the BLUE for β_1 in the large model is $\sigma^2(X_{11}'X_{11} + X_{21}'X_{21})^{-1}$, which is obviously smaller than $\mathrm{Var}(\widehat{\beta}_1)$ in the small model. This is because the large model has more observations (or information) and no nuisance parameters to be estimated.

It is often argued that more observations need to introduce more independent variables. In such a case, the large model can be described as

$$\mathrm{E}\begin{bmatrix}\mathbf{y}_1\\ \mathbf{y}_2\end{bmatrix} = \begin{bmatrix}X_{11} & 0\\ X_{21} & X_{22}\end{bmatrix}\begin{bmatrix}\beta_1\\ \beta_2\end{bmatrix}, \tag{1.3}$$

where X_{22} is of $p_2 \times k_2$ and β_2 is a k_2-vector. Note that X_{12}=0 since the small model holds. In this circumstance, $\theta = [\beta_1', \sigma^2]'$ that is a parameter vector of interest, while $\phi = \beta_2$. Let $(\widetilde{\beta}_1, \widetilde{\beta}_2)$ denote the BLUE in the regression model (1.3). It will be shown that

$$\mathrm{Var}(\widetilde{\beta}_1) = \sigma^2\left(\begin{bmatrix}X_{11} & 0\\ X_{21} & X_{22}\end{bmatrix}'\begin{bmatrix}X_{11} & 0\\ X_{21} & X_{22}\end{bmatrix}\right)^{11} \le \sigma^2(X_{11}'X_{11})^{-1} = \mathrm{Var}(\widehat{\beta}_1), \tag{1.4}$$

where A^{11} denotes the (1,1) block of the inverse matrix A^{-1}. Consequently the large model permits better inference than the small one.

In this article, we shall show that a result similar to (1.4) holds for a wide variety of statistical models. The estimation methods treated here are the methods of maximum likelihood and minimum chi-square due to Ferguson [5]. The results are applied to moment structure models, including a factor analysis model in Section 3. In Section 4, we discuss the relationship of our results to well-known inequalities.

2. MAIN RESULTS

First we consider the method of maximum likelihood. Let the distribution of $\mathbf{y}_1$ permit the density function $f(y_1|\theta)$; let $(\mathbf{y}_1, \mathbf{y}_2)$ permit $g(y_1, y_2|\theta, \phi)$. The first one is a small model; the latter is a large model. Assume that the model parameters are identified in both models and that these density functions satisfy the usual regularity conditions to derive the following asymptotic results. The purpose here is to show that the asymptotic covariance matrix of the maximum likelihood estimator (MLE) for θ based on the large model is smaller than that based on the small model.

The Fisher information matrices in the small and large models will be written as

$$i_{\theta\theta} = \mathrm{E}\left[\left(\frac{f_\theta}{f}\right)\left(\frac{f_\theta}{f}\right)'\right],$$

$$I = \begin{bmatrix} I_{\theta\theta} & I_{\theta\phi} \\ I_{\phi\theta} & I_{\phi\phi} \end{bmatrix} = \mathrm{E}\left(\begin{bmatrix} \dfrac{g_\theta}{g} \\ \dfrac{g_\phi}{g} \end{bmatrix}\begin{bmatrix} \dfrac{g_\theta}{g} \\ \dfrac{g_\phi}{g} \end{bmatrix}'\right).$$

Here we have written $f_\theta = \frac{\partial}{\partial\theta}f(\mathbf{y}_1|\theta)$, $g_\theta = \frac{\partial}{\partial\theta}g(\mathbf{y}_1, \mathbf{y}_2|\theta, \phi)$, and $g_\phi = \frac{\partial}{\partial\phi}g(\mathbf{y}_1, \mathbf{y}_2|\theta, \phi)$. As is well-known, the asymptotic covariance matrix of the MLE is given by the inverse of the Fisher information matrix. As a consequence, the inequality to be shown is

$$i_{\theta\theta}^{-1} \geq I^{11} = \left(I_{\theta\theta} - I_{\theta\phi}I_{\phi\phi}^{-1}I_{\phi\theta}\right)^{-1}. \tag{2.1}$$

A result (b) on page 331 in Rao [6] mentions that the information matrix based on the observed vector $(\mathbf{y}_1, \mathbf{y}_2)$ is not less than that based on any function of $(\mathbf{y}_1, \mathbf{y}_2)$. This result implies that the difference

$$\begin{bmatrix} I_{\theta\theta} & I_{\theta\phi} \\ I_{\phi\theta} & I_{\phi\phi} \end{bmatrix} - \begin{bmatrix} i_{\theta\theta} & 0 \\ 0 & 0 \end{bmatrix}$$

is non-negative definite, which further implies that

$$(I_{\theta\theta} - i_{\theta\theta}) - I_{\theta\phi}I_{\phi\phi}^{-1}I_{\phi\theta}$$

is also non-negative. Thus, (2.1) follows.

A direct proof that does not use Rao's result will be given in the APPENDIX.

Next we shall consider the method of minimum chi-square (Ferguson [5]) to estimate parameters. Let $\mathbf{y}_1$ and $\mathbf{y}_2$ be statistics depending on a sample size n such that

$$\sqrt{n}\begin{bmatrix} \mathbf{y}_1 - m_1(\theta) \\ \mathbf{y}_2 - m_2(\theta, \phi) \end{bmatrix} \xrightarrow{L} N(0, V),$$

where $m_1(\theta)$ and $m_2(\theta, \phi)$ are the asymptotic expectations of $\mathbf{y}_1$ and $\mathbf{y}_2$, and $V = \begin{bmatrix} V_{11} & V_{12} \\ V_{21} & V_{22} \end{bmatrix}$ is the asymptotic covariance matrix of $[\mathbf{y}_1', \mathbf{y}_2']'$. Notice that some elements of θ may be unrelated to $m_2(\theta, \phi)$.

According to Ferguson [5], the estimator $\widehat{\theta}$ based on the small model $\mathbf{y}_1$ is defined by a solution to

$$\min_\theta (\mathbf{y}_1 - m_1(\theta))' \widehat{V}_{11}^{-1} (\mathbf{y}_1 - m_1(\theta))$$

with $\widehat{V}_{11}$ a consistent estimator for V_{11}. Let $\Delta_{11} = \dfrac{\partial m_1(\theta)}{\partial \theta'}$. The asymptotic covariance matrix of $\widehat{\theta}$ is then represented as $(\Delta_{11}' V_{11}^{-1} \Delta_{11})^{-1}$. In the same manner, the estimator $(\widetilde{\theta}, \widetilde{\phi})$ based on the large model is defined through

$$\min_{(\theta,\phi)} \begin{bmatrix} \mathbf{y}_1 - m_1(\theta) \\ \mathbf{y}_2 - m_2(\theta, \phi) \end{bmatrix}' \widehat{V}^{-1} \begin{bmatrix} \mathbf{y}_1 - m_1(\theta) \\ \mathbf{y}_2 - m_2(\theta, \phi) \end{bmatrix}, \tag{2.2}$$

where $\widehat{V}$ is also a consistent estimator for V. Let $\Delta_{21} = \dfrac{\partial m_2(\theta,\phi)}{\partial \theta'}$ and $\Delta_{22} = \dfrac{\partial m_2(\theta,\phi)}{\partial \phi'}$. The asymptotic covariance matrix of $\widetilde{\theta}$ is given by

$$\left(\begin{bmatrix} \Delta_{11} & 0 \\ \Delta_{21} & \Delta_{22} \end{bmatrix}' V^{-1} \begin{bmatrix} \Delta_{11} & 0 \\ \Delta_{21} & \Delta_{22} \end{bmatrix} \right)^{11}. \tag{2.3}$$

Thus, what to prove here is

$$(\Delta_{11}' V_{11}^{-1} \Delta_{11})^{-1} \geq \left(\begin{bmatrix} \Delta_{11} & 0 \\ \Delta_{21} & \Delta_{22} \end{bmatrix}' V^{-1} \begin{bmatrix} \Delta_{11} & 0 \\ \Delta_{21} & \Delta_{22} \end{bmatrix} \right)^{11}. \tag{2.4}$$

The proof is very simple. Since V^{-1} can be written as $\begin{bmatrix} V_{11}^{-1} & 0 \\ 0 & 0 \end{bmatrix}$ plus a nonnegative definite matrix, substitution of this into (2.3) leads to

$$\left(\begin{bmatrix} \Delta_{11}' V_{11}^{-1} \Delta_{11} & 0 \\ 0 & 0 \end{bmatrix} + M \right)^{11}$$

for some nonnegative definite matrix $M = \begin{bmatrix} M_{11} & M_{12} \\ M_{21} & M_{22} \end{bmatrix}$. Thus, we have

$$\begin{aligned} \begin{bmatrix} \Delta_{11}' V_{11}^{-1} \Delta_{11} + M_{11} & M_{12} \\ M_{21} & M_{22} \end{bmatrix}^{11} &= (\Delta_{11}' V_{11}^{-1} \Delta_{11} + M_{11} - M_{12} M_{22}^{-1} M_{21})^{-1} \\ &\leq (\Delta_{11}' V_{11}^{-1} \Delta_{11})^{-1}, \end{aligned}$$

since $M_{11} - M_{12} M_{22}^{-1} M_{21}$ is nonnegative definite. This proves (2.4).

Consequently, it is seen that the larger model always makes better inference concerning θ as long as the above two estimation methods are employed. These results trivially extend to estimators based on a wider class of discrepancy functions in view of the equivalence of minimum chi-square and other minimum discrepancy functions (Shapiro [7]) and the asymptotic equivalence of MLE and other generalized least squares estimators in the context of covariance structure analysis (Browne [8]).

The inequality in (1.4) is obtained as a corollary of the result in (2.4) when $\Delta_{ij} = X_{ij}$ and $V = \sigma^2 I_{p_1+p_2}$.

It is important and interesting to investigate when the strict inequalities hold in (2.1) and (2.4). It would be possible to establish a corresponding result on the problem of testing statistical hypotheses. These are left as the future work.

3. EXAMPLES

This work was motivated by Bentler and Chou's [9] paper, which showed numerically that the asymptotic covariances of estimates for the factor loadings in a mean and covariance structure model are substantially smaller than those in the usual factor analysis model which is a covariance structure model.

Let S be the unbiased sample covariance matrix based on a sample of size $N = n+1$. In a covariance structure model, the population covariance matrix is represented by $\Sigma(\theta)$, that is

$$\mathrm{E}[S] = \Sigma(\theta). \tag{3.1}$$

One makes inference concerning θ based only on S. The method of maximum likelihood under the normality minimizes

$$F_{WL}(S, \Sigma(\theta)) = \log|\Sigma(\theta)| - \log|S| + \mathrm{tr}[\Sigma(\theta)^{-1}S] - p.$$

In covariance structure analysis, the method of minimum chi-square due to Ferguson is called the asymptotically distribution-free (ADF) method, which was introduced by Browne ([10], [11]) and developed by Bentler and Dijkstra [12]. The ADF method uses the fact that

$$\sqrt{n}(\mathrm{v}(S) - \mathrm{v}(\Sigma(\theta))) \xrightarrow{L} N(0,\ V_{11}),$$

where $\mathrm{v}(S)$ denotes a p^*-vector consisting of the distinct elements of a symmetric matrix S with $p^* = p(p+1)/2$ and V_{11} is a $p^* \times p^*$ matrix involving the fourth-order moments of observed variables.

Let $\bar{\mathbf{x}}$ be the sample mean vector. A mean and covariance structure model assumes

$$\begin{aligned} \mathrm{E}[S] &= \Sigma(\theta) \\ \mathrm{E}[\bar{\mathbf{x}}] &= \mu(\theta, \phi), \end{aligned} \tag{3.2}$$

where θ is a parameter of interest. Thus S is a small model; $(S, \bar{\mathbf{x}})$ is a large model. Under the normality assumption, the density functions of S and $(S, \bar{\mathbf{x}})$ are, respectively, $W(s \mid \Sigma(\theta)/n, n)$ and $W(s \mid \Sigma(\theta)/n, n) \cdot N(\bar{x} \mid \mu(\theta, \phi), \Sigma(\theta)/n)$. The method of maximum likelihood is equivalent to minimizing

$$F_{WL}(S, \Sigma(\theta)) + (\bar{\mathbf{x}} - \mu(\theta, \phi))'\Sigma(\theta)^{-1}(\bar{\mathbf{x}} - \mu(\theta, \phi)).$$

See Bentler ([13] page 226).

In the general case, we know that

$$\sqrt{n}\left(\begin{bmatrix} \mathrm{v}(S) \\ \bar{\mathbf{x}} \end{bmatrix} - \begin{bmatrix} \mathrm{v}(\Sigma(\theta)) \\ \mu(\theta, \phi) \end{bmatrix}\right) \xrightarrow{L} N(0,\ V),$$

where $V_{22} = \Sigma(\theta)$ and V_{12} is a matrix of the third-order moments. The estimator for (θ, ϕ) is obtained via (2.2). Ferguson's method for this problem was first implemented by Muthén [14]. The results obtained in the previous section show that the estimators for θ in the mean and covariance structure model (3.2) are better than those in the covariance structure model (3.1).

A mean and covariance structure model is naturally introduced from a usual factor analysis model in which

$$\mathbf{x} = \Lambda(\lambda)\mathbf{f} + \mathbf{u}, \tag{3.3}$$

where $\mathbf{f}$ and $\mathbf{u}$ are common and unique factors satisfying

$$\begin{aligned} \mathrm{E}[\mathbf{f}] = 0, \quad \mathrm{E}[\mathbf{u}] = 0, \quad \mathrm{Cov}[\mathbf{f}, \mathbf{u}] = 0, \\ \mathrm{Var}[\mathbf{f}] = \Phi, \quad \text{and} \quad \mathrm{Var}[\mathbf{u}] = \Psi. \end{aligned} \tag{3.4}$$

Then we have

$$\mathrm{Var}[\mathbf{x}] = \Lambda(\lambda)\Phi\Lambda(\lambda)' + \Psi \;\; (= \Sigma(\theta), \text{ say}). \tag{3.5}$$

The common factor $\mathbf{f}$ sometimes has a nonzero mean vector μ_f. In such a case, the mean of the observable vector is

$$\mathrm{E}[\mathbf{x}] = \Lambda(\lambda)\mu_f. \tag{3.6}$$

The model defined in (3.5) and (3.6) is a mean and covariance structure model (see also Browne [15]). The results obtained previously show that the mean and covariance structure model makes better inference than the covariance structure model (3.5) as far as estimation of $\theta = (\lambda, \Phi, \Psi)$ is concerned. What Bentler and Chou [9] demonstrated numerically is that the standard errors of the $\widehat{\lambda}$ in the mean and covariance structure model are substantially smaller than those in the covariance structure model in the ADF setting. Thus, the result obtained here provides a theoretical basis for their observation.

Muthén [14] discussed the ADF estimation in a multiple-group mean and covariance structure model, where in the g-th population the observed random p-vector is of the form:

$$\mathbf{x}^{(g)} = \mu^{(g)} + \Lambda^{(g)}\mathbf{f}^{(g)} + \mathbf{u}^{(g)}$$

and the structural equation is

$$\mathbf{f}^{(g)} = \alpha^{(g)} + B^{(g)}\mathbf{f}^{(g)} + \zeta^{(g)}$$

for $g = 1, \cdots, G$. Here $\mu^{(g)}$ is a p-vector of location parameters, $\alpha^{(g)}$ is a k-vector of intercepts, $B^{(g)}$ is a $k \times k$ matrix of slopes, $\zeta^{(g)}$ is a k-vector of residuals (for more details, see Muthén [14]). In such a case, let us write

$$\begin{aligned} \mathrm{E}[\mathbf{x}^{(g)}] &= \mu^{(g)}(\theta, \phi^{(g)}), \\ \mathrm{Var}[\mathbf{x}^{(g)}] &= \Sigma^{(g)}(\theta, \phi^{(g)}) \end{aligned}$$

for $g = 1, \cdots, G$, where θ is a common parameter vector across the G populations and $\phi^{(g)}$ appears only in the g-th population.

Consider the following two estimation methods: One is to estimate $(\theta, \phi^{(1)}, \cdots, \phi^{(G)})$ simultaneously based on the pooled samples from the G populations (this is a large model); the other is to estimate $(\theta, \phi^{(g)})$ by using the only sample from the g-th population (this is a small model) for each g. For the second case, we get G estimators for θ. Then, the results obtained here prove that the asymptotic covariance matrix of $\widehat{\theta}$ in the first method is smaller than that of each of the $\widehat{\theta}$'s in the second one. Furthermore, the asymptotic covariance matrix of $\widehat{\phi}^{(g)}$ in the first method is also smaller than that in the second one.

The next example is concerned with an estimation method in a factor analysis model, which was developed by Mooijaart [16], especially for nonnormal observations. He made use of the third-order moments as well as the second-order moments, and applied Ferguson's method to estimate parameters. In the factor analysis model described in (3.3) and (3.4), we further assume that $\mathbf{f}$ and $\mathbf{u}$ are independently distributed. In such a case, the third-order moments of $\mathbf{x}$ can be expressed as

$$\mathrm{E}[\mathbf{x} \otimes \mathbf{x} \otimes \mathbf{x}] = (\Lambda(\lambda) \otimes \Lambda(\lambda) \otimes \Lambda(\lambda))\, \Phi_3 + \Psi_3 \tag{3.7}$$

where $\Phi_3 = \mathrm{E}[\mathbf{f} \otimes \mathbf{f} \otimes \mathbf{f}]$ and $\Psi_3 = \mathrm{E}[\mathbf{u} \otimes \mathbf{u} \otimes \mathbf{u}]$. Let $m_3(\theta, \phi)$ be a column vector consisting of nonduplicated elements of (3.7), where $\phi = (\Phi_3, \Psi_3)$, and let s_3 be a vector of the sample third-order moments corresponding to $m_3(\theta, \phi)$. The estimation method of Mooijaart [16] is to minimize

$$\begin{bmatrix} \mathrm{v}(S) - \mathrm{v}(\Sigma(\theta)) \\ s_3 - m_3(\theta, \phi) \end{bmatrix}' \widehat{V}^{-1} \begin{bmatrix} \mathrm{v}(S) - \mathrm{v}(\Sigma(\theta)) \\ s_3 - m_3(\theta, \phi) \end{bmatrix}. \tag{3.8}$$

Here V is a weight matrix involving up to the 6th-order moments. For more details, see Mooijaart's paper.

In this example, again, it follows that the standard error of $\widehat{\theta}$ based on the large model (3.5) and (3.7) is smaller than that based on the small model (3.5) above.

The final example is two factor analysis models with the same number of factors in which observable variates in one factor model are part of those in the other model, that is, the large model, $[\mathbf{y}_1', \mathbf{y}_2']'$, has the following covariance structure:

$$\mathrm{Var}\left(\begin{bmatrix} \mathbf{y}_1 \\ \mathbf{y}_2 \end{bmatrix}\right) = \begin{bmatrix} \Lambda_1(\lambda) \\ \Lambda_2(\lambda, \gamma) \end{bmatrix} \Phi \begin{bmatrix} \Lambda_1(\lambda) \\ \Lambda_2(\lambda, \gamma) \end{bmatrix}' + \begin{bmatrix} \Psi_1 & 0 \\ 0 & \Psi_2 \end{bmatrix}.$$

The small model, $\mathbf{y}_1$, is then

$$\mathrm{Var}(\mathbf{y}_1) = \Lambda_1(\lambda)\Phi\Lambda_1(\lambda)' + \Psi_1.$$

Several authors have investigated effects of the augmentation of the number of variables in factor analysis, including Bentler and Kano [17], Kano [18], [19], [20], Schneeweiss [21], Williams [22] and others. Kano and Shapiro [23] showed that the asymptotic covariances of the MLE for Ψ_1 decrease as observable variates increase in number. The results obtained here apply to this situation, so that it follows that the asymptotic covariances of the estimators for (λ, Φ, Ψ_1) in the large model are smaller than those in the small model. Therefore, this result includes Kano and Shapiro's as a special case.

4. DISCUSSION

We should distinguish the situation developed in this article from the one which has been often discussed, in which case one model is obtained by replacing with the true value some of the parameters in the other model, that is, one compares between

$$g(y|\theta, \phi_0) \tag{4.1}$$

and

$$g(y|\theta, \phi), \tag{4.2}$$

where ϕ_0 is the true value of ϕ. In the model (4.1), ϕ does not have to be estimated. In such a situation, the asymptotic covariances of the MLE for θ based on the model (4.1) have been shown to be smaller than those on (4.2) by using a well-known inequality:

$$I_{\theta\theta}^{-1} \leq \begin{bmatrix} I_{\theta\theta} & I_{\theta\phi} \\ I_{\phi\theta} & I_{\phi\phi} \end{bmatrix}^{11} = \left(I_{\theta\theta} - I_{\theta\phi} I_{\phi\phi}^{-1} I_{\phi\theta}\right)^{-1}. \tag{4.3}$$

More generally, Altham [24] compared between the MLEs for (θ, ϕ) based on the two models $g(y|\theta,\phi)$ and $g(y|\theta(u),\phi(u))$, where u is a more basic parameter vector and the structures $\theta(u)$ and $\phi(u)$ are known, and showed that the model using the structure permits better inference. In the analysis of covariance structures, this fact was noted by Bentler and Mooijaart [25].

We should notice that the situation treated in this article is different from the above one because in our setting the numbers of observable variates are different between two models to be compared whereas Altham has assumed that they are the same.

The result in Rao [6] that was referred to in Section 2 is closely related to the current work. Suppose that the small model $\mathbf{y}_1$ is dependent on ϕ as well as θ, and let $i = \begin{bmatrix} i_{\theta\theta} & i_{\theta\phi} \\ i_{\phi\theta} & i_{\phi\phi} \end{bmatrix}$ be the Fisher information matrix based on $\mathbf{y}_1$. Then the result in Rao guarantees that $i \leq I$, which implies that $I^{11} \leq i^{11}$, provided that i is nonsingular. The large model yields better estimation. In our setting, unfortunately, the distribution of $\mathbf{y}_1$ is *unrelated* to ϕ, so that the information matrix i is singular. Thus, our results are not a straightforward corollary of Rao's, while the theoretical development here may be minimal. Some practical implications of our results (2.1) and (2.4) are interesting and useful. As shown in Section 3, there are many examples to which the result can be applied. For instance, even when one is not interested in the mean vector but is interested in the covariance structure only, the result obtained here suggests that use of the sample mean could make the inference on the covariance structure more accurate. For another thing, computational output for the standard errors of $\widehat{\theta}$ in a large model should probably be smaller than that in a small model (this is not always true, however); therefore, if the relation does not hold, one should doubt whether the analysis has been made properly. Thus, it is worthwhile noting explicitly the superiority of large models for estimation of θ.

Consider the following inequalities

$$I_{\theta\theta}^{-1} \leq I^{11} \leq i_{\theta\theta}^{-1}.$$

The first inequality is already known as described in (4.3), and the second one is the one we have developed here. These inequalities show that the MLE for θ using the model $g(y_1, y_2|\theta, \phi)$ is better than the MLE based on $f(y_1|\theta)$, and the MLE based on $g(y_1, y_2|\theta, \phi_0)$ is the best.

The corresponding inequalities involving estimators by the method of minimum chi-square are

$$\left(\begin{bmatrix} \Delta_{11} \\ \Delta_{21} \end{bmatrix}' V^{-1} \begin{bmatrix} \Delta_{11} \\ \Delta_{21} \end{bmatrix}\right)^{-1} \leq \left(\begin{bmatrix} \Delta_{11} & 0 \\ \Delta_{21} & \Delta_{22} \end{bmatrix}' V^{-1} \begin{bmatrix} \Delta_{11} & 0 \\ \Delta_{21} & \Delta_{22} \end{bmatrix}\right)^{11} \tag{4.4}$$
$$\leq (\Delta_{11}' V_{11}^{-1} \Delta_{11})^{-1}.$$

Note that the matrix in the far left side in (4.4) represents the asymptotic covariance matrix of the estimator based on (2.2) with a known ϕ_0. It turns out that the best is the estimator using $(m_1(\theta), m_2(\theta, \phi_0))$, the second is the one based on $(m_1(\theta), m_2(\theta, \phi))$, and the worst is based on $m_1(\theta)$.

Recall the multiple regression model described in Section 1. The far left quantity in (4.4) corresponds to the model (1.2), whereas the middle and the right side in (4.4) correspond to the models (1.3) and (1.1) as already noted in Section 2.

Finally, we have assumed the models are correctly specified and the sample is large enough. Generally speaking, a larger model will require more samples than a smaller model in order that the asymptotic results are relevant. Also, in practical situations, it is computationally more difficult to fit a larger model. Nevertheless, these results are quite informative.

ACKNOWLEDGMENT

This work is supported in part by USPHS grants DA00017 and DA01070. The authors would like to thank the referee for pointing out Rao's result which helps to simplify the proof of (2.1).

APPENDIX

We give a simple and direct alternative proof of (2.1).

Since $\mathbf{y}_1$ is the vector of marginal variates of $(\mathbf{y}_1, \mathbf{y}_2)$, we note that

$$f(y_1|\theta) = \int g(y_1, y_2|\theta, \phi) dy_2,$$

from which it follows that

$$\int g_\phi(y_1, y_2|\theta, \phi) dy_2 = \frac{\partial}{\partial \phi} f(y_1|\theta) = 0.$$

From these relations, we can express the conditional expectations of $\frac{g_\phi}{g}$ and $\frac{f_\phi}{f}$ given $\mathbf{y}_1 = y_1$ as follows:

$$\mathrm{E}\left[\frac{g_\phi}{g}\middle|\mathbf{y}_1 = y_1\right] = \int \frac{g_\phi}{g}\frac{g}{f} dy_2 = \frac{1}{f}\int g_\phi dy_2 = 0, \tag{A1}$$

$$\mathrm{E}\left[\frac{g_\theta}{g}\middle|\mathbf{y}_1 = y_1\right] = \frac{1}{f}\int g_\theta dy_2 = \frac{1}{f}\frac{\partial}{\partial\theta}\int g dy_2 = \frac{f_\theta}{f}. \tag{A2}$$

By Schwartz' inequality, we can evaluate the following conditional expectation as

$$\begin{aligned} &\mathrm{E}\left[\left(\frac{g_\theta}{g} - I_{\theta\phi}I_{\phi\phi}^{-1}\frac{g_\phi}{g}\right)\left(\frac{g_\theta}{g} - I_{\theta\phi}I_{\phi\phi}^{-1}\frac{g_\phi}{g}\right)'\middle|\mathbf{y}_1\right] \\ &\qquad \geq \mathrm{E}\left[\frac{g_\theta}{g} - I_{\theta\phi}I_{\phi\phi}^{-1}\frac{g_\phi}{g}\middle|\mathbf{y}_1\right]\mathrm{E}\left[\frac{g_\theta}{g} - I_{\theta\phi}I_{\phi\phi}^{-1}\frac{g_\phi}{g}\middle|\mathbf{y}_1\right]' \\ &\qquad = \left(\frac{f_\theta}{f}\right)\left(\frac{f_\theta}{f}\right)'. \end{aligned} \tag{A3}$$

The last equality holds in view of (A1) and (A2). Note that

$$I_{\theta\theta} - I_{\theta\phi}I_{\phi\phi}^{-1}I_{\phi\theta} = \mathrm{E}\left[\left(\frac{g_\theta}{g} - I_{\theta\phi}I_{\phi\phi}^{-1}\frac{g_\phi}{g}\right)\left(\frac{g_\theta}{g} - I_{\theta\phi}I_{\phi\phi}^{-1}\frac{g_\phi}{g}\right)'\right].$$

Taking expectation of both sides in (A3) in terms of $\mathbf{y}_1$ and using the above equality lead to (2.1). *Q.E.D.*

REFERENCES

1. Y. Fujikoshi, *Ann. Inst. Statist. Math.* **34**, 523–530 (1982).
2. Y. Fujikoshi, P. R. Krishnaiah, and J. Schmidhammer, in: *Advances in Multivariate Statistical Analysis*, A. K. Gupta (Ed.), pp. 45–61. D. Reidel Publishing Company (1987).
3. R. A. Wijsman, *J. Multi. Anal.* **18**, 169–177 (1986).
4. C. R. Rao, in: *Multivariate Analysis*, P. R. Krishnaiah (Ed.), pp. 87–103. Academic Press, New York (1966).
5. T. Ferguson, *Ann. Math. Statist.* **29**, 1046–1062 (1958).
6. C. R. Rao, *Linear Statistical Inference and Its Applications.* 2nd ed. Wiley, New York (1973).
7. A. Shapiro, *South African Statist. J.* **19**, 73–81 (1985).
8. M. W. Browne, *South African Statist. J.* **8**, 1–24 (1974).
9. P. M. Bentler and C. -P. Chou, Paper presented at Psychometric meetings, Los Angeles, CA (1989, July).
10. M. W. Browne, in: *Topics in Applied Multivariate Analysis*, D. M. Hawkins (Ed.), pp. 72–141. Cambridge University Press, Cambridge, England (1982).
11. M. W. Browne, *British J. Math. Statist. Psychol.* **37**, 62–83 (1984).
12. P. M. Bentler and T. Dijkstra, in: *Multivariate Analysis* **VI**, P. R. Krishnaiah (Ed.), pp. 9–42, North Holland, Amsterdam, The Netherlands (1985).
13. P. M. Bentler, *EQS Structural Equations Program Manual.* BMDP Statistical Software, Los Angeles (1989).
14. B. Muthén, *British J. Math. Statist. Psychol.* **42**, 55–62 (1989).
15. M. W. Browne, in: *Statistical Analysis of Measurement Error Models and Applications (Contemporary Mathematics, 112)*, P. J. Brown and W. A. Fuller (Eds.), pp. 211–225. American Mathematical Society, Providence, RI (1990).
16. A. Mooijaart, *Psychometrika* **50**, 323–342 (1985).
17. P. M. Bentler and Y. Kano, *Multi. Behav. Res.* **25**, 67–74 (1990).
18. Y. Kano, *Statist. Probab. Lett.* **2**, 241–244 (1984).
19. Y. Kano, *Ann. Inst. Statist. Math.* **39**, 57–68 (1986).
20. Y. Kano, *British J. Math. Statist. Psychol.* **39**, 221–227 (1986).
21. H. Schneeweiss, in: *Statistical Analysis of Measurement Error Models and Applications (Contemporary Mathematics, 112)*, P. J. Brown and W. A. Fuller (Eds.), pp. 33–40. American Mathematical Society, Providence, RI (1990).
22. J. S. Williams, *Psychometrika* **43**, 293–306 (1978).
23. Y. Kano and A. Shapiro, *South African Statist. J.* **21**, 131–139 (1987).
24. P. M. E. Altham, *J. Roy. Statist. Soc., Ser. B* **46**, 118-119 (1984).
25. P. M. Bentler and A. Mooijaart, *Psychol. Bull.* **106**, 315–317 (1989).

Stat. Sci. & Data Anal., pp. 197-209
K. Matsusita *et al.* (Eds)

Sensitivity Analysis in Covariance Structure Analysis: A Numerical Investigation in the Case of Confirmatory Factor Analysis

SHINGO WATADANI and YUTAKA TANAKA
Sanyo Gakuen Junior College, Hirai, Okayama 700, Japan
Department of Statistics, Okayama University, Tsushima, Okayama 700, Japan

Abstract. Some influence functions have been derived for sensitivity analysis in covariance structure analysis by Tanaka, Watadani and Moon [1]. A numerical investigation is performed using these influence functions with the emphasis on the evaluation of the goodness of approximation and on the influence on model selection.

Key words: covariance structure analysis, confirmatory factor analysis, influence function.

1. INTRODUCTION

Statistical methods which belong to covariance structure analysis (CSA) have a common fundamental assumption that the covariance matrix Σ of observable variables is expressed as a function of a set of parameters $\underline{\theta} = (\theta_1, \dots \theta_m)^T$, i.e. $\Sigma = \Sigma(\underline{\theta})$. The parameter vector $\underline{\theta}$ is estimated by minimizing a function called discrepancy function or fitting function $G\big(S, \Sigma(\underline{\theta})\big)$, which measures the discrepancy between the sample covariance matrix S and the reproduced covariance matrix $\Sigma(\underline{\theta})$. Discrepancy function is given by

$$G\big(S, \Sigma(\underline{\theta})\big) = \mathrm{tr}(S - \Sigma)^2/2, \tag{1}$$

for the unweighted least squares method,

$$G\big(S, \Sigma(\underline{\theta})\big) = \mathrm{tr}(I - S^{-1}\Sigma)^2/2, \tag{2}$$

for the generalized least squares method, and

$$G\big(S, \Sigma(\underline{\theta})\big) = \mathrm{tr}(\Sigma^{-1}S) - \log|\Sigma^{-1}S| - p, \tag{3}$$

for the maximum likelihood method.

In Tanaka, Watadani and Moon [1] some influence functions have been derived and used for sensitivity analysis in CSA including confirmatory factor analysis (CFA). In the present paper we perform a numerical investigation focusing on the evaluation of the goodness of approximation and on the influence on model selection in CFA.

2. INFLUENCE FUNCTIONS IN COVARIANCE STRUCTURE ANALYSIS

In general, from the minimization problem of a discrepancy function $G(S, \Sigma(\underline{\theta}))$ which is assumed to be differentiable with respect to $\underline{\theta}$, we can obtain a set of determinating equations for the estimate $\hat{\underline{\theta}}$ such that

$$\underline{g}\left(\underline{s}, \hat{\underline{\theta}}\right) = \underline{0}, \tag{4}$$

where $\underline{g}$ is defined as an $m \times 1$ vector-valued function $\underline{g}(\underline{s}, \underline{\theta}) = \partial G(\underline{s}, \underline{\theta})/\partial\underline{\theta}$, and $\underline{s}$ is defined as a $p^* \times 1$ vector $\underline{s} = \mathrm{vech}(S) = (s_{11}, \ldots, s_{p1}; s_{22}, \ldots, s_{p2}; \ldots; s_{pp})^T$, where $p^* = p(p+1)/2$. Here, we consider a small perturbation on the data which makes a change of the sample covariance matrix $S \to S + {\scriptstyle\Delta} S$ (in vector notation $\underline{s} \to \underline{s} + {\scriptstyle\Delta}\underline{s}$), and try to evaluate the corresponding change of the estimate $\hat{\underline{\theta}} \to \hat{\underline{\theta}} + {\scriptstyle\Delta}\underline{\theta}$. For this purpose we replace $\underline{s}$ and $\hat{\underline{\theta}}$ in **(4)** by $\underline{s} + {\scriptstyle\Delta}\underline{s}$ and $\hat{\underline{\theta}} + {\scriptstyle\Delta}\underline{\theta}$, respectively, and obtain

$$\underline{g}\left(\underline{s} + {\scriptstyle\Delta}\underline{s}, \hat{\underline{\theta}} + {\scriptstyle\Delta}\underline{\theta}\right) = \underline{0}. \tag{5}$$

Applying Taylor expansion in the neighborhood of $\left(\underline{s}, \hat{\underline{\theta}}\right)$ and neglecting the higher order terms, the approximate change ${\scriptstyle\Delta}\underline{\theta}$ is obtained by

$${\scriptstyle\Delta}\underline{\theta} \cong -\left[\frac{\partial \underline{g}\left(\underline{s}, \hat{\underline{\theta}}\right)}{\partial \underline{\theta}^T}\right]^{-1} \left[\frac{\partial \underline{g}\left(\underline{s}, \hat{\underline{\theta}}\right)}{\partial \underline{s}^T}\right] {\scriptstyle\Delta}\underline{s}, \tag{6}$$

corresponding to an arbitrary change ${\scriptstyle\Delta}\underline{s}$ of the sample covariance matrix. Considering the perturbation on the empirical cdf $\hat{F}_n \to (1-\varepsilon)\hat{F}_n + \varepsilon\hat{\delta}(\underline{x}_i)$, where $\hat{\delta}(\underline{x}_i)$ is the empirical cdf of only one individual $\underline{x}_i$, the so-called empirical influence function (EIF) for the estimate $\hat{\underline{\theta}}$ is defined as the derivative of $\hat{\underline{\theta}}$ with respect to ε. Thus, from **(6)** the EIF for $\hat{\underline{\theta}}$ is obtained as

$$EIF\left(\underline{x}_i; \hat{\underline{\theta}}\right) = -\left[\frac{\partial \underline{g}\left(\underline{s}, \hat{\underline{\theta}}\right)}{\partial \underline{\theta}^T}\right]^{-1} \left[\frac{\partial \underline{g}\left(\underline{s}, \hat{\underline{\theta}}\right)}{\partial \underline{s}^T}\right] EIF(\underline{x}_i; \underline{s}). \tag{7}$$

The *EIF* for the sample covariance matrix S, which is defined as the sum of squares and products matrix divided by n, is given by

$$EIF(\underline{x}_i;\, \underline{s}) \;=\; \mathrm{vech}\big[(\underline{x}_i - \bar{\underline{x}})(\underline{x}_i - \bar{\underline{x}})^T - S\big], \tag{8}$$

in vector notation. In the similar manner the corresponding theoretical influence functions (TIF) for $\underline{\theta}$ is obtained as

$$TIF(\underline{x};\, \underline{\theta}) \;=\; -\left[\frac{\partial \underline{g}\,(\underline{\sigma}, \underline{\theta})}{\partial \underline{\theta}^T}\right]^{-1} \left[\frac{\partial \underline{g}\,(\underline{\sigma}, \underline{\theta})}{\partial \underline{s}^T}\right] TIF(\underline{x};\, \underline{\sigma}), \tag{9}$$

and the TIF for $\underline{\sigma}$ is given by

$$TIF(\underline{x};\, \underline{\sigma}) \;=\; \mathrm{vech}\big[(\underline{x} - \underline{\mu})(\underline{x} - \underline{\mu})^T - \Sigma\big], \tag{10}$$

where Σ and $\underline{\sigma}$ are the population counterparts of S and $\underline{s}$, and $\underline{\theta}$ is defined as the solution of **(4)** with $\underline{s}$ replaced by $\underline{\sigma}$.

As other sample versions of influence functions, there are the sample influence function (SIF) and the deleted-EIF. The latter, which is abbreviated by $EIF_{(i)}$, is the EIF at $\underline{x}_i$ based on the sample without the i-th individual. The SIF for $\hat{\underline{\theta}}$ is defined as

$$SIF\left(\underline{x}_i;\, \hat{\underline{\theta}}\right) \;=\; -(n-1)\left(\hat{\underline{\theta}}_{(i)} - \hat{\underline{\theta}}\right), \tag{11}$$

and the deleted-EIF for $\hat{\underline{\theta}}_{(i)}$ and $\underline{s}_{(i)}$ are obtained by

$$EIF_{(i)}\left(\underline{x}_i;\, \hat{\underline{\theta}}_{(i)}\right) \;=\; -\left[\frac{\partial \underline{g}\left(\underline{s}_{(i)}, \hat{\underline{\theta}}_{(i)}\right)}{\partial \underline{\theta}^T}\right]^{-1} \left[\frac{\partial \underline{g}\left(\underline{s}_{(i)}, \hat{\underline{\theta}}_{(i)}\right)}{\partial \underline{s}^T}\right] EIF_{(i)}\left(\underline{x}_i;\, \underline{s}_{(i)}\right) \tag{12}$$

and

$$EIF_{(i)}\left(\underline{x}_i;\, \underline{s}_{(i)}\right) \;=\; \mathrm{vech}\left[\left(\underline{x}_i - \bar{\underline{x}}_{(i)}\right)\left(\underline{x}_i - \bar{\underline{x}}_{(i)}\right)^T - S_{(i)}\right], \tag{13}$$

respectively, where $\bar{\underline{x}}_{(i)}$, $S_{(i)}$, $\underline{s}_{(i)}$ and $\hat{\underline{\theta}}_{(i)}$ indicate mean vector, the sample covariance matrix, its vector notation and the estimate of $\underline{\theta}$ based on the sample without the i-th individual.

3. INFLUENCE MEASURES

1) Influence on the estimate $\hat{\underline{\theta}}$

To summarize the vector-valued influence function (IF) for $\hat{\underline{\theta}}$ into a scalar-valued measure we shall use Euclidean norm $\|(n-1)^{-1}IF\|$ or, if the estimated asymptotic covariance matrix acov $\left(\hat{\underline{\theta}}\right)$ is available, generalized Cook's distance defined as

$$CD = (n-1)^{-2} IF^T \left[\text{acov}(\hat{\underline{\theta}})\right]^{-1} IF. \tag{14}$$

Multiplier $(n-1)^{-2}$ is attached because $-(n-1)^{-1}IF$ corresponds to $\hat{\underline{\theta}}_{(i)} - \hat{\underline{\theta}}$. For the maximum likelihood or generalized least squares estimate $\hat{\underline{\theta}}$, acov $\left(\hat{\underline{\theta}}\right)$ is given by

$$\text{acov}\left(\hat{\underline{\theta}}\right) = \frac{2}{n}\left[\frac{\partial^2 G(S, \Sigma(\hat{\underline{\theta}}))}{\partial\underline{\theta}\partial\underline{\theta}^T}\right]^{-1}, \tag{15}$$

provided that appropriate assumptions are satisfied (see, e.g. Bollen [2]).

2) Influence on the precision of the estimate

We define a summarized measure CVR by analogy with $COVRATIO$ in regression diagnostics (see Belsley, Kuh and Welsch [3]), as

$$CVR = \left|\text{acov}\left(\hat{\underline{\theta}}_{(i)}\right)\right| \Big/ \left|\text{acov}\left(\hat{\underline{\theta}}\right)\right|. \tag{16}$$

If $CVR < 1$, the precision of the estimate becomes better when the i-th individual is omitted.

3) Influence on the goodness of fit

In the case of maximum likelihood the likelihood ratio statistic X^2 for the goodness of fit changes by omitting the i-th individual as

$$\begin{aligned} \Delta X^2 &= X^2_{(i)} - X^2 \\ &= (n-1)G\left(S_{(i)}, \Sigma\left(\hat{\underline{\theta}}_{(i)}\right)\right) - nG\left(S, \Sigma(\hat{\underline{\theta}})\right). \end{aligned} \tag{17}$$

If $\Delta X^2 < 0$ the goodness of fit becomes better when the i-th individual is omitted.

4) Influence on the model selection

Akaike's information criterion (AIC) can be effectively used as a criterion for selecting models (Akaike [4]). It is defined by

$$AIC = -2(\text{log maximum likelihood}) + 2(\text{number of free parameters}). \tag{18}$$

Let $AIC(\Sigma(\underline{\theta}))$ and $AIC(S)$ be the AIC for the assumed model $\Sigma(\underline{\theta})$ and for the model without any structure on the covariance matrix, respectively. Then, we have

$$AIC(\Sigma(\underline{\theta})) - AIC(S) = X^2 - 2\phi(\Sigma(\underline{\theta})), \tag{19}$$

where $\phi(\Sigma(\underline{\theta})) = p(p+1)/2 - m$ is the difference of the numbers of free parameters for these two models, which is equal to degrees of freedom for X^2. Therefore, once ΔX^2 is obtained, it is easy to compute ΔAIC, since the essential part of $AIC(S)$ is given by $\log|S|$ and $\Delta(\log|S|)$ is easily obtained.

4. APPROXIMATION OF THE ESTIMATE $\underline{\hat{\theta}}_{(i)}$ BASED ON $(n-1)$ INDIVIDUALS

For the evaluation of the SIF and the $EIF_{(i)}$ it is needed to compute the estimate $\underline{\hat{\theta}}_{(i)}$ based on the sample without the i-th individual. Since it requires high computing cost to compute $\underline{\hat{\theta}}_{(i)}$ accurately, we consider to use some approximation formulae.

1) Linear approximation using the EIF

Using the EIF in place of the SIF in **(11)**, we obtain a linear approximation of $\underline{\hat{\theta}}_{(i)}$ as

$$\underline{\hat{\theta}}_{(i)} \cong \underline{\tilde{\theta}}_{(i)} \equiv \underline{\hat{\theta}} - (n-1)^{-1} EIF\left(\underline{x}_i; \underline{\hat{\theta}}\right). \tag{20}$$

2) One-step approximation

From the expansion of the left-hand side of **(5)**

$$\underline{g}\left(\underline{s} + \Delta\underline{s}, \underline{\hat{\theta}} + \Delta\underline{\theta}\right) = \underline{g}\left(\underline{s} + \Delta\underline{s}, \underline{\hat{\theta}}\right) + \left[\frac{\partial \underline{g}\left(\underline{s} + \Delta\underline{s}, \underline{\hat{\theta}}\right)}{\partial \underline{\theta}^T}\right] \Delta\underline{\theta} + o(\|\Delta\underline{\theta}\|), \tag{21}$$

in the neighborhood of $(\underline{s} + \Delta\underline{s}, \underline{\hat{\theta}})$, an approximate estimate is obtained as

$$\underline{\hat{\theta}}_{(i)} \cong \underline{\hat{\theta}}^1_{(i)} \equiv \underline{\hat{\theta}} - \left[\frac{\partial \underline{g}\left(\underline{s}_{(i)}, \underline{\hat{\theta}}\right)}{\partial \underline{\theta}^T}\right]^{-1} \underline{g}\left(\underline{s}_{(i)}, \underline{\hat{\theta}}\right). \tag{22}$$

This equation is equivalent to the so-called one-step approximation (see Cook and Weisberg [5], p. 182) which corresponds to a single step of Newton's method with the initial value $\underline{\hat{\theta}}$.

3) Improvement of the linear approximation by additional one-step

To improve the approximation we can consider the second order Taylor expansion. The formulation is straightforward, but it is complicated and high computing cost is required. As a better approach we apply a single step of Newton's method with the initial value $\underline{\tilde{\theta}}_{(i)}$ which is given by the linear approximation, and obtain

$$\underline{\hat{\theta}}_{(i)} \cong \underline{\hat{\theta}}^2_{(i)} = \underline{\tilde{\theta}}_{(i)} - \left[\frac{\partial \underline{g}\left(\underline{s}_{(i)}, \underline{\tilde{\theta}}_{(i)}\right)}{\partial \underline{\theta}^T}\right]^{-1} \underline{g}\left(\underline{s}_{(i)}, \underline{\tilde{\theta}}_{(i)}\right). \tag{23}$$

5. APPROXIMATION OF THE SECOND DERIVATIVES OF G

The derivatives of G with respect to the elements of $\underline{\theta}$ are given by Jöreskog [6] as

$$\frac{\partial G}{\partial \theta_i} = \mathrm{tr}\left[W(\Sigma - S)W\frac{\partial \Sigma}{\partial \theta_i}\right], \tag{24}$$

where W is defined by $W = I$ in the unweighted least squares method, $W = S^{-1}$ in the generalized least squares method and $W = \Sigma^{-1}$ in the maximum likelihood method. The second derivatives of G are obtained as

$$\begin{aligned}\frac{\partial^2 G}{\partial \theta_i \partial \theta_j} &= \mathrm{tr}\left[W\frac{\partial \Sigma}{\partial \theta_i}W\frac{\partial \Sigma}{\partial \theta_j}\right] + 2\,\mathrm{tr}\left[W(\Sigma - S)\frac{\partial W}{\partial \theta_i}\frac{\partial \Sigma}{\partial \theta_j}\right] \\ &\quad + \mathrm{tr}\left[W(\Sigma - S)W\frac{\partial^2 \Sigma}{\partial \theta_i \partial \theta_j}\right]\end{aligned} \tag{25}$$

and

$$\frac{\partial^2 G}{\partial \theta_i \partial s_{jk}} = -\,\mathrm{tr}\left[WE^*_{jk}W\frac{\partial \Sigma}{\partial \theta_i}\right] + 2\,\mathrm{tr}\left[W(\Sigma - S)\frac{\partial W}{\partial s_{jk}}\frac{\partial \Sigma}{\partial \theta_i}\right] \tag{26}$$

where E^*_{jk} is a $p \times p$ matrix with 1's in the (j,k) and (k,j) elements and 0's in the other elements. Sometimes the second derivatives **(25)** and **(26)** are evaluated only by the first terms neglecting the terms containing $\Sigma - S$. We investigate numerically the effects for this approximation on various measures of influence discussed in section 3.

6. APPLICATION TO CONFIRMATORY FACTOR ANALYSIS

In confirmatory factor analysis the $p \times p$ covariance matrix Σ is expressed in terms of the $p \times q$ loading matrix Λ, the $q \times q$ correlation matrix Φ of the common factors and the $p \times p$ diagonal unique variance matrix Ψ as

$$\Sigma = \Lambda\Phi\Lambda^T + \Psi. \tag{27}$$

The parameters to be estimated are the elements of Λ, Φ and Ψ except for those specified with restrictions on the model. The first derivatives of Σ with respect to the parameters are given as

$$\begin{aligned}\frac{\partial \Sigma}{\partial \lambda_{ir}} &= E_{ir}\Phi\Lambda^T + \Lambda\Phi E_{ri}, \\ \frac{\partial \Sigma}{\partial \phi_{rs}} &= \Lambda E^*_{rs}\Lambda^T, \quad \frac{\partial \Sigma}{\partial \psi_{ii}} = E^*_{ii},\end{aligned} \tag{28}$$

and the second derivatives of Σ with respect to parameters are given as

$$\frac{\partial^2 \Sigma}{\partial \lambda_{ir} \partial \lambda_{js}} = \phi_{rs} E_{ij}^*, \quad \frac{\partial^2 \Sigma}{\partial \lambda_{ir} \partial \phi_{st}} = E_{ir} E_{st}^* \Lambda^T + \Lambda E_{st}^* E_{ri},$$
$$\frac{\partial^2 \Sigma}{\partial \lambda_{iu} \partial \psi_{jj}} = 0, \quad \frac{\partial^2 \Sigma}{\partial \phi_{rs} \partial \phi_{tu}} = 0, \quad \frac{\partial^2 \Sigma}{\partial \phi_{rs} \partial \psi_{jj}} = 0, \quad \frac{\partial^2 \Sigma}{\partial \psi_{ii} \partial \psi_{jj}} = 0. \qquad \textbf{(29)}$$

Here we consider the case of maximum likelihood. Then the first and second derivatives of the discrepancy function G given in **(3)** can be derived by using **(25)**, **(26)**, **(28)** and **(29)** as given in Tanaka, Watadani and Moon [1]. When the approximation discussed in section 5 is used, the second derivatives of G with respect to the parameters are given by

$$\frac{\partial^2 G}{\partial \lambda_{ir} \partial \lambda_{js}} \cong 2\left(\sigma^{ij}\gamma_{rs} + \eta_{is}\eta_{jr}\right), \quad \frac{\partial^2 G}{\partial \lambda_{ir} \partial \phi_{st}} \cong 2\left(\xi_{is}\beta_{rt} + \xi_{it}\beta_{rs}\right),$$
$$\frac{\partial^2 G}{\partial \lambda_{ir} \partial \psi_{jj}} \cong 2\sigma^{ij}\eta_{jr}, \quad \frac{\partial^2 G}{\partial \phi_{rs} \partial \phi_{tu}} \cong 2\left(\alpha_{rt}\alpha_{su} + \alpha_{ru}\alpha_{st}\right), \qquad \textbf{(30)}$$
$$\frac{\partial^2 G}{\partial \phi_{rs} \partial \psi_{ii}} \cong 2\xi_{ir}\xi_{is}, \quad \frac{\partial^2 G}{\partial \psi_{ii} \partial \psi_{jj}} \cong \left(\sigma^{ij}\right)^2,$$

where

$$\begin{aligned}
\Xi &= (\xi_{ir}) = \Sigma^{-1}\Lambda, \\
\mathrm{H} &= (\eta_{ir}) = \Sigma^{-1}\Lambda\Phi = \Xi\Phi, \\
\mathrm{A} &= (\alpha_{rs}) = \Lambda\Sigma^{-1}\Lambda = \Lambda^T\Xi, \\
\mathrm{B} &= (\beta_{rs}) = \Phi\Lambda^T\Sigma^{-1}\Lambda = \Phi\mathrm{A}, \\
\Gamma &= (\gamma_{rs}) = \Phi\Lambda^T\Sigma^{-1}\Lambda\Phi = \mathrm{B}\Phi.
\end{aligned} \qquad \textbf{(31)}$$

This approximation reduces the computing cost by approximately one-half.

7. NUMERICAL EXAMPLE

We have applied our method of sensitivity analysis to the open/closed book data (Mardia, Kent and Bibby [7], Table 1.2.1). The set of data consists of $n = 88$ individuals (students) and $p = 5$ variables (tests). We assume a two-factor model in confirmatory factor analysis, where the two factors may be correlated. We also assume, from the result of exploratory factor analysis (see, Mardia et al. [7]), that the first factor is mainly related to the first two variables and that the second factor is mainly related to the last three variables.

The maximum likelihood estimates of parameters are shown in Table 1, where asterisks ($*$) indicate that the parameters with them are fixed by constraints. The goodness of fit statistic ($\mathrm{X}^2 = 2.0966$) shows that the assumed model fits the data well.

Table 1
Estimates of parameters with their standard errors (in parentheses)
$(n = 88, p = 5, q = 2)$

$\hat{\underline{\lambda}}_1$	$\hat{\underline{\lambda}}_2$	$\hat{\phi}_{12}$	diag $\hat{\Psi}$
0.7007(0.1048)	0.0000*		0.5090(0.1030)
0.7898(0.1043)	0.0000*		0.3763(0.1042)
0.0000*	0.9255(0.0873)	0.8176(0.0711)	0.1434(0.0658)
0.0000*	0.7740(0.0947)		0.4009(0.0778)
0.0000*	0.7254(0.0968)		0.4738(0.0848)

$X^2 = 2.0966$; with 4 degrees of freedom.

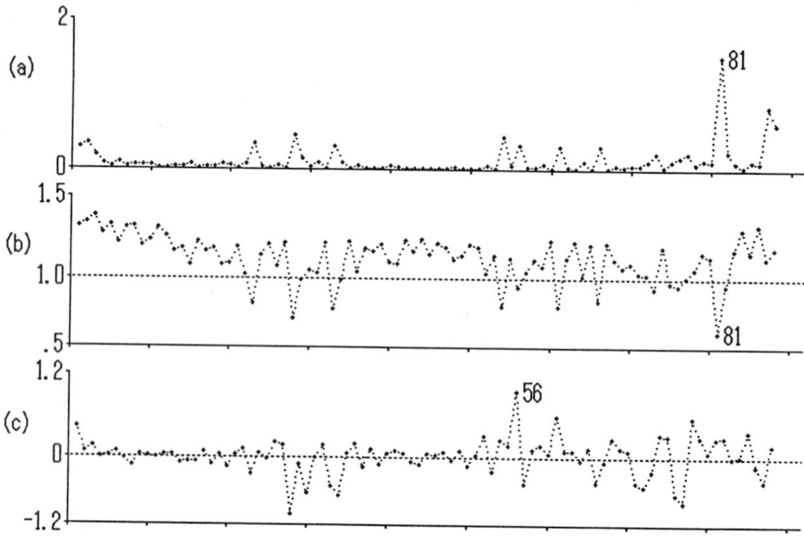

Figure 1 Index plots of (a) CD, (b) CVR and (c) ΔX^2 based on $\tilde{\underline{\theta}}_{(i)}$.

To investigate the influence of each individual we have computed three influence measures CD, CVR, and ΔX^2 based on the exact estimate $\hat{\underline{\theta}}_{(i)}$ (or SIF) and the linear approximation $\tilde{\underline{\theta}}_{(i)}$ (or EIF). Figure 1 shows index plots of these influence measures based on $\tilde{\underline{\theta}}_{(i)}$. From these index plots we can say that there is an highly influential individual, number 81, and that, when this individual is omitted, the goodness of fit becomes a little worse while the precision becomes much better. Table 2 shows the

Table 2
Estimates of parameters with their standard errors (in parentheses) based on the sample without the 81-th individual
($n = 87, p = 5, q = 2$)

$\hat{\underline{\lambda}}_1$	$\hat{\underline{\lambda}}_2$	$\hat{\phi}_{12}$	diag $\hat{\Psi}$
0.6557(0.1023)	0.0000*		0.5325(0.0992)
0.7384(0.0979)	0.0000*		0.3499(0.0908)
0.0000*	0.9377(0.0866)	0.8786(0.0673)	0.1322(0.0626)
0.0000*	0.7737(0.0950)		0.4129(0.0780)
0.0000*	0.7236(0.0971)		0.4876(0.0854)

$X^2 = 2.4512$; with 4 degrees of freedom.

estimates of parameters for the data without the individual number 81. Actually, as we expected from Figure 1, the goodness of fit has become worse and the precision has become better by omitting this individual. The effects on the estimates are not small, but within $\pm SE$.

Figure 2 shows three scatter diagrams of the influence measures based on the linear approximation $\tilde{\underline{\theta}}_{(i)}$ versus those based on the exact estimate $\hat{\underline{\theta}}_{(i)}$. These scatter diagrams show that the linear approximation based on the EIF is so good that we can use the approximate measures instead of the exact ones for detecting influential individuals.

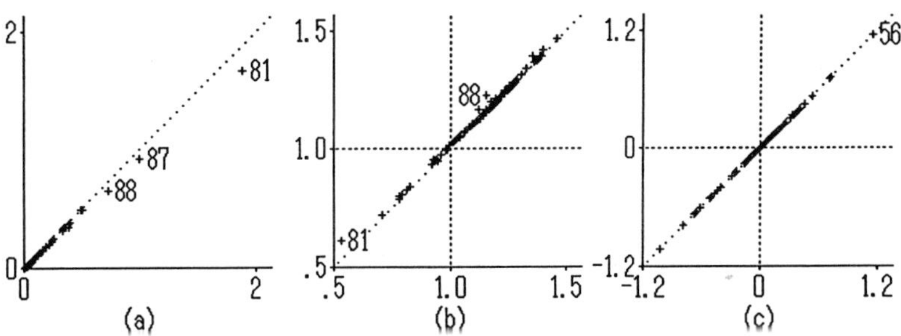

Figure 2 Scatter diagrams of (a) CD, (b) CVR and (c) ΔX^2 based on $\tilde{\underline{\theta}}_{(i)}$ (vertical) versus $\hat{\underline{\theta}}_{(i)}$ (horizontal).

We have also computed the influence measures based on the one-step approximation $\hat{\underline{\theta}}^1_{(i)}$ and the additional one-step after the linear approximation $\hat{\underline{\theta}}^2_{(i)}$. Figure 3

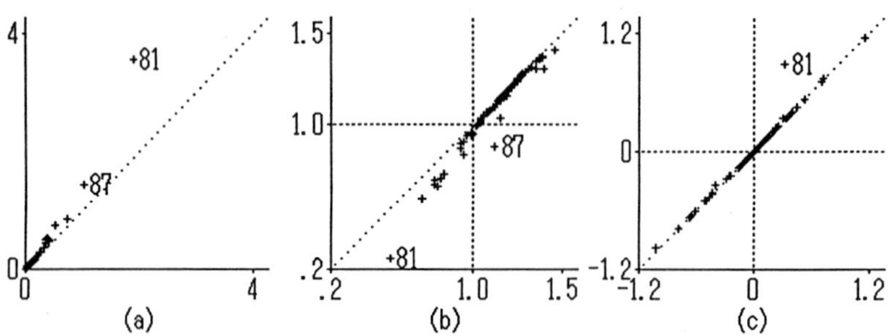

Figure 3 Scatter diagrams of (a) CD, (b) CVR and (c) ΔX^2 based on $\hat{\underline{\theta}}^1_{(i)}$ (vertical) versus $\hat{\underline{\theta}}_{(i)}$ (horizontal).

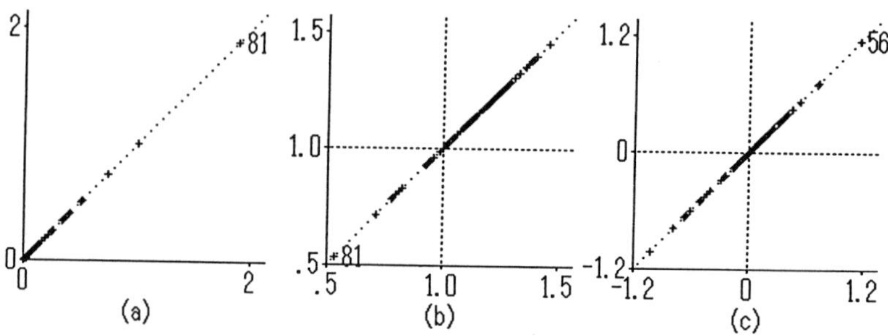

Figure 4 Scatter diagrams of (a) CD, (b) CVR and (c) ΔX^2 based on $\hat{\underline{\theta}}^2_{(i)}$ (vertical) versus $\hat{\underline{\theta}}_{(i)}$ (horizontal).

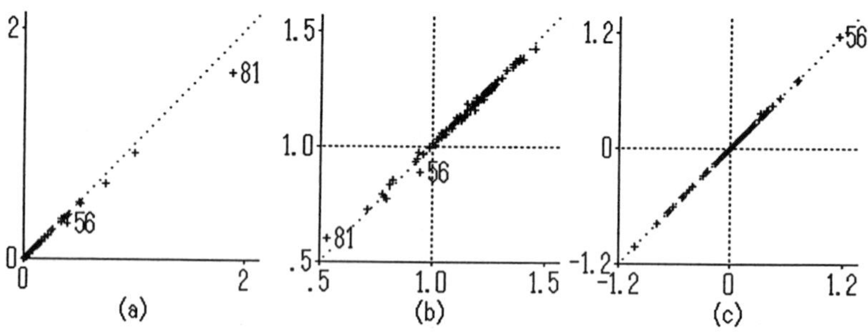

Figure 5 Scatter diagrams of (a) CD, (b) CVR and (c) ΔX^2 based on approximated $\tilde{\underline{\theta}}_{(i)}$ (vertical) versus $\hat{\underline{\theta}}_{(i)}$ (horizontal).

shows scatter diagrams of influence measures based on $\hat{\underline{\theta}}^1_{(i)}$ versus the exact estimate $\hat{\underline{\theta}}_{(i)}$, and Figure 4 shows those based on $\hat{\underline{\theta}}^2_{(i)}$ versus $\hat{\underline{\theta}}_{(i)}$. It is found, from Figure 2 and 3, that the linear approximation is better than the one-step approximation, and, from figure 2 and 4, that the approximation has improved much by the additional one-step.

To investigate the effect of the approximation in the second derivatives of G, we have computed the influence measures based on $\tilde{\underline{\theta}}_{(i)}$ with the second derivatives approximated by **(30)**. Figure 5 shows scatter diagrams of the influence measures based on $\tilde{\underline{\theta}}_{(i)}$ using the approximate second derivatives versus those based on $\hat{\underline{\theta}}_{(i)}$. The comparison of Figure 2 and Figure 5 reveals that the effect of this approximation is small and almost negligible.

Table 3
Estimates of parameters with their standard errors (in parentheses) for the second best model
($n = 88, p = 5, q = 2$)

$\hat{\underline{\lambda}}_1$	$\hat{\underline{\lambda}}_2$	$\hat{\phi}_{12}$	diag $\hat{\Psi}$
0.7027(0.1045)	0.0000*		0.5062(0.1025)
0.7875(0.1040)	0.0000*		0.3798(0.1032)
0.2682(0.1578)	0.6827(0.1569)	0.7432(0.0943)	0.1898(0.0601)
0.0000*	0.8063(0.0954)		0.3499(0.0806)
0.0000*	0.7530(0.0969)		0.4329(0.0842)

$X^2 = 0.1639$; with 3 degrees of freedom.

Next, we shall study the influence on model selection. To do this we consider other possible two-factor models as rival models. Among them we show the results for the model which is the second best following the original one. It is a model in which the first factor is related to the first three variables and the second factor is related to the last three variables. The estimated parameters in this model are shown in Table 3. The comparison of the AIC values shows that this rival model ($AIC = 1070.3147$) is a little worse than the original model ($AIC = 1070.2475$) for the whole sample. Figure 6 shows the index plots of $AIC\left(\Sigma(\tilde{\underline{\theta}}_{(i)})\right) - AIC(S_{(i)})$. From these index plots, it is found that, though the original model is better than the other for the whole sample, this rival model sometimes takes a smaller AIC value than the original model when we omit one of the individuals such as numbers 1, 56, 61 and 78. Particularly, the order of the AIC values is clearly interchanged for the sample without the 56-th individual. Also it is

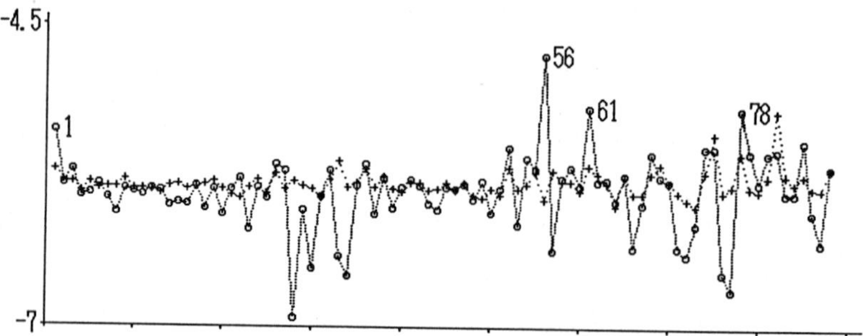

Figure 6 Index plots of $AIC\left(\Sigma(\tilde{\underline{\theta}}_{(i)})\right) - AIC(S_{(i)})$ (o: the original model, +: the rival model).

noticed that the variation of the value of $AIC\left(\Sigma(\tilde{\underline{\theta}}_{(i)})\right) - AIC(S_{(i)})$ (which is equivalent to the variation of the X^2 values) in the original model is larger than in the rival model. It may have connection with the fact that X^2 for the original model follows the chi-squared distribution with 4 degrees of freedom, while X^2 for the rival model follows the same type of distribution with 3 degrees of freedom, when both models fit the data well. In the above we used $AIC\left(\Sigma(\tilde{\underline{\theta}}_{(i)})\right) - AIC(S_{(i)})$ instead of $AIC\left(\Sigma(\tilde{\underline{\theta}}_{(i)})\right)$. The reason why $AIC\left(\Sigma(\tilde{\underline{\theta}}_{(i)})\right) - AIC(S_{(i)})$ can be used is that, since $AIC\left(\Sigma(\tilde{\underline{\theta}}_{(i)})\right)$ can be decomposed into two parts, $AIC\left(\Sigma(\tilde{\underline{\theta}}_{(i)})\right) - AIC(S_{(i)})$ and $AIC(S_{(i)})$, where the latter part does not depend on the assumed model, the result of model selection does not change by using $AIC\left(\Sigma(\tilde{\underline{\theta}}_{(i)})\right) - AIC(S_{(i)})$ instead of $AIC\left(\Sigma(\tilde{\underline{\theta}}_{(i)})\right)$. Also it may be worth to note that it was difficult to find the difference between the two models from the index plots of $AIC\left(\Sigma(\tilde{\underline{\theta}}_{(i)})\right)$, because the variation of $AIC(S_{(i)})$ ($SE = 3.436$) masked the variation of $AIC\left(\Sigma(\tilde{\underline{\theta}}_{(i)})\right) - AIC(S_{(i)})$ ($SE = 0.3207$ for the original model and $SE = 0.1084$ for the rival model).

8. CONCLUDING REMARKS

From the above numerical investigation we can say the followings:

1) The *EIF* can be used for detecting influential individuals instead of the *SIF*, which has the clear "leave-one-out" interpretation.
2) The linear approximation using the *EIF* is at least as good as the so-called one-step approximation. The former is better than the latter from the computational

point of view.

3) The additional one-step approximation is effective for the improvement of approximation.
4) The terms containing $(\Sigma - S)$ in the second derivatives of G have very small effects on the results. The neglection of these terms will be helpfull in complicated models, because the part of computing the second derivatives of G is the most troublesome in the development of the program.
5) We should be careful in model selection based on AIC, because there is a possibility that the model with the minimum AIC value varies by the influence of a few individuals.

Acknowledgement

The authors express deep thanks to the reviewer for valuable comments.

REFERENCES

1. Y. Tanaka, S. Watadani and S. H. Moon, Influence in covariance structure analysis: with an application to confirmatory factor analysis, *Comm. Statist.* **A20**, 3805–3821 (1991).

2. H. A. Bollen, *Structural Equations with Latent Variables*. John Wiley & Sons (1989).

3. D. A. Belsley, E. Kuh and R. E. Welsch, *Regression Diagnostics: Identifying Influential Data and Sources of Collinearity*. John Wiley & Sons. (1980).

4. H. Akaike, Factor analysis and AIC, *Psychometrika*, **52**, 317–332 (1973).

5. D. R. Cook and S. Weisberg, *Residuals and Influence in Regression*. Chapman and Hall. (1982).

6. K. G. Jöreskog Structural analysis of covariance and correlation matrices, *Psychometrika*, **43**, 443–477 (1978).

7. K. V. Mardia, J. T. Kent and J. M. Bibby, *Multivariate Analysis*. Academic Press. (1979).

Stat. Sci. & Data Anal., pp. 211-223
K. Matsusita *et al.* (Eds)

Likelihood Ratio Tests for Means and Covariances with Incomplete Multinormal Observations

SERGE B. PROVOST
Department of Statistical and Actuarial Sciences, The University of Western Ontario, London, Ontario, Canada N6A 3M1

Abstract. We consider a monotonic sample ($\mathbf{y}_j$, $j = 1, \ldots, n_1$) which may be defined as follows: whenever a subvector $\mathbf{y}_{ij}$ is missing from the observation vector $\mathbf{y}_j$, then all the subvectors $\mathbf{y}_{ab}$ for which $a \geq i$ and $b \geq j$ will also be missing. For such a sample, the observations vectors form a triangular array. We derive likelihood ratio statistics for testing hypotheses on mean vectors and covariance matrices on the basis of a monotonic normal sample. We also give the asymptotic distributions of these test statistics.

Keywords: Missing data, monotonic samples, likelihood ratio test, maximum likelihood estimates, asymptotic distribution.

1. INTRODUCTION

When analyzing the variables involved in certain experiments, it may happen that some of the observation vectors may be incomplete. This might occur by design, for example when the cost of collecting data on some of the variables is prohibitive, or by accident, for example when the data collected on certain variables are missing or unreliable due to the methodology or the recording equipment used. It may also happen that some respondents are unwilling to answer certain items on a survey questionnaire or that researchers who did not observe exactly the same set of variables combine their data.

Inference problems connected with various statistical models developed for handling missing data have been studied by Wilks[1], Lord[2], Edgett[3], Rao[4], Anderson[5], Trawinski and Bargmann[6], Mathai[7], Afifi and Elashoff[8,9,10,11], Smith and Pfaffenberger[12], Morrison[13], Eaton and Kariya[14,15,16] , Bhargava[17], Little[18], Rubin[19], Radhakrishnan[20], Srivastava[21], and Provost[22], among others. An important reference in connection with the analysis of experiments is Dodge[23].

It is assumed in this paper that the complete and incomplete observation vectors form a monotonic sample as defined in Section 2. The likelihood function of a monotonic sample from a multivariate normal population is maximized in Section 3. The likelihood ratio statistic for testing the independence of q subvectors on the basis of a monotonic sample is derived in Section 4 where the asymptotic distribution of the test statistic is also given . Likelihood ratio tests for means and covariances are respectively derived in Sections 5 and 6.

2. DEFINITION OF A MONOTONIC SAMPLE

Let

$$Y_{(\ell)} = \begin{pmatrix} Y_1 \\ Y_2 \\ \vdots \\ Y_\ell \end{pmatrix} \overset{ind}{\sim} N_{p_{(\ell)}}(\mu_{(\ell)}, \Sigma_{(\ell)}) ,$$

$$\begin{aligned} &\ell = 1, \dots, r; \quad Y_i : p_i \times 1 , \; i = 1, \dots, \ell ; \\ &Y_{(\ell)} : p_{(\ell)} \times 1, \quad p_{(\ell)} = p_1 + p_2 + \cdots + p_\ell ; \end{aligned} \tag{2.1}$$

that is, the $Y_{(\ell)}$'s are independently distributed as $p_{(\ell)}$–variate normal vectors with mean vectors $\mu'_{(\ell)} = (\mu'_1, \dots, \mu'_\ell)$ and positive definite covariance matrices $\Sigma_{(\ell)} \equiv (\Sigma_{kj})$, with $\Sigma_{kj} : p_k \times p_j, 1 \le k, j \le \ell$, $\ell = 1, \dots, r$.

Let

$$Y = Y_{(r)} \; , \; p = p_{(r)} \; , \; \mu = \mu_{(r)} \; \text{ and } \; \Sigma = \Sigma_{(r)} \; . \tag{2.2}$$

Let $(y_1, \dots, y_{n_r}), (y_{n_r+1}, \dots, y_{n_{r-1}}), \dots, (y_{n_3+1}, \dots, y_{n_2})$, and $(y_{n_2+1}, \dots, y_{n_1})$ be independent random samples from $Y_{(r)}, Y_{(r-1)}, \dots, Y_{(2)}$, and $Y_{(1)}$, respectively, where $Y_{(\ell)}, \ell = 1, \dots, r$, is defined in (2.1), and let each observation vector be partitioned as in (2.1):

$$y_j = \begin{pmatrix} y_{1j} \\ y_{2j} \\ \vdots \\ y_{\ell j} \end{pmatrix} \qquad \text{for } \; j = n_{\ell+1} + 1, \dots, n_\ell \; ; \; n_\ell > p_\ell \ge 1 \; ;$$

$$\ell = 1, \dots, r \; ; \quad n_{r+1} = 0 \; ; \tag{2.3}$$

the n_1 vectors y_j defined in (2.3) are referred to as a monotonic sample of size $(n_1, n_2, \dots, n_r)$, on $p_{(1)}, p_{(2)}, \dots, p_{(r)}$ subvectors $[n_1 > n_2 > \cdots > n_r > p = p_{(r)} > p_{(r-1)} > \cdots > p_{(2)} > p_{(1)}]$ from the multivariate normal distribution $N_p(\mu, \Sigma)$.

Monotonic samples exhibit the following triangular pattern:

$$\begin{array}{ccccccccccccc}
y_{11} & \cdots & y_{1n_r} & \cdots & y_{1n_{r-1}} & \cdots & \cdots & y_{1n_3} & \cdots & y_{1n_2} & \cdots & y_{1n_1} \\
y_{21} & \cdots & y_{2n_r} & \cdots & y_{2n_{r-1}} & \cdots & \cdots & y_{2n_3} & \cdots & y_{2n_2} & & \\
y_{31} & \cdots & y_{3n_r} & \cdots & y_{3n_{r-1}} & \cdots & \cdots & y_{3n_3} & & & & \\
\cdots & \cdots & \cdots & \cdots & \cdots & \cdots & \cdots & & & & & \\
\cdots & \cdots & \cdots & \cdots & \cdots & \cdots & & & & & & \\
y_{r-1,1} & \cdots & y_{r-1,n_r} & \cdots & y_{r-1,n_{r-1}} & & & & & & & \\
y_{r,1} & \cdots & y_{r,n_r} & & & & & & & & &
\end{array} \tag{2.4}$$

Definition 1. A sample $\{y_j, \; j = 1, \dots, n_1\}$ from a p-variate population is called a monotonic sample if whenever a subvector y_{ij} is missing from the observation vector y_j, then all the subvectors y_{ab} for which $a \ge i$ and $b \ge j$ will also be missing.

3. MAXIMIZATION OF THE LIKELIHOOD FUNCTION

Let

$$f_{(\ell)}(\boldsymbol{y}_{(\ell)}) = f_{(\ell)}(\boldsymbol{y}_1, \boldsymbol{y}_2, \ldots, \boldsymbol{y}_\ell) = \frac{\exp\{-\frac{1}{2}(\boldsymbol{y}_{(\ell)} - \boldsymbol{\mu}_{(\ell)})'\Sigma_{(\ell)}^{-1}(\boldsymbol{y}_{(\ell)} - \boldsymbol{\mu}_{(\ell)})\}}{(2\pi)^{p_{(\ell)}/2}|\Sigma_{(\ell)}|^{\frac{1}{2}}} \tag{3.1}$$

denote the joint normal density of the random vectors $\boldsymbol{Y}_1, \ldots, \boldsymbol{Y}_\ell$ defined in (2.1), $\ell = 1, \ldots, r$. It is assumed that the mean vectors and the covariance matrices are unknown.

Denoting the likelihood function of $(\boldsymbol{y}_{n_{\ell+1}+1}, \ldots, \boldsymbol{y}_{n_\ell})$ by

$$L_{(\ell)}(\boldsymbol{\mu}_{(\ell)}, \Sigma_{(\ell)}) = \prod_{j=n_{\ell+1}+1}^{n_\ell} f_{(\ell)}(\boldsymbol{y}_j) \ , \ \ell = 1, \ldots, r \ ; \ n_{r+1} = 0 \ ,$$

where $f_{(\ell)}(\cdot)$ is given in (3.1), the likelihood function of the monotonic sample defined in (2.3) can be written as

$$L_1(\boldsymbol{\mu}, \Sigma) = \prod_{l=1}^{r} L_{(\ell)}(\boldsymbol{\mu}_{(\ell)}, \Sigma_{(\ell)}) \ . \tag{3.2}$$

However this representation of the likelihood function does not lend itself to the derivation of maximum likelihood estimates (m.l.e.'s).

A more convenient representation of the overall likelihood functions can be obtained by expressing the joint density in terms of a marginal and conditional densities as follows:

$$\begin{aligned} f_{(r)}(\boldsymbol{y}_{(r)}) = f_{(r)}(\boldsymbol{y}_1, \boldsymbol{y}_2, \ldots, \boldsymbol{y}_r) =& f_{(1)}(\boldsymbol{y}_{(1)}) \cdot f_{2.(1)}(\boldsymbol{y}_2|\boldsymbol{y}_{(1)}) \cdot f_{3.(2)}(\boldsymbol{y}_3|\boldsymbol{y}_{(2)}) \\ & \cdots f_{r-1.(r-2)}(\boldsymbol{y}_{r-1}|\boldsymbol{y}_{(r-2)}) \cdot f_{r.(r-1)}(\boldsymbol{y}_r|\boldsymbol{y}_{(r-1)}) \end{aligned} \tag{3.3}$$

where $f_{(1)}(\boldsymbol{y}_{(1)})$ given in (3.1) is the marginal density of $\boldsymbol{Y}_{(1)}$ and for $k = 2, \ldots, r$, $f_{k.(k-1)}(\boldsymbol{y}_k|\boldsymbol{y}_{(k-1)})$ denotes the conditional normal density of $\boldsymbol{Y}_k$ given $\boldsymbol{Y}_{(k-1)} = \boldsymbol{y}_{(k-1)}$ where $\boldsymbol{y}_{(k-1)}$ is a fixed real vector having $p_{(k-1)}$ components. The conditional mean vector and covariance matrix of $\boldsymbol{Y}_\ell$ given $\boldsymbol{Y}_{(\ell-1)} = \boldsymbol{y}_{(\ell-1)}$ are respectively

$$\boldsymbol{\mu}_{\ell\cdot(\ell-1)} = \boldsymbol{\mu}_\ell + \Sigma^*_{\ell(\ell-1)}(\boldsymbol{y}_{(\ell-1)} - \boldsymbol{\mu}_{(\ell-1)}) = \boldsymbol{\mu}^*_\ell + \Sigma^*_{\ell(\ell-1)}\boldsymbol{y}_{(\ell-1)} \tag{3.4}$$

and

$$\Sigma_{\ell\ell.(\ell-1)} = \Sigma_{\ell\ell} - \Sigma^*_{\ell(\ell-1)}\Sigma_{(\ell-1)\ell} \ , \ \ell = 2, \ldots, r \ , \tag{3.5}$$

where

$$\Sigma_{(\ell-1)\ell} = \begin{pmatrix} \Sigma_{1\ell} \\ \Sigma_{2\ell} \\ \vdots \\ \Sigma_{\ell-1,\ell} \end{pmatrix} \ , \quad \Sigma_{\ell(\ell-1)} = \Sigma'_{(\ell-1)\ell} \ , \tag{3.6}$$

$$\Sigma^*_{\ell(\ell-1)} = \Sigma_{\ell(\ell-1)}\Sigma^{-1}_{(\ell-1)} \ , \ \boldsymbol{\mu}^*_\ell = \boldsymbol{\mu}_\ell - \Sigma^*_{\ell(\ell-1)}\boldsymbol{\mu}_{(\ell-1)} \ , \quad \ell = 2, \ldots, r \ , \tag{3.7}$$

(see Anderson [24, p.36], equations (5) and (3) for example).

Letting

$$f_{1.(0)}(\boldsymbol{y}_1|\boldsymbol{y}_{(0)}) = f_{(1)}(\boldsymbol{y}_{(1)}) \ , \ \boldsymbol{\mu}_{(0)} = \mathbf{0} \ , \ \boldsymbol{\mu}_{1.(0)} = \boldsymbol{\mu}_{(1)} = \boldsymbol{\mu}_1,$$
$$\Sigma_{1(0)} \equiv O \quad \text{and} \quad \Sigma_{11.(0)} = \Sigma_{(1)} = \Sigma_{11} \ , \tag{3.8}$$

one can write

$$f_{(r)}(\boldsymbol{y}_{(r)}) = \prod_{\ell=1}^{r} f_{\ell\cdot(\ell-1)}(\boldsymbol{y}_\ell|\boldsymbol{y}_{(\ell-1)})$$
$$= \prod_{\ell=1}^{r} \frac{\exp\{-\frac{1}{2}(\boldsymbol{y}_\ell - \boldsymbol{\mu}_{\ell\cdot(\ell-1)})'\Sigma_{\ell\ell\cdot(\ell-1)}^{-1}(\boldsymbol{y}_\ell - \boldsymbol{\mu}_{\ell\cdot(\ell-1)})\}}{(2\pi)^{p_\ell/2}|\Sigma_{\ell\ell.(\ell-1)}|^{\frac{1}{2}}} \ . \tag{3.9}$$

Letting

$$\boldsymbol{y}_{(\ell-1)j} = \begin{pmatrix} \boldsymbol{y}_{1j} \\ \boldsymbol{y}_{2j} \\ \vdots \\ \boldsymbol{y}_{\ell-1,j} \end{pmatrix} \ , \ j = 1,\ldots,n_\ell \ , \ \ell = 2,\ldots,r \ , \tag{3.10}$$

we denote the likelihood function of $\boldsymbol{y}_{\ell j}$ given $\boldsymbol{y}_{(\ell-1)j}$, for $j = 1,\ldots,n_\ell$, by

$$L_{\ell.(\ell-1)}(\boldsymbol{\mu}_{\ell.(\ell-1)} \ , \ \Sigma_{\ell\ell.(\ell-1)}) = \prod_{j=1}^{n_\ell} f_{\ell.(\ell-1)}(\boldsymbol{y}_{\ell j}|\boldsymbol{y}_{(\ell-1)j}) \ , \ \text{for } \ell = 1,2.,\ldots\ r, \tag{3.11}$$

where

$$L_{1.(0)}(\boldsymbol{\mu}_{1.(0)} \ , \ \Sigma_{11.(0)}) = \prod_{j=1}^{n_1} f_{(1)}(y_{1j}) \ ,$$
$$\boldsymbol{\mu}_{(1).0} = \boldsymbol{\mu}_{(1)} = \boldsymbol{\mu}_1 \ , \ \Sigma_{11.(0)} = \Sigma_{(1)} = \Sigma_{11} \ , \tag{3.12}$$

the conditional density $f_{\ell.(\ell-1)}(\cdot)$ is given in (3.9) and $\boldsymbol{\mu}_{\ell.(\ell-1)}$ and $\Sigma_{\ell\ell.(\ell-1)}$ are respectively given in (3.4) and (3.5). The overall likelihood function of the monotonic sample defined in (2.3) can then be expressed as follows in view of (3.3)

$$L_2(\boldsymbol{\mu}_{\ell.(\ell-1)} \ , \ \Sigma_{\ell\ell.(\ell-1)}; \ \ \ell = 1,\ldots,r) = \prod_{\ell=1}^{r} L_{\ell.(\ell-1)}(\boldsymbol{\mu}_{\ell.(\ell-1)} \ , \ \Sigma_{\ell\ell.(\ell-1)}) \ . \tag{3.13}$$

There is a one–to–one mapping between the set of parameters

$$M_1 = \{\boldsymbol{\mu}, \Sigma\} = \{\boldsymbol{\mu}_{(r)} \ , \ \Sigma_{(r)}\} = \{\boldsymbol{\mu}_j \ , \ \Sigma_{kj} \ , \ j = 1,\ldots,r; \ \ k \le j\} \ , \tag{3.14}$$

and

$$M_2 = \bigcap_{\ell=1}^{r} M_2^\ell = \bigcap_{\ell=1}^{r} \{\boldsymbol{\mu}_\ell^* \ , \Sigma_{\ell(\ell-1)}^* \ , \ \Sigma_{\ell\ell.(\ell-1)}\} \tag{3.15}$$

where

$$\boldsymbol{\mu}_\ell^* = \boldsymbol{\mu}_\ell - \Sigma_{\ell(\ell-1)}^* \boldsymbol{\mu}_{(\ell-1)} \ ; \tag{3.16}$$
$$\boldsymbol{\mu}_1^* = \boldsymbol{\mu}_1 \ , \ \Sigma_{1(0)}^* = O \text{ (the null matrix)} \quad \text{and} \quad \Sigma_{11.(0)} = \Sigma_{11},$$

and $\Sigma_{\ell\ell.(\ell-1)}$ and $\Sigma^*_{\ell(\ell-1)}$ are defined in (3.5) and (3.7) respectively. This is easily seen by noticing that if we fix the parameters of either set M_1 or M_2, the parameters of the other set will be uniquely determined. Then, noting that $L_1(\boldsymbol{\mu}, \Sigma) \equiv L_1(M_1)$ and that $L_2(\boldsymbol{\mu}_{\ell.(\ell-1)}, \Sigma_{\ell\ell.(\ell-1)};\ \ell = 1, \ldots, r) \equiv L_2(M_2)$ and using Lemma 3.2.3 of Anderson [24]:

"Let $f(\Theta)$ be a real-valued function defined on a set S and let ϕ be a single-valued function, with a single-valued inverse, on S to a set S^*; that is to each $\Theta \epsilon S$ there corresponds a unique $\Theta^* \epsilon S^*$. Let $g(\Theta^*) = f(\phi^{-1}(\Theta^*))$. Then if $f(\Theta)$ attains a maximum at $\Theta = \Theta_0$, $g(\Theta^*)$ attains a maximum at $\Theta^* = \Theta_0^* = \phi(\Theta_0)$. If the maximum of $f(\Theta)$ at Θ_0 is unique, so is the maximum of $g(\Theta^*)$ at Θ_0^*."

one can maximize the likelihood function of the monotonic sample L_2 given in (3.13) with respect to the set of parameters M_2 which is equivalent to maximizing L_1 given in (3.2) with respect to the set M_1.

We now consider the following subset of the monotonic sample defined in (2.3)

$$Y_{(\ell)} = \begin{pmatrix} \boldsymbol{y}_{1,1} & \cdots & \boldsymbol{y}_{1,n_\ell} \\ \vdots & & \vdots \\ \boldsymbol{y}_{\ell,1} & \cdots & \boldsymbol{y}_{\ell,n_\ell} \end{pmatrix} \equiv \boldsymbol{y}_{(\ell)1}, \ldots, \boldsymbol{y}_{(\ell)n_\ell} \tag{3.17}$$

$$Y_{(\ell)} : p_{(\ell)} \times n_\ell\ ;\ \boldsymbol{y}_{(\ell)j} : p_{(\ell)} \times 1\ ,\ j = 1, \ldots, n_\ell;\ \ \ell = 1, \ldots, r.$$

We note that $L_{\ell\cdot(\ell-1)}$ is given in (3.11) in terms of $Y_{(\ell)}$ since

$$\boldsymbol{y}'_{(\ell)j} = (\boldsymbol{y}'_{(\ell-1)j}, \boldsymbol{y}'_{\ell j})\ ,\ j = 1, \ldots, n_\ell\ . \tag{3.18}$$

Let

$$\bar{\mathbf{y}}_{(\ell)} = \sum_{j=1}^{n_\ell} \boldsymbol{y}_{(\ell)j}/n_\ell = \begin{pmatrix} \sum_{j=1}^{n_\ell} \boldsymbol{y}_{(\ell-1)j}/n_\ell \\ \sum_{j=1}^{n_\ell} \boldsymbol{y}_{\ell j}/n_\ell \end{pmatrix} \equiv \begin{pmatrix} \bar{\mathbf{y}}^0_{(\ell-1)} \\ \bar{\mathbf{y}}_\ell \end{pmatrix} \tag{3.19}$$

$$S_{(\ell)} = \sum_{j=1}^{n_\ell} (\boldsymbol{y}_{(\ell)j} - \bar{\mathbf{y}}_{(\ell)})(\boldsymbol{y}_{(\ell)j} - \bar{\mathbf{y}}_{(\ell)})' \equiv \begin{pmatrix} S_{(\ell)11} & S_{(\ell)12} \\ S_{(\ell)21} & S_{(\ell)22} \end{pmatrix} \tag{3.20}$$

where

$$S_{(\ell)11} = \sum_{j=1}^{n_\ell} (\boldsymbol{y}_{(\ell-1)j} - \bar{\boldsymbol{y}}^0_{(\ell-1)})(\boldsymbol{y}_{(\ell-1)j} - \bar{\boldsymbol{y}}^0_{(\ell-1)})'$$

$$S_{(\ell)12} = \sum_{j=1}^{n_\ell} (\boldsymbol{y}_{(\ell-1)j} - \bar{\boldsymbol{y}}^0_{(\ell-1)})(\boldsymbol{y}_{\ell j} - \bar{\boldsymbol{y}}_\ell)' = S'_{(\ell)21}$$

$$S_{(\ell)22} = \sum_{j=1}^{n_\ell} (\boldsymbol{y}_{\ell j} - \bar{\mathbf{y}}_\ell)(\boldsymbol{y}_{\ell j} - \bar{\mathbf{y}}_\ell)'$$

$$S_{(\ell)} : p_{(\ell)} \times p_{(\ell)}\ ;\ \ S_{(\ell)11} : p_{(\ell-1)} \times p_{(\ell-1)}\ \ ;\ S_{(\ell)12} : p_{(\ell-1)} \times p_\ell\ ;\ \ S_{(\ell)22} : p_\ell \times p_\ell\ .$$

Since $\boldsymbol{y}_{(\ell)1}, \ldots, \boldsymbol{y}_{(\ell)n_\ell}$ is a simple random sample from $N_{p_{(\ell)}}(\boldsymbol{\mu}_{(\ell)}, \Sigma_{(\ell)})$, the maximum likelihood estimates of the parameters are

$$\hat{\boldsymbol{\mu}}_{(\ell)} = \bar{\mathbf{y}}_{(\ell)} \quad \text{and} \quad \hat{\Sigma}_{(\ell)} = S_{(\ell)}/n_\ell \,. \tag{3.21}$$

Let

$$\hat{\boldsymbol{\mu}}_{(\ell)} = \begin{pmatrix} \hat{\boldsymbol{\mu}}_{(\ell-1)} \\ \hat{\boldsymbol{\mu}}_\ell \end{pmatrix} = \begin{pmatrix} \bar{\mathbf{y}}^0_{(\ell-1)} \\ \bar{\mathbf{y}}_\ell \end{pmatrix} \tag{3.22}$$

and

$$\hat{\Sigma}_{(\ell)} = \begin{pmatrix} \hat{\Sigma}_{(\ell-1)} & \hat{\Sigma}_{(\ell-1)\ell} \\ \hat{\Sigma}_{\ell(\ell-1)} & \hat{\Sigma}_{\ell\ell} \end{pmatrix} = \begin{pmatrix} S_{(\ell)11}/n_\ell & S_{(\ell)12}/n_\ell \\ S_{(\ell)21}/n_\ell & S_{(\ell)22}/n_\ell \end{pmatrix} , \tag{3.23}$$

where $\bar{\mathbf{y}}^0_{(\ell-1)}, \bar{\mathbf{y}}_\ell$, and $S_{(\ell)ij}, i,j = 1,2$, are defined in (3.19) and (3.20).

Noting that the mapping between the following sets of parameters is one–to–one

$$P_1 = \begin{pmatrix} \boldsymbol{\mu}_{(\ell)} \\ \Sigma_{(\ell)} \end{pmatrix} \equiv \begin{pmatrix} \boldsymbol{\mu}_{(\ell-1)} \\ \boldsymbol{\mu}_\ell \\ \Sigma_{(\ell-1)} \\ \Sigma_{\ell(\ell-1)} \\ \Sigma_{\ell\ell} \end{pmatrix}$$

and

$$P_2 = \begin{pmatrix} \boldsymbol{\mu}_{(\ell-1)} \\ \boldsymbol{\mu}^*_\ell = \boldsymbol{\mu}_\ell - \Sigma^*_{\ell(\ell-1)}\boldsymbol{\mu}_{(\ell-1)} \\ \Sigma_{(\ell-1)} \\ \Sigma^*_{\ell(\ell-1)} = \Sigma_{\ell(\ell-1)}\Sigma^{-1}_{(\ell-1)} \\ \Sigma_{\ell\ell\cdot(\ell-1)} = \Sigma_{\ell\ell} - \Sigma^*_{\ell(\ell-1)}\Sigma_{(\ell-1)\ell} \end{pmatrix} , \tag{3.24}$$

we can apply Corollary 3.2.1 of Anderson (1984) to the effect that the maximum likelihood estimators of functions of parameters are those functions of the maximum likelihood estimators of those parameters provided the transformation is one–to–one.

The maximum likelihood estimates of the parameters belonging to the set P_2 are therefore obtained as functions of the maximum likelihood estimates of the parameters belonging to the set P_1 given in (3.22) and (3.23):

$$\hat{\boldsymbol{\mu}}_{(\ell-1)} = \bar{\mathbf{y}}^0_{(\ell-1)} \quad ; \quad \hat{\Sigma}_{(\ell-1)} = S_{(\ell)11}/n_\ell \quad ; \tag{3.25}$$

$$\hat{\Sigma}^*_{\ell(\ell-1)} = \hat{\Sigma}_{\ell(\ell-1)}\hat{\Sigma}^{-1}_{(\ell-1)} = \frac{S_{(\ell)21}}{n_\ell} \cdot n_\ell(S_{(\ell)11})^{-1} = S_{(\ell)21}(S_{(\ell)11})^{-1} \; ;$$

$$\hat{\boldsymbol{\mu}}^*_\ell = \hat{\boldsymbol{\mu}}_\ell - \hat{\Sigma}^*_{\ell(\ell-1)}\hat{\boldsymbol{\mu}}_{(\ell-1)} = \bar{\mathbf{y}}_\ell - S_{(\ell)21}(S_{(\ell)11})^{-1}\bar{\mathbf{y}}^0_{(\ell-1)} \equiv \bar{\mathbf{y}}_\ell - S^*_{(\ell)21}\bar{\mathbf{y}}^0_{(\ell-1)}$$

$$\hat{\Sigma}_{\ell\ell\cdot(\ell-1)} = \hat{\Sigma}_{\ell\ell} - \hat{\Sigma}^*_{\ell(\ell-1)}\hat{\Sigma}_{(\ell-1)\ell} = \frac{S_{(\ell)22} - S_{(\ell)21}(S_{(\ell)11})^{-1}S_{(\ell)12}}{n_\ell} \equiv \frac{S_{(\ell)22.1}}{n_\ell}$$

for $\ell = 2,\ldots,r$, and $\hat{\boldsymbol{\mu}}_{(1)} = \bar{\mathbf{y}}_{(1)} = \bar{\mathbf{y}}_1$, $\hat{\Sigma}_{(1)} = S_{(1)}/n_1 \equiv S_{(1)22.1}/n_1$; $\bar{\mathbf{y}}_{(1)}$, $\bar{\mathbf{y}}_\ell$, $\bar{\mathbf{y}}^0_{(\ell-1)}$ and $S_{(\ell)}, S_{(\ell)ik}$, $i,k = 1,2$ being given in (3.19) and (3.20).

Now noting that the likelihood function L_2 given in (3.13) is a function of the parameters belonging to the parameters space M_2 as defined in (3.15), we can state the following theorem.

Theorem 1. On the basis of the monotonic sample of size $n_1, \ldots, n_r$ from $N_p(\boldsymbol{\mu}, \Sigma)$ defined in (2.3), the maximum likelihood estimates of the parameters $\{\boldsymbol{\mu}_\ell^*, \Sigma_{\ell(\ell-1)}, \Sigma_{\ell\ell\cdot(\ell-1)}\}$ belonging to the parametric space M_2 defined in (3.15) are available from (3.25) for $\ell = 1, 2, \ldots, r$. Furthermore these maximum likelihood estimates uniquely determine those of $\boldsymbol{\mu}$ and Σ.

Since the likelihood function L_2 defined in (3.13) is completely determined by the parameters belonging to the set M_2 for a given monotonic sample, we can maximize L_2 by substituting the m.l.e.'s of the parameters in M_2 for the corresponding parameters in L_2.

Consider $L_{\ell\cdot(\ell-1)}(\boldsymbol{\mu}_{\ell\cdot(\ell-1)}, \Sigma_{\ell\ell\cdot(\ell-1)})$ defined in (3.11) which, in view of (3.8) and (3.3), is equal to

$$(2\pi)^{-n_\ell p_\ell/2} |\Sigma_{\ell\ell\cdot(\ell-1)}|^{-n_\ell/2} \exp\{-\phi_\ell/2\} \tag{3.26}$$

where, in view of (3.9) and (3.4),

$$\begin{aligned}
\phi_\ell &= tr[\sum_{j=1}^{n_\ell} [\boldsymbol{y}_{\ell j} - (\boldsymbol{\mu}_\ell^* + \Sigma_{\ell(\ell-1)}^* \boldsymbol{y}_{(\ell-1)j})]' \\
&\qquad \Sigma_{\ell\ell\cdot(\ell-1)}^{-1} [\boldsymbol{y}_{\ell j} - (\boldsymbol{\mu}_\ell^* + \Sigma_{\ell(\ell-1)}^* \boldsymbol{y}_{(\ell-1)j})]] \\
&= tr[\, \Sigma_{\ell\ell\cdot(\ell-1)}^{-1} \sum_{j=1}^{n_\ell} [\boldsymbol{y}_{\ell j} - (\boldsymbol{\mu}_\ell^* + \Sigma_{\ell(\ell-1)}^* \boldsymbol{y}_{(\ell-1)j})] \qquad (3.27) \\
&\qquad [\boldsymbol{y}_{\ell j} - (\boldsymbol{\mu}_\ell^* + \Sigma_{\ell(\ell-1)}^* \boldsymbol{y}_{(\ell-1)j})]']
\end{aligned}$$

(using the fact that $tr(ABC) = tr(BCA)$). Now substituting the m.l.e.'s for the parameters, one has that

$$\begin{aligned}
\phi_\ell^* &= tr(n_\ell (S_{(\ell)22.1})^{-1} \sum_{j=1}^{n_\ell} [\boldsymbol{y}_{\ell j} - (\bar{\mathbf{y}}_\ell - S_{(\ell)21}^* \bar{\mathbf{y}}_{(\ell-1)}^0 + S_{(\ell)21}^* \boldsymbol{y}_{(\ell-1)j})] \\
&\qquad [\boldsymbol{y}_{\ell j} - (\bar{\mathbf{y}}_\ell - S_{(\ell)21}^* \bar{\mathbf{y}}_{(\ell-1)}^0 + S_{(\ell)21}^* \boldsymbol{y}_{(\ell-1)j})]') \qquad (3.28) \\
&= tr(n_\ell (S_{(\ell)22.1})^{-1} [S_{(\ell)22} - S_{(\ell)21}^* S_{(\ell)12} - S_{(\ell)21} S_{(\ell)12}^* \\
&\qquad + S_{(\ell)21}^* S_{(\ell)11} S_{(\ell)12}^*]) \qquad (3.29) \\
&= tr(n_\ell I_{p_\ell}) = n_\ell p_\ell
\end{aligned}$$

where $S_{(\ell)12}^*$ is the transpose of $S_{(\ell)21}^*$ and I_{p_ℓ} denotes the identity matrix of order p_ℓ. The maximum value attained by the expression in (3.26) is therefore

$$\max_{M_2^\ell}(L_{\ell.(\ell-1)}) = (2\pi e/n_\ell)^{-n_\ell p_\ell/2} |S_{(\ell)22.1}|^{-n_\ell/2} \tag{3.30}$$

for $\ell = 2, \ldots, r$ where $S_{(\ell)22.1}$ is given in (3.25).

Similarly, one obtains

$$\max_{M_2^1} L_{1.0} = L_1(\boldsymbol{\mu}_{(1)}, \Sigma_{(1)}) = (2\pi e/n_1)^{-n_1 p_1/2} |S_{(1)}|^{-n_1/2} \tag{3.31}$$

where $S_{(1)}$ is given in (3.20).

In view of (3.13) and (3.15), the maximum of the overall likelihood function of the monotonic sample specified by (2.3) is

$$\max_{M_2} L_2 = \prod_{\ell=1}^{r} \max_{M_2^\ell}(L_{\ell\cdot(\ell-1)}) \ . \tag{3.32}$$

Writing $S_{(1)}$ as $S_{(1)22.1}$ as in (3.25), we can now state the following theorem.

Theorem 2. The likelihood function of the monotonic sample specified by (2.3) attains its maximum at

$$\prod_{\ell=1}^{r}\{(2\pi e/n_\ell)^{-n_\ell p_\ell/2}|S_{(\ell)22.1}|^{-n_\ell/2}\} = \max_{M_2}(L_2) \tag{3.33}$$

where $S_{(\ell)22.1}$ is defined in (3.25).

4. LIKELIHOOD RATIO TEST OF INDEPENDENCE

Let

$$\boldsymbol{Y}^0 = \begin{pmatrix} {}_1\boldsymbol{Y} \\ {}_2\boldsymbol{Y} \\ \vdots \\ {}_q\boldsymbol{Y} \end{pmatrix} \sim N_p(\boldsymbol{\mu}^0 \ , \ \Sigma^0) \ , \ \Sigma^0 > O \tag{4.1}$$

be a p-dimensional normal vector partitioned into q subvectors ${}_\alpha\boldsymbol{Y}$ having ${}_\alpha p$ components with $p = {}_1p + {}_2p + \cdots + {}_qp$. The mean vector $\boldsymbol{\mu}^0$ and the covariance matrix Σ^0 are partitioned similarly:

$$\boldsymbol{\mu}^0 = \begin{pmatrix} {}_1\boldsymbol{\mu} \\ {}_2\boldsymbol{\mu} \\ \vdots \\ {}_q\boldsymbol{\mu} \end{pmatrix} \ , \quad \Sigma^0 = \begin{pmatrix} {}_{11}\Sigma & \cdots & {}_{1q}\Sigma \\ \vdots & & \vdots \\ {}_{q1}\Sigma & \cdots & {}_{qq}\Sigma \end{pmatrix} \equiv ({}_{\alpha\beta}\Sigma) \ . \tag{4.2}$$

Let

$${}_\alpha\boldsymbol{Y} = \begin{pmatrix} {}_\alpha Y_1 \\ {}_\alpha Y_2 \\ \vdots \\ {}_\alpha Y_{{}_\alpha r} \end{pmatrix} \ , \quad {}_\alpha\boldsymbol{\mu} = \begin{pmatrix} {}_\alpha\mu_1 \\ {}_\alpha\mu_2 \\ \vdots \\ {}_\alpha\mu_{{}_\alpha r} \end{pmatrix} \tag{4.3}$$

and

$${}_{\alpha\alpha}\Sigma = \begin{pmatrix} {}_{\alpha\alpha}\Sigma_{11} & \cdots & {}_{\alpha\alpha}\Sigma_{1{}_\alpha r} \\ \vdots & & \vdots \\ {}_{\alpha\alpha}\Sigma_{{}_\alpha r1} & \cdots & {}_{\alpha\alpha}\Sigma_{{}_\alpha r{}_\alpha r} \end{pmatrix} \ , \ 1 \le \alpha \le q \tag{4.4}$$

with

$${}_\alpha\mu_i : {}_\alpha p_i \times 1 \ , \ {}_{\alpha\alpha}\Sigma_{kj} : {}_\alpha p_k \times {}_\alpha p_j \ ; \quad 1 \le k, \ j \le {}_\alpha r \ , \ \alpha = 1, \ldots, q \ .$$

Treated in this section is the problem of testing the mutual independence of the q subvectors ${}_\alpha\boldsymbol{Y}$. The null hypothesis can be stated equivalently as follows:

$$H_0 : {}_{\alpha\beta}\Sigma = O \ , \ 1 \le \alpha, \beta \le q \ , \ \alpha < \beta \ . \tag{4.5}$$

The likelihood ratio test statistic is derived on the basis of a monotonic random sample of size $(n_1, n_2, \ldots, n_r)$ on $\boldsymbol{Y}' = (\boldsymbol{Y}_1', \ldots, \boldsymbol{Y}_r')$ exactly defined as in (2.3). [Note that it may be necessary to reorder the original vectors or their components in order to obtain the triangular pattern of a monotonic sample on $p_{(1)}, \ldots, p_{(r-1)}, p_{(r)}$ components with $p_{(r)} > p_{(r-1)} > \ldots > p_{(1)}$]. We wish to test for the independence of q sets of components from $\boldsymbol{Y}$. Those sets are denoted by ${}_1\boldsymbol{Y}, {}_2\boldsymbol{Y}, \ldots, {}_q\boldsymbol{Y}$ and form the vector $\boldsymbol{Y}^0$ of the possibly reordered components of $\boldsymbol{Y}$. We note that as long as the components of ${}_\alpha\boldsymbol{Y}$'s are ordered as they were in $\boldsymbol{Y}$, the observations on each ${}_\alpha\boldsymbol{Y}, \alpha = 1, \ldots, q$, constitute a monotonic sample.

Letting ${}_{\alpha\alpha}\Sigma \equiv_\alpha \Sigma$, $\alpha = 1, \ldots, q$ and using for each ${}_\alpha\boldsymbol{Y}$ the notation developed in Sections 2 and 3 with the addition of the subscript α to the left of each symbol, we have that the observation vectors on each ${}_\alpha\boldsymbol{Y}$ form a monotonic sample of size $({}_\alpha n_1, {}_\alpha n_2, \ldots, n_{{}_\alpha r})$ on ${}_\alpha p_{(1)}, {}_\alpha p_{(2)}, \ldots, p_{({}_\alpha r)}$ subvectors from $N_{{}_\alpha p}({}_\alpha\boldsymbol{\mu}, {}_\alpha\Sigma)$. From Theorem 2, the likelihood function of the monotonic sample on ${}_\alpha Y$ attains its maximum at

$$\prod_{\ell=1}^{{}_\alpha r} (2\pi e/{}_\alpha n_\ell)^{-({}_\alpha n_\ell)({}_\alpha p_\ell)/2} |{}_\alpha S_{(\ell)22.1}|^{-{}_\alpha n_\ell/2} = \max_{{}_\alpha M_2}({}_\alpha L_2) \tag{4.6}$$

for $\alpha = 1, \ldots, q$.

Under H_0 (the independence assumption), the maximum value of the overall likelihood is therefore

$$\prod_{\alpha=1}^{q} \max_{{}_\alpha M_2}({}_\alpha L_2) \equiv \max(L_2^0) \ . \tag{4.7}$$

The maximum value of the likelihood function for the monotonic sample on $\boldsymbol{Y}$ with respect to all real vectors $\boldsymbol{\mu}$ and positive definite symmetric matrix Σ in the parametric space M_1 (or equivalently M_2) is given by (3.33) since the same notation developed in Sections 2 and 3 is used.

The likelihood ratio statistic is therefore

$$\lambda = \frac{\max(L_2^0)}{\max_{M_2}(L_2)} \tag{4.8}$$

where $\max(L_2^0)$ is given in (4.7) and $\max_{M_2}(L_2)$ is given in (3.33).

We reject H_0 as defined in (4.5) when $\lambda \le \lambda(\alpha)$ where λ is given in (4.8) and $\lambda(\alpha)$ is a constant depending on α, the significance level of the test.

Asymptotically, $-2\ln\lambda$ is distributed as a central chi-square variable with p^* degrees of freedom where

$$p^* = \sum_{\alpha=1}^{q-1} \sum_{\beta=\alpha+1}^{q} {}_\alpha p \ {}_\beta p \tag{4.9}$$

is the number of parameters that are specified under the null hypothesis.

5. LIKELIHOOD RATIO TEST FOR COVARIANCE MATRICES

Using the notation of Section 2, we obtain the likelihood ratio statistic to test

$$\begin{aligned} H_0^1 \ &: \ \Sigma = V \ , \text{ a given positive definite matrix,} \\ &\quad \text{or equivalently, } \Sigma_{kj} = V_{kj} \ , \ 1 \le k \le j \le r \ , \\ &\quad \text{and } \boldsymbol{\mu} \text{ is assumed to be unknown} \end{aligned} \tag{5.1}$$

vs

$$H_A \ : \ \boldsymbol{\mu} \text{ and } \Sigma \text{ are unknown.}$$

From the relationships in (3.25), we see that when $\Sigma = V$, the m.l.e.'s of the mean subvectors are

$$\hat{\boldsymbol{\mu}}_{(\ell-1)} = \bar{\mathbf{y}}^0_{(\ell-1)}$$

and

$$\begin{aligned} \hat{\boldsymbol{\mu}}^*_\ell &= \hat{\boldsymbol{\mu}}_\ell - V^*_{\ell(\ell-1)} \hat{\boldsymbol{\mu}}_{(\ell-1)} \\ &= \bar{\mathbf{y}}_\ell - V_{\ell(\ell-1)} V^{-1}_{(\ell-1)} \bar{\mathbf{y}}^0_{(\ell-1)} \ , \quad \ell = 2, \ldots, r \end{aligned} \tag{5.2}$$

with

$$\hat{\boldsymbol{\mu}}_1 = \bar{\mathbf{y}}_{(1)} = \bar{\mathbf{y}}_1$$

where the subscripts and the superscripts for V have the same meaning as those associated with Σ .

Substituting these m.l.e.'s in (3.27), one has

$$\begin{aligned} \phi_\ell^{(1)} &= tr\Big[V^{-1}_{\ell\ell.(\ell-1)} \sum_{j=1}^{n_\ell} \Big[\mathbf{y}_{\ell j} - \Big(\hat{\boldsymbol{\mu}}_\ell - V^*_{\ell(\ell-1)} \hat{\boldsymbol{\mu}}_{(\ell-1)} + V^*_{\ell(\ell-1)} \mathbf{y}_{(\ell-1)j}\Big)\Big] \\ &\qquad \Big[\mathbf{y}_{\ell j} - \Big(\hat{\boldsymbol{\mu}}_\ell - V^*_{\ell(\ell-1)} \hat{\boldsymbol{\mu}}_{(\ell-1)} + V^*_{\ell(\ell-1)} \mathbf{y}_{(\ell-1)j}\Big)\Big]'\Big] \\ &= tr\Big[V^{-1}_{\ell\ell.(\ell-1)} \sum_{j=1}^{n_\ell} \Big[\mathbf{y}_{\ell j} - \sum_{j=1}^{n_\ell} \frac{\mathbf{y}_{\ell j}}{n_\ell} - V^*_{\ell(\ell-1)} \Big(\mathbf{y}_{(\ell-1)j} - \sum_{j=1}^{n_\ell} \frac{\mathbf{y}_{(\ell-1)j}}{n_\ell}\Big)\Big] \\ &\qquad \Big[\mathbf{y}_{\ell j} - \sum_{j=1}^{n_\ell} \frac{\mathbf{y}_{\ell j}}{n_\ell} - V^*_{\ell(\ell-1)} \Big(\mathbf{y}_{(\ell-1)j} - \sum_{j=1}^{n_\ell} \frac{\mathbf{y}_{(\ell-1)j}}{n_\ell}\Big)\Big]'\Big] \\ &= tr\Big[V^{-1}_{\ell\ell.(\ell-1)} \Big\{S_{(\ell)22} - V^*_{\ell(\ell-1)} S_{(\ell)12} - S_{(\ell)21} V^*_{(\ell-1)\ell} \\ &\qquad + V^*_{\ell(\ell-1)} S_{(\ell)11} V^*_{(\ell-1)\ell}\Big\}\Big] \ . \end{aligned} \tag{5.3}$$

Hence, under H_0^1, the maximum value of $L_{\ell.(\ell-1)}$ is

$$\max \ L^{(1)}_{\ell.(\ell-1)} = (2\pi)^{-n_\ell p_\ell/2} \, |V_{\ell\ell.(\ell-1)}|^{n_\ell/2} \exp\Big\{-\phi_\ell^{(1)}/2\Big\}, \quad \ell = 2, \ldots, r$$

and

$$\max\ L^{(1)}_{\ell.(0)} \;=\; \max\ L^{(1)}_{1} \;=\; (2\pi)^{-n_\ell p_\ell/2}\,|V_{11}|^{-n_1/2}\exp\left\{-trV_{11}^{-1}S_{(1)}\right\}\,. \tag{5.4}$$

Then, under H_0^1 , the likelihood function of the monotonic sample attains its maximum value at

$$\prod_{\ell=1}^{r}\max L^{(1)}_{\ell.(\ell-1)} \;\equiv\; \max L^{(1)}\,. \tag{5.5}$$

The likelihood ratio statistic for testing H_0^1 vs H_A is therefore

$$\frac{\max L^{(1)}}{\max\limits_{M_2} L_2} \;=\; \lambda_1 \tag{5.6}$$

where $\max\limits_{M_2} L_2$ is defined in (3.32). The quantity $-2\ln\lambda_1$ is asymptotically distributed as a chi-square variate having d degrees of freedom where

$$d \;=\; p(p+1)/2 \tag{5.7}$$

and $p = p_1 + \cdots + p_r$.

6. LIKELIHOOD RATIO TEST FOR MEAN VECTORS

We derive the likelihood ratio statistic to test

H_0^2 : $\boldsymbol{\mu} = \mathbf{m}$, a given mean vector, (6.1)
or equivalently, $(\boldsymbol{\mu}_1', \ldots, \boldsymbol{\mu}_r') \;=\; (\mathbf{m}_1', \ldots, \mathbf{m}_r')$,
and Σ is assumed to be an unknown positive definite covariance matrix

vs

H_A : $\boldsymbol{\mu}$ and Σ are unknown.

Under H_0^2, the m.l.e. of $\Sigma_{(\ell)}$ is

$$\hat{\Sigma}_{(\ell)} \;=\; \frac{s_{(\ell)}}{n_\ell} \;=\; \begin{pmatrix} s_{(\ell)11}/n_\ell & s_{(\ell)12}/n_\ell \\ s_{(\ell)21}/n_\ell & s_{(\ell)22}/n_\ell \end{pmatrix} \tag{6.2}$$

where $s_{(\ell)}$ is defined as $S_{(\ell)}$ in (3.20) except that $\bar{\mathbf{y}}_{(\ell)}$ is replaced by $\boldsymbol{\mu}_{(\ell)}$ in $s_{(\ell)}$.

From the one-to-one relationship between the sets P_1 and P_2 defined in (3.24), it follows that the m.l.e.'s of $\Sigma^*_{\ell.(\ell-1)}$ and $\Sigma_{\ell.(\ell-1)}$ are respectively

$$\hat{\Sigma}^*_{\ell(\ell-1)} \;=\; s_{(\ell)21}s_{(\ell)11}^{-1} \;\equiv\; s^*_{\ell(\ell-1)}$$

and

$$\hat{\Sigma}_{\ell\ell.(\ell-1)} \;=\; \frac{s_{(\ell)22} \;-\; s_{(\ell)21}s_{(\ell)11}^{-1}s_{(\ell)12}}{n_\ell}$$

$$\equiv\; \frac{s_{(\ell)22.1}}{n_\ell}\,, \qquad \ell = 2, \ldots, r, \tag{6.3}$$

with $\hat{\Sigma}_{(1)} = s_{(1)}/n_1$.

Furthermore

$$\hat{\mu}_\ell^* = \mathbf{m}_\ell - s^*_{\ell(\ell-1)}\mathbf{m}_{(\ell-1)} . \tag{6.4}$$

Substituting these m.l.e.'s in (3.27), one has

$$\begin{aligned}
\phi_\ell^{(2)} &= tr\Big[n_\ell \left(s_{(\ell)22.1}\right)^{-1} \\
&\quad \sum_{j=1}^{n_\ell}\left[\mathbf{y}_{\ell j} - \left(\mathbf{m}_\ell - s^*_{\ell(\ell-1)}\mathbf{m}_{(\ell-1)} + s^*_{\ell(\ell-1)}\mathbf{y}_{(\ell-1)j}\right)\right] \\
&\quad \left[\mathbf{y}_{\ell j} - \left(\mathbf{m}_\ell - s^*_{\ell(\ell-1)}\mathbf{m}_{(\ell-1)} + s^*_{\ell(\ell-1)}\mathbf{y}_{(\ell-1)j}\right)\right]'\Big] \\
&= tr\Big[n_\ell \left(s_{(\ell)22.1}\right)^{-1} \Big(s_{(\ell)22} - s^*_{\ell(\ell-1)}s_{(\ell)12} \\
&\quad - s_{(\ell)21}s^*_{(\ell-1)\ell} + s_{(\ell)21}s^{-1}_{(\ell)11}s_{(\ell)11}s^*_{(\ell-1)\ell}\Big)\Big] \\
&= tr\left(n_\ell I_{p_\ell}\right) = n_\ell p_\ell, \qquad \ell = 1, 2, \ldots, r .
\end{aligned} \tag{6.5, 6.6}$$

Then, under H_0^2, the likelihood function of the monotonic sample attains its maximum value at

$$\begin{aligned}
\prod_{\ell=1}^{r} \max L^{(2)}_{\ell.(\ell-1)} &= \prod_{\ell=1}^{r} (2\pi)^{-n_\ell p_\ell/2}\left|s_{(\ell)22.1}\right|^{-n_\ell/2}\exp\{-n_\ell p_\ell/2\} \\
&\equiv \max L^{(2)}
\end{aligned} \tag{6.7}$$

in view of (3.26) and (6.6).

The likelihood ratio statistic for testing H_0^2 vs H_A is therefore

$$\frac{\max L_{(2)}}{\max\limits_{M_2} L_2} = \prod_{\ell=1}^{r}\left(|S_{(\ell)22.1}|/|s_{(\ell)22.1}|\right)^{n_\ell/2} \equiv \lambda_2 \tag{6.8}$$

in view of (6.7), (3.30) and (3.32), and $-2\ln\lambda_2$ is asymptotically distributed as a chi-square variate with p degrees of freedom.

Acknowledgements

This research was supported by the Natural Sciences and Engineering Research Council of Canada. Thanks are also due to the referee for several valuable suggestions.

REFERENCES

1. S.S. Wilks, *Ann. Math. Statist.*, **3**, 163–195 (1932).
2. F.M. Lord, *J. Amer. Statist. Assoc.*, **50**, 870–876 (1955).
3. G.L.Edgett, *J. Amer. Statist. Assoc.*, **51**, 122–131 (1956).
4. C.R. Rao, *J. Roy. Statist. Soc. Ser. B*, **19**, 259–264 (1956).
5. T.W. Anderson, *J. Amer. Statist. Assoc.*, **52**, 200–203 (1957).

6. I.M. Trawinski and R.E. Bargmann, *Ann. Math. Statist.*, **35**, 647–657 (1964).
7. A.M. Mathai, *Trabajos de Estadistica*, **17**, 59–83 (1966).
8. A.A. Afifi and R.M. Elashoff, *J. Amer. Statist. Assoc.*, **61**, 595–604 (1966).
9. A.A. Afifi and R.M. Elashoff, *J. Amer. Statist. Assoc.*, **62**, 10–29 (1967).
10. A.A. Afifi and R.M. Elashoff, *J. Amer. Statist. Assoc.*, **64**, 337–358 (1969a).
11. A.A. Afifi and R.M. Elashoff, *J. Amer. Statist. Assoc.*, **64**, 359–365 (1969b).
12. W.B. Smith and R.C. Pfaffenberger, *Biometrics*, **26**, 625–639 (1970).
13. D.F. Morrison, *J. Amer. Statist. Assoc.*, **66**, 602–604 (1971).
14. M.L. Eaton and T. Kariya, Testing for independence with additional information. Technical Report # 238. University of Minnesota (1974).
15. M.L. Eaton and T. Kariya, Tests on means with additional information. Technical Report # 243. University of Minnesota (1975).
16. M.L. Eaton and T. Kariya, *Ann. Statist.*, **11**, 654–665 (1983).
17. R.P. Bhargava, *Ann. Inst. Statist. Math.*, **27**, 327–339 (1975).
18. R.J.A. Little, *Biometrika*, **63**, 593–604 (1976).
19. D.B. Rubin, *Biometrika*, **63**, 581–592 (1976).
20. R. Radhakrishnan, *Comm. Statist. Theory Methods*, **11**, 941–955 (1982).
21. M.S. Srivastava, *Comm. Statist. Theory Methods*, **14**,775–792 (1985).
22. S.B. Provost, *Comm. Statist. Theory Methods,* **17**, 1763–1765 (1988).
23. Y. Dodge, *Analysis of Experiments with Missing Data*, Wiley, New York (1985).
24. T.W. Anderson, *An Introduction to Multivariate Statistical Analysis*, Wiley, New York (1984).

Stat. Sci. & Data Anal., pp. 225-235
K. Matsusita *et al.* (Eds)

Maximally orthogonally invariant higher order moments and their application to testing elliptically-contouredness

AKIMICHI TAKEMURA

Faculty of Economics, University of Tokyo, Bunkyo, Tokyo, 113 Japan

Abstract. Consider joint higher order moments of a p dimensional random vector. The moments of a given order form a symmetric tensor. Now transform the random vector by an orthogonal matrix. This orthogonal transformation induces a transformation on the higher order moment tensors. The main purpose of this paper is to derive two forms of maximal invariant of this induced transformation. Based on invariant multivariate skewness, we give a geometric interpretation of a test of multivariate normality proposed by Isogai[1]. We also show that the test has the same null distribution under any multivariate elliptically contoured distribution (left orthogonally invariant distribution) and hence can be used to test elliptically-contouredness.

1. INTRODUCTION AND NOTATION

Let $X = (X_1, \dots, X_p)'$ be a p dimensional random vector. For simplicity assume $E(X) = 0, \mathrm{Var}(X) = E(XX') = I_p$. Let

$$\mu_{abc} = \mu_{abc}(X) = E(X_a X_b X_c) \tag{1}$$

be the joint third order moment (about the mean) of X . $\{\mu_{abc}\}$ form a symmetric tensor of degree 3

$$\boldsymbol{\mu}_3 = \boldsymbol{\mu}_3(X) = \{\mu_{abc}\}_{a,b,c=1,\dots,p} \tag{2}$$

Similarly denote k-th order joint moment of X by $\mu_{a_1 \dots a_k} = \mu_{a_1 \dots a_k}(X) = E(X_{a_1} \cdots X_{a_k})$ and symmetric tensor of degree k by $\boldsymbol{\mu}_k = \{\mu_{a_1,\dots,a_k}\}$.

Let $G = (g_{ab}) \in \mathcal{O}(p)$ be a $p \times p$ orthogonal matrix and consider the transformation $X \mapsto Y = GX$. This transformation induces a transformation of third order moments as follows.

$$\boldsymbol{\mu}_3(X) \mapsto \boldsymbol{\mu}_3(Y) = \{\mu_{abc}(Y)\}, \quad \mu_{abc}(Y) = \sum_{a',b',c'} g_{aa'} g_{bb'} g_{cc'} \mu_{a'b'c'}(X). \tag{3}$$

This can be regarded as an action of $\mathcal{O}(p)$ on the space $\mathcal{S}_3$ of symmetric tensor of degree 3. Similarly $\mathcal{O}(p)$ acts on the space $\mathcal{S}_k$ of $\boldsymbol{\mu}_k$. A natural question here is to obtain a maximal invariant of this transformation. Two forms of maximal invariant are given in Section 2.

We mention here why orthogonally invariant moments are considered. Let $\Sigma = E(ZZ')$ be the variance matrix of a random vector Z and let $X = \Sigma^{-1/2}Z$. Then $\mathrm{Var}(X) = I$. However $\Sigma^{-1/2}$ is determined only up to an arbitrary orthogonal matrix. Therefore from invariance viewpoint we should only consider orthogonally invariant moments of X, when X is obtained by standardizing the variance matrix of Z. In Section 3 we consider a test of multivariate normality and elliptically-contouredness (more precisely, left orthogonal invariance) with mean vector and variance matrix unknown. To test the distributional hypothesis it is natural first to standardize original observation vectors by subtracting the mean vector and multiplying by $S^{-1/2}$ where S is the sample variance matrix. If a test is based on sample moments of the standardized observation vectors, then only orthogonally invariant moments should be used.

2. EXPRESSION OF MAXIMAL INVARIANTS

Here we present two different expressions of maximal invariant of the transformation (3). For clarity of exposition we discuss third order moments in detail and give generalizations to higher order moments.

Our first approach to find maximal invariants is to generalize the definition of multivariate skewness by Malkovich and Afifi[6]. Let a unit vector g_1 ($g_1'g_1 = 1$) be defined

$$E(g_1'X)^3 = \max_{g'g=1} E(g'X)^3.$$

Next we move to the orthogonal complement of $\mathrm{Span}\{g_1\}$ and let a unit vector g_2 $(g_2'g_1 = 0)$ be defined by

$$E(g_2'X)^3 = \max_{g'g=1, g'g_1=0} E(g'X)^3.$$

Similarly $g_j \in \mathrm{Span}(g_1, \ldots, g_{j-1})^{\perp}$ is defined by

$$E(g_j'X)^3 = \max_{g'g=1, g'g_1=0, \ldots, g'g_{j-1}=0} E(g'X)^3 \tag{4}$$

Let $G' = (g_1, \ldots, g_p)$ be orthogonal and $Y = GX$. We claim that

$$\mu_{aab}(Y) \equiv 0, \ a < b \tag{5}$$

and the remaining third order moments of Y form a maximal invariant.

Theorem 2.1. *Consider the set of* $\mu_3(X)$ *such that the maximum of (4) is uniquely attained for* $j = 1, \ldots, p$. *On this subset of* S_3 *a maximal invariant of the transformation (3) is given by*

$$\mu_{aaa}(Y), a = 1, \ldots, p, \ \mu_{aab}(Y), a > b, \ \mu_{abc}(Y), a < b < c. \tag{6}$$

Proof. Let g and $g + \Delta g$ be unit vectors with Δg infinitesimal. From $0 = (g + \Delta g)'(g + \Delta g) - g'g = 2g'\Delta g + (\Delta g)'\Delta g$ we find that

$$g' \frac{\Delta g}{\|\Delta g\|} \to 0 \qquad (\Delta g \to 0).$$

Namely on the unit sphere the tangent vectors are orthogonal to the vector on the sphere. Otherwise the direction of $\Delta g / \|\Delta g\|$ is free. Since g_1 maximizes $E(g'X)^3$, we have

$$E[((g_1 + \Delta g)'X)^3] \leq E(g_1'X)^3, \qquad \forall \Delta g$$

or

$$3E[((\Delta g / \|\Delta g\|)'X)(g_1'X)^2] + o(\|\Delta g\|) \leq 0 \quad (\Delta g \to 0) \tag{7}$$

Writing $\tilde{g} = \Delta g / \|\Delta g\|$ and letting $\Delta g \to 0$ we see that for any $\tilde{g}$ such that $\tilde{g}'g_1 = 0$

$$E[(\tilde{g}'X)(g_1'X)^2] = 0. \tag{8}$$

In terms of transformed vector Y this is equivalent to $\mu_{11b}(Y) = 0, \quad b = 2, \ldots, p$.

We can apply the same reasoning regarding g_2. Let $g_2 + \Delta g$ be a unit vector such that $(g_2 + \Delta g)'g_1 = 0$. Then as $\Delta g \to 0$ the tangent vector $\tilde{g} = \Delta g / \|\Delta g\|$ satisfies

$$\tilde{g}'g_2 = 0, \ \tilde{g}'g_1 = 0 \qquad (\Delta g \to 0)$$

but otherwise the direction of $\tilde{g}$ is free. Therefore reasoning as above we see that for all $\tilde{g}$ such that $\tilde{g}'g_i = 0, i = 1, 2$

$$E[(\tilde{g}'X)(g_2'X)^2] = 0. \tag{9}$$

In terms of Y we have $\mu_{22b}(Y) = 0$, $b = 3, \ldots, p$. Now arguing similarly for $g_3, g_4, \ldots$ we obtain (5).

Now let H be orthogonal and $X^* = HX$. Obviously the above maximization procedure leads to the same Y for X and for X^*, because we assume that the maximization is unique. Hence the nonzero moments of Y are invariant. Conversely suppose that maximization for X and X^* respectively are attained by orthogonal matrices G and H and assume that the resulting nonzero moments are the same. Since other moments are identically equal to zero, we have

$$\mu_{abc}(GX) = \mu_{abc}(HX^*), \qquad \forall a, b, c.$$

This implies $\mu_{abc}(H'GX) = \mu_{abc}(X^*)$, i.e. , $H'GX$ and X^* have the same set of third order moments. This proves the theorem. ∎

Remark 2.1. As is made clear in the statement of Theorem 2.1, we can not well define the maximal invariant for the entire space $\mathcal{S}_3$ of symmetric tensor of degree 3. If maximizing g_j is not uniquely determined, then the subsequent vectors ($g_{j+1}, \ldots, g_p$) will in general depend on the choice of g_j . This contrasts with the case of the spectral decomposition of covariance matrix. For the spectral decomposition, characteristic roots are uniquely determined even if multiple roots exist. Note however that for almost all symmetric tensor of degree 3 (w.r.t. the Lebesgue measure $\prod_{a \le b \le c} d\mu_{abc}$) the maximization is unique.

Remark 2.2. Note that $\mathcal{O}(p)$ is $p(p-1)/2$ dimensional and in Theorem 2.1 we have the same number of moments $\mu_{aab}, a < b,$ eliminated from the maximal invariants.

Generalization of the above approach for moments of order $k > 3$ is immediate. Define g_1 by

$$E(g_1'X)^k = \max_{g'g=1} E(g'X)^k$$

and define $g_2, \ldots, g_p$ inductively by

$$E(g_j'X)^k = \max_{g'g=1, g'g_1=0, \ldots, g'g_{j-1}=0} E(g'X)^k. \tag{10}$$

Let G and Y be defined as in the case of third order moments. Then we have

Corollary 2.1. *Consider the set of $\mu_k(X)$ such that the maximum of (9) is uniquely attained (up to the sign of the maximizing vector) for $j = 1, \ldots, p$. On this subset of $\mathcal{S}_k$*

$$\mu_{a \cdots ab}(Y) \equiv 0, \quad \forall a < b, \tag{11}$$

and the remaining k -th order moments of Y form a maximal invariant.

Remark 2.3. If k is even, the sign of the maximizing vector g_j is not uniquely defined. However this nonuniqueness is trivial and the sign of g_j may be determined by an appropriate way.

Our second definition of maximal invariant is based on the following geometric consideration. Let $V_1 = XX'X$. V_1 is obtained from X by cubing the length of X without changing the direction of X . Let a unit vector g_1 be defined by

$$g_1 = E(V_1)/\,\|E(V_1)\|\,.$$

Note that g_1 is determined in terms of third order moments of X . In fact $E(V_1)$ is written as

$$E(V_1) = \begin{pmatrix} \mu_{111} + \mu_{122} + \cdots + \mu_{1pp} \\ \mu_{211} + \mu_{222} + \cdots + \mu_{2pp} \\ \vdots \\ \mu_{p11} + \mu_{p22} + \cdots + \mu_{ppp} \end{pmatrix} \tag{12}$$

Once g_1 is given we move to the orthogonal complement of g_1 and repeat the above procedure. Let $M_1 = \text{Span}(g_1)^{\perp}$ and let P_{M_1} be the orthogonal projector onto M_1 . Define $\tilde{X} = P_{M_1}X$. We now cube the length of $\tilde{X}$ and define

$$g_2 = E(V_2)/\,\|E(V_2)\|\,, \quad \text{where} \quad V_2 = \tilde{X}\tilde{X}'\tilde{X} \tag{13}$$

Similarly $g_j \in M_{j-1}$ is defined by

$$g_j = E(V_j)/\,\|E(V_j)\|\,, \quad V = \tilde{X}\tilde{X}'\tilde{X}, \quad \tilde{X} = P_{M_{j-1}}X \tag{14}$$

where $M_{j-1} = \text{Span}(g_1, \ldots, g_{j-1})^{\perp}$. Let $G' = (g_1, \ldots, g_p)$ and $Y = GX$.

For the case of third order moments we can show that

$$\mu_{aab}(Y) \equiv 0, \; b \geq a+2, \; \mu_{a,a,a+1}(Y) \equiv -\sum_{b=a+1}^{p} \mu_{b,b,a+1}(Y). \tag{15}$$

and we obtain the following result.

Theorem 2.2. *Consider the set of $\boldsymbol{\mu}_3(X)$ such that $E(V_j) \neq 0, j = 1, \ldots, p$. On this subset of S_3 a maximal invariant of the transformation (3) is given by* $\{\mu_{aaa}(Y), a = 1, \ldots, p, \; \mu_{aab}(Y), a > b, \; \mu_{abc}(Y), a < b < c\}$.

Note that the components of the maximal invariant are the same in Theorem 2.1 and Theorem 2.2.

Proof. Since $Y'Y = X'X$ is a scalar, the i-th element of $YY'Y$ is $Y_i(Y'Y)$. Its expectation is

$$E[Y_i(Y'Y)] = E[g_i'X(X'X)] = g_i'E(V_1) = 0, \qquad i = 2, \ldots, p$$

This is equivalent to

$$\mu_{11i}(Y) + \mu_{22i}(Y) + \cdots + \mu_{ppi}(Y) = 0, \qquad i = 2, \ldots, p. \tag{16}$$

Now let $\tilde{X} = P_{M_{j-1}}X$ and for $i > j$ consider $g_i'\tilde{X}\tilde{X}'\tilde{X}$. Note that $Y_i = g_i'X = g_i'\tilde{X}$ since $g_i \in M_{j-1}$. Furthermore $\tilde{X}'\tilde{X} = Y_j^2 + Y_{j+1}^2 + \cdots + Y_p^2$. Therefore

$$E[Y_i(Y_j^2 + Y_{j+1}^2 + \cdots + Y_p^2)] = g_i'E(\tilde{X}\tilde{X}'\tilde{X}) = g_i'E(V_j) = 0. \tag{17}$$

This is equivalent to

$$\mu_{jji}(Y) + \mu_{j+1,j+1,i}(Y) + \cdots + \mu_{ppi}(Y) = 0, \qquad i = j+1, \ldots, p. \tag{18}$$

From (16) and (18) pick relations for a fixed i. Then

$$\begin{aligned}
0 &= \mu_{ppi}(Y) + \cdots + \mu_{i-1,i-1,i}(Y), \\
0 &= \mu_{ppi}(Y) + \cdots + \mu_{i-1,i-1,i}(Y) + \mu_{i-2,i-2,i}(Y), \\
&\cdots \\
0 &= \mu_{ppi}(Y) + \cdots + \mu_{i-1,i-1,i}(Y) + \cdots + \mu_{22i}(Y), \\
0 &= \mu_{ppi}(Y) + \cdots + \mu_{i-1,i-1,i}(Y) + \cdots + \mu_{22i}(Y) + \mu_{11i}(Y).
\end{aligned} \tag{19}$$

Solving these equations we obtain

$$\begin{aligned}
\mu_{11i}(Y) &= \cdots = \mu_{i-2,i-2,i}(Y) = 0 \\
\mu_{i-1,i-1,i}(Y) &= -\sum_{j=i}^{p} \mu_{jji}(Y)
\end{aligned} \tag{20}$$

This is equivalent to (15). Therefore $\mu_{aab}(Y), a < b$, are either 0 or can be expressed by other moments in a simple manner. Hence these moments can be eliminated. The rest of the proof showing that the remaining moments of Y form a maximal invariant is totally similar to that of Theorem 2.1 and omitted. ∎

Generalizing the second approach to moments of order higher than 3 are not as straightforward as in the first approach. First consider odd moments. Let $l = 2k+1$. Instead of cubing the length of the random vector here we consider raising the length to the l-th power. Let $V_1 = (X'X)^k X$ and $g_1 = E(V_1)/\|E(V_1)\|$. As in the case of third order moments define for $j = 2, \ldots p$

$$g_j = E(V_j)/\|E(V_j)\|, \quad V_j = (\tilde{X}'\tilde{X})^k \tilde{X}, \quad \tilde{X} = P_{M_{j-1}} X \tag{21}$$

and define G and Y accordingly. Corresponding to (17) we now have

$$E[Y_i(Y_j^2 + Y_{j+1}^2 + \cdots + Y_p^2)^k] = 0, \qquad \forall i > j \tag{22}$$

Hence $\mu_{a\cdots ab}(Y), a < b,$ can be expressed by other moments. Therefore we have the following result.

Corollary 2.2. *Consider the set of* $\mu_{2k+1}(X)$ *such that* $E(V_j) \neq 0, j = 1, \ldots, p$ *in (21). On this subset of* S_{2k+1} *moments of the form* $\mu_{a\cdots ab}(Y), a < b,$ *can be expressed in terms of other moments by (22) and the remaining moments of* Y *form a maximal invariant.*

Our definition of maximal invariant for odd order moments can not be directly applied for even order moments. For even k we consider

$$\Psi = E(\|X\|^{k-2} XX'). \tag{23}$$

Let the spectral decomposition of Ψ be

$$\Psi = G'DG, \tag{24}$$

where G is orthogonal and $D = \text{diag}(d_1, \ldots, d_p)$, $d_1 \geq \cdots \geq d_p$ is a diagonal matrix with characteristic roots of Ψ as diagonal elements. Let $Y = GX$. Then $E(\|Y\|^{k-2} YY') = D$, hence

$$E[(Y_1^2 + \cdots + Y_p^2)^{k-2} Y_i Y_j] = 0, \qquad i < j. \tag{25}$$

From this relation we can express $\mu_{a\cdots ab}(Y), a < b,$ in terms of other moments of Y. Therefore we have the following result for even order moments.

Proposition 2.1. *Let* k *be even and let* G *be the orthogonal matrix in the spectral decomposition of* Ψ *in (24). Let* $Y = GX$. *Then moments of the form* $\mu_{a\cdots ab}(Y), a < b$ *can be expressed in terms of other moments of* Y *and the remaining moments of* Y *form a maximal invariant on the subset of* S_k *where the roots of* Ψ *are all distinct.*

3. A TEST OF LEFT ORTHOGONAL INVARIANCE

The first definition of the maximal invariant in the previous section is related to the test of multivariate normality proposed by Malkovich and Afifi[6]. Here we propose a test of multivariate normality based on the first direction g_1 of the second definition of maximal invariant. Our test is essentially the same as a test based on " $\mathrm{tr}\, S_2$ " proposed by Isogai[1],[2]. The derivation below gives a clear geometric interpretation and motivation of the test statistic. Another interpretation of $\mathrm{tr}\, S_2$ is discussed in Isogai[3].

Furthermore it will be shown that the null distribution of the test statistic is the same for any so called "left orthogonally invariant" distribution. Hence our test can be regarded as a test of left orthogonal invariance.

Let $Z = (Z_1, \ldots, Z_n)'$ be $n \times p$ observation matrix and let $\bar{Z} = \sum_{t=1}^n Z_t/n$, $S = \sum_{t=1}^n (Z_t - \bar{Z})(Z_t - \bar{Z})'/n$. Let $X_t = S^{-1/2}(Z_t - \bar{Z}), t = 1, \ldots, n$, where $S = S^{1/2}S^{1/2\prime}$ and $S^{-1/2} = (S^{1/2})^{-1}$. Let $X = (X_1, \ldots, X_n)' : n \times p$. X is the standardized observation matrix, whose columns have mean 0 and unit variance. Note that $X = X(Z)$ can be written as

$$X(Z) = (Z - 1_n \bar{Z}')S^{-1/2\prime} \tag{26}$$

and $X'X = nI_p$. For the sake of definiteness of our argument we here take $S^{1/2} \in G_T$, where G_T is the group of lower triangular matrices with positive diagonal elements. $S^{1/2}$ is obtained by the Cholesky decomposition of S.

The null hypothesis we consider here is that Z has an left orthogonally invariant distribution in the following sense (see Section 3.3 of Kariya and Sinha[4]:

$$H_0 : \mathcal{L}[Z - 1_n\mu'] = \mathcal{L}[Q(Z - 1_n\mu')], \qquad \forall Q \in \mathcal{O}(n), \tag{27}$$

where $\mathcal{L}[Z]$ denotes the distribution of Z, $\mu \in R^p$, $1_n = (1, \ldots, 1)' \in R^n$, and $\mathcal{O}(n)$ denotes the group of $n \times n$ orthogonal matrices. We also assume that Z is of rank r with probability 1.

Left orthogonal invariance is a natural notion of elliptically-contouredness for $n \times p$ random matrix. It can be easily shown that left orthogonal invariance together with independence of rows of Z imply multivariate normality of rows of Z. Therefore generalization from multivariate normality to left orthogonal invariance is achieved only by giving up the independence of the rows of Z. (See discussion in Sec.1.1 of Kariya and Sinha[4].)

The test statistic we consider is

$$U = \frac{1}{n}\sum_{t=1}^{n} X_t X_t' X_t. \tag{28}$$

U is a sample quantity corresponding to V_1 of the last section. Now we consider the null distribution of U.

For $b \in R^p$ and $A \in G_T$ consider the location and scale transformation of Z: $Z \mapsto ZA' + 1_n b'$. Then $\bar{Z} \mapsto A\bar{Z} + b, S \mapsto ASA'$. Now

$$\begin{aligned} X(ZA' + 1_n b') &= (ZA' - 1_n \bar{Z}'A')(ASA')^{-1/2\prime} \\ &= (Z - 1_n \bar{Z}')A'A'^{-1}S^{-1/2\prime} = X(Z). \end{aligned} \tag{29}$$

(we used the lower triangularity of $S^{1/2}$) and $X(Z)$ is seen to be invariant with respect to this transformation. Now Corollary 3.1 of Kariya and Sinha[4] guarantees that the null distribution of X is the same for all left orthogonally invariant distributions. The distribution of X can be described as follows. Let V be the Stiefel manifold of $M = (\text{span} 1_n)^\perp \subset R^n$ defined as follows

$$V = \{Y : n \times p \mid Y'Y = I_p, 1_n'Y = 0\} \tag{30}$$

Let $\tilde{\mathcal{O}}(n) = \{G | G \in \mathcal{O}(n), G1_n = 1_n\}$. Note that $X/\sqrt{n} \in V$ and

$$\mathcal{L}[X] = \mathcal{L}[QX], \ \forall Q \in \tilde{\mathcal{O}}(n). \tag{31}$$

By the uniquness of the Haar invariant distribution, we see that the distribution of X is the Haar invariant distribution on V.

Now $XH \in V, \forall H \in \mathcal{O}(p)$ and $\mathcal{L}[XH] = \mathcal{L}[QXH], \forall Q \in \tilde{\mathcal{O}}(n)$. This implies that the Haar invariant distribution is right orthogonally invariant as well, i.e., XH is again distributed according to the Haar invariant distribution on V for any $H \in \mathcal{O}(p)$.

So far we have assumed $S^{-1/2}$ to be lower triangular for simplicity. We now argue that the distribution of X does not depend on the choice of $S^{-1/2}$ as long as $S^{1/2}$ is chosen based on S alone. Consider the conditional distribution of X given S. It is easy to see that this conditional distribution satisfies (31) and hence is again the Haar invariant distribution on V. Now $S^{-1/2\prime}$ is determined only up to multiplication by an orthogonal matrix from the right, namely, $S^{-1/2\prime}H$ can be used instead of $S^{-1/2\prime}$ for any $H \in \mathcal{O}(p)$. Now $\mathcal{L}[XH|S] = \mathcal{L}[X|S]$ as long as H is based only on S. Considering the unconditional distribution of X we see that the distribution of X does not depend on the choice of $S^{-1/2}$.

We summarize the argument in the following theorem.

Theorem 3.1. *Suppose that the decomposition of* $S = S^{1/2}S^{1/2\prime}$ *is based only on* S *and let* $X = (Z - 1_n\bar{Z}')S^{-1/2\prime}$. *Under the null hypothesis (27) the distribution* *of X is the same for all left orthogonally invariant distributions. The distribution of X is the Haar invariant distribution on the Stiefel manifold V of (30).*

In view of this result from now on we assume that Z is multivariate normal $Z \sim N_{np}(0, I_n \otimes I_p)$ without loss of generality. As mentioned above we consider a test based on $U \in R^p$ of (28). Since the distribution of X under H_0 is right orthogonally invariant, the distribution of U is orthogonally invariant or spherically symmetric. Therefore it is natural to take the rejection region as

$$U'U > c. \tag{32}$$

We now express $U'U$ in terms of moments (around the mean) of the original observation vectors $Z_t, t = 1, \ldots, n$.

$$\begin{aligned} U'U &= \frac{1}{n^2}[S^{-1/2}\sum_{t=1}^{n}(Z_t - \bar{Z})'S^{-1}(Z_t - \bar{Z})(Z_t - \bar{Z})] \\ &\qquad [S^{-1/2}\sum_{u=1}^{n}(Z_u - \bar{Z})'S^{-1}(Z_u - \bar{Z})(Z_u - \bar{Z})]' \\ &= \frac{1}{n^2}\sum_{t,u}(Z_t - \bar{Z})'S^{-1}(Z_t - \bar{Z})(Z_t - \bar{Z})'S^{-1}(Z_u - \bar{Z}) \\ &\qquad (Z_u - \bar{Z})'S^{-1}(Z_u - \bar{Z}) \\ &= \frac{1}{n^2}\sum_{a,a',b,b',c,c'=1}^{p}\sum_{t,u=1}^{n} S^{aa'}(Z_{ta} - \bar{Z}_a)(Z_{ta'} - \bar{Z}_{a'}) \\ &\qquad S^{bb'}(Z_{tb} - \bar{Z}_b)(Z_{ub'} - \bar{Z}_{b'})S^{cc'}(Z_{uc} - \bar{Z}_a)(Z_{uc'} - \bar{Z}_{c'}) \\ &= \sum_{a,a',b,b',c,c'} m_{aa'b}m_{cc'b'}S^{aa'}S^{bb'}S^{cc'} \end{aligned} \tag{33}$$

where $S^{-1} = (S^{ab})$ and $m_{abc} = (1/n)\sum_{t=1}^{n}(Z_{ta} - \bar{Z}_a)(Z_{tb} - \bar{Z}_b)(Z_{tc} - \bar{Z}_c)$ is the sample third order moment of $Z_t, t = 1, \ldots, n$. This show that $U'U$ coincides with $\operatorname{tr} S_2$ proposed in Isogai[1],[2]. We have now shown that this statistic is a length squared of a geometrically meaningful random vector U .

Although the exact distribution of $U'U$ is complicated, the asymptotic null distribution of U and $U'U$ (as $n \to \infty$) is simple. In view of the fact that $S \xrightarrow{p} I$ and the fact that (when the population variance matrix $\Sigma = I$) $\sqrt{n}m_{abc}$'s are asymptotically normally and mutually independently distributed as

$$\begin{aligned} &\mathcal{L}[\sqrt{n}m_{aaa}] \to N(0, 6) \\ &\mathcal{L}[\sqrt{n}m_{aab}] \to N(0, 2), \quad a < b \end{aligned} \tag{34}$$

(see Kendall and Stuart[5], p.340), we see that the asymptotic distribution of $nU'U$ is the same as

$$n \sum_{a,b,c} m_{aab} m_{ccb} = \sum_{b} (\sqrt{n} \sum_{a} m_{aab})^2. \tag{35}$$

Now

$$\mathcal{L}[\sqrt{n} \sum_{a} m_{aab}] \rightarrow N(0, 6 + 2(p-1)) = N(0, 2(p+2)) \tag{36}$$

Hence we obtain the following result.

Theorem 3.1. *Under the null hypothesis (27) $nU'U/[2(p+2)]$ is asymptotically distributed according to χ^2 distribution with p degrees of freedom.*

This result under multivariate normality was already obtained in Isogai[1],[2]. Here we extended the result for left orthogonally invariant distributions. Now since U has spherically symmetric distribution, as a corollary we have

Corollary 3.1. *Under the null hypothesis (27) $\sqrt{n}U$ is asymptotically distributed as $N_p(0, 2(p+2)I_p)$.*

REFERENCES

1. T. Isogai, On measures of multivariate skewness and kurtosis. *Mathematica Japonica,* **28,** 251-261 (1983).
2. T. Isogai, Measure of multivariate skewness and kurtosis and tests for multivariate normality. Special report in the annual meeting of Japan mathematical association (1986).
3. T. Isogai, On using influence function for testing multivariate normality. *Ann.Inst.Stat. Math.,* **41,** 169-186 (1989).
4. T. Kariya and B.K. Sinha, *Robustness of Statistical Tests.* Academic Press (1989).
5. M.G. Kendall and A. Stuart, *The advanced Theory of Statistics, Vol. 1.* (4th ed.), Griffin (1977).
6. J.F. Malkovich and A.A. Afifi, On tests for multivariate normality. *J.A.S.A.,* **68,** 713-718 (1973).

Stat. Sci. & Data Anal., pp. 237-245
K. Matsusita *et al.* (Eds)

Asymptotic Theory for the Concentrated Matrix Langevin Distributions on the Grassmann Manifold

YASUKO CHIKUSE

Kagawa University, 2-1 Saiwai-cho, Takamatsu-shi, Kagawa-ken, Japan 760

Abstract. The Grassmann manifold $G_{k,m-k}$ consists of k-dimensional linear subspaces $\mathcal{V}$ in R^m, and, to each $\mathcal{V}$ in $G_{k,m-k}$, corresponds a unique $m \times m$ orthogonal projection matrix P, idempotent of rank k, onto $\mathcal{V}$. Let $P_{k,m-k}$ denote the set of all such orthogonal projection matrices. The matrix Langevin distribution, denoted by $L^{(P)}(m,k;F)$, on $P_{k,m-k}$ has the pdf proportional to $\exp(\operatorname{tr} FP)$, where F is an $m \times m$ positive semi-definite matrix with the svd $F = \Gamma\Delta\Gamma'$. We derive asymptotic expansions, for large concentration Δ, for distributions of matrix statistics and of their related functions in connection with testing problems on the $L^{(P)}(m,k;F)$ distributions.

Key words: Grassmann manifold, set of orthogonal projection matrices, concentrated matrix Langevin distributions, (noncentral) Wishart distributions, generalized noncentral Laguerre polynomials in multiple matrices, zonal and invariant polynomials.

1. INTRODUCTION

The Grassmann manifold $G_{k,m-k}$ consists of k-dimensional linear subspaces $\mathcal{V}$ in R^m. To each $\mathcal{V}$ in $G_{k,m-k}$, corresponds a unique $m\times m$ orthogonal projection matrix P, idempotent of rank k, onto $\mathcal{V}$. If k column vectors of an $m \times k$ matrix Y are orthonormal ($Y'Y = I_k$, i.e., Y belongs to the Stiefel manifold $V_{k,m}$) and span $\mathcal{V}$, then we have $YY' = P$. Let $P_{k,m-k}$ denote the set of all such orthogonal projection matrices.

Letting $(\underset{\sim}{y}_{k+1}, \ldots, \underset{\sim}{y}_m)$ be the orthogonal complement of $Y = (\underset{\sim}{y}_1, \ldots, \underset{\sim}{y}_k)$, an invariant measure on $G_{k,m-k}$ is given by the differential form (James [1]) $\bigwedge_{j=1}^{m-k} \bigwedge_{i=1}^{k} \underset{\sim}{y}'_{k+j}\, d\underset{\sim}{y}_i$ in terms of the exterior products ($\bigwedge$), where, for any matrix X $(= (x_{ij}))$, dX $(= (dx_{ij}))$ denotes the matrix of differentials. See, e.g., Muirhead [2] for the use of exterior products. Rewriting the above differential form, we have the invariant measure on $P_{k,m-k}$ (see Chikuse and Watson [3])

$$(dP) = \bigwedge_{j=1}^{m-k} \bigwedge_{i=1}^{k} \underset{\sim}{y}'_i\, dP \underset{\sim}{y}_{k+j}. \tag{1.1}$$

Let $[dP]$ denote the normalized invariant measure $(dP)/\left[\pi^{k(m-k)/2}\Gamma_k(k/2)/\Gamma_k(m/2)\right]$ of unit mass on $P_{k,m-k}$, where $\Gamma_k(a) = \pi^{k(k-1)/4} \prod_{j=1}^{k} \Gamma(a-(j-1)/2)$. A more detailed discussion of manifolds may be found in Farrell [4] and James [1].

Chikuse and Watson [3] discuss population distributions on $P_{k,m-k}$, and suggest a general form of probability density functions (pdf's) which is expressed in terms of a hypergeometric function ${}_rF_s$ of matrix argument (see Appendix). The case $r = s$ yields the "matrix Langevin distribution", denoted by $L^{(P)}(m, k; F)$, whose pdf with respect to $[dP]$ is given by

$$\mathrm{etr}(FP)/{}_1F_1(k/2; m/2; F), \tag{1.2}$$

where $\mathrm{etr}\, A = \exp(\mathrm{tr}\, A)$, and F is an $m \times m$ positive semi-definite matrix with a restriction imposed to ensure the identifiability of F, e.g., $\mathrm{tr}\, F$ being a fixed positive number. Let the singular value decomposition of F, with p being the rank of F, be

$$F = \Gamma\Delta\Gamma', \quad \text{where } \Gamma \in V_{p,m}\,, \text{ and } \Delta = \mathrm{diag}(\lambda_1, \ldots, \lambda_p),\ \lambda_1 \geq \cdots \geq \lambda_p > 0. \tag{1.3}$$

When $p \geq k$, writing $\Gamma = (\Gamma^{(1)} \vdots \Gamma^{(2)})$, $\Gamma^{(1)}$ being $m \times k$, it is seen that $P_0 = \Gamma^{(1)}\Gamma^{(1)\prime}$ is the mode of the distribution and gives $\max_P \mathrm{tr}\, FP = \mathrm{tr}\, FP_0 = \sum_{j=1}^{k} \lambda_j$; P_0 is unique if $\lambda_k > \lambda_{k+1}$ in (1.3). For $p < k$, the mode is not unique, with $\max_P \mathrm{tr}\, FP = \mathrm{tr}\, \Delta$. The λ_j are the parameters of concentration. This is an analogue of Downs' [5] distribution on the Stiefel manifold. See also Watson [6] for a discussion of the $L^{(P)}(m, k; F)$ distribution.

This paper investigates asymptotic behaviors of statistics in connection with testing problems on the $L^{(P)}(m, k; F)$ distribution, when Δ is (i.e., all of the λ_j are) large. Let P be a random matrix on $P_{k,m-k}$, having the $L^{(P)}(m, k; F)$ distribution, where Γ and Δ are known. We are concerned with an asymptotic expansion, for large Δ, for the distribution of the matrix variate $(m \geq k + p)$

$$V = 2\Delta^{1/2}(I_p - \Gamma' P \Gamma)\Delta^{1/2}, \tag{1.4}$$

where $A^{1/2}$ is defined as the unique square root of a positive semi-definite matrix A.

Assuming $p = k$; hence $\Gamma\Gamma'$ is the "unique" mode of the distribution, we consider the problem of testing the null hypothesis

$$H_0 : \Gamma = \Gamma_0, \quad \text{a given matrix in } V_{k,m}, \tag{1.5}$$

against a sequence of local alternative hypotheses,

$$H_1 : \Gamma = (\Gamma_0 + \Psi\Delta^{-1/2})(I_k + \Delta^{-1/2}\Psi'\Psi\Delta^{-1/2})^{-1/2}, \tag{1.6}$$

for any $m \times k$ matrix Ψ such that $\Psi'\Gamma_0 = 0$, when Δ is known and large. When we have a random sample $P_1, \ldots, P_n$ from the $L^{(P)}(m, k; F)$ distribution, $n\bar{P} = \sum_{j=1}^{n} P_j$ is sufficient for the parameter F. We shall derive an asymptotic expansion, for large Δ, for the distribution of the matrix statistic

$$S = 2n\Delta^{1/2}(I_k - \Gamma_0' \bar{P} \Gamma_0)\Delta^{1/2}, \tag{1.7}$$

under the alternative hypothesis (1.6) $(m \geq 2k)$. We also discuss asymptotic theory for a multi-sample test problem. The traces of the matrix statistics may be useful univariate statistics for the testing problems, and asymptotic expansions for their distributions are obtained.

In Appendix, we briefly present results on zonal and invariant polynomials in multiple matrices, and the generalized noncentral Laguerre polynomials in multiple matrices. These results are utilized for the derivation in this paper.

It may be worth noting related works. Asymptotic theory for large sample sizes and for high dimension on $P_{k,m-k}$ has been investigated by Chikuse and Watson [3] and Chikuse [7], respectively. There exists a considerably large literature of orientation statistical analysis on $V_{k,m}$ in recent years; see e.g., Chikuse [8], Downs [5], Khatri and Mardia [9], and the references cited in these papers.

2. ASYMPTOTIC EXPANSIONS FOR THE $L^{(P)}(m,k;F)$ DISTRIBUTION

2.1. An Asymptotic Expansion for a Random Matrix

We consider the matrix variate V defined by (1.4), when Γ and Δ are known and Δ is large. The characteristic function (c.f.) of V is, for a $p\times p$ symmetric matrix T,

$$\Phi_V(T)=E(\operatorname{etr} iTV)=\operatorname{etr}(2i\Delta T)\,{}_1F_1{}^{-1}(\Delta)\,{}_1F_1(\Delta(I_p-2iT)),\tag{2.1}$$

where we use the notation ${}_1F_1(A)={}_1F_1(k/2;m/2;A)$ throughout this paper. Utilizing the asymptotic expansion for the ${}_1F_1$ function, for large Δ, in terms of the ${}_2F_0$ function (Constantine and Muirhead [10, Theorem 3.2]), and then expanding the ${}_2F_0$ functions (see (A.1)), we obtain

$$\begin{aligned}\Phi_V(T)&=|I_p-2iT|^{-b}{}_2F_0{}^{-1}(a,b;\Delta^{-1})\,{}_2F_0(a,b;\Delta^{-1}(I_p-2iT)^{-1})\\&=|I_p-2iT|^{-b}\Big\{1-abC_{(1)}(\Delta^{-1})+abC_{(1)}(\Delta^{-1}(I_p-2iT)^{-1})\\&\quad+\sum_{\lambda\vdash 2}[(ab)^2-(a)_\lambda(b)_\lambda/2]C_\lambda(\Delta^{-1})-(ab)^2C_{(1)}(\Delta^{-1})C_{(1)}(\Delta^{-1}(I_p-2iT)^{-1})\\&\quad+\sum_{\lambda\vdash 2}(a)_\lambda(b)_\lambda C_\lambda(\Delta^{-1}(I_p-2iT)^{-1})/2+O(\Delta^{-3})\Big\},\\&\qquad\qquad\text{with } a=(p-k+1)/2 \text{ and } b=(m-k)/2.\end{aligned}\tag{2.2}$$

The pdf of V is given by the inversion formula

$$a_p\int_{\mathcal{S}_p}\operatorname{etr}(-iVT)\Phi_V(T)(dT),\quad\text{with } a_p=2^{p(p-1)/2}/(2\pi)^{p(p+1)/2},\tag{2.3}$$

where the integration is over the space $\mathcal{S}_p$ of all $p\times p$ real symmetric matrices. Assuming $m\geq k+p$, we can evaluate the integral

$$\begin{aligned}\mathcal{I}_\lambda&=a_p\int_{\mathcal{S}_p}\operatorname{etr}(-iVT)|I_p-2iT|^{-b}C_\lambda(\Delta^{-1}(I_p-2iT)^{-1})(dT)\\&=(2i)^{-p(p+1)/2}a_p|\Delta|^{b-(p+1)/2}\operatorname{etr}(-V/2)\\&\qquad\times\int_{\mathcal{R}(Z)>0}\operatorname{etr}(\Delta^{-1/2}V\Delta^{-1/2}Z/2)|Z|^{-b}C_\lambda(Z^{-1})(dZ)\end{aligned}\tag{2.4}$$

$$=w_p(V;2b,I_p)C_\lambda(\Delta^{-1}V/2)/(b)_\lambda,\tag{2.5}$$

where $w_p(\cdot\,; f, I_p)$ denotes the pdf of the Wishart $W_p(f, I_p)$ distribution. Here, we made the transformation $Z = \Delta^{1/2}(I_p - 2iT)\Delta^{1/2}$, and utilized the inverse Laplace transform of the zonal polynomial (Constantine [11, (9)]) to obtain (2.5) from (2.4), where the integration is over all $p \times p$ matrices of the form $Z = \mathcal{R}(Z) + i\mathcal{I}(Z)$ for fixed positive definite matrix $\mathcal{R}(Z)$ and arbitrary real symmetric matrix $\mathcal{I}(Z)$.

Thus, we establish

Theorem 2.1. *Let $P \in P_{k,m-k}$ be distributed as $L^{(P)}(m, k; F)$ with the svd $F = \Gamma\Delta\Gamma'$ ($m \geq k + p$). Then, for large Δ, the pdf of the matrix variate V defined by (1.4) has the expansion*

$$w_p(V; m-k, I_p)\Big\{1 - ab \operatorname{tr}\Delta^{-1} + a \operatorname{tr}\Delta^{-1}V/2 + \sum_{\lambda \vdash 2}[(ab)^2 - (a)_\lambda(b)_\lambda/2]C_\lambda(\Delta^{-1}) - a^2b(\operatorname{tr}\Delta^{-1})\operatorname{tr}\Delta^{-1}V/2 + \sum_{\lambda \vdash 2}(a)_\lambda C_\lambda(\Delta^{-1}V)/8 + O(\Delta^{-3})\Big\}, \tag{2.6}$$

where a and b are given in (2.2).

2.2. Asymptotic Expansions for Testing the Mode

For the matrix statistic S defined by (1.7), considered in connection with the test ((1.5), (1.6)) when $p = k$, we note that $S = \sum_{j=1}^{n} W_j$, and $\Phi_S(T) = \prod_{j=1}^{n} \Phi_{W_j}(T)$, where $W_j = 2\Delta^{1/2}(I_k - \Gamma_0'P_j\Gamma_0)\Delta^{1/2}$, $j = 1, \ldots, n$. Hence, let us first consider the matrix $W = 2\Delta^{1/2}(I_k - \Gamma_0'P\Gamma_0)\Delta^{1/2}$, where $P \in P_{k,m-k}$ has the $L^{(P)}(m, k; F)$ distribution. The c.f. of W is, for a $k \times k$ symmetric matrix T,

$$\Phi_W(T) = \operatorname{etr}(2i\Delta T)\,{}_1F_1^{-1}(\Delta)\,{}_1F_1(E), \quad \text{with } E = \Gamma\Delta\Gamma' - 2i\Gamma_0\Delta^{1/2}T\Delta^{1/2}\Gamma_0'. \tag{2.7}$$

We choose an $m \times (m-k)$ matrix Γ_1 such that $(\Gamma_0 \vdots \Gamma_1) \in O(m)$, and write $\Psi = \Gamma_1\Psi_1$, with Ψ_1 being $(m-k) \times k$. Then, the alternative hypothesis (1.6) can be rewritten as

$$\Gamma = \Gamma_0 + \Gamma_0A_0 + \Gamma_1A_1, \tag{2.8}$$

where

$$A_0 = -\Delta^{-1/2}\Psi_1'\Psi_1\Delta^{-1/2}/2 + 3\Delta^{-1/2}\Psi_1'\Psi_1\Delta^{-1}\Psi_1'\Psi_1\Delta^{-1/2}/8 + O(\Delta^{-3}),$$

and

$$A_1 = \Psi_1\Delta^{-1/2} - \Psi_1\Delta^{-1}\Psi_1'\Psi_1\Delta^{-1/2}/2 + O(\Delta^{-5/2}). \tag{2.9}$$

We can rewrite E, with (2.8) substituted, as

$$E = (\Gamma_0 \vdots \Gamma_1)E^*(\Gamma_0 \vdots \Gamma_1)', \quad \text{with } E^* = \begin{bmatrix} \Delta^{1/2}(I_k - 2iT)\Delta^{1/2} + A_{11} & \vdots & A_{12} \\ \cdots & & \cdots \\ A_{12}' & \vdots & A_{22} \end{bmatrix}, \tag{2.10}$$

where

$$A_{11} = \Delta A_0' + A_0\Delta + A_0\Delta A_0' \ (= O(1)),$$
$$A_{12} = \Delta A_1' + A_0\Delta A_1' \ (= O(\Delta^{1/2})), \quad \text{and} \quad A_{22} = A_1\Delta A_1' \ (= O(1)). \tag{2.11}$$

We utilize the asymptotic expansion for the ${}_1F_1$ function of the argument matrix E^* of the form (2.10) for large Δ (Constantine and Muirhead [10, (3.4)], Muirhead [12, (7.2)]), and the Kummer relation (James [13, (51)]). Combining with the previous asymptotic expansion for ${}_1F_1(\Delta)$ for large Δ, in (2.7), in view of the expansion $\log|I+X| = -\sum_{j=1}^{\infty} \operatorname{tr}(-X)^j/j$ for "small" X, we obtain, after some algebra,

$$\begin{aligned}\Phi_W(T) = |U|^{(m-k)/2} \operatorname{etr}[\Omega(U-I_k)/2]\{1 + (m-k)\operatorname{tr}\Delta^{-1}[(m-k)I_k + \Omega]/4 \\ - (m-k)\operatorname{tr}\Sigma^{1/2}(\Delta^{-1})(I_k - U)\Sigma^{1/2}(\Delta^{-1})/8 \\ + \operatorname{tr}\Lambda(\Delta^{-1})[\Omega^{1/2}(I_k-U)\Omega^{1/2}]^2/8 + O(\Delta^{-2})\},\end{aligned} \tag{2.12}$$

where

$$U = (I_k - 2iT)^{-1}, \qquad \Omega = 2\Psi'\Psi,$$
$$\Sigma(\Delta^{-1}) = 2(1+m-k)\Delta^{-1} + \Delta^{-1}\Omega + \Omega\Delta^{-1}, \tag{2.13}$$

and

$$\Lambda(\Delta^{-1}) = \Omega^{1/2}\Delta^{-1}\Omega^{-1/2} + \Omega^{-1/2}\Delta^{-1}\Omega^{1/2}.$$

Thus, inverting $\Phi_S(T) = \Phi_W^n(T)$, in view of (A.2), (A.3), with $X = \Omega^{1/2}(I_k - U)\Omega^{1/2}$ and $A = \Lambda(\Delta^{-1})$, and (A.4), we establish

Theorem 2.2. *Under the alternative hypothesis (1.6), for large Δ (with n fixed), the pdf of the matrix statistic S defined by (1.7) has the expansion*

$$\begin{aligned}w_k(S;(m-k)n, I_k; n\Omega)\{1 + (m-k)n\operatorname{tr}\Delta^{-1}[(m-k)I_k + \Omega]/4 \\ - L^u_{(1)}(S/2;\Sigma^{1/2}(\Delta^{-1});\Omega)/4 + n\sum_{\lambda\vdash 2}\sum_{\phi\in\lambda\cdot(1)} a_\phi^{\lambda,(1)} \\ \times L^u_{\lambda,(1);\phi}(S/2;\Omega^{1/2},\Lambda^{1/2}(\Delta^{-1});\Omega)/8((m-k)n/2)_\lambda + O(\Delta^{-2}),\}\end{aligned} \tag{2.14}$$

where $u = [(m-k)n - k - 1]/2$, $w_k(\cdot\,;(m-k)n, I_k; n\Omega)$ and $L^u_{(1)}$, $L^u_{\lambda,(1);\phi}$ are the pdf of the noncentral Wishart $W_k((m-k)n, I_k; n\Omega)$ distribution and the associated generalized noncentral Laguerre polynomials, respectively (see Appendix), and the coefficients $a_\phi^{\lambda,(1)}$ and the matrices Ω, $\Sigma(\Delta^{-1})$ and $\Lambda(\Delta^{-1})$ are given by (A.2), (A.3) and (2.13), respectively.

Suppose we have a random sample $P_{i1}, \ldots, P_{in_i}$ of size n_i from the $L^{(P)}(m,k;F_i)$ distribution, with the svd $F_i = \Gamma_i\Delta_i\Gamma_i'$, $\Gamma_i \in V_{k,m}$, and $\Delta_i > 0$, $k \times k$ diagonal, for $i = 1, \ldots, q$. We consider the multi-sample problem of testing the null hypothesis

$$H_{0q} : \Gamma_i = \Gamma_{i0}, \quad \text{a given matrix in } V_{k,m}, \text{ for } i = 1, \ldots, q, \tag{2.15}$$

against a sequence of local alternative hypotheses,

$$H_{1q}:\Gamma_i=(\Gamma_{i0}+\Psi_i\Delta_i^{-1/2})(I_k+\Delta_i^{-1/2}\Psi_i'\Psi_i\Delta_i^{-1/2})^{-1/2}, \tag{2.16}$$

for any $m\times k$ matrices Ψ_i such that $\Psi_i'\Gamma_{i0}=0$, $i=1,\dots,q$, when all of the Δ_i are known and large. Putting $\bar{P}_i=\sum_{j=1}^{n_i}P_{ij}/n_i$, $i=1,\dots,q$, we may consider the matrix statistic

$$\bar{S}=\sum_{i=1}^{q}S_i,\quad \text{where } S_i=2n_i\Delta_i^{1/2}(I_k-\Gamma_{i0}'\bar{P}_i\Gamma_{i0})\Delta_i^{1/2}. \tag{2.17}$$

By a similar method, we establish

Theorem 2.3. *Under the alternative hypothesis (2.16), for large Δ_i (with n_i fixed), $i=1,\dots,q$, the pdf of the matrix statistic $\bar{S}$ defined by (2.17) has the expansion*

$$\begin{aligned}&w_k(\bar{S};(m-k)\bar{n},I_k;\bar{\Omega})\Big\{1+(m-k)\sum_{i=1}^{q}n_i\operatorname{tr}\Delta_i^{-1}[(m-k)I_k+\Omega_i]/4\\&-L_{(1)}^{\bar{u}}(\bar{S}/2;\bar{\Sigma}^{1/2}(\bar{\Delta}^{-1});\bar{\Omega})/4+\sum_{i=1}^{q}n_i\sum_{\lambda\vdash 2}\sum_{\phi\in\lambda\cdot(1)}a_\phi^{\lambda,(1)}\\&\times L_{\lambda,(1);\phi}^{\bar{u}}(\bar{S}/2;\Omega_i^{1/2},\Lambda_i^{1/2}(\Delta_i^{-1});\bar{\Omega})/8((m-k)\bar{n}/2)_\lambda+O(\Delta_i^{-2})\Big\},\end{aligned} \tag{2.18}$$

where

$$\begin{aligned}&\bar{n}=\sum_{i=1}^{q}n_i,\quad \bar{\Omega}=\sum_{i=1}^{q}n_i\Omega_i,\quad \bar{u}=[(m-k)\bar{n}-k-1]/2,\\&\bar{\Sigma}(\bar{\Delta}^{-1})=\sum_{i=1}^{q}n_i\Sigma_i(\Delta_i^{-1})/\bar{n},\end{aligned} \tag{2.19}$$

and $\Sigma_i(\Delta_i^{-1})$ and $\Lambda_i(\Delta_i^{-1})$ are defined similarly to $\Sigma(\Delta^{-1})$ and $\Lambda(\Delta^{-1})$, respectively, given in (2.13), with Δ_i and Ω_i replacing Δ and Ω, respectively.

The c.f. $\Phi_w(t)$ $(=Ee^{it\operatorname{tr}W})$ of $w=\operatorname{tr}W$ is given by $\Phi_W(tI_k)$, which is

$$\begin{aligned}\Phi_w(t)=(1-2it)^{-k(m-k)/2}\operatorname{etr}[((1-2it)^{-1}-1)\Omega/2]\{1+a_0(\Delta^{-1})\\+(1-2it)^{-1}a_1(\Delta^{-1})+(1-2it)^{-2}a_2(\Delta^{-1})+O(\Delta^{-2})\},\end{aligned} \tag{2.20}$$

where

$$\begin{aligned}&a_0(\Delta^{-1})=\operatorname{tr}\Delta^{-1}(\Omega^2-(m-k)I_k)/4,\\&a_1(\Delta^{-1})=\operatorname{tr}[(m-k)(1+m-k)\Delta^{-1}+(m-k)\Delta^{-1}\Omega-\Omega\Delta^{-1}\Omega-\Delta^{-1}\Omega^2]/4,\end{aligned} \tag{2.21}$$

and

$$a_2(\Delta^{-1})=\operatorname{tr}\Delta^{-1}\Omega^2/4.$$

Thus, we have the expansion for the c.f. $\Phi_s(t)\ (= \prod_{j=1}^{n} \Phi_{w_j}(t))$ of $s = \operatorname{tr} S$, $S = \sum_{j=1}^{n} W_j$, and utilizing the inversion formula for the noncentral χ^2 distribution, we obtain an asymptotic expansion for the pdf of s; and similarly that for $\bar{s} = \operatorname{tr} \bar{S}$. We establish

Theorem 2.4. *Under the alternative hypotheses (1.6) and (2.16), for large Δ (with n fixed) and Δ_i (with n_i fixed), $i = 1, \ldots, q$, the pdf's of $s = \operatorname{tr} S$ and $\bar{s} = \operatorname{tr} \bar{S}$ have the expansions*

$$\chi^2_{k(m-k)n}(s; n\omega)[1 + na_0(\Delta^{-1})] + \chi^2_{k(m-k)n+2}(s; n\omega)na_1(\Delta^{-1}) + \chi^2_{k(m-k)n+4}(s; n\omega)na_2(\Delta^{-1}) + O(\Delta^{-2}), \tag{2.22}$$

and

$$\chi^2_{k(m-k)\bar{n}}(\bar{s}; \bar{\omega})\left[1 + \sum_{i=1}^{q} n_i a_0^{(i)}(\Delta_i^{-1})\right] + \chi^2_{k(m-k)\bar{n}+2}(\bar{s}; \bar{\omega}) \sum_{i=1}^{q} n_i a_1^{(i)}(\Delta_i^{-1}) + \chi^2_{k(m-k)\bar{n}+4}(\bar{s}; \bar{\omega}) \sum_{i=1}^{q} n_i a_2^{(i)}(\Delta_i^{-1}) + O(\Delta_i^{-2}), \tag{2.23}$$

respectively. Here, $\chi^2_f(\cdot\,; \omega)$ denotes the pdf of the noncentral χ^2_f distribution, $\omega = \operatorname{tr} \Omega$, $a_j(\Delta^{-1})$, $j = 0, 1, 2$, are defined by (2.21), $\bar{n} = \sum_{i=1}^{q} n_i$, $\bar{\omega} = \sum_{i=1}^{q} n_i \operatorname{tr} \Omega_i$, and the $a_j^{(i)}(\Delta^{-1})$ are defined similarly to the $a_j(\Delta^{-1})$, with Δ_i and Ω_i replacing Δ and Ω, respectively.

It is noted here that asymptotic expansions under the null hypotheses H_0 ((1.5)) and H_{0q} ((2.15)) are readily obtained by putting $\Psi = 0$ and the $\Psi_i = 0$ in Theorems 2.2–2.4.

APPENDIX

The zonal polynomials $C_\lambda(X)$ in a $p \times p$ symmetric matrix X were defined by the theory of group representations of the real linear group $Gl(p, R)$ of $p \times p$ real nonsingular matrices on the vector space of homogeneous polynomials of degree l on the space of $p \times p$ symmetric matrices (James [13] and Constantine [14]). Here, $[2\lambda]$ indexes each irreducible representation, where λ is an ordered partition of l, i.e., $\lambda = (l_1, \ldots, l_p)$, $l_1 \geq \cdots \geq l_p \geq 0$, $\sum_{j=1}^{p} l_j = l$, this being denoted by $\lambda \vdash l$. The $C_\lambda(X)$, $\lambda \vdash l = 0, 1, \ldots$, are invariant under the transformation $X \to HXH'$, $H \in O(p)$, where $O(p)$ is the orthogonal group of $p \times p$ orthonormal matrices, and constitute a basis of the space of all homogeneous symmetric polynomials in the latent roots of X. The polynomial $(\operatorname{tr} X)^l$ then has a unique decomposition $(\operatorname{tr} X)^l = \sum_\lambda C_\lambda(X)$, where the sum $\sum_\lambda$ ranges over all $\lambda \vdash l$; note that $\operatorname{tr} X = C_{(1)}(X)$.

The hypergeometric function ${}_rF_s(a_1, \ldots, a_r; b_1, \ldots, b_s; X)$ of symmetric matrix argument X has a serial representation

$$\sum_{l=0}^{\infty} \sum_{\lambda} [(a_1)_\lambda \cdots (a_r)_\lambda / (b_1)_\lambda \cdots (b_s)_\lambda\, l!] C_\lambda(X), \tag{A.1}$$

where the a_i and b_j are real or complex constants, and $(a)_\lambda = \prod_{j=1}^{p}(a-(j-1)/2)_{l_j}$, with $(a)_l = a(a+1)\cdots(a+l-1)$.

The invariant polynomials $C_\phi^{\lambda,\rho}(X,Y)$ in two $p\times p$ symmetric matrix arguments X and Y were defined (Davis [15, 16]), extending the zonal polynomials, by the theory of group representations of $Gl(p,R)$ on the vector space of polynomials in two arguments, homogeneous of degree l, r on the space of $p\times p$ symmetric matrices. $C_\phi^{\lambda,\rho}(X,Y)$ is defined when the irreducible representation indexed by $[2\phi]$ ($\phi \vdash (l+r)$) occurs (possibly with multiplicity greater than one) in the decomposition of the Kronecker product $[2\lambda]\otimes[2\rho]$ of the irreducible representations indexed by $[2\lambda]$ and $[2\rho]$ into a direct sum of irreducible representations of $Gl(p,R)$; this being denoted by $\phi\in\lambda\cdot\rho$. The $C_\phi^{\lambda,\rho}(X,Y)$ are invariant under the simultaneous transformations $X\to HXH'$, $Y\to HYH'$, $H\in O(p)$. We define $C_\lambda^{\lambda,0}(X,Y)=C_\lambda(X)$, $C_\rho^{0,\rho}(X,Y)=C_\rho(Y)$. The $C_\phi^{\lambda,\rho}$ polynomials have been extended to the invariant polynomials in more than two matrix arguments by Chikuse [17]. The polynomials have been tabulated up to three matrices and the first few degrees by Davis [15, 18].

We may write

$$\operatorname{tr} X^2A = C_{(1)}(X^2A) = \sum_{\lambda\vdash 2}\sum_{\phi\in\lambda\cdot(1)} a_\phi^{\lambda,(1)} C_\phi^{\lambda,(1)}(X,A), \tag{A.2}$$

for suitable coefficients $a_\phi^{\lambda,(1)}$, where the sum $\sum_{\phi\in\lambda\cdot(1)}$ ranges over all $[2\phi]$ occurring in $[2\lambda]\otimes[2(1)]$. Referring to the tables of the invariant polynomials yields

$$\begin{aligned}\operatorname{tr} X^2A &= C_{(3)}^{(2),(1)}(X,A) - C_{(2,1)}^{(2),(1)}(X,A)/6\\ &\quad - 5^{1/2}C_{(2,1)}^{(1^2),(1)}(X,A)/12 + C_{(1^3)}^{(1^2),(1)}(X,A)/4.\end{aligned} \tag{A.3}$$

Extending the Chikuse and Davis' [19] generalized (central) Laguerre polynomials in multiple matrices (generalization of the Laguerre polynomials in one matrix due to Herz [20] and Constantine [11]) to the "noncentral" case, Chikuse [21] defined and discussed the generalized noncentral Laguerre polynomials in multiple matrices, associated with the joint distribution of multiple noncentral Wishart distributions. Here, we present results for the case involving the invariant polynomials in two matrix arguments, which are relevant for the derivation in this paper. The generalized noncentral Laguerre polynomials $L_{\lambda,\rho;\phi}^{u}(X;A,B;\Omega)$, in $p\times p$ symmetric matrix argument X, two $p\times p$ constant matrices A, B, and a $p\times p$ symmetric noncentral matrix Ω, are defined as the orthogonal polynomials associated with the noncentral Wishart $W_p(2u+p+1, I_p/2;\Omega)$ distribution. The following *generalized noncentral multivariate Rodrigues formula* is of great use in multivariate distribution theory:

$$\begin{aligned}&L_{\lambda,\rho;\phi}^{(n-p-1)/2}(X/2;A,B;\Omega)w_p(X;n,I_p;\Omega)/(n/2)_\lambda\\ &\quad= \left[2^{p(p-1)/2}/(2\pi)^{p(p+1)/2}\right]\int_{S_p}\operatorname{etr}(-iXT)|I_p-2iT|^{-n/2}\\ &\qquad\times \operatorname{etr}\frac{1}{2}\Omega[(I_p-2iT)^{-1}-I_p]C_\phi^{\lambda,\rho}\big(A(I_p-(I_p-2iT)^{-1})A',BB'\big)(dT),\end{aligned} \tag{A.4}$$

where the integration is over the space $\mathcal{S}_p$ of all $p \times p$ real symmetric matrices.

Acknowledgment
The author is grateful to the referee for helpful comments.

REFERENCES

1. A.T. James, *Ann. Math. Statist.* **25**, 40–75 (1954).
2. R.J. Muirhead, *Aspects of Multivariate Statistical Theory.* Wiley, New York (1982).
3. Y. Chikuse and G.S. Watson, *Research Paper* No.43, Kagawa University (1992).
4. R.H. Farrell, *Multivariate Calculation.* Springer-Verlag, Berlin (1985).
5. T.D. Downs, *Biometrika* **59**, 665–676 (1972).
6. G.S. Watson, Personal communication (1991).
7. Y. Chikuse, *Research Paper* No.48, Kagawa University (1992).
8. Y. Chikuse, *J. Multivariate Analysis* **39**, 270–283 (1991).
9. C.G. Khatri and K.V. Mardia, *J. Roy. Statist. Soc. Ser. B* **39**, 95–106 (1977).
10. A.G. Constantine and R.J. Muirhead, *J. Multivariate Analysis* **6**, 369–391 (1976).
11. A.G. Constantine, *Ann. Math. Statist.* **37**, 215–225 (1966).
12. R.J. Muirhead, *Ann. Statist.* **6**, 5–33 (1978).
13. A.T. James, *Ann. Math. Statist.* **35**, 475–501 (1964).
14. A.G. Constantine, *Ann. Math. Statist.* **34**, 1270–1285 (1963).
15. A.W. Davis, *Ann. Inst. Statist. Math.* **A31**, 465–485 (1979).
16. A.W. Davis, in: *Multivariate Analysis,* P.R. Krishnaiah (Ed.), vol V, pp.287–299. North-Holland, Amsterdam (1980).
17. Y. Chikuse, in: *Multivariate Statistical Analysis,* R.P. Gupta (Ed.), pp.53–68. North-Holland, Amsterdam (1980).
18. A.W. Davis, *Ann. Inst. Statist. Math.* **A33**, 297–313 (1981).
19. Y. Chikuse and A.W. Davis, *Ann. Inst. Statist. Math.* **A38**, 109–122 (1986).
20. C.S. Herz, *Ann. Math.* **61**, 474–523 (1955).
21. Y. Chikuse, *Research Paper* No.47, Kagawa University (1992).

Stat. Sci. & Data Anal., pp. 247-252
K. Matsusita *et al.* (Eds)

On Canonical Correlations and the Degrees of Non-Orthogonality in the Three-way Layout

JULIE BÉRUBÉ, ROBERT E. HARTWIG and GEORGE P. H. STYAN

Dept. of Statistics, University of Michigan, 419 South State Street, 1444 Mason Hall, Ann Arbor, MI 48109, USA
Dept. of Mathematics, Box 8205, North Carolina State University, Raleigh, NC 27695-8205, USA
Dept. of Mathematics & Statistics, McGill University, 805 ouest, rue Sherbrooke, Montréal, QC, Canada H3A 2K6

Abstract

We study four kinds of canonical correlations associated with the usual three-way layout of experimental design. We define the degree t of non-orthogonality as the number of positive canonical correlations strictly less than 1 and show that t may be viewed as the degree of non-idempotence or the degree of asymmetry of the product of two symmetric idempotent matrices. We obtain several new inequalities and equalities for these canonical correlations and for the various associated degrees of non-orthogonality.

Key words: Canonical correlations, degree of asymmetry, degree of disconnectedness, degree of non-idempotence, degree of non-orthogonality, experimental design, idempotent matrices, projectors, rank equalities and inequalities, three-way layout.

INTRODUCTION AND PRELIMINARIES

We consider the usual three-way layout of experimental design with white noise

$$\mathcal{E}(\mathbf{y}) = \mathbf{X}_1\boldsymbol{\alpha} + \mathbf{X}_2\boldsymbol{\beta} + \mathbf{X}_3\boldsymbol{\tau} = (\mathbf{X}_1 : \mathbf{X}_2 : \mathbf{X}_3)\begin{pmatrix}\boldsymbol{\alpha}\\ \boldsymbol{\beta}\\ \boldsymbol{\tau}\end{pmatrix} = \mathbf{X}\boldsymbol{\gamma};\ \mathcal{D}(\mathbf{y}) = \sigma^2\mathbf{I}, \qquad (1)$$

where the vectors $\boldsymbol{\alpha}$, $\boldsymbol{\beta}$, and $\boldsymbol{\tau}$ consist of the row, column and treatment effects, respectively. The matrices $\mathbf{X}_1$, $\mathbf{X}_2$, $\mathbf{X}_3$ are the $n \times r$, $n \times c$ and $n \times v$ "design matrices" identifying the correspondence between the elements of $\mathbf{y}$ and, respectively, the rows, columns and treatments, of the three-way layout; the $n \times (r+c+v)$ partitioned matrix $\mathbf{X} = (\mathbf{X}_1 : \mathbf{X}_2 : \mathbf{X}_3)$, therefore, is the design matrix for the full three-way layout. Since exactly one treatment is applied to each observation which appears in precisely one row and one column, we have $\mathbf{X}_1\mathbf{e}^{(r)} = \mathbf{X}_2\mathbf{e}^{(c)} = \mathbf{X}_3\mathbf{e}^{(v)} = \mathbf{e}^{(n)}$, where $\mathbf{e}^{(a)}$ is the $a \times 1$ vector of ones, $a = r, c, v, n$.

We consider the following four different types of canonical correlations:

(i) $\rho_h^{(i.j)}$ between $\mathbf{X}_i'\mathbf{y}$ and $\mathbf{X}_j'\mathbf{y}$,

(ii) $\rho_h^{(i.jk)}$ between $\mathbf{X}_i'\mathbf{y}$ and $(\mathbf{X}_j : \mathbf{X}_k)'\mathbf{y}$,

(iii) $\rho_h^{(i,j\,|\,k)}$ between $\mathbf{X}_i'\mathbf{y}$ and $\mathbf{X}_j'\mathbf{M}_k\mathbf{y}$,

(iv) $\rho_h^{(i\,|\,k,j\,|\,k)}$ between $\mathbf{X}_i'\mathbf{M}_k\mathbf{y}$ and $\mathbf{X}_j'\mathbf{M}_k\mathbf{y}$, where $i \neq j, i \neq k, j \neq k$; $i, j, k = 1, 2, 3$; $h = 1, \ldots, m$. The upper limits m of the index h denote the numbers of positive canonical correlations; we write $m = 0$ when the associated vectors are uncorrelated. There are fifteen kinds of canonical correlations ρ_h in all, three each of types (i), (ii) and (iv), plus six of type (iii).

The "residual matrix" $\mathbf{M}$ in (iii) and (iv) may be defined by $\mathbf{M} = \mathbf{I} - \mathbf{X}\mathbf{X}^+$, with $\mathbf{X}^+$ the Moore-Penrose inverse of $\mathbf{X}$, and so $\mathbf{M}$ is the orthogonal projector on the null space $\mathcal{N}(\mathbf{X}')$. The "hat matrix" $\mathbf{H} = \mathbf{X}(\mathbf{X}'\mathbf{X})^-\mathbf{X}' = \mathbf{X}\mathbf{X}^+ = \mathbf{I} - \mathbf{M}$ associated with the design matrix $\mathbf{X}$ is the orthogonal projector on the range or column space $\mathcal{C}(\mathbf{X})$.

Canonical correlations of types (i), (ii), and (iv) were studied in some detail in [1], while those of type (iii) were introduced in [2]; further details are given in [3]. Canonical correlations of type (i), since the third factor is ignored, apply directly to two-way layouts or subdesigns; cf. e.g., [4]. As in [1, 4] we define, for each of the fifteen kinds of canonical correlations,

u = the number of canonical correlations *equal* to 1

t = the number of *nonzero* canonical correlations *strictly less* than 1.

The number of *nonzero* canonical correlations m is then

$$m = u + t. \quad \mathbf{(2)}$$

We will say that the corresponding layout is "orthogonal" whenever $t = 0$, and "connected" whenever $u = 1$ for canonical correlations of types (i) and (ii) and $u = 0$ for canonical correlations of types (iii) and (iv). We may thus interpret, and define, t as the "degree of non-orthogonality" and u as the "degree of disconnectedness".

Whenever we need to be more specific, however, we add subscripts to m, u and t, corresponding to the superscripts in (i), (ii), (iii) and (iv), e.g., for type (i) we would write $m_{i.j} = u_{i.j} + t_{i.j}$; when the dispersion matrix $\mathcal{D}(\mathbf{y}) = \sigma^2\mathbf{I}$, the cross-covariance matrix between $\mathbf{X}_i'\mathbf{y}$ and $\mathbf{X}_j'\mathbf{y}$ is $\sigma^2\mathbf{X}_i'\mathbf{X}_j$, and the equation $m_{i.j} = u_{i.j} + t_{i.j}$ may be written as the following rank formula:

$$\operatorname{rank}(\mathbf{X}_i'\mathbf{X}_j) = \operatorname{rank}(\mathbf{X}_i) + \operatorname{rank}(\mathbf{X}_j) - \operatorname{rank}(\mathbf{X}_i : \mathbf{X}_j) + \operatorname{rank}(\mathbf{X}_i'\mathbf{M}_j\mathbf{M}_i\mathbf{X}_j), \quad \mathbf{(3)}$$

where $u_{i.j} = \operatorname{rank}(\mathbf{X}_i) + \operatorname{rank}(\mathbf{X}_j) - \operatorname{rank}(\mathbf{X}_i : \mathbf{X}_j)$ and $t_{i.j} = \operatorname{rank}(\mathbf{X}_i'\mathbf{M}_j\mathbf{M}_i\mathbf{X}_j)$.

Equation **(3)** is a special case of the following more general result established in [5]:

LEMMA 1. *For complex matrices* $\mathbf{A}$, $p \times h$ *and* $\mathbf{B}$, $p \times k$, *the rank of the product*

$$\operatorname{rank}(\mathbf{A}^*\mathbf{B}) = \operatorname{rank}(\mathbf{A}) + \operatorname{rank}(\mathbf{B}) - \operatorname{rank}(\mathbf{A}:\mathbf{B}) + \operatorname{rank}(\mathbf{A}^*\mathbf{M}_\mathbf{B}\mathbf{M}_\mathbf{A}\mathbf{B}), \quad \mathbf{(4)}$$

where $\mathbf{A}^*$ *is the conjugate transpose of* $\mathbf{A}$, $(\mathbf{A}:\mathbf{B})$ *is the* $p \times (h + k)$ *partitioned matrix with* $\mathbf{A}$ *placed next to* $\mathbf{B}$, *and* $\mathbf{M}_\mathbf{X}$ *is the orthogonal projector on the null space* $\mathcal{N}(\mathbf{X}')$, $\mathbf{X} = \mathbf{A}, \mathbf{B}$.

Proof. In view of (2.10) and (2.35) in [6], we have

$$\operatorname{rank}(\mathbf{A}^*\mathbf{M}_\mathbf{B}\mathbf{M}_\mathbf{A}\mathbf{B}) = \operatorname{rank}(\mathbf{A}^*\mathbf{M}_\mathbf{B}\mathbf{H}_\mathbf{A}\mathbf{B}) = \operatorname{rank}(\mathbf{M}_\mathbf{B}\mathbf{H}_\mathbf{A}\mathbf{B}) = \operatorname{rank}(\mathbf{B} : \mathbf{H}_\mathbf{A}\mathbf{B}) - \operatorname{rank}(\mathbf{B}). \quad \mathbf{(5)}$$

Since elementary column operations do not change rank and $\mathbf{B} - \mathbf{H}_\mathbf{A}\mathbf{B} = \mathbf{M}_\mathbf{A}\mathbf{B}$, it follows that

$$\text{rank}(\mathbf{B} : \mathbf{H_A B}) = \text{rank}(\mathbf{M_A B} : \mathbf{H_A B}) = \\ \text{rank}(\mathbf{M_A B}) + \text{rank}(\mathbf{H_A B}) = \text{rank}(\mathbf{A} : \mathbf{B}) - \text{rank}(\mathbf{A}) + \text{rank}(\mathbf{A^*B}). \quad (6)$$

Combining **(6)** with **(5)** yields **(4)** . ❐

From this lemma we easily obtain the following corollary:

COROLLARY 1. *Let* **A** *be* $n \times n$ *so that* $\mathbf{A}^2 = \mathbf{0}$. *Then*

$$2\text{rank}(\mathbf{A}) = \text{rank}(\mathbf{A} : \mathbf{A^*}) = \text{rank}(\mathbf{AA^*} + \mathbf{A^*A}) = \text{rank}(\mathbf{A} + \mathbf{A^*}) = \text{rank}(\mathbf{A} - \mathbf{A^*}). \quad (7)$$

RESULTS

Our first result is the following theorem:

THEOREM 1. *The degree* $t_{i.j}$ *of non-orthogonality between factor i and factor j satisfies*

$$\begin{aligned} t_{i.j} &= \text{rank}[\mathbf{H}_i\mathbf{H}_j - (\mathbf{H}_i\mathbf{H}_j)^2] \\ &= \text{rank}(\mathbf{H}_i\mathbf{H}_j - \mathbf{H}_i\mathbf{H}_j\mathbf{H}_i) \\ &= \text{rank}(\mathbf{H}_{ij} - \mathbf{H}_i - \mathbf{H}_j + \mathbf{H}_i\mathbf{H}_j) \\ &= \tfrac{1}{2}\,\text{rank}(\mathbf{H}_i\mathbf{H}_j - \mathbf{H}_j\mathbf{H}_i), \end{aligned} \quad (8)$$

where the "hat" matrix $\mathbf{H}_k = \mathbf{X}_k(\mathbf{X}_k'\mathbf{X}_k)^-\mathbf{X}_k' = \mathbf{X}_k\mathbf{X}_k^+$ *is the orthogonal projector on the column space* $\mathcal{C}(\mathbf{X}_k)$, $k = i, j$, *and* $\mathbf{H}_{ij}$ *is the orthogonal projector on* $\mathcal{C}(\mathbf{X}_i : \mathbf{X}_j)$.

Proof: Since $t_{i.j}$ is the rank of the cross-covariance matrix between the vectors of adjusted totals, cf. [4], we see that $t_{i.j} = \text{rank}(\mathbf{X}_i'\mathbf{M}_j\mathbf{M}_i\mathbf{X}_j)$, where $\mathbf{M}_i = \mathbf{I} - \mathbf{H}_i$, and so $t_{i.j} = \text{rank}(\mathbf{H}_i\mathbf{M}_j\mathbf{M}_i\mathbf{H}_j) = \text{rank}(\mathbf{H}_i\mathbf{M}_j\mathbf{H}_i\mathbf{H}_j) = \text{rank}[\mathbf{H}_i\mathbf{H}_j - (\mathbf{H}_i\mathbf{H}_j)^2]$; this proves the first equality in **(8)**.

As $\text{rank}(\mathbf{H}_j\mathbf{M}_i\mathbf{H}_j) = \text{rank}(\mathbf{H}_j\mathbf{M}_i)$ it follows that $t_{i.j} = \text{rank}(\mathbf{H}_i\mathbf{H}_j\mathbf{M}_i\mathbf{H}_j) = \text{rank}(\mathbf{H}_i\mathbf{H}_j\mathbf{M}_i) = \text{rank}(\mathbf{H}_i\mathbf{H}_j - \mathbf{H}_i\mathbf{H}_j\mathbf{H}_i)$, which establishes the second equality in **(8)**.

To prove the third equality in **(8)** we note that

$$\mathbf{H}_{ij} - \mathbf{H}_i - \mathbf{H}_j + \mathbf{H}_i\mathbf{H}_j = (\mathbf{H}_{ij} - \mathbf{H}_i)(\mathbf{H}_{ij} - \mathbf{H}_j) = \mathbf{H}_{j|i}\,\mathbf{H}_{i|j},$$

where $\mathbf{H}_{i|j}$ is the "adjusted" hat matrix

$$\mathbf{H}_{i|j} = \mathbf{H}_{ij} - \mathbf{H}_j = \mathbf{M}_j\mathbf{X}_i(\mathbf{X}_i'\,\mathbf{M}_j\mathbf{X}_i)^-\mathbf{X}_i'\mathbf{M}_j. \quad (9)$$

It then follows that

$$\text{rank}(\mathbf{H}_{ij} - \mathbf{H}_i - \mathbf{H}_j + \mathbf{H}_i\mathbf{H}_j) = \\ \text{rank}(\mathbf{H}_{j|i}\,\mathbf{H}_{i|j}) = \text{rank}[(\mathbf{H}_{j|i}\,\mathbf{H}_{i|j})'] = \text{rank}(\mathbf{H}_{i|j}\mathbf{H}_{j|i}) = \text{rank}(\mathbf{X}_i'\mathbf{M}_j\mathbf{M}_i\mathbf{X}_j') = t_{i.j}.$$

To prove the fourth, and last, equality in **(8)**, we write

$$\mathbf{T}_{ij} = \mathbf{H}_i\mathbf{H}_j - \mathbf{H}_i\mathbf{H}_j\mathbf{H}_i = \mathbf{H}_i\mathbf{H}_j\mathbf{M}_i$$

and observe that $\mathbf{T}_{ij}^2 = \mathbf{0}$. Since $\mathbf{H}_i\mathbf{H}_j - \mathbf{H}_j\mathbf{H}_i = \mathbf{T}_{ij} - \mathbf{T}_{ij}'$ it follows from **(7)** that

$$\text{rank}(\mathbf{H}_i\mathbf{H}_j - \mathbf{H}_j\mathbf{H}_i) = \text{rank}(\mathbf{T}_{ij} - \mathbf{T}_{ij}') = 2\,\text{rank}(\mathbf{T}_{ij}) = 2\,t_{i.j}$$

which completes the proof. ❐

Whenever $t_{i.j} = 0$ we say that the factors i and j are weakly orthogonal, cf. [1, 7]. From Theorem 1 and Lemma 1 we immediately obtain:

COROLLARY 2. *The factors i and j are weakly orthogonal, i.e., $t_{i.j} = 0$, if and only if any one of the following six conditions holds:*

$$\mathbf{H}_i\mathbf{H}_j = \mathbf{H}_j\mathbf{H}_i\,,$$

$$(\mathbf{H}_i\mathbf{H}_j)^2 = \mathbf{H}_i\mathbf{H}_j,$$

$$\mathbf{H}_i\mathbf{H}_j = \mathbf{H}_i\mathbf{H}_j\mathbf{H}_i,$$

$$\mathbf{H}_{i|j}\mathbf{H}_{j|i} = \mathbf{0}\,,$$

$$\mathbf{H}_{ij} = \mathbf{H}_i + \mathbf{H}_j - \mathbf{H}_i\mathbf{H}_j\,,$$

$$\text{rank}(\mathbf{H}_{ij}) = \text{rank}(\mathbf{H}_i) + \text{rank}(\mathbf{H}_j) - \text{rank}(\mathbf{H}_i\mathbf{H}_j). \tag{10}$$

Some 45 algebraic characterizations for the commutativity of two orthogonal projectors were established in [8]; cf. also [9, 10].

Whenever weak pairwise orthogonality holds, certain equalities among the other canonical correlations follow. Our next theorem lists some of these equalities. Parts (a) and (b) were announced (without proofs) in [1], while part (c) is new.

THEOREM 2. *Let the factors i and j be weakly orthogonal, i.e., $t_{i.j} = 0$, and let k be a third factor. Then*

(a) $\{\rho^{(i.jk)}\} = \{\rho^{(i|j,\,k|j)}\} + \{u_{i.j} \text{ ones}\}$,

(b) $\{\rho^{(j.ik)}\} = \{\rho^{(j|i,\,k|i)}\} + \{u_{i.j} \text{ ones}\}$, *and*

(c) $\{\rho^{(i,\,k|j)}\} = \{\rho^{(i|j,\,k|j)}\}$ *and* $\{\rho^{(j,\,k|i)}\} = \{\rho^{(j|i,\,k|i)}\}\}$.

Proof: (a) The set of canonical correlations between factor i and the other two factors j and k is

$$\{\rho^{(i.jk)}\} = \{\text{ch}^{1/2}[\mathbf{X}_i'\mathbf{X}_{jk}(\mathbf{X}_{jk}'\mathbf{X}_{jk})^{-}\mathbf{X}_{jk}'\mathbf{X}_i(\mathbf{X}_i'\mathbf{X}_i)^{-}]\} = \{\text{ch}^{1/2}(\mathbf{H}_i\mathbf{H}_{jk})\}, \tag{11}$$

the set of positive square roots of the characteristic roots of $\mathbf{H}_i\mathbf{H}_{jk}$, where $\mathbf{H}_{jk}$ is the orthogonal projector on $\mathcal{C}(\mathbf{X}_j : \mathbf{X}_k)$. Following **(9)** we have $\mathbf{H}_{jk} = \mathbf{H}_j + \mathbf{H}_{k|j}$ and so

$$\{\rho^{(i.jk)}\} = \{\text{ch}^{1/2}[\mathbf{H}_i(\mathbf{H}_j + \mathbf{H}_{k|j})]\}. \tag{12}$$

We now consider the characteristic polynomial of $\mathbf{H}_i(\mathbf{H}_j + \mathbf{H}_{k|j})$

$$|\lambda\mathbf{I} - \mathbf{H}_i(\mathbf{H}_j + \mathbf{H}_{k|j})| = |\lambda\mathbf{I} - \mathbf{H}_i\mathbf{H}_j| \cdot |\mathbf{I} - (\lambda\mathbf{I} - \mathbf{H}_i\mathbf{H}_j)^{-1}\mathbf{H}_i\mathbf{H}_{k|j}| \tag{13}$$

for all $\lambda \neq \mathrm{ch}(\mathbf{H}_i\mathbf{H}_j) = \rho_h^{(i.j)}$; $h = 1, \ldots, m_{i.j}$. We can simplify the last term in **(13)** using the commutativity of the projectors $\mathbf{H}_i$ and $\mathbf{H}_j$ (since $t_{i.j} = 0$, cf. Corollary 2) and so

$$(\lambda\mathbf{I} - \mathbf{H}_i\mathbf{H}_j)\mathbf{H}_i\mathbf{H}_{k|j} = \lambda\mathbf{H}_i\mathbf{H}_{k|j} - \mathbf{H}_i\mathbf{H}_j\mathbf{H}_i\mathbf{H}_{k|j} = \lambda\mathbf{H}_i\mathbf{H}_{k|j} - \mathbf{H}_i\mathbf{H}_j\mathbf{H}_{k|j} = \lambda\mathbf{H}_i\mathbf{H}_{k|j}.$$

The characteristic polynomial **(13)** is, therefore, equivalent to

$$|\lambda\mathbf{I} - \mathbf{H}_i\mathbf{H}_j| \cdot |\mathbf{I} - (1/\lambda)\mathbf{H}_i\mathbf{H}_{k|j}| = (\lambda - 1)^{u_{i.j}}\lambda^{n-u_{i.j}}\lambda^{-n}|\lambda\mathbf{I} - \mathbf{H}_i\mathbf{H}_{k|j}| \tag{14}$$

for $\lambda \neq 0$ and $\lambda \neq \rho_h^{(i.j)}$; $h = 1, \ldots, m_{i.j}$. We recall that the nonzero eigenvalues of $\mathbf{H}_i\mathbf{H}_j$ are the squares of the nonzero canonical correlations between $\mathbf{X}_i'\mathbf{y}$ and $\mathbf{X}_j'\mathbf{y}$ and are all equal to 1 since $\mathbf{H}_i\mathbf{H}_j$ is idempotent whenever $t_{i.j} = 0$, cf. Corollary 2. Moreover,

$$\mathbf{H}_i\mathbf{H}_{k|j} = (\mathbf{H}_i - \mathbf{H}_i\mathbf{H}_j)\mathbf{H}_{k|j} = (\mathbf{H}_{ij} - \mathbf{H}_j)\mathbf{H}_{k|j} = \mathbf{H}_{i|j}\mathbf{H}_{k|j}, \tag{15}$$

since then $\mathbf{H}_i - \mathbf{H}_i\mathbf{H}_j = \mathbf{H}_{ij} - \mathbf{H}_j$. The right-hand side of **(14)** then equals

$$\left(\frac{\lambda - 1}{\lambda}\right)^{u_{i.j}}|\lambda\mathbf{I} - \mathbf{H}_{i|j}\mathbf{H}_{k|j}|,$$

which establishes (a). The proof is similar for (b), and (c) follows at once from the equality in **(15)** and the fact that $t_{i.j} = t_{j.i}$. ❒

If one factor, say j, is orthogonal to each of the other two factors, i.e., $t_{i.j} = t_{j.k} = 0$, then a further set of equalities between certain canonical correlations is obtained.

THEOREM 3. *Let* $t_{i.j} = t_{j.k} = 0$. *Then*

(a) $\{\rho^{(i.jk)}\} = \{\rho^{(k.ij)}\} + \{u_{i.j} - u_{j.k} \text{ ones}\}$ *if* $u_{i.j} \geq u_{j.k}$,

(b) $\{\rho^{(j.ik)}\} = \{\rho^{(i|k,\, j|k)}\} + \{u_{j.k} \text{ ones}\} = \{\rho^{(j|i,\, k|i)}\} + \{u_{i.j} \text{ ones}\}$.

If in addition $u_{i.j} = u_{j.k}$, *then*

(c) $\{\rho^{(i.jk)}\} = \{\rho^{(k.ij)}\}$ *and*

(d) $\{\rho^{(i|k,\, j|k)}\} = \{\rho^{(j|i,\, k|i)}\} = \{\rho^{(i,\, j|k)}\} = \{\rho^{(j,\, k|i)}\}$.

Proof: To prove (a), we start with **(14)**, i.e.,

$$\left(\frac{\lambda - 1}{\lambda}\right)^{u_{i.j}}|\lambda\mathbf{I} - \mathbf{H}_i\mathbf{H}_{k|j}|.$$

This characteristic polynomial is equivalent to

$$\left(\frac{\lambda - 1}{\lambda}\right)^{u_{i.j}}|\lambda\mathbf{I} - \mathbf{H}_i(\mathbf{H}_k - \mathbf{H}_j\mathbf{H}_k)| = \left(\frac{\lambda - 1}{\lambda}\right)^{u_{i.j}}|\lambda\mathbf{I} + \mathbf{H}_j\mathbf{H}_k - \mathbf{H}_{ij}\mathbf{H}_k|, \tag{16}$$

since $t_{i.j} = 0$ is equivalent to $\mathbf{H}_i\mathbf{H}_j = \mathbf{H}_i + \mathbf{H}_j - \mathbf{H}_{ij}$ (cf. Corollary 2). We may rewrite **(16)** as

$$\left(\frac{\lambda-1}{\lambda}\right)^{u_{i.j}} |\lambda\mathbf{I} - \mathbf{H}_{ij}\mathbf{H}_k| \cdot |\mathbf{I} + (\lambda\mathbf{I} - \mathbf{H}_{ij}\mathbf{H}_k)^{-1}\mathbf{H}_j\mathbf{H}_k|. \tag{17}$$

The last term in **(17)** simplifies since $(\lambda\mathbf{I} - \mathbf{H}_{ij}\mathbf{H}_k)\mathbf{H}_j\mathbf{H}_k = (\lambda - 1)\mathbf{H}_j\mathbf{H}_k$, using the commutativity of $\mathbf{H}_j$ and $\mathbf{H}_k$. We may rewrite **(17)**, therefore, as

$$\left(\frac{\lambda-1}{\lambda}\right)^{u_{i.j}} |(\lambda-1)\mathbf{I} + \mathbf{H}_j\mathbf{H}_k| \cdot |\lambda\mathbf{I} - \mathbf{H}_{ij}\mathbf{H}_k| = \left(\frac{\lambda-1}{\lambda}\right)^{u_{i.j}-u_{j.k}} |\lambda\mathbf{I} - \mathbf{H}_{ij}\mathbf{H}_k|,$$

which proves (a). The equalities in (b) can be obtained directly from Theorem 2, while the equalities in (c) and (d) follow at once from (a) and (b). ❐

Acknowledgements

The authors are very grateful to an anonymous referee for several very helpful suggestions. The research of the first and third authors was supported in part by the Natural Sciences and Engineering Research Council of Canada and by the Fonds pour la Formation de Chercheurs et l'Aide à la Recherche du Gouvernement du Québec.

REFERENCES

[1] G. P. H. Styan, Canonical correlations in the three-way layout, in: *Pacific Statistical Congress* (I. S. Francis, B. F. J. Manly and F. C. Lam, Eds.). North-Holland, Amsterdam, 433–438 (1986).

[2] K. J. Worsley, G. P. H. Styan and J. Bérubé, Genstat ANOVA efficiency factors and canonical efficiency factors for non-orthogonal designs. *Genstat Newsletter*, **26**, 11–21 (1991).

[3] J. Bérubé, Some properties of three-way layouts, MSc thesis. Dept. of Mathematics and Statistics, McGill University, Montréal (1991).

[4] D. Latour and G. P. H. Styan, Canonical correlations in the two-way layout, in: *Proc. First International Tampere Seminar on Linear Statistical Models and their Applications* (T. Pukkila and S. Puntanen, Eds.). Dept. of Mathematical Sciences, University of Tampere, 225–243 (1985).

[5] J. K. Baksalary and G. P. H. Styan, Around a formula for the rank for a matrix product, with some statistical applications, in: *Graphs, Matrices and Designs: A Festschrift for Norman J. Pullman* (R. S. Rees, Ed.). Marcel Dekker, New York, in press (1992). [Report A-244, Dept. of Mathematical Sciences, University of Tampere (1990).]

[6] G. Marsaglia and G. P. H. Styan, Equalities and inequalities for ranks of matrices. *Linear and Multilinear Algebra*, **2**, 269-292 (1974).

[7] C. G. Khatri and K. R. Shah, Orthogonality in multiway classifications. *Linear Algebra and its Applications*, **82**, 215–224 (1986).

[8] J. K. Baksalary, Algebraic characterizations and statistical implications of the commutativity of orthogonal projectors, in: *Proc. Second International Tampere Conference in Statistics* (T. Pukkila and S. Puntanen, Eds.). Dept. of Mathematical Sciences, University of Tampere, 113–142 (1987).

[9] K. Nordström, Commutativity of orthogonal projectors: a lattice-theoretic approach, in: *A Spectrum of Statistical Thought: Essays in Statistical Theory, Economics, and Population Genetics in Honour of Johan Fellman* (G. Rosenqvist, K. Juselius, K. Nordström, and J. Palmgren, Eds.). Svenska Handelshögskolan, Helsinki, 175–187 (1991).

[10] C. R. Rao and H. Yanai, General definition and decomposition of projectors and some applications to statistical problems. *Journal of Statistical Planning and Inference*, **3**, 1–17 (1979).

Stat. Sci. & Data Anal., pp. 253-264
K. Matsusita *et al.* (Eds)

Partial Canonical Correlations Associated with the Inverse and Some Generalized Inverses of a Partitioned Dispersion Matrix

HARUO YANAI and SIMO PUNTANEN
Research Division, The National Center for University Entrance Examination, 2-19-23 Komaba, Meguro-ku, Tokyo 153, Japan
Department of Mathematical Sciences, University of Tampere, P.O. Box 607, SF-33101 Tampere, Finland

Abstract. Given three sets of random variables $X_1'y$, $X_2'y$, and $X_3'y$, we first develop partial canonical correlations between $X_1'y$ and $X_2'y$ after elimination of $X_3'y$ in terms of the elements of the inverse of the joint dispersion matrix $\mathbf{V}$ of all the variables. Secondly, the result is extended to the case where $\mathbf{V}$ is singular by means of extensive use of both the orthogonal and oblique projectors.

Key words: Projector, orthogonal projector, generalized inverse, canonical correlation, partial canonical correlation.

1. INTRODUCTION

Jewell and Bloomfield [1] showed that the canonical correlations between two random vectors $\mathbf{x}_1$ and $\mathbf{x}_2$, whose joint positive definite dispersion matrix is

$$\mathbf{V} = \begin{pmatrix} \mathbf{V}_{11} & \mathbf{V}_{12} \\ \mathbf{V}_{21} & \mathbf{V}_{22} \end{pmatrix} = \begin{pmatrix} \mathbf{X}_1'\mathbf{X}_1 & \mathbf{X}_1'\mathbf{X}_2 \\ \mathbf{X}_2'\mathbf{X}_1 & \mathbf{X}_2'\mathbf{X}_2 \end{pmatrix}, \tag{1.1}$$

are precisely the same as the canonical correlations associated with the inverse of the dispersion matrix; i.e., $\mathbf{V}^{-1}$, where $\mathbf{V}^{-1}$ is partitioned according to (1.1).

Further, Latour, Puntanen and Styan [2] extended the results to a singular $\mathbf{V}$, using a certain class of symmetric reflexive generalized inverses of $\mathbf{V}$. Related studies were made also by Khatri [3]. Recently, Baksalary, Puntanen, and Yanai [4] considered the problem by making extensive use of the projection operator, and established a series of results characterizing the situations when all canonical correlations are less than one.

We extend these earlier results to the situation where the variables are partitioned into $\mathbf{x}_1$, $\mathbf{x}_2$, and $\mathbf{x}_3$ and the associated dispersion matrix is

$$\mathbf{V} = \begin{pmatrix} \mathbf{V}_{11} & \mathbf{V}_{12} & \mathbf{V}_{13} \\ \mathbf{V}_{21} & \mathbf{V}_{22} & \mathbf{V}_{23} \\ \mathbf{V}_{31} & \mathbf{V}_{32} & \mathbf{V}_{33} \end{pmatrix}, \tag{1.2}$$

and study the partial canonical correlations based on (1.2), its inverse and a generalized inverse.

2. PRELIMINARY ALGEBRAIC RESULTS

Let $\mathbb{R}^n$ denote the vector space of real n-tuples and let $\mathbb{R}^{n \times m}$ stand for the set of $n \times m$ matrices with real elements. Symbols $\mathcal{S}(\mathbf{A})$, $r(\mathbf{A})$, $\mathbf{A}'$, $\mathbf{A}^-$ denote the column space, the rank, the transpose, and a generalized inverse of $\mathbf{A}$, respectively. The notation $(\mathbf{A} : \mathbf{B})$ means a partitioned matrix comprising of $\mathbf{A}$ and $\mathbf{B}$, and $\mathbf{I}_n$ stands for the identity matrix of order n. Further, let $\mathcal{S}(\mathbf{A})^\perp$ denote the orthocomplement of $\mathcal{S}(\mathbf{A})$ (with respect to the standard inner product).

Moreover, the symbols $\mathbf{P}_\mathbf{A}$, $\mathbf{Q}_\mathbf{A}$, $\mathbf{P}_{(\mathbf{A} : \mathbf{B})}$, and $\mathbf{Q}_{(\mathbf{A} : \mathbf{B})}$ denote the orthogonal projectors onto $\mathcal{S}(\mathbf{A})$, $\mathcal{S}(\mathbf{A})^\perp$, $\mathcal{S}(\mathbf{A} : \mathbf{B})$, and $\mathcal{S}(\mathbf{A} : \mathbf{B})^\perp$, respectively. The following lemma on the decomposition of these projectors appears to be useful for proving the main theorems and lemmas to be considered in later sections.

Lemma 1. *For any* $\mathbf{A} \in \mathbb{R}^{n \times m}$ *and* $\mathbf{B} \in \mathbb{R}^{n \times r}$, *the following decompositions hold:*

(a) $\mathbf{P}_{(\mathbf{A} : \mathbf{B})} = \mathbf{P}_\mathbf{A} + \mathbf{P}_{\mathbf{B}/\mathbf{A}}$,

where $\mathbf{P}_{\mathbf{B}/\mathbf{A}}$ *is the orthogonal projector onto* $\mathcal{S}(\mathbf{Q}_\mathbf{A}\mathbf{B})$, *i.e.,*

$$\mathbf{P}_{\mathbf{B}/\mathbf{A}} = \mathbf{Q}_\mathbf{A}\mathbf{B}(\mathbf{B}'\mathbf{Q}_\mathbf{A}\mathbf{B})^-\mathbf{B}'\mathbf{Q}_\mathbf{A};$$

(b) $\mathbf{Q}_{(\mathbf{A} : \mathbf{B})} = \mathbf{Q}_{\mathbf{B}/\mathbf{A}}\mathbf{Q}_\mathbf{A}$, *where* $\mathbf{Q}_{\mathbf{B}/\mathbf{A}} = \mathbf{I}_n - \mathbf{P}_{\mathbf{B}/\mathbf{A}}$.

Proof. For a proof of (a), see e.g. Theorem 6 of Rao and Yanai [5]. Using (a), the proof of (b) follows immediately, since

$$\mathbf{Q}_{(\mathbf{A} : \mathbf{B})} = \mathbf{I}_n - \mathbf{P}_{(\mathbf{A} : \mathbf{B})} = \mathbf{Q}_\mathbf{A} - \mathbf{P}_{\mathbf{B}/\mathbf{A}} = \mathbf{Q}_\mathbf{A} - \mathbf{Q}_\mathbf{A}\mathbf{P}_{\mathbf{B}/\mathbf{A}} = \mathbf{Q}_\mathbf{A}\mathbf{Q}_{\mathbf{B}/\mathbf{A}}$$

and $\mathbf{Q}_{\mathbf{B}/\mathbf{A}}\mathbf{Q}_\mathbf{A} = \mathbf{Q}_\mathbf{A}\mathbf{Q}_{\mathbf{B}/\mathbf{A}}$, due to symmetry of $\mathbf{Q}_{(\mathbf{A} : \mathbf{B})}$. □

In Lemma 2 we collect some results concerning the matrix

$$\mathbf{P}_{\mathbf{A}\cdot\mathbf{B}} = \mathbf{A}(\mathbf{A}'\mathbf{Q}_{\mathbf{B}}\mathbf{A})^{-}\mathbf{A}'\mathbf{Q}_{\mathbf{B}}. \tag{2.1}$$

It is worth noting that

$$\mathrm{r}(\mathbf{P}_{\mathbf{A}\cdot\mathbf{B}}) = \mathrm{r}(\mathbf{Q}_{\mathbf{B}}\mathbf{A}) \tag{2.2}$$

for all choices of $(\mathbf{A}'\mathbf{Q}_{\mathbf{B}}\mathbf{A})^{-}$ but $\mathbf{P}_{\mathbf{A}\cdot\mathbf{B}}$ is invariant with respect to the choice of $(\mathbf{A}'\mathbf{Q}_{\mathbf{B}}\mathbf{A})^{-}$ if and only if $\mathcal{S}(\mathbf{A}') \subset \mathcal{S}(\mathbf{A}'\mathbf{Q}_{\mathbf{B}})$, i.e., $\mathcal{S}(\mathbf{A}) \cap \mathcal{S}(\mathbf{B}) = \{\mathbf{0}\}$; see Rao and Mitra ([6], pp. 21 and 43). Some further properties are given in the following; see Rao and Yanai ([5], Note 6) and Baksalary, Puntanen, and Yanai ([4], Lemma 1):

Lemma 2. *The following four statements are equivalent:*

(a) $\mathcal{S}(\mathbf{A}) \cap \mathcal{S}(\mathbf{B}) = \{\mathbf{0}\}$, i.e., $\mathcal{S}(\mathbf{A} : \mathbf{B}) = \mathcal{S}(\mathbf{A}) \oplus \mathcal{S}(\mathbf{B})$,

(b) $\mathbf{P}_{\mathbf{A}\cdot\mathbf{B}}$ *is the (oblique) projector onto* $\mathcal{S}(\mathbf{A})$ *along* $\mathcal{S}(\mathbf{B}) \oplus \mathcal{S}(\mathbf{A} : \mathbf{B})^{\perp}$,

(c) $\mathbf{P}_{(\mathbf{A}\,:\,\mathbf{B})} = \mathbf{P}_{\mathbf{A}\cdot\mathbf{B}} + \mathbf{P}_{\mathbf{B}\cdot\mathbf{A}}$,

(d) $\mathcal{S}(\mathbf{P}_{\mathbf{A}\cdot\mathbf{B}})$ *is invariant with respect to the choice of* $(\mathbf{A}'\mathbf{Q}_{\mathbf{B}}\mathbf{A})^{-}$,

Furthermore,

(e) $\mathcal{S}(\mathbf{A} : \mathbf{B} : \mathbf{C}) = \mathcal{S}(\mathbf{A}) \oplus \mathcal{S}(\mathbf{B}) \oplus \mathcal{S}(\mathbf{C}) \Leftrightarrow$

$$\mathbf{P}_{(\mathbf{A}\,:\,\mathbf{B}\,:\,\mathbf{C})} = \mathbf{P}_{\mathbf{A}\cdot(\mathbf{B}\,:\,\mathbf{C})} + \mathbf{P}_{\mathbf{B}\cdot(\mathbf{A}\,:\,\mathbf{C})} + \mathbf{P}_{\mathbf{C}\cdot(\mathbf{A}\,:\,\mathbf{B})}.$$

For a general discussion on projectors and generalized inverses, see also Yanai [7].

Lemma 3. *The following three statements are equivalent:*

(a) $\mathcal{S}(\mathbf{A} : \mathbf{B} : \mathbf{C}) = \mathcal{S}(\mathbf{A}) \oplus \mathcal{S}(\mathbf{B}) \oplus \mathcal{S}(\mathbf{C})$,

(b) $\mathcal{S}(\mathbf{Q}_{\mathbf{C}}\mathbf{A} : \mathbf{Q}_{\mathbf{C}}\mathbf{B}) = \mathcal{S}(\mathbf{Q}_{\mathbf{C}}\mathbf{A}) \oplus \mathcal{S}(\mathbf{Q}_{\mathbf{C}}\mathbf{B})$, $\mathcal{S}(\mathbf{A}) \cap \mathcal{S}(\mathbf{C}) = \{\mathbf{0}\}$, *and* $\mathcal{S}(\mathbf{B}) \cap \mathcal{S}(\mathbf{C}) = \{\mathbf{0}\}$,

(c) $\mathcal{S}(\mathbf{A} : \mathbf{B} : \mathbf{C}) = \mathcal{S}(\mathbf{A} : \mathbf{B}) \oplus \mathcal{S}(\mathbf{C})$ *and* $\mathcal{S}(\mathbf{A}) \cap \mathcal{S}(\mathbf{B}) = \{\mathbf{0}\}$.

Moreover, if any of the conditions above is satisfied, then

(d) $\mathcal{S}(\mathbf{A}) \cap \mathcal{S}(\mathbf{B}) = \{\mathbf{0}\}$, $\mathcal{S}(\mathbf{A}) \cap \mathcal{S}(\mathbf{C}) = \{\mathbf{0}\}$, *and* $\mathcal{S}(\mathbf{B}) \cap \mathcal{S}(\mathbf{C}) = \{\mathbf{0}\}$.

Proof. By the definition of the direct sum, statement (a) implies the pairwise disjointness of $\mathcal{S}(\mathbf{A})$, $\mathcal{S}(\mathbf{B})$, and $\mathcal{S}(\mathbf{C})$, i.e., (a) $\Rightarrow$ (d). To prove that (a) implies (b), we first notice that (a) is equivalent to $\mathrm{r}(\mathbf{A} : \mathbf{B} : \mathbf{C}) = \mathrm{r}(\mathbf{A}) + \mathrm{r}(\mathbf{B}) + \mathrm{r}(\mathbf{C})$. Hence, in view of decomposition

$$\mathcal{S}(\mathbf{A} : \mathbf{B} : \mathbf{C}) = \mathcal{S}(\mathbf{Q_C A} : \mathbf{Q_C B}) \oplus \mathcal{S}(\mathbf{C}), \tag{2.3}$$

we get

$$\mathrm{r}(\mathbf{A}) + \mathrm{r}(\mathbf{B}) = \mathrm{r}(\mathbf{Q_C A} : \mathbf{Q_C B}) \leq \mathrm{r}(\mathbf{Q_C A}) + \mathrm{r}(\mathbf{Q_C B}) \leq \mathrm{r}(\mathbf{A}) + \mathrm{r}(\mathbf{B}),$$

and so $\mathrm{r}(\mathbf{Q_C A} : \mathbf{Q_C B}) = \mathrm{r}(\mathbf{Q_C A}) + \mathrm{r}(\mathbf{Q_C B})$, i.e., $\mathcal{S}(\mathbf{Q_C A} : \mathbf{Q_C B}) = \mathcal{S}(\mathbf{Q_C A}) \oplus \mathcal{S}(\mathbf{Q_C B})$. Hence we have proved that (a) $\Rightarrow$ (b). To go the other way, we observe that decomposition (2.3) and the two latter conditions of (b) imply that $\mathrm{r}(\mathbf{A} : \mathbf{B} : \mathbf{C}) = \mathrm{r}(\mathbf{A}) + \mathrm{r}(\mathbf{B}) + \mathrm{r}(\mathbf{C})$ which means that (b) $\Rightarrow$ (a). The equivalence of (a) and (c) is easy to establish. ❑

Observe, in view of the symmetry of the condition (a), that each of the conditions (b) and (c) may be supplemented by the counterparts with the roles of **A**, **B**, and **C** interchanged.

3. CANONICAL CORRELATIONS

Throughout this paper, we will consider **y** as an n-dimensional random vector with the dispersion matrix proportional to the identity matrix of order n. Further, let $\mathbf{X}_1 \in \mathbb{R}^{n\times p}$ and $\mathbf{X}_2 \in \mathbb{R}^{n\times q}$. Then the joint dispersion matrix of $\mathbf{X}_1'\mathbf{y}$ and $\mathbf{X}_2'\mathbf{y}$ is proportional to the matrix **V** given in (1.1).

The ith largest canonical correlation between two random vectors $\mathbf{X}_1'\mathbf{y}$ and $\mathbf{X}_2'\mathbf{y}$ is denoted as $\mathrm{cc}_i(\mathbf{X}_1'\mathbf{y}, \mathbf{X}_2'\mathbf{y})$, and the set of all corresponding positive canonical correlations as $\{\mathrm{cc}(\mathbf{X}_1'\mathbf{y}, \mathbf{X}_2'\mathbf{y})\}$. The ith largest eigenvalue of a matrix **A** (having real eigenvalues) is denoted as $\mathrm{ch}_i(\mathbf{A})$ and the set of all nonzero eigenvalues as $\{\mathrm{nzch}(\mathbf{A})\}$.

Let the inverse of the positive definite **V**, defined in (1.1), be partitioned as

$$\mathbf{V}^{-1} = \begin{pmatrix} \mathbf{V}^{11} & \mathbf{V}^{12} \\ \mathbf{V}^{21} & \mathbf{V}^{22} \end{pmatrix} = (\mathbf{V}^{(1)} : \mathbf{V}^{(2)}). \tag{3.1}$$

If **V** is singular, then we consider a corresponding partitioned generalized inverse of **V**. Further, let $\mathbf{P}_i$ and $\mathbf{Q}_i$ denote the orthogonal projectors onto $\mathcal{S}(\mathbf{X}_i)$ and $\mathcal{S}(\mathbf{X}_i)^\perp$ ($i = 1, 2$), respectively. Then straightforward calculation proves the following lemma.

Lemma 4. *Let* $\mathbf{V}^{-1}$ *be partitioned as in* (3.1) *and let*

$$\mathbf{F} = (\mathbf{X}_1 : \mathbf{X}_2)\mathbf{V}^{(1)} \quad \textit{and} \quad \mathbf{G} = (\mathbf{X}_1 : \mathbf{X}_2)\mathbf{V}^{(2)}.$$

Then

$$\mathbf{F} = \mathbf{Q}_2\mathbf{X}_1(\mathbf{X}_1'\mathbf{Q}_2\mathbf{X}_1)^{-1} \quad \textit{and} \quad \mathbf{G} = \mathbf{Q}_1\mathbf{X}_2(\mathbf{X}_2'\mathbf{Q}_1\mathbf{X}_2)^{-1}.$$

Lemma 5. *Let* $\mathbf{V}$ *be positive definite and let* $\mathbf{F}$ *and* $\mathbf{G}$ *be as in Lemma 4. Then the following six sets are identical:*

(a) $\{\mathrm{cc}^2(\mathbf{X}_1'\mathbf{y}, \mathbf{X}_2'\mathbf{y})\}$, (d) $\{\mathrm{nzch}[\mathbf{V}_{11}^{-1}\mathbf{V}_{12}\mathbf{V}_{22}^{-1}\mathbf{V}_{21}]\}$,

(b) $\{\mathrm{cc}^2(\mathbf{F}'\mathbf{y}, \mathbf{G}'\mathbf{y})\}$, (e) $\{\mathrm{nzch}[(\mathbf{V}^{11})^{-1}\mathbf{V}^{12}(\mathbf{V}^{22})^{-1}\mathbf{V}^{21}]\}$,

(c) $\{\mathrm{cc}^2(\mathbf{X}_1'\mathbf{Q}_2\mathbf{y}, \mathbf{X}_2'\mathbf{Q}_1\mathbf{y})\}$, (f) $\{\mathrm{nzch}(\mathbf{P}_1\mathbf{P}_2)\}$.

Results of Lemma 5 are quoted from Jewell and Bloomfield [1], Styan ([8], Theorem 2.5), and Baksalary, Puntanen, and Yanai ([4], Lemma 5). It may be mentioned that Lemma 4 immediately gives the proof for the equality of the sets in (b) and (c).

Consider now the situation when $\mathbf{V}$ might be singular. Then by Theorem 2.5 in Styan [8], we know that the positive canonical correlations between $\mathbf{X}_1'\mathbf{Q}_2\mathbf{y}$ and $\mathbf{X}_2'\mathbf{Q}_1\mathbf{y}$ are all less than one (denote by t their number) and are identical with those positive canonical correlations between $\mathbf{X}_1'\mathbf{y}$ and $\mathbf{X}_2'\mathbf{y}$ which are not equal to one; briefly denoted

$$\mathrm{cc}_i(\mathbf{X}_1'\mathbf{Q}_2\mathbf{y}, \mathbf{X}_2'\mathbf{Q}_1\mathbf{y}) < 1 \text{ for all } i = 1, \ldots, t, \tag{3.2}$$

and

$$\{\mathrm{cc}(\mathbf{X}_1'\mathbf{Q}_2\mathbf{y}, \mathbf{X}_2'\mathbf{Q}_1\mathbf{y})\} = \{\mathrm{nucc}(\mathbf{X}_1'\mathbf{y}, \mathbf{X}_2'\mathbf{y})\}, \tag{3.3}$$

where "nucc" refers to those canonical correlations which are less than one. Note that if $\mathbf{V}$ is positive definite, then there are no unit canonical correlations between $\mathbf{X}_1'\mathbf{y}$ and $\mathbf{X}_2'\mathbf{y}$ and hence the equality between the sets (a) and (c) in Lemma 5 holds for all canonical correlations.

Furthermore, let $(\mathbf{V}^{(1)} : \mathbf{V}^{(2)}) = \mathbf{V}^-$ be a generalized inverse of $\mathbf{V}$, partitioned according to (3.1), and let

$$\mathbf{F} = (\mathbf{X}_1 : \mathbf{X}_2)\mathbf{V}^{(1)} \quad \text{and} \quad \mathbf{G} = (\mathbf{X}_1 : \mathbf{X}_2)\mathbf{V}^{(2)}. \tag{3.4}$$

Then the joint dispersion matrix of $\mathbf{F'y}$ and $\mathbf{G'y}$ is

$$\begin{pmatrix} \mathbf{F'F} & \mathbf{F'G} \\ \mathbf{G'F} & \mathbf{G'G} \end{pmatrix} = (\mathbf{V}^-)'\mathbf{V}\mathbf{V}^-.$$

Following Baksalary, Puntanen, and Yanai [4], we note that a matrix $\mathbf{V}^{\sim}$ is a symmetric reflexive (and thus nonnegative definite) generalized inverse of $\mathbf{V}$ if and only if it admits the representation $(\mathbf{V}^-)'\mathbf{V}\mathbf{V}^-$ for some generalized inverse $\mathbf{V}^-$. Hence it follows that considering the canonical correlations associated with a symmetric reflexive generalized inverse of $\mathbf{V}$ is equivalent to considering the canonical correlations between $\mathbf{F'y}$ and $\mathbf{G'y}$, where $\mathbf{F}$ and $\mathbf{G}$ are defined in (3.4), with a suitably chosen $\mathbf{V}^- = (\mathbf{V}^{(1)} : \mathbf{V}^{(2)})$. Without going to further details, we reproduce two useful lemmas from Baksalary, Puntanen, and Yanai [4].

Lemma 6. *Let*

$$\mathbf{F} = (\mathbf{X}_1 : \mathbf{X}_2)\mathbf{V}^{(1)} \quad \textit{and} \quad \mathbf{G} = (\mathbf{X}_1 : \mathbf{X}_2)\mathbf{V}^{(2)},$$

where $(\mathbf{V}^{(1)} : \mathbf{V}^{(2)})$ *is a generalized inverse of* $\mathbf{V}$, *partitioned according to* (3.1). *Then*

(a) $\mathbf{X}_1'\mathbf{G} = \mathbf{0}$ *if and only if* $\mathcal{S}(\mathbf{G}) = \mathcal{S}(\mathbf{Q}_1\mathbf{X}_2)$,

(b) $\mathbf{X}_2'\mathbf{F} = \mathbf{0}$ *if and only if* $\mathcal{S}(\mathbf{F}) = \mathcal{S}(\mathbf{Q}_2\mathbf{X}_1)$.

Lemma 7. *Let* $\mathbf{F}$ *and* $\mathbf{G}$ *be defined as in Lemma 6. Then the following three statements are equivalent:*

(a) *the canonical correlations between* $\mathbf{X}_1'\mathbf{y}$ *and* $\mathbf{X}_2'\mathbf{y}$ *are all less than one,*

(b) $\mathbf{X}_1\mathbf{F}'\mathbf{X}_1 = \mathbf{X}_1,\ \mathbf{X}_1\mathbf{F}'\mathbf{X}_2 = \mathbf{0},\ \mathbf{X}_2\mathbf{G}'\mathbf{X}_2 = \mathbf{X}_2,\ \mathbf{X}_2\mathbf{G}'\mathbf{X}_1 = \mathbf{0},$

(c) $\mathbf{X}_1\mathbf{F}' = \mathbf{P}_{1\cdot 2},\ \mathbf{X}_2\mathbf{G}' = \mathbf{P}_{2\cdot 1}.$

Corresponding conditions can be expressed for the canonical correlations between $\mathbf{F'y}$ *and* $\mathbf{G'y}$ *by interchanging the roles of* $\mathbf{X}_1, \mathbf{X}_2$ *and* $\mathbf{F}, \mathbf{G}$. *Further, the following four statements are equivalent:*

(d) *the canonical correlations between* $\mathbf{X}_1'\mathbf{y}$ *and* $\mathbf{X}_2'\mathbf{y}$ *and between* $\mathbf{F'y}$ *and* $\mathbf{G'y}$ *are all less than one,*

(e) $\mathbf{X}_1'\mathbf{G} = \mathbf{0}$ *and* $\mathbf{X}_2'\mathbf{F} = \mathbf{0}$,

(f) $\mathcal{S}(\mathbf{G}) = \mathcal{S}(\mathbf{Q}_1\mathbf{X}_2)$ *and* $\mathcal{S}(\mathbf{F}) = \mathcal{S}(\mathbf{Q}_2\mathbf{X}_1)$,

(g) $\{\mathrm{cc}(\mathbf{X}_1'\mathbf{y}, \mathbf{X}_2'\mathbf{y})\} = \{\mathrm{cc}(\mathbf{X}_1'\mathbf{Q}_2\mathbf{y}, \mathbf{X}_2'\mathbf{Q}_1\mathbf{y})\} = \{\mathrm{cc}(\mathbf{F}'\mathbf{y}, \mathbf{G}'\mathbf{y})\}$.

4. PARTIAL CANONICAL CORRELATIONS

Consider an $n\times m$ matrix $\mathbf{X}$ which is partitioned as

$$\mathbf{X} = (\mathbf{X}_1 : \mathbf{X}_2 : \mathbf{X}_3).$$

Then the joint dispersion matrix of $\mathbf{X}_1'\mathbf{y}$, $\mathbf{X}_2'\mathbf{y}$, and $\mathbf{X}_3'\mathbf{y}$ is proportional to the matrix

$$\mathbf{V} = \begin{pmatrix} \mathbf{V}_{11} & \mathbf{V}_{12} & \mathbf{V}_{13} \\ \mathbf{V}_{21} & \mathbf{V}_{22} & \mathbf{V}_{23} \\ \mathbf{V}_{31} & \mathbf{V}_{32} & \mathbf{V}_{33} \end{pmatrix} = \begin{pmatrix} \mathbf{X}_1'\mathbf{X}_1 & \mathbf{X}_1'\mathbf{X}_2 & \mathbf{X}_1'\mathbf{X}_3 \\ \mathbf{X}_2'\mathbf{X}_1 & \mathbf{X}_2'\mathbf{X}_2 & \mathbf{X}_2'\mathbf{X}_3 \\ \mathbf{X}_3'\mathbf{X}_1 & \mathbf{X}_3'\mathbf{X}_2 & \mathbf{X}_3'\mathbf{X}_3 \end{pmatrix} = \mathbf{X}'\mathbf{X}. \tag{4.1}$$

Let a generalized inverse of $\mathbf{V}$ be partitioned similarly as

$$\mathbf{V}^- = \begin{pmatrix} \tilde{\mathbf{V}}^{11} & \tilde{\mathbf{V}}^{12} & \tilde{\mathbf{V}}^{13} \\ \tilde{\mathbf{V}}^{21} & \tilde{\mathbf{V}}^{22} & \tilde{\mathbf{V}}^{23} \\ \tilde{\mathbf{V}}^{31} & \tilde{\mathbf{V}}^{32} & \tilde{\mathbf{V}}^{33} \end{pmatrix} = (\mathbf{V}^{(1)} : \mathbf{V}^{(2)} : \mathbf{V}^{(3)})\,. \tag{4.2}$$

Furthermore, define

$$\mathbf{F} = \mathbf{X}\mathbf{V}^{(1)},\ \ \mathbf{G} = \mathbf{X}\mathbf{V}^{(2)},\ \text{ and }\ \mathbf{H} = \mathbf{X}\mathbf{V}^{(3)}. \tag{4.3}$$

We observe that $\mathrm{r}(\mathbf{F} : \mathbf{G} : \mathbf{H}) = \mathrm{r}(\mathbf{V}\mathbf{V}^-) = \mathrm{r}(\mathbf{V})$. Hence we have $\mathcal{S}(\mathbf{F} : \mathbf{G} : \mathbf{H}) = \mathcal{S}(\mathbf{X})$, and so

$$\mathbf{P}_{(\mathbf{F}\,:\,\mathbf{G}\,:\,\mathbf{H})} = \mathbf{P}_{123}. \tag{4.4}$$

In (4.4) we have used the symbol $\mathbf{P}_{123}$ for the orthogonal projector onto $\mathcal{S}(\mathbf{X})$. Similarly, the notations $\mathbf{P}_{ij}$ and $\mathbf{Q}_{ij}$ mean the orthogonal projectors onto $\mathcal{S}(\mathbf{X}_i : \mathbf{X}_j)$ and $\mathcal{S}(\mathbf{X}_i : \mathbf{X}_j)^\perp$, respectively.

By the partial canonical correlations between $\mathbf{X}_1'\mathbf{y}$ and $\mathbf{X}_2'\mathbf{y}$ after elimination of $\mathbf{X}_3'\mathbf{y}$, we mean the canonical correlations between $\mathbf{X}_1'\mathbf{Q}_3\mathbf{y}$ and $\mathbf{X}_2'\mathbf{Q}_3\mathbf{y}$. These kinds of canonical correlations were considered in some detail by Styan [9] for the special case where the column vector of ones belongs to each of the three column spaces $\mathcal{S}(\mathbf{X}_1)$, $\mathcal{S}(\mathbf{X}_2)$, $\mathcal{S}(\mathbf{X}_3)$; see also [10, 11]. Let us first consider the situation when $\mathbf{V}$, defined in (4.1), is nonsingular.

Theorem 1. *Let a positive definite* $\mathbf{V}$ *and its inverse be partitioned as in* (4.1) *and* (4.2), *respectively, and let* $\mathbf{F}$, $\mathbf{G}$, *and* $\mathbf{H}$ *be defined in* (4.3). *Then the following five sets are identical:*

(a) $\{\mathrm{cc}^2(\mathbf{F}'\mathbf{y}, \mathbf{G}'\mathbf{y})\}$,

(b) $\{\mathrm{cc}^2(\mathbf{X}_1'\mathbf{Q}_{23}\mathbf{y}, \mathbf{X}_2'\mathbf{Q}_{13}\mathbf{y})\}$,

(c) $\{\mathrm{cc}^2(\mathbf{X}_1'\mathbf{Q}_3\mathbf{y}, \mathbf{X}_2'\mathbf{Q}_3\mathbf{y})\}$,

(d) $\{\mathrm{nzch}[\mathbf{V}_{11\cdot 3}^{-1}\mathbf{V}_{12\cdot 3}\mathbf{V}_{22\cdot 3}^{-1}\mathbf{V}_{21\cdot 3}]\}$, *where* $\mathbf{V}_{ij\cdot 3} = \mathbf{V}_{ij} - \mathbf{V}_{i3}\mathbf{V}_{33}^{-1}\mathbf{V}_{3j} = \mathbf{X}_i'\mathbf{Q}_3\mathbf{X}_j$,

(e) $\{\mathrm{nzch}[(\tilde{\mathbf{V}}^{11})^{-1}\tilde{\mathbf{V}}^{12}(\tilde{\mathbf{V}}^{22})^{-1}\tilde{\mathbf{V}}^{21}]\}$.

Proof. (a) $\Leftrightarrow$ (b): Using Lemma 4, we have $\mathbf{F} = \mathbf{Q}_{23}\mathbf{X}_1(\mathbf{X}_1'\mathbf{Q}_{23}\mathbf{X}_1)^{-1}$ and $\mathbf{G} = \mathbf{Q}_{13}\mathbf{X}_2(\mathbf{X}_2'\mathbf{Q}_{13}\mathbf{X}_2)^{-1}$, thus establishing the proof.

(b) $\Leftrightarrow$ (c): Let $\mathbf{Q}_{i/j}$ be the orthogonal projector onto $\mathcal{S}(\mathbf{Q}_j\mathbf{X}_i)^{\perp}$. Then by Lemma 1(b), we have $\mathbf{Q}_{23} = \mathbf{Q}_3\mathbf{Q}_{2/3}$ and $\mathbf{Q}_{13} = \mathbf{Q}_3\mathbf{Q}_{1/3}$, thus establishing

$$\begin{aligned}\{\mathrm{cc}(\mathbf{F}'\mathbf{y}, \mathbf{G}'\mathbf{y})\} &= \{\mathrm{cc}(\mathbf{X}_1'\mathbf{Q}_{23}\mathbf{y}, \mathbf{X}_2'\mathbf{Q}_{13}\mathbf{y})\}\\ &= \{\mathrm{cc}(\mathbf{X}_1'\mathbf{Q}_3\mathbf{Q}_{2/3}\mathbf{y}, \mathbf{X}_2'\mathbf{Q}_3\mathbf{Q}_{1/3}\mathbf{y})\}\\ &= \{\mathrm{cc}(\mathbf{X}_1'\mathbf{Q}_3\mathbf{y}, \mathbf{X}_2'\mathbf{Q}_3\mathbf{y})\}. \qquad (4.5)\end{aligned}$$

The last equality in (4.5) follows by replacing the role of $\mathbf{X}_i$ by $\mathbf{Q}_3\mathbf{X}_i$ ($i = 1, 2$) in (3.3), noting that $\mathbf{Q}_{i/3}$ is the orthogonal projector onto $\mathcal{S}(\mathbf{Q}_3\mathbf{X}_i)^{\perp}$, and observing that since $\mathbf{V}$ is positive definite there are no unit canonical correlations between $\mathbf{X}_1'\mathbf{Q}_3\mathbf{y}$ and $\mathbf{X}_2'\mathbf{Q}_3\mathbf{y}$.

The proof of (c) $\Leftrightarrow$ (d) is easy to establish. To prove (d) $\Leftrightarrow$ (e), we first note that

$$\begin{pmatrix} \tilde{\mathbf{V}}^{11} & \tilde{\mathbf{V}}^{12} \\ \tilde{\mathbf{V}}^{21} & \tilde{\mathbf{V}}^{22} \end{pmatrix} = \begin{pmatrix} \mathbf{V}_{11\cdot 3} & \mathbf{V}_{12\cdot 3} \\ \mathbf{V}_{21\cdot 3} & \mathbf{V}_{22\cdot 3} \end{pmatrix}^{-1},$$

and use the result (d) $\Leftrightarrow$ (e) of Lemma 5. ❑

It may be worth noting that one method for obtaining $\{cc(\mathbf{X}_1'\mathbf{Q}_3\mathbf{y}, \mathbf{X}_2'\mathbf{Q}_3\mathbf{y})\}$ is to use the property

$$\{nucc(\mathbf{X}_1 : \mathbf{X}_3)'\mathbf{y}, (\mathbf{X}_2 : \mathbf{X}_3)'\mathbf{y}]\} = \{cc(\mathbf{X}_1'\mathbf{Q}_3\mathbf{y}, \mathbf{X}_2'\mathbf{Q}_3\mathbf{y})\},$$

which can be proved using (3.3) since

$$\mathbf{Q}_{13}(\mathbf{X}_2 : \mathbf{X}_3) = \mathbf{Q}_{13}\mathbf{X}_2 = \mathbf{Q}_{1/3}\mathbf{Q}_3\mathbf{X}_2 \quad \text{and} \quad \mathbf{Q}_{23}(\mathbf{X}_1 : \mathbf{X}_3) = \mathbf{Q}_{23}\mathbf{X}_1 = \mathbf{Q}_{2/3}\mathbf{Q}_3\mathbf{X}_1.$$

Next we consider the situation where $\mathbf{V}$ may be singular.

Lemma 8. *Let* $\mathbf{F}$, $\mathbf{G}$, *and* $\mathbf{H}$ *be defined as in* (4.3). *Then*

(a) $\mathbf{P}_{123} = \mathbf{X}_1\mathbf{F}' + \mathbf{X}_2\mathbf{G}' + \mathbf{X}_3\mathbf{H}'$

$\qquad = \mathbf{F}\mathbf{X}_1' + \mathbf{G}\mathbf{X}_2' + \mathbf{H}\mathbf{X}_3' = \mathbf{P}_{(\mathbf{F} : \mathbf{G} : \mathbf{H})}$;

(b) $\mathbf{F}'(\mathbf{X}_2 : \mathbf{X}_3) = \mathbf{0} \Leftrightarrow \mathcal{S}(\mathbf{F}) = \mathcal{S}(\mathbf{Q}_{23}\mathbf{X}_1) = \mathcal{S}(\mathbf{Q}_{2/3}\mathbf{Q}_3\mathbf{X}_1)$,

$\mathbf{G}'(\mathbf{X}_1 : \mathbf{X}_3) = \mathbf{0} \Leftrightarrow \mathcal{S}(\mathbf{G}) = \mathcal{S}(\mathbf{Q}_{13}\mathbf{X}_2) = \mathcal{S}(\mathbf{Q}_{1/3}\mathbf{Q}_3\mathbf{X}_2)$,

$\mathbf{H}'(\mathbf{X}_1 : \mathbf{X}_2) = \mathbf{0} \Leftrightarrow \mathcal{S}(\mathbf{H}) = \mathcal{S}(\mathbf{Q}_{12}\mathbf{X}_3) = \mathcal{S}(\mathbf{Q}_{1/2}\mathbf{Q}_2\mathbf{X}_3)$.

The proof of (a) follows immediately from (4.3) and (4.4). Further, notice that the result (b) is a particular case of Lemma 6. Next we state two lemmas which follow from a slight modification of Lemma 7.

Lemma 9. *The following three statements are equivalent:*

(a) *the canonical correlations between* $\mathbf{X}_1'\mathbf{y}$ *and* $\mathbf{X}_3'\mathbf{y}$, *between* $\mathbf{X}_2'\mathbf{y}$ *and* $\mathbf{X}_3'\mathbf{y}$, *and between* $\mathbf{X}_1'\mathbf{Q}_3\mathbf{y}$ *and* $\mathbf{X}_2'\mathbf{Q}_3\mathbf{y}$, *are all less than one*;

(b) $\mathbf{X}_1\mathbf{F}'\mathbf{X}_1 = \mathbf{X}_1,\ \mathbf{X}_1\mathbf{F}'\mathbf{X}_2 = \mathbf{0},\ \mathbf{X}_1\mathbf{F}'\mathbf{X}_3 = \mathbf{0}$,

$\mathbf{X}_2\mathbf{G}'\mathbf{X}_1 = \mathbf{0},\ \mathbf{X}_2\mathbf{G}'\mathbf{X}_2 = \mathbf{X}_2,\ \mathbf{X}_2\mathbf{G}'\mathbf{X}_3 = \mathbf{0}$,

$\mathbf{X}_3\mathbf{H}'\mathbf{X}_1 = \mathbf{0},\ \mathbf{X}_3\mathbf{H}'\mathbf{X}_2 = \mathbf{0},\ \mathbf{X}_3\mathbf{H}'\mathbf{X}_3 = \mathbf{X}_3$;

(c) $\mathbf{X}_1\mathbf{F}' = \mathbf{P}_{1\cdot 23}$, $\mathbf{X}_2\mathbf{G}' = \mathbf{P}_{2\cdot 13}$, *and* $\mathbf{X}_3\mathbf{H}' = \mathbf{P}_{3\cdot 12}$, *where* $\mathbf{P}_{i\cdot jk}$ *is defined according to* (2.1).

The equivalence between (a) and (b) in Lemma 9 is an immediate consequence of Lemma 9(a) and Lemma 3. Similarly, the following lemma is established.

Lemma 10. *The following three statements are equivalent:*

(a) *the canonical correlations between* $\mathbf{F'y}$ *and* $\mathbf{H'y}$, *between* $\mathbf{G'y}$ *and* $\mathbf{H'y}$, *and between* $\mathbf{F'Q_H y}$ *and* $\mathbf{G'Q_H y}$ *are all less than one;*

(b) $\mathbf{F X_1' F = F}, \quad \mathbf{F X_1' G = 0}, \quad \mathbf{F X_1' H = 0},$

$\mathbf{G X_2' F = 0}, \quad \mathbf{G X_2' G = G}, \quad \mathbf{G X_2' H = 0},$

$\mathbf{H X_3' F = 0}, \quad \mathbf{H X_3' G = 0}, \quad \mathbf{H X_3' H = H};$

(c) $\mathbf{F X_1' = P_{F\cdot(G\,:\,H)}}, \quad \mathbf{G X_2' = P_{G\cdot(F\,:\,H)}}, \quad \mathbf{H X_3' = P_{H\cdot(F\,:\,G)}}.$

From Lemmas 9 and 10, the following theorem is established.

Theorem 2. *Consider the following statements:*

(a) *the canonical correlations between* $\mathbf{X_1'y}$ *and* $\mathbf{X_3'y}$, *between* $\mathbf{X_2'y}$ *and* $\mathbf{X_3'y}$, *and between* $\mathbf{X_1'Q_3y}$ *and* $\mathbf{X_2'Q_3y}$ *are all less than one;*

(b) *the canonical correlations between* $\mathbf{F'y}$ *and* $\mathbf{H'y}$, *between* $\mathbf{G'y}$ *and* $\mathbf{H'y}$, *and between* $\mathbf{F'Q_H y}$ *and* $\mathbf{G'Q_H y}$ *are all less than one;*

(c) $\mathbf{F'(X_2 : X_3) = 0}, \ \mathbf{G'(X_1 : X_3) = 0},$ *and* $\mathbf{H'(X_1 : X_2) = 0};$

(d) $\mathcal{S}(\mathbf{F}) = \mathcal{S}(\mathbf{Q_{23}X_1}), \ \mathcal{S}(\mathbf{G}) = \mathcal{S}(\mathbf{Q_{13}X_2}),$ *and* $\mathcal{S}(\mathbf{H}) = \mathcal{S}(\mathbf{Q_{12}X_3});$

(e) $\{\mathrm{cc}(\mathbf{F'y}, \mathbf{G'y})\} = \{\mathrm{cc}(\mathbf{X_1'Q_3y}, \mathbf{X_2'Q_3y})\} = \{\mathrm{cc}(\mathbf{X_1'Q_{23}y}, \mathbf{X_2'Q_{13}y})\},$

$\{\mathrm{cc}(\mathbf{G'y}, \mathbf{H'y})\} = \{\mathrm{cc}(\mathbf{X_2'Q_1y}, \mathbf{X_3'Q_1y})\} = \{\mathrm{cc}(\mathbf{X_2'Q_{13}y}, \mathbf{X_3'Q_{12}y})\},$

$\{\mathrm{cc}(\mathbf{F'y}, \mathbf{H'y})\} = \{\mathrm{cc}(\mathbf{X_1'Q_2y}, \mathbf{X_3'Q_2y})\} = \{\mathrm{cc}(\mathbf{X_1'Q_{23}y}, \mathbf{X_3'Q_{12}y})\};$

(f) $\mathbf{P_{1\cdot 23}' = P_{F\cdot(G\,:\,H)}}, \ \mathbf{P_{2\cdot 13}' = P_{G\cdot(F\,:\,H)}},$ *and* $\mathbf{P_{3\cdot 12}' = P_{H\cdot(F\,:\,G)}}.$

Then we have:

$$\text{(a) and (b)} \Leftrightarrow \text{(c)} \Leftrightarrow \text{(d)} \Leftrightarrow \text{(e)} \Leftrightarrow \text{(f)}.$$

Proof. First we prove that (a) & (b) $\Rightarrow$ (c). If (a) and (b) hold, then from Lemmas 8 and 9 it follows that

$$\mathbf{F X_1' F = F}, \quad \mathbf{X_1 F' X_2 = 0}, \quad \mathbf{X_1 F' X_3 = 0}, \tag{4.6}$$

$$\mathbf{G X_2' G = G}, \quad \mathbf{X_2 G' X_1 = 0}, \quad \mathbf{X_2 G' X_3 = 0}, \tag{4.7}$$

$$\mathbf{H X_3' H = H}, \quad \mathbf{X_3 H' X_1 = 0}, \quad \mathbf{X_3 H' X_2 = 0}. \tag{4.8}$$

Premultiplying the second and third equations in (4.6), (4.7), and (4.8) by $\mathbf{F}'$, $\mathbf{G}'$, and $\mathbf{H}'$, respectively, and using the first equations, we obtain (c). On the other hand, Lemma 8(a) implies that $\mathbf{X_1 F' X_1 = X_1}$, $\mathbf{X_2 G' X_2 = X_2}$, and $\mathbf{X_3 H' X_3 = X_3}$, and hence (c) implies Lemma 9(b), i.e., (c) implies (a). The proof of (c) $\Rightarrow$ (b) is similar. The equivalence (c) $\Leftrightarrow$ (d) is a particular case of Lemma 6. To prove that (d) $\Rightarrow$ (e), we observe that (d) implies

$$\{\mathrm{cc}(\mathbf{F'y}, \mathbf{G'y})\} = \{\mathrm{cc}(\mathbf{X_1' Q_3 Q_{2/3} y}, \mathbf{X_2' Q_3 Q_{1/3} y})\}.$$

Further, by (3.3) we have

$$\{\mathrm{cc}(\mathbf{X_1' Q_3 Q_{2/3} y}, \mathbf{X_2' Q_3 Q_{1/3} y})\} = \{\mathrm{nucc}(\mathbf{X_1' Q_3 y}, \mathbf{X_2' Q_3 y})\}. \tag{4.9}$$

But since (d) holds then (a) also holds and there are no unit canonical correlations between $\mathbf{X_1' Q_3 y}$ and $\mathbf{X_2' Q_3 y}$. This shows that in (4.9) we can replace "nucc" by "cc" and so $\{\mathrm{cc}(\mathbf{F'y}, \mathbf{G'y})\} = \{\mathrm{cc}(\mathbf{X_1' Q_3 y}, \mathbf{X_2' Q_3 y})\}$. The other parts of (e) follow similarly. Hence we proved that (d) $\Rightarrow$ (e). Using the equivalence of (e) and (f) of Lemma 7, we observe that (e) $\Rightarrow$ (d).The proof of (d) $\Rightarrow$ (f) is obvious. Finally we show that (f) $\Rightarrow$ (d). If (f) holds then

$$\mathbf{P'_{1\cdot 23}} = \mathbf{Q_{23} X_1 [(X_1' Q_{23} X_1)^-]' X_1'} = \mathbf{F (F' Q_{(G\,:\,H)} F)^- F' Q_{(G\,:\,H)}}, \tag{4.10}$$

where $(\mathbf{X_1' Q_{23} X_1})^-$ and $(\mathbf{F' Q_{(G\,:\,H)} F})^-$ are some fixed generalized inverses. Using (2.2) and the first equality in (4.10) we observe that that $\mathcal{S}(\mathbf{P'_{1\cdot 23}}) = \mathcal{S}(\mathbf{Q_{23} X_1})$ so that the column space $\mathcal{S}(\mathbf{P'_{1\cdot 23}})$ is actually invariant for all choices of $(\mathbf{X_1' Q_{23} X_1})^-$. Therefore, due to the second equality in (4.10), also $\mathcal{S}(\mathbf{P_{F\cdot(G\,:\,H)}})$ is invariant with respect to the choice of $(\mathbf{F' Q_{(G\,:\,H)} F})^-$ which by Lemma 2 means that $\mathcal{S}(\mathbf{P_{F\cdot(G\,:\,H)}}) = \mathcal{S}(\mathbf{F})$. Hence we have proved that $\mathcal{S}(\mathbf{Q_{23} X_1}) = \mathcal{S}(\mathbf{F})$. The other parts of the implication (f) $\Rightarrow$ (d) can be proved similarly. ❑

From Theorem 2, it is interesting to note that the canonical correlations between $\mathbf{F'y}$ and

$\mathbf{G'y}$ turn out to be the partial canonical correlations between $\mathbf{X}_1'\mathbf{y}$ and $\mathbf{X}_2'\mathbf{y}$ after elimination of $\mathbf{X}_3'\mathbf{y}$ even when the dispersion matrix in (4.1) fails to be nonsingular provided that the "rank-additivities" $r(\mathbf{X}_1 : \mathbf{X}_2 : \mathbf{X}_3) = r(\mathbf{X}_1) + r(\mathbf{X}_2) + r(\mathbf{X}_3)$ and $r(\mathbf{F} : \mathbf{G} : \mathbf{H}) = r(\mathbf{F}) + r(\mathbf{G}) + r(\mathbf{H})$ hold.

Acknowledgements

The authors are very grateful to Professor Jerzy K. Baksalary for helpful discussions regarding this research area at the International Workshop on Linear Models, Experimental Designs, and Related Matrix Theory, held at the University of Tampere in August 1990. This reseach was completed when the second author was a Senior Researcher Fellow in the Academy of Finland.

REFERENCES

1. N. P. Jewell and P. Bloomfield, *Ann. Statist.*, **11**, 837-847 (1983).
2. D. Latour, S. Puntanen and G. P. H. Styan, in: *Proceedings of the Second International Tampere Conference in Statistics* , T. Pukkila and S. Puntanen, (Eds.), pp. 541-553. Dept. of Mathematical Sciences, University of Tampere, Tampere (1987).
3. C. G. Khatri, *J. Multivariate Anal.*, **34**, 211-226 (1990).
4. J. K. Baksalary, S. Puntanen and H. Yanai, *Linear Algebra Appl.*, **176**, 61-74 (1992).
5. C. R. Rao and H. Yanai, *J. Statist. Plann. Inference*, **3**, 1-17 (1979).
6. C. R. Rao and S. K. Mitra, *Generalized Inverse of Matrices and Its Applications*. Wiley, New York (1971).
7. H. Yanai, *Comp. Statist. Data Anal.*, **10**, 251-260 (1990).
8. G. P. H. Styan, in: *Proceedings of the First International Tampere Seminar on Linear Statistical Models and their Applications.*, T. Pukkila and S. Puntanen, (Eds.), pp. 37-75. Dept. of Mathematical Sciences, University of Tampere, Tampere (1985).
9. G. P. H. Styan, in: *Pacific Statistical Congress*, I. S. Francis, B. F. J. Manly and F. C. Lam (Eds.), pp. 433-438. North Holland, Amsterdam (1986).
10. J. Bérubé, *Some properties of three-way layouts*, MSc Thesis. Dept. of Mathematics and Statistics. McGill University, Montréal (1991).
11. K. J. Worsley, G. P. H. Styan and J. Bérubé, *Genstat Newsletter*, **26**, 11-21 (1991).

Stat. Sci. & Data Anal., pp. 265-276
K. Matsusita *et al.* (Eds)

Approximation to the Upper Percentiles of T^2_{max}-type Statistics

TAKASHI SEO and MINORU SIOTANI

Department of Mathematics, Faculty of Science, Hiroshima University, Higashi-Hiroshima 724, Japan
Natural Sciences Division, Meisei University, Ōme City, Tokyo 198, Japan

Abstract. T^2_{max}-type statistics defined as the maximum of the square of distances between two points in a random sample drawn from a multivariate distribution are useful in simultaneous statistical inferences. To make statistical inferences on comparisons of several treatments, we need the upper percentiles of T^2_{max}-type statistics. However, it is formidable that the exact sample distributions of T^2_{max}-type statistics are obtained under the assumption of multivariate normality. In this paper, the modified second approximation method is applied to evaluate the upper percentiles of T^2_{max}-type statistics. The estimated percentiles are investigated by Monte Carlo studies and the valid ranges of parameters for the formulae of the modified second approximation are discussed.

Key words: Asymptotic Expansion Formulae, Bonferroni's Inequalities, Hotelling's T^2, Modified Second Approximation, Monte Carlo Studies, Simultaneous Confidence Intervals, T^2_{max}-type Statistics.

1. INTRODUCTION

T^2_{max}-type statistics defined as the maximum of the square of distances between two points in a random sample drawn from a multivariate distribution are useful in some problems in simultaneous statistical inference. The simultaneous confidence interval estimation based on the union-intersection principle was discussed by Roy and Bose[1]. However, it is formidable that the exact sample distributions of T^2_{max}-type statistics are obtained under the assumption of multivariate normality. Siotani[2], [3] discussed how to obtain the approximate percentiles of T^2_{max} by using Bonferroni's inequalities, which have been reproduced in Krishnaiah[4]. Krishnaiah[5] also discussed the testing procedure using T^2_{max} and their optimum properties. The simultaneous multivariate procedures using Hotelling's T^2 statistics have been given by Jensen[6]. The numerical examination on the accuracy of the modified second approximate percentiles of T^2_{max}-type statistics were given for the multivariate studentized range case by Seo and Siotani[7]. This paper concerns with the case of T^2_{max}-type statistics with a control and the modified second approximate percentiles of this statistic are given by the much simplified formulae of Siotani's results after a great deal of computation. The estimated percentiles are investigated by Monte Carlo studies and the valid ranges of parameters for the formulae of the modified second approximation are discussed. The simultaneous comparisons among mean vectors using these statistics are also discussed.

2. T^2_{max}-TYPE STATISTICS AND MODIFIED SECOND APPROXIMATION METHOD

2.1 T^2_{max}-type Statistics

Let $y_1, y_2, \cdots, y_N$ be N (column) random vectors having $N_p(\mathbf{0}, \gamma\Sigma)$ $(\gamma > 0)$ and let $\delta\Sigma$ $(\gamma > |\delta|)$ be a covariance matrix of y_i and y_j $(i \neq j)$. Siotani[2], [8] considered a general statistic having the form

$$T^2_{max} = \max_{1 \le i \le N}\{y_i' S^{-1} y_i\}, \tag{2.1}$$

where S is an unbiasedly estimated matrix of Σ with ν d.f., which is independent of y's. These type statistics are useful in simultaneous statistical inferences.

As the special cases of T^2_{max}-type statistics, the multivariate studentized range R_{max} is defined as the positive square root of the maximum of generalized studentized distances between any two points in random sample drawn from a multivariate distribution, that is, the positive square root of

$$R^2_{max} = \max_{i<j}\{(\boldsymbol{x}_i - \boldsymbol{x}_j)' S^{-1} (\boldsymbol{x}_i - \boldsymbol{x}_j)\}, \tag{2.2}$$

where $\boldsymbol{x}_i, i = 1, \cdots N$ are N independent random vectors distributed as $N_p(\boldsymbol{\xi}, \Lambda)$. When $p = 1$, R_{max} reduces the ordinary studentized range in the univariate case; $R_{max} = (x_{max} - x_{min})/\ell$, where ℓ is the standard deviation. A short note on the background and the discussion of R_{max} are derived by Seo and Siotani[7].

In the similar way, let $\boldsymbol{x}_0$ be a control variate which is independent of $\boldsymbol{x}_i$'s and has the same distribution with that of $\boldsymbol{x}_i$. Then the maximum of the squared studentized deviates from a control is defined by

$$T^2_{max \cdot c} = \max_i\{D_i^2\}, \qquad D_i^2 = (\boldsymbol{x}_i - \boldsymbol{x}_0)' S^{-1} (\boldsymbol{x}_i - \boldsymbol{x}_0). \tag{2.3}$$

It is easily seen that these statistics defined in (2.2) and (2.3) are the special cases of T^2_{max}-type statistics. To make statistical inferences on comparisons of several treatments, we need the percentiles of these statistics. However, it is formidable that the exact sample distributions of T^2_{max}-type statistics are obtained even under the assumption of multivariate normality. The modified second approximation method is explained to evaluate the upper percentiles of these statistics in the next subsection.

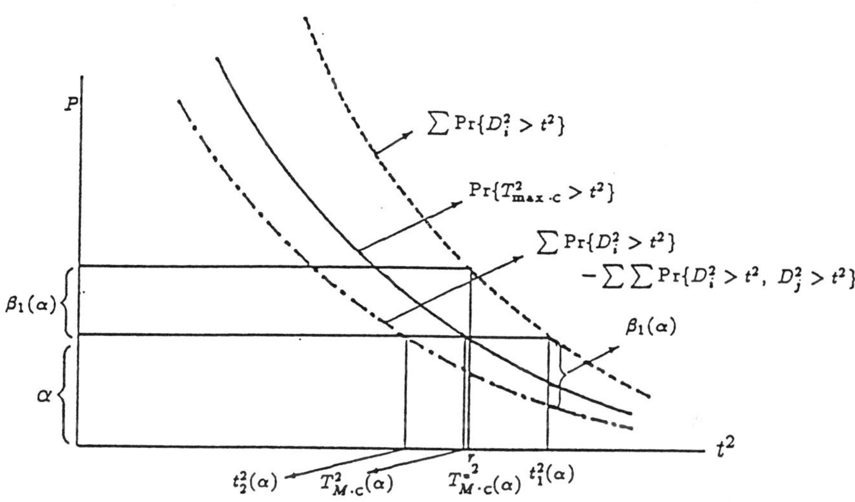

Figure 2.1. Illustration of the modified second approximation

2.2 Modified Second Approximation Method

The modified second approximation method for the special case $T^2_{\max\cdot C}$ treated in this paper is described below. The method is based on inclusion-exclusion formula and associated Bonferroni's inequalities for $\Pr\{T^2_{\max\cdot C} > t^2\}$; Let us define N events

$$E \equiv [T^2_{\max\cdot C} > t^2], \qquad E_i \equiv [D_i^2 > t^2], \quad i = 1, 2, \cdots, N. \tag{2.4}$$

Then we have

$$\Pr(E) = \sum_{i=1}^{N} \Pr(E_i) - \sum_{i<j} \Pr(E_i, E_j) + \cdots + (-1)^{N+1} \Pr(E_1, E_2, \cdots, E_N) \tag{2.5}$$

and

$$\sum_{i=1}^{N} \Pr(E_i) > \Pr(E) > \sum_{i=1}^{N} \Pr(E_i) - \sum_{i<j} \Pr(E_i, E_j). \tag{2.6}$$

Let $t_1^2(\alpha)$ be the first approximation to the exact upper $100\alpha\%$ point $T^2_{M\cdot C}(\alpha)$ of $T^2_{\max\cdot C}$, which is the solution of the equation

$$\Pr\{D_i^2 > t_1^2(\alpha)\} = \alpha/N. \tag{2.7}$$

Then it follows from (2.6) that

$$\alpha > \Pr\{T^2_{\max\cdot C} > t_1^2(\alpha)\} > \alpha - \beta_1(\alpha), \tag{2.8}$$

where

$$\beta_1(\alpha) = \frac{N(N-1)}{2}\Pr\{D_i^2 > t_1^2(\alpha),\ D_j^2 > t_1^2(\alpha)\}. \tag{2.9}$$

Obviously $t_1^2(\alpha)$ is an overestimate of $T^2_{M\cdot C}(\alpha)$.

Referring to the Figure 2.1, *the (singly) modified second approximation* is defined by $T^{*2}_{M\cdot C}(\alpha)$ satisfying

$$\Pr\{D_i^2 > T^{*2}_{M\cdot C}(\alpha)\} = \{\alpha + \beta_1(\alpha)\}/N. \tag{2.10}$$

There is no information whether this method gives an overestimate or underestimate of $T^2_{M\cdot C}(\alpha)$, although $T^{*2}_{M\cdot C}(\alpha)$ is smaller than $t_1^2(\alpha)$.

If the modified second approximation is still an overestimate with significant excess, this method may be repeated once more to obtain an accurate approximation. $T^{**2}_{M\cdot C}(\alpha)$ which is called *the doubly modified second approximation* is satisfying

$$\Pr\{D_i^2 > T^{**2}_{M\cdot C}(\alpha)\} = \{\alpha + \beta_2(\alpha)\}/N \tag{2.11}$$

where

$$\beta_2(\alpha) = \frac{N(N-1)}{2}\Pr\{D_i^2 > T^{*2}_{M\cdot C}(\alpha),\ D_j^2 > T^{*2}_{M\cdot C}(\alpha)\}. \tag{2.12}$$

The accuracy of approximations depends of course on the values of parameters, α, p, N and ν. The doubly modified second approximation procedure is possibly needed when the distribution has a long right tail and/or when α is pretty small like 0.001. In order to evaluate $T^{*2}_{M\cdot C}(\alpha)$, we need to know $t_1(\alpha)$ and $\beta_1(\alpha)$ which will be done in the next section.

3. MODIFIED SECOND APPROXIMATION $T^{*2}_{M\cdot C}(\alpha)$

Note that $\boldsymbol{y}_i = \boldsymbol{x}_i - \boldsymbol{x}_0 \sim N_p(\boldsymbol{0}, 2\Lambda)$, $T_i^2 = \frac{1}{2}\boldsymbol{y}_i' S^{-1}\boldsymbol{y}_i = \frac{1}{2}D_i^2$ is Hotelling's T^2 statistic with ν d.f. and hence $[(\nu - p + 1)/p\nu]T_i^2$ is distributed according to the F distribution with p and $\nu - p + 1$ d.f.'s for i $(i = 1, \cdots, N)$. Hence the first approximation $t_1^2(\alpha)$ is obtained from (2.7) as

$$t_1^2(\alpha) = [2\nu p/(\nu - p + 1)]F_{p,\nu-p+1}(\alpha/N). \tag{3.1}$$

The exact evaluation of $\beta_1(\alpha)$ is quite complicated; hence we adopted the large sample approximation based on asymptotic expansion, which is due to Siotani[2], in complicated, time- and money-consuming form, and some checks for the accuracy were done only for $p = 1, 2$. After a great deal of computation using the reduction formulae for the density function of the χ^2-distribution with respect to its number of d.f., Siotani's inconvenient formula is here much simplified as shown below.

3.1 Asymptotic expansion for $\Pr\{D_1^2 > t_1^2(\alpha), D_2^2 > t_1^2(\alpha)\}$

The following notations are used:

$$\chi^2 \equiv \chi^2(\alpha/N;p), \quad \eta \equiv \frac{2}{3}\chi^2, \quad \sum \equiv \sum_{j=0}^{\infty},$$

$$g_a(\eta) \equiv \frac{1}{\Gamma(a)}\eta^{a-1}e^{-\eta} \ (a>0), \quad G_a(\eta) \equiv \int_\eta^\infty g_a(t)\,dt,$$

$$g_{(p/2)-1}(\eta) \equiv -\frac{1}{2\sqrt{\pi}}\eta^{-3/2}e^{-\eta} \ \text{ for } \ p=1; \ \equiv 0 \ \text{ for } \ p=2,$$

$$L(\eta) \equiv (\frac{3}{4})^{\frac{1}{2}p}\sum \frac{\Gamma(\frac{1}{2}p+j)}{j!\Gamma(\frac{1}{2}p)}(\frac{1}{4})^j G^2_{\frac{1}{2}p+j}(\eta),$$

$$H(\eta) \equiv (\frac{3}{4})^{\frac{1}{2}p}\sum \frac{\Gamma(\frac{1}{2}p+j)}{j!\Gamma(\frac{1}{2}p)}(\frac{1}{4})^j g_{\frac{1}{2}p+j}(\eta)G_{\frac{1}{2}p+j}(\eta),$$

$$H^*(\eta) \equiv (\frac{3}{4})^{\frac{1}{2}p}\sum \frac{\Gamma(\frac{1}{2}p+j)}{j!\Gamma(\frac{1}{2}p)}(\frac{1}{4})^j g_{\frac{1}{2}p-1+j}(\eta)G_{\frac{1}{2}p+j}(\eta),$$

$$H(\eta;j^r) \equiv (\frac{3}{4})^{\frac{1}{2}p}\sum \frac{\Gamma(\frac{1}{2}p+j)}{j!\Gamma(\frac{1}{2}p)}j^r(\frac{1}{4})^j g_{\frac{1}{2}p+j}(\eta)G_{\frac{1}{2}p+j}(\eta), \qquad r=1,2,3,$$

$$Q(\eta) \equiv (\frac{3}{4})^{\frac{1}{2}p}\sum \frac{\Gamma(\frac{1}{2}p+j)}{j!\Gamma(\frac{1}{2}p)}(\frac{1}{4})^j g^2_{\frac{1}{2}p+j}(\eta),$$

$$Q(\eta;d_j) \equiv (\frac{3}{4})^{\frac{1}{2}p}\sum \frac{\Gamma(\frac{1}{2}p+j)}{j!\Gamma(\frac{1}{2}p)}d_j\,(\frac{1}{4})^j g^2_{\frac{1}{2}p+j}(\eta),$$

$$d_j = j,\ j^2,\ \frac{1}{p+2j},\ \frac{1}{p+2+2j}.$$

Then we have

$$\Pr\{D_1^2 > t_1^2(\alpha), D_2^2 > t_1^2(\alpha)\} = A_0 + \nu^{-1}A_1 + \nu^{-2}A_2 + O(\nu^{-3}), \tag{3.2}$$

where

$$A_0 = L(\eta),$$

$$A_1 = \frac{2}{9}\chi^4 H(\eta) - \frac{4}{3}\chi^2 H(\eta;j) + \frac{8}{9}\chi^4 Q(\eta) - \frac{8}{9}(p-1)\chi^4 Q(\eta;\frac{1}{p+2j}),$$

$$A_2 = b_{20}H(\eta) + b_{21}H(\eta;j) + b_{22}H(\eta;j^2) + b_{23}H(\eta;j^3) + b_{24}H^*(\eta) + c_{20}Q(\eta) + c_{21}Q(\eta;j) + c_{22}Q(\eta;j^2) + c_{23}Q(\eta;\frac{1}{p+2j}) + c_{24}Q(\eta;\frac{1}{p+2+2j}),$$

coefficients b_{2i}'s and c_{2i}'s being

$$b_{20} = \frac{1}{324}\chi^2[4\chi^6 + 8(3p-11)\chi^4 + 6(9p^2 - 11p - 2)\chi^2 + 27p^2(p-2)],$$

$$b_{21} = \frac{1}{54}\chi^2[8(3p+1)\chi^2 + 3p(3p-8)],$$

$$b_{22} = \frac{1}{9}\chi^2(6\chi^2 - 3p + 8),$$

$$b_{23} = -\frac{2}{3}\chi^2, \qquad b_{24} = -\frac{1}{9}\chi^4(\chi^2 + p)^2,$$

$$c_{20} = \frac{1}{81}\chi^4[49\chi^4 + 12(8p-13)\chi^2 + 18(2p-1)],$$

$$c_{21} = -\frac{76}{27}\chi^6, \qquad c_{22} = \frac{28}{9}\chi^4,$$

$$c_{23} = \frac{4}{81}(p-1)\chi^4[4(p-4)\chi^4 - 12(2p-1)\chi^2 - 9(p-1)],$$

$$c_{24} = -\frac{16}{81}(p^2-1)\chi^8.$$

Now multiplying (3.2) by $N(N-1)/2$ to obtain $\beta_1(\alpha)$, the modified second approximation $T^{*2}_{M\cdot C}(\alpha)$ to $T^2_{M\cdot C}(\alpha)$ is computed by (2.10), i.e.,

$$\Pr\{D_i^2 > T^{*2}_{M\cdot C}(\alpha)\} = \{\alpha + \beta_1(\alpha)\}/N,$$

which is

$$T^{*2}_{M\cdot C}(\alpha) = [2\nu p/(\nu - p + 1)]F_{p,\nu-p+1}[\{\alpha + \beta_1(\alpha)\}/N].$$

The asymptotic expansion formulae up to the term of order ν^{-2} are used instead of $\beta_1(\alpha)$. The program for computing $T^{*2}_{M\cdot C}(\alpha)$ is written by FORTRAN. Subroutines are taken from JSA-1972 in Yamauti(ed.)[9]. To compute $L(\eta)$, $H(\eta)$, etc., the summation is taken, for each case, over the terms up to the term of order 10^{-12}.

4. NUMERICAL EXAMINATION ON THE ACCURACY OF $T^{*2}_{M\cdot C}(\alpha)$

In the same way as in Seo and Siotani[7] the Monte Carlo numerical experiments are made, first to compare the simulated values with the modified second approximation in Section 3, secondly to examine the valid ranges of parameters.

4.1. Simulated densities of $T^2_{\max\cdot C}$ and simulated values $\overline{T}^2_{M\cdot C}(\alpha)$ of $T^2_{M\cdot C}(\alpha)$

Based on normal random vectors generated from $N_p(\mathbf{0}, I_p)$, 10,000 of $T^2_{\max\cdot C}$ are computed for each set parameters (α, p, N, ν). Then the upper $100\alpha\%$ point $T^2_{M\cdot C}(\alpha)$ is estimated with aid of fine histogram. This process is repeated 100 times to obtain 100 estimates $T^2_{Mi}(\alpha)$, $i = 1, 2, \cdots, 100$ of $T^2_{M\cdot C}(\alpha)$. We then calculated the average $\overline{T}^2_{M\cdot C}(\alpha)$ and the standard deviation of the average, which is denoted by $S.D. \equiv S_{\overline{T}^2_{M\cdot C}(\alpha)}$.

To give the image of the distribution of $T^2_{\max\cdot C}$, the simulated densities calculated based on 10,000 values of $T^2_{\max\cdot C}$ are given in the form of histogram. Table 4.1 gives the simulated values $\overline{T}^2_{M\cdot C}(\alpha)$ and 10 times its standard deviation $10(S.D.)$ for $p = 3, 5$; $\alpha = 0.10, 0.05, 0.01, 0.001$; $N = 2, 5, 10, 15$; $\nu = 10, 20, 40$. It may be seen from the magnitude of the standard deviation $S.D.$ of $\overline{T}^2_{M\cdot C}(\alpha)$ that the simulated values $\overline{T}^2_{M\cdot C}(\alpha)$ are good enough for using as the reference values in the examination of precision of $T^{*2}_{M\cdot C}(\alpha)$, except the cases of $p = 3, 5$, $\nu = 10$, $\alpha = 0.001$.

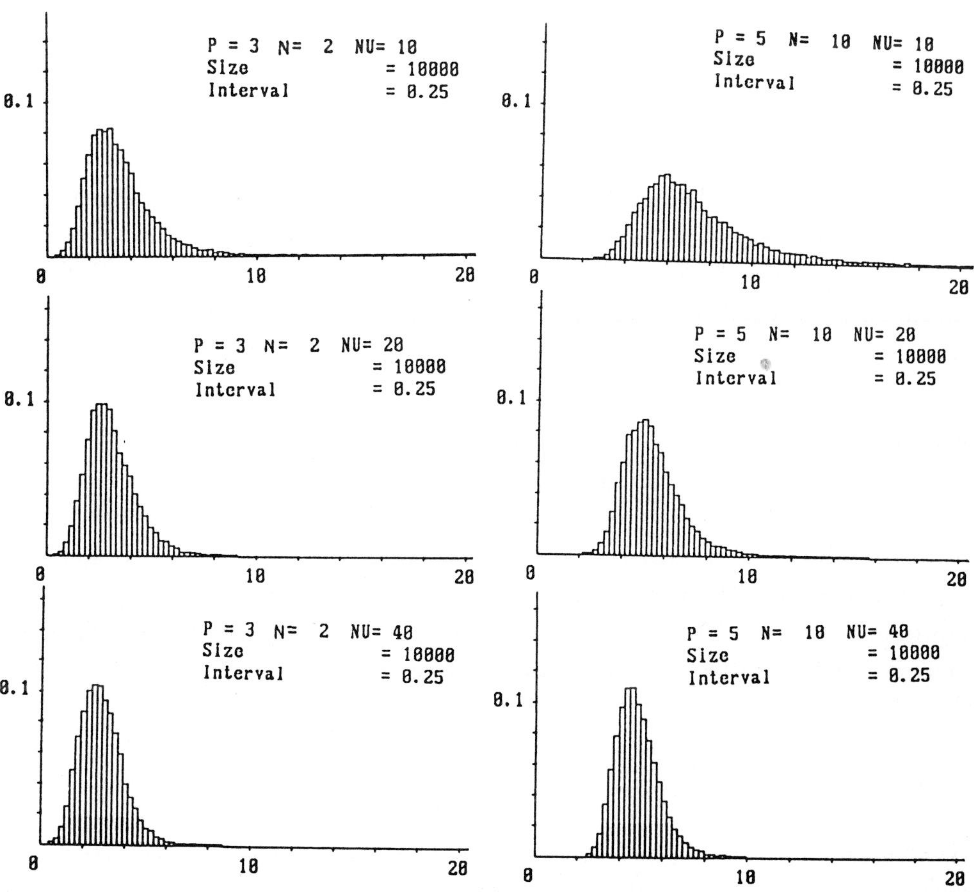

Figure 4.1. Examples of the simulated densities of $T^2_{\max\cdot C}$ in the form of histogram

As seen from Figure 4.1, the simulated density-histograms for those exceptional cases have long tails, which may make the standard deviation large. To obtain more accurate simulated value $\overline{T}^2_{M\cdot C}(\alpha)$, the repetition more than 100 may have been needed.

4.2 Examination on the precision of $T^{*2}_{M\cdot C}(\alpha)$ and the valid ranges of parameters

There are the following two main sources for errors of $T^{*2}_{M\cdot C}(\alpha)$:

(1) the asymptotic expansion used in evaluation of $\beta_1(\alpha)$,
(2) the modified second approximation method itself.

The error due to first source (1) is expected to decrease in magnitude as ν increases. For the second source (2), the validity of the method depends on the values of parameters, particularly on values of p and ν.

There is no information whether the method gives an overestimate or underestimate of $T^2_{M\cdot C}(\alpha)$. In this subsection, examinations on overall errors are made as well as on the valid ranges of parameters. For this purpose, the following values are calculated:

$$e_i \equiv T^{*2}_{M\cdot C}(\alpha) - \overline{T}^2_{M\cdot C}(\alpha), \quad re_i \equiv [T^{*2}_{M\cdot C}(\alpha) - \overline{T}^2_{M\cdot C}(\alpha)]/\overline{T}^2_{M\cdot C}(\alpha), \qquad i = 3, 5,$$

which were listed in Table 4.1 with the modified second approximation $T^{*2}_{M\cdot C}(\alpha)$. The tendency of the results in Table 4.1 is quite similar to that of Seo and Siotani[7] in which the upper percentiles of the multivariate studentized range. For $p = 8$, the same examination was done at several selected values of parameters and we obtained also the similar results as in Seo and Siotani[7].

4.3 The doubly modified second approximation

First note that more than half values of e_i or re_i, $i = 3, 5$ in Table 4.1 have minus sign. This means that $T^{*2}_{M\cdot C}(\alpha)$'s corresponding to those minus e_i's or re_i's are underestimate, so the doubly modified second approximation procedure is not effective to use. For $T^{*2}_{M\cdot C}(\alpha)$'s corresponding to positive e_i's or re_i's, the doubly modified second approximation procedure has a good effect on the approximation, so the doubly modified approximation procedure will be used only for the cases with $N = 2$.

5. AN APPLICATION

As an application of the use of $T^2_{\max}$-type statistics, we consider the construction of simultaneous confidence intervals for the comparisons among means of several treatments with a control.

Table 4.1. Values of $\overline{T}^2_{M\cdot c}(\alpha)$, $10(S.D.)$, $T^{*2}_{M\cdot c}(\alpha)$ and their Errors and Relative Errors

$p=3$		$\alpha=0.10$				$\alpha=0.05$			
		N				N			
ν		2	5	10	15	2	5	10	15
	$\overline{T}^2_{M\cdot c}$	28.645	38.460	45.991	50.108	38.760	50.854	59.662	64.286
		(0.435)	(0.508)	(0.556)	(0.665)	(0.739)	(0.841)	(0.909)	(1.081)
10	$T^{*2}_{M\cdot c}$	28.964	38.694	45.597	49.157	39.023	50.815	59.172	63.470
	e_3	0.319	0.234	−0.394	−0.951	0.263	−0.039	−0.490	−0.816
	re_3	0.011	0.006	−0.009	−0.019	0.007	−0.001	−0.008	−0.013
	$\overline{T}^2_{M\cdot c}$	20.254	25.957	30.311	32.759	25.688	32.035	36.898	39.518
		(0.206)	(0.252)	(0.269)	(0.340)	(0.362)	(0.376)	(0.438)	(0.488)
20	$T^{*2}_{M\cdot c}$	20.360	26.020	30.106	32.330	25.744	32.000	36.548	39.033
	e_3	0.106	0.063	−0.205	−0.429	0.056	−0.035	−0.350	−0.485
	re_3	0.005	0.002	−0.007	−0.013	0.002	−0.001	−0.010	−0.012
	$\overline{T}^2_{M\cdot c}$	17.435	21.883	25.260	27.326	21.528	26.320	29.911	32.110
		(0.171)	(0.203)	(0.218)	(0.251)	(0.239)	(0.280)	(0.311)	(0.355)
40	$T^{*2}_{M\cdot c}$	17.517	21.961	25.170	26.947	21.604	26.315	29.746	31.656
	e_3	0.082	0.078	−0.090	−0.379	0.076	−0.005	−0.165	−0.454
	re_3	0.005	0.004	−0.004	−0.014	0.004	−0.000	−0.006	−0.014

$p=3$		$\alpha=0.01$				$\alpha=0.001$			
		N				N			
ν		2	5	10	15	2	5	10	15
	$\overline{T}^2_{M\cdot c}$	69.778	88.922	102.696	111.667	141.942	176.257	204.330	220.444
		(2.237)	(2.427)	(3.068)	(3.552)	(14.145)	(15.617)	(18.807)	(15.845)
10	$T^{*2}_{M\cdot c}$	70.007	88.182	101.146	107.853	142.396	175.771	200.191	213.163
	e_3	0.229	−0.740	−1.550	−3.814	0.454	−0.486	−4.139	−7.281
	re_3	0.003	−0.008	−0.015	−0.034	0.003	−0.003	−0.021	−0.034
	$\overline{T}^2_{M\cdot c}$	39.630	47.611	53.458	56.828	63.589	74.340	82.452	87.195
		(1.068)	(1.018)	(1.096)	(1.243)	(4.096)	(3.826)	(5.267)	(5.168)
20	$T^{*2}_{M\cdot c}$	39.693	47.476	53.234	56.423	64.156	74.576	82.473	86.942
	e_3	0.063	−0.135	−0.224	−0.405	0.567	0.236	0.021	−0.253
	re_3	0.002	−0.003	−0.004	−0.007	0.009	0.003	0.000	−0.003
	$\overline{T}^2_{M\cdot c}$	31.350	36.663	40.890	43.342	46.276	52.591	57.288	60.404
		(0.627)	(0.643)	(0.742)	(0.698)	(2.242)	(2.256)	(2.329)	(2.452)
40	$T^{*2}_{M\cdot c}$	31.433	36.749	40.694	42.925	46.680	52.857	57.537	60.235
	e_3	0.083	0.086	−0.196	−0.417	0.404	0.266	0.249	−0.169
	re_3	0.003	0.002	−0.005	−0.010	0.009	0.005	0.004	−0.003

Table 4.1. Continued

$p=5$		$\alpha=0.10$				$\alpha=0.05$			
		N				N			
ν		2	5	10	15	2	5	10	15
10	$\overline{T}^2_{M\cdot c}$	68.334 (1.082)	92.735 (1.318)	111.082 (1.724)	123.094 (1.672)	94.516 (1.998)	125.138 (2.354)	150.172 (2.950)	165.183 (3.125)
	$T^{*2}_{M\cdot C}$	69.011	93.595	111.310	120.469	95.170	126.697	149.448	161.223
	e_5	0.677	0.860	−0.228	−2.625	0.654	1.559	−0.724	−3.960
	re_5	0.010	0.009	−0.002	−0.022	0.007	0.012	−0.005	−0.024
20	$\overline{T}^2_{M\cdot c}$	34.594 (0.398)	43.095 (0.421)	49.868 (1.022)	53.343 (0.402)	42.949 (0.574)	52.450 (0.637)	59.858 (1.434)	63.788 (0.641)
	$T^{*2}_{M\cdot C}$	34.573	43.249	49.510	52.917	42.831	52.445	59.445	63.273
	e_5	−0.021	0.154	−0.358	−0.426	−0.118	−0.005	−0.413	−0.515
	re_5	−0.001	0.004	−0.007	−0.008	−0.003	−0.000	−0.006	−0.008
40	$\overline{T}^2_{M\cdot c}$	26.966 (0.303)	32.594 (0.246)	36.981 (0.259)	39.535 (0.266)	32.354 (0.417)	38.281 (0.331)	42.908 (0.401)	45.684 (0.363)
	$T^{*2}_{M\cdot C}$	26.893	32.723	36.914	39.232	32.220	38.340	42.786	45.263
	e_5	−0.073	0.129	−0.067	−0.303	−0.134	0.059	−0.122	−0.421
	re_5	−0.003	0.004	−0.002	−0.008	−0.004	0.002	−0.003	−0.010

$p=5$		$\alpha=0.01$				$\alpha=0.001$			
		N				N			
ν		2	5	10	15	2	5	10	15
10	$\overline{T}^2_{M\cdot c}$	183.309 (7.496)	237.675 (8.745)	281.670 (10.630)	311.426 (13.056)	427.373 (49.268)	541.342 (62.869)	645.752 (83.788)	719.273 (74.998)
	$T^{*2}_{M\cdot C}$	184.354	240.058	280.859	302.289	431.649	557.065	652.556	704.636
	e_5	1.045	2.383	−0.811	−9.137	4.276	15.723	6.804	−14.637
	re_5	0.006	0.010	−0.003	−0.030	0.010	0.028	0.010	−0.021
20	$\overline{T}^2_{M\cdot c}$	64.559 (1.480)	76.850 (1.664)	86.224 (3.098)	91.044 (1.922)	102.561 (5.442)	118.913 (6.458)	134.804 (10.914)	140.042 (7.193)
	$T^{*2}_{M\cdot C}$	64.372	76.500	85.517	90.528	102.935	119.641	132.412	139.688
	e_5	0.187	−0.350	−0.707	−0.516	0.374	0.728	−2.392	−0.354
	re_5	0.003	−0.005	−0.008	−0.005	0.004	0.006	−0.018	−0.003
40	$\overline{T}^2_{M\cdot c}$	45.167 (0.859)	51.580 (0.834)	56.869 (0.910)	59.957 (0.983)	64.114 (2.793)	72.100 (3.137)	77.718 (3.038)	81.862 (2.889)
	$T^{*2}_{M\cdot C}$	44.860	51.681	56.744	59.613	64.285	72.160	78.139	81.595
	e_5	−0.307	0.101	−0.125	−0.344	0.171	0.060	0.421	−0.267
	re_5	−0.007	0.002	−0.002	−0.006	0.003	0.001	0.005	−0.003

Note. $\overline{T}^2_{M\cdot C} \equiv \overline{T}^2_{M\cdot C}(\alpha)$, the figures in the brackets : (the standard deviation of $\overline{T}^2_{M\cdot C}$)×10.

Suppose there are k treatments Π_i, $i = 1, \cdots, k$, a control Π_0 and their corresponding populations Π_i and Π_0 are normal, $N_p(\boldsymbol{\mu}_i, \Sigma)$, $i = 1, \cdots, k$ and $N_p(\boldsymbol{\mu}_0, \Sigma)$, respectively. On the basis of random sample of the same size m drawn independently from those populations, we calculate

$$\text{sample means: } \overline{\boldsymbol{x}}_i, \ i = 1, \cdots, k, \ \text{and } \overline{\boldsymbol{x}}_0,$$

$$\text{pooled sum of squares: } (k+1)(m-1)S = \sum_{i=0}^{k}\sum_{r=1}^{m}(\boldsymbol{x}_{ir} - \overline{\boldsymbol{x}}_i)(\boldsymbol{x}_{ir} - \overline{\boldsymbol{x}}_i)'.$$

Note that S has $(k+1)(m-1)$ degrees of freedom. The simultaneous confidence intervals on

$$\begin{gathered} \boldsymbol{a}'(\boldsymbol{\mu}_i - \boldsymbol{\mu}_0) \text{ for all nonzero, nonstochastic } \boldsymbol{a} \\ \text{and for all } i = 1, \cdots k \end{gathered} \tag{5.1}$$

are given by

$$\boldsymbol{a}'(\overline{\boldsymbol{x}}_i - \overline{\boldsymbol{x}}_0) - H^{\frac{1}{2}} \leq \boldsymbol{a}'(\boldsymbol{\mu}_i - \boldsymbol{\mu}_0) \leq \boldsymbol{a}'(\overline{\boldsymbol{x}}_i - \overline{\boldsymbol{x}}_0) + H^{\frac{1}{2}} \tag{5.2}$$

$$\text{for all } \boldsymbol{a}(\neq \boldsymbol{0}) \text{ and all } i = 1, \cdots, k,$$

$$H = \frac{2}{m}\tilde{T}^2_{M\cdot C}(\alpha)\boldsymbol{a}'S^{-1}\boldsymbol{a} \tag{5.3}$$

where $\tilde{T}^2_{M\cdot C}(\alpha)$ is the upper 100α percent point of

$$\tilde{T}^2_{\max\cdot C} \equiv \max_i\{(\frac{m}{2})(\overline{\boldsymbol{x}}_i - \overline{\boldsymbol{x}}_0 - \boldsymbol{\mu}_i + \boldsymbol{\mu}_0)'S^{-1}(\overline{\boldsymbol{x}}_i - \overline{\boldsymbol{x}}_0 - \boldsymbol{\mu}_i + \boldsymbol{\mu}_0)\}.$$

Note that $\boldsymbol{y}_i = \sqrt{m}(\boldsymbol{x}_i - \boldsymbol{\mu}_i)$, $i = 1, \cdots, k$ and $\Lambda = \Sigma$, $\nu = (k+1)(m-1)$. Then $\boldsymbol{y}_1$, $\boldsymbol{y}_2$, $\cdots$, $\boldsymbol{y}_k$ and $\boldsymbol{y}_0$ are independent normal variates having $N_p(\boldsymbol{0}, \Sigma)$ and

$$2\tilde{T}^2_{\max\cdot C} = \max_i\{(\boldsymbol{y}_i - \boldsymbol{y}_0)'S^{-1}(\boldsymbol{y}_i - \boldsymbol{y}_0)\} = T^2_{\max\cdot C}.$$

Hence

$$\tilde{T}^2_{M\cdot C}(\alpha) = \frac{1}{2}T^2_{M\cdot C}(\alpha).$$

Thus the modified second estimate $\tilde{T}^{*2}_{M\cdot C}(\alpha)$ of $\tilde{T}^2_{M\cdot C}(\alpha)$ can be obtained by

$$\tilde{T}^{*2}_{M\cdot C}(\alpha) = \frac{1}{2}T^{*2}_{M\cdot C}(\alpha).$$

Consequently, using $\tilde{T}^{*2}_{M\cdot C}(\alpha)$ instead of $\tilde{T}^2_{M\cdot C}(\alpha)$ in (5.3), the simultaneous confidence intervals in (5.2) have a simultaneous confidence coefficient approximately equal to $(1-\alpha)$.

In the similar way the simultaneous confidence intervals on all pairwise comparisons among mean vectors are given by using the modified second approximation for $R_{\max}$ by Seo and Siotani[7].

Acknowledgements

The authors would like to thank Professor Y. Fujikoshi, Hiroshima University, and the referee for their helpful comments. The authors also wish to thank Miss M. Takahashi for her help in numerical work.

REFERENCES

1. S. N. Roy and R. C. Bose, *Ann. Math. Stat.*, **24**, 513–536 (1953).
2. M. Siotani, *Ann. Inst. Stat. Math.*, **10**, 183–208 (1959).
3. M. Siotani, *Bulletin of the International Statistical Institute*, **38**, 591–599 (1961).
4. P. R. Krishnaiah, *in: Handbook of Statistics, Vol. I,* P. R. Krishnaiah (Ed.), pp. 745–971 North-Holland, Amsterdam (1980).
5. P. R. Krishnaiah, *in: Multivariate Analysis–II,* P. R. Krishnaiah (Ed.), pp. 121–143. Academic Press, New York (1969).
6. D. R. Jensen, *Biometrics*, **28**, 39–53 (1972).
7. T. Seo and M. Siotani, to appear in *J. Japan Statist. Soc.*, (1992).
8. M. Siotani, *Ann. Inst. Stat. Math.*, **11**, 167–182 (1960).
9. Z. Yamauti(Ed.), *Statistical Tables and Formulas with Computer Applications JSA-1972*, Japanese Standards Association (1972).

Stat. Sci. & Data Anal., pp. 277-290
K. Matsusita *et al.* (Eds)

Curvature Measures in Data Analysis

ROBERT E. KASS and ELIZABETH H. SLATE
Department of Statistics, Carnegie Mellon University, Pittsburgh, PA 15213

Abstract We show how the curvature measures used to study nonlinearity in nonlinear regression may be generalized to other models. We first briefly review some roles for curvature both in nonlinear regression (Bates and Watts[1]) and in general asymptotic inference based on α-connections (Amari[2]). We then indicate the relationship between these two separate areas of research and from this define data-analytical curvature measures for exponential family nonlinear models.

Key Words: α-connections; curved exponential family; exponential family nonlinear model; parameter-effects curvature; statistical curvature.

1. INTRODUCTION

Research concerning the effects of certain kinds of "curvature" on large-sample inference developed in two quite separate areas of statistics. One had to do with nonlinearity of nonlinear regression surfaces (Beale[3]; Bates and Watts[1]). The second began as an attempt to understand better the way exponential families play a special role in statistical theory (Efron[4]; Amari[2],[5]). Though it seemed natural to consider possible relationships between these two distinct lines of research, the methodological use of doing so was only fully articulated recently in a paper by Kass and Slate[6]. The emphasis in Kass and Slate[6] is on data analytical tools. Here, we will instead concentrate on the underlying geometrical concepts.

To summarize background material, we present the most basic facts from the geometry of large-sample Maximum Likelihood Estimation in Section 2.1, and from the geometry of nonlinear regression in Section 2.2. In Section 3, definitions of the α-connections introduced by Amari[2] are provided for completeness. We then show in Section 4 how the two areas of research may be brought together. First, by examining the formal relationship of the curvatures calculated in nonlinear regression to those computed in the general context of asymptotic inference. Here, we follow Kass[7]. Then, it may be observed that the "parameter-effects array", which is fundamental in the nonlinear regression analysis of Bates and Watts[1] and others, is a connection coefficient array and, thus, the α-connections could furnish a formal framework for generalization. This was implicit in Kass[8], but the methodological consequences were not investigated. We summarize the resulting data analysis techniques of Kass and Slate[6] in Section 5, and add a few further remarks in Section 6.

Notation:

In general, if β is a parameterization of a regular parametric family of densities and $\ell(\beta)$ is the loglikelihood function, then we will write the Fisher information matrix as $i(\beta)$ and define

$$g_{\lambda\mu} = i(\beta)_{\lambda\mu} = E(\frac{\partial \ell}{\partial \beta_\lambda}\frac{\partial \ell}{\partial \beta_\mu}).$$

The elements of the inverse of this matrix will be written

$$g^{\lambda\mu} = [i(\beta)^{-1}]_{\lambda\mu}. \tag{1.1}$$

2. GEOMETRY OF ASYMPTOTIC INFERENCE

2.1. Maximum likelihood estimation

An important class of models for geometrical analysis of large-sample statistical theory begins with densities belonging to a k-dimensional regular exponential family,

$$p(y \mid \eta) = \exp\{y\eta - \psi(\eta) + b(y)\} \tag{2.1}$$

where y and η are in R^k. If η may be written in terms of an m-dimensional parameter θ as $\eta = \eta(\theta)$, with the parameter space Θ being an open subset of R^m, (2.1) becomes

$$p(y \mid \theta) = \exp\{y\eta(\theta) - \psi(\eta(\theta)) + b(y)\}. \tag{2.2}$$

If the mapping $\theta \to \eta(\theta)$ from Θ into the natural parameter space N is one-to-one and infinitely differentiable, with full-rank Jacobian and an infinitely differentiable inverse mapping, then the family defined by (2.2) is an m-dimensional *curved exponential family.* See Kass[7] for examples, discussion, and references.

Efron[4] defined the *statistical curvature* of a one-dimensional family and showed some of the ways in which it is relevant to large-sample inference. Amari[2],[5] generalized and extended Efron's work using a differential-geometrical framework to handle multidimensional families. Perhaps the most basic way to think of statistical curvature is as a measure of the insufficiency of the MLE. This comes from considering the *information loss* of an estimator T, which is $ni(\theta) - i^T(\theta)$, where $ni(\theta)$ is the information matrix based on the full sample and $i^T(\theta)$ is the information provided by the estimator (the information matrix computed from the distribution of T instead of the full sample). This quantity would be zero if T were sufficient (as is the MLE for an exponential family) and would become infinite as $n \to \infty$ if T were inefficient. In one dimension, the limiting relative information loss of the MLE is

$$\lim i(\theta)^{-1}(ni(\theta) - i^T(\theta)) = \gamma^2. \tag{2.3}$$

Thus, the greater the statistical curvature, the more information is lost by the MLE when it is not sufficient.

In reviewing the main results from the work of Efron and others, Kass[7] emphasized scalar curvatures, partly to lead toward relationships with curvatures used in nonlinear regression. In particular, as a generalization of statistical curvature γ, Kass used

$$\gamma^2 = \sum_{a,b,c,d} g^{ac} g^{bd} \cdot < (\partial_{ab}\eta)_N, (\partial_{cd}\eta)_N >_{\eta}, . \tag{2.4}$$

where g^{ab} is defined by (1.1) and $(\cdot)_N$ indicates a normal component of the vector, computed with respect to the inner product defined by the $n \times n$ Fisher information matrix $i(\eta(\theta))$. With this definition of γ, the limiting relative information loss in the multidimensional case satisfies

$$\lim tr(i(\theta)^{-1}[ni(\theta) - i^T(\theta)]) = \gamma^2. \tag{2.5}$$

Statistical curvature is also related to the deficiency of an estimator and to large deviations of an estimator. In addition, it may be interpreted as the approximate coefficient of variation of observed information (fluctuating, asymptotically, about expected information) and provides the rate of convergence of the Fisher scoring algorithm (see Kass[7] for references).

Several other curvature measures are of interest. By permuting the indices in the inverse information matrices appearing in (2.4), we obtain an alternative reduction of the three-way array $(\partial_{ab}\eta)_N$,

$$m^2\bar{\gamma}^2 = \sum_{a,b,c,d} g^{ab} g^{cd} < (\partial_{ab}\eta)_N, (\partial_{cd}\eta)_N >$$

which will be interpreted below. Then, substituting the mean-value parameterization μ of the exponential family in place of η, and using the tangential components instead of the normal components, we have

$$\omega^2 = \sum_{a,b,c,d} g^{ac} g^{cd} < (\partial_{ab}\mu)_T, (\partial_{cd}\mu)_T >_{\mu(\theta)} \qquad (2.6)$$

and

$$m^2 \bar{\omega}^2 = \sum_{a,b,c,d} g^{ab} g^{cd} < (\partial_{ab}\mu)_T, (\partial_{cd}\mu)_T >_{\mu(\theta)} .$$

The latter is directly related to the asymptotic relative bias of the MLE, while ω appears in the formula for the limiting risk of the MLE (see Kass[7]).

2.2. Nonlinear Regression

The normal nonlinear regression model begins with a model function $f(\theta, x)$ for $\theta \in \Theta \subseteq R^m$ and then

$$Y_i = \eta_i(\theta) + \epsilon_i$$

with $\eta_i(\theta) = f(\theta, x_i)$ and $\epsilon_i \sim N(0, \sigma^2)$, independently, for $i = 1, \ldots, n$. Letting $\eta = (\eta_1, \ldots, \eta_n)$, $\eta(\theta)$ is an m-dimensional surface in R^n. The vector θ is estimated by the least-squares estimator $\hat{\theta}$, which is also the MLE, and approximate inference is carried out by proceeding as if the model were linear based on the linear approximation to the regression surface $\eta(\theta)$ at the least-squares fitted value $\eta(\hat{\theta})$. When the surface is approximately linear, the normality-based theory of ordinary least squares leads to confidence intervals and regions that have approximately correct coverage probabilities. When the surface is substantially nonlinear, the coverage probabilities will be faulty. Thus, measures of nonlinearity have been developed in the hope of identifying problematic situations.

The nonlinearity measures of Beale[3] and Bates and Watts[1] are summaries of second derivatives of the regression surface, which would be zero if the surface were exactly linear. The second derivatives, appropriately standardized, are decomposed into components that are tangential to the surface (at $\eta(\hat{\theta})$) and components that are normal to the surface. The standardization ensures certain invariance properties of the resulting measures. The measures based on the normal components become parameterization-invariant and thus can not be decreased by a transformation of parameters (such as taking logs of positive parameters). The nonlinearity measures based on tangential components *do* depend on nonlinear transformations, and thus indicate the possible improvement in inferences following reparameterization (again, as when logarithms are used). On the other hand, these measures are defined so as to remain affine invariant. This is important because an affine transformation of θ should not change the degree to which the surface is judged approximately linear (or $\hat{\theta}$ is judged approximately normal). We now provide the definitions of some of these measures.

The measure of parameterization-invariant nonlinearity proposed by Beale[3] is an average curvature among certain curves that slice through the regression surface at the least-squares fitted value $\eta(\hat{\theta})$. The curves have the form $c_v(t) = \eta(\hat{\theta} + tv)$ at $c(0) = \eta(\hat{\theta})$, where v is a nonzero vector in R^m, and have been called "lifted lines" by Bates and Watts[1]. Each such curve has a second derivative vector which decomposes into normal and tangential components $c_v''(0) = c_v''(0)_N + c_v''(0)_T$, and the curvature of the curve is

$$\kappa_N(v) = \| c_v'(0) \|^{-2} \cdot \| c_v''(0)_N \| . \tag{2.7}$$

This curvature may be considered a measure of curvature of the surface at $\eta(\hat{\theta})$ in the direction of the tangent vector $c_v'(0)$. Beale's measure is then

$$\gamma_{RMS}^2 = \frac{1}{S_m} \int_S (\kappa_N(v))^2 \cdot dS \tag{2.8}$$

where the integral is over the sphere $\{v : \| c_v'(0) \| = 1\}$, and $S_m = \pi^{\frac{m}{2}} / \Gamma(\frac{m}{2})$, the surface area of the unit sphere in R^m.

The tangential-component analogues begin with

$$\kappa_T(v) = \| c_v'(0) \|^{-2} \cdot \| (c_v''(0))_T \| \tag{2.9}$$

which may be averaged over the unit sphere, as in (2.8),

$$\omega_{RMS}^2 = \frac{1}{S_m} \int_S (\kappa_T(v))^2 \cdot dS. \tag{2.10}$$

An important computational simplification involves a transformation that has three steps. Let TM be the tangent plane to the surface $\eta(\theta)$ at $\hat{\theta}$, and $V = D\eta(\hat{\theta})$. Geometrically, the first step is to rotate the n-dimensional Euclidean coordinate system so that the first m coordinates span TM, and last $n - m$ are orthogonal to it. Algebraically this is accomplished via the QR-decomposition, with $QR = V$. The second step is to introduce a linear transformation of the parameter θ such that the vector derivatives of η with respect to the components of the new parameter ϕ coincide with the orthonormal basis for TM in the rotated coordinate system. This transformation is

$$\phi = R_1(\theta - \hat{\theta}) \tag{2.11}$$

where R_1 is the $m \times m$ upper portion of the $n \times m$ matrix R. Finally, the surface in these new coordinates and parameterization becomes

$$\xi(\phi) = Q^T(\eta(\hat{\theta} + R^{-1}\phi) - \eta(\hat{\theta}))$$

and the second derivative array A is defined by its components

$$a_{\lambda ab} = \frac{\partial \xi_\lambda}{\partial \phi_a \partial \phi_b}(\hat{\phi}) \tag{2.12}$$

with the index λ ranging over the m tangential components followed by the $n - m$ normal components as indicated in the notation

$$A = A^T | A^N.$$

We note that this definition does not include the normalizing factor $s \cdot m^{1/2}$, where s is the residual root-mean-square, used by Bates and Watts[1]. The curvature measures may be computed easily from the A^T and A^N arrays, using formulae given by Bates and Watts. We return to these below.

An empirical finding of several authors is that the parameterization-invariant measures are usually small. This means that, in practice, reparameterization can often substantially improve approximate inferences. This also leads to increased attention on the array A^T and resulting tangential-component measures such as ω_{RMS}. Bates and Watts[9] referred to the ideal parameterization as providing a "uniform coordinate system" for the regression surface, just as the Euclidean rectangular grid provides a uniform coordinate system in ordinary linear regression. They noted that large elements of the A^T array indicate departures from this nice uniformity that ensures good inferences.

3. α-CONNECTIONS

The α-connection coefficients are defined by

$$\overset{1}{\Gamma}_{\lambda\mu\nu}(\beta) = E(\frac{\partial^2 \ell}{\partial \beta_\lambda \partial \beta_\mu} \frac{\partial \ell}{\partial \beta_\nu})$$

$$\overset{-1}{\Gamma}_{\lambda\mu\nu}(\beta) = \overset{1}{\Gamma}_{\lambda\mu\nu}(\beta) + E(\frac{\partial \ell}{\partial \beta_\lambda} \frac{\partial \ell}{\partial \beta_\mu} \frac{\partial \ell}{\partial \beta_\nu})$$

and

$$\overset{\alpha}{\Gamma}_{\lambda\mu\nu}(\beta) = \frac{1-\alpha}{2} \overset{-1}{\Gamma}_{\lambda\mu\nu}(\beta) + \frac{1+\alpha}{2} \overset{1}{\Gamma}_{\lambda\mu\nu}(\beta).$$

Note that $\overset{1}{\Gamma}_{\lambda\mu\nu}$ and $\overset{-1}{\Gamma}_{\lambda\mu\nu}$ correspond to $\alpha = 1$ and $\alpha = -1$, respectively; these are the *exponential* and *mixture* connection coefficients. The $\overset{\alpha}{\Gamma}_{\lambda\mu\nu}$ coefficients are in covariant form. An alternative form includes one contravariant (or "raised") index and is defined by

$$\overset{\alpha}{\Gamma}{}^{\nu}_{\lambda\mu}(\beta) = \sum_{\kappa} g^{\nu\kappa}\, \overset{\alpha}{\Gamma}_{\lambda\mu\kappa}(\beta).$$

4. RELATIONSHIPS

4.1. Curvature measures in nonlinear regression

From the definition of the normal nonlinear regression model, the vector $Y = (Y_1, \ldots, Y_n)$ has a distribution belonging to the n-dimensional multivariate normal family $N(\eta, \sigma^2 \cdot I_n)$, where $\eta = (\eta_1, \ldots, \eta_n)$ is restricted to lie on an m-dimensional surface specified by $\eta = \eta(\theta)$. Thus, for each fixed σ, this family is the subfamily of the n-dimensional normal location family that consists of distributions with location parameter restricted by $\eta = \eta(\theta)$. Assuming regularity of the surface, therefore, for each fixed σ, the normal nonlinear regression model is a curved exponential family and general measures discussed in Section 2 may be then be applied in this special case. Since geometrical analysis usually ignores σ and considers only the nonlinear regression surface, let us do this here and assume σ is known.

There are two scalar summaries of the A^N array that play an important role in the geometry of surfaces. They are analogous to two forms of squared trace for a matrix h, $tr(hh^T)$ and $tr(h)^2$. Including a factor of σ^2 in their definition, they are

$$\gamma^2 = \sigma^2 \cdot \sum_{\lambda=m+1}^{n} \sum_{b,c} (a_{\lambda bc})^2 \tag{4.1}$$

and

$$\bar{\gamma}^2 = m^{-2}\sigma^2 \cdot \sum_{\lambda=m+1}^{n} \Big(\sum_{b} a_{\lambda bb}\Big)^2 \tag{4.2}$$

where λ is summed over the $n - m$ normal components. Here we should perhaps remark that the use of the factor m in the definition of $\bar{\gamma}$ but not γ occurs as a notational quirk because the latter is σ times what is usually called the *mean curvature* of the surface. In differential geometry, for $\sigma = 1$, the *Riemannian scalar curvature* is $r = m^2\bar{\gamma}^2 - \gamma^2$.

What is especially interesting is that the notation here and in Section 2.1 is not accidental: the quantities γ and $\bar{\gamma}$ here are special cases of the quantities defined there. In particular, γ is the *statistical curvature.*

The root-mean-squared curvature of Section 2.2 may now be rewritten in terms of these summaries of the A array according to the relation

$$m(m+2) \cdot \gamma_{RMS}^2 = m^2\bar{\gamma}^2 + 2\gamma^2. \tag{4.3}$$

(The corresponding equation in Kass[7] omitted the factor $m(m+2)$.)

As in (4.1) and (4.2), we define

$$\omega^2 = \sigma^2 \cdot \sum_{\lambda=1}^{m} \sum_{b,c} (a_{\lambda bc})^2 \tag{4.4}$$

and

$$\bar{\omega}^2 = m^{-2} \sigma^2 \cdot \sum_{\lambda=1}^{m} (\sum_{b} a_{\lambda bb})^2 \tag{4.5}$$

where λ is now summed over the m tangential components. We then have the tangential-component analogue of (2.8),

$$m(m+2) \cdot \omega_{RMS}^2 = m^2 \bar{\omega}^2 + 2\omega^2. \tag{4.6}$$

The quantities ω, $\bar{\omega}$, and ω_{RMS}, which are computed using (2.12), are invariant to affine transformations of the parameter space. Again, the point is that ω and $\bar{\omega}$ have interesting interpretations mentioned at the end of Section 2.1. Equation (4.6) helps connect the two distinct lines of research mentioned in sections 2.1 and 2.2.

4.2. Interpretation of α-connections

In the case of nonlinear regression, using the parameterization ϕ appearing in the definition (2.12), the α-connection coefficients become nothing other than components of the A^T array,

$$a_{cab} = \overset{\alpha}{\Gamma}{}^{\,c}_{ab}(\hat{\phi}).$$

This suggests the possibility of generalizing the A^T array by using the α-connection coefficients, and using α-connections to compute curvature measures. But is this an interesting idea for data analysis? What would be the interpretation of such an array?

In general, when connection coefficients satisfy

$$\Gamma^a_{bc}(\beta) = 0,$$

$a = 1, \ldots, m$; $b = 1, \ldots, m$; $c = 1, \ldots, m$ for some coordinate system β it means that, in terms of the geometry of the connection, the coordinate system is "uniform" in the sense used by Bates and Watts[9] noted at the end of Section 2.2. Departures from this "uniformity" would be reflected by large curvatures computed in the given geometry. Thus, if the special coordinate system β that possesses this "uniformity" is of interest, departures from it will be meaningful, and curvatures will potentially be useful.

In the case of the α-connections, it does turn out that for certain values of α the resulting parameterizations β are inherently interesting because of their statistical properties. This comes from the characterization by Hougaard[10] of various parameterizations using a second-order differential equation. He considered the case of one-parameter curved exponential families. Kass[8] showed that Hougaard's equation could be rewritten and generalized using the α-connections in terms of a parameter β of an m-dimensional family

$$\overset{\alpha}{\Gamma}{}^{a}_{bc}(\beta) = 0,$$

$a = 1,\ldots,m$; $b = 1,\ldots,m$; $c = 1,\ldots,m$. If the equations hold for $\alpha = 1$, 0, -1/3, -1 then β is, respectively, the natural parameterization, the asymptotic skewness-reducing parameterization, the variance-stabilizing parameterization, the asymptotic bias-reducing parameterization. These values of α occur in the work of Amari[2],[5]. In addition, and more importantly for the methodology of Kass and Slate[6], if the equations hold for $\alpha = \frac{1}{3}$ then β is the parameterization in which

$$E\left(\frac{\partial^3 \ell}{\partial\theta_a \partial\theta_b \partial\theta_c}\right) = 0$$

for all a, b, c. This parameterization is of special interest because the loglikelihood function in terms of β becomes approximately quadratic, so the likelihood function should be approximately normal.

When β satisfies these equations, it thus provides a "uniform coordinate system" in the sense of Bates and Watts[9] with respect to the α-connection geometry rather than the usual Euclidean geometry. Therefore, curvatures calculated in one of these geometries will be of interest. For example, large curvatures in the $\alpha = \frac{1}{3}$ geometry will indicate departures from "uniformity" representing deviations away from the "quadratic loglikelihood parameterization" (in which the expected third derivatives of the loglikelihood vanish). Thus, it becomes statistically meaningful to use the α-connection array in place of the A^T array for families other than nonlinear regression models. Simply put, this is the proposal made by Kass and Slate[6].

5. CURVATURE MEASURES FOR EXPONENTIAL FAMILY NONLINEAR MODELS

As we have said at the end of the previous section, we may interpret and compute curvatures from $\overset{\alpha}{\Gamma}{}^{c}_{ab}(\hat{\phi})$ by analogy with the interpretation of, and curvatures based on, A^T. Here, we specialize this idea to an important class of models.

Consider a set of densities

$$p^{(1)}(z \mid \nu, \sigma) = \exp\{[z\nu(\theta) - \psi(\nu)]/\sigma^2 + b(z,\sigma)\}$$

that, for fixed σ, form a one-dimensional regular exponential family with natural parameter ν. Taking n copies of such a family and writing $y = (y_1, ..., y_n)$ with y_i replacing z (for $i = 1, \ldots, n$) and $\eta = (\eta_1, ..., \eta_n)$ with η_i replacing ν (for $i = 1, \ldots, n$), the product family of densities $p(y \mid \eta, \sigma) = \Pi_{i=1}^{n} p^{(1)}(y_i \mid \eta_i, \sigma)$ is, for fixed σ, a regular exponential family of order n. We let the natural parameter space of this family be denoted by N. When $\eta_i = f(\theta; x_i)$ for some function f, parameter vector θ and explanatory variables x_i ($i = 1, \ldots, n$) we obtain an *exponential family nonlinear model*, which will have densities of the form

$$p(y \mid \eta, \sigma) = \prod_{i=1}^{n} \exp\{[y_i\eta_i(\theta) - \psi(\eta_i)]/\sigma^2 + b(y_i, \sigma)\}. \tag{5.1}$$

Generalized linear models (McCullagh and Nelder, 1989) are exponential family nonlinear models, and when the dispersion parameter σ is known, exponential family nonlinear models become curved exponential families. We assume the parameter space Θ is an open subset of R^m and $\theta \to \eta(\theta)$ is an embedding (i.e., the mapping from Θ into N is one-to-one and infinitely differentiable, with full-rank Jacobian and an infinitely differentiable inverse mapping).

As in the case of nonlinear regression, analysis of these models may be based on the embedding $\theta \to \eta(\theta)$, ignoring the presence of the parameter σ and effectively treating it as if it were known. The reason is that the estimating equations for θ and σ are separable and the multidimensional parameter θ is usually the parameter of interest while σ is a unidimensional nuisance parameter; once $\hat{\theta}$ is found, an estimate of σ may be obtained immediately from the deviance, and inference about θ proceeds by straightforward modification of the case in which σ is known. Thus, the geometry of these models we will consider will ignore σ, treating it as if it were known. We therefore write

$$\mathcal{Q}_0 = \{p(\cdot \mid \eta(\theta), \sigma) : \theta \in \Theta\},$$

and take

$$\mathcal{Q} = \{p(\cdot \mid \eta, \sigma) : \eta \in N\}$$

to be the unrestricted exponential dispersion model, speaking of θ and η as parameterizations for these models, ignoring σ. By construction, $\mathcal{Q}$ is an n-fold product of a one-dimensional regular exponential family with itself, and we let $\mathcal{Q}^{(1)}$ denote this one-dimensional exponential family.

Now suppose we have a parameterization τ of $\mathcal{Q}^{(1)}$. We can use this parameterization for each of the n copies of $\mathcal{Q}^{(1)}$ comprising $\mathcal{Q}$ to define a parameterization $\zeta = (\zeta_1, \ldots, \zeta_n)$ for $\mathcal{Q}$, which we will call the *product parameterization* based on τ. We then obtain an m-dimensional surface $\zeta(\theta)$ in R^n that represents $\mathcal{Q}_0$. The simple way to understand the method of Kass and Slate[6] is that we compute curvatures of the surface $\zeta(\theta)$ in very much the same way curvatures of a regression surface are computed in nonlinear regression. We carry out the steps leading to (2.12) using ζ in place of η, with one modification. In an exponential family nonlinear model, the variances of $\hat{\zeta}_1, \ldots, \hat{\zeta}_n$ may be inhomogeneous. To accommodate this, we weight the observations according to the Fisher information matrix $i(\zeta)$. Thus, letting $G = i(\zeta(\hat{\theta}))$, we begin with the decomposition

$$QR = G^{1/2} D\zeta(\hat{\theta}) \tag{5.2}$$

and take R_1 to be the upper $m \times m$ part of R. We then use (2.11) with this new definition of R_1, and write the surface in the rotated coordinate system as

$$\xi(\phi) = Q^T G^{1/2}(\zeta(\hat{\theta} + R_1^{-1}\phi) - \zeta(\hat{\theta})).$$

With this new definition of $\xi(\theta)$ we now apply (2.12) getting the computational formula

$$A_\zeta = [Q^T G^{1/2}][L^T D^2 \zeta(\hat{\theta}) L] \tag{5.3}$$

where $L = R_1^T$, the brackets again indicating multiplication of the three-way arrays, and the subscript ζ being is used to indicate explicit dependence on the choice of ζ. This new A-array may be decomposed according to the first m and last $n - m$ values of the index k as

$$A_\zeta = A_\zeta^T | A_\zeta^N,$$

representing tangential and normal components with respect to the inner product based on the Fisher information matrix G. We summarize, as follows.

METHOD FOR EXPONENTIAL FAMILY NONLINEAR MODELS. *For the product parameterization ζ of $\mathcal{Q}$ constructed as described above from a parameterization τ of $\mathcal{Q}^{(1)}$, we consider the array A_ζ to be the generalization of the A array. Once A_ζ is substituted for A, the calculation of curvatures proceeds exactly as in the nonlinear* *regression setting and we use the measures (4.1), (4.2), (4.4), and (4.5). Furthermore, if we define γ_{RMS} and ω_{RMS} analogously to the way they are defined in nonlinear regression, we again have the computational formulae (2.8) and (4.6).*

We have not yet said how this description is related to the α-connections or why it is reasonable. This comes from the following.

RESULT. For each α there exists a parameterization τ of $\mathcal{Q}^{(1)}$, unique up to an affine transformation, such that when the A_ζ^T array is computed by the method above in terms of the product parameterization ζ of $\mathcal{Q}$ defined from τ, the connection coefficients in terms of ϕ satisfy

$$\overset{\alpha}{\Gamma}{}^{c}_{ab}(\hat{\phi}) = a_{cab}.$$

This result says that when ζ is determined from α (via τ), the method above will result in curvature measures that are calculated with respect to the α-connection geometry. We have concentrated here on the A_ζ^T array, but a similar statement could be made about the A_ζ^N array. As far as the importance of this method is concerned, we believe it mainly consists in the use of the $\alpha = \frac{1}{3}$ connection, which leads to $E(\ell'''(\tau)) = 0$ (and, incidentally, $\ell'''(\hat{\tau}) = 0$), and thus its curvatures measure departures from what we may call the *quadratic loglikelihood parameterization.* As Slate[12] shows for natural exponential families having quadratic variance functions (NEF-QVF families), the likelihood is remarkably close to normal in terms of τ, often being satisfactorily normal for sample sizes as small as 1 or 2. Thus, if the curvatures computed in terms of the resulting A_ζ ($\alpha = \frac{1}{3}$) are small, we would expect the likelihood to be approximately normal. Expressions for the quadratic loglikelihood parameterization in the NEF-QVF families may be found in Slate[13].

Investigations in Slate[13] and examples in Kass and Slate[7] lead us to believe the method will be useful in extending nonlinearity measures used in nonlinear regression to the more general class of exponential family nonlinear models.

6. CLOSING COMMENTS

We have shown how curvature measures of nonlinearity used in nonlinear regression may be generalized to apply to other models. The mathematical technique is to base the curvature measures on α-connections. We focused on the important class of exponential family nonlinear models, where the generalization is conceptually clearest and computationally most easily carried out. Though the method is more general, we emphasized the use of studying departures from the "quadratic loglikelihood parameterization", using the α-connection with $\alpha = \frac{1}{3}$. Additional details and further justification may be found in Kass and Slate[6].

We did not have space here to consider the problem from a pure likelihood or Bayesian point of view. In fact, there are similar measures based on the observed third derivatives of the loglikelihood function (or log posterior) defined in Kass and Slate[6]. One of these is particularly easy to compute using interactive Bayesian methods given in Tierney, Kass, and Kadane[13] and implemented in LISP-STAT (Tierney[14]. Further discussion concerning diagnostics and the role of reparameterization in Bayesian analysis may be found in Kass and Slate[15].

REFERENCES

1. Bates, D. M. and Watts, D. G. (1980). Relative curvature measures of nonlinearity, *J. Royal Statist. Soc.*, B, **42**:1-25.

2. Amari, S.-I. (1982). Differential geometry of curved exponential families – curvatures and information loss, *Ann. Statist.* **10**: 357-387.

3. Beale, E.M.L. (1960). Confidence regions in non-linear estimation, *J. Royal Statist. Soc.* B, **22**: 41-88.

4. Efron, B. (1975). Defining the curvature of a statistical problem (with applications to second-order efficiency). *Ann. Statist.* **3**, 1189-1242.

5. Amari, S.-I. (1985). Differential-geometric methods in statistics. *Lecture Notes in Statistics*, **28**, New York: Springer.

6. Kass, R.E. and Slate, E.H. (1993). Some diagnostics of maximum likelihood and posterior non-normality. *Ann. Statist.*, to appear.

7. Kass, R.E. (1989). The geometry of asymptotic inference (with discussion), *Statist. Sci.*, **4**: 188-234.

8. Kass, R.E. (1984). Canonical parameterizations and zero parameter-effects curvature, *J. Royal Statist. Soc.* B, **46**: 86-92.

9. Bates, D.M. and Watts, D.G. (1981). Parameter transformations for improved approximate confidence regions in nonlinear least squares. *Ann. Statist.*, **9**, 1152-1167.

10. Hougaard, P. (1982). Parameterizations of non-linear models. *J. Royal Statist. Soc. B*, **44**, 244-252.

11. McCullagh, P. and Nelder, J.A. (1983). Generalized Linear Models, New York: Chapman & Hall.

12. Slate, E.H. (1991). "Reparameterization of Statistical Models," Ph.D. Thesis, Department of Statistics, Carnegie Mellon University.

13. Kass, Tierney, L. and Kadane, J.B. (1991). "Asymptotic evaluation of integrals arising in Bayesian statistics," in *Computing Science and Statistics: Proceedings of the 22nd Symposium on the Interface*, eds. C. Page and R. LePage, pp. 38-42.

14. Tierney, L. (1991). "LISP-STAT" John Wiley & Sons, New York.

15. Kass, R.E. and Slate, E.H. (1992) "Reparameterization and diagnostics of posterior non-normality," *Bayesian Statistics 4*, edited by J.O. Berger, J.M. Bernardo, A.P. Dawid, D.V. Lindley and A.F.M. Smith, Oxford University Press.

Stat. Sci. & Data Anal., pp. 291-299
K. Matsusita *et al.* (Eds)

Statistical Inference from Observations with Censoring and Grouping for Exponential Families

SHINTO EGUCHI

Department of Mathematics, Faculty of Sciences, Shimane University, Mastue 690 Japan

Abstract. This paper discusses the maximum likelihood method for a censoring and grouping model from a differential geometric viewpoint. In an independent contingency table with grouping the *EM* algorithm is obtained and several numerical examples are given. A simple exponential model is presented to recover the loss of information causing from grouping and censoring.

Keywords: Censoring and grouping model, contingency table, differential geometric method, *EM* algorithm, exponential family, information loss.

1. INTRODUCTION

This paper is concerned with a problem of statistical inference based on incomplete observations via grouping and censoring, see Blight[1]. A sample space is assumed to be partitioned into an uncensored subspace and m subspaces which are censored except for reporting which space the observation is in. Thus only the counts of the grouped subspace is observed if the observation falls into one of censored subspaces. We concentrate on the case where the original sample can be reduced to a sufficient statistic, so that the order of statistic is inflated by m because of grouping to m subspaces. The differential geometric aspects of statistical inference are discussed in the light of the embedding of the resulting parametric model into a higher-order exponential family. Furthermore we discuss the *EM* algorithm for obtaining the maximum likelihood estimate.

Section 2 discusses the censoring and grouping model in two-way contingency tables. A simple derivation of the *EM* algorithm is presented by the Lagrange multiplier method. Section 3 gives an example in which the information loss via censoring and grouping is recovered by designing the number of grouping with the sample size.

2. GROUPING AND CENSORING MODEL

We assume that the underlying distribution is of exponential type, or a random vector X has a distribution with density

$$p(x:\theta) = \exp[x'\theta - \psi(\theta)]$$

Here the k-parameter is called the natural parameter, while the expectation parameter η is defined by the relation $\eta = E(\boldsymbol{X})$, so that $\eta = (\partial/\partial\theta)\psi(\theta)$. For a partition $\Omega_0 \cup \cdots \cup \Omega_m$ of the sample space Ω we suppose that the actual observation is complete only on the subspace Ω_0 and is censored on other spaces $\Omega_1 \cup \cdot \cdot \cdot \cup \Omega_m$ except for observing the number of the subspace included the observation. Based on N observations $\{(\boldsymbol{x}_\alpha) : 1 \leq \alpha \leq N\}$ the sufficient statistic $\boldsymbol{x} = \Sigma_\alpha \boldsymbol{x}_\alpha / N$ is reduced to the actual sufficient statistic $\boldsymbol{y} = (\boldsymbol{x}_0, p_1, \cdots, p_m)$, where $\boldsymbol{x}_0$ is the mean vector of uncensored observations and p_i is the observed frequency of counts falling into the i-th space Ω_i. Thus the log-likelihood function is decomposed into $l(\theta) = l_1(\theta) + l_2(\theta)$, where

$$l_1(\theta) = \bar{x}_0'\theta - \bar{p}_1\psi_0(\theta) \quad \text{and} \quad l_2(\theta) = \sum\nolimits^m{}_{i=0} p_i\psi_i(\theta) - \psi(\theta),$$

where

$$\psi_i(\theta) = \log \int_{\Omega_i} \exp(\boldsymbol{x}'\theta)\, d\mu(\boldsymbol{x}).$$

Here $l_1(\theta)$ is based on the uncensored observations in Ω_0 and $l_2(\theta)$ is the multinominal log-likelihood with cell probabilities $\pi_i(\theta) = \exp[\psi_i(\theta) - \psi(\theta)]$ from the observed frequencies (p_i) by grouping the partition. Hence the parametric family can be regarded as a subfamily embedded into an k+m order exponential family by regarding cell probabilities (π_i) as m free parameters in addition to the original parameter θ. The information matrix $I_\theta(\boldsymbol{y})$ with $\boldsymbol{y}$ is given by

$$I_\theta(\boldsymbol{y}) = \partial\partial'\psi(\theta) - \Sigma^m{}_{i=1}\pi_i(\theta)\partial\partial'\psi_i(\theta),$$

which is decomposed into

$$I_\theta(\boldsymbol{x}_0) = \pi_0(\theta)\partial\partial'\psi_0(\theta)$$

and

$$I_\theta(p) = \partial\partial'\psi(\theta) - \sum_{i=0}^{m} \pi_i(\theta)\partial\partial'\psi_i(\theta),$$

where $p = (p_1, \cdots, p_m)'$ and ' denotes matrix transpose. It is shown in the appendix that the information matrix with p can be rewritten as

$$I_\theta(p) = \sum_{i=0}^{m} \pi_i(\theta)(\eta_i - \eta)(\eta_i - \eta) \tag{1.2}$$

which is the variance matrix of a discreet variable Z with probability function defined by $\text{Prob}(Z = \eta_i) = \pi_i(\theta)$ $(i = 0, \cdots, m)$. In this way the variance matrix $\partial\partial'\psi(\theta)$ of x, say $I_\theta(x)$, is reduced to $I_\theta(p)$ via grouping the partition. The relative loss of information in reducing from the unobservable statistic x to the actual statistic y is given by

$$\Delta(\theta) = I_\theta(x)^{-1/2} I_\theta(y) I_\theta(x)^{1/2}$$

$$= [\partial\partial'\psi(\theta)]^{-1/2}[\Sigma^m{}_{j=1}\, \pi_j(\theta)\partial\partial'\psi_j(\theta)]\, [\partial\partial'\psi(\theta)]^{-1/2},$$

where $A^{1/2}$ denotes the Gramian square root for A positive semidefinite. The exponential embedding curvature is defined as

$$H^{(e)}(\theta) = E\{(\partial\partial' l + I_\theta(y)) I_\theta(y)^{-1} (\partial\partial' l + I_\theta(y))\}$$

$$- [E\partial\partial' l \otimes \partial l]\{I_\theta(y) \otimes I_\theta(y)\}^{-1}[E\partial l \otimes \partial\partial' l]$$

see page 49-54 Amari[4], where E denotes the expectation and $\otimes$ the Kronecker product. Note that the skewness tensor vanishes everywhere and that the exponential connection is reduced to that for the multinominal log-likelihood $l_2(\theta)$. Thus we get

$$H^{(e)}(\theta) = \sum_{j=1}^{m} \pi_j\, \partial\partial'\psi_j\, I_y^{-1}\, \partial\partial'\psi_j - \sum_{j=1}^{m} \pi_j\, \partial\partial'\psi_j\, I_y^{-1} \sum_{j=1}^{m} \pi_j\, \partial\partial'\psi_j$$

$$- \sum_{i=0}^{m} \pi_i\, \partial\partial'\psi_i \otimes (\eta_i - \eta)\, (I_y \otimes I_y)^{-1} \sum_{i=0}^{m} (\eta_i - \eta) \otimes (\pi_i\, \partial\partial'\psi_i).$$

In general this expresses the lower bound for the limiting loss of information in reducing from y to a first-order efficient estimator, which the maximum likelihood estimator (*mle*) attains.

Exponential Distribution

Specifically we discuss a type I censorship model for a exponential variable X with density

$\theta\exp(-\theta x)$ on a space of positive numbers. Let X be assumed top be censored if $X > c$ with a fixed time c. The log-likelihood function is thus

$$l(\theta) = -\theta\bar{x}_0 + \bar{p}_0 \log\theta - \bar{p}_1 c\theta,$$

from which it follows that the information is

$$I_y(\theta) = (1 - \exp(-c\theta))\theta^{-2}$$

and that the embedding curvature is

$$H^{(e)}(\theta) = \frac{e^{-c\theta}}{\theta^2} - \frac{c^2 e^{-c\theta}}{(1 - e^{-c\theta})^2}$$

Consequently we observe that $I_y(\theta)$ rapidly converges to $I_x(\theta)$ with exponential order and also $H^{(e)}(\theta)$ vanishes as c increases.

We now consider the *EM* algorithm for obtaining the *mle*, see Blight[1] and Dempster et al. [3]. With a one-to-one transformation of parameters the *EM* algorithm has the same rule as the *mle* at each iteration step. Hence the convergence property is also parametrization invariant. he conjugate form consists of the E-step: $\eta = \bar{x}_0 + \Sigma\bar{p}_j\, \eta_j(\theta)$ and the M-step by $\theta = (\partial\psi)^{-1}(\eta)$, where $\eta_j(\theta) = \partial\psi_j(\theta)$. Accordingly one iteration forms as

$$H(\eta) = \bar{x}_0 + \Sigma\bar{p}_j\, \eta_j((\partial\psi)^{-1}(\eta)).$$

Note that

$$\partial^* H(\eta) = \Sigma\bar{p}_j\, \partial\partial\, \psi_j\, (\partial\partial'\, \psi(\theta))^{-1}.$$

A sufficient condition for the global convergence of the iteration process $\{H^k(\eta_0)\}_{k\geq 1}$ is that the spectral radius of $\partial^* H(\eta)$ is less than 1 for all η, see Ortega and Rheinboldt[4].

The *mle* $\hat{\theta}$ is characterized as a fixed point of the mapping H, which is guaranteed to be unique under the sufficient condition. The statistic $\bar{p}_j\, \eta_j(\hat{\theta})$ is regarded as the predictor for the missing value

$$\frac{1}{N} \sum_{x_\alpha \in \Omega_j} x_\alpha$$

for $j = 1,\dots, m$. Consequently the ideal statistic $\bar{x}$ is predicted as $\partial\eta(\hat{\theta})$, or $\bar{x}_0 + \Sigma\bar{p}_j\ \eta_j(\hat{\theta})$ with variance matrix $I_x(\theta)\ I_y(\theta)^{-1}I_x(\theta)$. The estimated embedding curvature based on the *mle* $\hat{\theta}$ is given by

$$H^{(e)}(\theta) = \sum_{j=1}^{m} \hat{\pi}_j \partial\partial'\hat{\psi}_j \hat{I}_y^{-1} \partial\partial'\hat{\psi}_j - \sum_{j=1}^{m} \hat{\pi}_j \partial\partial'\hat{\psi}_j \hat{I}_y^{-1} \sum_{j=1}^{m} \hat{\pi}_j \partial\partial'\hat{\psi}_j,$$

where the hat denotes the evaluation of θ as $\hat{\theta}$.

3. INDEPENDENT CONTINGENCY TABLE

We discuss a model with censoring and grouping in contingency tables. Let R *and* C be two independent response variable with m+1 levels and n+1 levels, respectively. Thus the $(m+1)(n+1)$ observed counts x_{ij} for the evens $\{R = i, C = j\}$ $(0 \le i \le m,\ 0 \le j \le n)$, so that the log-likelihood function is written as

$$l^*(\boldsymbol{p},\boldsymbol{q}) = \sum_{i=0}^{m} x_{i\cdot} \log p_i + \sum_{j=0}^{n} x_{\cdot j} \log q_j,$$

where $x_{i\cdot} = \Sigma_j x_{ij}$, $x_{\cdot j} = \Sigma_i x_{ij}$ with the marginal cell probability vectors (p_i) and (q_j).

We now consider the case of grouping and the observations via partition

$$K(0) \cup \dots \cup K(r)$$

of the sample space $\{0, \dots, m\} \times \{0, \dots, n\}$. Here we can observe only the counts $\boldsymbol{y} = (y_a)$ for the events $\{(R,C) \in K(a)\}$ $(0 \le a \le r)$ and the likelihood based on $\boldsymbol{y}$ becomes $l(\boldsymbol{p},\boldsymbol{q}) = \Sigma y_a \log \pi_a$ with

$$\pi_a = \Sigma_{(i,j) \in K(a)}\, p_i\, q_j.$$

If the partition is columnwise designed, or $K(a) = \{(a,j) \mid 0 \le j \le n\}$ $(a = 0, \dots, m)$, then the *mle* of p_i is $y_{i\cdot}$ and all q_j's are unestimable. Hereafter we assume all p_i's and q_j's to be estimable. A sufficient condition is that the Jacobi matrix

$$\left[\frac{\partial \pi_a}{\partial p_i}, \frac{\partial \pi_a}{\partial q_j}\right]_{1 \le a \le r,\ 0 \le i \le m,\ 0 \le j \le n}$$

is of rank $m+n$ for all $\boldsymbol{p}$ and $\boldsymbol{q}$, which implies that the parametric model is regarded as a curved exponential subfamily in multinominal distributions with r grouped cells.

By the use of Lagrange's method we give the conjugate form of the *EM* algorithm: The differentiation of the objective function $l + \lambda\,(1 - \Sigma_0{}^m p_i) + \mu\,(1 - \Sigma_0{}^n q_j)$ with p_i and q_j leads to

$$\sum_{a \in P(i)} \sum_{j \in J_a(i)} \frac{y_a}{\pi_a} q_j = \lambda \qquad (i = 0, \ldots, m) \tag{2.1}$$

and

$$\sum_{a \in Q(j)} \sum_{i \in I_a(j)} \frac{y_a}{\pi_a} p_i = \mu \qquad (j = 0, \ldots, n)\,,$$

where $J_a(i) = \{j \mid (i,j) \in K(a)\}$, $P(i) = \{a \mid J_a(i) \neq \phi\}$, $I_a(j) = \{i \mid (i,j) \in K(a)\}$ and $Q(i) = \{a \mid I_a(i) \neq \phi\}$. By multiplying both sides of equation (2.1) with p_i for $i = 0, \ldots, m$ we get the sum

$$\lambda = \lambda \Sigma_0{}^m\, p_i = x_{..} \qquad \text{(the total count)}$$

and similarly $\mu = x_{..}$. The conjugate form $H(\boldsymbol{p},\boldsymbol{q}) = (p_i{}^*, q_j{}^*)_{i,j}$ is given by

$$p_i^* = \sum_{a \in P(i)} \sum_{j \in J_a(i)} \frac{y_a}{x_{..}} \frac{p_i q_j}{\pi_a} \qquad \text{and} \qquad q_i^* = \sum_{a \in Q(i)} \sum_{i \in I_a(j)} \frac{y_a}{x_{..}} \frac{p_i q_j}{\pi_a}.$$

We can proceed a similar discussion for a case of multiway table.

Example. Let a 3×3 table be grouped into 6 categories as follows:

$$K(1) = \{\,(1,1)\},\ K(2) = \{(1,2),\ (2,1)\},\ K(3) = \{(3,1),(2,2)\},\ K(4) = \{(13)\}$$

$$K(5) = \{(2,3),(3,2)\} \quad \text{and} \quad K(6) = \{(3,3)\}.$$

Then for the observed frequency y_a falling into $K(a)$ $(a = 1, \ldots, 6)$ the *EM* algorithm is given by

$$p_1^* = y_1 + y_2 \frac{p_1 q_2}{p_1 q_2 + p_2 q_1} + y_4$$

$$p_2^* = y_2 \frac{p_1 q_2}{p_1 q_2 + p_2 q_1} + y_3 \frac{p_3 q_2}{p_3 q_1 + p_2 q_2} + y_5 \frac{p_2 q_3}{p_3 q_1 + p_2 q_2}$$

$$q_1^* = y_1 + y_2 \frac{p_2 q_1}{p_1 q_2 + p_2 q_1} + y_3 \frac{p_3 q_1}{p_3 q_1 + p_2 q_2}$$

and

$$q_2^* = y_2\frac{p_1q_2}{p_1q_2+p_2q_1} + y_3\frac{p_2q_2}{p_1q_2+p_2q_1}$$

with $p_3^* = 1 - p_1^* - p_2^*$ and $q_3^* = 1 - q_1^* - q_2^*$. In this way the form is simple, see the following table for a numerical performance.

Table. Numerical examples

Case 1	
the observed	y = (.008, .046, .118, .064, .302, .464)
true value	(p_1, p_2) = (.100, .200), (q_1, q_2) = (.100, .200)
complete *mle*	(p_1, p_2) = (.094, .220), (q_1, q_2) = (.102, .224)
incomplete *mle*	(p_1, p_2) = (.090, .257), (q_1, q_2) = (.106, .189)
iteration number	1409
Case 2	
the observed	y = (.368, .218, .086, .220, .080, .030)
true value	(p_1, p_2) = (.700, .200), (q_1, q_2) = (.500, .200)
complete *mle*	(p_1, p_2) = (.724, .170), (q_1, q_2) = (.508, .184)
incomplete *mle*	(p_1, p_2) = (.697, .208), (q_1, q_2) = (.528, .159)
iteration number	86
Case 3	
the observed	y = (.054, .100, .026, .548, .208, .066)
true value	(p_1, p_2) = (.700, .200), (q_1, q_2) = (.100, .100)
complete *mle*	(p_1, p_2) = (.680, .236), (q_1, q_2) = (.084, .108)
incomplete *mle*	(p_1, p_2) = (.682, .241), (q_1, q_2) = (.079, .109)
iteration number	17
Case 4	
the observed	y = (.082, .188, .210, .116, .262, .144)
true value	(p_1, p_2) = (.300, .300), (q_1, q_2) = (.300, .300)
complete *mle*	(p_1, p_2) = (.288, .328), (q_1, q_2) = (.302, .298)
incomplete *mle*	(p_1, p_2) = (.275, .386), (q_1, q_2) = (.293, .284)
iteration number	194

Here the observed vales are simulated by 500 random numbers from the respective true values and the iteration rule is decided to be stopped until the sum of absolute values of the difference between the present ad the next step is less than 10^{-6}. *The stating point of* (p_1, p_2, q_1, q_2) is adopted as (.333, .333, .3333, .333) *for all the cases.*

According to the table we observe that the *EM* algorithm enjoys a mild convergence to the incomplete *mles* for all the cases. In the case 1 the iteration number may seem to be larger but the actual time to be convergent is small because one iteration form is quite simple.
Alternatively the Newton-Raphthon algorithm may become complicated with subroutine of the inverse matrix.

4. EXPONENTIAL DISTRIBUTION WITH GROUPING

We discuss the loss of information because of censoring and grouping, which is given by (1.3) when the number of m of grouping increases as the size N of sample. A partition is given to make the loss of information vanish as follows. Let X have an exponential distribution with the hazard ratio θ. Define a grouping model with a partition $\{I_i\}_{i\in T}$ of $[0,\infty)$ with $[t_i, t_{i+1})$. For some j, the observation is complete only on I_j and elsewhere there is grouping. The likelihood function is therefore

$$l(\theta) = \exp[\, y_0 - \sum_{i\neq j} p_i\{t_i\theta - \log(1 - \exp((t_i - t_{i+1})\theta))\}]$$

and hence the *EM* algorithm is gin by the conjugate form

$$H(\eta) = y_0 - \Sigma_{i\neq j}\, \{p_i\, t_i - \eta) - (t_i - t_{i+1})/\{\eta(1 - \exp[-(t_i - t_{i+1})\theta])\}$$

with the expectation parameter η with relation to $\eta = -1/\theta$. As a specific case of partition $\{I_i^m\}_{1\leq i\leq m}$ we define as $I_i^m = [i/\sqrt{m},\ (i+1)/\sqrt{m})$ for $i \leq m-1$ and $I_m^m = [(m+1)/\sqrt{m},\ \infty)$. The cell probability in I_i^m is thus

$$\pi_{im}(\theta) = \begin{cases} \exp[-i\theta/\sqrt{m}](1 - exp[-\theta/\sqrt{m}]) & (i \leq m-1) \\ \exp[-\theta/\sqrt{m} - \theta\sqrt{m}] & (i = m) \end{cases}$$

and the cumulant transfration is

$$\psi_{im}(\theta) = \log \int_{I_i^m} \frac{1}{\theta} \exp[-x\theta]\, dx$$

$$= \begin{cases} -i\theta/\sqrt{m} - \log(1 - exp[-\theta/\sqrt{m}])/\theta & (i \leq m-1) \\ \exp[-\theta/\sqrt{m} - \theta\sqrt{m}] & (i = m) \end{cases}.$$

Consequently the limiting loss of information is given by

$$\Delta(\theta) = \lim_{N\to\infty} \Sigma_1^m\, \pi_{im}(\theta)\, \partial\partial'\, \psi_{im}(\theta)$$

which vanishes for al θ if $m = O(N^\delta)$ with constant $\delta > 0$. Thus we can conclude that the *mle* based on grouped observations recover the efficiency relative to the complete observation case.

REFERENCES

1.Blight, B.J. (1970), Biometrika 57, 389-395.

2. Amari, S. (1985), Lecture Note in Statistics 28, Springer-verlag, New York.

3. Dempster, A.P., N.M. Laird and D.B. Rubin (1977), Jour. Roy. Statist. Soc. B 39, 1-38.

4. Ortega, J.M. and W.C. Rheinboldt (1974), Academic Press, New York.

Appendix

Proof of (2.1). By definition,

$$\exp[\psi(\theta)] = \Sigma_0{}^m \exp[\psi_i(\theta)]$$

of which first and second-order differentiation leads to

$$\eta = \Sigma_0{}^m \pi_i(\theta)\eta_i$$

and

$$\partial\partial' \psi(\theta) = \partial\eta' = \Sigma_0{}^m \pi_i(\theta)\partial\partial' \psi_i(\theta) + \Sigma_1{}^m \pi_j(\theta)\eta_j\eta_j' - \eta\eta' .$$

The substitution of the last equation (1.1) leads to (1.2).

Stat. Sci. & Data Anal., pp. 301-311
K. Matsusita *et al.* (Eds)

Pitman Closeness of the Equivariant Shrinkage Estimators of the Normal Variance

NARIAKI SUGIURA
Department of Mathematics, University of Tsukuba, Tsukuba, Ibaraki 305 Japan

Abstract. It is shown that the best affine equivariant estimator for estimating the variance in a normal linear model under squared error loss or under entropy loss, is closer than every non-truncated equivariant shrinkage estimator. Pitman's probability of closeness of Brewster-Zidek estimator with respect to the best affine equivariant estimator is shown to be considerably smaller than 1/2 except for large sample sizes. It is also shown by numerical example that the best affine equivariant estimator is not always closer than the Stein estimator, which is a truncated scale-equivariant shrinkage estimator.

Key Words : Variance of normal population, Best affine equivariant estimator, Stein estimator, Brewster-Zidek estimator, Squared error loss, Entropy loss.

1. INTRODUCTION

A canonical form of normal linear model is given by random observable vector $X(r \times 1)$ having normal distribution $N(\mu, \sigma^2 I_r)$ and the error sum of squares S where S/σ^2 has chi-square distribution with n degrees of freedom. X and S are assumed to be independently distributed. Let $L(d,\sigma^2)$ be a loss function for estimating σ^2, based on X and S with unknown mean vector μ in r dimensional euclidean space. In this paper, we shall only consider squared error loss $L_s(d,\sigma^2)= (d/\sigma^2-1)^2$ or entropy loss $L_e(d,\sigma^2)=d/\sigma^2 - \log(d/\sigma^2) - 1$ as a loss function. The best affine equivariant estimator(=best scalar multiple of S) is known to be $d_{BAE}(S)= S/(n+2)$ under L_s and $d_{BAE}^*(S)=S/n$ under L_e. Stein estimator is given by $d_{Stein}(X,S) = \min\{ d_{BAE}(S), (S+X'X)/(n+r+2)\}$ under L_s, which is better than $d_{BAE}(S)$ as was proved by Stein[1]. Similarly under L_e, it is given by $d_{Stein}^*(X,S) = \min \{d_{BAE}^*(S), (S+X'X) /(n+r)\}$ which is better than $d_{BAE}^*(S)$. Any equivariant estimator for the transformation $(X,S)\rightarrow(cHX, c^2S)$ for any nonzero real number c and any r x r orthogonal matrix H is expressed by $d(X,S)=d_{BAE}(S)\phi(T)$, where $\phi(T)$ is any function of $T=S/(S+X'X)$. We shall call $d_{BAE}(S)\phi(T)$ a nontruncated equivariant shrinkage estimator if $P(0<\phi(T)<1)=1$ and a truncated equivariant shrinkage estimator if $P(0<\phi(T)\leq 1)=1$ and $P(\phi(T)=1) > 0$. Obviously Stein estimator is a truncated equivariant shrinkage estimator.

In this paper we shall show that the Pitman probability of closeness of any given nontruncated equivariant shrinkage estimator $d_{BAE}(S)\phi(T)$ with respect to $d_{BAE}(S)$ defined by $P[L(d_{BAE}(S)\phi(T), \sigma^2) < L(d_{BAE}(S),\sigma^2)]$ is less than 1/2 for all σ^2, n and r. If $P[L(d_0(X,S),\sigma^2) < L(d(X,S), \sigma^2)] >1/2$ holds for two nontruncated equivariant shrinkage estimators $d_0(X,S)$ and $d(X,S)$, we shall say that the estimator $d_0(X,S)$ is closer than $d(X,S)$ under loss function L. This slightly extended notion of Pitman[2] nearness were used by Sugiura[3] and Peddada[4] among others. Brewster-Zidek estimator is a nontruncated equivariant shrinkage estimator, which is better than

$d_{BAE}(S)$ under L_s or better than $d_{BAE}^*(S)$ under L_e as was proved by Brewster and Zidek[5] and has the following expression due to Rukhin[6] under L_s:

$$\phi(T) = 1 - \frac{2}{n+r+2} \frac{T^{n/2+1}(1-T)^{r/2}}{\int_T^1 z^{n/2+1}(1-z)^{r/2-1} dz} \tag{1.1}$$

which is equivalently written by

$$= 1 - \frac{r}{n+r+2} \frac{T^{n/2+1}}{{}_2F_1(\frac{r}{2};-\frac{n+2}{2};\frac{r+2}{2};1-T)}. \tag{1.2}$$

The last expression is useful for numerical computation, where ${}_2F_1$ stands for the hypergeometric function defined generally by

$$ {}_pF_q(a_1,..a_p;b_1,...,b_q;x) = \sum_{k=0}^{\infty} \frac{(a_1)_k \cdots (a_p)_k}{(b_1)_k \cdots (b_q)_k} \frac{x^k}{k!} \quad (p \le q+1) \tag{1.3}$$

with $(a)_k = a(a+1)...(a+k-1)$. The hypergeometric function ${}_0F_1$ and ${}_1F_1$ will also be used in this paper. Under L_e, Brewster-Zidek estimator has a form $d_{BAE}^*(S)\phi^*(T)$ where

$$\phi^*(T) = 1 - \frac{2}{n+r} \frac{T^{n/2}(1-T)^{r/2}}{\int_T^1 z^{n/2}(1-z)^{r/2-1} dz} \tag{1.4}$$

$$= 1 - \frac{r}{n+r} \frac{T^{n/2}}{{}_2F_1(\frac{r}{2};-\frac{n}{2};\frac{r+2}{2};1-T)}. \tag{1.5}$$

Our numerical examples show that Pitman probability of closeness of Brewster-Zidek estimator with respect to $d_{BAE}(S)$ is considerably smaller than 1/2 under L_s, but that of $d_{BAE}^*(S)$ under L_e is not so small. For truncated equivariant shrinkage estimator, the Pitman probability of closeness of $d_{BAE}(S)\phi(T)$ with respect to $d_{BAE}(S)$ should be defined by

$$P[L(d_{BAE}(S)\phi(T),\sigma^2) < L(d_{BAE}(S),\sigma^2)] + \frac{1}{2} P[\phi(T)=1] \tag{1.6}$$

as is noted by Nayak[7] and Kubokawa[8]. This is not always less than 1/2, which is provided by Stein estimators. Our numerical examples show that it is less than 1/2 in most cases but slightly larger than 1/2 for moderately large λ under L_s and that it is larger than 1/2 in most cases but slightly less than 1/2 for small λ under L_e. Kubokawa[8,9] proved that the Pitman probability of closeness of the Stein type truncated estimator with respect to the closest scalar multiple of S(=S/med (χ_n^2)) is always larger than 1/2, where med(χ_n^2) stands for the median of the chi-square distribution with n degrees of freedom, in a more general problem of estimating the generalized variance. Recent paper by Maatta and Casella[10] on estimating the variance of a normal population may be noted. It is known by them or Sugiura and Konno[11,12] that the reduction of risk of these improved estimators is substantial, when r is large.

2. PITMAN CLOSENESS OF EQUIVARIANT SHRINKAGE ESTIMATOR

The better estimators than the best affine equivariant estimator in the sense of minimizing risk, were found within a class of equivariant estimators given by $d_{BAE}(S)\phi(T)$ as is stated in Section 1. However this is not the case in obtaining closer estimators within a class of nontruncated equivariant shrinkage estimators. One possible explanation of this rather unusual phenomenon is, as Rao[13] noted, that mean squared error puts too much weight for the large deviation of the estimator from the true value of σ^2, whereas probability of closeness is concerned with concentration of the estimator in the neighbourhoods of the true value and needs no shrinking of the estimator. The result is shown in the following slightly extended form:

Theorem 2.1. *Suppose that* $0<c\leq 1/med(\chi_n^2)$. *Then (a) under* L_s, *an affine equivariant estimator cS is closer than any nontruncated equivariant shrinkage estimators* $cS\phi(T)$ *with* $0<\phi(T)<1$. *(b) Under* L_e, *an affine equivariant estimator cS is closer than any nontruncated equivariant shrinkage estimators* $cS\phi(T)$ *with* $0<\phi(T)<1$.

Proof. It suffices to show the result under L_s. The Pitman probability of closeness of $cS\phi(T)$ with respect to cS is written by

$$\begin{aligned} P[(\frac{cS\phi(T)}{\sigma^2}-1)^2<(\frac{cS}{\sigma^2}-1)^2] &= P[(1-\phi(T))\{2-\frac{cS}{\sigma^2}(1+\phi(T))\}<0] \\ &= P[\frac{S}{\sigma^2}>\frac{2}{c(1+\phi(T))}] \\ &< P[S/\sigma^2>1/c] \\ &\leq \frac{1}{2}. \end{aligned}$$

Khattree[14] discussed the case of $\phi(T)$=const. in Theorem 2.1 in multivariate situation.

Corollary 2.1. (a)*Under* L_s, *the estimator* $d_{BAE}(S)$ *is closer than any nontruncated equivariant shrinkage estimators* $d_{BAE}(S)\phi(T)$ *with* $0<\phi(T)<1$.

(b) Under L_e , *the estimator* $d_{BAE}^*(S)$ *is closer than any nontruncated equivariant shrinkage estimators* $d_{BAE}^*(S)\phi(T)$ *with* $0<\phi(T)<1$.

Proof. Since $d_{BAE}(S)=S/(n+2)$ and $d_{BAE}^*(S)=S/n$, it suffices to note that $n+2>med(\chi_n^2)$ and $n>med(\chi_n^2)$ respectively, which are true by the mode-median-mean inequality for the chi-square distribution with n degrees of freedom by Groeneveld and Meeden[15] for instance.

Put $\lambda=\mu'\mu/\sigma^2$. Then we get the following useful expression of the Pitman probability of closeness for numerical computation:

Theorem 2.2. *(a) Under L_s, Pitman probability of closeness of the Brewster-Zidek estimator with respect to the BAE estimator is given by*

$$P_{BZ:BAE} = e^{-\lambda/2} \int_0^1 \frac{t^{n/2-1}(1-t)^{r/2-1}}{B(n/2,r/2)} \int_{a(t)}^{\infty} \frac{v^{(n+r)/2-1}e^{-v/2}}{\Gamma((n+r)/2)2^{(n+r)/2}} {}_0F_1(\frac{r}{2};\frac{\lambda}{4}v(1-t))dvdt \tag{2.1}$$

where $a(t)=2(n+2)/\{t(1+\phi(t))\}$; $\phi(T)$ is defined by (1.1) and ${}_0F_1$ stands for the hypergeometric function defined by (1.3).
(b) Under L_e, Pitman probability of closeness of the Brewster-Zidek estimator with respect to the BAE estimator is given by (2.1), where $a(t)=-n\,[\log\phi^*(t)]/\{t(1-\phi^*(t))\}$ and $\phi^*(t)$ are defined by (1.4).

Proof. Noting that the probability of closeness of Brewster-Zidek estimator with respect to BAE estimator is given by $P(V>2(n+2)/\{T(1+\phi(T))\})$ under Ls and $P(V>-n[\log\phi^*(T)]/\{T(1-\phi^*(T))\})$ under L_e, where $V=(S+X'X)/\sigma^2$ and that V and $T=S/(S+X'X)$ has the joint density function

$$e^{-\lambda/2}\sum_{k=0}^{\infty}\frac{(\lambda/2)^k}{k!}\frac{v^{(n+r)/2+k-1}e^{-v/2}}{\Gamma((n+r)/2+k)2^{(n+r)/2+k}}\frac{t^{n/2-1}(1-t)^{r/2+k-1}}{B(n/2,r/2+k)} \tag{2.2}$$

which is a special case of Shorrock and Zidek[16] or Sugiura and Konno[11], we get the desired expression.

Tables II and III give the value of $P_{BZ:BAE}$ under L_s and under L_e respectively based on Theorem 2.2, in which double integral need to be evaluated. We applied Takahashi-Mori[17]'s double exponential transformation formulas for each line integrals. They are given by trapezoidal rule after transformation of the domain of integration to infinite interval $(-\infty,+\infty)$. For integration on finite interval (a,b), it is given by

$$\int_a^b f(x)dx = \int_{-1}^{1} f(\frac{b-a}{2}y+\frac{b+a}{2})dy$$

$$= \frac{b-a}{2}\int_{-\infty}^{+\infty} f(\frac{b-a}{2}\phi(t)+\frac{b+a}{2})\phi'(t)dt$$

$$\sim \frac{b-a}{2}h\sum_{i=-\infty}^{+\infty} f(\frac{b-a}{2}\phi(ih)+\frac{b+a}{2})\phi'(ih), \tag{2.3}$$

where $\phi(t)=\tanh\{\frac{\pi}{2}\sinh(t)\}$ with jacobian $\phi'(t)=\frac{\pi}{2}\cosh(t)/\{\cosh(\frac{\pi}{2}\sinh(t)\}^2$. For half-infinite interval (a,∞), it is given by

$$\int_a^{\infty} f(x)dx = \int_0^{\infty} f(y+a)dy$$

$$= \int_{-\infty}^{+\infty} f(\phi(t)+a)\phi'(t)dt$$

$$\sim h \sum_{i=-\infty}^{+\infty} f(\phi(ih)+a)\phi'(ih), \qquad (2.4)$$

where $\phi(t)=\exp\{\frac{\pi}{2}\sinh(t)\}$ and $\phi'(t)=\frac{\pi}{2}\cosh(t)\exp\{\frac{\pi}{2}\sinh(t)\}$. These formulas are obtained by Takahasi and Mori[17] and are known to be very powerful unless the integrand oscilates so strong. The optimality of the transformations was disscused by them. By varying mesh size h we can check the precision of this approximation. To check the numerical accuracy of the values of $P_{BZ:BAE}$, we have computed

$$Q_{BZ:BAE} = e^{-\lambda/2} \int_0^1 \frac{t^{n/2-1}(1-t)^{r/2-1}}{B(n/2,r/2)} \int_0^{a(t)} \frac{v^{(n+r)/2-1}e^{-v/2}}{\Gamma((n+r)/2)2^{(n+r)/2}}\, {}_0F_1(\tfrac{r}{2};\tfrac{\lambda}{4}v(1-t))dvdt \qquad (2.5)$$

separately by the same double exponential formulas and check the identity $P_{BZ;BAE}+Q_{BZ:BAE}=1$ for n=1~50, r=1,5 and λ=0~20. We take h=.2~.1 and ih=-3~3 in double exponential formula for most cases. Typical examples of them are listed in Table I and the results are similar for other cases, from which it is quite likely that $P_{BZ:BAE}$ is correct at least up to three decimals.

Table I

Accuracy of $P_{BZ:BAE}+Q_{BZ:BAE}$

		r=1					r=5		
n	λ	$P_{BZ:BAE}$	$Q_{BZ:BAE}$	P+Q	n	λ	$P_{BZ:BAE}$	$Q_{BZ:BAE}$	P+Q
1	0	.06771	.93248	1.0002	1	0	.05310	.94668	.9998
	5	.07570	.92360	.9993		5	.06239	.93782	1.0002
	20	.08142	.91887	1.0003		20	.07454	.92534	.9999
10	0	.26100	.73900	1.0000	10	0	.22259	.77741	1.0000
	5	.27585	.72414	1.0000		5	.25076	.74924	1.0000
	20	.28445	.71557	1.0000		20	.27941	.72061	1.0000
50	0	.38181	.61823	1.0000	50	0	.35428	.64575	1.0000
	5	.39098	.60904	1.0000		5	.37566	.62435	1.0000
	20	.39562	.60428	.9999		20	.39397	.60591	.9999

Including other examples, we can say that the approximations to $P_{BZ:BAE}+Q_{BZ:BAE}$ are poorest when n=1.

From Table II we can see that the probability of closeness is a monotonically increasing function of λ, but the values are much smaller than 1/2 under L_s and that not so much smaller under L_e in Table III. The limiting probabilites as λ tends to infinity are given by $P(\chi_n^2>n+2)$ where χ_n^2 stands for the chi-square varaiate with n degrees of freedom. This is because X'X tends to infinity in probability, as λ tends to infinity and thus T tends to zero, $\phi(T)$ in (1.1) tends to 1 in probability which implies that the probability of closeness $P(S/\sigma^2>2(n+2)/(1+\phi(T)))$ tends to $P(\chi_n^2>n+2)$. It is easily seen that $\lim_{n\to\infty} P(\chi_n^2>n+2)=1/2$. Moreover we can show that the probability of closeness tends to 1/2 as n tends to infinity for fixed λ. For this purpose we shall put $Z=\sqrt{n/2}\,(S/n-1)$ which has asymptotically standard normal distribution for large n and write $T=S/(S+X'X)=1-X'X/n+O_p(n^{-3/2})$. It follows that $\phi(T)$ in (1.1) has an expansion

$$1-\frac{r}{n+r+2}R(X'X)(1+O_p(n^{-1/2})),$$

where

$$R(X'X)=\frac{(X'X)^{r/2-1}e^{-X'X/2}}{\int_0^{X'X}u^{r/2-1}e^{-u/2}du}. \tag{2.6}$$

Table II

Probability of closeness of Brewster-Zidek estimator w.r.t. the BAE estimator under L_s

n	r	λ	prob.	n	r	λ	prob.	n	r	λ	prob.	n	r	λ	prob.
1	1	0	.068	1	5	0	.053	5	1	0	.195	5	5	0	.160
		1	.070			1	.055			1	.199			1	.166
		5	.076			5	.062			5	.210			5	.185
		10	.079			10	.068			10	.216			10	.199
		20	.081			20	.075			20	.220			20	.212
		∞	.083			∞	.083			∞	.221			∞	.221
10	1	0	.261	10	5	0	.223	20	1	0	.320	20	5	0	.285
		1	.265			1	.230			1	.324			1	.291
		5	.276			5	.251			5	.333			5	.312
		10	.282			10	.267			10	.338			10	.326
		20	.284			20	.279			20	.340			20	.337
		∞	.285			∞	.285			∞	.341			∞	.341
				50	1	0	.382	50	5	0	.354				
						1	.384			1	.360				
						5	.391			5	.376				
						10	.394			10	.386				
						20	.396			20	.394				
						∞	.396			∞	.396				

Hence the probability of closeness can be evaluated as

$$P(S/\sigma^2>2(n+2)/(1+\phi(T)))=P(Z>\frac{\sqrt{2n}}{2(n+r+2)}R(X'X))+O(\frac{1}{n})$$

$$=1-E[\Phi(\frac{\sqrt{2n}}{2(n+r+2)}R(X'X))]+O(\frac{1}{n})$$

$$= \frac{1}{2} - \frac{\sqrt{n}}{2\pi(n+r+2)} E[\, R(X'X)] + O(\frac{1}{n}),$$

which tends to 1/2 as n tends to infinity.

Under entropy loss we can similarly prove that $\lim_{\lambda \to \infty} P^{(e)}_{BZ:BAE} = P(\chi_n^2 > n)$, which is less than 1/2 and $\lim_{n \to \infty} P^{(e)}_{BZ:BAE} = 1/2$. The convergence to 1/2 is faster in Table III than in Table II.

Table III

Probability of closeness of Brewster-Zidek estimator w.r.t. the BAE estimator under L_e

n	r	λ	prob.	n	r	λ	prob.	n	r	λ	prob.	n	r	λ	prob.
1	1	0	.253	1	5	0	.208	5	1	0	.373	5	5	0	.311
		1	.260			1	.214			1	.380			1	.322
		5	.279			5	.234			5	.398			5	.353
		10	.289			10	.250			10	.407			10	.377
		20	.298			20	.267			20	.413			20	.398
		∞	.317			∞	.317			∞	.417			∞	.417
10	1	0	.408	10	5	0	.354	20	1	0	.434	20	5	0	.391
		1	.414			1	.364			1	.439			1	.399
		5	.428			5	.393			5	.449			5	.424
		10	.435			10	.414			10	.455			10	.441
		20	.440			20	.432			20	.458			20	.454
		∞	.440			∞	.440			∞	.458			∞	.458
				50	1	0	.461	50	5	0	.428				
						1	.461			1	.434				
						5	.468			5	.452				
						10	.472			10	.463				
						20	.473			20	.471				
						∞	.473			∞	.473				

By the same argument as in Theorem 2.2, we can get the following theorem:

Theorem 2.3. *(a) Under L_s, Pitman probability of closeness of Stein estimator with respect to the BAE estimator defined by (1.4) with $\phi(T) = \min\{1, \frac{n+2}{(n+r+2)T}\}$ is given by*

$$e^{-\lambda/2}\{\frac{1}{2}\int_0^b \frac{t^{n/2-1}(1-t)^{r/2-1}}{B(n/2,r/2)}\, {}_1F_1(\frac{n+r}{2};\frac{r}{2};\frac{\lambda(1-t)}{2})dt$$

$$+ \int_b^1 \frac{t^{n/2-1}(1-t)^{r/2-1}}{B(n/2,r/2)} \int_{a(t)}^{\infty} \frac{v^{(n+r)/2-1}e^{-v/2}}{\Gamma((n+r)/2)2^{(n+r)/2}}\, {}_0F_1(\frac{r}{2};\frac{\lambda}{4}v(1-t))dvdt\}, \tag{2.7}$$

where $b=(n+2)/(n+r+2)$; $a(t)=2/\{t/(n+2)+1/(n+r+2)\}$ and ${}_0F_1$ stands for the hypergeometric function defined by (1.2).

(b) Under L_e, Pitman probability of closeness of Stein estimator with respect to the BAE estimator defined by (1.4) with $\phi(T) = \min\{1, \frac{n}{(n+r)T}\}$ is given by (2.7), where b=n/(n+r) and $a(t)= \{n \log(t/b)\}/(t-b)$.

The first term of (2.7) stands for the probability $P_1=P[\phi(T)=1]/2$ and the second term stands for $P_2=P[L(d_{BAE}(S)\phi(T),\sigma^2)<L(d_{BAE}(S),\sigma^2)]$ under L_s. Based on Theorem 2.3 we can compute the probability of closeness of Stein estimator given in the following Tables V and VI. To check the numerical accuracy of our double integration based on double exponential formulas by Takahasi and Mori[17], we first computed the values of P_1 by double exponential formula and compared with those by Mathematica for Macintosh II, which are shown in fourth column in Table IV in parentheses. This is a univariate integral and can easily be computed by Mathematica for Macintosh II. The agreement between two methods of integration are excellent and our numerical values are reliable up to six digits. Then we computed the values of each double integrals

$$P_2 = e^{-\lambda/2} \int_b^1 \frac{t^{n/2-1}(1-t)^{r/2-1}}{B(n/2,r/2)} \int_{a(t)}^{\infty} \frac{v^{(n+r)/2-1}e^{-v/2}}{\Gamma((n+r)/2)2^{(n+r)/2}} {}_0F_1(\tfrac{r}{2};\tfrac{\lambda}{4}v(1-t))dvdt \tag{2.8}$$

$$Q_2 = e^{-\lambda/2} \int_b^1 \frac{t^{n/2-1}(1-t)^{r/2-1}}{B(n/2,r/2)} \int_0^{a(t)} \frac{v^{(n+r)/2-1}e^{-v/2}}{\Gamma((n+r)/2)2^{(n+r)/2}} {}_0F_1(\tfrac{r}{2};\tfrac{\lambda}{4}v(1-t))dv\}dt \tag{2.9}$$

seperately by double exponential formula and the sum is equal to

$$P_2 + Q_2 = e^{-\lambda/2} \int_b^1 \frac{t^{n/2-1}(1-t)^{r/2-1}}{B(n/2,r/2)} {}_1F_1(\tfrac{n+r}{2};\tfrac{r}{2};\tfrac{\lambda}{2}(1-t))dt \tag{2.10}$$

the integrand of which is equal to that of P_1. When n=r=1 and λ=0, we get P_1=1/3 and P_2+Q_2= 1/3, which agrees with the values in Table IV. The sum P_2+Q_2 and the values of RHS of (2.10) by Mathematica for Macintosh II are shown in the last column in Table IV, from which we can

Table IV

Accuracy of P_1 and P_2+Q_2 by two methods of integration

n	r	λ	P1 (Mathematica)	P2	Q2	P2+Q2(Mathematica)
1	1	0	.333333(.333333)	.054893	.278441	.333334(.333333)
1	1	10	.497169(.497170)	.002347	.003313	.005660(.005660)
1	1	20	.499946(.499947)	.00006712	.00003840	.0001055(.0001055)
10	5	0	.277321(.277321)	.175182	.270177	.445359(.445357)
10	5	10	.480073(.480073)	.026578	.013296	.03987(.03985)
10	5	20	.498674(.498674)	.002184	.000466	.002650(.002651)
50	1	0	.165760(165760)	.274042	.394227	.6683(.6685)
50	1	10	.492621(.492623)	.0071426	.0075996	.01474(.01475)
50	1	20	.49995(.49987)	.0001338	.0001170	.0002507(.0002510)

say that our appoximate values of numerical integration for P_2 are likely to be reliable at least up to three decimals. We can see from Table V that (i) When r>1, BAE estimator is closer than Stein estimator, for small λ, but for large λ, the Stein estimator is closer than BAE estimator, though the probability of closeness is larger than half only slightly. (ii) When r=1, BAE estimator is closer than Stein estimator. We did not have larger values of $P_{S:BAE}$ than 1/2 for other values of λ not cited in Table V. For the other cases examined it is less than half under L_S.

Table V

Probability of closeness of Stein estimator w.r.t. the BAE estimator under L_S

n	r	λ	P1	P2	Pit.Cl.	n	r	λ	P1	P2	Pit.Cl.
1	1	0	.333	.055	.388	2	1	0	.276	.090	.367
		1	.390	.042	.431			1	.350	.068	.418
		5	.479	.012	.491			5	.470	.019	.489
		10	.497	.002	.500			10	.496	.004	.500
		20	.500	.000	.500			20	.500	.000	.500
5	1	0	.218	.150	.369	10	1	0	.191	.196	.388
		1	.308	.111	.419			1	.287	.143	.430
		5	.459	.030	.489			5	.454	.037	.491
		10	.494	.005	.499			10	.494	.006	.500
		20	.500	.000	.500			20	.500	.000	.500
20	1	0	.176	.236	.412	50	1	0	.166	.274	.440
		1	.275	.170	.445			1	.267	.195	.463
		5	.450	.042	.493			5	.448	.047	.495
		10	.493	.007	.500			10	.493	.007	.500
		20	.500	.000	.500			20	.500	.000	.500
1	2	0	.387	.052	.439	2	2	0	.333	.085	.418
		1	.420	.043	.463			1	.381	.070	.450
		5	.479	.017	.497			5	.469	.027	.496
		10	.496	.005	.501			10	.494	.007	.501
		20	.500	.000	.500			20	.500	.000	.500
5	2	0	.267	.141	.407	10	2	0	.231	.184	.415
		1	.331	.113	.443			1	.303	.145	.448
		5	.454	.041	.495			5	.446	.049	.495
		10	.492	.010	.501			10	.490	.011	.501
		20	.500	.000	.500			20	.500	.000	.500
20	2	0	.209	.221	.430	50	2	0	.195	.256	.451
		1	.286	.171	.457			1	.273	.196	.470
		5	.441	.055	.496			5	.437	.060	.497
		10	.489	.011	.501			10	.489	.012	.500
		20	.500	.000	.500			20	.500	.000	.500
1	5	0	.428	.051	.480	2	5	0	.385	.083	.468
		1	.442	.047	.490			1	.408	.076	.484
		5	.476	.029	.505			5	.462	.044	.506
		10	.492	.013	.505			10	.488	.019	.506
		20	.499	.002	.501			20	.499	.002	.501
5	5	0	.320	.136	.455	10	5	0	.277	.175	.453
		1	.356	.121	.476			1	.321	.153	.474
		5	.442	.065	.507			5	.430	.077	.507
		10	.483	.025	.507			10	.480	.027	.507
		20	.499	.003	.501			20	.499	.002	.501
20	5	0	.247	.209	.457	50	5	0	.225	.242	.467
		1	.297	.179	.476			1	.278	.203	.481
		5	.421	.084	.505			5	.415	.088	.503
		10	.479	.027	.505			10	.478	.026	.504
		20	.499	.002	.501			20	.499	.002	.500

Table VI

Probability of closeness of Stein estimator w.r.t. the BAE estimator under L_e

n	r	λ	P1	P2	Pit.Cl.	n	r	λ	P1	P2	Pit.Cl.
1	1	0	.250	.229	.479	2	1	0	.211	.264	.476
		1	.318	.188	.506			1	.293	.208	.501
		5	.446	.076	.522			5	.445	.071	.517
		10	.487	.021	.509			10	.490	.016	.506
		20	.499	.002	.501			20	.500	.001	.500
5	1	0	.182	.297	.479	10	1	0	.170	.313	.483
		1	.275	.224	.499			1	.268	.230	.499
		5	.446	.065	.511			5	.446	.062	.508
		10	.491	.012	.503			10	.492	.011	.502
		20	.500	.000	.500			20	.500	.000	.500
20	1	0	.165	.323	.487	50	1	0	.161	.331	.492
		1	.265	.234	.499			1	.263	.236	.499
		5	.446	.059	.505			5	.446	.057	.503
		10	.492	.010	.502			10	.492	.009	.501
		20	.500	.000	.500			20	.500	.000	.500
1	2	0	.289	.225	.514	2	2	0	.250	.256	.506
		1	.331	.202	.533			1	.305	.221	.526
		5	.429	.114	.542			5	.428	.106	.534
		10	.474	.048	.522			10	.479	.036	.515
		20	.496	.008	.504			20	.498	.003	.502
5	2	0	.216	.284	.499	10	2	0	.201	.297	.498
		1	.284	.234	.517			1	.275	.237	.512
		5	.431	.093	.523			5	.432	.085	.517
		10	.484	.024	.509			10	.486	.020	.506
		20	.499	.001	.501			20	.500	.001	.500
20	2	0	.193	.305	.498	50	2	0	.188	.311	.498
		1	.270	.238	.508			1	.267	.238	.505
		5	.433	.079	.512			5	.434	.074	.508
		10	.487	.017	.504			10	.488	.015	.502
		20	.500	.001	.500			20	.500	.000	.500
1	5	0	.313	.235	.553	2	5	0	.284	.261	.545
		1	.339	.226	.565			1	.313	.246	.559
		5	.398	.175	.573			5	.394	.173	.567
		10	.440	.113	.554			10	.448	.096	.554
		20	.478	.043	.521			20	.488	.024	.512
5	5	0	.250	.281	.531	10	5	0	.233	.289	.522
		1	.289	.256	.545			1	.279	.256	.534
		5	.398	.153	.551			5	.402	.136	.539
		10	.461	.067	.528			10	.468	.051	.519
		20	.495	.009	.505			20	.497	.005	.502
20	5	0	.222	.293	.515	50	5	0	.214	.295	.509
		1	.272	.253	.525			1	.268	.248	.516
		5	.406	.122	.528			5	.408	.109	.518
		10	.473	.040	.513			10	.475	.032	.508
		20	.498	.003	.501			20	.499	.002	.501

Under L_e, it is larger than 1/2 except for the cases of small λ. (iii) The probability of Stein estimator being equal to the BAE estimator is monotonically increasing with respect to λ, whereas the probability of $L(d_{Stein},\sigma^2) < L(d_{BAE},\sigma^2)$ is a monotonically decreasing function of λ. (iv) It may be remarked that probability of Stein estimator being equal to $d_{BAE}(S)$ is very high when λ is not in the neighborhood of zero under L_s.

Acknowledgements

The author is grateful to the referee for his careful reading of the earlier version of the paper with valuable comments and corrections.

REFERENCES

1. C. Stein, *Ann. Inst. Statist. Math.*, **16**, 155-160 (1964).
2. E.J.G. Pitman, *Proc. Camb. Phil. Soc.*, **33**, 212-222 (1937).
3. N. Sugiura, *J.Japan Statist. Soc.*, **14**, 145-155 (1984).
4. S.D. Peddada, *Amer. Statistician*, **39**, 298-299 (1985).
5. J.F. Brewster, and J.V. Zidek, *Ann. Statist.*, **2**, 21-38 (1974).
6. A.L. Rukhin, *J. Amer. Statist. Assoc.*, **82**, 925-928 (1987).
7. T.K. Nayak, *J. Statist. Plan. Inf.*, **24**, 259-268 (1990).
8. T. Kubokawa, *Commun. Statist.-Theor. Meth.*, **20**, 3499-3523 (1991).
9. T. Kubokawa, *Canadian J. Statist.*, **18**, 59-62 (1990).
10. J.M. Maatta and G. Casella, *Statistical Science*, **5**, 90-120 (1990).
11. N. Sugiura, and Y. Konno, in: *Advances in Multivariate Statistical Analysis*, A.K.Gupta(Ed.), 353-371. D.Reidel, Holland (1987).
12. N. Sugiura, and Y. Konno, *Ann. Inst. Statis. Math.*, **40**, 329-341 (1988).
13. C.R. Rao, in: *Statistics and Related Topics*, M. Csörgö, D.A. Dowson, J.N.K.Rao and Md.E.Saleh(Eds.), 123-143. North Holland, Amsterdam (1981).
14. R. Khattree, *Commun. Statist.-Theor. Meth.* **16**, 263-274 (1987).
15. R.A. Groeneveld, and G. Meeden, *Amer. Statistician* **31**, 120-121 (1977).
16. R.W. Shorrock, and J.V. Zidek, *Ann. Statist.* **4**, 629-638 (1976).
17. H. Takahasi and M. Mori, *Publ. RIMS, Kyoto Univ.* **9**, 721-741 (1974).

Stat. Sci. & Data Anal., pp. 313-324
K. Matsusita *et al.* (Eds)

Normal Approximation to the Distribution of the Sample Mean in the Exponential Family

RYUEI NISHII and TAKEMI YANAGIMOTO

Hiroshima University, Hiroshima 730, Japan
The Institute of Statistical Mathematics, Tokyo 106, Japan

Abstract. Consider the testing problem on the mean against the one-sided alternative in the exponential family. Relating to the test statictics, we discuss the normalizing transformation of the sample mean and the signed log-likelihood ratio. Their asymptotic expansions are obtained up to $O(n^{-2})$.

Key words: Asymptotic expansion, exponential dispersion model, exponential family, normalizing transformation, signed-likelihood ratio.

1. INTRODUCTION

Let $a(x)e^{\theta x-\kappa(\theta)}$ be a probability density function (pdf) in the exponential family, where $a(x)$ and $\kappa(\theta)$ are known functions. The one-dimensional natural parameter θ varies in a set of $\Theta \subset \mathbf{R}^1$. Suppose that a random variable X has this pdf. Then, its mean μ is given by $\kappa'(\theta)$. Using the mean parameter, we rewrite the pdf as

$$f(x;\mu) = a(x)\exp\{b(\mu)x - \kappa(b(\mu))\}, \quad \mu \in b(\Theta), \tag{1.1}$$

where $\theta = b(\mu)$ is the inverse function of $\kappa'(\theta) = \mu$. Note that the mean is strictly increasing in θ. The term $\kappa(\theta)$ is called a cumulant generator because i-th cumulants of X are given by $d^i\kappa(\theta)/d\theta^i$.

Suppose that μ_0 is a given constant, and put $\theta_0 = b(\mu_0)$. Then we consider to test the hypothesis H against the one-sided alternative K such that

$$H : \mu = \mu_0 \quad (\text{or} \quad \theta = \theta_0) \quad \text{vs} \quad K : \mu > \mu_0 \quad (\text{or} \quad \theta > \theta_0). \tag{1.2}$$

Suppose X_1, X_2, ..., X_n be a random sample of size n from the pdf $f(x,\mu)$. One may use a test statistic $n^{1/2}(\overline{X} - \mu)/\kappa''(\theta)^{1/2}$, where $\overline{X}$ is the sample mean. The standard normal distribution will be available in this case. However, this convergence has error with order $O(n^{-1/2})$, and one can not improve the convergence rate by adjusting the mean and/or variance.

In this note, we discuss two approaches to improve this convergence rate. Section 2 deals with the normalizing transformation of $\overline{X}$. Section 3 concerns with the signed log-likelihood ratio. Applications to the exponential dispersion model and numerical examples are given in Section 4.

2. NORMALIZING TRANSFORMATION

Let $g(x)$ be a smooth function. Then it is well known that $U = n^{1/2}\{g(\overline{X}) - g(\mu)\}/\{\kappa''(\theta)^{1/2}g'(\mu)\}$ is asymptotically normally distributed with mean 0 and variance 1. The asymptotic expansion is given by

$$\Pr[U - n^{-1/2}g''(\mu)/\{2g'(\mu)\} \leq u]$$
$$= \Phi(u) - n^{-1/2}\{3g''(\mu)/g'(\mu) + \kappa'''(\theta)/\kappa''(\theta)^2\}\phi(u)/6 + O(n^{-1}), \quad (2.1)$$

where $\Phi(u)$ is the standard normal distribution function and $\phi(u) = \Phi'(u)$. If g is a solution of the differential equation $3g''(\kappa'(\theta))/g'(\kappa'(\theta)) + \kappa'''(\theta)/\kappa''(\theta)^2 = 0$, then, the term with the $O(n^{-1/2})$ in (2.1) vanishes. Its solution, say $T(\kappa'(\theta))$, is called the normalizing transformation. See [1] and [2] for a general procedure for finding the normalizing transformation. Specifically, we obtain $T(\kappa'(\theta)) = \int \kappa''(\theta)^{2/3}d\theta$. Let

$$V = n^{1/2}\{T(\overline{X}) - T(\mu)\}/\{\kappa''(\theta)^{1/2}T'(\mu)\}. \quad (2.2)$$

All the derivatives of $T(\kappa'(\theta))$ are given in terms of those of $\kappa'(\theta)$. Consequently, the general expansion on U yields the following theorem:

Theorem 2.1. Let $\overline{X}$ be a sample mean and T be the normalizing transformation of the family of pdf's (1.1). Then, (i) the asymptotic mean and variance of V in (2.2) are respectively given by $m_{NT} + O(n^{-5/2})$ and $s_{NT}^2 + O(n^{-3})$ with

$$m_{NT} = -n^{-1/2}\kappa_3/6 - n^{-3/2}(21\kappa_3^3 - 20\kappa_3\kappa_4 + 3\kappa_5)/72, \quad (2.3)$$

and

$$s_{NT}^2 = 1 + n^{-1}(3\kappa_3^2 - 2\kappa_4)/6 + n^{-2}(5551\kappa_3^4 - 7728\kappa_3^2\kappa_4 + 1539\kappa_3\kappa_5 + 1170\kappa_4^2 - 162\kappa_6)/1944, \quad (2.4)$$

where $\kappa_i = \{\kappa''(\theta)\}^{-i/2}\kappa^{(i)}(\theta)$.

(ii) The asymptotic expansion is

$$\Pr\{(V - m_{NT})/s_{NT} \le v\} = \Phi(v) - [n^{-1}(4\kappa_3{}^2 - 3\kappa_4)H_3/216$$
$$+n^{-3/2}\{(134\kappa_3{}^3 - 144\kappa_3\kappa_4 + 27\kappa_5)H_2/324 + (40\kappa_3{}^3 - 45\kappa_3\kappa_4 + 9\kappa_5)H_4/1620\} \tag{2.5}$$
$$+n^{-2}(e_1H_3 + e_2H_5 + e_3H_7)]\phi(v) + O(n^{-5/2}),$$

where $H_i = \phi(v)^{-1}(-d/dv)^i\phi(v)$ are Hermite polynomials and

$$\begin{cases} e_1 = (21871\kappa_3{}^4 - 32016\kappa_3{}^2\kappa_4 + 7200\kappa_3\kappa_5 + 4896\kappa_4{}^2 - 864\kappa_6)/31104, \\ e_2 = (24\kappa_3{}^4 - 36\kappa_3{}^2\kappa_4 + 8\kappa_3\kappa_5 + 6\kappa_4{}^2 - \kappa_6)/720, \\ e_3 = (16\kappa_3{}^4 - 4\kappa_3{}^2\kappa_4 + 9\kappa_4{}^2)/93312. \end{cases} \tag{2.6}$$

3. SIGNED LOG-LIKELIHOOD RATIO

The second approach is based on the log likelihood ratio statistic $LR \equiv \log\{L(\overline{X})/L(\mu_0)\}$, where $L(\mu)$ is the likelihood $\prod f(x_i;\mu)$. Then, the signed log likelihood ratio defined by

$$W = \mathrm{sgn}(\overline{X} - \mu)\sqrt{2LR} = \mathrm{sgn}(\overline{X} - \mu)\sqrt{2n\{b(X)\overline{X} - b(\mu)\overline{X} - \kappa(b(\overline{X})) + \kappa(b(\mu))\}} \tag{3.1}$$

is asymptotically normally distributed.

Theorem 3.1. The asymptotic bias and variance of W are respectively given by $m_{LR} + O(n^{-5/2})$ and $s_{LR}{}^2 + O(n^{-3})$ with

$$m_{LR} = -n^{-1/2}\kappa_3/6 - n^{-3/2}(50\kappa_3{}^3 - 45\kappa_3\kappa_4 + 6\kappa_5)/240, \tag{3.2}$$

$$s_{LR}{}^2 = 1 + n^{-1}(14\kappa_3{}^2 - 9\kappa_4)/36 + n^{-2}(1300\kappa_3{}^4 - 1725\kappa_3{}^2\kappa_4 + 240\kappa_4{}^2 + 324\kappa_3\kappa_5 - 30\kappa_6)/720. \tag{3.3}$$

We standardize the signed log likelihood W in the following two ways:

$$W_1 = (W - m_{LR})/s_{LR} \quad \text{and} \quad W_2 = W/\sqrt{s_{LR}{}^2 + m_{LR}{}^2}. \tag{3.4}$$

The first standardization may be found in Appendix C of [3] up to $O(n^{-1})$. The second is due to the correction such that ${W_2}^2$ has mean one, i.e., ${s_{LR}}^2 + {m_{LR}}^2$ is the Bartlett correction factor.

Theorem 3.2. The standardized signed log likelihoods defined in (3.4) have the expansions

$$\Pr\{W_1 \le v\} = \Phi(v) + [c_1 n^{-3/2} H_2 + c_2 n^{-2} H_3]\phi(v) + O(n^{-5/2}), \quad (3.5)$$

$$\Pr\{W_2 \le v\} = \Phi(v)+[(\kappa_3 n^{-1/2}/6+d_1 n^{-3/2})+d_2 n^{-3/2} H_2+d_3 n^{-2} H_3]\phi(v)+O(n^{-5/2}), \quad (3.6)$$

where

$$\begin{cases} c_1 &= (1660{\kappa_3}^3 - 1755\kappa_3\kappa_4 + 324\kappa_5)/6480, \\ c_2 &= (9320{\kappa_3}^4 - 13500{\kappa_3}^2\kappa_4 + 2025{\kappa_4}^2 + 3024\kappa_3\kappa_5 - 360\kappa_6)/25920, \\ d_1 &= (125{\kappa_3}^3 - 120\kappa_3\kappa_4 + 18\kappa_5)/720, \\ d_2 &= (550{\kappa_3}^3 - 585\kappa_3\kappa_4 + 108\kappa_5)/2160, \\ d_3 &= (695{\kappa_3}^4 - 978{\kappa_3}^2\kappa_4 + 135{\kappa_4}^2 + 216\kappa_3\kappa_5 - 24\kappa_6)/1728. \end{cases} \quad (3.7)$$

The expansion (3.5) shows that W_1 is asymptotically standard normally distributed with error $O(n^{-3/2})$. Hence the signed log likelihood ratio is superior to the normalizing transformation. Barndorff-Nielsen in [4] and [5] proved this point in the general setting. His standardization of W, say W^*, is essentially based on the saddlepoint approximation. W^* can be recognized as a sort of polynomial normalizing transformation, in other words, W is adjusted not by constants but by a random variable.

The expansions for the distributions of ${W_1}^2$ and ${W_2}^2$ are easily derived from (3.5) and (3.6).

Corollary 3.1. The squared variates ${W_1}^2$ and ${W_2}^2$ have the expansion:

$$\Pr\{{W_1}^2 \le w\} = \Psi(w) + 2c_2 n^{-2}(w^2 - 3w)\psi(w) + O(n^{-3}), \quad (3.8)$$

$$\Pr\{{W_2}^2 \le w\} = \Psi(w) + 2d_3 n^{-2}(w^2 - 3w)\psi(w) + O(n^{-3}), \quad (3.9)$$

where $\Psi(w)$ denotes the distribution function of a chi-squared distribution with one degree of freedom, $\psi(w) = \Psi'(w)$, and c_2 and d_3 are found in (3.7).

This corollary shows that both the statistics ${W_1}^2$ and ${W_2}^2$ have asymptotically the chi-squared distribution with one degree of freedom. Their approximation errors are of order

$O(n^{-2})$, though W_2 converges to $N(0,1)$ with error $O(n^{-1/2})$. This is a quite interesting fact.

4. APPLICATION TO THE EXPONENTIAL DISPERSION MODEL

A family of pdf's of the form $a(x;\lambda)e^{\lambda\{\theta x-\kappa(\theta)\}}$ is called an exponential dispersion (ED) model, where $\lambda(>0)$ denotes a concentration parameter. See, [6] for a review on this family. The above discussion is valid also for the ED model. In this case, λ plays a similar role of a sample size n if λ is known, since the ED model is reproductive with respect to λ. Accordingly, the formulae in Theorems 2.1, 3.1 and 3.2 still hold for λ in place of n.

As an example, consider the ED model whose cumulant generator $\kappa(\theta)$ is given by $\kappa_\alpha(\theta) = (\alpha-1)\alpha^{-1}\{\theta/(\alpha-1)\}^\alpha$ if $\alpha \in (0,1)$; $= -\log(-\theta)$ if $\alpha = 0$. It is known that such a family exists and is called the ED model with power variance function. An exact form of the density function is seen in, e.g., [7]. This family will be denoted by $\mathrm{ED}^{(\alpha)}$. The mean function of $\mathrm{ED}^{(\alpha)}$ is given by $\mu = \{\theta/(\alpha-1)\}^{\alpha-1}$. For examples, $\mathrm{ED}^{(0)}$ coicides with a family of gamma distributions, and the $\mathrm{ED}^{(1/2)}$ coincides with that of inverse gaussian distributions.

Suppose that a random variable X follows a pdf in $\mathrm{ED}^{(\alpha)}$ with the concentration parameter λ for $0 \le \alpha < 1$. [8] showed that the normalizing transformation of X is given by the Box-Cox transformation $X^{(1-2\alpha)/(3-3\alpha)}$. After adjusting its asymptotic mean and variance for a large λ, we obtain the following statistic

$$V_\alpha = \begin{cases} (3-3\alpha)(1-2\alpha)^{-1}\lambda^{1/2}\{(X/\mu)^{(1-2\alpha)/(3-3\alpha)} - 1\}/\mu^{\alpha/(2-2\alpha)} & \text{if } \alpha \ne 1/2, \\ \lambda^{1/2}\log(X/\mu)/\mu^{3/2} & \text{if } \alpha = 1/2. \end{cases}$$

On the other hand, the signed log likelihood ratio is obtained as

$$W_\alpha = \begin{cases} \mathrm{sgn}(X-\mu)\sqrt{2\lambda\alpha^{-1}(\alpha-1)\{(\alpha-1)X^{\alpha/(\alpha-1)} - \alpha\mu^{1/(\alpha-1)}X + \mu^{\alpha/(\alpha-1)}\}} & \text{if } \alpha \ne 0, \\ \mathrm{sgn}(X-\mu)\sqrt{2\lambda\{X/\mu - 1 - \log(X/\mu)\}} & \text{if } \alpha = 0. \end{cases} \tag{4.2}$$

Calculating the derivatives of $\kappa_\alpha(\theta)$, we have

Theorem 4.1. Let X be a random variable with the pdf in the ED with the cumulant generator $(\alpha-1)\alpha^{-1}\{\theta/(\alpha-1)\}^\alpha$ with $0 \le \alpha < 1$, and let $\lambda_* = \lambda/\mu^{\alpha/(1-\alpha)}$, where $\mu = \{\theta/(\alpha-1)\}^{\alpha-1}$ and λ denote the mean and the concentration parameter respectively. Then,

(i) the asymptotic bias and variance of V_α in (4.1) are respectively given by $m_{\alpha,NT} + O(\lambda^{-5/2})$ and $s_{\alpha,NT}{}^2 + O(\lambda^{-3})$ with

$$m_{\alpha,NT} = \lambda_*{}^{-1/2}(\alpha-2)/\{6(1-\alpha)\} + \lambda_*{}^{-3/2}\alpha(\alpha-2)(4\alpha-5)/\{72(1-\alpha)^3\},$$

$$s_{\alpha,NT}{}^2 = 1+\lambda_*{}^{-1}\alpha(\alpha-2)/\{6(1-\alpha)^2\}+\lambda_*{}^{-2}(\alpha-2)(370\alpha^3-477\alpha^2-66\alpha+52)/\{1944(1-\alpha)^4\}.$$

(ii) The asymptotic expansion of the distribution function is

$$\Pr\{(V_\alpha - m_{\alpha,NT})/s_{\alpha,NT} \le v\} = \Phi(v) - [\lambda_*{}^{-1}(\alpha-2)(\alpha+1)H_3/\{216(1-\alpha)^2\}$$
$$+\lambda_*{}^{-3/2}\{(\alpha-2)(17\alpha^2-5\alpha-4)/\{324(1-\alpha)^3\}H_2+(\alpha-2)(\alpha+1)(2\alpha-1)H_4/\{810(1-\alpha)^3\}$$
$$+\lambda_*{}^{-2}\{(\alpha-2)(1087\alpha^3-714\alpha^2-396\alpha+136)H_3/\{31104(1-\alpha)^4\}-\alpha(\alpha-2)(\alpha+1)H_5/\{720(1-\alpha)^3\}$$
$$+(\alpha-2)^2(\alpha+1)^2H_7/\{93312(1-\alpha)^4\}\}]\phi(v) + O(\lambda^{-5/2}).$$

(iii) The asymptotic bias and variance of the log likelihood ratio (4.2) are respectively given by $m_{\alpha,LR} + O(\lambda^{-5/2})$ and $s_{\alpha,LR}{}^2 + O(\lambda^{-3})$ with

$$m_{\alpha,LR} = \lambda_*{}^{-1/2}(\alpha-2)/\{6(1-\alpha)\} + \lambda_*{}^{-3/2}(\alpha-2)(11\alpha^2-17\alpha+2)/\{240(1-\alpha)^3\},$$

$$s_{\alpha,LR}{}^2 = 1+\lambda_*{}^{-1}(\alpha-2)(5\alpha-1)/\{36(1-\alpha)^2\}$$
$$+\lambda_*{}^{-2}(\alpha-2)(109\alpha^3-201\alpha^2+54\alpha+4)/\{720(1-\alpha)^4\}.$$

(iv) The asymptotic expansion of the distribution function is

$$\Pr\{(W_\alpha - m_{\alpha,LR})/s_{\alpha,LR} \le v\}$$
$$= \Phi(v) - [\lambda_*{}^{-3/2}(\alpha-2)(229\alpha^2-133\alpha-2)H_2/\{6480(1-\alpha)^3\}$$
$$+\lambda_*{}^{-2}(\alpha-2)(509\alpha^3-516\alpha^2+69\alpha+14)H_3/\{25920(1-\alpha)^4\}]\phi(v) + O(\lambda^{-5/2}),$$

$$\Pr\{W_\alpha/\sqrt{s_{\alpha,LR}{}^2 + m_{\alpha,LR}{}^2} \le v\}$$
$$= \Phi(v) - [\lambda_*{}^{-1/2}(\alpha-2)/\{6(1-\alpha)\} + \lambda_*{}^{-3/2}\{(\alpha-2)(23\alpha^2-26\alpha-4)/\{720(1-\alpha)^3\}$$
$$+(\alpha-2)(73\alpha^2-31\alpha-14)H_2/\{2160(1-\alpha)^3\}\}$$
$$+\lambda_*{}^{-2}(\alpha-2)(2\alpha-1)(22\alpha^2-19\alpha-2)H_3/\{1728(1-\alpha)^4\}]\phi(v) + O(\lambda^{-5/2}).$$

Example. Gamma distributions

Putting $\alpha = 0$, we get the expansion on the gamma distribution. Let X follow a gamma distribution with the pdf $\lambda^{-\lambda}\Gamma(\lambda)^{-1}x^{\lambda-1}e^{\lambda\{-x/\mu-\log\mu\}}$. Then its normalizing transformation is given by the cubic root transformation $V_0 = 3\lambda^{1/2}\{(X/\mu)^{1/3} - 1\}$, which is known as Wilson-Hilferty transformation. The signed log likelihood ratio is obtained as $W_0 = \text{sgn}(X-\mu)\sqrt{2\lambda\{X/\mu - 1 - \log(X/\mu)\}}$. The asymptotic expansions of the distribution functions are

$$\Pr\{(V_0 + \lambda^{-1/2}/3)/\sqrt{1 - 13\lambda^{-2}/243} \le v\}$$
$$= \Phi(v) - [-\lambda^{-1}H_3/108 + \lambda^{-3/2}(2H_2/81 + H_4/405)$$
$$+\lambda^{-2}(-17H_3/1944 + H_5/23328)]\phi(v) + O(\lambda^{-5/2}),$$

$$\Pr\{(W_0 + \lambda^{-1/2}/3 + \lambda^{-3/2}/60)/\sqrt{1 + \lambda^{-1}/18 - \lambda^{-2}/90} \le v\}$$
$$= \Phi(v) - [\lambda^{-3/2}H_2/1620 - 7\lambda^{-2}H_3/6480]\phi(v) + O(\lambda^{-5/2}),$$

$$\Pr\{W_0/\sqrt{1 + \lambda^{-1}/6} \le v\}$$
$$= \Phi(v) - [-\lambda^{-1/2}/3 + \lambda^{-3/2}(1/90 + 7H_2/540) - \lambda^{-2} \times 17H_3/432]\phi(v) + O(\lambda^{-5/2}).$$

As a numerical example, we consider the gamma distribution with $\lambda = n$ and $\mu = n$. This distribution is obtained by the sum of a random sample of size n from the exponential distribution. Figures 1.1, 1.2, 2.1 and 2.2 show that the difference between the exact distribution function and the approximated distribution functions when $n = 1$ (exponential distribution). The first order approximation mean $\Phi(v_0)$ with $v_0 = 3n^{1/2}\{(x/n)^{1/3} - 1\}$ and $\Phi(w_0)$ with $w_0 = \text{sgn}(x-n)\sqrt{2\lambda\{x/n - 1 - \log(x/n)\}}$, the second order approximation mean $\Phi(v_0 + n^{-1/2}/3)$ and $\Phi(w_0 + n^{-1/2}/3)$ and the third order approximation mean $\Phi(v_0 + n^{-1/2}/3) - n^{-1}H_3/108$ and $\Phi((w_0 + n^{-1/2}/3)/\sqrt{1 + n/18})$. The x-axis is standardized by the mean n and the variance n as $(x - n)/\sqrt{n}$.

Both the first order approximations are very poor. However, the approximation with second or third order works well. As a whole, the signed log likelihood ratio is superior to the cubic root transformation.

Acknowledgments
The authors are very grateful for the referee's valuable comments and careful reading.

REFERENCES

1. S. Konishi, *Biometrika* **68**, 647-651 (1981).
2. N. Niki and S. Konishi, *Ann. Inst. Statist. Math.* **38**, 371-383 (1986).
3. P. McCullagh and J.A. Nelder, *Generalized linear models.* Second Ed. Chapman and Hall, London (1989).
4. O.E. Barndorff-Nielsen, *Biometrika* **73**, 307-322 (1986).
5. O.E. Barndorff-Nielsen, *Scand. J. Statist.* **17**, 157-160 (1990).
6. B. Jørgensen, *J. R. Statist. Soc. B* **49**, 127-162 (1987).
7. P. Hougaard, *Biometrika* **73**, 387-96 (1986). (Its correction *Biometrika* **75**, 395, 1988)
8. R. Nishii, *Tech. Repo. #279, Statist. Research Group, Hiroshima Univ.* (1990).

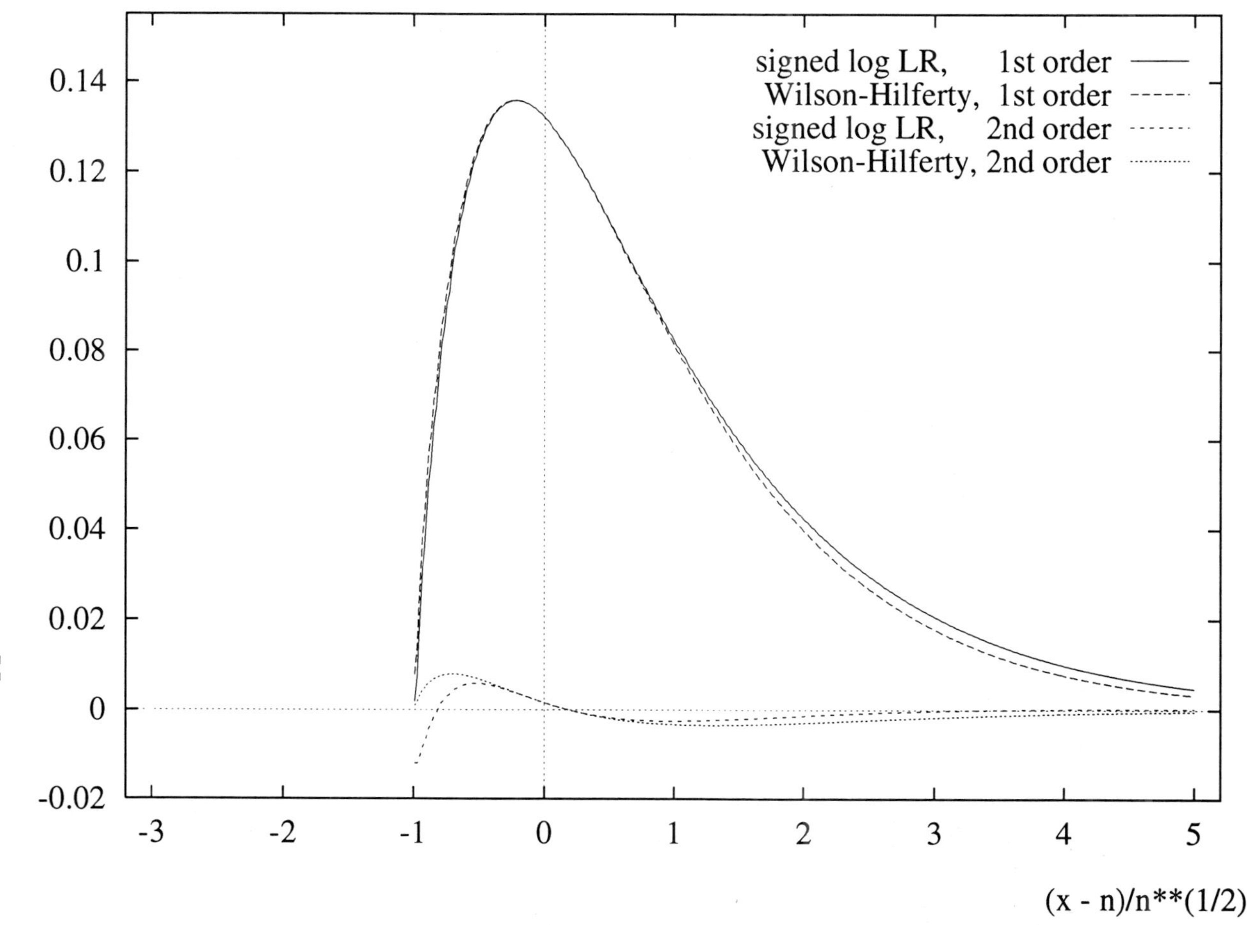

Figure 1.1. APPROXIMATION OF SUM FROM EXPONENTIAL DISTRIBUTION, n = 1

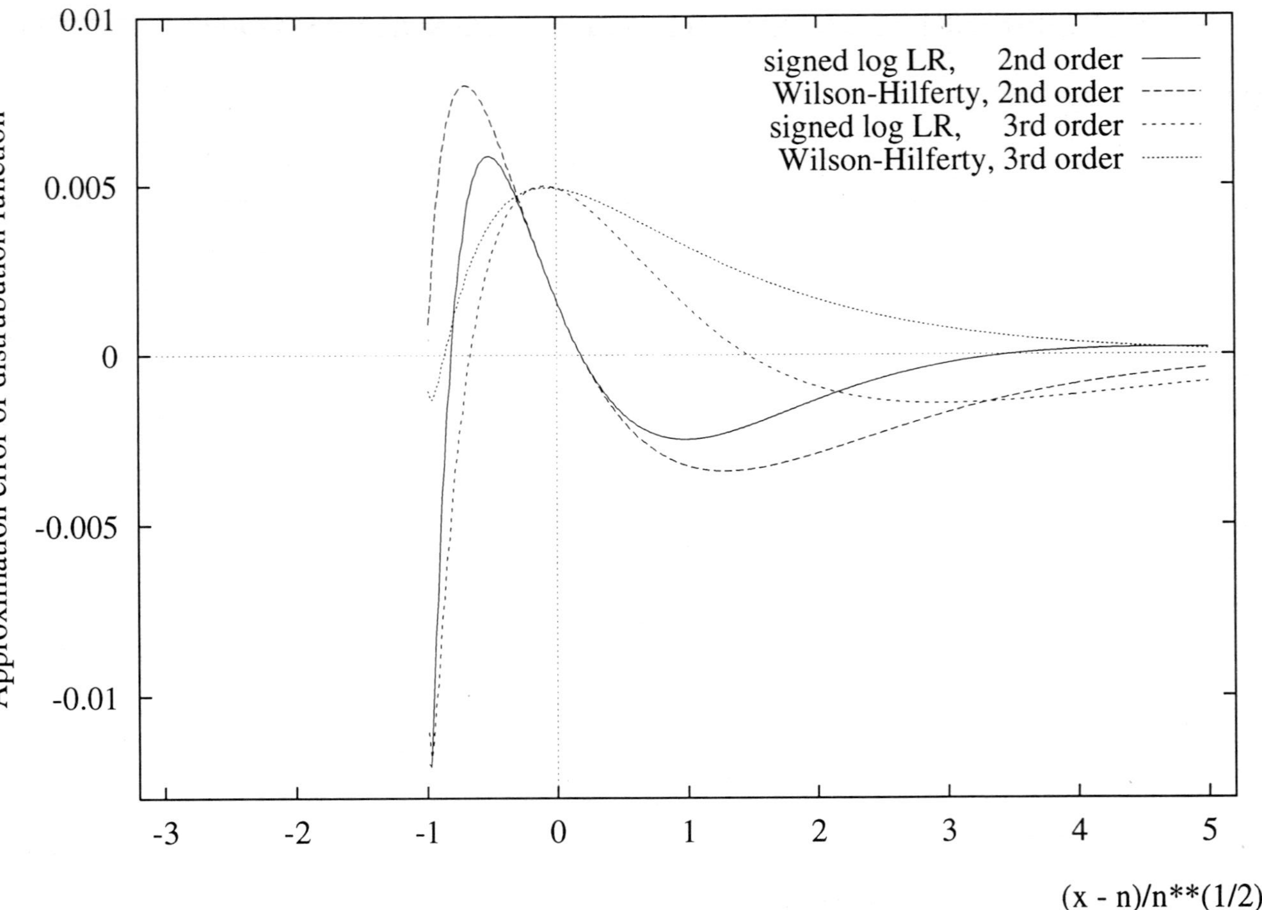

Figure 1.2. APPROXIMATION OF SUM FROM EXPONENTIAL DISTRIBUTION, n = 1

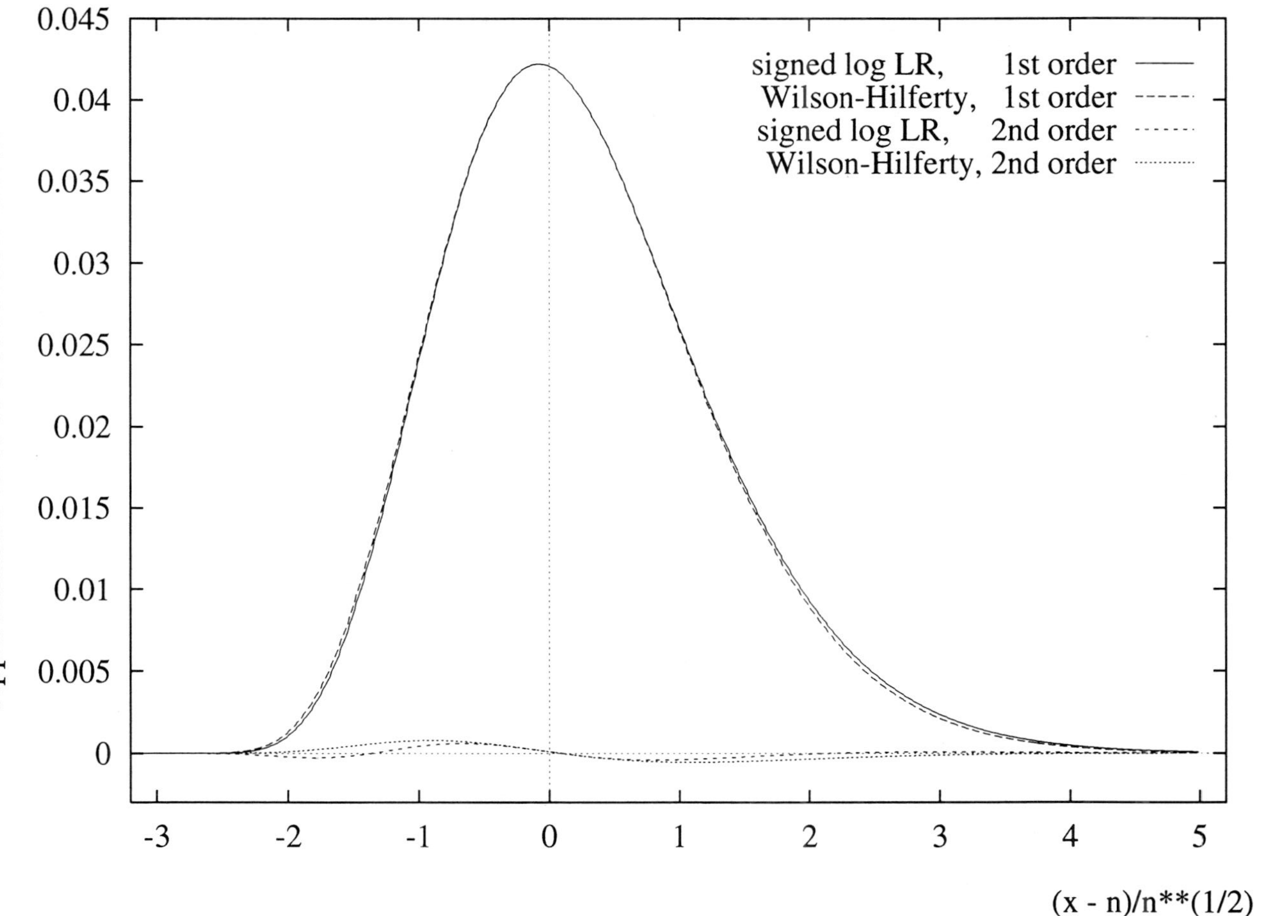

Figure 2.1. APPROXIMATION OF SUM FROM EXPONENTIAL DISTRIBUTION, $n = 10$

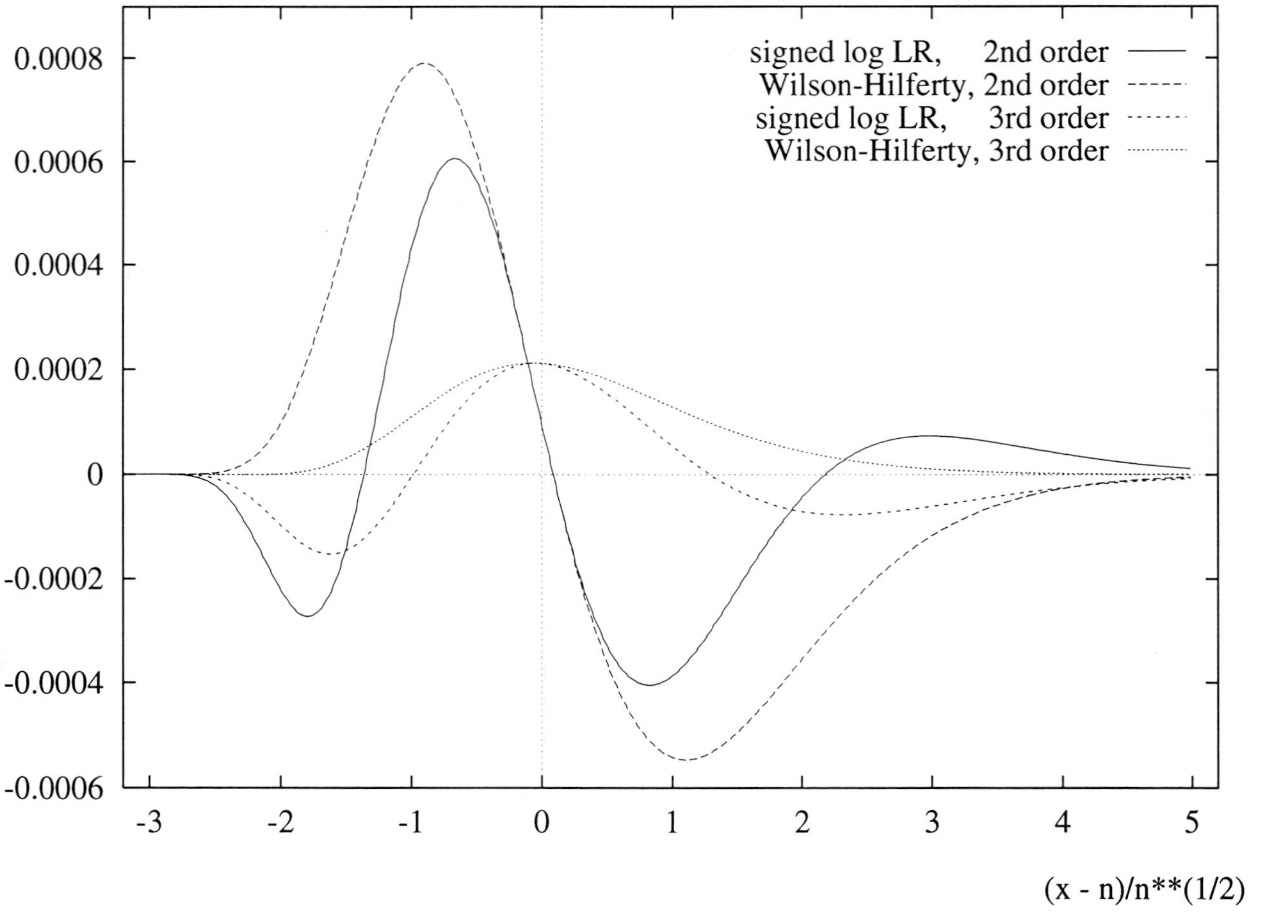

Figure 2.2. APPROXIMATION OF SUM FROM EXPONENTIAL DISTRIBUTION, n = 10

Stat. Sci. & Data Anal., pp. 325-336
K. Matsusita *et al.* (Eds)

A New Approach to Asymptotic Distributions of Maximum Likelihood Ratio Statistics

JIAN ZHANG and GUOYING LI
Institute of Systems Science, Academia Sinica, Beijing 100080

Abstract. In this paper, an empirical process method is suggested for deriving the asymptotic distribution of the maximum likelihood ratio statistics (MLRS) when the true parameter value may be not on the common boundary of Ω_0 and Ω_A. The conditions required are very mild. Consequently, we are able to answer the question about the asymptotic distribution of MLRS which was put forward by Cox and Hinkley [1]. Some examples are given. An application in testing the reliability of a series or parallel system is provided.

Key words: Maximum likelihood ratio statistics, asymptotic distributions, empirical processes.

1. INTRODUCTION

Let $(\mathcal{X}, B_{\mathcal{X}})$ be a sample space. Let $\{P_\varphi : \varphi \in \Omega\}$ be a family of parametric distributions on this space and μ be a σ-finite measure on $(\mathcal{X}, B_{\mathcal{X}})$. Assume that $\{P_\varphi : \varphi \in \Omega\}$ is uniformly absolutely continuous with respect to μ. Therefore, there exist Radon-Nikodym derivatives $f(x,\varphi) = dP_\varphi(x)/d\mu, \varphi \in \Omega$, where Ω is a subset of some metric space. Let $Y_1, \cdots, Y_n$ be the observations of the population $f(x,\theta)$ with $\theta \in \Omega$ being unknown. It is desired to test the null hypothesis H_0: $\theta \in \Omega_0$, a subset of Ω; versus the alternative $H_A : \theta \in \Omega_A$, another subset of Ω. The maximum likelihood ratio statistic (MLRS), W, defined by $\exp\{W/2\} = \sup_{\varphi\in\Omega_A} \prod_{i=1}^{n} f(Y_i,\varphi)/ \sup_{\varphi\in\Omega_0} \prod_{i=1}^{n} f(Y_i,\varphi)$ are often applied to construct the maximum likelihood ratio test (MLRT), which is especially useful in the cases with composite hypotheses. The asymptotic properties were extensively investigated in literature. Cox and Hinkley [1] gave an excellent summary up to then. A more recent work was provided by Self and Liang [2]. Set

$$l(\varphi, y) = \begin{cases} \log f(y,\varphi), & \text{when} \quad f(y,\varphi) \neq 0, \\ -\infty, & \text{otherwise.} \end{cases}$$

Suppose that $l(\varphi, y)$ has limit (can be infinity) for φ on the boundary of Ω. Define

$$l(\varphi, y) = \lim_{\Omega\ni\tau\to\varphi} l(\tau, y).$$

For convenience, we define $(-\infty + \infty) = 0$, $(-\infty + a) = -\infty, a \in \mathbb{R}^1$. Suppose that the true parameter is θ, and that θ_0 and θ_A are the minimum points of $E_\theta[l(\varphi, Y) - l(\theta, Y)]$

over Ω_0 and Ω_A respectively. Denote the maximum likelihood estimator (MLE) of θ by $\widehat{\theta}$, and the MLE restricted in turn to Ω_0 and Ω_A by $\widehat{\theta}_0$ and $\widehat{\theta}_A$. To derive the asymptotic distributions of MLRS under the null and alternative hypotheses, the approach used so far are classical; and based on the asymptotic behaviours of the MLE $\widehat{\theta}, \widehat{\theta}_0$ and $\widehat{\theta}_A$ and the Taylor series expansion or the analogues. Here, we make three remarks.

(a). Classical methods rely heavily on the consistency of $\widehat{\theta}, \widehat{\theta}_0$, and $\widehat{\theta}_A$. (cf. Cox and Hinkley [1], p. 332-334). However, in some cases, $\widehat{\theta}_0$ or $\widehat{\theta}_A$ is not consistent even if $\widehat{\theta}$ is consistent. Th following is an example.

Example 1. Let Y follow a p-dimensional normal distribution $N_p(\theta, I_p)$ with unknown mean θ and unit covariance matrix I_p. Let $(Y_1, \cdots, Y_n)$ be a i.i.d. sample from Y. Let $\Omega_0 = \{\varphi \in \mathbb{R}^p : \|\varphi\| \leq 1\}, \Omega_A = \{\varphi \in \mathbb{R}^p : \|\varphi\| > 1\}$ and $\Omega = \mathbb{R}^p$. Then, by definition,

$$W = n \inf_{\Omega_0}\{\|\overline{Y} - \varphi\|^2\} - n \inf_{\Omega_A}\{\|\overline{Y} - \varphi\|^2\},$$

where $\overline{Y} = \frac{1}{n}\sum_{i=1}^n Y_i$.

$$\widehat{\theta}_A = \begin{cases} \overline{Y}, & \text{when} \quad \|\overline{Y}\| > 1, \\ \dfrac{\overline{Y}}{\|\overline{Y}\|}, & \text{when} \quad 0 < \|\overline{Y}\| \leq 1, \\ \text{any } \varphi \in \mathbb{R}^p & \text{with} \quad \|\varphi\| = 1, \quad \text{otherwise.} \end{cases}$$

$$\widehat{\theta}_0 = \begin{cases} \overline{Y}, & \text{when} \quad \|\overline{Y}\| \leq 1, \\ \dfrac{\overline{Y}}{\|\overline{Y}\|}, & \text{otherwise.} \end{cases}$$

When $\theta = 0 \in \Omega_0$, then $\widehat{\theta} = \widehat{\theta}_0 \to 0$ and $\sqrt{n}\overline{Y} \sim N_p(0, I_p)$. Consequently, $\widehat{\theta}_A$ converges weakly to a uniform distribution on p-dimensional unit sphere. Hence, it is impossible for $\widehat{\theta}_A$ to converge to any constant by probability.

(b). For composite hypotheses, it is usually assumed that θ, θ_0 and θ_A are inner points of, respectively, Ω, Ω_0 and Ω_A (Cox and Hinkley [1], p. 335). But, it is quite often that θ_0 and θ_A can not be both inner points in turn to Ω_0 and Ω_A. Additionally, in some cases, the limit distributions of W are very difficult to derive by classical methods (cf. Cox and Hinkley [1], p. 337).

(c). In classical approaches, pretty strong assumptions are made on $l(\varphi,Y)$,Ω,Ω_0 and Ω_A. For example, at least the first two continuous derivatives of $l(\varphi,Y)$ are needed, thus some common distributions, such as the double exponential distribution, are excluded.

This paper presents a new approach based on empirical processes, particularly the theorem in Zhang and Cheng [3]. The asymptotic behavior of W is obtained under very mild conditions without any derivatives of $l(\varphi, y)$ involved. In addition, the open problem raised by Cox and Hinkley [1] p. 337 can be solved by our results.

Let $P_n g(\cdot)$ denote $\frac{1}{n}\sum_{i=1}^n g(Y_i)$, $Pg(Y)$ denote the expectation of $g(Y)$ and "$\xrightarrow{d}$" denote the convergence by distribution. Throughout the paper, the notations above are adopted.

2. MAIN RESULTS

Let $\overline{C}$ be the closure of the set C, and P_θ be the probability measure corresponding to $f(\cdot,\varphi)$. Let Ω_1 is a subset of Ω. For any $\theta \in \Omega$ fixed, put $g(\varphi,y) = l(\varphi,y) - l(\theta,y)$, $\mathcal{G}_\theta = \{g(\varphi,y) : \varphi \in \overline{\Omega}_1\}$.

Theorem 2.1. Suppose that $\Omega \subset \mathbb{R}^p$, Ω_1 is bounded, and that there exist $\alpha > 0$, $\delta > 0$ and a nonnegative function $M(\cdot)$, satisfying $E_\theta M(Y) < +\infty$, such that for any $\varphi, \tau \in \overline{\Omega}_1$ with $\|\varphi - \tau\| \leq \delta$,

$$|l(\varphi,y) - l(\tau,y)|^2 \leq M(y)\|\varphi - \tau\|^\alpha. \tag{2.1}$$

Then,

$$\sqrt{n}(\sup_{\varphi\in\overline{\Omega}_1} P_n g(\varphi,\cdot) - \sup_{\varphi\in\overline{\Omega}_1} P_\theta g(\varphi,\cdot)) \xrightarrow{d} \sup_{\varphi\in B_{1\theta}} W_\theta(g(\varphi,\cdot)),$$

where $B_{1\theta} = \{\varphi \in \overline{\Omega}_1 : E_\theta g(\varphi,Y) = \sup_{\tau\in\overline{\Omega}_1} E_\theta g(\tau,Y)\}$, W_θ is the P_θ-bridge indexed by $\mathcal{G}_\theta$(cf. Pollard [4],p.149).

Corollary 1. Assume that, for $\theta \in \Omega$ there exist $m_\theta(>0)$ and a neighborhood of $B_{1\theta}$, say $O(B_{1\theta}, m_\theta) \triangleq \{\varphi : \inf_{\tau\in B_{1\theta}} \|\varphi - \tau\| \leq m_\theta\}$ such that

$$E_\theta\{\sup_{\substack{\varphi\notin O(B_{1\theta},m_\theta)\\ \varphi\in\overline{\Omega}}} g(\varphi,Y)\} < 0.$$

Suppose that there exist $\alpha > 0, \delta > 0$ and a nonnegative function $M(Y)$ with $E_\theta M(Y) < +\infty$ such that for any $\varphi,\tau \in \Omega, \|\varphi - \tau\| \leq \delta$, (2.1) holds. Then, the probability of event that the confidence set $C = \{\theta \in \Omega : \sup_{\varphi\in\overline{\Omega}} P_n g(\varphi,\cdot) < b/\sqrt{n}\}$ covers the false parameter value $\tau(\neq\theta)$ approximately equals

$$P_\theta\{\sup_{\varphi\in B_{1\tau}} W(g(\varphi,\cdot)) < \sqrt{n}(b/\sqrt{n} - \sup_{\varphi\in\overline{\Omega}} P_\theta g(\varphi,\cdot))\}.$$

In the following,let $\Omega_1 = \Omega_0 \cup \Omega_A$, $B_0 = \{\varphi \in \overline{\Omega}_0\text{:}\ E_\theta g(\varphi,Y) = \sup_{\tau\in\overline{\Omega}_0} E_\theta g(\tau,Y)\}$ and $B_A = \{\varphi \in \overline{\Omega}_A\text{:}\ E_\theta g(\varphi,Y) = \sup_{\tau\in\overline{\Omega}_A} E_\theta g(\tau,Y)\}$.

Theorem 2.2. Suppose that $\Omega \subset \mathbb{R}^p$, Ω_1 is bounded, and inequality (2.1) holds. Then

$$\begin{aligned}&\sqrt{n}(W/n - 2\sup_{\varphi\in\Omega_A} E_\theta g(\varphi,Y) + 2\sup_{\varphi\in\Omega_0} E_\theta g(\varphi,Y))\\ &\xrightarrow{d} 2\sup_{\varphi\in B_A} W_\theta(g(\varphi,\cdot)) - 2\sup_{\varphi\in B_0} W_\theta(g(\varphi,\cdot))\end{aligned} \tag{2.2}$$

Remark 2.1. It is obvious that if $B_A \neq B_0$, the limit distribution in (2.2) is nondegenerate. Especially, if $B_0 = \{\theta_0\}, B_A = \{\theta_A\}$ and $\theta_0 \neq \theta_A$, then (2.2) results a normal distribution. When $B_A = B_0$, the conclusion of Theorem 2.1 is $W/\sqrt{n} \to 0$ in probability; however in most cases the limit distribution of W has previously derived.

Remark 2.2. The result on p.335 of Cox and Hinkely [1] follows immediately from this theorem, because in that case the true parameter is $\theta_0, B_0 = \{\theta_0\}, B_A = \{\theta_A\}$ and $\theta_0 \neq \theta_A$.

The following proposition is the direct application of Theorem 2.2 and easy to prove.

Proposition 1. Suppose there is an exponential family $f(x;\varphi) = dP_\varphi(x)/d\mu = C(\varphi)\exp[\sum_{i=1}^k Q_i(\varphi)T_i(x)]h(x)$, $\varphi \in \overline{\Omega}_1 \subset \mathbb{R}^p$. Assume that $\log C(\varphi)$ and $Q_i(\varphi)$, $1 \le i \le k$, all satisfy Lipschitz's condition with an exponent α, that is, there exist two positive constants D and δ_0 such that for φ, $\tau \in \overline{\Omega}$ with $\|\varphi - \tau\| \le \delta$,

$$\max\{|\log(C(\varphi)/C(\tau))|, |Q_i(\varphi) - Q_i(\tau)| : 1 \le i \le k\} \le D\|\varphi - \tau\|^\alpha.$$

Assume that $E_\theta \max_{1 \le i \le k} T_i^2(X) < +\infty, \log(C(\varphi))$ and $Q_i(\varphi), 1 \le i \le k$, are all bounded on $\overline{\Omega}_1$. Then

$$\begin{aligned} &[l(\varphi; X) - l(\tau, X)]^2 \\ &\le D^2[(K+1) + (K+1)K \max_{1\le i\le k} T_i^2(X)]\|\varphi - \tau\|^{2\alpha}. \end{aligned} \tag{2.3}$$

Furthermore, if Ω_1 is bounded, then the conclusion of Theorem 2.2 remains true. In particular, define $Q(\varphi) = (Q_1(\varphi),\cdots,Q_k(\varphi))^\tau$, $T(x) = (T_1(x),\cdots,T_k(x))^\tau$, and assume that $Q(\varphi)$ has the first derivative on $\overline{\Omega}_1$, and that

$$\begin{aligned} &\sup_{\varphi\in\overline{\Omega}_1} \max_{\substack{1\le i\le k \\ 1\le j\le p}} \left|\frac{\partial Q_i(\varphi)}{\partial \varphi_j}\right| < +\infty, \\ &E_\theta \max_{1\le i\le k} T_i^2(X) < +\infty, \end{aligned}$$

with $\varphi = (\varphi_1,\cdots,\varphi_p)$. Then, (2.3) holds with $\alpha = 1$.

The assumption that "Ω_1 is bounded" can be removed. Before that, we need some more notations. Set $d(\varphi, A) = \inf_{\tau\in A}\|\varphi - \tau\|(\varphi \in \mathbb{R}^p, A \subset \mathbb{R}^p), \xi(\varphi, Y, \rho) = \sup\{g(\tau, Y) : \tau \in \overline{\Omega}_1, \|\tau - \varphi\| \le \rho\}, \zeta(\varphi, Y, \rho) = \sup\{g(\tau, Y) : \tau \in \overline{\Omega}_1, \|\tau - \varphi\| > \rho\}, B_n = \{\varphi \in \overline{\Omega}_1 : \sum_{i=1}^n g(\varphi, Y_i) = \sup_{\tau\in\overline{\Omega}} \sum_{i=1}^n g(\tau, Y_i)\}, B_\theta = \{\varphi \in \overline{\Omega}_1 : E_\theta g(\varphi, Y) = \sup_{\tau\in\overline{\Omega}_1} E_\theta g(\tau, Y)\}, \Omega_a = \{\varphi \in \Omega : d(\varphi, B_\theta) < a\}$, and $\mathcal{G}_a = \{g(\varphi, y) : \varphi \in \Omega_a\}$. Recall that the true parameter $\theta \in \Omega$ is fixed. We assume the following conditions.

C1. For each $\varphi \in \overline{\Omega}_1$ and $\rho > 0$, $\xi(\varphi, Y, \rho)$ is a measurable function of y;

C2. For every $\varphi \in \overline{\Omega}_1$, there exist $\rho > 0$, $r_0 > 0$ such that

$$E_\theta|\xi(\varphi, Y, \rho)| < +\infty, E_\theta|\zeta(\varphi, Y, r_0)| < +\infty;$$

C3. $\lim_{\varphi\in\Omega, d\to+\infty} g(\varphi, y) = -\infty$, a.s. P_θ;

C4. $B_\theta \ne \emptyset$, and there exists an $a > 0$ such that for every $\varphi \in \Omega_a$,

$$E_\theta|g(\varphi, Y)| < +\infty;$$

C5. There exists a set V with P_θ -probability one such that for each $y \in V, l(\varphi, y)$ is continuous on $\overline{\Omega}_1$.

Theorem 2.3. Suppose that C1–C5 hold. Then: (i) B_θ is closed and bounded, (ii) $\sup_{\varphi\in B_\theta} d(\varphi, B_n) \to 0$ a.s. Furthermore, if there exist $a > 0, \alpha > 0, \delta > 0, \varphi \in B_\theta$ and a nonnegative function $M(\cdot)$ with $E_\theta M(Y) < +\infty$ such that for any $\varphi, \tau \in \Omega_a, \|\varphi - \tau\| \le \delta$, (2.1) holds, then the conclusion (2.2) in Theorem 2.2 remains true.

Remark 2.3. Conclusion (ii) is a generalization of the consistency result of MLE in Wald [5].

Remark 2.4. No matter where the true parameter θ is, and no matter whether $\overline{\Omega}_0 \cap \overline{\Omega}_A = \emptyset$, and whether $B_0 \cap B_A = \emptyset$, the last two theorems remain true. Therefore, the last two theorems have solved the open problem in Cox and Hinkley ([1],p.337) about the asymptotic property of W (i.e. W' in their term) for $\overline{\Omega}_0 \cap \overline{\Omega}_A = \emptyset$, $E_\theta g(\varphi, Y)$ being unimodal, and then θ_A being a boundary point of Ω_A.

Remark 2.5. Condition (2.1) on $l(\varphi, y)$ is much weaker than those in previous work.

3. PROOFS OF MAIN RESULTS

The proofs of the three theorems in Section 2 are all based on the following lemma.

Lemma 3.1(Zhang and Cheng [3]). Let π be a bounded subset of a metric space $\mathcal{G}$ with a metric d. Let X be random variable which follows probability distribution P. Let distribution functional $V(t) = V(t, P)$ be continuous on π with respect to the metric d. Let $V_n(t) = V_n(t, P_n)$ be a measurable sample functional for each $t \in \pi$ such that $\sup_{t\in\pi} V_n(t)$ is also measurable, where P_n is the empirical distribution of P. Set

$$B \triangleq \{t \in \pi : V(t) = \sup_{s\in\pi} V(s)\},$$

$$S_n(t) = \sqrt{n}(V_n(t) - V(t)), \quad n \geq 1, t \in \pi.$$

If there is a process $\{S(t) : t \in \pi\}$ with continuous sample paths such that

$$\sup_{t\in\pi} |S_n(t) - S(t)| \to 0, \quad \text{a.s.} \quad (P),$$

then,

$$\sqrt{n}(\sup_{t\in\pi} V_n(t) - \sup_{t\in\pi} V(t)) \to \sup_{t\in B} S(t), \quad \text{a.s.} \quad (P).$$

Take Theorem 2.1 as an example. Set

$$\begin{aligned} \pi &= \overline{\Omega}_1, \\ V(t) &= E_\theta g(t, Y), \\ B &= B_1, \quad \overline{V}_n(t) = P_n g(t, Y), \\ \overline{S}_n(t) &= \sqrt{n}(\overline{V}_n(t) - V(t)), t \in \overline{\Omega}_1, \end{aligned}$$

Then

$$\sqrt{n}(\sup_{\pi\in\overline{\Omega}_1} P_n g(\pi, \cdot) - \sup_{\pi\in\overline{\Omega}_1} P_\theta g(\pi, \cdot)) = \sqrt{n}(\sup_{t\in\pi} \overline{V_n}(t) - \sup_{t\in\pi} V(t)). \tag{2.4}$$

To prove Theorem 2.1, we first show

$$\overline{S_n} \xrightarrow{d} \overline{S},$$

where $\overline{S} = \{\overline{S}(t) : \quad t \in \pi\}$ is a Gaussian process. Take advantage of Representation Theorem (cf. Pollard [4],p.71), we can find two processes S_n and S such that

$$\sup |S_n(t) - S(t)| \rightarrow \quad 0, \quad \text{a.s.} \quad (P).$$

Thus, the limit distribution of (2.4) follows by Lemma 3.1 immediately.

Proof of Theorem 2.1: By (2.1), it follows that $\mathcal{G}_\theta$ is permissible and that

$$N_2(\delta, \rho_{P_n}, \mathcal{G}_\theta) \leq N_2(\frac{\delta^{2/\alpha}}{(P_n M(Y))^{1/\alpha}}, \|\cdot\|, \overline{\Omega}_1), \tag{2.5}$$

where $\|\cdot\|$ is the Euclidean norm, ρ_{P_n} is a seminorm on $L^2(P)$ defined by $\rho_{P_n}(f) = (P_n f^2)^{1/2}$ and $N_2(\varepsilon, \rho_{P_n}, \mathcal{G}_\theta)$ and $N_2(\varepsilon, \|\cdot\|, \overline{\Omega}_1)$ are the covering numbers (cf. Pollard [4], p.143) of $\mathcal{G}_\theta$ and $\overline{\Omega}_1$ respectively. Note that $\overline{\Omega}_1$ is a bounded closed subset of a p-dimensional Euclidean space, by (2.5) it is easy to prove that there are positive constants $c_0(\theta)$ and c_1, such that

$$\limsup_n N_2(\delta, \rho_{P_n}, \mathcal{G}_\theta) \leq c_0(\theta)\delta^{-c_1}, \quad \text{a.s.} \quad (P_\theta).$$

Hence, as $\delta \rightarrow 0$,

$$\lim_{n\rightarrow+\infty} \sup J_2(\delta, P_n, \mathcal{G}_\theta) \leq \int_0^\delta [2\log(c_0^2(\theta)u^{-2c_1-1})]^{1/2} du \rightarrow 0,$$

where

$$J_2(\delta, P, \mathcal{F}) = \int_0^\delta [2\log(N_2(u, \rho_p, \mathcal{F})^2/u)]du.$$

This implies that, for any $\zeta > 0$ and $\varepsilon > 0$, there exists $r > 0$ such that

$$\lim_{n\rightarrow+\infty} \sup P_\theta\{J_2(r, P_n, \mathcal{G}_\theta) > \zeta\} < \varepsilon.$$

By Equicontinuity Lemma (cf. Pollard [4],p.150) and Central Limit Theorem (cf. Pollard [4],p.154), there exists a P_θ -bridge, say W_θ, indexed by $\mathcal{G}_\theta$ with uniformly continuous sample paths with respect to the seminorm ρ_{P_θ} such that

$$\{\sqrt{n}(P_n - P)h : h \in \mathcal{G}_\theta\} \xrightarrow{d} \{W_\theta(h) : h \in \mathcal{G}_\theta\},$$

where $\rho_{P_\theta} = (P_\theta f^2)^{1/2}$. Set

$$\mathcal{H}_\theta \triangleq \{q(\cdot) : q(\cdot) \text{ is continuous uniformly on } \mathcal{G}_\theta \text{ with respect to } \rho_{P_\theta}\}.$$

For any q_1 and q_2 lying in $\mathcal{H}_\theta$, define

$$d(q_1, q_2) = \sup_{h\in\mathcal{G}_\theta} |q_1(h) - q_2(h)|,$$

then d is a metric on $\mathcal{H}_\theta$ and the sample paths of W_θ belong to $\mathcal{H}_\theta$ almost surely. By the continuity of $q(\cdot)$ in $\mathcal{H}_\theta$ and the assumptions in Theorem 2.1, it follows that $\rho_{P_\theta}(g(\varphi;\cdot) - g(\tau)\cdot)) \rightarrow 0(\|\varphi - \tau\| \rightarrow 0)$ and $q(h(\cdot))$ is continuous on $\overline{\Omega}_1$ provided that $h(\cdot) \in \mathcal{G}_\theta$. Consequently, $\mathcal{H}_\theta$ is separable and has a countable dense subset. $\{\overline{S}_n(t) : t \in \pi\} \xrightarrow{d}$

$\{W_\theta(g(t,\cdot)): \quad t \in \pi\}$. By Representation Theorem, there exist two processes $\{S_n(t): t \in \pi\}$ and $\{S(t): t \in \pi\}$ which follow the same distributions as those of $\{\overline{S}_n(t): t \in \pi\}$ and $\{W_\theta(g(t,\cdot)): \quad t \in \pi\}$, respectively, and satisfy

$$\sup_{t\in\pi} |S_n(t) - S(t)| \to 0, \quad \text{a.s.} \quad (P_\theta).$$

Set

$$V_n(t) = S_n(t)/\sqrt{n} + V(t), t \in \pi.$$

Then, by Lemma 3.1, we have

$$\sqrt{n}(\sup_{t\in\pi} V_n(t) - \sup_{t\in\pi} V(t)) \to \sup_{\in\in B_{1\theta}} S(t), \quad \text{a.s.} \quad (P_\theta).$$

Consequently,

$$\sqrt{n}(\sup_{t\in\pi} \overline{V_n}(t) - \sup_{t\in\pi} V(t)) \xrightarrow{d} \sup_{t\in B_{1\theta}} \overline{S}(t).$$

The proof of Theorem 2.1 is completed.

The proof of Theorem 2.2 is very similar to that of Theorem 2.1, in which we only note that, by a way similar to the proof of Lemma 3.1 (cf. Zheng and Cheng [3]), we can prove

$$\begin{aligned}&\sqrt{n}(\sup_{t\in\overline{\Omega}_A} V_n(t) - \sup_{t\in\overline{\Omega}_A} V(t)) - \sqrt{n}(\sup_{t\in\overline{\Omega}_0} V_n(t) - \sup_{t\in\overline{\Omega}_0} V(t))\\ &\to \sup_{t\in B_A} S(t) - \sup_{t\in B_0} S(t), \ \text{a.s.} \ (P_\theta).\end{aligned}$$

The remainder of this section is devoted to proving Theorem 2.3. First, we give two lemmas whose proofs are similar to that of Wald [5].

Lemma 3.2. Under condition C1–C5, we have

$$\lim_{\rho\to o^+} E\theta[\xi(\varphi, Y, \rho)] = E_\theta g(\varphi, Y), \quad \varphi \in \overline{\Omega}_1,$$

$$\lim_{r\to+\infty} E_\theta[\zeta(\varphi, Y, r)] = -\infty, \quad \varphi \in \overline{\Omega}_1,$$

and $E_\theta g(\varphi, Y)$ is continuous on $\overline{\Omega}_1$ with respect to φ.

Lemma 3.3 Assume that, for each $\theta \in \Omega$,

(I) $\lim_{\rho\to o^+} E_\theta[\xi(\varphi, Y, \rho)] = E_\theta g(\varphi, Y), \varphi \in \overline{\Omega}_1$;

(II) $\lim_{r\to+\infty} E_\theta[\zeta(\varphi, Y, r)] = -\infty, \varphi \in \overline{\Omega}_1$;

(III) $B_\theta \neq \emptyset$, and $\quad E_\theta|g(\varphi, Y)| \leq +\infty, \varphi \in B_\theta$.

Then, for arbitrary $\theta \in \Omega, \delta > 0$, $\theta_0 \in B_\theta$ and any closed subset D of $\overline{\Omega}_1 \backslash \Omega_a$, we have

$$P_\theta\{\lim_{n\to+\infty} \sup_{\varphi\in D} \prod_{i=1}^{n} f(Y_i, \varphi) / \prod_{i=1}^{n} f(Y_i, \theta_0) = 0\} = 1.$$

Proof of Theorem 2.3: Analogously to Wald [5], we conclude, applying Lemmas 3.2 and 3.3, that

$$\sup_{\varphi \in B_\theta} d(\varphi, B_n) \to 0, \quad \text{a.s.} \quad (P_\theta),$$

and that B_θ is a bounded closed set. Consequently, Ω_a is bounded. Then, by Theorem 2.2, conclusion (2.2) in Theorem 2.2 remains true.

4. DISCUSSION AND EXAMPLES

Example 4.1. (Continuation of Example 1.1). Assume that Y has the p-dimensional normal distribution $N_p(\theta, I_p)$ with mean $\theta \in \mathbb{R}^p$ and unit covariance matrix I_p. Let $Y_1, \cdots, Y_n$ be n i.i.d. observations of Y. Let the null hypothesis, H_0, assert $\theta \in \Omega_0$ and the alternative hypothesis H_A assert $\theta \in \Omega_A$, where Ω_0 is a closed set and Ω_A is an open set. Then, we have

$$W = \inf_{\varphi \in \Omega_0} \{\sum_{i=1}^n \|Y_i - \varphi\|^2\} - \inf_{\varphi \in \Omega_A} \{\sum_{i=1}^n \|Y_i - \varphi\|^2\},$$

$$l(\varphi, Y) = -1/2\|Y - \varphi\|^2 - p/2 \log(2\pi).$$

It is easy to see that the assumptions in Theorem 2.3 hold here. Let B_0 and B_A be defined in Section 2. By Theorem 2.3,

$$\begin{aligned} &\sqrt{n}(W/n - \inf_{\varphi \in \Omega_0} E_\theta \|Y - \varphi\|^2 + \inf_{\varphi \in \Omega_A} \|Y - \varphi\|^2) \\ &\xrightarrow{d} 2 \sup_{B_0} W_\theta(g(\varphi, \cdot)) - 2 \sup_{B_A} W_\theta(g(\varphi, \cdot)), \end{aligned} \tag{4.1}$$

Take $\Omega_0 = \{\varphi : \|\varphi\| \leq 1\}$ and $\Omega_A = \{\varphi : \|\varphi\| > 1\}$. Then, it is direct to show that $B_0 = \{\theta\}$, $B_A = \{\frac{\theta}{\|\theta\|}\}$ when $\theta \neq 0$ and $\|\theta\| < 1$.

Thus, the limit distribution in (4.1) is nondegenerate normal distribution if $0 < \|\theta\| < 1$. When $\theta = 0$, we have $B_0 = \{0\}$, $B_A = \{\varphi : \|\varphi\| = 1\}$ and, in this case, the limit distribution in (4.1) is equal that of an infimum of a Gaussian process indexed by p-dimensional sphere which is nondegenerate. When $\|\theta\| > 1$, $B_0 = \{\frac{\theta}{\|\theta\|}\}$, $B_A = \{\theta\}$ and the limit distribution in (4.1) is a nondegenerate normal distribution. The asymptotic behavior above can not be derived by the previously existing results on MLRT. However, for $\Omega_0 = \{\varphi : \|\varphi\| \leq 1\}$ and $\Omega_A = \{\varphi : \|\varphi\| > 1\}$, it can be obtained by the analytic method directly. Note that

$$W = \begin{cases} n(\|\overline{Y}\| - 1)^2, & \text{when } \|\overline{Y}\| > 1, \\ -n(\|\overline{Y}\| - 1)^2, & \text{otherwise;} \end{cases}$$

where $\overline{Y} = n^{-1} \sum_{i=1}^n Y_i$. By some analytic calculations, we can conclude that:

(1) if the true parameter value $\|\theta\| > 1$, then

$$P_\theta(\|\overline{Y}\| \leq 1) \to 0, \ (n \to +\infty).$$

$$\sqrt{n}(W/n - (\|\theta\| - 1)^2) \xrightarrow{d} 2\frac{\theta^\tau}{\|\theta\|} N_p(0, I_p)(\|\theta\| - 1);$$

(2) if $\|\theta\| < 1$ and $\|\theta\| \neq 0$, then

$$P_\theta(\|\overline{Y}\| > 1) \to 0,$$

$$n^{1/2}(W/n + (\|\theta\| - 1)^2) \xrightarrow{d} 2\frac{\theta^\tau}{\|\theta\|} N_p(0, I_p)(1 - \|\theta\|);$$

(3) if $\|\theta\| = 0$, then

$$P_\theta(\|\overline{Y}\| > 1) \to 0,$$

$$n^{1/2}(W/n + 1) \xrightarrow{d} 2\|N_p(0, I_p)\|;$$

(4) if $\|\theta\| = 1$, then

$$W \xrightarrow{d} \text{sign}(\theta^\tau N_p(0, I_p)) \cdot (\theta^\tau N_p(0, I_p))^2,$$

where sign $(x) = 1$, when $x \geq 0$; -1, otherwise. For example, if $\|\theta\| = 1$, then

$$\begin{aligned}
& P_\theta\{W < d\} \\
&= P_\theta\{n((\|\overline{Y}\| - 1)^2 - (\|\theta\| - 1)^2) < d, \|\overline{Y}\| > 1\} \\
&+ P_\theta\{n(-(\|\overline{Y}\| - 1)^2 \\
&+ (\|\theta\| - 1)^2 < d), \|\overline{Y}\|| \leq 1\} \\
&= P_\theta\{\frac{\theta^\tau}{\|\theta\|} n^{1/2}(\overline{Y} - \theta) n^{1/2}(\|\overline{Y}\| - 1) < d, \|\overline{Y}\| - \|\theta\| > 0\} \\
&+ P_\theta\{\frac{\theta^\tau}{\|\theta\|} n^{1/2}(\overline{Y} - \theta) n^{1/2}(-\|\overline{Y}\| + 1) < d, \|\overline{Y}\| - \|\theta\| \leq 0\} \\
&+ o(1) \\
&= P_\theta\{(\theta^\tau n^{1/2}(\overline{Y} - \theta))^2 < d, \theta^\tau \cdot n^{1/2}(\overline{Y} - \theta) > 0\} \\
&+ P_\theta\{-(\theta^\tau n^{1/2}(\overline{Y} - \theta))^2 < d, \theta^\tau n^{1/2}(\overline{Y} - \theta) \leq 0) + o(1) \\
&\to P_\theta\{\text{sign}(\theta^\tau N_p(0, I_p)) \cdot (\theta^\tau N_p(0, I_p))^2 < d\},
\end{aligned}$$

that is,

$$W \xrightarrow{d} \text{sign}(\theta^\tau N_p(0, I_p)) \cdot (\theta^\tau N_p(0, I_p))^2.$$

For the case of $\|\theta\| = 1$, Theorem 2.3 gives a degenerate limit distribution of $W/\sqrt{n}$, while the analytic method concludes a nondegenerate limit distribution of W. However, this is a very special case. In general, when $B_0 = B_A$, Theorems 2.2 and 2.3 can only yield a degenerate limit distribution of $W/\sqrt{n}$. Can we derive a nondegenerate limit distribution of $c_n W$ for some constants c_n by the empirical process method ? This problem deserves to study further.

Now we give three examples which are not included in Proposition 1, but conditions (2.1) and C1-C5 still hold.

Example 4.2. Let $f(\varphi, x) = \dfrac{c(\varphi)}{\|x - \varphi\|^2 + 1}$, $\quad c(\varphi) = (\displaystyle\int \frac{1}{\|x - \varphi\|^2 + 1} dx)^{-1}$, $\varphi \in \mathbb{R}^p$, $x \in \mathbb{R}^p$. Then, for any $\tau \in \mathbb{R}^p$,

$$|l(\varphi, x) - l(\tau, x)| \leq \frac{(2\|x\| + \|\varphi\| + \|\tau\|)}{\|x - \varphi\|^2 + 1} \cdot \|\varphi - \tau\|.$$

Example 4.3. Double exponential distribution, $f(x,\varphi) = c(\varphi)\exp(-|x-\varphi|)$, $\varphi \in \mathbb{R}^p$, $x \in \mathbb{R}^p$, where $|\cdot|$ is the block metric defined by $|x-\varphi| = \sum_{i=1}^p |x_i - \varphi_i|$. It is obvious that, for all φ and τ,

$$|l(\varphi,x) - l(\tau,x)| = ||x-\varphi| - |x-\tau|| \le |\varphi - \tau|.$$

The next example satisfies that $P_\theta l(\varphi, Y) = -\infty, P_\theta|g(\varphi,Y)| < +\infty, \varphi \in \mathbb{R}^p$.

Example 4.4. Let $\varphi \in \mathbb{R}^p$, $x \in \mathbb{R}^p$ and

$$f(x,\varphi) = \frac{c(\varphi)}{(\|x-\varphi\|+1)(\log(\|x-\tau\|+2))^{1+\delta}}, 1 > \delta > 0.$$

It is easy to show that, for any φ and τ,

$$|l(\varphi,x) - l(\tau,x)| \le (1 + \frac{1}{2\log 2})\|\varphi - \tau\|.$$

and condition C1–C5 hold for $\Omega_1 = \mathbb{R}^p$.

The last example arises from the reliability design.

Example 4.5. In the reliability design, we often come across the problem of testing the reliability of a series or parallel system, that is, we want to test whether the reliability of a system achieves the specified value and, then, to give its confidence region. For convenience, we only discuss the case of a parallel system composed of two partial systems. Let the life variables of these two partial systems be X and Y, respectively, which are independent of each other and have normal distributions (The mathematical treatment is the same for the case that the life variables have the logarithmic normal distributions), that is, $(X,Y) \sim N((\mu_1,\mu_2)^\tau, \begin{pmatrix} \sigma_1^2 & 0 \\ 0 & \sigma_2^2 \end{pmatrix})$. Then, the lifetimes of X and Y are

$$L_1 = \mu_1 + \Phi^{-1}(1-\alpha)\sigma_1,$$

and

$$L_2 = \mu_2 + \Phi^{-1}(1-\alpha)\sigma_2$$

respectively, where Φ is the standard normal distribution function and $0 < \alpha < 1$ is the probability level of life. Usually α take a value of 0.85, 0.90, 0.95 or 0.99. The problem of the reliability testing can be described as follows.

The null hypothesis $H_0 : L_1 > L_0$ or $L_2 > L_0$ and both σ_1 and σ_2 are positive; the alternative hypothesis $H_A : 0 \le L_1 \le L_0$ and $0 \le L_2 \le L_0$ and both σ_1 and σ_2 are positive; where L_0 is a specified tolerance lower limit for the reliability lifetime. Set

$$\begin{aligned} \Omega_0 = \ & \{(\mu_1,\mu_2,\sigma_1,\sigma_2) : \mu_1 + \Phi^{-1}(1-\alpha)\sigma_1 > L_0 \\ & \text{or} \quad \mu_2 + \Phi^{-1}(1-\alpha)\sigma_2 > L_0, \quad \sigma_1 > 0, \quad \sigma_2 > 0\}, \end{aligned}$$

$$\begin{aligned} \Omega_A = \ & \{(\mu_1,\mu_2,\sigma_1,\sigma_2) : 0 \le \mu_1 + \Phi^{-1}(1-\alpha)\sigma_1 \le L_0 \\ & \text{or} \quad 0 \le \mu_2 + \Phi^{-1}(1-\alpha)\sigma_2 \le L_2, \quad \sigma_1 > 0, \quad \sigma_2 > 0\}. \end{aligned}$$

Let $(X_1, Y_2), \cdots, (X_n, Y_n)$ be n i.i.d. observations of (X, Y). Then,

$$\begin{aligned}
W &= \{-\inf_{\Omega_A} + \inf_{\Omega_0}\}\{\sum_{i=1}^n (\frac{x_i - \mu_1}{\sigma_1})^2 + \sum_{i=1}^n (\frac{Y_i - \mu_2}{\sigma_2})^2 + n(\log \sigma_1^2 + \log \sigma_2^2)\} \\
\overline{\sigma}_1 &= 1/2\Phi^{-1}(1-\alpha)(\mu_1 - L_0) + [((\Phi^{-1}(1-\alpha))^2/4 + 1)(\mu_1 - L_0)^2 + \sigma_1^2]^{1/2}, \\
\overline{\mu}_1 &= L_0 - \Phi^{-1}(1-\alpha)\overline{\sigma}_1, \\
\overline{\sigma}_2 &= 1/2\Phi^{-1}(1-\alpha)(\mu_2 - L_0) + [((\Phi^{-1}(1-\alpha))^2/4 + 1)(\mu_2 - L_0) + \sigma_2^2]^{1/2}, \\
\overline{\mu}_2 &= L_0 - \Phi^{-1}(1-\alpha)\overline{\sigma}_2.
\end{aligned}$$

Let true parameter value of (X, Y) be $(\mu_1, \mu_2, \sigma_1, \sigma_2)$ and $\overline{X} = n^{-1}\sum_{i=1}^n X_i$, $\overline{S}_X^2 = n^{-1}\sum_{i=1}^n (X_i - \overline{X})^2$, $\overline{Y} = n^{-1}\sum_{i=1}^n Y_i$, $\overline{S}_Y^2 = n^{-1}\sum_{i=1}^n (Y_i - \overline{Y})^2$. Define

$$\begin{aligned}
h(x, \mu, v, \sigma, \tau) &= (\frac{x - \mu}{\sigma})^2 + \log \sigma^2 + (\frac{y - v}{\tau})^2 + \log \tau^2, \\
\mathcal{G} &= \{h(\cdot, \mu, v, \sigma, \tau) : (\mu, v, \sigma, \tau) \in \mathbb{R}^2 \times (\mathbb{R}^+)^2\}, \\
\overline{W} &= -\{\log \overline{S}_X^2 + \log \overline{S}_Y^2 + 1\} \\
&\quad + \min\{(\frac{\overline{S}_X^2 + (\overline{X} - \overline{\mu}_1)^2}{\overline{\sigma}_1^2} + \log \overline{\sigma}_1^2 + \log \overline{S}_Y^2), \\
&\quad (\frac{\overline{S}_Y^2 + (\overline{Y} - \overline{\mu}_2)^2}{\overline{\sigma}_2^2} + \log \overline{\sigma}_2^2 + \log \overline{S}_X^2)\}, \\
\Psi_0 &= \{(\psi_1, \psi_3, \psi_2, \psi_4) : \\
&\quad \frac{\sigma_1^2 + (\mu_1 - \psi_1)^2}{\psi_2^2} + \frac{\sigma_2^2 + (\mu_1 - \psi_3)^2}{\psi_4^2} + \log \psi_2^2 + \log \psi_4^2 \\
&= \min[\frac{\sigma_1^2 + (\mu_1 - \overline{\mu}_1)^2}{\overline{\sigma}_1^2} + \log \sigma_2^2 + 1 + \log \overline{\sigma}_1^2, \\
&\quad \frac{\sigma_2^2 + (\mu_2 - \overline{\mu}_2)^2}{\overline{\sigma}_2^2} + 1 + \log \overline{\sigma}_2^2 + \log \sigma_1^2]\}, \\
B_0 &= \Psi_0 \cap \{(\overline{\mu}_1, \mu_2, \overline{\sigma}_1, \sigma_2), \quad (\mu_1, \overline{\mu}_2, \sigma_1, \overline{\sigma}_2)\}.
\end{aligned}$$

If $(\mu_1, \mu_2, \sigma_1, \sigma_2) \in \Omega_A$, then, by Theorem 2.3 we have

$$\sqrt{n}(W/n - \overline{W}) \xrightarrow{d} \underline{W}(h_1) - \sup_{h \in B_0} \underline{W}(h),$$

where $h_1(x) = h(x, \mu_1, \mu_2, \sigma_1, \sigma_2)$ and $\underline{W}$ is a P-bridge indexed by $\mathcal{G}$. The details of proof for this result appeared in Zhang [6]. For $(\mu_1, \mu_2, \sigma_1, \sigma_2)$ lying in the common boundary of Ω_0 and Ω_A or being the inner point of Ω_0, the limit distribution of W has been presented in Zhang [6].

At the end of this paper, we point out that for the case of the parameter space Ω with an infinite dimension, if the covering integral satisfies

$$\int_0^\delta [2\log[N(u, d, \Omega)^2/u]^{1/2} du \to 0 \ (\delta \to o^+),$$

then, Theorem 2.2 and Theorem 2.3 still hold with a suitable modification.

Acknowledgments

This research was supported in part by National Natural Science Foundation of China.

REFERENCES

1. D. Cox and D. Hinkley, *Theoretical Statistics*, Chapman and Hall , London (1974).
2. S. Self and K. Liang, *JASA*, **82**, 605-610 (1987).
3. J. Zhang and P. Cheng, *Journal of Systems Science and Mathematical Science*, **9**, 370-382 (1989).
4. D. Pollard, *Convergence of Stochastic Processes*, Springer-Verlag, New York (1984).
5. A. Wald, *Ann. Math. Statist.*, **20**, 595-601 (1949).
6. J. Zhang, *Technical Report*, Institute of Systems Science, Academia Sinica, Beijing (1991).

Stat. Sci. & Data Anal., pp. 337-343
K. Matsusita *et al.* (Eds)

Test of homogeneity of parameters

TAKESI HAYAKAWA
Hitotsubashi University, Kunitachi, Tokyo, Japan

Abstract. The paper deals with the asymptotic expansions of the distribution of the likelihood ratio criterion for testing homogeneity of parameters of k population and of the power function of it.

Key words: Likelihood ratio test, homogeneity, asymptotic expansion, power function.

1. INTRODUCTION

Let $x_i' = (x_{i1}, \cdots, x_{in_i})$, $i = 1, 2, \cdots, k$ be a random sample of the i-th population with a probability density function (p.d.f.) $f(x|\theta_i)$ which depends on an unknown parameter θ_i. The problem considered here is to test a hypothesis of homogeneity of parameters

$$H : \theta_1 = \theta_2 = \cdots = \theta_k \quad (= \theta, \text{say}),$$

against the alternative

$$K : \text{violation of at least one equality.}$$

The likelihood ratio criterion (LRC) for H against K would be expressed as

$$\lambda = \prod_{i=1}^{k} \prod_{\alpha=1}^{n_i} \frac{f(x_{i\alpha}|\tilde{\theta})}{f(x_{i\alpha}|\hat{\theta}_i)}, \tag{1}$$

where $\hat{\theta}_i$ is the maximum likelihood estimator (m.l.e.) of θ_i based on n_i observations x_i under K, and $\tilde{\theta}$ is the m.l.e. of θ under H based on $n = \sum_{i=1}^{k} n_i$ observations $x_1, x_2, \cdots, x_k$, respectively. Hayakawa [1] considered this problem under a condition that there was a linear relation between $\tilde{\theta}$ and $\hat{\theta}_1, \hat{\theta}_2, \cdots, \hat{\theta}_k$ such that

$$\tilde{\theta} = \rho_1 \hat{\theta}_1 + \rho_2 \hat{\theta}_2 + \cdots + \rho_k \hat{\theta}_k , \tag{2}$$

where $\rho_i = n_i/n \ (> 0)$, $\sum_{i=1}^{k} \rho_i = 1$.

The purpose of this paper is to study the asymptotic behavior of $-2 \log \lambda$ under various situations without this restriction (2).

Defining the log-likelihood function by

$$L_i(\theta_i) = \sum_{\alpha=1}^{n_i} \log f(x_{i\alpha}|\theta_i), \quad i = 1, 2, \cdots, k ,$$

we have the statistic $2\log\lambda = 2\sum_{i=1}^{k}\{L_i(\tilde{\theta}) - L_i(\hat{\theta}_i)\}$. The following notations and conventions will be used. We assume that each $L_i(\theta_i)$ should be regular with respect to θ_i-derivatives.

(i) $$y_i^{(l)} = n_i^{-l/2}\frac{\partial^l L_i(\theta_i)}{\partial\theta_i^l}, \quad l = 1,2,\cdots, \quad z_i \equiv y_i^{(1)}$$

$$v_i = \sqrt{n_i}(\hat{\theta}_i - \theta_i), \qquad w = \sqrt{n}(\tilde{\theta} - \theta)$$

$$m_{(r_1^{\alpha_1},\ldots,r_l^{\alpha_l})}(\theta_i) = \int\left\{\frac{\partial^{r_1}\log f(x|\theta_i)}{\partial\theta_i^{r_1}}\right\}^{\alpha_1}\cdots\left\{\frac{\partial^{r_l}\log f(x|\theta_i)}{\partial\theta_i^{r_l}}\right\}^{\alpha_l} f(x|\theta_i)dx$$

(ii) The following equalities hold for $m_{(\cdot)}$'s by the regularity condition.

$$\begin{aligned} m_{(2)}(\theta_i) &+ m_{(1^2)}(\theta_i) = 0 \\ m_{(3)}(\theta_i) &+ 3m_{(21)}(\theta_i) + m_{(1^3)}(\theta_i) = 0 \\ m_{(4)}(\theta_i) &+ 4m_{(31)}(\theta_i) + 3m_{(2^2)}(\theta_i) + 6m_{(21^2)}(\theta_i) + m_{(1^4)}(\theta_i) = 0 \end{aligned} \tag{3}$$

(iii) Any function evaluated at $\theta = \hat{\theta}$ will be denoted by attaching the symbol $\wedge$.

2. NULL DISTRIBUTION OF LRC

Expanding $L_i(\tilde{\theta})$ in Taylor series at $\tilde{\theta} = \hat{\theta}_i$, and noting $L_i^{(1)}(\hat{\theta}_i) = 0$ as $\hat{\theta}_i$ is the m.l.e. of θ_i, we have

$$\begin{aligned} 2\log\lambda &= \sum_{i=1}^{k} L_i^{(2)}(\hat{\theta}_i)(\tilde{\theta} - \hat{\theta}_i)^2 + \frac{1}{3}\sum_{i=1}^{k} L_i^{(3)}(\hat{\theta}_i)(\tilde{\theta} - \hat{\theta}_i)^3 \\ &\quad + \frac{1}{12}\sum_{i=1}^{k} L_i^{(4)}(\hat{\theta}_i)(\tilde{\theta} - \hat{\theta}_i)^4 + \cdots, \end{aligned} \tag{4}$$

where $L_i^{(l)}(\cdot)$ is the l-th derivative of $L_i(\cdot)$. Expanding $\frac{\partial L_i(\tilde{\theta})}{\partial\theta}$ at θ, we have

$$\begin{aligned} 0 &= \sum_{i=1}^{k}\left.\frac{\partial L_i(\theta)}{\partial\theta}\right|_{\theta=\tilde{\theta}} \\ &= \sum_{i=1}^{k}\frac{\partial L_i(\theta)}{\partial\theta} + \sum_{i=1}^{k}\frac{\partial^2 L_i(\theta)}{\partial\theta^2}(\tilde{\theta} - \theta) + \frac{1}{2}\sum_{i=1}^{k}\frac{\partial^3 L_i(\theta)}{\partial\theta^3}(\tilde{\theta} - \theta)^2 \\ &\quad + \frac{1}{6}\sum_{i=1}^{k}\frac{\partial^4 L_i(\theta)}{\partial\theta^4}(\tilde{\theta} - \theta)^3 + \cdots \\ &= \sqrt{n}\sum\sqrt{\rho_i}z_i + \sqrt{n}\sum\rho_i y_i^{(2)}\cdot w + \frac{1}{2}\sqrt{n}\sum\rho_i\sqrt{\rho_i}y_i^{(3)}\cdot w^2 \\ &\quad + \frac{1}{6}\sqrt{n}\sum\rho_i^2 y_i^{(4)}\cdot w^3 + \cdots . \end{aligned}$$

Solving the above equation with respect to w, we have

$$\begin{aligned} w &= -\frac{\sum\sqrt{\rho_i}z_i}{\sum\rho_i y_i^{(2)}} - \frac{1}{2}\frac{\sum\rho_i\sqrt{\rho_i}y_i^{(3)}}{\{\sum\rho_i y_i^{(2)}\}^3}(\sum\sqrt{\rho_i}z_i)^2 \\ &\quad - \frac{1}{2}\frac{\{\sum\rho_i\sqrt{\rho_i}y_i^{(3)}\}^2}{\{\sum\rho_i y_i^{(2)}\}^5}(\sum\sqrt{\rho_i}z_i)^3 + \frac{1}{6}\frac{\sum\rho_i^2 y_i^{(4)}}{\{\sum\rho_i y_i^{(2)}\}^4}(\sum\sqrt{\rho_i}z_i)^3 + o_p\left(\frac{1}{n}\right). \end{aligned}$$

Similary we have

$$v_i = -\frac{z_i}{y_i^{(2)}} - \frac{1}{2}\frac{y_i^{(3)}}{(y_i^{(2)})^3}z_i^2 - \frac{1}{2}\frac{(y_i^{(3)})^2}{(y_i^{(2)})^5}z_i^3 + \frac{1}{6}\frac{y_i^{(4)}}{(y_i^{(2)})^4}z_i^3 + o_p\left(\frac{1}{n}\right).$$

Thus we have

$$\begin{aligned}
\sqrt{n}(\tilde{\theta} - \hat{\theta}_i) &= \frac{1}{\sqrt{\rho_i}}\frac{z_i}{y_i^{(2)}} - \frac{\sum\sqrt{\rho_j}z_j}{\sum \rho_j y_j^{(2)}} + \frac{1}{2\sqrt{\rho_i}}\frac{y_i^{(3)}}{\{y_i^{(2)}\}^3}z_i^2 \\
&\quad - \frac{1}{2}\frac{\sum \rho_j\sqrt{\rho_j}y_j^{(3)}}{(\sum \rho_j y_j^{(2)})^3}(\sum \rho_j z_j)^2 \\
&\quad + \frac{1}{2\sqrt{\rho_i}}\frac{(y_i^{(3)})^2}{\{y_i^{(2)}\}^5}z_i^3 - \frac{1}{2}\frac{(\sum \rho_j\sqrt{\rho_j}y_j^{(3)})^2}{(\sum \rho_j y_j^{(2)})^5}(\sum\sqrt{\rho_j}z_j)^3 \\
&\quad - \frac{1}{6\sqrt{\rho_i}}\frac{y_i^{(4)}}{\{y_i^{(2)}\}^4}z_i^3 + \frac{1}{6}\frac{\sum \rho_j^2 y_j^{(4)}}{(\sum \rho_j y_j^{(2)})^4}(\sum\sqrt{\rho_j}z_j)^3 + o_p\left(\frac{1}{n}\right) \qquad (5)
\end{aligned}$$

Inserting (5) in (4), the LRC is asymptotically expanded under the null hypothesis as follows.

$$\begin{aligned}
2\log\lambda &= l_0 + l_1 + l_2 + o_p(1/n)\,, \qquad (6)\\
l_0 &= \sum_{i=1}^{k}\rho_i y_i^{(2)}\left\{\frac{z_i}{\sqrt{\rho_i}y_i^{(2)}} - \frac{\sum\sqrt{\rho_j}z_j}{\sum \rho_j y_j^{(2)}}\right\}^2, \\
l_1 &= \frac{1}{3}\sum_{i=1}^{k}\frac{y_i^{(3)}z_i^3}{(y_i^{(2)})^3} - \frac{1}{3}\sum_{i=1}^{k}\rho_i\sqrt{\rho_i}y_i^{(3)}\frac{(\sum\sqrt{\rho_i}z_i)^3}{(\sum \rho_i y_i^{(2)})^3}, \\
l_2 &= \frac{1}{4}\sum_{i=1}^{k}\frac{(y_i^{(3)})^2 z_i^4}{(y_i^{(2)})^5} - \frac{1}{4}\frac{(\sum \rho_i\sqrt{\rho_i}y_i^{(3)})^2(\sum\sqrt{\rho_j}z_j)^4}{(\sum \rho_i y_i^{(2)})^5} \\
&\quad - \frac{1}{12}\sum\frac{y_i^{(4)}z_i^4}{(y_i^{(2)})^4} + \frac{1}{12}\sum \rho_i^2 y_i^{(4)}\frac{(\sum\sqrt{\rho_i}z_i)^4}{(\sum \rho_i y_i^{(2)})^4}
\end{aligned}$$

To find the moment generating function $M(t)$ of $-2\log\lambda$, we need to use the Edgeworth type A series expansion of the joint probability density function of z_i, $y_i^{(2)}$, $y_i^{(3)}$ and $y_i^{(4)}$, $i = 1, 2, \cdots, k$, which would be stated in the following lemma.

Lemma. The joint probability density function of z_i, $y_i^{(2)}$, $y_i^{(3)}$ and $y_i^{(4)}$, $i = 1, 2, \cdots, k$ is expressed up to order $O(1/n)$ as,

$$f = f_0\left[1 + \frac{F_1}{\sqrt{n}} + \frac{F_2}{n}\right] + o\left(\frac{1}{n}\right)\,, \qquad (7)$$

where

$$\begin{aligned}
f_0 &= \prod_{i=1}^{k}(2\pi m_{(1^2)}(\theta_i))^{-1/2}\exp\{-z_i^2/2m_{(1^2)}(\theta_i)\}\prod_{l=2}^{4}\delta_{li} \\
F_1 &= \frac{1}{6}\sum_{i=1}^{k}\frac{1}{\sqrt{\rho_i}}m_{(1^3)}(\theta_i)H_3(z_i) - \sum_{i=1}^{k}\frac{1}{\sqrt{\rho_i}}m_{(21)}(\theta_i)H_1(z_i)d_{2i}^{(1)}
\end{aligned}$$

$$
\begin{aligned}
F_2 &= \frac{1}{2}\sum_{i=1}^{k}\frac{1}{\rho_i}\{m_{(2^2)}(\theta_i) - m_{(2)}^2(\theta_i)\}d_{2i}^{(2)} \\
&\quad - \frac{1}{2}\sum_{i=1}^{k}\frac{1}{\rho_i}\{m_{(21^2)}(\theta_i) - m_{(2)}(\theta_i)m_{(1^2)}(\theta_i)\}H_2(z_i)d_{2i}^{(1)} \\
&\quad - \sum_{i=1}^{k}\frac{1}{\rho_i}m_{(31)}(\theta_i)H_1(z_i)d_{3i}^{(1)} + \frac{1}{2}\sum_{i=1}^{k}\frac{1}{\rho_i}m_{(21^2)}(\theta_i)H_2(z_i)d_{2i}^{(2)} \\
&\quad + \frac{1}{24}\sum_{i=1}^{k}\frac{1}{\rho_i}\{m_{(1^4)}(\theta_i) - 3m_{(1^2)}^2(\theta_i)\}H_4(z_i) \\
&\quad - \frac{1}{6}\sum_{i=1}^{k}\frac{1}{\rho_i}m_{(21)}(\theta_i)m_{(1^3)}(\theta_i)H_4(z_i)d_{2i}^{(1)} \\
&\quad + \frac{1}{72}\sum_{i=1}^{k}\frac{1}{\rho_i}m_{(1^3)}(\theta_i)H_6(z_i) \\
&\quad + \frac{1}{2}\sum_{i\neq j}\frac{1}{\sqrt{\rho_i\rho_j}}m_{(21)}(\theta_i)m_{(21)}(\theta_j)H_1(z_i)H_1(z_j)d_{2i}^{(1)}d_{2j}^{(1)} \\
&\quad - \frac{1}{6}\sum_{i\neq j}\frac{1}{\sqrt{\rho_i\rho_j}}m_{(21)}(\theta_i)m_{(1^3)}(\theta_j)H_1(z_i)H_3(z_j)d_{2i}^{(1)} \\
&\quad + \frac{1}{72}\sum_{i\neq j}\frac{1}{\sqrt{\rho_i\rho_j}}m_{(1^3)}(\theta_i)m_{(1^3)}(\theta_j)H_3(z_i)H_3(z_j)\,, \\
\delta_{li} &= \delta(y_i^{(l)} - m_{(l)}(\theta_i)/n_i^{(l-2)/2}) \\
d_{li}^{(r)} &\equiv \delta^{(r)}(y_i^{(l)} - m_{(l)}(\theta_i)/n_i^{(l-2)/2})/\delta(y_i^{(l)} - m_{(l)}(\theta_i)/n_i^{(l-2)/2}),
\end{aligned}
$$

and $\delta^{(r)}$ is the r-th derivative of Dirac delta-function δ. δ enjoys the following properties.

$$
\begin{aligned}
\delta(x-a) &= 0\,, \qquad x \neq a \\
\int \delta(x-a)dx &= 1 \\
\int h(\cdots,x,\cdots)\delta(x-a)dx &= h(\cdots,a\,,\cdots) \\
\int h(\cdots,x,\cdots)\delta^{(r)}(x-a)dx &= (-1)^r\left[\frac{\partial^r h}{\partial x^r}\right]_{x=a}
\end{aligned}
$$

$H_r(z)$ is defined by

$$
\frac{d^n}{dx^n}\exp\left(-\frac{x^2}{2\alpha}\right) = (-1)^n H_n(x)\exp\left(-\frac{x^2}{2\alpha}\right)
$$

The moment generating function $M(t)$ of $-2\log\lambda$ under H is expressed as

$$
\begin{aligned}
M(t) &= E[\exp(-2t\log\lambda)] \\
&= \int\cdots\int \exp\{-2t\log\lambda\}f\cdot\Pi dz_i dy_i^{(2)}dy_i^{(3)}dy_i^{(4)}\,,
\end{aligned}
$$

where all $m_{(\ \ )}(\theta_i)$'s are replaced by $m_{(\ \ )}(\theta)$ in f. Carrying out the integration with respect to z_i, $y_i^{(2)}$, $y_i^{(3)}$, $y_i^{(4)}$, $i = 1, 2, \cdots, k$ and using the regularity conditions, we have after some lengthy calculation

$$
M(t) = (1-2t)^{-(k-1)/2}\left[1 + \frac{A}{n}\left\{\frac{1}{1-2t} - 1\right\} + o\left(\frac{1}{n}\right)\right]\,,
$$

$$A = \left[\frac{1}{8m_{(1^2)}^2}\{m_{(2^2)} - 2m_{(21^2)} - m_{(1^4)}\} + \frac{1}{24m_{(1^2)}^3}\{m_{(3)}m_{(21)} - 8m_{(21)}m_{(1^3)} - 5m_{(3)}m_{(1^3)}\}\right]\left(\sum\frac{1}{\rho_i} - 1\right),$$

$m_{(\cdot)}$ denotes $m_{(\cdot)}(\theta)$, respectively. Inverting $M(t)$, the distribution function of $-2\log\lambda$ is expressed asymptotically as

$$P\{-2\log\lambda \le x|H\} = P_{k-1} + \frac{A}{n}\{P_{k+1} - P_{k-1}\} + o\left(\frac{1}{n}\right), \tag{8}$$

where $P_f = P\{\chi_f^2 \le x\}$.

Inclusion of $\sum 1/\rho_i$ in (8) shows that the homogeneity of sample size and small number of populations would be necessary for having a good approximation to the limiting chi-square distribution. By taking

$$c = 1 - \frac{2A}{n(k-1)} + o\left(\frac{1}{n}\right), \tag{9}$$

we have

$$P\{-2c\log\lambda \le x|H\} = P_{k-1} + o\left(\frac{1}{n}\right).$$

c is Bartlett correction factor[2].

Example For the cases (i) $N(\theta,1)$ and (ii) $\log N(\theta,1)$, the correction factor becomes $c = 1 + o(1/n)$. It is natural because the exact distribution of $-2\log\lambda$ for these cases under the null hypothesis H is nothing but a chi-square one with $k-1$ degrees of freedom. For the cases of

(iii) Laplace distribution with pdf

$$\exp\{-|x|/\theta\}/2\theta, \quad -\infty < x < \infty. \quad \theta > 0$$

(iv) Gamma distribution with pdf

$$\exp\{-x/\theta\}x^{a-1}/\Gamma(a)\theta^a, \quad x \ge 0,\ a > 0,\ \theta > 0$$

(v) Weibull distribution with pdf

$$mx^{m-1}\exp\{-x^m/\theta\}/\theta, \quad x \ge 0,\ m > 0,\ \theta > 0$$

(vi) Double exponential distribution with pdf

$$\exp\{-\exp(x)/\theta + x\}/\theta, \quad x \ge 0,\ \theta > 0,$$

the correction factor becomes

$$c = 1 - \left(\sum_{i=1}^{k}\frac{1}{\rho_i} - 1\right)/3n(k-1) + o(1/n), \tag{10}$$

which is the correction factor due to Bartlett [2] for the test of homogeneity of variances of k normal populations.

(vii) Langevin distribution with pdf

$$\exp(\theta\mu' x)/a_p(\theta), \quad \theta > 0, \quad x,\mu \in S_p = \{y \mid y \in R^p,\ y'y = 1\}$$

where μ is a known vector and

$$a_p(\theta) = \frac{2\pi^{\frac{p}{2}}}{\Gamma\left(\frac{p}{2}\right)}\, {}_0F_1\left(\frac{p}{2}; \frac{\theta^2}{4}\right)$$

with ${}_0F_1$ denoting the Bessel function. The correction factor is

$$c = 1 - \frac{1}{n(k-1)}\left\{\frac{5}{12}\frac{(A'')^2}{(A')^3} - \frac{1}{4}\frac{A'''}{(A')^2}\right\}\left(\sum_{i=1}^{k}\frac{1}{\rho_i} - 1\right)$$

and

$$A(\theta) = \frac{d}{d\theta}\log a_p(\theta), \quad A' = \frac{dA}{d\theta}, \text{ etc.}$$

3. LRC UNDER LOCAL ALTERNATIVES

In this section we consider the asymptotic behavior of LRC under the sequence of Pitman's local alternatives K_n,

$$K_n : \theta_i = \theta + \varepsilon_i/\sqrt{n_i}\,, \quad i = 1, 2, \cdots, k.$$

It should be noted that w is expanded under K_n as

$$\begin{aligned} \sqrt{n}(\tilde{\theta} - \theta) &= -\frac{1}{\sum \rho_i y_i^{(2)}} \cdot \sum \sqrt{\rho_i}(z_i - \varepsilon_i y_i^{(2)}) - \frac{1}{2}\sum \sqrt{\rho_i}\varepsilon_i^2 y_i^{(3)} \cdot \frac{1}{\sum \rho_i y_i^{(2)}} \\ &\quad - \sum \rho_i \varepsilon_i y_i^{(3)} \cdot \frac{\sum \sqrt{\rho_i}(z_i - \varepsilon_i y_i^{(2)})}{(\sum \rho_i y_i^{(2)})^2} \\ &\quad - \frac{1}{2}\sum \rho_i \sqrt{\rho_i} y_i^{(3)} \cdot \frac{\{\sum \sqrt{\rho_i}(z_i - \varepsilon_i y_i^{(2)})\}^2}{(\sum \rho_i y_i^{(2)})^3} + o_p(1/\sqrt{n}) \end{aligned} \tag{11}$$

and LRC is expanded under K_n as

$$\begin{aligned} 2\log\lambda &= \bar{l}_0 + \bar{l}_1 + o_p(1/\sqrt{n}), \\ \bar{l}_0 &= \sum_{i=1}^{k}\frac{1}{y_i^{(2)}}(z_i - \varepsilon_i y_i^{(2)})^2 - \frac{\{\sum \sqrt{\rho_i}(z_i - \varepsilon_i y_i^{(2)})\}^2}{\sum \rho_i y_i^{(2)}}, \\ \bar{l}_1 &= \sum_{i=1}^{k} y_i^{(3)} \cdot \frac{z_i(z_i - \varepsilon_i y_i^{(2)})\varepsilon_i}{(y_i^{(2)})^2} - \sum_{i=1}^{k}\sqrt{\rho_i} y_i^{(3)}\varepsilon_i^2 \cdot \frac{\sum \sqrt{\rho_i}(z_i - \varepsilon_i y_i^{(2)})}{\sum \rho_i y_i^{(2)}} \\ &\quad - \sum_{i=1}^{k}\rho_i \varepsilon_i y_i^{(3)} \cdot \frac{\{\sum \sqrt{\rho_i}(z_i - \varepsilon_i y_i^{(2)})\}^2}{(\sum \rho_i y_i^{(2)})^2} \\ &\quad + \frac{1}{3}\sum_{i=1}^{k} y_i^{(3)} \cdot \frac{(z_i - \varepsilon_i y_i^{(2)})^3}{(y_i^{(2)})^3} - \frac{1}{3}\sum_{i=1}^{k}\rho_i\sqrt{\rho_i} y_i^{(3)} \cdot \frac{\{\sum \sqrt{\rho_i}(z_i - \varepsilon_i y_i^{(2)})\}^3}{(\sum \rho_i y_i^{(2)})^3} \end{aligned}$$

To handle the moment generating function $\bar{M}(t)$ of $-2\log\lambda$ under K_n, we expand all functions $m_{(1^2)}(\theta_i)$ at θ as

$$m_{(1^2)}(\theta_i) = m_{(1^2)}(\theta) - \frac{1}{\sqrt{n}}\frac{\varepsilon_i}{\sqrt{\rho_i}}(m_{(3)}(\theta) + m_{(21)}(\theta)) + o\left(\frac{1}{\sqrt{n}}\right).$$

Then f of (7) is expressed as

$$f = \prod_{i=1}^{k} \frac{1}{\sqrt{2\pi m_{(1^2)}}} \exp\left\{-\frac{1}{2}\frac{z_i^2}{m_{(1^2)}}\right\} \left[1 + \frac{1}{\sqrt{n}}\bar{F}_1 + o\left(\frac{1}{\sqrt{n}}\right)\right] \tag{12}$$

$$\bar{F}_1 = \frac{1}{6} m_{(1^3)} \sum_{i=1}^{k} \frac{1}{\sqrt{\rho_i}} H_3(z_i) - m_{(21)} \sum_{i=1}^{k} \frac{1}{\sqrt{\rho_i}} H_1(z_i) \bar{d}_{2i}^{(1)} - \frac{1}{2m_{(1^2)}} (m_{(3)} + m_{(21)}) \sum_{i=1}^{k} \frac{\varepsilon_i}{\sqrt{\rho_i}} (z_i^2 - m_{(1^2)}),$$

where $\bar{d}_{2i}^{(1)} = \delta^{(1)}(y_i^{(2)} - m_{(2)}(\theta)) \,/ \delta(y_i^{(2)} - m_{(2)}(\theta))$ and all $m_{(\cdot)}$'s are $m_{(\cdot)}(\theta)$, respectively. Then the moment generating function $\bar{M}(t)$ of $-2\log\lambda$ under K_n is expressed as

$$\bar{M}(t) = (1-2t)^{-(k-1)/2} \exp\left[\frac{t}{1-2t} m_{(1^2)} \cdot \left\{\sum \varepsilon_i^2 - \left(\sum \sqrt{\rho_i}\varepsilon_i\right)^2\right\}\right] \cdot \left[1 + \frac{1}{\sqrt{n}} \sum_{l=0}^{2} \bar{a}_l (1-2t)^{-l} + o(1/\sqrt{n})\right]. \tag{13}$$

where

$$\bar{a}_2 = \frac{1}{6} m_{(1^3)} (\pi_1 - 3\pi_2 + 2\pi_3)$$

$$\bar{a}_1 = \frac{1}{2} m_{(21)} \pi_1 + \frac{1}{2} (m_{(1^3)} - m_{(21)}) \pi_2 - \frac{1}{2} m_{(1^3)} \pi_3$$

$$\bar{a}_0 = -\frac{1}{6} (m_{(1^3)} + 3m_{(21)}) \pi_1 + \frac{1}{2} m_{(21)} \pi_2 + \frac{1}{6} m_{(1^3)} \pi_3$$

$$\pi_1 = \sum_{i=1}^{k} \frac{\varepsilon_i^3}{\sqrt{\rho_i}}, \quad \pi_2 = \sum_{i=1}^{k} \varepsilon_i^2 \sum_{i=1}^{k} \sqrt{\rho_i}\varepsilon_i, \quad \pi_3 = \left(\sum_{i=1}^{k} \sqrt{\rho_i}\varepsilon_i\right)^3$$

Inversion of $\bar{M}(t)$ gives the asymptotic expansion of the power function of $-2\log\lambda$ under the Pitman's local alternatives as

$$P\{-2\log\lambda \le x | K_n\} = \bar{P}_{k-1}(\Delta^2) + \frac{1}{\sqrt{n}} \sum_{i=0}^{2} \bar{a}_l \bar{P}_{k+2l-1}(\Delta^2) + o\left(\frac{1}{\sqrt{n}}\right),$$

where $\bar{P}_f(\Delta^2) = P\{\chi_f^2(\Delta^2) \le x\}$ and $\chi_f^2(\Delta^2)$ is a non-central chi-square random variable with the non-centrality parameter $\Delta^2 = \frac{1}{2} m_{(1^2)} \sum_{i=1}^{k} \{\varepsilon_i - \sqrt{\rho_i}(\sum_{j=1}^{k} \sqrt{\rho_j}\varepsilon_j)\}^2$ and f degrees of freedom. This shows that $-2\log\lambda$ is locally unbiased because of the homogeneity of parameter ε_i's.

REFERENCES

1. Takesi Hayakawa, in: *Essays in Probability and Statistics* (Prof. J. Ogawa Memorial Volume), pp. 265-285. Shinko Tsusho, Co. Ltd. (1976).
2. M.S. Bartlett, *Proc. Roy. Soc., London,* **A160**, 268-282 (1937).

Stat. Sci. & Data Anal., pp. 345-356
K. Matsusita *et al.* (Eds)

Determining the No-Observed-Adverse-Effect Level in Continuous Response

YASUKI KIKUCHI*, TAKASHI YANAGAWA** and HARUTOSHI NISHIYAMA**
** Department of Genaral Education, Sasebo College of Technology, Sasebo 857-11, Japan*
*** Department of Mathematics, Kyushu University 33, Fukuoka 812, Japan*

Abstract The determination of the no-observed-adverse-effect level (NOAEL) is discussed when observed response are from normal distribution. Three tests which incorporate order restriction, and may be used for the determination of the NOAEL are critically examined, including the Williams test, the Dunnett type multiple comparison test, and a new test proposed herein. Tables of the critical values for the Dunnett type multiple comparison test and the new test are given. An alternative method based on the Akaike Information Criterion (AIC) is also developed. It is shown that these testing procedures must be cautiously applied in deciding the NOAEL; that the procedure based on the AIC performs well.

Key words : AIC procedure, Dunnett type multiple comparison test, no-observed-adverse-effect level, order restriction, pooling step-up test, toxicology, Williams test

1 INTRODUCTION

The no-observed-adverse-effect level (NOAEL) is the highest experimental dose at which increase in adverse toxicological endpoint is zero or otherwise acceptably small. The concept of the NOAEL is central to systematic assesment of risk from toxicants, as currently practiced. For example, inclusion of the NOAEL value in the report of laboratory experiments is recommended by the Pharmaceutical Affairs Bureau, Japanese government (GLP, 1989). Also the U.S.EPA utilizes the NOAEL in setting regulatory levels for exposure to non-cancerous toxic substances (U.S.EPA [1], [2]).

An experiment with continuous response data is described by the number of experimental subjects at risk (n_i), average response of interest (X_i), and the exposure level (d_i), for i=0, 1,..., k as given in Table 1. The subscript zero refers to the control group, making $d_0=0$; otherwise the dose values are arbitrary, subject to order $0=d_0<d_1<\cdots<d_k$.

Table 1.
Continuous Response Data

Dose Level	$d_0=0$	<	d_1	<	d_2	<	$\cdots$	<	d_i	<	$\cdots$	<	d_k
Number of Subjects	n_0		n_1		n_2		$\cdots$		n_i		$\cdots$		n_k
Average Response	X_0		X_1		X_2		$\cdots$		X_i		$\cdots$		X_k

We suppose in this paper that X_i's are independently distributed, X_i is distributed as $N(\mu_i, \sigma^2/n_i)$, and that an unbiased estimator S^2 of σ^2 is distributed as $\sigma^2\chi^2_\nu/\nu$ independently of X_i, where ν is the degrees of freedom of χ^2_ν. The corresponding problem in categorical response data has been discussed in the twin paper (Yanagawa et al. [3]).

One statistical approach that may be employed for deciding the NOAEL is to test the hypothesis of no difference in the true mean between the control group and a treatment group, paring the control group for a test with each treatment group sequentially. In this paper, we study first the behavior of the three tests which incorporate order restriction, and may be used for the determination of the NOAEL, including the Williams test, the Dunnett type multiple comparison test, and a new test proposed herein. An alternative method based on the Akaike Information Criterion (AIC) is also developed. Simulation studies are carried out to compare these procedures. A new procedure implementing the Akaike Information Criterion (AIC) is shown to work well. These three tests and the AIC-procedure are applied to the data of hemoglobin in blood discussed in Yoshimura [4].

2 THE TESTS FOR THE NOAEL

Suppose that μ_i are monotonically ordered with $\mu_0 \le \mu_1 \le \cdots \le \mu_k$. Alternatively, one could suppose that $\mu_0 \ge \mu_1 \ge \cdots \ge \mu_k$. Let $\widehat{\mu}_i$ and $\widehat{\mu}_i^*$ be the maximum likelihood estimators (mle's) of μ_i subject to the constraints $\mu_0 \le \mu_1 \le \cdots \le \mu_k$ and $\mu_1 \le \mu_2 \le \cdots \le \mu_k$, respectively, constructed by the pool-adjacent-violators algolithm (PAVA) (Barlow et al. [5]). Let δ denote the largest value d_i such that $\mu_0 = \mu_i$. In this section, we consider the Williams test, the Dunnett type multiple comparison test, and a new test not using the PAVA estimator. These testing procedures assign the NOAEL a dose value based on the sample data.

2.1 Williams Test

Initially, the null hypothesis $H_0^k : \mu_0 = \mu_1 = \cdots = \mu_k$ is tested against $H_a^k : \mu_0 \le \mu_1 \le \cdots \le \mu_k$ $(\mu_0 < \mu_k)$. If H_0^k is not rejected, then the estimated NOAEL takes the value d_k, i.e., $\widehat{\delta} = d_k$; if it is rejected, $\widehat{\delta}$ is d_0, $d_1, \ldots$, d_{k-2} or d_{k-1}, as determined by the subsequent test. Thus we could write $H_0^k : \delta = d_k$ vs. $H_a^k : \delta < d_k$. If H_0^k is rejected, then $H_0^{k-1} : \mu_0 = \mu_1 = \cdots = \mu_{k-1}$ should be tested conditionally on having rejected H_0^k. The alternative hypothesis can be written as $H_a^{k-1} : \mu_0 \le \mu_1 \le \cdots \le \mu_{k-1}$ $(\mu_0 < \mu_{k-1})$. Equivalently, the second test is of $H_0^{k-1} : \delta = d_{k-1}$ vs. $H_a^{k-1} : \delta < d_{k-1}$. If H_0^{k-1} is not rejected, then $\widehat{\delta} = d_{k-1}$; otherwise, the third hypothesis H_0^{k-2} is tested conditionally on having rejected H_0^k and H_0^{k-1}. This procedure is continued until one of the null hypotheses is not rejected or the hypothesis $H_0^1 : \mu_0 = \mu_1$ is tested against $H_a^1 : \mu_0 < \mu_1$ conditionally on having rejected $H_0^k, \ldots$, H_0^3 and $H_0^2 : \mu_0 = \mu_1 = \mu_2$. The test statistic employed in testing $H_0^i : \mu_0 = \mu_1 = \cdots = \mu_i$ vs. $H_a^i : \mu_0 \le \mu_1 \le \cdots \le \mu_i$ $(\mu_0 < \mu_i)$ is $T_i^* = (\widehat{\mu}_i^* - X_0)[2S^2/n]^{-1/2}$, where $n = n_0 = n_1 = \cdots = n_k$. The critical values of this test, $t_{i,\nu}(\alpha)$; $i = 1, 2, \ldots, k$, are given in Williams [6] when $n_0 = n_1 = \cdots = n_k$, and Williams [7] when $n_0 > n_1 = \cdots = n_k$. For a specified test size α, the test rejects H_0^i if $t_i^* \ge t_{i,\nu}(\alpha)$. Although error is reported in computing the critical values (Brown [8]), we find that these values provide good approximation to the nominal values.

One may replace $\widehat{\mu}_i^*$ with $\widehat{\mu}_i$ in the Williams test. The justification of this replacement is discussed in Appendix 1.

2.2 The Dunnett type Multiple Comparison Test

Alternatively we may incorporate the order restriction into the Dunnett type multiple comparison test (Dunnett [9]) for the NOAEL. With the PAVA estimator $\widehat{\mu}_i$, this leads to the statistic $T_i = (\widehat{\mu}_i - \widehat{\mu}_0)[2S^2/n]^{-1/2}$, where $n=n_0=n_1=\cdots=n_k$. For a specified test size α, this test first selects the critical point $d_{k,\nu}(\alpha)$ such that

$$Pr[T_k < d_{k,\nu}(\alpha)] = 1 - \alpha,$$

under $\mu_0=\mu_1=\cdots=\mu_k$, from which the estimated NOAEL, $\widehat{\delta}$, is determined according to:

if $t_1 \geq d_{k,\nu}(\alpha)$, then $\widehat{\delta} = d_0$,
if $t_1 < d_{k,\nu}(\alpha)$ and $t_2 \geq d_{k,\nu}(\alpha)$, then $\widehat{\delta} = d_1$,
$\vdots$
if $t_{k-1} < d_{k,\nu}(\alpha)$ and $t_k \geq d_{k,\nu}(\alpha)$, then $\widehat{\delta} = d_{k-1}$,
if $t_k < d_{k,\nu}(\alpha)$, then $\widehat{\delta} = d_k$.

The distribution of T_k under $\mu_0=\mu_1=\cdots=\mu_k$ for k=2, 3 and 4 is given in Appendix 2. Using this distribution, the exact critical points $d_{k,\nu}(\alpha)$ for k=2, 3 and 4; α=0.05; and ν=3(1)30, 32(2)50, 60, 120 and ∞ are computed and given in Appendix 3. Note that Williams [10] gives the critical points of this test based on the limiting distribution as $k\to\infty$, and Marcus [11] gives those values when $\nu\to\infty$.

2.3 A Pooling Step-Up Test

This test does not use the PAVA estimator, but uses the original response X_i, and pools X_0, $X_1,\ldots$, so far as no significant difference is detected between dose levels d_0 and d_1, between d_0 and d_2, and so on, to increase the power of the test. Initially the null hypothesis $H_0^1 : \mu_0=\mu_1$ is tested against $H_a^1 : \mu_0<\mu_1$. If H_0^1 is rejected, then the estimated NOAEL takes the value $\widehat{\delta} = d_0$; if it is not rejected, then $\widehat{\delta}$ is $d_1, d_2,\ldots, d_{k-1}$ or d_k, as determined by the subsequent test. If H_0^1 is not rejected, then $H_0^2 : \mu_0=\mu_2$ should be tested conditionally on having not rejected H_0^1. The alternative hypothesis can be written as $H_a^2 : \mu_0<\mu_2$. If H_0^2 is rejected, then $\widehat{\delta} = d_1$; otherwise, the third hypothesis H_0^3 is tested conditionally on having not rejected H_0^1 and H_0^2, and this procedure is continued until one of the null hypotheses is rejected or the hypothesis $H_0^k : \mu_0=\mu_k$ is tested against $H_a^k : \mu_0<\mu_k$ conditionally on having not rejected $H_0^i : \mu_0=\mu_i$ (i=1, 2,..., k−1). The test statistic employed in testing $H_0^i : \mu_0=\mu_i$ vs. $H_a^i : \mu_0<\mu_i$ is $\widetilde{T}_i = (X_i - \widetilde{X}_{i-1})[S^2\{1/(n_0 + n_1+\cdots+n_{i-1})+(1/n_i)\}]^{-1/2}$, where $\widetilde{X}_{i-1} = (n_0X_0+\cdots+n_{i-1}X_{i-1})/(n_0+n_1+\cdots+n_{i-1})$ is the pooled mean. For a specified test size α, the test rejects H_0^i if $\widetilde{t}_i \geq p_{k,i,\nu}(\alpha)$.

We decide $p_{k,i,\nu}(\alpha)$, the critical point for the i-th test, to satisfy

$$Pr[\widehat{\delta} < d_k] = \alpha, \tag{2.1}$$

under $\mu_0=\mu_1=\cdots=\mu_k$. Note that the Dunnett type test is made to satisfy this equation. Suppose that the test size at each stage is equal, say α', then

$$Pr[\widetilde{T}_i \geq p_{k,i,\nu}(\alpha) \mid \widetilde{T}_1 < p_{k,1,\nu}(\alpha), \widetilde{T}_2 < p_{k,2,\nu}(\alpha), \cdots, \widetilde{T}_{i-1} < p_{k,i-1,\nu}(\alpha)] = \alpha'$$

and

$$Pr[\widehat{\delta} < d_k] = 1 - Pr[\widetilde{T}_1 < p_{k,1,\nu}(\alpha), \widetilde{T}_2 < p_{k,2,\nu}(\alpha), \cdots, \widetilde{T}_k < p_{k,k,\nu}(\alpha)]$$
$$= 1 - (1 - \alpha')^k.$$

Therefore from (2.1) the test size α' is given by

$$\alpha' = 1 - (1 - \alpha)^{1/k}.$$

It is shown that

$$Pr[\widetilde{T}_1 < p_{k,1,\nu}(\alpha), \widetilde{T}_2 < p_{k,2,\nu}(\alpha), \cdots, \widetilde{T}_m < p_{k,m,\nu}(\alpha)]$$
$$= \int_0^{\infty} \prod_{i=1}^{m} \Phi\left(p_{k,i,\nu}(\alpha)[\chi_\nu^2/\nu]^{1/2}\right) g(\chi_\nu^2) d\chi_\nu^2,$$

where Φ and $g(\chi_\nu^2)$ is the distribution function of $N(0,1)$ and the density function of χ_ν^2, respectively. Therefore $p_{k,i,\nu}(\alpha)$ are obtained sequentially by one-variate numerical integration. The table for the exact critical point $p_{k,i,\nu}(\alpha)$ for k=2, 3 and 4; α=0.05; and ν=3(1)30, 32(2)50, 60, 120 and ∞ is given in Appendix 4. Note that this test and table can be used when the sample sizes are not equal.

2.4 Power Comparison by Simulation

Denote the Williams test, the Dunnett type test, and the pooling step-up test, by *W*-test, *D*-test, and *PS*-test, respectively, for simplicity. To study the power of the tests, we consider the probabilities of the correct decision in Case 1 ($\mu_0=\mu_1=\mu_2$), Case 2 ($\mu_0=\mu_1\leq\mu_2$), and Case 3 ($\mu_0\leq\mu_1=\mu_2$). Selecting k=2, $n_0=n_1=n_2=10$, $\mu_0=0$, $\sigma^2=1$, we generated 2000 sets of $\{x_{ij}\}$ (j=1,..., n_i; i=0,..., k) for Case 1 ($\mu_0=\mu_1=\mu_2$), Case 2 ($\mu_0=\mu_1\leq\mu_2$), and Case 3 ($\mu_0\leq\mu_1=\mu_2$) with $\mu_2-\mu_0=c\sigma$ (c=0(0.1)1.5), and size α=0.05.

The values of the estimated probability of the correct decision of the NOAEL in Case 1, i.e., test size for those three tests were, 0.948 for *W*-test, 0.949 for *D*-test, and 0.956 for *PS*-test, and these were essentially equal. Figure 1 shows the estimated probability, $Pr[\widehat{\delta} = d_1 \mid \mu_0=\mu_1\leq\mu_2]$, of the correct decision of the NOAEL in Case 2. The horizontal axis represents the value of c and the vertical axis represents the probability of the correct decision. The order of the magnitude of the estimated $Pr[\widehat{\delta} = d_1 \mid \mu_0=\mu_1\leq\mu_2]$ was as follows:

D-test > *W*-test > *PS*-test for $0 \leq c \leq 0.5$,
D-test > *PS*-test > *W*-test for $0.6 \leq c \leq 1.2$,
PS-test > *D*-test > *W*-test for $1.3 \leq c \leq 1.5$.

Figure 2 shows the estimated probability, $Pr[\widehat{\delta} = d_0 \mid \mu_0\leq\mu_1=\mu_2]$, of the correct decision of the NOAEL in Case 3. The order of this probability was as follows:

PS-test > *W*-test > *D*-test for $0 \leq c \leq 0.4$,
W-test > *PS*-test > *D*-test for $0.5 \leq c \leq 0.8$,
W-test > *D*-test > *PS*-test for $0.9 \leq c \leq 1.5$.

In summary, the three tests are competing. More specifically, if one is interested in Case 2, the Dunnett type multiple comparison test could be suggested; on the other hand, if one is interested in Case 3, the Williams test could be recommended. The pooling step-up test competes well with the other tests. Note that the Williams test and the Dunnett type test are essentially for the case of equal sample sizes, whereas the pooling step-up test may be used for any sample sizes.

2.5 Criticism of the Statistical Tests

When $n_0=n_1=n_2=10$, $c=0.7$, and the test size is 0.05, it is shown that the maximum probability of the correct decision of these tests in Case 2 and Case 3 are less than 0.5. The smaller values of the probabilities of the correct decision indicate the larger the doses selected as the NOAEL, showing that the smaller the sample sizes, the larger the values selected as the NOAEL. This behavior of the statistical tests is unacceptable for deciding the NOAEL since smaller sample tends to make the dose levels appear safer. In most toxicological studies, the size of the sample is strictly restricted and we must increase the size of the tests, for example from 5% to 15%, to increase the power. However, it is not easy to show how much test size we must increase. We may circumvent this difficulty by the use of the Akaike Information Criterion (AIC).

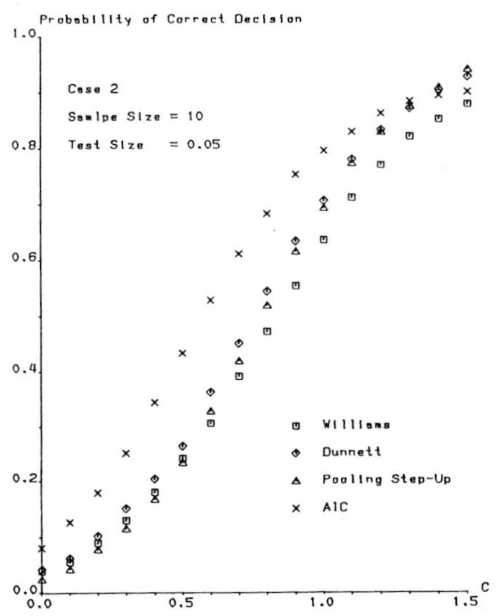

Figure 1. The estimated $Pr[\hat{\delta}=d_1|\mu_0=\mu_1\leq\mu_2]$

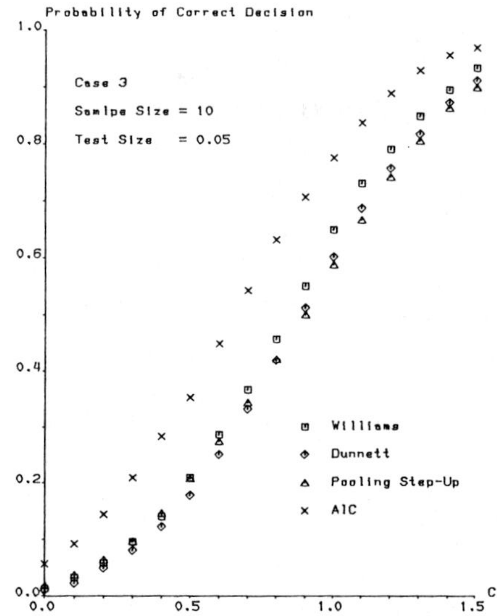

Figure 2. The estimated $Pr[\hat{\delta}=d_0|\mu_0\leq\mu_1=\mu_2]$

3 APPLICATION OF THE AIC

For simplicity we illustrate this method when k=2. Let γ_1=μ_1−μ_0 and γ_2=μ_2−μ_0. Note that

$$\mu_0 = \mu_1 = \mu_2 \text{ if and only if } \gamma_1 = \gamma_2 = 0,$$
$$\mu_0 = \mu_1 \leq \mu_2 \text{ if and only if } \gamma_1 = 0,\ \gamma_2 \geq 0,$$
$$\mu_0 \leq \mu_1,\ \mu_0 \leq \mu_2 \text{ if and only if } \gamma_1 \geq 0,\ \gamma_2 \geq 0,$$

and that the order restriction $\mu_0 \leq \mu_1 \leq \mu_2$ is equivalent to $\gamma_1 \geq 0$ and $\gamma_2 - \gamma_1 \geq 0$.

The log likelihood is

$$l(\mu_0, \gamma_1, \gamma_2, \sigma^2) = -\frac{n_0 + n_1 + n_2}{2}\{\log(2\pi) + \log\sigma^2\} - \frac{1}{2\sigma^2}Q(\mu_0, \gamma_1, \gamma_2),$$

where

$$Q(\mu_0, \gamma_1, \gamma_2) = \sum_{j=1}^{n_0}(x_{0j} - \mu_0)^2 + \sum_{j=1}^{n_1}\{x_{1j} - (\mu_0 + \gamma_1)\}^2 + \sum_{j=1}^{n_2}\{x_{2j} - (\mu_0 + \gamma_2)\}^2.$$

Put

$$L(\mu_0, \gamma_1, \gamma_2, \sigma^2) = 2 \times l(\mu_0, \gamma_1, \gamma_2, \sigma^2)$$
$$- 2 \times (\text{number of parameters involved in the likelihood}),$$

which is the log likelihood function penalized by the number of parameters involved in the model. Let $\widehat{\mu}_0$, $\widehat{\gamma}_1$, $\widehat{\gamma}_2$ and $\widehat{\sigma}^2$ be the mle's of μ_0, γ_1, γ_2 and σ^2 which maximize $L(\mu_0, \gamma_1, \gamma_2, \sigma^2)$, respectively. Then $L(\widehat{\mu}_0, \widehat{\gamma}_1, \widehat{\gamma}_2, \widehat{\sigma}^2)$ measures the goodness-of-fit of the four-parameter model to the data. Next we suppose that γ_1=0 is known and that μ_0, γ_2 and σ^2 are the parameters in the model. Let $\widehat{\mu}_0(\gamma_1{=}0)$, $\widehat{\gamma}_2(\gamma_1{=}0)$ and $\widehat{\sigma}^2(\gamma_1{=}0)$ be the mle's of μ_0, γ_2 and σ^2, under γ_1=0, respectively. Then $L(\widehat{\mu}_0(\gamma_1{=}0), 0, \widehat{\gamma}_2(\gamma_1{=}0), \widehat{\sigma}^2(\gamma_1{=}0))$ is a measure of goodness-of-fit of the three-parameter model to the data. The penalized likelihood function has been established to measure the goodness-of-fit by adjusting the number of the parameters involved in the model. Thus if $L(\widehat{\mu}_0, \widehat{\gamma}_1, \widehat{\gamma}_2, \widehat{\sigma}^2) < L(\widehat{\mu}_0(\gamma_1{=}0), 0, \widehat{\gamma}_2(\gamma_1{=}0), \widehat{\sigma}^2(\gamma_1{=}0))$, we may select the three-parameter model. This idea of the model selection is first proposed by Akaike [12], and is widely known as the Akaike Information Criterion (AIC).

We take into account the order restriction $\gamma_1 \geq 0$ and $\gamma_2 - \gamma_1 \geq 0$. The mle's of γ_1 and γ_2 could violate this restriction, and if this is the case the AIC is no more a justifiable measure of model selection. Thus we introduce switchings to modify the AIC and propose the following general procedure for the determination of the NOAEL.

Step 1 : Establish the full parameter $(\mu_0, \gamma_1, \ldots, \gamma_k, \sigma^2)$ model.
Step 2 : Obtain the mle's $\widehat{\gamma}_i$ of γ_i involved in the model.
Step 3 : If $\widehat{\gamma}_{i+1} - \widehat{\gamma}_i \leq 0$ for all i, i=0, 1,..., k−1, decide $\widehat{\delta} = d_k$.
If $\widehat{\gamma}_{i+1} - \widehat{\gamma}_i \geq 0$ for all i, i=0, 1,..., k−1, goto Step 5,
otherwise proceed to Step 4.
Step 4 : Let i_0 be the smallest i such that $\widehat{\gamma}_{i+1} - \widehat{\gamma}_i < 0$.
Reestablish the model $(\mu_0, \gamma_1, \ldots, \gamma_k, \sigma^2)$ with $\gamma_{i_0+1} = \gamma_{i_0}$
and return to Step 2.

Step 5 : Let A_j (j=1, 2,..., m) be the group of $\hat{\gamma}_i$'s generated by PAVA such that $a < b$ if $a \in A_j$, $b \in A_{j'}$ for $j < j'$ and $a = b$ if $a \in A_j$ and $b \in A_j$. We plug the A_j into the set of γ_i's and the penalized likelihood L by $L(\gamma_{A_1},\ldots,\gamma_{A_m})$. Obtain the maximum values of $L_0=L(\gamma_{A_1},\ldots,\gamma_{A_m})$, $L_1=L(0,\gamma_{A_2},\ldots,\gamma_{A_m})$, $L_2=L(0,0,\gamma_{A_3},\ldots,\gamma_{A_m}),\ldots$, and $L_m=L(0,0,\ldots,0)$, respectively, with respect to the parameters γ_{A_i}'s involved in each function.

Step 6 : If L_0 is the largest, decide $\hat{\delta} = d_0$

If L_i is the largest, decide $\hat{\delta} = d_{i_0}$, where $i_0 = \max\{j : \hat{\gamma}_j \in A_i\}$.

Because of the many switchings introduced in the above procedure, we carefully studied the behavior of the procedure based on the AIC by simulation. Figure 1 and 2 show the estimated probabilities of the correct decision for the AIC procedure which have been obtained by the same framework of the simulation as the statistical tests. These figures show that the AIC procedure provides much larger probabilities of the correct decision, in Case 2 and Case 3, than those procedures based on the statistical tests. For example, the probabilities of the correct decision with the AIC, the Williams test, the Dunnett type multiple comparison and pooling step-up test in Case 2 when c=0.7 are given, respectively, by 0.611, 0.392, 0.452 and 0.420; and the corresponding values in Case 3 are 0.543, 0.367, 0.332 and 0.344.

The estimated probability of the correct decision with the AIC procedure in Case 1 is 0.863. This indicates that, if the test sizes of the three statistical tests had been chosen at $1-0.863 \approx 0.14$, those procedures would have given the similar values of the correct decision as the AIC procedure in Case 2 and Case 3. Whereas it is not easy to determine the most appropriate test size of statistical tests for the situation in the present paper, the results of the simulation studies show that the procedure based on the AIC takes balance the probabilities of the correct decision in Case 1, 2 and 3, and provides better results for the determination of the NOAEL than the procedures based on the statistical tests.

4 AN APPLICATION

We now present a sample application of the methods described in this paper. The data are from Yoshimura [4]. Male rats were randomly selected into four groups of size 10, and were administered four different levels of dose for 35 days. Table 2 shows the amount of hemoglobin in blood of animals which is represented from Yoshimura [4].

Table 2.
Amount of Hemoglobin in Blood (mg/ml)

Dose d_j (mg/kg)	Hemoglobin in Blood (mg/ml)									
0	153	153	152	156	158	151	151	150	148	157
5	158	152	152	152	151	151	157	147	155	146
10	153	146	138	152	140	146	156	142	147	153
20	137	139	141	141	143	133	147	144	151	156

We suppose in this example that μ_i are monotonically ordered with $\mu_0 \geq \mu_1 \geq \mu_2 \geq \mu_3$, and use the statistics with reversing the signs of data. From the table we have $x_0 = -152.9$, $x_1 = -152.1$, $x_2 = -147.3$, $x_4 = -143.2$, $s^2 = 969.5/36$, and $\nu = 36$. Thus we take $\widehat{\mu}_i = \widehat{\mu}_i^* = x_i$ for $i = 0,1,2,3$. Taking test size = 0.05, the NOAEL in each procedure is computed as follows.

The Williams Test

$t_3^* = (\widehat{\mu}_3^* - x_0)[2s^2/10]^{-1/2} = 4.180$, $t_2^* = 2.413$, $t_1^* = 0.345$.
$t_3^* > t_{3,36}(0.05) = 1.79$, $t_2^* > t_{2,36}(0.05) = 1.77$, $t_1^* < t_{1,36}(0.05) = 1.69$,
deciding $\widehat{\delta} = d_1$.

The Dunnett type Multiple Comparison Test

$t_3 = (\widehat{\mu}_3 - \widehat{\mu}_0)[2s^2/10]^{-1/2} = 4.180$, $t_2 = 2.413$, $t_1 = 0.345$.
$t_1 < d_{3,36}(0.05) = 1.875 < t_2$, deciding $\widehat{\delta} = d_1$.

The Pooling Step-Up Test

$\widetilde{x}_1 = (x_0 + x_1)/2 = -152.5$, $\widetilde{x}_2 = (x_0 + x_1 + x_2)/3 = -150.8$.
$\widetilde{t}_1 = (x_1 - x_0)[2s^2/10]^{-1/2} = 0.345$, $\widetilde{t}_2 = (x_2 - \widetilde{x}_1)[2s^2/10]^{-1/2} = 2.587$,
$\widetilde{t}_3 = (x_3 - \widetilde{x}_2)[2s^2/10]^{-1/2} = 3.993$.
$\widetilde{t}_1 < p_{3,1,36}(0.05) = 2.205$, $\widetilde{t}_2 > p_{3,2,36}(0.05) = 2.202$, deciding $\widehat{\delta} = d_1$.

The AIC procedure

$\widehat{\gamma}_1 = x_1 - x_0 = 0.8$, $\widehat{\gamma}_2 = x_2 - x_0 = 5.6$, $\widehat{\gamma}_1 = x_3 - x_0 = 9.7$.
Since $\widehat{\gamma}_1 > 0$, $\widehat{\gamma}_2 - \widehat{\gamma}_1 > 0$, and $\widehat{\gamma}_3 - \widehat{\gamma}_2 > 0$, we compare $L(\widehat{\mu}_0, \widehat{\gamma}_1, \widehat{\gamma}_2, \widehat{\gamma}_3, \widehat{\sigma}^2)$, $L(\widehat{\mu}_0(\gamma_1{=}0), 0, \widehat{\gamma}_2(\gamma_1{=}0), \widehat{\gamma}_3(\gamma_1{=}0), \widehat{\sigma}^2(\gamma_1{=}0))$, $L(\widehat{\mu}_0(\gamma_1{=}\gamma_2{=}0), 0, 0, \widehat{\gamma}_3(\gamma_1{=}\gamma_2{=}0), \widehat{\sigma}^2(\gamma_1{=}\gamma_2{=}0))$, and $L(\widehat{\mu}_0(\gamma_1{=}\gamma_2{=}\gamma_3{=}0), 0, 0, 0, \widehat{\sigma}^2(\gamma_1{=}\gamma_2{=}\gamma_3{=}0))$. The values of these likelihoods are obtained respectively by -251.03, -249.16, -253.96, and -264.63. Since the value of $L(\widehat{\mu}_0(\gamma_1{=}0), 0, \widehat{\gamma}_2(\gamma_1{=}0), \widehat{\gamma}_3(\gamma_1{=}0), \widehat{\sigma}^2(\gamma_1{=}0))$ is the largest among these, the NOAEL is decided to be d_1. Therefore, in this example, the result by the AIC procedure agrees with those of three testing procedures with the test size 0.05.

Acknowledgement

The authors are greateful to the referee for his/her valuable comments and suggestions.

References

1. U. S. Environmental Protection Agency, Reference Dose (RfD): Description and Use in Health Risk Assessment, Integrated Risk Information System (IRIS): Appendix A. Intra-Agency Reference Dose Work Group, Office of Health and Environmental Assessment, Washington, D.C. (1986).
2. U. S. Environmental Protection Agency, Interim methods for Development of Inhalation Reference Doses. Office of Health and Environmental Assessment, Washington, D.C. (1989).
3. T. Yanagawa, Y. Kikuchi and K.G. Brown, Statistical issues of the no-observed-adverse-effect level in categorical response. *Environmental Health Perspectives*, in press.
4. I. Yoshimura, *Statistical analysis of toxicology and drug effect* (in Japanese), Saienteisuto-sha, Tokyo (1987).
5. R.E. Barlow, D.J. Bartholomew, J.M. Bremmer and H.D. Brunk, *Statistical Inference under Order*

Restriction: The Theory and Application of Isotonic Regression, Wiley, New York (1972).

6. D.A. Williams, A test for differences between treatment means when several dose levels are compared with a zero dose control. *Biometrics*, **27**, 103-117 (1971).
7. D.A. Williams, The comparison of several dose levels with a zero control. *Biometrics*, **28**, 519-531 (1972).
8. K.G. Brown, Williams' test. in : *The Encyclopedia of Statistical Sciences*, **vol. 9**, pp. 633-635, Wiley, New York (1988).
9. C.W. Dunnett, A multiple comparisons procedure for comparing several treatments with a control. *J. Amer. Statist. Ass.*, **50**, 1096-1121 (1955).
10. D.A. Williams, Some inference procedures for monotonically ordered normal means. *Biometrika*, **64**, 9-14 (1977).
11. R. Marcus, The powers of some tests of the equality of normal means against an ordered alternative. Biometrika, **63**, 177-183 (1976).
12. H. Akaike, Information theory and an extension of the maximum likelihood principle, *2nd Inter. Symp. on Information Theory*, B.N. Petrov and F. Csaki (Eds.), Akademiai Kiado, Budapest, pp. 267-281 (1973).

Appendix 1 An Equivalence of Two Statistics in the Williams Type

Proposition

The test statistics $T_i' = (\widehat{\mu}_i - X_0)[2S^2/n]^{-1/2}$ and $T_i^* = (\widehat{\mu}_i^* - X_0)[2S^2/n]^{-1/2}$ are equivalent, if T_i' is positive.

proof

Let A_j (j=1,...,r) be the groups of X_i (i=1,...,k) generated by the PAVA. Then (A_1,..., A_r) is a partition of (X_0, X_1,..., X_k). It is clear that $X_0 \in A_1$. Furthermore, if X_i belongs to A_j, then $\widehat{\mu}_i = \text{Av}(A_j)$, where $\text{Av}(A_j) = \sum n_i x_i / \sum n_i$ and the summations are taken over i such that $X_i \in A_j$. First we prove the case when $i = 1$. If $X_1 \in A_1$, then $\widehat{\mu}_1 = \widehat{\mu}_0 = \text{Av}(A_1)$ since $X_0 \in A_1$. On the other hand $\widehat{\mu}_1 = \widehat{\mu}_0 = \min_{0 \le v \le k}(\sum_{l=0}^{v} n_l X_l / \sum_{l=0}^{v} n_l) \le X_0$, conflicting with $T_1' > 0$. Thus $X_1 \notin A_1$ but $X_1 \in A_2$, namely $\widehat{\mu}_0 = X_0$ and $\widehat{\mu}_1$ does not contain X_0. Therefore $\widehat{\mu}_1 = \widehat{\mu}_1^*$. In general, if $\widehat{\mu}_{i_1} > \widehat{\mu}_{i_2}$, there exist two integers j_1 and j_2 such that $1 \le j_2 < j_1 \le r$ and $\widehat{\mu}_{i_1} = \text{Av}(A_{j_1})$ and $\widehat{\mu}_{i_2} = \text{Av}(A_{j_2})$. Therefore $\widehat{\mu}_{i_1}$ does not contain X_i such that $X_i \notin A_{j_1}$. Since $j_1 \neq 1$ and $X_0 \in A_1$, $\widehat{\mu}_{i_1}$ does not contain X_0. From this if $\widehat{\mu}_i > X_0$, there could be the cases such that $\widehat{\mu}_i \ge \widehat{\mu}_{i-1} \ge \cdots \ge \widehat{\mu}_1 > X_0 = \widehat{\mu}_0$, $\widehat{\mu}_i \ge \widehat{\mu}_{i-1} \ge \cdots \ge \widehat{\mu}_2 > X_0 \ge \widehat{\mu}_1$, ..., or $\widehat{\mu}_i > X_0 \ge \widehat{\mu}_{i-1} \ge \cdots \ge \widehat{\mu}_1$, but $\widehat{\mu}_i$ does not contain X_0 and therefore $\widehat{\mu}_i = \widehat{\mu}_i^*$ in any case.

Appendix 2 Distribution of T_k

Let $f_k(\boldsymbol{u})$ be the density function of k-variate t-distribution, i.e.,

$$f_k(\boldsymbol{u}) = \frac{\Gamma((\nu+k)/2)}{(\pi\nu)^{k/2}\Gamma(\nu/2)|R|^{1/2}} \left(1+\nu^{-1}\boldsymbol{u}'R^{-1}\boldsymbol{u}\right)^{-(\nu+k)/2},$$

where $\boldsymbol{u} = (u_1, \ldots, u_k)'$, and R is the k by k matrix with diagonal elements 1 and non-diagonal elements 0.5. Then the distribution of T_k is given as follows:

$$\begin{aligned} Pr[T_2 = 0] &= 1/3, \\ Pr[T_3 = 0] &= 1/4, \\ Pr[T_4 = 0] &= 1/5; \end{aligned}$$

and for $t > 0$,

$$Pr[T_2 \geq t] = 1 - \int_{-\infty}^{t} \int_{2(u_2-t)}^{2t-u_2} f_2(\boldsymbol{u}) du_1 du_2,$$

$$\begin{aligned} Pr[T_3 \geq t] = 1 &- \int_{-\infty}^{t} \int_{-\infty}^{u_3-t} \int_{3(u_3-t)-u_2}^{3t-(u_2+u_3)} f_3(\boldsymbol{u}) du_1 du_2 du_3 \\ &- \int_{-\infty}^{t} \int_{u_3-t}^{u_3} \int_{2(u_3-t)}^{3t-(u_2+u_3)} f_3(\boldsymbol{u}) du_1 du_2 du_3 \\ &- \int_{-\infty}^{t} \int_{u_3}^{2t-u_3} \int_{u_2+u_3-2t}^{3t-(u_2+u_3)} f_3(\boldsymbol{u}) du_1 du_2 du_3, \end{aligned}$$

$$\begin{aligned} Pr[T_4 \geq t] = 1 &- \int_{-\infty}^{t} \int_{-\infty}^{u_4-t} \int_{-\infty}^{2(u_4-t)-u_3} \int_{4(u_4-t)-(u_2+u_3)}^{4t-(u_2+u_3+u_4)} f_4(\boldsymbol{u}) du_1 du_2 du_3 du_4 \\ &- \int_{-\infty}^{t} \int_{-\infty}^{u_4-t} \int_{2(u_4-t)-u_3}^{2u_4-u_3} \int_{2(u_4-t)}^{4t-(u_2+u_3+u_4)} f_4(\boldsymbol{u}) du_1 du_2 du_3 du_4 \\ &- \int_{-\infty}^{t} \int_{-\infty}^{u_4-t} \int_{2u_4-u_3}^{3t-(u_3+u_4)} \int_{2(u_2+u_3+u_4)/3-2t}^{4t-(u_2+u_3+u_4)} f_4(\boldsymbol{u}) du_1 du_2 du_3 du_4 \\ &- \int_{-\infty}^{t} \int_{u_4-t}^{u_4} \int_{-\infty}^{u_4-t} \int_{3(u_4-t)-u_2}^{4t-(u_2+u_3+u_4)} f_4(\boldsymbol{u}) du_1 du_2 du_3 du_4 \\ &- \int_{-\infty}^{t} \int_{u_4-t}^{u_4} \int_{u_4-t}^{2u_4-u_3} \int_{2(u_4-t)}^{4t-(u_2+u_3+u_4)} f_4(\boldsymbol{u}) du_1 du_2 du_3 du_4 \\ &- \int_{-\infty}^{t} \int_{u_4-t}^{u_4} \int_{2u_4-u_3}^{3t-(u_3+u_4)} \int_{2(u_2+u_3+u_4)/3-2t}^{4t-(u_2+u_3+u_4)} f_4(\boldsymbol{u}) du_1 du_2 du_3 du_4 \\ &- \int_{-\infty}^{t} \int_{u_4}^{2t-u_4} \int_{-\infty}^{(u_3+u_4)/2-t} \int_{3(u_3+u_4)/2-u_2-3t}^{4t-(u_2+u_3+u_4)} f_4(\boldsymbol{u}) du_1 du_2 du_3 du_4 \\ &- \int_{-\infty}^{t} \int_{u_4}^{2t-u_4} \int_{(u_3+u_4)/2-t}^{(u_3+u_4)/2} \int_{u_3+u_4-2t}^{4t-(u_2+u_3+u_4)} f_4(\boldsymbol{u}) du_1 du_2 du_3 du_4 \\ &- \int_{-\infty}^{t} \int_{u_4}^{2t-u_4} \int_{(u_3+u_4)/2}^{3t-(u_3+u_4)} \int_{2(u_2+u_3+u_4)/3-2t}^{4t-(u_2+u_3+u_4)} f_4(\boldsymbol{u}) du_1 du_2 du_3 du_4. \end{aligned}$$

Appendix 3 Upper α points $d_{k,\nu}(\alpha)$ of the distribution of T_k

for the Dunnett type Multiple Comparison Test (α=0.05)

ν	k=1	k=2	k=3	k=4
3	2.353	2.702	2.847	2.925
4	2.132	2.404	2.512	2.569
5	2.015	2.250	2.340	2.387
6	1.943	2.156	2.236	2.276
7	1.895	2.093	2.166	2.202
8	1.860	2.047	2.115	2.149
9	1.833	2.013	2.078	2.110
10	1.812	1.987	2.049	2.079
11	1.796	1.965	2.025	2.054
12	1.782	1.948	2.006	2.034
13	1.771	1.934	1.990	2.017
14	1.761	1.921	1.977	2.003
15	1.753	1.911	1.965	1.991
16	1.746	1.902	1.955	1.980
17	1.740	1.894	1.946	1.971
18	1.734	1.887	1.938	1.963
19	1.729	1.880	1.932	1.956
20	1.725	1.875	1.925	1.949
21	1.721	1.870	1.920	1.944
22	1.717	1.865	1.915	1.938
23	1.714	1.861	1.910	1.933
24	1.711	1.857	1.906	1.929
25	1.708	1.854	1.902	1.925
26	1.706	1.850	1.899	1.921
27	1.703	1.848	1.896	1.918
28	1.701	1.845	1.893	1.915
29	1.699	1.842	1.890	1.912
30	1.697	1.840	1.887	1.909
32	1.694	1.836	1.883	1.904
34	1.691	1.832	1.879	1.900
36	1.688	1.829	1.875	1.896
38	1.686	1.826	1.872	1.893
40	1.684	1.823	1.869	1.890
42	1.682	1.821	1.866	1.887
44	1.680	1.818	1.864	1.885
46	1.679	1.816	1.862	1.882
48	1.677	1.815	1.860	1.880
50	1.676	1.813	1.858	1.878
60	1.671	1.806	1.851	1.871
120	1.658	1.790	1.833	1.852
∞	1.645	1.774	1.815	1.834

Appendix 4 Upper α points $p_{k,i,\nu}(\alpha)$ of the distribution of $\widetilde{T}_k$ for the Pooling Step-Up Test (α=0.05)

	k=2 α'=2.532×10^{-2}		k=3 α'=1.695×10^{-2}			k=4 α'=1.274×10^{-2}			
ν	i=1	i=2	i=1	i=2	i=3	i=1	i=2	i=3	i=4
3	3.166	3.019	3.716	3.549	3.411	4.146	3.965	3.814	3.686
4	2.764	2.681	3.169	3.080	3.002	3.474	3.381	3.299	3.227
5	2.560	2.505	2.897	2.840	2.788	3.146	3.088	3.035	2.987
6	2.438	2.397	2.736	2.695	2.658	2.954	2.913	2.875	2.841
7	2.356	2.324	2.630	2.599	2.570	2.828	2.797	2.768	2.742
8	2.298	2.272	2.555	2.530	2.507	2.739	2.715	2.692	2.670
9	2.254	2.233	2.499	2.479	2.459	2.673	2.653	2.635	2.617
10	2.221	2.202	2.456	2.439	2.422	2.623	2.606	2.590	2.575
11	2.194	2.178	2.422	2.407	2.392	2.582	2.568	2.554	2.541
12	2.172	2.157	2.394	2.380	2.368	2.550	2.537	2.525	2.513
13	2.153	2.141	2.370	2.359	2.347	2.523	2.511	2.501	2.490
14	2.138	2.126	2.351	2.340	2.330	2.500	2.490	2.480	2.471
15	2.125	2.114	2.334	2.324	2.315	2.480	2.471	2.462	2.454
16	2.113	2.103	2.320	2.311	2.302	2.463	2.455	2.447	2.439
17	2.103	2.094	2.307	2.299	2.291	2.449	2.441	2.434	2.427
18	2.094	2.086	2.296	2.288	2.281	2.436	2.429	2.422	2.415
19	2.087	2.079	2.286	2.279	2.272	2.424	2.418	2.411	2.405
20	2.080	2.072	2.277	2.271	2.264	2.414	2.408	2.402	2.396
21	2.073	2.066	2.269	2.263	2.257	2.405	2.399	2.394	2.388
22	2.068	2.061	2.262	2.256	2.251	2.397	2.391	2.386	2.381
23	2.062	2.056	2.256	2.250	2.245	2.389	2.384	2.379	2.374
24	2.058	2.052	2.250	2.245	2.239	2.382	2.377	2.373	2.368
25	2.053	2.048	2.244	2.239	2.234	2.376	2.371	2.367	2.363
26	2.049	2.044	2.239	2.235	2.230	2.370	2.366	2.362	2.357
27	2.046	2.041	2.235	2.230	2.226	2.365	2.361	2.357	2.353
28	2.042	2.037	2.231	2.226	2.222	2.360	2.356	2.352	2.348
29	2.039	2.034	2.227	2.222	2.218	2.355	2.352	2.348	2.344
30	2.036	2.032	2.223	2.219	2.215	2.351	2.348	2.344	2.340
32	2.031	2.027	2.216	2.213	2.209	2.343	2.340	2.337	2.334
34	2.026	2.022	2.210	2.207	2.204	2.337	2.334	2.331	2.328
36	2.022	2.018	2.205	2.202	2.199	2.331	2.328	2.325	2.322
38	2.018	2.015	2.201	2.198	2.195	2.325	2.323	2.320	2.318
40	2.015	2.012	2.197	2.194	2.191	2.321	2.318	2.316	2.313
42	2.012	2.009	2.193	2.190	2.188	2.316	2.314	2.312	2.309
44	2.010	2.007	2.190	2.187	2.184	2.313	2.310	2.308	2.306
46	2.007	2.004	2.186	2.184	2.182	2.309	2.307	2.305	2.303
48	2.005	2.002	2.184	2.181	2.179	2.306	2.304	2.302	2.300
50	2.003	2.000	2.181	2.179	2.177	2.303	2.301	2.299	2.297
60	1.995	1.992	2.171	2.169	2.167	2.291	2.290	2.288	2.286
120	1.974	1.973	2.146	2.145	2.144	2.262	2.261	2.261	2.260
∞	1.955	1.955	2.121	2.121	2.121	2.234	2.234	2.234	2.234

Stat. Sci. & Data Anal., pp. 357-364
K. Matsusita *et al.* (Eds)

Asymptotics on the Statistics for a Family of Non-Regular Distributions

MASAFUMI AKAHIRA

Institute of Mathematics, University of Tsukuba, Tsukuba, Ibaraki 305, Japan

Abstract. It is presented that a set of the extreme statistics and an ancillary statistic is second order asymptotically sufficient in some sense. It is also shown that the asymptotic density of the extreme statistics is given up to the second order. Further their applications are discussed.

Key words: Extreme statistics, ancillary statistic, second order asymptotic sufficiency, one-directional family, asymptotic density.

1. INTRODUCTION

Asymptotic sufficiency was discussed by LeCam [1] under the regularity conditions. In non-regular cases the asymptotic sufficiency was studied by Akahira [2], Weiss [3] and Mita [4] among others.

In the location parameter situation where a density function has a bounded support, it is shown in Akahira [5] that the maximum probability estimator (MPE) is second order asymptotically efficient in some sense. Then the MPE is constructed from the extreme statistics and the maximum likelihood estimator. So, it seems to be interesting to consider the second order asymptotic sufficiency in such cases. In this paper, for one-directional family of distributions including such cases, it is presented that the second order asymptotically sufficient statistic is given by a set of the extreme statistics and an ancillary statistic (see also Akahira [6]). Further, it is shown that the asymptotic density of the extreme statstics is given up to the second order, i.e. the order n^{-1}, where n is a size of the sample. Finally their applications are discussed.

2. NOTATIONS AND ASSUMPTIONS

Suppose that $X_1, \ldots, X_n$ are independent and identically distributed random variables with a density function $f(x,\theta)$ with respect to a σ-finite measure μ, where θ belongs to a parameter spase Θ which is assumed to be an open set of $\mathbf{R}^1$. For each $\theta \in \Theta$, let $A(\theta)$ be a support of $f(x,\theta)$, that is, $A(\theta) = \{x : f(x,\theta) > 0\}$. The determination of $A(\theta)$ is not unique in so far as any null set may be added to it, but in the sequel we take one fixed determination of $A(\theta)$ for all θ, which satisfies the following conditions.
(A.1) For any disjoint points θ_1 and θ_2 in Θ neither

$$A(\theta_1) \supset A(\theta_2) \quad \text{nor} \quad A(\theta_1) \subset A(\theta_2).$$

(A.2) For any θ_1, θ_2 and θ_3 with $\theta_1 < \theta_2 < \theta_3$

$$A(\theta_1) \cap A(\theta_3) \subset A(\theta_1) \cap A(\theta_2),$$
$$A(\theta_1) \cap A(\theta_3) \subset A(\theta_2) \cap A(\theta_3).$$

The above conditions are used to define the one-directional family of distributions in the paper by Akahira and Takeuchi [7]. Further we assume the following conditions.
(A.3) For almost all $x[\mu]$, $f(x,\theta)$ is twice differentiable in θ.

$$(\text{A.4})_1 \quad \int_{A(\theta+\Delta\theta)-A(\theta)} f(x,\theta+\Delta\theta)d\mu = a_{1\pm}(\theta)\Delta\theta + a_{2\pm}(\theta)(\Delta\theta)^2 + o((\Delta\theta)^2) \quad \text{as} \quad \Delta\theta \to 0 \pm 0,$$

$$(\text{A.4})_2 \quad \int_{A(\theta)-A(\theta+\Delta\theta)} f(x,\theta)d\mu = b_{1\pm}(\theta)\Delta\theta + b_{2\pm}(\theta)(\Delta\theta)^2 + o((\Delta\theta)^2) \quad \text{as} \quad \Delta\theta \to 0 \pm 0,$$

$$(\text{A.4})_3 \quad \int_{(A(\theta+\Delta\theta)-A(\theta+2\Delta\theta))\cap(A(\theta)-A(\theta+2\Delta\theta))} f(x,\theta+\Delta\theta)d\mu = b_{1\pm}(\theta)\Delta\theta + c_{2\pm}(\theta)(\Delta\theta)^2 + o((\Delta\theta)^2) \quad \text{as} \quad \Delta\theta \to 0 \pm 0,$$

where + and − signs should be read consistently here.
(A.5) For each $\theta \in \Theta$,

$$0 < I(\theta) = \int_{A(\theta)} \frac{1}{f(x,\theta)} \left\{ \frac{\partial f(x,\theta)}{\partial \theta} \right\}^2 d\mu < \infty.$$

(A.6) There exist a positive number ε and a μ−integrable function $G_\theta(x)$ on $A(\theta)$, independent of $\Delta\theta$, such that

$$|f(x,\theta+2\Delta\theta) - 2f(x,\theta+\Delta\theta) + f(x,\theta)|/(\Delta\theta)^2 \le G_\theta(x)$$

for all $x \in A(\theta+2\Delta\theta) \cap A(\theta+\Delta\theta) \cap A(\theta)$ with $|\Delta\theta| \le \varepsilon$.

(A.7) For any disjoint points θ_1 and θ_2 in Θ, $\mu(A(\theta_1) - A(\theta_2)) > 0$.

(A.8) For each $\theta \in \Theta$

$$a_{1+}(\theta) - b_{1+}(\theta) = a_{1-}(\theta) - b_{1-}(\theta),$$
$$c_{2+}(\theta) - a_{2+}(\theta) - 2b_{2+}(\theta) = c_{2-}(\theta) - a_{2-}(\theta) - 2b_{2-}(\theta).$$

(A.9) For $v < 0 < u$

$$\int_{(A(\theta)-A(\theta+un^{-1}))\cap(A(\theta)-A(\theta+vn^{-1}))} f(x,\theta)d\mu = o(n^{-2}).$$

Under the above conditions we shall discuss the second order asymptotic sufficiency of the statistics and obtain the asymptotic density of the extreme statistics up to the second order.

3. SECOND ORDER ASYMPTOTICS ON THE EXTREME STATISTICS

It is presented that the second order asymptotically sufficient statistic is given by a set of the extreme statistics and an ancillary statistic (see also Akahira [6]). It is shown that the asymptotic density of the extreme statistics is given up to the second order, i.e. the order n^{-1}. We define $\underline{\theta}$ and $\overline{\theta}$ by

$$\underline{\theta} = \inf\{\theta | X_i \in A(\theta),\ i=1,\dots,n\} \quad \text{and} \quad \overline{\theta} = \sup\{\theta | X_i \in A(\theta),\ i=1,\dots,n\},$$

respectively. We put

$$Z_1(\theta) = \frac{1}{\sqrt{n}} \sum_{i=1}^{n} \left\{ \frac{\partial}{\partial\theta} \log f(X_i,\theta) - \alpha(\theta) \right\}$$

for $\underline{\theta} < \theta < \overline{\theta}$, where $\alpha(\theta) = b_{1+}(\theta) - a_{1+}(\theta)$. We also set $\beta(\theta) = 2\{c_{2+}(\theta) - a_{2+}(\theta) - 2b_{2+}(\theta)\} - I(\theta)$ and $\hat{\theta}^* = (\underline{\theta} + \overline{\theta})/2$. Then we have the following theorems.

Theorem 3.1. *Assume that the conditions (A.1) to (A.8) hold. Then the statistic* $(\overline{\theta}, \underline{\theta}, Z_1(\hat{\theta}^*))$ *is second order asymptotically sufficient in the sense that the joint density of* $(X_1, \ldots, X_n)$ *can be factorized up to the second order as*

$$\prod_{i=1}^{n} f(x_i, \theta) = \prod_{i=1}^{n} f(x_i, \hat{\theta}^*)\{g_\theta(\overline{\theta}, \underline{\theta}, Z_1(\hat{\theta}^*)) + o_p(1/n)\},$$

where

$$g_\theta(\overline{\theta}, \underline{\theta}, Z_1(\hat{\theta}^*)) = \Big[\exp\Big\{-\alpha(\hat{\theta}^*)n(\hat{\theta}^* - \theta) - (1/\sqrt{n})Z_1(\hat{\theta}^*)n(\hat{\theta}^* - \theta) + (1/n)\beta(\hat{\theta}^*)(n(\hat{\theta}^* - \theta))^2\Big\}\Big]\chi_{(\underline{\theta}, \overline{\theta})}(\theta)$$

with the indicator $\chi_{(\underline{\theta}, \overline{\theta})}(\theta)$ *of the open interval* $(\underline{\theta}, \overline{\theta})$.

The proof is omitted since it is given in Akahira [6].

Theorem 3.2. *Assume that the conditions (A.1), (A.2), (A.3),* $(A.4)_2$ *and (A.9) hold. Then the second order asymptotic density* $f_{\overline{\theta}, \underline{\theta}}(u, v)$ *of* $(\overline{\theta}, \underline{\theta})$ *is given by*

$$f_{\overline{\theta}, \underline{\theta}}(u, v) = \begin{cases} = e^{-\{b_{1+}(\theta)u + b_{1-}(\theta)v\}}\Big[-b_{1+}(\theta)b_{1-}(\theta) + \Big\{b_{1+}(\theta)b_{1-}(\theta) \\ \quad - 2\big(b_{1+}(\theta)b_{2-}(\theta)v + b_{2+}(\theta)b_{1-}(\theta)u\big) + b_{1+}(\theta)b_{1-}(\theta)\big(b_{2+}(\theta)u^2 \\ \quad + b_{2-}(\theta)v^2 + (1/2)\big(b_{1+}(\theta)u + b_{1-}(\theta)v\big)^2 - 2\big(b_{1+}(\theta)u \\ \quad + b_{1-}(\theta)v\big)\big)\Big\}n^{-1} + o(n^{-1})\Big] & \text{for } v < 0 < u, \\ = 0 & \text{otherwise.} \end{cases}$$

The proof is given in section 5.

Remark 3.1. As is easily seen from Theorem 3.2, the statistics $\overline{\theta}$ and $\underline{\theta}$ are asymptotically independent, but not so up to the second order.

4. APPLICATIONS

The higher order asymptotic efficiency of the maximum probability estimator is discussed by Akahira [5] in the following location parameter situation. Let $X_1, \ldots, X_n$ be a sequence of independent and identically distributed real random variables with the density $f_0(x - \theta)$ with respect to the Lebesgue measure. We assume the following conditions.

(C.1) $f_0(x) > 0$ for $a < x < b$,

$f_0(x) = 0$ for $x \leq a,\ x \geq b$,

where both a and b are finite.

(C.2) $f_0(x)$ is twice continuously differentiable in the open interval (a,b) and

$$\lim_{x\to a+0} f_0(x) = \lim_{x\to b-0} f_0(x) = c, \quad \lim_{x\to b-0} f_0'(x) = -\lim_{x\to a+0} f_0'(x) = h,$$

where c is a positive constant and h is a non-positive constant.

(C.3) $f_0(x)$ is symmetric around $x = (a+b)/2$.

(C.4)

$$0 < \int_a^b \{f_0'(x)\}^2/f_0(x)d\mu < \infty.$$

In order to discuss the higher order asymptotic efficiency, it is necessary to obtain the bound for the asymptotic distribution of higher order asymptotically median unbiased estimators, under the conditions (C.1) to (C.4). Then it is convenient to apply Theorems 3.1 and 3.2 for getting the bound. It is easily seen that the support $A(\theta)$ is given by the interval $(a+\theta, b+\theta)$. Assume that the conditions (C.1) to (C.4) hold. Then it follows that the conditions implies (A.1) to (A.5), (A.7) to (A.9). From (C.1), (C.2) and $(A.4)_2$ it follows that $b_{1+}(\theta) = c$, $b_{1-}(\theta) = -c$, $b_{2+}(\theta) = b_{2-}(\theta) = -h/2$, $\underline{\theta} = \max_{1\le i\le n} X_i - b$ and $\overline{\theta} = \min_{1\le i\le n} X_i - a$. From Theorem 3.1 we see that the statistic $(\overline{\theta}, \underline{\theta}, Z_1(\hat{\theta}^*))$ is second order asymptotically sufficient, where $Z_1(\hat{\theta}^*) = (1/\sqrt{n})\sum_{i=1}^n \{(\partial/\partial\theta)\log f(X_i - \hat{\theta}^*)\}$ with $\hat{\theta}^* = (\min_{1\le i\le n} X_i + \max_{1\le i\le n} X_i - a - b)/2$, provided that the condition (A.6) holds in addition to (C.1) to (C.4). From Theorem 3.2 we see that the asymptotic density of $(\overline{\theta}, \underline{\theta})$ is given by

$$f_{\overline{\theta},\underline{\theta}}(u,v) = \begin{cases} c^2 e^{-c(u-v)}\Big[1 - \Big\{1 + \dfrac{h}{c}(u-v) - \dfrac{h}{2}(u^2+v^2) + \dfrac{c^2}{2}(u-v)^2 \\ \qquad - 2c(u-v)\Big\}n^{-1} + o(n^{-1})\Big] & \text{for } v < 0 < u, \\ 0 & \text{otherwise,} \end{cases}$$

which coincides with the result of Akahira [5].

5. PROOF OF THEOREM 3.2.

Putting

$$\underline{\theta}_i = \inf\{\theta | X_i \in A(\theta)\} \quad \text{and} \quad \overline{\theta}_i = \sup\{\theta | X_i \in A(\theta)\} \quad (i = 1, \ldots, n),$$

we have $\underline{\theta} = \max_{1\le i\le n} \underline{\theta}_i$ and $\overline{\theta} = \min_{1\le i\le n} \overline{\theta}_i$. Then the joint distribution function $F_\theta(u,v)$ of $(\overline{\theta}, \underline{\theta})$ is given by

$$\begin{aligned}
F_\theta(u,v) &= P_\theta^n\left\{n(\underline{\theta}-\theta) \le v, n(\overline{\theta}-\theta) \le u\right\} \\
&= P_\theta^n\left\{\max_{1\le i\le n} \underline{\theta}_i \le \theta + vn^{-1}, \min_{1\le i\le n} \overline{\theta}_i \le \theta + un^{-1}\right\} \\
&= P_\theta^n\left\{\max_{1\le i\le n} \underline{\theta}_i \le \theta + vn^{-1}\right\} \\
&\qquad - P_\theta^n\left\{\max_{1\le i\le n} \underline{\theta}_i \le \theta + vn^{-1}, \min_{1\le i\le n} \overline{\theta}_i > \theta + un^{-1}\right\} \\
&= \left(P_\theta\left\{\underline{\theta}_1 \le \theta + vn^{-1}\right\}\right)^n - \left(P_\theta\left\{\underline{\theta}_1 \le \theta + vn^{-1}, \overline{\theta}_1 > \theta + un^{-1}\right\}\right)^n \qquad (5.1)
\end{aligned}$$

for $v < 0 < u$, since $\underline{\theta}_1, \dots, \underline{\theta}_n$ are mutually independent and $\overline{\theta}_1, \dots, \overline{\theta}_n$ are so. It is easily seen that, except for $v < 0 < u$, the density function of $(\overline{\theta}, \underline{\theta})$ is equal to 0. We also obtain

$$\begin{aligned}
\left(P_\theta\left\{\underline{\theta}_1 \le \theta + vn^{-1}\right\}\right)^n &= \left(1 - P_\theta\left\{\underline{\theta}_1 > \theta + vn^{-1}\right\}\right)^n \\
&= \left(1 - P_\theta\left\{A(\theta + vn^{-1})^c\right\}\right)^n.
\end{aligned}$$

Indeed, it is clear that the first equality holds. On the second one, when $X_1 \in A(\theta)$, it is enough to prove that the event $\{\underline{\theta}_1 > \theta + vn^{-1}\}$ is equivalent to the event $\{X_1 \in A(\theta + vn^{-1})^c\}$. If $\underline{\theta}_1 > \theta + vn^{-1}$, then $X_1 \notin A(\theta + vn^{-1})$ since $X_1 \in A(\theta + vn^{-1})$ implies $\underline{\theta} + vn^{-1} \ge \underline{\theta}_1 = \inf\{\theta | X_1 \in A(\theta)\}$. To show the converse assertion, assume that $\theta_1 \le \theta + vn^{-1}$. If $\underline{\theta}_1 = \theta + vn^{-1}$, then $X_1 \in A(\theta + vn^{-1})$, which leads to a contradiction. If $\underline{\theta}_1 < \theta + vn^{-1}$, then there exists a $\theta^* \in \Theta$ such that $\theta^* < \theta + vn^{-1}$ and $X_1 \in A(\theta^*)$, hence, by (A.2), $X_1 \in A(\theta + vn^{-1})$, since $\theta^* < \theta + vn^{-1} < \theta$, which reaches a contradiction. Thus the converse claim is obtained. From $(A.4)_2$ we have

$$\begin{aligned}
\left(P_\theta\left\{\underline{\theta}_1 \le \theta + vn^{-1}\right\}\right)^n &= \left\{1 - \int_{A(\theta)-A(\theta+vn^{-1})} f(x,\theta)d\mu\right\}^n \\
&= e^{-b_{1-}(\theta)v}\left[1 - \left\{b_{2-}(\theta) + \frac{1}{2}b_{1-}^2(\theta)\right\}v^2 n^{-1} + o(n^{-1})\right]. \qquad (5.2)
\end{aligned}$$

We also have

$$\begin{aligned}
&P_\theta\left\{\underline{\theta}_1 \le \theta + vn^{-1}, \overline{\theta}_1 > \theta + un^{-1}\right\} \\
=&1 - P_\theta\{\underline{\theta}_1 > \theta + vn^{-1}\} - P_\theta\{\overline{\theta}_1 \le \theta + un^{-1}\} + P_\theta\{\underline{\theta}_1 > \theta + vn^{-1}, \overline{\theta}_1 \le \theta + un^{-1}\} \\
=&1 - P_\theta\left(A(\theta + vn^{-1})^c\right) - P_\theta\left(A(\theta + un^{-1})^c\right) + P_\theta\left(A(\theta + vn^{-1})^c \cap A(\theta + un^{-1})^c\right)
\end{aligned}$$

$$=1 - P_\theta\left(A(\theta+vn^{-1})^c \cup A(\theta+un^{-1})^c\right)$$
$$=P_\theta(A(\theta+vn^{-1}) \cap A(\theta+un^{-1}))$$
$$=G_\theta(u,v) \quad \text{(say)}. \tag{5.3}$$

From (5.1), (5.2) and (5.3) we obtain

$$F_\theta(u,v) = e^{-b_{1-}(\theta)v}\left[1 - \left\{b_{2-}(\theta) + \frac{1}{2}b_{1-}^2(\theta)\right\}v^2 n^{-1} + o(n^{-1})\right] - \{G_\theta(u,v)\}^n,$$

hence

$$\frac{\partial^2 F_\theta(u,v)}{\partial u \partial v} = -n(n-1)\{G_\theta(u,v)\}^{n-2}\frac{\partial G_\theta}{\partial u}\frac{\partial G_\theta}{\partial v} - n\{G_\theta(u,v)\}^{n-1}\frac{\partial^2 G_\theta(u,v)}{\partial u \partial v}. \tag{5.4}$$

Since, by $(A.4)_2$,

$$P_\theta(A(\theta) - A(\theta+un^{-1})) = \int_{A(\theta)-A(\theta+un^{-1})} f(x,\theta)d\mu$$
$$= b_{1+}(\theta)un^{-1} + b_{2+}(\theta)u^2n^{-2} + o(n^{-2}),$$
$$P_\theta(A(\theta) - A(\theta+vn^{-1})) = \int_{A(\theta)-A(\theta+vn^{-1})} f(x,\theta)d\mu$$
$$= b_{1-}(\theta)vn^{-1} + b_{2-}(\theta)v^2n^{-2} + o(n^{-2})$$

for $v < 0 < u$, it follows from (A.9) that

$$G_\theta(u,v) = P_\theta(A(\theta) \cap A(\theta+vn^{-1}) \cap A(\theta+un^{-1}))$$
$$= 1 - P_\theta(A(\theta) - A(\theta+un^{-1})) - P_\theta(A(\theta) - A(\theta+vn^{-1}))$$
$$+ P_\theta((A(\theta) - A(\theta+un^{-1})) \cap (A(\theta) - A(\theta+vn^{-1})))$$
$$= 1 - \{b_{1+}(\theta)u + b_{1-}(\theta)v\}n^{-1} - \{b_{2+}(\theta)u^2 + b_{2-}(\theta)v^2\}n^{-2} + o(n^{-2}), \tag{5.5}$$

hence

$$\frac{\partial G_\theta(u,v)}{\partial u} = -b_{1+}(\theta)n^{-1} - 2b_{2+}(\theta)un^{-2} + o(n^{-2}), \tag{5.6}$$
$$\frac{\partial G_\theta(u,v)}{\partial v} = -b_{1-}(\theta)n^{-1} - 2b_{2-}(\theta)vn^{-2} + o(n^{-2}), \tag{5.7}$$
$$\frac{\partial^2 G_\theta(u,v)}{\partial u \partial v} = o(n^{-2}). \tag{5.8}$$

From (5.4), (5.6), (5.7) and (5.8) we obtain

$$
\begin{aligned}
\frac{\partial^2 F_\theta(u,v)}{\partial u \partial v} &= -n(n-1)\{G_\theta(u,v)\}^{n-2}\frac{\partial G_\theta(u,v)}{\partial u}\frac{\partial G_\theta(u,v)}{\partial v} + o(n^{-1}) \\
&= -n(n-1)\big[1 - \{b_{1+}(\theta)u + b_{1-}(\theta)v\}n^{-1} - \{b_{2+}(\theta)u^2 + b_{2-}(\theta)v^2\}n^{-2} \\
&\quad + o(n^{-2})\big]^{n-2}\{b_{1+}(\theta)n^{-1} + 2b_{2+}(\theta)un^{-2} + o(n^{-2})\} \\
&\quad \cdot \{b_{1-}(\theta)n^{-1} + 2b_{2-}(\theta)vn^{-2} + o(n^{-2})\} + o(n^{-1}) \\
&= -\big[\exp(n-2)\log(1 - (b_{1+}(\theta)u + b_{1-}(\theta)v)n^{-1} \\
&\quad - (b_{2+}(\theta)u^2 + b_{2-}(\theta)v^2)n^{-2} + o(n^{-2}))\big]n(n-1)\big\{b_{1+}(\theta)b_{1-}(\theta)n^{-2} \\
&\quad + 2b_{1+}(\theta)b_{2-}(\theta)vn^{-3} + 2b_{2+}(\theta)b_{1-}(\theta)un^{-3} + o(n^{-3})\big\} \\
&= -\Bigg[\exp\Bigg\{-(b_{1+}(\theta)u + b_{1-}(\theta)v) - (b_{2+}(\theta)u^2 + b_{2-}(\theta)v^2)n^{-1} \\
&\quad - \frac{1}{2}(b_{1+}(\theta)u + b_{1-}(\theta)v)^2n^{-1} + 2(b_{1+}(\theta)u + b_{1-}(\theta)v)n^{-1} + o(n^{-1})\Bigg\}\Bigg] \\
&\quad \cdot \big\{b_{1+}(\theta)b_{1-}(\theta) + 2b_{1+}(\theta)b_{2-}(\theta)vn^{-1} + 2b_{2+}(\theta)b_{1-}(\theta)un^{-1} \\
&\quad - b_{1+}(\theta)b_{1-}(\theta)n^{-1} + o(n^{-1})\big\} \\
&= e^{-\{b_{1+}(\theta)u + b_{1-}(\theta)v\}}\Bigg[-b_{1+}(\theta)b_{1-}(\theta) + \Bigg\{b_{1+}(\theta)b_{1-}(\theta) \\
&\quad - 2(b_{1+}(\theta)b_{2-}(\theta)v + b_{2+}(\theta)b_{1-}(\theta)u) \\
&\quad + b_{1+}(\theta)b_{1-}(\theta)\Bigg(b_{2+}(\theta)u^2 + b_{2-}(\theta)v^2 \\
&\quad + \frac{1}{2}(b_{1+}(\theta)u + b_{1-}(\theta)v)^2 - 2(b_{1+}(\theta)u + b_{1-}(\theta)v)\Bigg)\Bigg\}n^{-1} + o(n^{-1})\Bigg]
\end{aligned}
$$

for $v < 0 < u$. This completes the proof.

REFERENCES

1. L. LeCam, *Proc. Third Berkeley Symp. on Math. Statist. and Prob.*, **1**, 129-156 (1956).
2. M. Akahira, *Rep. Univ. Electro-Comm.*, **27**, 125-128 (1976).
3. L. Weiss, *Selecta Statistica Canadiana*, **5**, 141-150 (1979).
4. H. Mita, *Tokyo J. Math.*, **2**, 323-335 (1979).
5. M. Akahira, *Ann. Inst. Statist. Math.*, **43**, 181-195 (1991).
6. M. Akahira, to appear in *Metron*.
7. M. Akahira and K. Takeuchi, *Ann. Inst. Statist. Math.*, **39**, 593-610 (1987).

Stat. Sci. & Data Anal., pp. 365-374
K. Matsusita *et al.* (Eds)

Error Bounds for Asymptotic Expansions of Some Distributions in a SUR Model

SATORU MUKAIHATA and YASUNORI FUJIKOSHI
Department of Mathematics, Faculty of Science, Hiroshima University, Higashi-Hiroshima 724, Japan

Abstract. In this paper we obtain error bounds for asymptotic expansions of the distributions of Zellner's estimator in a SUR model. Error bounds are derived by obtaining certain modifications of Fujikoshi and Shimizu [1]'s results. In our derivation it is essential to obtain appropriate approximations for some moments with error bounds.

Key words: Asymptotic expansions, distribution functions, error bounds, standardized and Studentized estimators, SUR model.

1. INTRODUCTION

In this paper we study error bounds for asymptotic expansions of the distribution of the Zellner's estimator in a SUR model. An error bound has been obtained by Kariya and Maekawa [2] for an asymptotic expansion of the distribution function of the standardized estimator. Our purpose is to obtain an improvement for their error bound and to derive an error bound for an asymptotic expansion of the distribution function of the Studentized estimator.

Our method is based on modifications of Fujikoshi and Shimizu [1]'s method. It is easily seen that the distribution function of the Studentized estimator as well as the standardized estimator can be expressed as the one of a scale mixture of the standard normal variable Z, i.e., $X = \sigma Z$, where σ is a positive-valued random variable and is independent of Z. It is known (see, e.g., Fujikoshi and Shimizu [1]) that the distribution function of X can be approximated with an error bound. In the use of the result it is necessary to evaluate the exact moments of $\sigma^{\pm 2}$. But it is generally difficult to obtain the exact ones in our cases. However, by deriving appropriate approximations for their moments with error bounds, we will obtain error bounds for asymptotic expansions of the distribution of X. In Section 2 we derive an extended result on two types of approximations for the distribution function of a scale mixture of the standard normal distribution function and their error bounds. In Section 3 the standardized and Studentized Zellner's estimators are expressed as a scale mixture of the standardized normal distribution function. Asymptotic expansions of the distribution function of these estimators and their error bounds are obtained in Section 4.

2. SCALE MIXTURE OF THE STANDARD NORMAL DISTRIBUTION

Here we consider a scale mixture of the standard normal variable, i.e.

$$X = \sigma Z, \tag{2.1}$$

where σ is a positive-valued random variable and is independent of Z. Let Φ and ϕ be the distribution and the probability density function of Z, respectively. Then the distribution function $F(x)$ of X can be expressed as

$$\begin{aligned} F(x) &= E_\sigma[\Phi(x/\sigma)] \\ &= E_\sigma[\Phi(x/\sqrt{\sigma^2})]. \end{aligned}$$

Two types of approximations for $F(x)$ have been proposed (cf. Fujikoshi and Shimizu [1]): For $\delta = -1, 1$,

$$\Phi_{\delta,k}(x) = E_\sigma[\Phi_{\delta,k}(x,\sigma)], \tag{2.2}$$

where

$$\Phi_{\delta,k}(x,\sigma) = \Phi(x) + \sum_{j=1}^{k-1} \frac{1}{j!} a_{\delta,j}(x)\phi(x)(\sigma^{2\delta} - 1)^j.$$

Here the coefficients are defined by

$$a_{\delta,j}(x) = \left(\frac{d}{ds}\right)^j \Phi(xs^{-\delta/2})\Big|_{s=1} \Big/ \phi(x),$$

or equivalently

$$\begin{aligned} a_{1,j}(x) &= -2^{-j} H_{2j-1}(x), \\ a_{-1,j}(x) &= (-1)^{j-1} 2^{-j} \{x^{2j-1} + \sum_{i=1}^{j-1} 1\cdot 3 \cdots (2i-1) \binom{j-1}{i} x^{2j-2i-1}\}, \end{aligned}$$

where $H_j(x)$ is the Hermite polynomial defined by $(\frac{d}{dx})^j \phi(x) = (-1)^j H_j(x)\phi(x)$.

When the exact moments of $\sigma^{2\delta}$ are not available, we cannot use the approximation $\Phi_{\delta,k}(x)$. However, we may consider to approximate $(\sigma^{2\delta}-1)^j$ by a simple variable, $m^*_{\delta,j}(\sigma^2)$, whose exact moments are available. Namely, suppose that for $j = 1, \cdots, k-1$,

$$(\sigma^{2\delta} - 1)^j = m^*_{\delta,j}(\sigma^2) + R_{\delta,j}(\sigma^2) \tag{2, 3}$$

and $E_\sigma[m^*_{\delta,j}(\sigma^2)]$ can be evaluated.

Theorem 2.1. *Let $X = \sigma Z$ be a scale mixture of the standard normal variable. Suppose that for a given positive integer k, $E[\sigma^{2k}] < \infty$, $E[\sigma^{-2k}] < \infty$ and $E_\sigma[m^*_{\delta,j}(\sigma^2)] < \infty, j = 1, \cdots, k-1$. Then, letting*

$$\Phi^*_{\delta,k}(x) = \Phi(x) + \sum_{j=1}^{k-1} \frac{1}{j!} a_{\delta,j}(x)\phi(x) E_\sigma[m^*_{\delta,j}(\sigma^2)], \tag{2.4}$$

it holds that

$$\begin{aligned}\sup_x |F(x) - \Phi^*_{\delta,k}(x)| &\le \frac{1}{k!}\bar{a}_{\delta,k} E_\sigma[(\sigma^2 \vee \sigma^{-2} - 1)^k] + \sum_{j=1}^{k-1} \frac{1}{j!}\bar{a}_{\delta,j} |E_\sigma[R_{\sigma,j}(\sigma^2)]| \\ &\le \frac{1}{k!}\bar{a}_{\delta,k} E_\sigma[|\sigma^2 - 1|^k + |\sigma^{-2} - 1|^k] + \sum_{j=1}^{k-1} \frac{1}{j!}\bar{a}_{\delta,j} E_\sigma[|R_{\delta,j}(\sigma^2)|],\end{aligned}$$

where $\bar{a}_{\delta,j} = \sup_x |a_{\delta,j}(x)\phi(x)|, j = 1, \cdots, k.$

Proof. It is known (cf. Fujikoshi and Shimizu [1]) that

$$\sup_x |F(x) - \Phi_{\delta,k}(x)| \le \frac{1}{k!}\bar{a}_{\delta,k} E_\sigma[(\sigma^2 \vee \sigma^{-2} - 1)^k]. \tag{2.5}$$

Therefore, the result follows from that

$$\Phi_{\delta,k}(x) = \Phi^*_{\delta,k}(x) + \sum_{J=1}^{k-1} \frac{1}{j!} a_{\delta,j}(x)\phi(x) E_\sigma[R_{\delta,j}(\sigma^2)].$$

We note that if the first $k-1$ moments of $\sigma^{2\delta}$ in Theorem 2.1 can be obtained, we can take $R_{\delta,j} = 0$ for $j = 1, \cdots, k-1$.

3. ZELLNER'S ESTIMATOR IN A SUR MODEL

Let

$$\boldsymbol{y} = A\boldsymbol{\beta} + \boldsymbol{u} \text{ with } \boldsymbol{u} \sim N(\boldsymbol{0}, \Omega) \tag{3.1}$$

be a linear normal regression model, where A is an $n \times k$ fixed matrix and $\Omega \in S(n)$. Here $S(n)$ denotes the set of all positive definite matrices of order n. We consider the two equations SUR model which has the following structure in (3.1):

$$A = \begin{pmatrix} A_1 & O \\ O & A_2 \end{pmatrix}, \quad \boldsymbol{y} = \begin{pmatrix} \boldsymbol{y}_1 \\ \boldsymbol{y}_2 \end{pmatrix}, \quad \boldsymbol{u} = \begin{pmatrix} \boldsymbol{u}_1 \\ \boldsymbol{u}_2 \end{pmatrix} \text{ and } \boldsymbol{\beta} = \begin{pmatrix} \boldsymbol{\beta}_1 \\ \boldsymbol{\beta}_2 \end{pmatrix}, \tag{3.2}$$

where $A_i : m \times k_i, \boldsymbol{y}_i : m \times 1, \boldsymbol{u}_i : m \times 1, \boldsymbol{\beta}_i : k_i \times 1, k_1 + k_2 = k$ and $n = 2m$. Further, the covariance matrix of $\boldsymbol{u}$ is of the form

$$\Omega = \mathrm{Cov}(\boldsymbol{u}) = \Sigma \otimes I_m, \text{ where } \Sigma = (\sigma_{ij}) \in S(2). \tag{3.3}$$

The UZE (Zellner's estimator with the unrestricted covariance matrix) in this case is given by

$$\hat{\boldsymbol{\beta}} = \hat{\boldsymbol{\beta}}(\hat{\Omega}) = (A'\hat{\Omega}^{-1}A)^{-1}A'\hat{\Omega}^{-1}\boldsymbol{y} \tag{3.4}$$

with $\hat{\Omega} = \left(\frac{1}{q}S\right) \otimes I_m$, where $S = Y'\{I_m - A^*(A^{*\prime}A^*)^- A^{*\prime}\}Y$, $Y = [\boldsymbol{y}_1, \boldsymbol{y}_2]$, $A^* = [A_1, A_2]$ and $q = m - \mathrm{rank}(A^*)$. As is well known, the conditional distribution of $\hat{\boldsymbol{\beta}}(\hat{\Omega})$ given $\hat{\Omega}$ is

$$N(\boldsymbol{\beta}, (A'\hat{\Omega}^{-1}A)^{-1}A'\hat{\Omega}^{-1}\Omega\hat{\Omega}^{-1}A(A'\hat{\Omega}^{-1}A)^{-1}),$$

and S is distributed as a Wishart distribution $W_2(\Sigma, q)$.

In this paper, we consider the distribution of a linear combination of $\hat{\boldsymbol{\beta}}$. More precisely, we consider the distributions of the standardized variable

$$X = \{\boldsymbol{a}'(\hat{\boldsymbol{\beta}} - \boldsymbol{\beta})\}/\Delta = (\tilde{D}/\Delta) \cdot \{\boldsymbol{a}'(\hat{\boldsymbol{\beta}} - \boldsymbol{\beta})\}/\tilde{D} = \sigma Z \tag{3.5}$$

and the Studentized variable

$$\tilde{X} = \boldsymbol{a}'(\hat{\boldsymbol{\beta}} - \boldsymbol{\beta})\}/D = (\tilde{D}/D) \cdot \{\boldsymbol{a}'(\hat{\boldsymbol{\beta}} - \boldsymbol{\beta})\}/\tilde{D} = \tilde{\sigma} Z, \tag{3.6}$$

where $\boldsymbol{a}$ is an $n \times 1$ constant vector and

$$\begin{aligned}
&\tilde{D} = \{\boldsymbol{a}'(A'\hat{\Omega}^{-1}A)^{-1}A'\hat{\Omega}^{-1}\Omega\hat{\Omega}^{-1}A(A'\hat{\Omega}^{-1}A)^{-1}\boldsymbol{a}\}^{1/2}, \\
&\Delta = \{\boldsymbol{a}'(A'\Omega^{-1}A)^{-1}\boldsymbol{a}\}^{1/2}, \quad D = \{\boldsymbol{a}'(A'\hat{\Omega}^{-1}A)^{-1}\boldsymbol{a}\}^{1/2}, \\
&Z = \{\boldsymbol{a}'(\hat{\boldsymbol{\beta}} - \boldsymbol{\beta})\}/\tilde{D}, \quad \sigma = \tilde{D}/\Delta \text{ and } \tilde{\sigma} = \tilde{D}/D.
\end{aligned}$$

Let $F(x)$ and $\tilde{F}(x)$ be the distribution functions of X and $\tilde{X}$, respectively. Then it is easily seen that

$$F(x) = E_S[\Phi(x/\sigma)] \text{ and } \tilde{F}(x) = E_S[\Phi(x/\tilde{\sigma})]. \tag{3.8}$$

In general, it is very difficult to obtain the exact distributions of X and $\tilde{X}$. Moreover, they will be very complicated even if they are obtained. We will obtain asymptotic expansions for $F(x)$ and $\tilde{F}(x)$ and their error bounds by using Theorem 2.1.

For a notational simplicity, let

$$P = \frac{1}{n}A'\Omega^{-1}A, \quad \bar{A} = \begin{pmatrix} \bar{A}_1 \\ \bar{A}_2 \end{pmatrix} = \frac{1}{\sqrt{n}}\Omega^{-1/2}AP^{-1/2}, \tag{3, 9}$$

and

$$\boldsymbol{b} = \{\boldsymbol{a}'P^{-1}\boldsymbol{a}\}^{-1/2}P^{-1/2}\boldsymbol{a}. \tag{3, 10}$$

Then $\bar{A}'\bar{A} = I_k$ and $\boldsymbol{b}'\boldsymbol{b} = 1$.

4. THE DISTRIBUTION OF ZELLNER'S ESTIMATOR

4.1. The distribution of X

Kariya and Maekawa [2] obtained error bounds for asymptotic expansions of the distribution function of X defined by (3.5). Their method is based on the use of a bound in terms of characteristic functions. By using Theorem 2.1, we can improve their error bounds as follows:

Theorem 4.1. *Let* $M = \bar{A}_1'\bar{A}_1 - \bar{A}_2'\bar{A}_2$, $N = \bar{A}_1'\bar{A}_2 + \bar{A}_2'\bar{A}_1$ *and*

$$\Phi^*(x) = \Phi(x) - \frac{1}{2(q+1)}\{2 - \boldsymbol{b}'(M^2 + N^2)\boldsymbol{b}\}\phi(x).$$

Then

$$\text{(i)}\ \sup_x |F(x) - \Phi^*(x)| \le \frac{8c_1}{(q-3)(q-5)} + \frac{4c_2}{(q+1)(q+3)} \quad (q > 5)$$

$$\text{(ii)}\ \sup_x |F(x) - \Phi(x)| \le c_3(q-3)^{-1} \quad (q > 3),$$

where $c_1 = \frac{1}{8}\sup_x |H_3(x)\phi(x)|, c_2 = \frac{1}{2}\sup_x |H_1(x)\phi(x)|$ *and* $c_3 = 2c_2$.

Proof. It has been essentially proved in Kariya and Maekawa [2] that $\sigma^2 \ge 1$,

$$E[\sigma^2 - 1] \le \frac{2}{(q-3)} \quad \text{for } q > 3,$$

$$E[(\sigma^2-1)^2] \le \frac{8}{(q-3)(q-5)} \quad \text{for } q > 5 \tag{4.1}$$

and

$$|E[\sigma^2 - 1] - \frac{1}{q+1}\{2 - b'(M^2+N^2)b\}| \le \frac{4}{(q+1)(q+3)}.$$

Therefore, the desired results are obtained from Theorem 2.1.

Remark. The results obtained by Kariya and Maekawa [2] are the ones obtained by replacing c_1, c_2 and c_3 by $(2\pi)^{-1}$, π^{-1} and $(2/\pi)^{1/2}$, respectively. The constants c_i's are computed (see Fujikoshi [3]) as

$$c_1 = 0.0688\cdots, c_2 = 0.1210\cdots, c_3 = 0.2420\cdots.$$

Therefore, our bounds are less than the halves of their bounds.

4.2. The distribution of $\tilde{X}$

In this section, we obtain asymptotic expansions for the distribution function $\tilde{F}(x)$ of $\tilde{X}$ defined by (3.6) and their error bounds. First, we state the main result.

Theorem 4.2. *Let* $\tilde{\Phi}^*(x) = \Phi(x) - (3/2)(q-3)^{-1}x\phi(x)$ *and* $\tilde{\varepsilon} = \sup_x |\tilde{F}(x) - \tilde{\Phi}^*(x)|$. *Then if* $q > 5$,

$$\text{(i)}\ \tilde{\varepsilon} \le d_1\{D_1(q) + D_2(q)\} + d_2 \cdot \frac{4(2q-3)}{(q-2)(q-3)},$$

$$\text{(ii)}\ \tilde{\varepsilon} \le d_3\{D_1(q) + D_2(q)\} + d_2[\frac{6}{q-3} + \frac{2}{q+1} + \{\frac{48}{(q+1)(q+3)(q+5)}\}^{1/2}],$$

$$\text{(iii)}\ \sup_x |\tilde{F}(x) - \Phi(x)| \le d_2[\{D_1(q)\}^{1/2} + \{D_2(q)\}^{1/2}],$$

where $d_i = c_i$ *in Theorem 4.1* $(i = 1, 2)$, $d_3 = \frac{1}{8}\sup_x |(x^3 + x)\phi(x)| = 0.0791\cdots$, $D_1(q) = 2(3q^2 + 11q - 30)/\{(q-2)(q-3)(q-5)\}$ *and* $D_2(q) = 6q^{-1}$.

Our proof is based on the use of Theorem 2.1, and hence we need to to obtain some inequalities as in (4.1) for $\tilde{\sigma}^2$ and $\tilde{\sigma}^{-2}$. Let

$$\tilde{S} = \frac{1}{q}\Sigma^{-1/2}S\Sigma^{-1/2}, \quad W = \tilde{S} \otimes I_m = \Omega^{-1/2}\hat{\Omega}\Omega^{-1/2},$$

$$C_j = \bar{A}'(W^{-1} - I_n)^j \bar{A} \ (j = 1, 2), \quad Q = (I_k + C_1)^{-1}, \\ J = Q(I_k + 2C_1 + C_2)Q. \tag{4.2}$$

Then from (3.7), (3.9), (3.10) and (4.2),

$$\begin{aligned} \tilde{\sigma}^2 &= \{\boldsymbol{b}'(\bar{A}'W^{-1}\bar{A})^{-1}\bar{A}'W^{-2}\bar{A}(\bar{A}'W^{-1}\bar{A})^{-1}\boldsymbol{b}\}/\{\boldsymbol{b}'(\bar{A}'W^{-1}\bar{A})^{-1}\boldsymbol{b}\} \\ &= \{\boldsymbol{b}'(I_k + C_1)^{-1}(I_k + 2C_1 + C_2)(I_k + C_1)^{-1}\boldsymbol{b}\}/\{\boldsymbol{b}'(I_k + C_1)^{-1}\boldsymbol{b}\} \\ &= (\boldsymbol{b}'J\boldsymbol{b})/(\boldsymbol{b}'Q\boldsymbol{b}) \\ &= \sigma^2(\boldsymbol{b}'Q\boldsymbol{b})^{-1} \end{aligned} \tag{4.3}$$

Lemma 4.1. *It holds that*

$$\begin{aligned} (i)\ \tilde{\sigma}^2 - 1 &= \boldsymbol{b}'C_1\boldsymbol{b} - R_1, \text{ where} \\ R_1 &= \boldsymbol{b}'C_1^2Q\boldsymbol{b} - \frac{\{\boldsymbol{b}'C_1Q\boldsymbol{b}\}^2}{\boldsymbol{b}'Q\boldsymbol{b}} - \frac{\sigma^2 - 1}{\boldsymbol{b}'Q\boldsymbol{b}}, \\ (ii)\ \tilde{\sigma}^{-2} - 1 &= -\boldsymbol{b}'C_1\boldsymbol{b} + R_{-1}, \text{ where} \\ R_{-1} &= \boldsymbol{b}'C_1^2Q\boldsymbol{b} + \boldsymbol{b}'Q\boldsymbol{b} \cdot (\sigma^{-2} - 1). \end{aligned}$$

Proof. Using the relations $\boldsymbol{b}'\boldsymbol{b} = 1$ and $Q = I_k - C_1Q$, we can easily obtain the desired results from (4.3).

Lemma 4.2. *Let $\lambda_1 \le \lambda_2$ be the latent roots of $\tilde{S}^{-1}$. Then*

$$(i)\ 1 \le \sigma^2 \le (\lambda_1 + \lambda_2)^2/(4\lambda_1\lambda_2),$$
$$(ii)\ \lambda_1 \le \tilde{\sigma}^2 \le \lambda_2.$$

Proof. Let ρ be a 2×2 orthogonal matrix such that $\rho\tilde{S}^{-1}\rho' = \text{diag}\{\lambda_1, \lambda_2\}$, and let

$$\tilde{A} = \begin{pmatrix} \tilde{A}_1 \\ \tilde{A}_2 \end{pmatrix} = (\rho \otimes I_m)\bar{A}, \quad B = \tilde{A}_2'\tilde{A}_2, \tag{4.4}$$

where $\tilde{A}_i : m \times k$. Then $\bar{A}'\bar{A} = I_k$ implies $I_k = \tilde{A}'\tilde{A} = \tilde{A}_1'\tilde{A}_1 + \tilde{A}_2'\tilde{A}_2$.

Note that ρ and so $\tilde{A}$ are random. Further, let $0 \le \eta_1 \le \cdots \le \eta_k$ be the latent roots of $B = \tilde{A}_2'\tilde{A}_2$ and let Ψ be a $k \times k$ orthogonal matrix such that

$$\Psi B \Psi' = \text{diag}\{\eta_1, \cdots, \eta_k\}. \tag{4.5}$$

Here we use arguments similar to the ones as in Kariya and Maekawa [2], who used in obtaining (4.1). Since $\tilde{A}_1'\tilde{A}_1 = I_k - B \ge 0, 0 \le \eta_i \le 1$ $(i = 1, \cdots, k)$. Then we can write C_j $(j = 1, 2)$ and $Q = (I_k + C_1)^{-1}$ as follows:

$$C_j = (\lambda_1 - 1)^j(I_k - B) + (\lambda_2 - 1)^j B \quad (j = 1, 2), \\ Q = \{\lambda_1(I_k - B) + \lambda_2 B\}^{-1}. \tag{4.6}$$

Hence from (4.5) and (4.6), $\Psi J\Psi' = \mathrm{diag}\{h(\eta_1), \cdots, h(\eta_k)\}$, where

$$h(\eta) = \frac{\lambda_1^2(1-\eta)+\lambda_2^2\eta}{\{\lambda_1(1-\eta)+\lambda_2\eta\}^2} \qquad (0 \le \eta \le 1).$$

The first result (i) follows from that

$$1 \le h(\eta) \le (\lambda_1+\lambda_2)^2/(4\lambda_1\lambda_2), \quad 0 \le \eta \le 1.$$

Next, from the fact that $Q = (I_k + C_1)^{-1} > 0$, it holds that

$$\alpha_1 \le \tilde{\sigma} = \boldsymbol{b}'J\boldsymbol{b}/(\boldsymbol{b}'Q\boldsymbol{b}) \le \alpha_2,$$

where α_1 and α_2 are the minimum and the maximum of the latent roots of $JQ^{-1} = (I_k + C_1)^{-1}(I_k + 2C_1 + C_2)$, respectively. Then from (4.5) and (4.6), $\Psi JQ^{-1}\Psi' = \mathrm{diag}\{h(\eta_1), \cdots, h(\eta_k)\}$, where in this case

$$h(\eta) = \frac{\lambda_1^2(1-\eta)+\lambda_2^2\eta}{\lambda_1(1-\eta)+\lambda_2\eta} \qquad (0 \le \eta \le 1).$$

It is easy to see that $\lambda_1 = h(0) \le h(\eta) \le h(1) = \lambda_2$, which proves the second result (ii).

Now, since $q\tilde{S} \sim W_2(I_2, q)$, we can write $\tilde{S}$ as

$$\tilde{S} = \frac{1}{q}\begin{pmatrix} t_{11}^2 & t_{11}t_{12} \\ t_{11}t_{12} & t_{12}^2 + t_{22}^2 \end{pmatrix}, \tag{4.7}$$

where $t_{ii}^2 \sim \chi^2_{q-i+1}$, $i = 1, 2, t_{12} \sim N(0,1)$ and t_{11}, t_{12} and t_{22} are mutually independent (see, e.g., Siotani, Hayakawa and Fujikoshi [4]). Then $\lambda_1 + \lambda_2$ and $\lambda_1\lambda_2$ are expressed as

$$\begin{aligned} \lambda_1 + \lambda_2 &= q(t_{11}^2 + t_{12}^2 + t_{22}^2)(t_{11}^2 t_{22}^2)^{-1} = 2V_1^{-1}(\frac{1}{2q}V_2)^{-1}, \\ \lambda_1\lambda_2 &= q^2(t_{11}^2 t_{22}^2)^{-1} = V_1^{-1}(\frac{1}{2q}V_2)^{-2}, \end{aligned} \tag{4.8}$$

where $V_1 = 4t_{11}^2 t_{22}^2 V_2^{-2} \sim B((q-1)/2, 1)$, $V_2 = t_{11}^2 + t_{12}^2 + t_{22}^2 \sim \chi^2_{2q}$ and V_1 is independent of V_2 (these results can be easily obtained). Further, it follows that

$$0 \le \sigma^2 - 1 \le (\lambda_1+\lambda_2)^2(4\lambda_1\lambda_2)^{-1} - 1 = (1-V_1)V_1^{-1} \tag{4.9}$$

from Lemma 4.2 and that $1 - V_1 \sim B(1, (q-1)/2)$.

Lemma 4.3. *Let C_i's, R_δ's and $\tilde{\sigma}^2$ be the ones defined in (4.2), Lemma 4.1 and (3.7), respectively. Then*

(*i*) $E(C_1) = \dfrac{3}{q-3}I_k, \quad q > 3.$

(*ii*) $E[|R_1|] \le \dfrac{4(2q-3)}{(q-2)(q-3)}, \quad q > 3.$

(*iii*) $E[|R_{-1}|] \le \dfrac{6}{q-3} + \dfrac{2}{q+1} + \{\dfrac{48}{(q+1)(q+3)(q+5)}\}^{1/2}, \quad q > 3.$

(*iv*) $E[(\tilde{\sigma}^2 - 1)^2] \le \dfrac{2(3q^2+11q-30)}{(q-2)(q-3)(q-5)}, \quad q > 5.$

(*v*) $E[(\tilde{\sigma}^{-2} - 1)^2] \le 6q^{-1}, \quad q > 1.$

Proof. (i) is immediately obtained from that $q\tilde{S} \sim W_2(I_2, q)$ and (4.2).

Next we prove (ii). From (4.5) and (4.6), $\Psi C_1^2 Q \Psi' = \mathrm{diag}\{h(\eta_1), \cdots, h(\eta_k)\}$, where

$$h(\eta) = \frac{\{(\lambda_1 - 1) + (\lambda_2 - \lambda_1)\eta\}^2}{\lambda_1 + (\lambda_2 - \lambda_1)\eta} \quad (0 \le \eta \le 1).$$

It is seen that $h(\eta)$ satisfies inequalities

$$0 \le h(\eta) \le \max\{(\lambda_1 - 1)^2 \lambda_1^{-1},\ (\lambda_2 - 1)^2 \lambda_2^{-1}\},$$

and therefore

$$0 \le \boldsymbol{b}' C_1^2 Q \boldsymbol{b} \le \max\{(\lambda_1 - 1)^2 \lambda_1^{-1},\ (\lambda_2 - 1)^2 \lambda_2^{-1}\}. \tag{4.10}$$

Further, from $Q = (I_k + C_1)^{-1} > 0$ we have

$$\alpha_1 \le \boldsymbol{b}' C_1 Q \boldsymbol{b} (\boldsymbol{b}' Q \boldsymbol{b})^{-1} \le \alpha_2,$$

where α_1 and α_2 are the minimum and the maximum of the latent roots of C_1, respectively. From (4.5) and (4.6), it follows that $\Psi C_1 \Psi' = \mathrm{diag}\{h(\eta_1), \cdots, h(\eta_k)\}$, where in this case $h(\eta) = \lambda_1 - 1 + (\lambda_2 - \lambda_1)\eta$ $(0 \le \eta \le 1)$. The function $h(\eta)$ satisfies

$$\lambda_1 - 1 = h(0) \le h(\eta) \le h(1) = \lambda_2 - 1.$$

Hence

$$\lambda_1 - 1 \le \boldsymbol{b}' C_1 Q \boldsymbol{b} (\boldsymbol{b}' Q \boldsymbol{b})^{-1} \le \lambda_2 - 1.$$

In a similar manner, we have

$$(\lambda_1 - 1)\lambda_1^{-1} \le \boldsymbol{b}' C_1 Q \boldsymbol{b} \le (\lambda_2 - 1)\lambda_2^{-1}$$

and therefore

$$0 \le \frac{(\boldsymbol{b}' C_1 Q \boldsymbol{b})^2}{\boldsymbol{b}' Q \boldsymbol{b}} \le \max\{(\lambda_1 - 1)^2 \lambda_1^{-1},\ (\lambda_2 - 1)^2 \lambda_2^{-1}\}. \tag{4.11}$$

Hence from (4.10), (4.11) and (4.8),

$$\begin{aligned} E\left[\left|\boldsymbol{b}' C_1^2 Q \boldsymbol{b} - \frac{(\boldsymbol{b}' C_1 Q \boldsymbol{b})^2}{\boldsymbol{b}' Q \boldsymbol{b}}\right|\right] &\le E[\max\{(\lambda_1 - 1)^2 \lambda_1^{-1},\ (\lambda_2 - 1)^2 \lambda_2^{-1}\}] \\ &\le E[(\lambda_1 + \lambda_2) + (\lambda_1^{-1} + \lambda_2^{-1})] - 4 \qquad (4.12) \\ &= \frac{6}{q - 3}. \end{aligned}$$

Furthermore, we have

$$0 \le \frac{\sigma^2 - 1}{\boldsymbol{b}' Q \boldsymbol{b}} = \frac{\boldsymbol{b}' Q (C_2 - C_1^2) Q \boldsymbol{b}}{\boldsymbol{b}' Q \boldsymbol{b}} \le (\sqrt{\lambda_2} - \sqrt{\lambda_1}\,)^2 \tag{4.13}$$

and therefore

$$E\left(\frac{\sigma^2-1}{b'Qb}\right) \leq E[\ \lambda_1+\lambda_2-2\sqrt{\lambda_1\lambda_2}\ ] = \frac{2q}{(q-2)(q-3)}. \tag{4.14}$$

Using (4.12) $\sim$ (4.14) and Lemma 4.1, we can obtain (ii).

Now, we prove (iii). In the same manner as the proof of (ii),

$$0 < \lambda_2^{-1} \leq b'Qb \leq \lambda_1^{-1} = \frac{V_2}{2q}(1+\sqrt{1-V_1})$$

from (4.8), and $0 \leq 1-\sigma^{-2} \leq 1-V_1$ from (4.9). Hence

$$0 \leq b'Qb\cdot(1-\sigma^{-2}) \leq \frac{V_2}{2q}(1-V_1)(1+\sqrt{1-V_1}),$$

and

$$\begin{aligned} E[b'Qb\cdot(1-\sigma^{-2})] &\leq E[(2q)^{-1}V_2(1-V_1)(1+\sqrt{1-V_1})] \\ &\leq \frac{2}{q+1}+\{\frac{48}{(q+1)(q+3)(q+5)}\}^{1/2} \end{aligned}$$

from the definition of V_i's. Further from (4.10) and (4.12),

$$E(b'C_1^2Qb) \leq E[\max\{(\lambda_1-1)^2\lambda_1^{-1},\ (\lambda_2-1)^2\lambda_2^{-1}\}] \leq \frac{6}{q-3}.$$

Using the above results and Lemma 4.1, we can obtain (iii).

Finally, (iv) and (v) are immediately obtained from Lemma 4.2 and (4.8).

Proof of Theorem 4.2. First from Lemmas 4.1(i) and 4.3(i),

$$\begin{aligned} \tilde{\varepsilon} &= \sup_x |\tilde{F}(x)-\Phi(x)+\frac{1}{2}x\phi(x)E(b'C_1b)| \\ &\leq \tilde{\varepsilon}_1+\tilde{\varepsilon}_2, \end{aligned}$$

where

$$\begin{aligned} \tilde{\varepsilon}_1 &= \sup_x |\tilde{F}(x)-\Phi(x)+\frac{1}{2}x\phi(x)E[(\tilde{\sigma}^2-1)]| \\ \tilde{\varepsilon}_2 &= \sup_x |\frac{1}{2}x\phi(x)E(R_1)|. \end{aligned}$$

Hence from Theorem 2.1 and Lemma 4.3, we can obtain (i). Similarly, it is shown that

$$\tilde{\varepsilon} \leq \bar{\varepsilon}_1+\bar{\varepsilon}_2,$$

where $\bar{\varepsilon}_1$ and $\bar{\varepsilon}_2$ are the the ones obtained from $\tilde{\varepsilon}_1$ and $\tilde{\varepsilon}_2$ by replacing $\tilde{\sigma}^2$ and R_1 by $\tilde{\sigma}^{-2}$ and R_{-1}, respectively. The inequality is obtained by using Lemma 4.2(ii) and 4.3(i). Again, using Theorem 2.1 and Lemma 4.3, we can prove (ii).

Further, from Theorem 2.1

$$\sup_x |\tilde{F}(x) - \Phi(x)| \le d_2\{E[|\tilde{\sigma}^2 - 1| + |\tilde{\sigma}^{-2} - 1|]\}$$
$$\le d_2[\{E[(\tilde{\sigma}^2 - 1)^2]\}^{1/2} + \{E[(\tilde{\sigma}^{-2} - 1)^2]\}^{1/2}]$$

and therefore (iii) follows from Lemma 4.3(iv) and (v).

The referee of this paper points out that the upper bound (i) in Theorem 4.2 is less than the upper bound (ii) in Theorem 4.2 for $q = 10(1) \sim 100$, based on numerical calculations. These upper bounds have been derived by using each of the two types of approximations in Theorem 4.1 and the inequalities (ii) $\sim$ (v) in Lemma 4.3. It may be noted that the upper bounds (ii) $\sim$ (v) in Lemma 4.3 are not the best ones. So, it is interesting to study certain improvements on the upper bounds (i) and (ii) in Theorem 4.2.

Acknowledgements

The authors wish to thank the referee for pointing out that the upper bound (i) in Theorem 4.2 is less than the upper bound (ii) in Theorem 4.2 for $q = 10(1) \sim 100$.

REFERENCES

1. Y. Fujikoshi and R. Shimizu, *Sugaku Expositions,* **3**, 75-96 (1990).
2. T. Kariya and K. Maekawa, *Ann. Inst. Statist. Math.,* **34**, 281-297 (1982).
3. Y. Fujikoshi, *Hiroshima Math. J.,* **17**, 309-324 (1987).
4. M. Siotani, T. Hayakawa and Y. Fujikoshi, *Modern Multivariate Statistical Analysis: A Graduate Course and Handbook,* American Science Press, Columbus, Ohio (1985).

Stat. Sci. & Data Anal., pp. 375-382
K. Matsusita *et al.* (Eds)

Second Order Asymptotic Bound for the Variance of Estimators for the Double Exponential Distribution

MASAFUMI AKAHIRA and KEI TAKEUCHI

Institute of Mathematics, University of Tsukuba, Tsukuba, Ibaraki 305, Japan

Research Center for Advanced Science and Technology, University of Tokyo
4-6-1 Komaba, Meguro-ku, Tokyo 156, Japan

Abstract. The Bhattacharyya type bound for the variance of unbiased estimators of a location parameter of the double exponential distribution is obtained, and also the loss of information of the maximum likelihood estimator based on the distribution rounded off is discussed.

Key words: Bhattacharyya type bound, unbiased estimator, loss of information, maximum likelihood estimator.

1. INTRODUCTION

The double exponential distribution with unknown location parameter case is first order regular but second order non-regular, in the sense that the density admits the first order differentiability with respect to the parameter but not the second order. From that property it follows that the first order asymptotic theory of regular estimation can be applied but in the second order, the situation becomes non-regular. For example, R.A. Fisher, already in the paper [1], noted that the loss of information of the maximum likelihood estimator is of order $\sqrt{n}$ instead of constant order as in the regular case. In Akahira and Takeuchi [2], [3], Akahira [4], Sugiura and Naing [5] among others, the second order (or more precisely next to the first order) properties of the maximum likelihood estimator and other estimators were discussed. In this paper we shall discuss the problem from a different point of view, and shed lights on the situation. Indeed, we shall obtain the Bhattacharyya type bound for the variance of unbiased estimators of a location parameter of the double exponential distribution and the loss of information of the maximum likelihood estimator based on the distribution rounded off.

2. THE BHATTACHARYYA TYPE BOUND FOR THE VARIANCE OF UNBIASED ESTIMATORS

Suppose that $X_1, \ldots, X_n$ are independent and identically distributed random variables with a double exponential (two-sided exponential) density

$$f(x-\theta) = (1/2)\exp(-|x-\theta|) \qquad (-\infty < x < \infty\ ;\ -\infty < \theta < \infty).$$

Then, for an unbiased estimator $\hat{\theta} = \hat{\theta}(X_1, \ldots, X_n)$ of the location parameter θ, we have the Cramér-Rao bound, i.e.

$$V_\theta(\hat{\theta}) \geq 1/nI(\theta) = 1/n,$$

where $V_\theta(\ \cdot\ )$ denotes the variance and

$$I(\theta) = E_\theta\left[\{(\partial/\partial\theta)\log f(X-\theta)\}^2\right] = E_\theta\left[\{\mathrm{sgn}(X-\theta)\}^2\right] = 1.$$

Since $f(x)$ is not twice differentiable, we can not further differentiate $\log f(x)$, hence the Bhattacharyya bound can not be obtained. However, we can obtain a similar bound as follows. Define

$$Z_1(\theta) = \frac{1}{\sqrt{n}}\sum_{i=1}^n \mathrm{sgn}(X_i - \theta).$$

Then, for an unbiased estimator $\hat{\theta}$ of θ, we have

$$\int\cdots\int Z_1(\theta)\hat{\theta}(\mathbf{x})\prod_{i=1}^n f(x_i-\theta)\prod_{i=1}^n dx_i = \frac{1}{\sqrt{n}} \tag{2.1}$$

where $\mathbf{x} = (x_1, \ldots, x_n)$. Indeed, we obtain

$$\int\cdots\int \hat{\theta}(\mathbf{x})\frac{\partial}{\partial\theta}\prod_{i=1}^n f(x_i-\theta)\prod_{i=1}^n dx_i = 1,$$

since the differentiation under the integral sign is allowed. Hence

$$\begin{aligned}
1 &= \int\cdots\int \hat{\theta}(\mathbf{x})\left\{\frac{\partial}{\partial\theta}\log\prod_{i=1}^n f(x_i-\theta)\right\}\prod_{i=1}^n f(x_i-\theta)\prod_{i=1}^n dx_i \\
&= \int\cdots\int \hat{\theta}(\mathbf{x})\left\{\sum_{i=1}^n\frac{\partial}{\partial\theta}\log f(x_i-\theta)\right\}\prod_{i=1}^n f(x_i-\theta)\prod_{i=1}^n dx_i \\
&= \int\cdots\int \sqrt{n}Z_1(\theta)\hat{\theta}(\mathbf{x})\prod_{i=1}^n f(x_i-\theta)\prod_{i=1}^n dx_i,
\end{aligned}$$

which implies that (2.1) holds. From (2.1) it follows that

$$\int \cdots \int Z_1(\theta + \Delta\theta)\hat{\theta}(\mathbf{x}) \prod_{i=1}^{n} f(x_i - \theta - \Delta\theta) \prod_{i=1}^{n} dx_i = \frac{1}{\sqrt{n}}$$

from which we have

$$\int \cdots \int \left\{ \frac{Z_1(\theta + \Delta\theta) - Z_1(\theta)}{\Delta\theta} + \sqrt{n} Z_1^2(\theta) \right\} \hat{\theta}(\mathbf{x}) \prod_{i=1}^{n} f(x_i - \theta) \prod_{i=1}^{n} dx_i = o(1). \quad (2.2)$$

Indeed, from (2.1) and (2.2) we have

$$\begin{aligned}
0 &= \int \cdots \int \left\{ Z_1(\theta + \Delta\theta)\hat{\theta}(\mathbf{x}) \prod_{i=1}^{n} f(x_i - \theta - \Delta\theta) - Z_1(\theta)\hat{\theta}(\mathbf{x}) \prod_{i=1}^{n} f(x_i - \theta) \right\} \prod_{i=1}^{n} dx_i \\
&= \int \cdots \int \{Z_1(\theta + \Delta\theta) - Z_1(\theta)\} \hat{\theta}(\mathbf{x}) \prod_{i=1}^{n} f(x_i - \theta) \prod_{i=1}^{n} dx_i \\
&\quad + \Delta\theta \int \cdots \int Z_1(\theta + \Delta\theta)\hat{\theta}(\mathbf{x}) \frac{\partial}{\partial\theta} \prod_{i=1}^{n} f(x_i - \theta) \prod_{i=1}^{n} dx_i + o(\Delta\theta) \\
&= \int \cdots \int \{Z_1(\theta + \Delta\theta) - Z_1(\theta)\} \hat{\theta}(\mathbf{x}) \prod_{i=1}^{n} f(x_i - \theta) \prod_{i=1}^{n} dx_i \\
&\quad + \Delta\theta \int \cdots \int Z_1(\theta')\hat{\theta}(\mathbf{x}) \frac{\partial}{\partial\theta} \prod_{i=1}^{n} f(x_i - \theta' + \Delta\theta) \prod_{i=1}^{n} dx_i + o(\Delta\theta) \\
&= \int \cdots \int \{Z_1(\theta + \Delta\theta) - Z_1(\theta)\} \hat{\theta}(\mathbf{x}) \prod_{i=1}^{n} f(x_i - \theta) \prod_{i=1}^{n} dx_i \\
&\quad + \Delta\theta \int \cdots \int Z_1(\theta')\hat{\theta}(\mathbf{x}) \left\{ \frac{\partial}{\partial\theta} \prod_{i=1}^{n} f(x_i - \theta') \right\} \prod_{i=1}^{n} dx_i + o(\Delta\theta) \\
&= \int \cdots \int \{Z_1(\theta + \Delta\theta) - Z_1(\theta)\} \hat{\theta}(\mathbf{x}) \prod_{i=1}^{n} f(x_i - \theta) \prod_{i=1}^{n} dx_i \\
&\quad + \Delta\theta \int \cdots \int Z_1(\theta)\hat{\theta}(\mathbf{x}) \left\{ \frac{\partial}{\partial\theta} \log \prod_{i=1}^{n} f(x_i - \theta) \right\} \prod_{i=1}^{n} f(x_i - \theta) \prod_{i=1}^{n} dx_i + o(\Delta\theta) \\
&= \int \cdots \int \{Z_1(\theta + \Delta\theta) - Z_1(\theta)\} \hat{\theta}(\mathbf{x}) \prod_{i=1}^{n} f(x_i - \theta) \prod_{i=1}^{n} dx_i \\
&\quad + \Delta\theta \int \cdots \int \sqrt{n}\hat{\theta}(\mathbf{x}) Z_1^2(\theta) \prod_{i=1}^{n} f(x_i - \theta) \prod_{i=1}^{n} dx_i + o(\Delta\theta),
\end{aligned}$$

which implies that (2.2) holds. Putting, for $\Delta\theta = O(1/\sqrt{n})$,

$$Z_2(\theta) = \frac{Z_1(\theta + \Delta\theta) - Z_1(\theta)}{\Delta\theta} + \sqrt{n} Z_1^2(\theta),$$

we have from (2.2)

$$\int \cdots \int Z_2(\theta)\hat{\theta}(\mathbf{x}) \prod_{i=1}^{n} f(x_i - \theta) \prod_{i=1}^{n} dx_i = o(1). \tag{2.3}$$

Note that $Z_2(\theta)$ depends on $\Delta\theta$, but for the sake of simplicity we omit $\Delta\theta$ from the expression. Then it follows from (2.1) and (2.3) that the Bhattacharyya type bound for the variance of unbiased estimators $\hat{\theta}$ of θ is given by

$$V_\theta(\hat{\theta}) \geq I^{11}(\theta)/n, \tag{2.4}$$

where $I^{11}(\theta)/n$ is the (1,1)-element of the matrix

$$\begin{pmatrix} nE_\theta[Z_1^2(\theta)] & \sqrt{n}E_\theta[Z_1(\theta)Z_2(\theta)] \\ \sqrt{n}E_\theta[Z_1(\theta)Z_2(\theta)] & E_\theta[Z_2^2(\theta)] \end{pmatrix}^{-1},$$

that is,

$$I^{11}(\theta) = \left[E_\theta[Z_1^2(\theta)] - \{E_\theta[Z_1(\theta)Z_2(\theta)]\}^2 / E_\theta[Z_2^2(\theta)]\right]^{-1} \tag{2.5}$$

(see, e.g., Zacks [6]). Since

$$Z_1(\theta) = \frac{1}{\sqrt{n}} \sum_{i=1}^{n} \operatorname{sgn}(X_i - \theta) = \frac{1}{\sqrt{n}} \sum_{i=1}^{n} X_i' \quad \text{(say)},$$

it follows that

$$E_\theta[Z_1^2(\theta)] = E_\theta\left[\left\{\frac{1}{\sqrt{n}} \sum_{i=1}^{n} X_i'\right\}^2\right] = \frac{1}{n} \sum_{i=1}^{n} E_\theta[X_i'^2] = 1. \tag{2.6}$$

Since

$$\begin{aligned} Z_2(\theta) &= \frac{1}{\sqrt{n}} \sum_{i=1}^{n} \frac{1}{\Delta\theta}\{\operatorname{sgn}(X_i - \theta - \Delta\theta) - \operatorname{sgn}(X_i - \theta)\} + \sqrt{n}Z_1^2(\theta) \\ &= \frac{1}{\sqrt{n}} \sum_{i=1}^{n} Y_i + \sqrt{n}Z_1^2(\theta) \qquad \text{(say)} \end{aligned}$$

and $E_\theta[Y_i] = -1 + (\Delta\theta/2) + o(\Delta\theta)$, $E_\theta[Y_i^2] = (2/\Delta\theta) - 1 + o(1)$, $E_\theta[X_i'Y_i] = -1 + (\Delta\theta/2) + o(\Delta\theta)$ $(i = 1, \ldots, n)$, it follows that

$$\begin{aligned} E_\theta[Z_1(\theta)Z_2(\theta)] &= E_\theta\left[\frac{1}{\sqrt{n}} \sum_{i=1}^{n} X_i' \left(\frac{1}{\sqrt{n}} \sum_{i=1}^{n} Y_i + \sqrt{n}Z_1^2(\theta)\right)\right] \\ &= \frac{1}{n}E_\theta\left[\sum_{i=1}^{n} X_i'Y_i + \sum_{i \neq j}\sum X_i'Y_j\right] + \frac{1}{n}E_\theta\left[\sum_i\sum_j\sum_k X_i'X_j'X_k'\right] \\ &= -1 + \frac{\Delta\theta}{2} + o(\Delta\theta). \end{aligned} \tag{2.7}$$

Since

$$E_\theta\left[(\sum_{i=1}^{n} Y_i)^2\right] = nE_\theta[Y_i^2] + n(n-1)\{E_\theta(Y_i)\}^2$$

$$= n^2 + \frac{2n}{\Delta\theta} + o\left(\frac{n}{\Delta\theta}\right),$$

$$E_\theta\left[Z_1^2(\theta)\sum_{i=1}^{n} Y_i\right] = \frac{1}{n}E_\theta\left[\left(\sum_{i=1}^{n} X_i'^2 + \sum_{i \neq j}\sum X_i' X_j'\right)\sum_{k=1}^{n} Y_k\right]$$

$$= nE_\theta\left[Y_i\right] = -n + \frac{n\Delta\theta}{2} + o(n\Delta\theta),$$

$$E_\theta\left[Z_1^4(\theta)\right] = \frac{1}{n^2}E_\theta\left[\left(\sum_{i=1}^{n} X_i'\right)^4\right] = \frac{1}{n^2}E_\theta\left[\sum_{i=1}^{n} X_i'^4 + 3\sum_{i \neq k}\sum X_i'^2 X_k'^2\right]$$

$$= \frac{1}{n^2}\{n + 3n(n-1)\} = 3 - \frac{2}{n},$$

it follows that

$$E_\theta\left[Z_2^2(\theta)\right] = n + \frac{2}{\Delta\theta} + 2n\left(-1 + \frac{\Delta\theta}{2}\right) + n\left(3 - \frac{2}{n}\right) + o\left(\frac{1}{n}\right)$$

$$= 2n + o(n). \tag{2.8}$$

From (2.4) to (2.8) we have the following theorem.

Theorem. *The Bhattacharyya type bound for the variance of unbiased estimators $\hat{\theta}$ of θ is given by*

$$\mathrm{Var}_\theta(\hat{\theta}) \geq \frac{1}{n}\left(1 + \frac{1}{2n} + o\left(\frac{1}{n}\right)\right),$$

i.e.

$$\mathrm{Var}_\theta(\sqrt{n}\hat{\theta}) \geq \left(1 + \frac{1}{2n} + o\left(\frac{1}{n}\right)\right). \tag{2.9}$$

This gives the second order bound for the variance of unbiased estimators of θ. But the second term of the right-hand side of (2.9) is of order n^{-1}, while, for all estimators thus far discussed, the second order in the asymptotic variance is of order $n^{-1/2}$, and it is a situation still to be investigated and explained. The bound (2.9) may not be sharp. The related results can be found in Akahira and Takeuchi [3].

3. THE LOSS OF INFORMATION OF THE ESTIMATOR BASED ON THE DISTRIBUTION ROUNDED OFF

In a practical situation, all data are given only up to some decimal unit, hence all "real" distributions are not strictly continuous but are discrete distributions up to rounding off. And rounding off incurs loss of information, which tends to zero as the width of the rounding goes to zero, so the continuous distribution can be considered as the limiting case, hence a practical approximation for the actual situation where the width of round-off is small enough. In the regular case we do not need to go beyond the above consideration, but in a non-regular case, we must take another aspect into consideration. When the data is rounded, the class of distributions is no longer irregular, admitting differentiation with respect to the parameter as many times as we want. Hence the "asymptotic deficiency" of the maximum likelihood estimator (MLE) and other estimators becomes of order $O(1)$, but not of higher order as in the limiting non-regular case. Therefore, if we restrict our attention to the class of the MLE and other similar regular types of estimators, these must be optimum "compromise" for the width of rounding between the loss of information in the data due to rounding and the loss of information (asymptotic deficiency) due to estimators.

In this section we round off the double exponential distribution, consider the loss of information by the distribution and the loss of information of the MLE, and deal with their sum as the loss of information of the MLE based on the distribution.

In the case when the density is given by $f(x-\theta) = (1/2)\exp(-|x-\theta|)$ $(-\infty < x < \infty;\ -\infty < \theta < \infty)$, we consider to round off the density as follows. Without loss of generality we assume that $0 \le \theta < h$. For any integer j we define $p_j(\theta)$ by

$$p_j(\theta) = \int_{jh}^{(j+1)h} f(x-\theta)dx.$$

Since

$$p_0(\theta) = \left(2 - e^{-\theta} - e^{\theta-h}\right)/2, \tag{3.1}$$

it follows that

$$p_0'(\theta) = \left(e^{-\theta} - e^{\theta-h}\right)/2 = (h-2\theta)/2 + o(h), \tag{3.2}$$

$$p_0''(\theta) = -\left(e^{-\theta} - e^{\theta-h}\right)/2 = -1 + (h/2) + o(h). \tag{3.3}$$

Then it follows from (3.1) to (3.3) that the amount $I_h(\theta)$ of information, $J_h(\theta)$ and $M_h(\theta)$ on the distribution rounded off are defined and calculated by

$$\begin{aligned} I_h(\theta) &= \sum_j \left\{ \frac{\partial}{\partial\theta} \log p_j(\theta) \right\}^2 p_j(\theta) \\ &= \sum_{j\neq 0} \left\{ \frac{\partial}{\partial\theta} \log p_j(\theta) \right\}^2 p_j(\theta) + \left\{ \frac{\partial}{\partial\theta} \log p_0(\theta) \right\}^2 p_0(\theta) \\ &= \sum_{j\neq 0} p_j(\theta) + \{p_0'(\theta)/p_0(\theta)\}^2 p_0(\theta) = 1 - 2\theta\left(1 - \frac{\theta}{h}\right) + o(h), \end{aligned} \tag{3.4}$$

$$\begin{aligned} J_h(\theta) &= \sum_j \left\{ \frac{\partial}{\partial\theta} \log p_j(\theta) \right\} \left\{ \frac{\partial^2}{\partial\theta^2} \log p_j(\theta) \right\} p_j(\theta) \\ &= \left\{ \frac{\partial}{\partial\theta} \log p_0(\theta) \right\} \left\{ \frac{\partial^2}{\partial\theta^2} \log p_0(\theta) \right\} p_0(\theta) \\ &= -1 + \frac{2\theta}{h} + o(1), \end{aligned} \tag{3.5}$$

$$\begin{aligned} M_h(\theta) &= \sum_j \left\{ \frac{\partial^2}{\partial\theta^2} \log p_j(\theta) + I_h(\theta) \right\}^2 p_j(\theta) \\ &= \sum_{j\neq 0} I_h^2(\theta) p_j(\theta) + \left\{ \frac{\partial^2}{\partial\theta^2} \log p_0(\theta) + I_h(\theta) \right\}^2 p_0(\theta) \\ &= I_h^2(\theta)(1 - p_0(\theta)) + \left\{ (p_0''(\theta)/p_0(\theta)) - (p_0'(\theta)/p_0(\theta))^2 + I_h(\theta) \right\}^2 p_0(\theta) \\ &= -1 + (2/h) + o(1). \end{aligned} \tag{3.6}$$

Since the amount $I(\theta)$ of information on the double exponential density $f(x-\theta)$ is equal to 1, the loss of information by rounding off is defined by $n\{I(\theta) - I_h(\theta)\}$, and the loss $D_h(\theta)$ of information (asymptotic deficiency) of the MLE by rounding off is also given by $\{I_h(\theta)M_h(\theta) - J_h^2(\theta)\}/I_h^2(\theta)$ (see Akahira [7]). Hence the loss of information of the MLE based on the distribution rounded off is given by

$$n\{I(\theta) - I_h(\theta)\} + D_h(\theta) = nI(\theta) - \{nI_h(\theta) - D_h(\theta)\}. \tag{3.7}$$

In order to get h minimizing it, we obtain h maximizing the average value of (3.7), i.e.

$$(1/h)\int_0^h \{nI_h(\theta) - D_h(\theta)\} d\theta, \tag{3.8}$$

since θ is considered to be random in the interval $[0, h)$. From (3.4), (3.5) and (3.6) we have

$$D_h(\theta) = \{I_h(\theta)M_h(\theta) - J_h^2(\theta)\}/I_h^2(\theta) = \frac{2}{h} + \frac{4\theta}{h} - \frac{4\theta^2}{h^2} + o(1),$$

hence

$$nI_h(\theta) - D_h(\theta) = \left(n - \frac{2}{h}\right)\left\{1 - 2\theta\left(1 - \frac{\theta}{h}\right)\right\}.$$

Then we obtain

$$(1/h)\int_0^h \{nI_h(\theta) - D_h(\theta)\}d\theta = \left(n - \frac{2}{h}\right)\left(1 - \frac{h}{3}\right) + o(1).$$

Hence it is seen that the value of h maximizing (3.8) is given by $\sqrt{6/n}$, and also its value of (3.8) is done by $n - (2\sqrt{6n}/3) + (2/3)$. Now, in the second term, a quantity of order $\sqrt{n}$ appears, unlike the regular case when it should be of constant order, and the coefficient is not still satisfactorily explained in more general contexts.

REFERENCES

1. R. A. Fisher, *Proc. Cambridge Philos. Soc.*, **22**, 700-725 (1925).
2. M. Akahira and K. Takeuchi, *Asymptotic Efficiency of Statistical Estimators : Concepts and Higher Order Asymptotic Efficiency,* Lecture Notes in Statistics 7, Springer, New York (1981).
3. M. Akahira and K. Takeuchi, *Austral. J. Statist.*, **32**, 281-291 (1990).
4. M. Akahira, *Ann. Inst. Statist. Math.*, **40**, 311-328 (1988).
5. N. Sugiura and M. T. Naing, *Comm. Statist., A - Theory Methods* **18**, 541-554 (1989).
6. S. Zacks, *The Theory of Statistical Inference,* Wiley, New York (1971).
7. M. Akahira, *The Structure of Asymptotic Deficiency of Estimators,* Queen's Papers in Pure and Applied Mathematics **75**, Queen's University Press, Kingston, Ontario, Canada (1986).

Stat. Sci. & Data Anal., pp. 383-394
K. Matsusita *et al.* (Eds)

Asymptotic Expansions for $E_\theta\{\min(t,m)\}$ and $E_\theta\{\bar{x}_{\min(t,m)}\}$

HAJIME TAKAHASHI
Hitotsubashi University, Kunitachi, Tokyo, Japan

Abstract. Let $x_1, x_2, \cdots$ be a sequence of independent and normally distributed random variables with unknown mean θ and variance 1. Let $t = \inf\{n \geq 1 : s_t \geq \sqrt{2a(n+c)}\}$ and $t(m) = \min\{t, m\}$. We shall obtain asymptotic expansions of $E_\theta\{t(m)\}$ and $E_\theta\{\bar{x}_{t(m)}\}$ as $a \to \infty$, for all θ in some neighborhood of $\theta_0 = [2a/m]$ of width $O(a^{-1/2})$.

Key words: Repeated significance tests, non-linear renewal theorem.

1. INTRODUCTION

Let P_θ be a probability measure under which $x_1, x_2, \cdots$ are independent and normally distributed random variables with unknown mean θ and variance 1. To test the hypothesis H_0: $\theta=0$ against the alternative H_1: $\theta>0$ sequentially, Armitage [1] suggests the following procedure (Repeated Significance Test). For any constants $a > 0$, $c \geq 0$ and positive integer m, we define the stopping time

$$t = \inf\{n \geq 1;\ s_n \geq \sqrt{2a(n+c)}\,\} \tag{1}$$

and $t(m) = \min\{t, m\}$, where $s_n = x_1 + \cdots + x_n$, $n \geq 1$ and $\inf\{\phi\} = \infty$. Then the RST rejects the hypothesis H_0 if and only if $t \leq m$. And, using the method of numerical integration, Armitage, McPherson and Row [2] obtained the tables of a, c and m for given significance level and then they calculated the power and the expected sample sizes at the specified θ values.

Siegmund ([3], [4]), on the other hand, utilized the non-linear renewal theory (cf. Woodroofe ([5], [6]), Lai and Siegmund [7]) to calculate the limit of these quantities as $m = m(a) \to \infty$, $a \to \infty$ in such a way that $2a/m$ is fixed ($= \theta_0^2$). Later Takahashi and Woodroofe [8], Woodroofe and Takahashi [9], Takahashi ([10], [11]), and Woodroofe and Keener [12] obtained second order asymptotic expansions for the error probabilities and expected sample sizes. And as a method of numerical approximation, these results give fairly accurate approximations for all $\theta \neq \theta_0$. But the problem still remains when θ is in some neighborhood of θ_0 of the width $O(a^{-1/2})$ as $a \to \infty$. Thus, we shall obtain the asymptotic expansions of $P_\theta\{t < m\}$, $E_\theta[t(m)]$ and $E_\theta[\bar{x}_{t(m)}]$, $\bar{x}_n = s_n/n$, for all $\theta = \theta_0(1 + u/\sqrt{2a})$ as $m \to \infty$, $a \to \infty$ with $2a/m = \theta_0^2$. Throughout the paper we shall let a goes to infinity through the integral multiple of $\theta_0^2/2$, so that $m = 2a/\theta_0^2$ are integer. Hence for each θ, we shall let $N = [2a/\theta^2]$ and

$$\rho = (2a/\theta^2) - [2a/\theta^2] = (2a/\theta^2) - N \tag{2}$$

the fractional part of the number $(2a/\theta^2)$, where $[b]$ denotes the largest integer not exceeding b. We shall refer the reader Siegmund [13], where closely related results are presented and proved in the line of Siegmund [3]. We have applied the different method and have obtained sharper results.

2. PRELIMINARY LEMMAS

For any $\theta > 0$, we let

$$R = s_t - \sqrt{2a(t+c)} \quad \text{and} \quad t^* = (\theta/2)(t-N)/\sqrt{N}. \tag{3}$$

The asymptotic joint distribution of (t^*, R) is closely related to the probability (Takahashi, [10])

$$\psi(\theta,r) = P_0\{s_k \le \theta k/2 - r, \ \forall k \ge 1\}, \ -\infty < r < \infty. \tag{4}$$

In addition, for i, $j = 0, 1, 2, \cdots$, we let

$$\nu_j^{(i)}(\theta) = (2/\theta)\int_0^\infty r^j \psi^{(i)}(\theta,r)dr, \tag{5}$$

where $\psi^{(i)}(\theta,r) = (\partial^i/\partial\theta^i)\psi(\theta,r)$. And we shall write $\nu_j(\theta)$ for $\nu_j^{(0)}(\theta)$.

Lemma 1. Let $x_k = (\theta/2)(k-N)/\sqrt{N}$, $k = 0, 1, \cdots$. Then for all $\theta = \theta_0(1+u/\sqrt{2a})$, as $a \to \infty$

$$\begin{aligned} P_\theta\{t=k\} &= P_\theta\{t^* = x_k\} = (\theta/2\sqrt{N})\phi(x_k) \\ &\times \Big[1 + (1/\sqrt{m})\big\{(1/\theta_0)(-2x_k + (1/2)x_k^3) + \nu_1(\theta_0)x_k + (1/2)c\theta_0 x_k\big\} \\ &+ (1/m)\big\{(-1/\theta_0^2) + (1/\theta_0)\nu_1(\theta_0) + \nu_1^{(1)}(\theta_0) - (1/2)\nu_2(\theta_0) \\ &\qquad + ((3/2) - (1/2)\theta_0\nu_1(\theta_0))\,c + \big((u/\theta_0)\nu_1(\theta_0) + u\nu_1^{(1)}(\theta_0) + cu\big)\,x_k \\ &+ \big((4/\theta_0^2) - (5/2\theta_0)\nu_1(\theta_0) - \nu_1^{(1)}(\theta_0) + (1/2)\nu_2(\theta_0) \\ &\qquad - ((9/4) - (1/2)\theta_0\nu_1(\theta_0))\,c\big)\,x_k^2 \\ &+ \big((-13/8\theta_0^2) + (1/2\theta_0)\nu_1(\theta_0) + (1/4)c\big)\,x_k^4 \\ &\qquad + (1/8\theta_0^2)x_k^6 + (1/2)c^2\theta_0^2(x_k^2 - 1)\big\} + o(m^{-1})\Big], \end{aligned} \tag{6}$$

and

$$\begin{aligned} &P_\theta\{t=k,\, R \in dr\} = \\ &(\theta/2\sqrt{N})\phi(x_k)\Big[\psi(\theta_0, r) + (1/\sqrt{m})\big\{(x_k^3/(2\theta_0) - x_k/\theta_0)\psi(\theta_0, r) \\ &+ (u - x_k)\psi^{(1)}(\theta_0, r) + r x_k \psi(\theta_0, r) + (1/2)c\theta_0 x_k \psi(\theta_0, r)\big\} + o(m^{-1/2})\Big]\, dr, \end{aligned} \tag{7}$$

where $\phi(x)$ denotes the standard normal density function.

Proof. The asymptotic expansion for $P_\theta\{t^* < x_k, R > r_0\}$ is given in Theorem 2.1 of Takahashi [10]. Since $P_\theta\{t^* = x_k\} = P_\theta\{t^* < x_{k+1}\} - P_\theta\{t^* < x_k\}$. (6) and (7) follows from the straightforward algebra.

Most of the analysis in this paper consists of calculating the summation of the form $(\theta/2\sqrt{N})\sum_{k=1}^m x_k^i \phi(x_k)$.

Lemma 2. For all $\theta = \theta_0(1 + u/\sqrt{2a})$, as $a \to \infty$

$$\begin{aligned}(\theta/2\sqrt{N})\sum_{k=1}^{m}\phi^{(i+1)}(x_k) &= \\ &\phi^{(i)}(u) + (1/\sqrt{m})((u^2/2\theta_0) + (1/4)\theta_0 + (1/2)\theta_0\rho)\phi^{(i+1)}(u) \\ &+ (1/m)\{(u/2)(1+3\rho)\phi^{(i+1)}(u) + (1/2)[((u^2/2\theta_0) + (1/2)\theta_0\rho)^2 \\ &+ (1/2)\theta_0((u^2/2\theta_0) + (1/2)\theta_0\rho) + (1/24)\theta_0^2]\phi^{(i+2)}(u)\}\cdot(1+o(1))\end{aligned} \tag{8}$$

for $i = -1, 0, 1, 2, \cdots$, where $\phi^{(i)}(u) = (\partial^i/\partial u^i)\phi(u)$ $i = 0, 1, 2, \cdots$ and $\phi^{(-1)}(u) = \Phi(u)$, the standard normal cumulative distribution function.

Proof. Since,

$$x_m = u + (1/\sqrt{m})((u^2/2\theta_0) + (1/2)\theta_0\rho) + (1/m)((3/2)u\rho) + o(a^{-1}) \quad \text{as} \quad a \to \infty.$$

For $i = -1, 0, 1, \cdots$ by Taylor's theorem

$$\begin{aligned}\phi^{(i)}(x_m) = \phi^{(i)}(u) &+ (1/\sqrt{m})[(u^2/2\theta_0) - (1/4)\theta_0 + (1/2)\theta_0\rho]\phi^{(i+1)}(u) \\ &+ (1/m)[((3/2)u\rho - u/2)\phi^{(i+1)}(u) \\ &+ ((u^4/8\theta_0^2) + (1/4)u^2\rho - u^2/8 + (1/8)\theta_0^2\rho^2 + (1/48)\theta_0^2)\phi^{(i+2)}(u^*)]\end{aligned} \tag{9}$$

where u^* denotes the intermediate point. On the other hand,

$$\phi^{(i)}(x_m) = \sum_{k=1}^{m}\int_{x_{k-1}}^{x_k}\phi^{(i+1)}(x)dx + \bar{o}(a^{-\infty}) \tag{10}$$

where $\bar{o}(a^{-\infty})$ denotes the term of the order of magnitude smaller than $a^{-\alpha}$ for every $\alpha > 0$ as $a \to \infty$. For each $k = 1, 2, \cdots$, we expand $\phi^{(i+1)}(x)$ inside the integral sign on the right hand side of (10) into Taylor series about x_k. And then integrating each term separately, we obtain

$$\begin{aligned}\phi^{(i)}(x_m) &= (\theta/2\sqrt{N})\left\{\sum_{k=1}^{m}\phi^{(i+1)}(x_k) - (\theta_0/4\sqrt{m})\sum_{k=1}^{m}\phi^{(i+2)}(x_k)\right. \\ &\left.+ (1/m)\left[(-u/2)\sum_{k=1}^{m}\phi^{(i+2)}(x_k) + (\theta_0^2/24)\sum_{k=1}^{m}\phi^{(i+3)}(x_k^*)\right]\right\} + \bar{o}(a^{-\infty})\end{aligned} \tag{11}$$

as $a \to \infty$ where x_k^* is an intermediate point.

It follows from (10) and (11) that as $a \to \infty$

$$\begin{aligned}(\theta/2\sqrt{N})\sum_{k=1}^{m}\phi^{(i+1)}(x_k) &= \\ &\phi^{(i)}(u) + (1/\sqrt{m})\{((u^2/2\theta_0) - (1/4)\theta_0 + (1/2)\theta_0\rho)\phi^{(i+1)}(u) \\ &+ (\theta_0/4)[(\theta/2\sqrt{N})\sum_{k=1}^{m}\phi^{(i+2)}(x_k)]\} + (1/m)\{((3/2)u\rho - u/2)\phi^{(i+1)}(u) \\ &+ ((u^4/8\theta_0) + (1/4)u^2\rho - u^2/8 + (1/8)\theta_0^2\rho^2 + (1/48)\theta_0^2)\phi^{(i+2)}(u^*) \\ &+ (u/2)[(\theta/2\sqrt{N})\sum_{k=1}^{m}\phi^{(i+2)}(x_k)] - (\theta_0^2/24)[(\theta/2\sqrt{N})\sum_{k=1}^{m}\phi^{(i+3)}(x_k)]\} + \bar{o}(a^{-\infty}).\end{aligned} \tag{12}$$

Repeating the same analysis to $(\theta/2\sqrt{N})\sum_{k=1}^{m}\phi^{(i+j)}(x_k)$ $(j = 2, 3, 4, \cdots)$, we obtain the lemma

Corollary 2.1. As $a \to \infty$

$$(\theta/2\sqrt{N})\sum_{k=1}^{m}\phi(x_k) = \Phi(u) + (1/\sqrt{m})[(u^2/2\theta_0) + \theta_0/4 + \theta_0\rho/2]\phi(u)$$
$$+(1/m)[(u/2)(1+3\rho)\phi(u) + (1/2)\{((u^2/2\theta_0) + \theta_0\rho/2)^2$$
$$+(\theta_0/2)((u^2/2\theta_0) + \theta_0\rho/2) + \theta_0^2/24\}(-u\phi(u))]\cdot(1+o(1))$$

$$(\theta/2\sqrt{N})\sum_{k=1}^{m}x_k\phi(x_k) = -\phi(u) + (1/\sqrt{m})[u^3/(2\theta_0) + \theta_0 u/4 + (\theta_0 u/2)\rho]\phi(u)$$
$$+(1/m)[(-u^6/8\theta_0^2) + ((1/8\theta_0^2) - 1/8)u^4 + (5/8 - \theta_0^2/48)u^2 + \theta_0^2/48$$
$$+\rho\{-u^4/4 + (7/4 - \theta_0^2/8)u^2 + \theta_0^2/8 + \rho(\theta_0^2/8)(1-u^2)\}]\phi(u)(1+o(1))$$

$$(\theta/2\sqrt{N})\sum_{k=1}^{m}x_k^2\phi(x_k) = \Phi(u) - u\phi(u) + (1/\sqrt{m})[(u^4/2\theta_0) + \theta_0 u^2/4 + (\theta_0\rho/2)u^2]\phi(u)$$
$$+(1/m)[((-u^7/8\theta_0^2) + ((1/4\theta_0^2) - 1/8)u^5 + (3/4 - \theta_0^2/48)u^3 + (\theta_0^2/24)u$$
$$+\rho\{-u^5/4 + (2 - \theta_0^2/8)u^3 + (\theta_0^2/4)u - \rho((\theta_0^2/8)u^3 - (\theta_0^2/4)u)\}]\phi(u)(1+o(1))$$

$$\begin{aligned}
(\theta/2\sqrt{N})\sum_{k=1}^{m}x_k^3\phi(x_k) &= -(u^2+2)\phi(u) + (1/\sqrt{m})[u^5/(2\theta_0) + \theta_0 u^3/4 \\
&\quad +\theta_0 u^3\rho/2]\phi(u)(1+o(1)) \\
(\theta/2\sqrt{N})\sum_{k=1}^{m}x_k^4\phi(x_k) &= 3\Phi(u) - (u^3+3u)\phi(u) + (1/\sqrt{m})[u^6/(2\theta_0) \\
&\quad +\theta_0 u^4/4 + \theta_0 u^4\rho/2]\phi(u)(1+o(1)) \\
(\theta/2\sqrt{N})\sum_{k=1}^{m}x_k^5\phi(x_k) &= -(u^4+4u^2+8)\phi(u)(1+o(1)) \\
(\theta/2\sqrt{N})\sum_{k=1}^{m}x_k^6\phi(x_k) &= 15\Phi(u) - (u^5+5u^3+15u)\phi(u)(1+o(1)) \\
(\theta/2\sqrt{N})\sum_{k=1}^{m}x_k^6\phi(x_k) &= -(u^6+6u^4+24u^2+48)\phi(u)(1+o(1))
\end{aligned}$$

3. ASYMPTOTIC EXPANSIONS

In this section we shall obtain higher order asymptotic expansions for $P_\theta\{t < m\}$, $E_\theta\{t(m)\}$ and $E_\theta\{\bar{x}_{t(m)}\}$ for all θ in the neighborhood of θ_0 of width $O(m^{-1/2})$.

Theorem 1. *For all* $\theta = \theta_0(1 + u/\sqrt{2a})$, *as* $a \to \infty$

$$\begin{aligned}
P_\theta\{t < m\} &= \Phi(u) + (1/\sqrt{m})[(-1/4)\theta_0 + 1/\theta_0 - \nu_1(\theta_0) + (1/2)\theta_0\rho - (1/2)\theta_0 c]\phi(u) \\
&\quad +(1/m)[-1/\theta_0^2 - (1/48)\theta_0^2 - (1/4)\theta_0\nu_1(\theta_0) - (1/2)\nu_2(\theta_0) \qquad (13)\\
&\quad +\rho(1/2 + (1/8)\theta_0^2 + (1/2)\theta_0\nu_1(\theta_0) - (1/8)\theta_0^2\rho) \\
&\quad +c(1/2 - (1/8)\theta_0^2 - (1/2)\theta_0\nu_1(\theta_0) + (1/4)\theta_0^2\rho - (1/2)\theta_0^2 c)] + o(1/m)
\end{aligned}$$

Proof. The theorem follows easily from Proposition 3 below.

Theorem 2. *For all $\theta = \theta_0(1+u/\sqrt{2a})$, as $a \to \infty$*

$$\begin{aligned}
E_\theta\{t(m)\} &= m - \sqrt{m}\{(2/\theta_0)(\phi(u)+u\Phi(u))\} \\
&+\{(3u/\theta_0^2)\phi(u)+[(3/\theta_0^2)u^2-1/\theta_0^2+(2/\theta_0)\nu_1(\theta_0)-\rho+c]\Phi(u)\} \\
&+(1/\sqrt{m})\{[(-4/\theta_0^3)u^2+(1/24)\theta_0+(2/\theta_0)\nu_1^{(1)}(\theta_0)-(1/\theta_0)\nu_2(\theta_0) \\
&+\rho((1/2\theta_0)u^2+\nu_1(\theta_0)-(1/4)\theta_0\rho) \\
&+c(2/\theta_0-\nu_1(\theta_0)+(1/2)\theta_0\rho-c\theta_0)]\phi(u) \\
&+[(-4/\theta_0^3)u^3-1/\theta_0^2+(2/\theta_0)\nu_1(\theta_0)-\rho+c]\Phi(u)\}+o(1/\sqrt{m}) \\
&= m-\sqrt{m}A+B+(1/\sqrt{m})C+o(1/\sqrt{m}) \quad \text{(Say)}.
\end{aligned} \tag{14}$$

Proof. Since $x_k = (\theta/2)(k-N)/\sqrt{N}$, we may write

$$\begin{aligned}
k = m &+ \sqrt{m}\{(2/\theta_0)x_k-(2/\theta_0)u\}+\{(3/\theta_0^2)u^2-(4u/\theta_0^2)x_k-\rho\} \\
&+ (1/\sqrt{m})\{(6/\theta_0^3)u^2x_k-(4/\theta_0^3)u^3-(\rho/\theta_0)x_k\}+o(1/\sqrt{m}), \\
&\qquad \text{as } a \to \infty.
\end{aligned} \tag{15}$$

It follows that

$$\begin{aligned}
E_\theta\{t(m)\} &= \sum_{k=1}^m kP_\theta\{t=k\}+mP_\theta\{t>m\} \\
&= \sum_{k=1}^m\{m+\sqrt{m}(2/\theta_0)(x_k-u)+((3/\theta_0^2)u^2-(4u/\theta_0^2)x_k-\rho) \\
&\quad +(1/\sqrt{m})(-4u^3/\theta_0^3+(6u^2/\theta_0^3-\rho/\theta_0)x_k)\}P_\theta\{t=k\} \\
&\quad +mP_\theta\{t>m\}+o(1/\sqrt{m}), \text{ as } a \to \infty.
\end{aligned} \tag{16}$$

From Lemma 1 and some algebra, the above equation becomes

$$\begin{aligned}
&m + (\theta/2\sqrt{N})\sum_{k=1}^m \phi(x_k)\{\sqrt{m}[(2/\theta_0)(x_k-u)] \\
&\quad +[(3/\theta_0^2)u^2-\rho-((2/\theta_0)u\nu_1(\theta_0)+cu)x_k+((-4/\theta_0^2+(2/\theta_0)\nu_1(\theta_0)+c)x_k^2 \\
&\qquad -(u/\theta_0^2)x_k^3+(1/\theta_0^2)x_k^4] \\
&\quad +(1/\sqrt{m})[\{(-4/\theta_0^3)u^3+(2/\theta_0^3)u-(2u/\theta_0^2)\nu_1(\theta_0)-(2u/\theta_0)\nu_1^{(1)}(\theta_0)+(u/\theta_0)\nu_2(\theta_0) \\
&\qquad +c(u\nu_1(\theta_0)-(3/\theta_0)u)\} \\
&\quad +\{(-2/\theta_0^3)+((1/\theta_0^2)u^2+2/\theta_0^2)\nu_1(\theta_0)+(2/\theta_0-(2/\theta_0)u^2)\nu_1^{(1)}(\theta_0) \\
&\quad -(1/\theta_0)\nu_2(\theta_0)+\rho(1/\theta_0-\nu_1(\theta_0)-(1/2)\theta_0c)+c((-1/2\theta_0)u^2+3/\theta_0-\nu_1(\theta_0))\}x_k \\
&\quad +\{(3u/\theta_0^2)\nu_1(\theta_0)+(4u/\theta_0)\nu_1^{(1)}(\theta_0)-(u/\theta_0)\nu_2(\theta_0)+c((9/2\theta_0)u-u\nu_1(\theta_0))\}x_k^2 \\
&\quad +\{(3/2\theta_0^3)u^2+(8/\theta_0^3)-(5/\theta_0^2)\nu_1(\theta_0)-(2/\theta_0)\nu_1^{(1)}(\theta_0)+(1/\theta_0)\nu_2(\theta_0)-\rho/(2\theta_0) \\
&\qquad -c((9/2\theta_0)-\nu_1(\theta_0))\}x_k^3 \\
&\quad +\{(5/4\theta_0^3)u-(u/\theta_0^2)\nu_1(\theta_0)-(c/2\theta_0)u\}x_k^4 + \{-13/(4\theta_0^3)+(1/\theta_0^2)\nu_1(\theta_0)+(c/2\theta_0)\}x_k^5 \\
&\quad -u/(4\theta_0^3)x_k^6+1/(4\theta_0^3)x_k^7+c^2\theta_0\{x_k^3-ux_k^2-x_k+u\}]+o(1/\sqrt{m})\}.
\end{aligned}$$

In the next step we invoke Lemma 2 to approximate each $(\theta/2\sqrt{N})\sum_{k=1}^m x_k^i\phi(x_k)$ $i = 0, 1,\cdots, 7$ up to the appropriate order. For example,

$$\begin{aligned}
&(\theta/2\sqrt{N})\sum_{k=1}^m \phi(x_k)(x_k-u) \\
&\quad = -\{\phi(u)+u\Phi(u)+(1/m)[(-1/2)\{((u^2/2\theta_0)+(1/2)\theta_0\rho)^2 \\
&\qquad +(1/2)\theta_0((u^2/2\theta_0)+(1/2)\theta_0\rho)+(1/24)\theta_0^2\}]\phi(u)(1+o(1))
\end{aligned}$$

The theorem follows after long and tedious calculations. We shall omit the detail.

Theorem 3. *For all $\theta = \theta_0(1 + u/\theta_0\sqrt{m})$, as $a \to \infty$*

$$
\begin{aligned}
E_\theta\{\bar{x}_{t(m)}\} &= \theta_0 + u/\sqrt{m} + (1/m)\{(2/\theta_0)\Phi(u)\} \\
&\quad +(1/\sqrt{m^3})\{[6/\theta_0^2 - (4/\theta_0)\nu_1(\theta_0) + \rho((1/2)\theta_0^2 + 2 + (1/2)u^2 \\
&\quad -(1/2)\theta_0^2\rho) - c]\phi(u) + [2u/\theta_0^2 + u\rho]\Phi(u)\} + o(1/\sqrt{m^3})\,. \\
&= \theta_0 + u/\sqrt{m} + (1/m)\Lambda + (1/\sqrt{m^3})\Xi + o(1/\sqrt{m^3}) \quad \text{(Say)}
\end{aligned}
\tag{17}
$$

Proof. We shall prove the theorem from the series of Propositions.

Proposition 1.

$$
\begin{aligned}
E_\theta\{\bar{x}_{t(m)}\} &= E_\theta\{[(1/t) - (1/m)]s_t I_{[t\le m]}\} + (\theta/m)E_\theta\{t(m)\}\,. \\
&= I + II \quad \text{(say)}
\end{aligned}
\tag{18}
$$

Proof. First we write

$$
E_\theta\{\bar{x}_{t(m)}\} = E_\theta\{\bar{x}_t I_{[t\le m]}\} + E_\theta\{\bar{x}_m I_{[t>m]}\}\,, \tag{19}
$$

where I_A denotes the indicator function of the set A. Now

$$
\begin{aligned}
E_\theta\{\bar{x}_m I_{[t>m]}\} &= E_\theta\{\bar{x}_m\} - E_\theta\{\bar{x}_m I_{[t\le m]}\} \\
&= \theta - E_\theta\{\bar{x}_m I_{[t\le m]}\}\,.
\end{aligned}
\tag{20}
$$

And,

$$
\begin{aligned}
m \cdot E_\theta\{\bar{x}_m I_{[t\le m]}\} &= E_\theta\{(s_m - s_t + s_t)I_{[t\le m]}\} \\
&= \sum_{k=1}^{m} E_\theta\{(s_m - s_k)I_{[t=k]}\} + E_\theta\{s_t I_{[t\le m]}\}
\end{aligned}
\tag{21}
$$

Since $(s_m - s_k)$ is independent of the event $[t = k]$ $(k \le m)$, the above equation becomes

$$
\begin{aligned}
&\sum_{k=1}^{m} E_\theta\{s_m - s_k\}P_\theta\{t = k\} + E_\theta\{s_t I_{[t\le m]}\} \\
&\quad = m\theta P_\theta\{t \le m\} - \theta[E_\theta\{t(m)\} - mP_\theta\{t > m\}] + E_\theta\{s_t I_{[t\le m]}\}\,.
\end{aligned}
\tag{22}
$$

The proposition follows from substitutions.

To analyse I, we first write

$$
\begin{aligned}
&E_\theta\{[(1/t) - (1/m)]s_t\, I_{[t\le m]}\} \\
&\quad = E_\theta\{[(1/t) - (1/m)]\,(\theta_0\sqrt{m}\sqrt{t}\sqrt{1 + (c/t)} + R)I_{[t\le m]}\}\,.
\end{aligned}
\tag{23}
$$

We then expand $1/t$ and $\sqrt{t}$ into "Taylor series" about $1/m$ and $\sqrt{m}$ respectively.

Proposition 2. As $a \to \infty$,

$$
\begin{aligned}
E_\theta\{[(1/t) - (1/m)]s_t\, I_{[t\le m]}\} &= (2/\sqrt{m})\left[uP_\theta\{t \le m\} - \sum_{k=1}^{m} x_k P_\theta\{t = k\}\right] \\
&\quad + (1/m)\left[-(u^2/\theta_0 - \theta_0\rho)P_\theta\{t \le m\} + (2/\theta_0)\sum_{k=1}^{m} x_k^2 P_\theta\{t = k\}\right] \\
&\quad + (1/\sqrt{m^3})[((1/\theta_0^2)u^3 + 2u\rho + cu)P_\theta\{t \le m\}
\end{aligned}
$$

$$+((-1/\theta_0^2)u^2 - \rho - c)\sum_{k=1}^{m} x_k P_\theta\{t=k\} \tag{24}$$
$$+(1/\theta_0^2)u\sum_{k=1}^{m} x_k^2 P_\theta\{t=k\} - (3/\theta_0^2)\sum_{k=1}^{m} x_k^3 P_\theta\{t=k\}$$
$$+(2/\theta_0)u\sum_{k=1}^{m}\int_0^\infty rP_\theta\{t=k,\ R\in dr\}$$
$$-(2/\theta_0)\sum_{k=1}^{m} x_k\int_0^\infty rP_\theta\{t=k,\ R\in dr\}] + o(1/\sqrt{m^3})$$

Proof. Let $B=\{|t-m|\le m^{11/20}\}$. Then on the set B, we have

$$\begin{aligned}\sqrt{t} = {}& \sqrt{m} - (1/\theta_0)(u-t^*) + (1/\sqrt{m})[(1/\theta_0)^2\{u^2-ut^*-(1/2)t^{*2}\}-(1/2)\rho]\\ &+(1/m)[(1/\theta_0)^3\{-u^3+u^2t^*+(1/2)ut^{*2}+(1/2)t^{*3}\}-(1/2\theta_0)\rho u]\\ &+(1/\sqrt{m})^3[(1/\theta_0)^4\{u^4-u^3t^*-(1/2)u^2t^{*2}-(1/2)ut^{*3}-(5/8)t^{*4}\}\\ &-(1/8)\rho^2-(1/4\theta_0^2)t^{*2}\rho]+O_p(a^{-35/20}),\end{aligned} \tag{25}$$

$$\begin{aligned}\sqrt{1+(c/t)} = {}& 1+(c/2m)+(1/\sqrt{m})^3(c/\theta_0)\{u-t^*\}\\ &+(1/m)^2(c/\theta_0^2)\{(1/2)u^2-2ut^*+2t^{*2}-(1/8)c\theta_0^2\}+O_p(a^{-47/20}),\end{aligned} \tag{26}$$

and

$$\begin{aligned}(1/t)-(1/m) = {}& (1/\sqrt{m})^3(2/\theta_0)\{u-t^*\} + (1/\sqrt{m})^4(1/\theta_0)^2\{(u^2-4ut^*+4t^{*2})+\theta_0^2\rho\}\\ &+(1/\sqrt{m})^5(1/\theta_0)^3\{(-2u^2t^*+8ut^{*2}-8t^{*3})+\rho(4\theta_0^3u-3\theta_0^2t^*)\}+O_p(a^{-56/20}).\end{aligned}$$

Note that $P_\theta\{B^c\}=\bar{o}(a^{-\infty})$ as $a\to\infty$, where B^c is the complement of the set B. And it is not difficult to show that $(t^{-1}-m^{-1})s_t$ is uniformly integrable (cf. Woodroofe [5]). It follows that

$$\begin{aligned}&E_\theta\{[(1/t)-(1/m)]s_tI_{[t\le m]}\} =\\ &\quad E_\theta\{[(2/\sqrt{m})\{u-t^*\}+(1/m)(1/\theta_0)\{2t^{*2}-u^2+\theta_0^2\rho\}\\ &\quad +(1/\sqrt{m})^3\{(1/\theta_0)^2(u^3-u^2t^*+ut^{*2}-3t^{*3})+\rho(2u-t^*)\\ &\quad\quad +c(u-t^*)+(2/\theta_0)(u-t^*)R\}]I_{[t\le m]}\}.\end{aligned} \tag{27}$$

It is straightforward to write down the expectation in terms of $P_\theta\{t=k\}$. Remember that $t^*=(\theta/2)(t-N)/\sqrt{N}$ becomes x_k when $t=k$, it follows

$$E_\theta\{t^{*i}I_{[t\le m]}\} = \sum_{k=1}^{m} x_k^i P_\theta\{t=k\},\quad i=0,1,2,3, \tag{28}$$

and

$$\begin{aligned}E_\theta\{t^{*i}RI_{[t\le m]}\} &= \sum_{k=1}^{m} x_k^i E_\theta\{RI_{[t=k]}\}\\ &= \sum_{k=1}^{m} x_k^i\int_0^\infty rP_\theta\{t=k,\ R\in dr\}\quad i=0,1.\end{aligned} \tag{29}$$

The proposition follows by substitutions.

By Lemmas 1 and 2, we obtain

Proposition 3.

$$\begin{aligned}P_\theta\{t \le m\} &= \Phi(u) + (1/\sqrt{m})\{(1/\theta_0) + (1/4)\theta_0 - \nu_1(\theta_0) + (\theta_0\rho/2) - (c/2)\theta_0\}\phi(u)\\ &+(1/m)\{-(1/\theta_0)^2 - (1/48)\theta_0^2 + (1/4)\theta_0\nu_1(\theta_0) - (1/2)\nu_2(\theta_0)\\ &\quad +\rho\{(1/2) + (\theta_0/2)\nu_1(\theta_0) - (\theta_0^2/8)(1+\rho)\}\\ &+c((1/2) + (1/8)\theta_0^2 - (1/2)\theta_0\nu_1(\theta_0) + (\rho\theta_0^2/4) - (c/2)\theta_0^2)\}u\phi(u) + o(1/m),\end{aligned} \tag{30}$$

$$\begin{aligned}\sum_{k=1}^m x_k P_\theta\{t=k\} =&\\ &-\phi(u) + (1/\sqrt{m})\{[(1/2\theta_0) + (1/4)\theta_0 - \nu_1(\theta_0) + (\rho/2)\theta_0 - (c/2)\theta_0]u\phi(u) \qquad (31)\\ &+[(-1/2\theta_0) + \nu_1(\theta_0) + (c/2)\theta_0]\Phi(u)\}\\ &+(1/m)\{[(1/48)\theta_0^2 - ((1/48)\theta_0^2 - (1/8) + (1/2\theta_0^2))u^2 + ((1/4)\theta_0 - (1/2\theta_0))u^2\nu_1(\theta_0)\\ &-(1/2)\nu_2(\theta_0)(1-u^2) + \nu_1^{(1)}(\theta_0)\\ &+c\{1 + ((1/4) + (1/8)\theta_0^2)u^2 - (1/2)\theta_0(u^2+1)\nu_1(\theta_0)\\ &+(\rho/4)\theta_0^2u^2 - (c/2)\theta_0^2(u^2+1)\}]\phi(u)\\ &+[(1/\theta_0)\nu_1(\theta_0) + \nu_1^{(1)}(\theta_0) + c]u\Phi(u)\} + o(1/m),\end{aligned}$$

$$\begin{aligned}\sum_{k=1}^m x_k^2 P_\theta\{t=k\} &= \Phi(u) - u\phi(u)\\ &+(1/\sqrt{m})\{(1/4)\theta_0u^2 - (u^2+2)\nu_1(\theta_0) + (\rho/2)\theta_0u^2\\ &-(c/2)\theta_0(u^2+2)\}\phi(u) + o(1/\sqrt{m}),\end{aligned} \tag{32}$$

$$\sum_{k=1}^m x_k^3 P_\theta\{t=k\} = -(u^2+2)\phi(u) + o(1), \tag{33}$$

$$\sum_{k=1}^m \int_0^\infty r P_\theta\{t=k,\, R\in dr\} = \nu_1(\theta_0)\Phi(u) + o(1), \tag{34}$$

and

$$\sum_{k=1}^m x_k \int_0^\infty r P_\theta\{t=k,\, R\in dr\} = -\nu_1(\theta_0)\phi(u) + o(1). \tag{35}$$

From Propositions 2 and 3, we obtain

Proposition 4. As $a \to \infty$

$$\begin{aligned}E_\theta\{[(1/t) - (1/m)]s_t I_{[t\le m]}\} &= (2/\sqrt{m})\{u\Phi(u) + \phi(u)\}\\ &+(1/m)\{(-1/\theta_0)u\phi(u) + [(-1/\theta_0)u^2 + (3/\theta_0) - 2\nu_1(\theta_0) + \rho\theta_0 - c\theta_0]\Phi(u)\}\\ &+(1/\sqrt{m})^3\{[(u^2/\theta_0^2) + ((6/\theta_0^2) - (1/24)\theta_0^2) - (4/\theta_0)\nu_1(\theta_0)\\ &\quad -2\nu_1^{(1)}(\theta_0) + \nu_2(\theta_0) + \rho(2 + (1/2)\theta_0^2 - (\rho/4)\theta_0^2 - \theta_0\nu_1(\theta_0))\\ &\quad +c(-3 + \theta_0\nu_1(\theta_0) - (\rho/2)\theta_0^2 + c\theta_0^2)]\phi(u)\\ &+[(1/\theta_0^2)u^3 + (1/\theta_0^2)u - (2/\theta_0)u\nu_1^{(1)}(\theta_0) + 2\rho u - cu]\Phi(u)\} + o((1/\sqrt{m})^3).\end{aligned} \tag{36}$$

To evaluate II, from Theorem 1 and algebra we have

Proposition 5.

$$
\begin{aligned}
(\theta/m)E_\theta\{t(m)\} &= \theta_0 + (1/\sqrt{m})\{u - 2(u\Phi(u) + \phi(u))\} \\
&+ (1/m)\{(1/\theta_0)u\phi(u) + [(1/\theta_0)(u^2-1) + 2\nu_1(\theta_0) - \rho\theta + c\theta_0]\Phi(u)\} \\
&+ (1/\sqrt{m})^3\{[(-u^2/\theta_0^2) + (1/24)\theta_0^2 + 2\nu_1^{(1)}(\theta_0) - \nu_2(\theta_0) \\
&+\rho((u^2/2) + \theta_0\nu_1(\theta_0) - (\rho/4)\theta_0^2) + c(2 - \theta_0\nu_1(\theta_0) + (\rho/2)\theta_0^2 - c\theta_0^2)]\phi(u) \\
&-[(-u^3/\theta_0^2) + (u/\theta_0^2) + 2u\nu_1^{(1)}(\theta_0) - \rho u + cu]\Phi(u)\} + o((1/\sqrt{m})^3)\,.
\end{aligned}
\tag{37}
$$

The theorem follows from Propositions 4 and 5.

4. SIMULATION RESULTS

We shall investigate the numerical accuracy of our asymptotic expansions. In tables 1 and 2 we compare the asymptotic expansion (13) with the simulation results for various a, m and c values. The first term on the right hand side of (13) alone constitutes the first order approximation to the power. And up to $1/\sqrt{m}$ terms and $1/m$ terms are called second and third order approximation respectively. Simulation results are based on 25,000 independent replication of Monte Carlo experiments using the random number generator RANN 2 in SSL II. Also the constants which appear in the asymptotic expansions are calculated numerically using Spitzer's formula and the results of Nabeya [14], see also Takahashi [11]. When $m = 148$, $\theta_0 = 0.284$, both second and third order approximation dominate the first order, but the differences are small and may be neglected irrespective to c values. For $m = 10$, $\theta_0 = 0.944$, we found that the second and third order approximation dominate the first order especially when $c = 4$.

Table 1.
Power of the test ($\sqrt{2a} = 3.45$ $m = 148$ ($\theta_0 = 0.284$))

θ	ρ	simulation	first order	second order	third order
0.275	0.39	0.5705	0.4584	0.5527	0.5562
		0.5405	0.4584	0.5342	0.5375
0.280	0.82	0.5879	0.4826	0.5793	0.5808
		0.5612	0.4826	0.5607	0.5621
0.283	0.62	0.6034	0.4972	0.5930	0.5933
		0.5826	0.4972	0.5744	0.5746
0.284	0.57	0.6101	0.5020	0.5977	0.5975
		0.5802	0.5020	0.5791	0.5789
0.285	0.54	0.6150	0.5069	0.6024	0.6018
		0.5861	0.5069	0.5838	0.5832
0.290	0.53	0.6377	0.5311	0.6263	0.6237
		0.6094	0.5311	0.6077	0.6053
0.295	0.77	0.6558	0.5552	0.6589	0.6463
		0.6356	0.5552	0.6325	0.6283

first row (c=0) second row (c=4)

Table 2.
Power of the test $(\sqrt{2a} = 2.986 \quad m = 10 \quad (\theta_0 = 0.944))$

θ	ρ	c	simulation	first order	second order	third order
0.935	0.20	0	0.5633	0.4883	0.5144	0.5162
		4	0.3145	0.4883	0.2763	0.2860
0.940	0.09	0	0.5664	0.4946	0.5143	0.5152
		4	0.3195	0.4946	0.2761	0.2806
0.943	0.03	0	0.5708	0.4984	0.5143	0.5145
		4	0.3222	0.4984	0.2760	0.2774
0.944	0.01	0	0.5728	0.4997	0.5143	0.5143
		4	0.3249	0.4997	0.2760	0.2763
0.945	0.98	0	0.5723	0.5009	0.5738	0.5738
		4	0.3191	0.5009	0.3356	0.3349
0.950	0.88	0	0.5841	0.5073	0.5739	0.5732
		4	0.3307	0.5073	0.3357	0.3305
0.954	0.80	0	0.5800	0.5123	0.5740	0.5727
		4	0.3334	0.5123	0.3359	0.3269

first row (c=0) second row (c=4)

In table 3 and 4 we shall compare first order= $m - \sqrt{m}A$, second order= $m - \sqrt{m}A + B$ and third order= $m - \sqrt{m}A + B + C/\sqrt{m}$ asymptotic expansion of $E\{t(m)\}$ (see equation (14)) with the simulation results.

Table 3.
Expected sample size $(\sqrt{2a} = 3.45 \quad m = 148)$

θ	simulation	first order	second order	third order
0.275	112.14	118.07	112.83	112.89
	117.19	118.07	114.84	116.82
0.280	110.86	115.61	110.71	111.02
	115.96	115.61	113.03	114.98
0.283	110.09	114.08	109.65	110.10
	114.36	114.08	111.95	113.89
0.284	109.34	113.56	109.29	109.78
	114.52	113.56	111.59	113.53
0.285	109.01	113.03	108.93	109.46
	113.99	113.03	111.23	113.18
0.290	106.82	110.32	107.03	107.76
	112.33	110.32	109.44	111.40
0.295	105.65	107.49	105.02	105.92
	110.11	107.49	107.67	109.67

first row (c=0) second row (c=4)

Table 4.
Expected sample size ($\sqrt{2a} = 2.986$ $m = 5$)

θ	simulation	first order	second order	third order
0.935	7.825	7.425	7.476	7.612
	9.366	7.425	9.430	8.418
0.940	7.779	7.373	7.499	7.633
	9.344	7.373	9.478	8.438
0.943	7.737	7.341	7.513	7.646
	9.347	7.341	9.507	8.450
0.944	7.751	7.331	7.518	7.650
	9.340	7.331	9.516	8.454
0.945	7.761	7.320	7.021	7.214
	9.334	7.320	9.025	8.257
0.950	7.703	7.267	7.039	7.235
	9.328	7.267	9.068	8.278
0.952	7.698	7.223	7.054	7.251
	9.321	7.223	9.103	8.295

first row (c=0) second row (c=4)

As in the previous case the asypmtotic expansion gives us a fairly accurate results for large and small m values. Since the difference between second and third order approximation is very small in table 3 and second order looks better in table 4, we may conclude that the second order approximation may be the best so far. Also tables 5 and 6 present the comparison of the asymptotic expansion of $E_\theta\{\bar{x}_{t(m)}\}$ and the simulation results, where the second order= $\theta + \Lambda/m$, and the third order= $\theta + \Lambda/m + \Xi/\sqrt{(m)^3}$.

Table 5.
Expected value of the sample mean. ($\sqrt{2a} = 3.45$ $m = 148$)

θ	simulation	second order	third order
0.275	0.3424	0.2968	0.3109
	0.3193	0.2968	0.3100
0.280	0.3492	0.3030	0.3176
	0.3241	0.3030	0.3168
0.283	0.3510	0.3067	0.3215
	0.3298	0.3067	0.3206
0.284	0.3550	0.3079	0.3228
	0.3297	0.3079	0.3219
0.285	0.3558	0.3092	0.3241
	0.3324	0.3092	0.3232
0.290	0.3638	0.3153	0.3307
	0.3377	0.3153	0.3298
0.295	0.3670	0.3215	0.3374
	0.3473	0.3215	0.3365

first row (c=0) second row (c=4)

Table 6.
Expected value of the sample mean. ($\sqrt{2a} = 2.986$ $m = 5$)

θ	simulation	second order	third order
0.935	1.1162	1.0384	1.0902
	0.9807	1.0384	1.0397
0.940	1.1237	1.0448	1.0940
	0.9890	1.0448	1.0435
0.943	1.1334	1.0486	1.0962
	0.9897	1.0486	1.0457
0.944	1.1315	1.0498	1.0969
	0.9937	1.0498	1.0465
0.945	1.1286	1.0511	1.1231
	0.9894	1.0511	1.0726
0.950	1.1410	1.0574	1.1281
	0.9961	1.0574	1.0776
0.954	1.1424	1.0625	1.1320
	1.0006	1.0625	1.0815

first row (c=0) second row (c=4)

Again the third order approximation dominate the others when m is large. But the second order approximation turn out to be the best for small m value and $c = 4$.

Acknowledgements
The author thanks both referee and Prof. Nabeya, who pointed out the existence of ρ in equation (2). Research Supported in part by the Ministry of Education of Japan under Grant-in-Aid for Scientific Research (C) 01530013. Hitotsubashi University.

REFERENCES

1. P. Armitage, *Sequential Medical Trials, 2nd ed.* Blackwell, Oxford (1975).
2. P. Armitage, C. K. McPherson and B. C. Row, *J. Roy. Statists. Soc. Ser.* **A 132** 235-244 (1969).
3. D. Siegmund, *Biometrika* **65** 177-189 (1977).
4. D. Siegmund, *Biometrika* **65** 341-349 (1978).
5. M. Woodroofe, *Ann. Probability* **4** 67-80 (1976a).
6. M. Woodroofe, *Biometrika* **63** 101-110 (1976b).
7. T.L. Lai and D. Siegmund, *Ann. Statists.* **5** 946-954 (1977).
8. H. Takahashi and M. Woodroofe, *Comm. Statists. A-Theory Methods* **10** 2113-2135 (1981).
9. M. Woodroofe and H. Takahashi, *Ann. Statists.* **10** 895-908 (1982).
10. H. Takahashi, *Ann. Statists.* **15** 278-295 (1987).
11. H. Takahashi, *J. of Japan Stat. Soc.* **20** 51-60 (1990).
12. M. Woodroofe and R. Keener, *Ann. Probability* **15** 102-114 (1987).
13. D. Siegmund, *Sequential Analysis*, 341-349. Springer-Verlag (1985).
14. S. Nabeya, *J. Statists. Comput. Simul.* **16** 223-240 (1983).

Stat. Sci. & Data Anal., pp. 395-405
K. Matsusita *et al.* (Eds)

Aspects of Goodness-of-Fit

MICHAEL A. STEPHENS
Simon Fraser University, Burnaby, B. C., Canada V5A 1S6

Abstract. In this article, two important methods of testing fit to a distribution are discussed and compared. They are the family of tests based on the empirical distribution function of a random sample, and the family based on plotting the order statistics against a suitable set of constants and examining the fit of a line through the plotted points. The two sets will be called EDF tests and Regression tests respectively.

Key words: Correlation tests; EDF tests; Probability plot; Regression tests; Tests for exponentiality; Tests for normality.

1. THE GOODNESS-OF-FIT PROBLEM

Suppose a random sample of n values $x_1, x_2, x_3, \ldots, x_n$ is given, and it is desired to test that the sample comes from the distribution $F(x;\theta)$. The parameter θ represents a vector of parameters in the distribution; they may all be known, so that the tested distribution is completely specified — this situation will be called Case 0 — or some or all of the parameters may have to be estimated from the sample. Thus a test might be required of the hypothesis that the sample comes from a normal distribution with mean μ and variance σ^2, or that a sample comes from a Gamma distribution with scale parameter β and shape parameter m. For the present, we assume the distribution $F(x;\theta)$ is continuous. We shall sometimes write the distribution as $F(x)$ for brevity.

2. EDF TESTS

The empirical distribution function of the sample is defined as follows:

$$F_n(x) = \begin{cases} 0, & x < x_{(1)}, \\ i/n, & x_{(i)} \leq x < x_{(i+1)},\ i = 1, 2, \ldots, n-1, \\ 1, & x \geq x_{(n)}. \end{cases}$$

The EDF thus represents, for any value of x, the fraction of the observations less than or equal to x; it clearly parallels $F(x)$, which gives the probability that an observation is less than x. In fact, by the Glivenko-Cantelli lemma, $|F_n(x) - F(x)| \to 0$ as $n \to \infty$. In 1933 Kolmogorov proposed a test based on the discrepancy $z_n(x) = F_n(x) - F(x)$, and Smirnov followed by proposing two related tests. The Kolmogorov-Smirnov tests, as they have come to be called, are defined as:

$$D_+ = \sup_x\{z_n(x)\};\ D_- = \sup_x\{-z_n(x)\}.$$

The statistic actually introduced by Kolmogorov was $D = \max(D_+, D_-)$. At about the same time, Cramér and von Mises were considering tests based on the integral of $z_n(x)$. The Cramér-von Mises family of statistics is

$$C = n \int_{-\infty}^{\infty} \{z_n(x)\}^2 \psi(x)\, dF(x),$$

where $\psi(x)$ is a weight function which can be used to vary the importance of different parts of the x-axis. Two commonly-used weight functions are $\psi(x) = 1$, giving the Cramér-von Mises statistic W^2, and $\psi(x) = \{F(x)[1 - F(x)]\}^{-1}$, giving the Anderson-Darling statistic A^2. In addition, W^2 can be modified to yield Watson's statistic U^2 given by

$$U^2 = n \int_{-\infty}^{\infty} \left\{F_n(x) - F(x) - \int_{-\infty}^{\infty} [F_n(x) - F(x)]\, dF(x)\right\}^2 dF(x).$$

2.1. Computing formulas

The definitions of these statistics look rather difficult to handle, but in fact very easy computing formulas exist. They are derived by means of the Probability Integral Transformation (PIT). This is the transformation

$$z = F(x; \theta).$$

It is well known that this transformation gives a variable z which is uniformly distributed between 0 and 1, written $U(0,1)$. If the Kolmogorov-Smirnov and Cramér-von Mises statistics are now calculated from the EDF of the z-values, with $F(z) = z$, the uniform distribution, it may easily be shown that the values are the same as those calculated from the original x-diagram. The z-diagram then gives the computing formulas following, with $z_{(i)} = F(x_{(i)}, \theta)$:

$$\begin{aligned}
D^+ &= \max[i/n - z_{(i)}];\\
D^- &= \max[z_{(i)} - (i-1)/n];\\
D &= \max(D^+, D^-);\\
W^2 &= \frac{1}{12n} + \sum_i \left\{z_{(i)} - \frac{2i-1}{2n}\right\}^2;\\
U^2 &= W^2 - n(\bar{z} - 0.5)^2 \text{ (where } \bar{z} = \textstyle\sum_i z_{(i)}/n);\\
A^2 &= -n - \frac{1}{n}\sum_i (2i-1)[\log z_{(i)} + \log(1 - z_{(n+1-i)})]. \qquad (1)
\end{aligned}$$

2.2. Estimated parameters

Suppose one or more components of θ are unknown, but are estimated by an efficient method from the sample values. These values are then inserted where necessary in the PIT above, and the statistics are calculated from the resulting z-values using the formulas (1). The unordered z-values are not now uniformly distributed; we describe them as super-uniform, because they almost always give much smaller values for the statistics, implying that the z-values are more evenly spaced than a genuine uniform sample.

2.3. Another transformation to uniformity

It is well-known that if events are occurring randomly in time, say at times $t_1, t_2, \ldots, t_n$, (the clock is started at time zero), and if the values are transformed by $z_{(i)} = t_{(i)}/t_{(n)}$, the set of $n-1$ values $z_{(i)}, i = 1, 2, \ldots, n-1$, will be distributed $U(0,1)$. An interesting set of events which gives superuniform $z_{(i)}$ are the ends of reigns (deaths or abdications) of the Kings and Queens of England, starting with time zero as the accession of William I in 1066 — it is hard to explain this phenomenon, even though it is obvious that successive reigns have lengths which are correlated: see Pearson [1].

2.4. Distribution theory

When the continuous distribution tested is completely specified (this is called Case 0), so that the test of fit becomes a test that the z-values are uniformly distributed, percentage points of the EDF statistics are either known exactly, or can be approximated very accurately. Details and tables are given by Stephens [2]. Furthermore, it is possible to modify the statistics so that only the asymptotic points need be tabulated. To do this, a modified form T^* of the EDF statistic T is used which is an easily calculated function of T and the sample size n. The resulting T^* is then compared with the asymptotic points of the test statistic. Modified forms and tables are given in Biometrika Tables for Statisticians, Vol II, Table 54, and also in Stephens [2].

When parameters are estimated efficiently (that is, with asymptotic variances given by the inverse of the Fisher information matrix), and used in the PIT to give the z-values from which the statistics are calculated, asymptotic percentage points can be calculated for statistics of the Cramér-von Mises family. These include W^2, U^2, and A^2. The asymptotic points depend on the distribution being tested , but not on location or scale parameters in the distribution; however, they do depend on shape parameters such as occur in the Gamma or Weibull or von Mises distributions. The points for finite n would be very difficult to calculate, and would have to be determined by Monte Carlo methods. Fortunately, for these statistics, the finite-n points converge very quickly to the asymptotic points, so that the latter may be used for practical purposes — a test with very small sample size would in any case have very little power.

For Kolmogorov-Smirnov statistics the distribution theory is more difficult. Again, points will not depend on the true values of location or scale parameters, but even asymptotic points are very difficult to calculate. Such tables as exist have usually been found by Monte Carlo methods. In addition, points for finite n do not converge rapidly to the asymptotic points for these statistics, so that it is necessary to give either the finite-n points (obtained by Monte Carlo) or modified forms, as was done for Case 0.

For both families of statistics, extensive tables of points are given by Stephens [2] for testing for the normal, exponential, Gamma, Weibull, extreme-value, von Mises and Cauchy distributions, so that the tests are available for practical use.

2.5. Power

The power of a test statistic will of course depend on several factors, including the size (or α-level) of the test, the sample size, and especially on the alternative to the tested distribution. Nevertheless, some general remarks can be made concerning the power of EDF statistics:

1. As two-sided omnibus tests (that is, tests against all alternatives, or at least a wide

range of alternatives), the Cramér-von Mises family is more powerful in general than the Kolmogorov-Smirnov family. This might be expected, as the former "tests" the hypothesized distribution all along the range of values of x, while the latter looks for a marked discrepancy between the EDF and the hypothesized $F(x)$, possibly only around one point.

2. For Case 0, there is a difference in power between the statistics, according to whether the alternative distribution is mostly a change in the location of the distribution, or a change in the scale. W^2 and A^2 will detect a change in location, and U^2 a change in scale. The Kolmogorov statistic D also detects a change in location.
3. If there is a change in location, *and the direction is known*, the statistics D^+ or D^- can be very powerful; however, if the wrong statistic is used, the power can easily be less than the α-level — that is, the test is biased. D^+ detects the situation where the true location is less than that tested, and D^- detects the opposite situation.
4. When parameters are estimated, the differences between the powers for these various types of alternative tend to fade, although the Cramér-von Mises family will still be better overall than the Kolmogorov-Smirnov tests.
5. On the whole, the recommended test statistic is the Anderson-Darling A^2; it is particularly effective in detecting outliers, that is, observations which are further into the tails than expected, and this is often the situation which the tester most wishes to detect.

Further details on all these statistics are given by Stephens [2]; a discussion of their use, and comparisons with other statistics, for the "observations random in time" situation described briefly above, is in Stephens [3].

3. REGRESSION TESTS

3.1. Introduction

For the second part of this paper, we describe another group of tests, to be called regression tests. They are based on a well-established and popular technique for testing fit to selected distributions, the *probability plot.* In regression tests, the order statistics $x_{(i)}$ of a sample are plotted on the vertical axis of a graph, against t_i, a set of constants which depend only on i, along the horizontal axis. (In the probability plot, the axes were reversed, but for convenience in introducing test statistics we keep them as above). The constants t_i are chosen so that the relationship between the $x_{(i)}$ and t_i is approximately a straight line. Historically, the linear relationship was often judged by eye, but more recently, test statistics have been developed, based on the parameters associated with the straight-line fit, when this is done by ordinary or generalised least squares.

Regression tests arise naturally when unknown parameters in the tested distribution $F(x;\theta)$ are location and scale parameters. Suppose $F(x;\theta)$ is $F_0(w)$, where $F_0(w)$ is a completely specified distribution and $w = (x-\alpha)/\beta$; then $\theta = (\alpha, \beta)$ with α a location parameter and β a scale parameter. A sample with order statistics $x_{(i)}$ can be derived from a set of values w from $F_0(w)$ with order statistics $w_{(i)}$, by the relationship

$$x_{(i)} = \alpha + \beta w_{(i)}, \qquad i = 1, \ldots, n. \tag{2}$$

An obvious example is the test for normality, where the density of w is given by $f(w) = (2\pi)^{-1/2}\exp(-w^2/2)$. Let $\Phi(w) = \int_{-\infty}^{w} f(t)\,dt$; then $F_0(w) = \Phi(w)$ and $F(x;\theta) = \Phi(w)$ with $w = (x-\mu)/\sigma$.

In the more general case, let $m_i = E(w_{(i)})$; then, from (2) we have

$$E(x_{(i)}) = \alpha + \beta m_i \tag{3}$$

and a plot of $x_{(i)}$ against m_i should be approximately a straight line with intercept α on the vertical axis and slope β. The values m_i are the most natural values to plot along the horizontal axis, but for most distributions they are difficult to calculate. Various authors have therefore proposed alternatives t_i which are convenient functions of i; then (3) can be replaced by the model

$$x_{(i)} = \alpha + \beta t_i + \epsilon_i \tag{4}$$

where ϵ_i is an "error" which only for $t_i = m_i$ will have mean zero.

It is then important to find a good method of testing how well the data fits the line (3) or (4). One way is simply to measure the correlation coefficient $r(x,t)$ between the paired sets $x_{(i)}$ and t_i. A second method is to estimate β using generalised least squares, and to compare this estimate with the estimate of scale given by the sample variance. We now examine these two procedures.

3.2. The correlation coefficient as test statistic

In discussing the correlation coefficient $r(x,t)$, we extend the usual meaning of correlation, and also that of variance and covariance, to apply to constants as well as random variables. Thus let x refer to the vector $x_{(1)}, \ldots, x_{(n)}$, and t to the vector $t_1, \ldots, t_n$; let $\bar{x} = \sum x_{(i)}/n$ and $\bar{t} = \sum t_i/n$, and define the sums

$$\begin{aligned} S(x,t) &= \sum (x_{(i)} - \bar{x})(t_i - \bar{t}) = \sum x_{(i)} t_i - n\bar{x}\bar{t} \\ S(x,x) &= \sum (x_{(i)} - \bar{x})^2 = \sum (x_i - \bar{x})^2 \\ S(t,t) &= \sum (t_i - \bar{t})^2. \end{aligned}$$

$S(x,x)$ will often be called S^2.

The correlation coefficient between x and t is

$$r(x,t) = \frac{S(x,t)}{[S(x,x)S(t,t)]^{1/2}}. \tag{5}$$

Statistics $r(x,m)$ or $r^2(x,m)$ are natural statistics for testing the fit of x to the model (3), since if a "perfect" sample is given, that is, a sample whose ordered values fall exactly at their expected values, $r(x,m)$ will be 1; more generally, the value of $r(x,m)$ can be interpreted as a measure of how closely the sample resembles a perfect sample. Tests based on $r(x,m)$, or equivalently on $r^2(x,m)$, will be one-tailed, with rejection of H_0 occurring only for low values of r.

However, as $n \to \infty$, $r^2(x,m) \to 0$ on H_0. A statistic which does have an asymptotic distribution is

$$Z(x,m) = n\{1 - r^2(x,m)\}. \tag{6}$$

Then $Z(x,m)$ is an equivalent statistic to r^2, based on the sum of squares of the residuals after the line (3) has been fitted. In common with many other goodness-of-fit statistics, for example chi-square and the EDF statistics, $Z(x,m)$ has the property that the larger it is, the worse the fit. Sarkadi [4] showed consistency of the test based on $r(x,m)$ for normality, and Gerlach [5] has shown consistency for correlation tests based on $r(x,m)$, or equivalently $Z(x,m)$, for a wide class of distributions including all the usual continuous distributions. This is to be expected, since, for large n, we can expect our sample to become perfect in the sense above. We can expect the consistency property to extend to $r(x,t)$ provided that t approaches m sufficiently rapidly for large samples.

3.3. The correlation test for the normal distribution

For the normal distribution $N(\mu,\sigma^2)$, $f(w)=(2\pi)^{-1/2}\exp(-w^2/2)$, with $w=(x-\mu)/\sigma$; thus $\alpha=\mu$ and $\beta=\sigma$, and the m_i are the expected values of standard normal order statistics. Equation (3) becomes

$$E(x_{(i)})=\mu+\sigma m_i. \tag{7}$$

For the normal distribution $\bar{m}=0$, and $r^2(x,m)$ can conveniently be written in vector notation. Let x be the vector $(x_{(1)},\ldots,x_{(n)})$, and let m be the vector $(m_1,\ldots,m_n)$; let primes, eg. x' and m', denote transposes of vectors or matrices.

Then

$$r^2(x,m)=\frac{(x'm)^2}{(m'm)S^2}. \tag{8}$$

The values of m_i required for the calculation of $r^2(x,m)$ have been well tabulated, and good computer programs are also available.

This statistic will later on be seen to be identical to W', the Shapiro-Francia statistic, so that, for testing normality, we shall refer to $r^2(x,m)$ also as W'. Tables for W' have been given by Shapiro and Francia [6].

In practice, it is easier to interpolate in tables of $Z(x,m)$ rather than $r^2(x,m)$, and Stephens [7] has produced tables for $Z(x,m)$ for both complete and censored samples. The null hypothesis that the sample comes from a normal distribution is rejected for large values of $Z(x,m)$.

De Wet and Venter [8] have proposed the use of the statistic $r(x,H)$, where $H_i=\Phi^{-1}\left(\frac{i}{n+1}\right)$. Use of H_i makes distribution theory easier, and de Wet and Venter have given the asymptotic null distribution of $Z(x,H)$. The H_i must be found numerically, using one of the excellent approximations available for $\Phi^{-1}(\cdot)$. The values of H_i and m_i are close in the middle of the sample, but are wider apart at the extremes. However, in Leslie *et. al.* [9] it is shown that $Z(x,H)$ and $Z(x,m)$ have the same null asymptotic distributions.

3.4. The Shapiro-Wilk procedure

We next turn to the second method of testing mentioned above, in which the parameters α and β in the model $x_{(i)}=\alpha+\beta m_i$ are estimated by generalised least squares. Using our previous notation, let $w_{(i)}$ be the order statistics from $F(w)$ with $\alpha=0$ and $\beta=1$; let $m_i=E(w_{(i)})$ as before, and let $E(w_{(i)}-m_i)(w_{(j)}-m_j)=V_{ij}$, the covariance of $w_{(i)}$ and $w_{(j)}$. Then let x be the column vector with components $x_{(1)},\ldots,x_{(n)}$, let m be a column vector with components $m_1,\ldots,m_n$, and let 1 be a column vector with each component equal to 1. Let V be the matrix with elements V_{ij}. The generalised least squares estimates of α and β are then

$$\hat{\alpha}=-m'Gx \text{ and } \hat{\beta}=1'Gx, \tag{9}$$

where

$$G=\frac{V^{-1}(1m'-m1')V^{-1}}{(1'V^{-1}1)(m'V^{-1}m)-(1'V^{-1}m)^2}. \tag{10}$$

For some distributions, for example the normal and exponential, these equations simplify considerably.

A method of testing fit has been proposed by Shapiro and Wilk [10,11] for testing normality and exponentiality. The procedure used is basically to compare the estimate of β^2 given by equation (9) with the estimate of β^2 given by the sample variance; the ratio of these estimates, multiplied by a constant, is taken as the test statistic. In the case of tests for normality, slight modifications of the first estimate of β^2 have also been suggested, since the estimate is complicated to calculate.

For the Shapiro-Wilk test for normality, α and β in (3) are μ and σ respectively; the estimates of these parameters given by (9) then become

$$\hat{\mu} = \bar{x} \text{ and } \hat{\sigma} = \frac{m'V^{-1}x}{m'V^{-1}m}.$$

The test statistic proposed by Shapiro and Wilk [10] is

$$W = \frac{\hat{\sigma}^2 R^4}{S^2 C^2} \tag{11}$$

where $S^2 = \sum(x_{(i)} - \bar{x})^2 = \sum(x_i - \bar{x})^2$, $R^2 = m'V^{-1}m$, and $C^2 = m'V^{-1}V^{-1}m$. The factors R^4 and C^2 ensure that W always takes values between 0 and 1.

Suppose the vector a is defined by $a = V^{-1}m/C$; then

$$W = \frac{(a'x)^2}{S^2} = \frac{(\sum_i a_i x_{(i)})^2}{S^2}.$$

In order to calculate W, the vector a is needed, and this in turn requires values of m and V^{-1}, derived from V. For values of n between 21 and 50, Shapiro and Wilk used approximations for the components a_i of a, and gave a table of values of a_i for sample sizes from $n = 3$ to 50. They also gave Monte Carlo points with which to make the test. The test is one-tailed: small values of W are significant.

A test similar to W, but for use with $n \geq 50$, was later suggested by Shapiro and Francia [6]. This is based on the observation of Gupta [12], who noted that the estimate $\hat{\sigma}$ is almost the same if V^{-1} is ignored in equation (10); the test statistic then given by Shapiro and Francia is

$$W' = \frac{(m'x)^2}{(m'm)^2 S^2}.$$

As has already been observed, this is equivalent to the sample correlation statistic $r^2(x, m)$.

3.5. Asymptotic equivalence of the Shapiro-Wilk and correlation statistics

Thus we have the remarkable result that the Shapiro-Wilk statistic *for testing normality* approaches the correlation coefficient $r^2(x, m)$. It is interesting to ask why this is so: why V^{-1} can be "ignored" when calculating W. Stephens [13] has shown heuristically that, for large n, m becomes an eigenvector of V, and $Vm \to \frac{1}{2}m$; then $V^{-1}m \to 2m$, $m'V^{-1}x \to 2m'x$, and $m'V^{-1}m \to 2m'm$. Hence $W \to W'$ because the factor 2 cancels in the numerator and denominator of W. The above results were proved rigorously by Leslie [14]. Stephens [13] also gives other asymptotic eigenvalues and eigenvectors of V.

4. POWER COMPARISONS

Shapiro and Wilk [10] gave power results for W, based on Monte Carlo studies. Unfortunately, the comparisons with EDF statistics were inaccurate — the EDF statistics were compared with Case 0 tables, and not the Case 3 tables to be used when the parameters μ and σ are estimated by $\bar{x}$ and s. Stephens [15] later gave comparisons based on the correct tables. These show that W is barely superior overall to EDF statistics, and especially only slightly superior to the Anderson-Darling A^2. Both statistics tend to have higher power than older statistics such as b_1 and b_2, the coefficients of skewness and kurtosis.

5. TWO QUESTIONS

The above results show that W, equivalent to the correlation coefficient $r^2(x, m)$, appears to give overall the most powerful omnibus test *for normality*. Two questions can then be asked:

(a) Will the Shapiro-Wilk procedure give good tests for other distributions?

(b) Will the correlation coefficient be successful for testing other distributions?

With reference to the first question, we first observe that for other distributions tested, the Shapiro-Wilk procedure will not necessarily lead to a statistic which is asymptotically equivalent to the correlation coefficient.

For the exponential distribution, where $F(x; \theta) = 1 - \exp\{-(x - \alpha)/\beta\}$, provided $x \geq \alpha$, (thus $F(w) = 1 - \exp(-w)$, and $\theta = (\alpha, \beta)$), the estimates in (3) become

$$\hat{\alpha} = x_{(1)} \text{ and } \hat{\beta} = \frac{n(\bar{x} - x_{(1)})}{(n-1)}.$$

The ratio $\hat{\beta}^2/S^2$, omitting some factors involving n, leads to the statistic

$$W_E = \frac{n(\bar{x} - x_{(1)})^2}{(n-1)S^2}.$$

Although this statistic has been proposed, and points given, for testing exponentiality (Shapiro and Wilk [11], Currie [16]), it has not proved powerful (Stephens [3]). Furthermore, it does not provide a *consistent* test, which means that there will be some distributions, not exponential, which would not be detected with power approaching 1, when the test for exponentiality is applied to large samples. Sarkadi [4] first pointed this out, by observing that W_E is equivalent, for large samples, to the coefficient of variation (CV) of the sample. To fix ideas, suppose α is known to be zero (this is frequently the case when the exponential distribution is used, although the discussion which follows is easily adapted to the case where α is not zero). Then, for large n, $x_{(1)} \to \alpha = 0$, and $W_E \to \bar{x}^2/S^2$. The coefficient of variation is $S^2/\bar{x}^2$, so that $W_E \to 1/\text{CV}$, and for large n, this is 1. However, many distributions have CV $= 1$, and a very large sample from one of these will have a W_E also approaching 1. The power of W_E will then approach a constant (less than 1) depending on the variance of W_E.

Spinelli and Stephens [17] have given power studies with samples taken from some other distributions with CV $= 1$, where the power of W_E is seen to diminish as the sample size n increases. Lockhart and Stephens [18] have explored the question of non-consistency further, and have shown that only for a very limited family of distributions, including the normal, does the Shapiro-Wilk procedure give a consistent test. Thus this technique of basing a test on the ratio of the regression estimate of scale to that given by the sample standard deviation cannot be recommended except for the normal case.

We now turn to the second question above. Since the correlation coefficient is powerful for testing normality, will it be equally successful for tests on other distributions? First, it should be emphasised that the most appropriate correlation is that between x and m: in the normal case, $H_i = \Phi^{-1}\{i/(n+1)\}$ was "sufficiently close" to m that the correlations $r(x, H)$ and $r(x, m)$ were approximately equal for large samples, and so had the same power. This is *not so* for other distributions. For example, for the exponential distribution, $m_i = \sum_{j=1}^{i}(n+1-j)^{-1}$, and $H_i = -\log\{1 - i/(n+1)\}$, and these are not close enough in the tails to give $r(x, H)$ as much power as $r(x, m)$. We can expect this result to be true also for other long-tailed distributions, and the question of when $r(x, H)$ can replace $r(x, m)$ for those distributions for which m is hard to calculate is itself an interesting research topic. See, for example, McLaren and Lockhart [19].

We therefore confine further discussion to the properties of $r(x, m)$. We have seen that, in contrast to W, $r(x, m)$ always gives a consistent test. However, McLaren and Lockhart [19] show that the asymptotic relative efficiency of correlation tests can be zero compared with EDF tests.

Stephens [20] adapted W_E to test exponentiality in the case where α is known; this can be compared with most power studies on other statistics which usually assume $\alpha = 0$. Stephens [3] gives some tables for comparison, and these demonstrate that the W statistics are in general less powerful than EDF statistics.

6. CENSORED DATA

One attraction of correlation statistics is the fact that the correlation coefficient is well-known to most applied statisticians, and the formula is very easy to calculate. This is true also for censored observations of types I or II, where missing observations are all at one end of the sample, often in the right-hand tail where higher values occur. Because of this appeal, Stephens [7], as was stated earlier, gives many tables of $Z(x, m) = n\{1 - r^2(x, m)\}$, or of the corresponding $Z(x, H)$, for use with right-censored data and for testing the exponential, Weibull and other distributions. EDF statistics have also been adapted for censored data, and formulas and tables for these statistics are given by Stephens [2]. For censored data, as for full samples, the statistics $Z(x, m)$ and $Z(x, H)$ may not be as powerful in general as EDF statistics; much depends on the influence of the tail observations which are lost by censoring. Finally, randomly censored data poses a unique problem in testing fit. The Kaplan-Meier estimate of $F(x)$ can be used for EDF statistics, and $r^2(x, m)$ can still be calculated if it is known *which* ordered observations have been lost, but in either case tables are difficult to provide. More work is needed on this topic.

7. TESTS FOR DISCRETE DISTRIBUTIONS

Until now, tests have been discussed only for continuous distributions. The correlation coefficient and the Shapiro-Wilk procedure do not adapt readily to discrete distributions, but EDF tests can be adapted. The technique is based on measures of discrepancy between the cumulative histograms of observed values and expected values (Pearson's χ^2 measures the discrepancy within each cell, and does not sum the observeds and expecteds). Pettitt and Stephens [21] gave some distribution theory for the Kolmogorov-Smirnov statistic for testing uniformity, and Freedman [22] discussed the Watson U^2 statistic, one of the Cramér-von Mises family, for the discrete uniform test. Recently, Lockhart and Stephens [23] have extended the test to include W^2 and A^2, and Spinelli and Stephens [24] develop a test for the Poisson distribution using Cramér-von Mises statistics. Power studies show these tests to be quite effective. In particular, for the test for normality, the EDF statistics will be more powerful than Pearson's χ^2 when the alternative is a trend in the cell probabilities — for example, to test that the probability of a defective item produced in a factory is the same each week, against the alternative that it decreases with time.

8. SUMMARY AND FINAL REMARKS

In this paper we have reviewed two important methods of testing fit — EDF statistics and regression methods based on the probability plot. Tests based on the EDF and those based on the correlation coefficient $r(x, m)$ are *consistent*, whereas those derived from use of the Shapiro-Wilk procedure are *not consistent* in general. The exception is the test for normality. For large samples, the correlation coefficient, however, can have low efficiency compared with EDF tests. For smaller samples, and for censored data, the situation is less clear, and more work is needed. Of course, other techniques for testing fit exist, based on Pearson's χ^2, on spacings, or on the empirical characteristic function. In general, for tests for continuous distributions, Pearson's χ^2 has low power compared with EDF statistics, due to the loss of information resulting from the grouping required. EDF statistics compare well with the other methods also, and, for overall testing against omnibus alternatives, these statistics are recommended. For specified *limited* alternatives, clearly other tests (for example the Likelihood Ratio test) can have good properties. Stephens [2,3,7] discusses these issues, but much more research can be done, both on mathematical aspects of the statistics, and on practical comparisons of tests.

Acknowledgement.
This work was supported by the Natural Science and Engineering Research Council of Canada and by the U. S. Office of Naval Research. The author is grateful to these agencies, and also for the invitation of the Program Committee of the Third Pacific Area Statistical Conference, for the invitation to present this paper and for their generous hospitality in Japan.

References

[1] E. S. Pearson. Comparison of tests for randomness of points on a line. *Biometrika* **50** 315–325 (1963).

[2] M. A. Stephens. Tests based on EDF statistics. Chapter 4 in *Goodness-of-fit techniques* (R.B. d'Agostino and M.A. Stephens, eds.). New York: Marcel Dekker (1986).

[3] M. A. Stephens. Tests for the exponential distribution. Chapter 10 in *Goodness-of-fit techniques* (R.B. d'Agostino and M.A. Stephens, eds.). New York: Marcel Dekker (1986).

[4] K. Sarkadi. The consistency of the Shapiro-Francia test. *Biometrika* **62** 445–450 (1975).

[5] B. Gerlach. A consistent correlation-type goodness-of-fit test; with application to the two-parameter Weibull distribution. *Math. Operationsforsch. Statist. Ser. Statist.* **10** 427–452 (1979).

[6] S. S. Shapiro and R. S. Francia. Approximate analysis of variance test for normality. *J. Amer. Statist. Assoc.* **67** 215–216 (1972).

[7] M. A. Stephens. Tests based on regression and correlation. Chapter 5 in *Goodness-of-fit techniques* (R.B. d'Agostino and M.A. Stephens, eds.). New York: Marcel Dekker (1986).

[8] T. De Wet and J. H. Venter. Asymptotic distribution of certain test criteria of normality. *South African Statist. J.* **6** 135–149 (1972).

[9] J. Leslie, S. Fotopoulos and M. A. Stephens. Asymptotic distribution of the Shapiro-Wilk W for testing normality. *Annals of Statistics* **14** 1497–1506 (1986).

[10] S. S. Shapiro and M. B. Wilk. An analysis of variance test for normality (complete samples). *Biometrika* **52** 591–611 (1965).

[11] S. S. Shapiro and M. B. Wilk. An analysis of variance test for the exponential distribution (complete samples). *Technometrics* **14** 355–370 (1972).

[12] A. K. Gupta. Estimation of the mean and standard deviation of a normal population from a censored sample. *Biometrika* **39** 266–273 (1952)

[13] M. A. Stephens. Asymptotic properties for covariance matrices of order statistics. *Biometrika* **62** 23–28 (1975).

[14] J. Leslie. Asymptotic properties and a new approximation for both the covariance matrix of normal order statistics and its inverse. *Colloq. Math. Soc. Janos Bolyai on Goodness of Fit* **45** (1984).

[15] M. A. Stephens. EDF statistics for goodness-of-fit and some comparisons. *J. Amer. Statist. Assoc.* **69** 730–737 (1974).

[16] I. D. Currie, The upper tail of the distribution of W-exponential. *Scand. J. Statist.* **7** 147–149 (1980).

[17] J. J. Spinelli and M. A. Stephens. Tests for exponentiality when origin and scale parameters are unknown. *Technometrics* **29** 471–476 (1987).

[18] R. A. Lockhart and M. A. Stephens. The non-consistency of the Shapiro-Wilk procedure. Research report, Dept. of Mathematics and Statistics, Simon Fraser University (1992).

[19] G. D. McLaren and R. A. Lockhart. On the asymptotic efficiency of certain tests of fit. *Canad. J. Statist.* **15** 159–167 (1987).

[20] M. A. Stephens. On the W test for exponentiality with origin known. *Technometrics* **20** 33–35 (1978).

[21] A. N. Pettitt and M. A. Stephens. The Kolmogorov-Smirnov goodness-of-fit statistic with dicrete and grouped data. *Technometrics* **19** 205–210 (1977).

[22] L. Freedman. Watson's U^2 statistic for discrete distributions. *Biometrika* **68** 708–711 (1981).

[23] R. A. Lockhart and M. A. Stephens. Cramér-von Mises statistics for discrete distributions. Research report, Dept. of Mathematics and Statistics, Simon Fraser University (1992).

[24] J. J. Spinelli and M. A. Stephens. EDF tests for the Poisson distribution. Research report, Dept. of Mathematics and Statistics, Simon Fraser University (1992).

Stat. Sci. & Data Anal., pp. 407-413
K. Matsusita *et al.* (Eds)

On the Central Limit Theorem in Hilbert Space with Application to U–Statistics

MADAN L. PURI and V.V. SAZONOV
Indiana University and Steklov Mathematical Institute
Bloomington, Indiana, U.S.A. and Moscow, Russia

Abstract. Estimates of the rate of convergence on balls in the Central Limit Theorem in Hilbert space are presented. An application to the study of the limit behavior of the distributions of Hilbert valued U–statistics is given.

Key words: Central Limit Theorem, Hilbert space, U-statistics.

Let $X_1, X_2, \ldots$ be independent and identically distributed (i.i.d.) random variables with values in a separable Hilbert space H. Assume that $E|X_1|^2 < \infty$ and, for simplicity, that $EX_1 = 0$. Denote by V the covariance operator corresponding to X_1, i.e. $(Vx, y) = E(X_1, x)(X_1, y)$ and let $\sigma_1^2 \geq \sigma_2^2 \geq \ldots$ be the eigenvalues of V. For $a \in H$ and $r \geq 0$ denote also $B_r(a) = \{x \in H : |x - a| < r\}$, $B_r = B_r(0)$. Recently in [1] the following theorem was proved.

THEOREM 1. *For any $a \in H$, $r \geq 0$, $n = 1, 2, \ldots$*

$$\Delta_n(a,r) = |P(S_n \in B_r(a)) - P(Y \in B_r(a))| \leq c \left(\prod_{i=1}^{6} \sigma_i^{-1} \right) \sigma^3 \beta (1 + |a|^3) n^{-\frac{1}{2}} \tag{1}$$

where c (here and below, with or without indices) is a positive constant and $\sigma^2 = E|X_1|^2$, $\beta = E|X_1|^3$.

Estimate **(1)** is a precise estimate. Indeed, from [2] and [3] it is easy to deduce that for any $c_0 > 0$ and $1 \geq \tau_1^2 \geq \ldots \geq \tau_\sigma^2$, there exist $a \in H$, $|a| > c_0$, and a probability measure P on H such that if $X_1, X_2, \ldots$ are i.i.d. random variables distributed as P, then $EX_1 = 0$, $\sigma_i^2 = \tau_i^2$, $i = 1, \ldots, 6$, $\beta < \infty$, and

$$\liminf_{n \to \infty} n^{\frac{1}{2}} \sup_{r \geq 0} \Delta_n(a,r) \geq c \left(\prod_{1}^{6} \sigma_i^{-1} \right) \sigma^3 \beta (1 + |a|^3). \tag{2}$$

The proof of **(1)** basically goes along the following lines. Assume that $\sigma = 1$ (this is not a restriction since the general case is reduced to this one if instead of X_i we consider $(E|X_i|^2)^{-\frac{1}{2}} X_i$, $i = 1, 2, \ldots$). We can also assume that

$$\left(\prod_{1}^{6} \sigma_i^{-1} \right) \beta n^{-\frac{1}{2}} \leq c_1, \tag{3}$$

since otherwise the theorem is obvious.

Note that if $X_i^{(1)} = X_i 1_{\{|X_i| \le n^{\frac{1}{2}}\}}$, $i = 1, 2, ..., n$, and $S'_n = n^{-\frac{1}{2}} \sum_{i=1}^{n} X_i^{(1)}$, then, as is easy to prove, for any Borel set E

$$\Delta_{n1}(a, r) = |P(S_n \in E) - P(S'_n \in E)| \le nP(|X_1| > n^{\frac{1}{2}}) \le \beta n^{-\frac{1}{2}}.$$

Thus it is enough to prove **(1)** with S_n replaced by S'_n.

Define $R = (\sigma_6^{-8} \beta^3 n)^{\frac{1}{8}}$. We will assume that $R \le n^{\frac{1}{2}}$; the opposite case is treated similarly but slightly differently. Note that if **(3)** is satisfied, then

$$P(B_R^c) \le \beta R^{-3} \le 1/2$$

if c_1 is small enough (which we assume). In this case, denoting P (resp. P_1) the probability measure corresponding to X_1 (resp. $X_1^{(1)}$) and defining P_2 as

$$P_2(A) = P(A \cap B_R)/P(B_R),$$

we have $P_1 \ge P_2/2$ and, hence,

$$P_1 = (P_2 + P_3)/2,$$

where P_3 is a probability measure.

Now let P_4 be the probability measure corresponding to the H-valued random variable

$$Z' = \zeta(Z - EZ) + Y' + EZ, \tag{4}$$

where ζ is a bounded real random variable such that $E\zeta = 0$, $E\zeta^2 = 1/2$ and $E\zeta^3 = 1$, Z is distributed as P_2, Y' is $(0, V_2/2)$ Gaussian, V_2 is the covariance operator of P_2, and ζ, Z, Y' are independent. It is easy to check that Z' has mean EZ, covariance operator V_2 and for any $h_1, h_2, h_3 \in H$

$$E(Z', h_1), (Z', h_2)(Z', h_3) = E(Z, h_1)(Z, h_2)(Z, h_3),$$

i.e. Z and Z' have the same moments of the first three orders, and in this sense they are "close". At the same time P_4, the distribution of Z', has a "solid" Gaussian component, the distribution of Y'. Denote

$$P_5 = (P_4 + P_3)/2$$

and let $X_i^{(5)}$, $i = 1, 2, ...$ be independent random variables with distribution P_5. Define $S''_n = n^{-\frac{1}{2}} \sum_1^n X_i^{(5)}$. The next step is to estimate

$$\Delta_{n2}(a, r) = |P(S'_n \in B_r(a)) - P(S''_n \in B_r(a))|.$$

If, for a measure μ on H, we denote $\bar{\mu}(\cdot) = \mu(n^{\frac{1}{2}} \cdot)$, $\mu^n = \mu * ... * \mu (n \text{ times})$, then, of course, the distribution of S'_n (resp. S''_n) can be written as $((\bar{P}_2 + \bar{P}_3)/2)^n$ (resp. $((\bar{P}_4 + \bar{P}_3)/2)^n$ and

$$\begin{aligned}\Delta_{n2}(a, r) &= \left| \left(((\bar{P}_2 + \bar{P}_3)/2)^n - ((\bar{P}_4 + \bar{P}_3)/2)^n \right) (B_r(a)) \right| \\ &\le 2^{-n} \left(\sum\nolimits_1 + \sum\nolimits_2 \right) \binom{n}{q} (\bar{P}_2^q - \bar{P}_4^q) * \bar{P}_3^{n-q}(B_r(a)) = I_1 + I_2,\end{aligned}$$

where $\sum_1$ is the summation over all integers q such that

$$|q-(n+1)/2| < (1/2)n^{7/10}$$

and $\sum_2$ is the summation over the remaining q, $0 \le q \le n$. Well known estimates of the tails of the binomial distribution give $I_2 \le cn^{-\frac{1}{2}}$ and one has to estimate I_1 composed of terms with q "not very different" from $(n+1)/2$. Consider a term

$$\nu_q = ((\bar{P}_2^q - \bar{P}_4^q) * \bar{P}_3^{n-q}(B_r(a)).$$

One can show that under our conditions the function $\bar{P}_4^q * \bar{P}_3^{n-q}(B_{r^{\frac{1}{2}}}(a))$ as a function of r is a distribution function with density bounded by $c\sigma_1^{-1}\sigma_2^{-1}$ (this is due to the solid Gaussian component in P_4). Thus by the Esséen inequality, for any $T > 0$

$$|\nu_q| \le c\left(\int_{|t|\le T} g_n(t)|t|^{-1}dt + \sigma_1^{-1}\sigma_2^{-1}T^{-1}\right),$$

where $g_n(t) = |f_{2n}(t) - f_{4n}(t)|$,

$$f_{kn}(t) = E\exp\left\{it\left|n^{-\frac{1}{2}}\left(\sum_1^q X_j^{(k)} + \sum_1^{n-q} X_j^{(3)}\right) - a\right|^2\right\},$$

and $X_j^{(k)}$ are independent of $X_j^{(3)}$, $k = 2, 4$. Then we choose an appropriately large T and for "large" $|t|$, $T_1 \le |t| \le T$, we use the inequality $|g_n(t)| \le |f_{2n}(t)| + |f_{4n}(t)|$ and estimate the integrals

$$\int_{T_1\le t\le T} |f_{kn}(t)|\ |t|^{-1}dt, \qquad k = 2, 4 \tag{5}$$

individually. For "small" t, we proceed as follows. Using the fact that $\bar{P}_2$ and $\bar{P}_4$ have the same moments of the first three orders, we can write

$$\begin{aligned}
&\left|\int_{|t|\le T_1} g_n(t)|t|^{-1}dt\right| \\
&\le \int_{|t|\le T_1} |t|^{-1}\left|\int \exp\{it|x-a|^2\}\,(\bar{P}_2^q - \bar{P}_4^q) * \bar{P}_3^{n-q}(dx)\right| dt \\
&\le \sum_{j=0}^{q-1}\int_{|t|\le T_1} |t|^{-1}\left|\iint f(1)\bar{P}_2^j * \bar{P}_4^{q-j-1} * \bar{P}_3^{n-q}(dx)(\bar{P}_2 - \bar{P}_4)(dy)\right| dt \\
&\le \frac{1}{6}\sum_{j=1}^{q-1}\int_{|t|\le T_1} |t|^{-1}\left|\int_0^1\iint f^{(4)}(\lambda)\bar{P}_2^j * \bar{P}_4^{q-j-1} * \bar{P}_3^{n-q}(dx)(\bar{P}_2 - \bar{P}_4)(dy)(1-\lambda)^3 d\lambda\right| dt
\end{aligned} \tag{6}$$

where $f(\lambda) = \exp\{it|x + \lambda y - a|^2\}$. Estimates of the integrals in (**5**) and in the right side of (**6**) are basically similar and use versions of the Götze–Yurinskii Lemma (see [4], [5]) which gives estimates of expressions like

$$|f_{kn}(t)|, \left|\int f^{(4)}(\lambda)\bar{P}_2^j * \bar{P}_4^{q-j-1} * \bar{P}_5^{n-q}(dx)\right|.$$

Next comes the estimation of

$$\Delta_{n3}(a,r) = |P(S_n'' \in B_r(a)) - P(Y \in B_r(a))|.$$

To this aim the following estimate is proved first. Let G be the $(0,V)$ Gaussian probability measure on $H, R = \bar{P}_5 - \bar{G}$ and let Y_1 be a $(0,U)$ Gaussian H–valued random variable with $trU \le 2$. Denote $\rho_1^2 \ge \rho_2^2 \ge ...$ the eigenvalues of U and let $e_1, e_2, ...$ be the corresponding eigenvectors. Then for any $r \ge 0$, $b \in H$

$$\left|\int P(|Y_1 - x - b| \le r)R(dx)\right| \le c\left(\prod_1^6 \rho_i^{-1}\right)\left(\beta + \sum_{i=2}^3 E|(X_1, \bar{b}_j)|^3\right) n^{-3/2} \quad \textbf{(7)}$$

where $\bar{b}_2 = \sum_5^6 (b, e_j)e_j$, $\bar{b}_3 = \sum_7^\infty (b, e_j)e_j$. The proof of this inequality is rather long and technical.

To estimate $\Delta_{n3}(a,r)$ we write

$$\Delta_{n3}(a,r) = \left|(\bar{P}_5^n - \bar{G}^n)(B_r(a))\right| \le \sum_{m=0}^{n-1} I_m,$$

where $I_m = |\bar{P}_5^m * \bar{G}^{n-m-1} * R(B_r(a))|$. If $m \le (n-1)/2$, we represent I_m as

$$I_m = \left|\int\int \bar{G}^{n-m-1}(B_r(a) - x - y)R(dx)\bar{P}_5^m(dy)\right|$$

and apply **(7)** with Y_1 $(0, ((n-m-1)/n)V)$ Gaussian (note that $(n-m-1)/n \ge 1/4$ if $n \ge 2$). If $m > (n-1)/2$ we write I_m as

$$I_m = \left|\int\int \bar{P}_5^m(B_r(a) - x - y)R(dx)\bar{G}^{n-m-1}(dy)\right|$$

and represent $\bar{P}_5^m$ as

$$\bar{P}_5^m = ((\bar{P}_4 + \bar{P}_3)/2)^m = \left(\sum\nolimits_1 + \sum\nolimits_2\right) 2^{-m} \binom{m}{\ell} \bar{P}_4^\ell * \bar{P}_3^{m-\ell}$$

where $\sum_1$ is the summation over all ℓ such that $|2\ell - m| < n^{7/10}$ and $\sum_2$ is the summation over all remaining $\ell \in \{0, ..., m\}$. Using known estimates of the tails of the binomial distribution we have $2^{-m}\sum_2 \binom{m}{\ell} \le cn^{-3/2}$, and hence

$$I_m \le 2^{-m} \sum\nolimits_1 \binom{m}{\ell} I_{m\ell} + cn^{-3/2},$$

where

$$\begin{aligned} I_{m\ell} &= \left|\int\int\int \bar{P}_4^\ell(B_r(a) - x - y - z)R(dx)\bar{G}^{n-m-1}(dy)\bar{P}_3^{m-\ell}(dz)\right| \\ &= \left|\int\int\int\int G_\ell(B_r(a) - x - y - z - u)R(dx)\bar{G}^{n-m-1}(dy)\bar{P}_3^{m-\ell}(dz)\bar{P}_6^\ell(du)\right|, \end{aligned}$$

since $\bar{P}_4^\ell = G_\ell * \bar{P}_6^\ell$, where G_ℓ is $(0, (\ell/2n)V_2)$ Gaussian and P_6 is the distribution of $\zeta(Z - EZ) + EZ$ in **(4)**. Observe that $\ell/n \ge c'$ if $n \ge c''$, and under condition **(3)**

$\sigma_{2j}^2 \geq \sigma_j^2$, $j = 1, ..., 6$, where $\sigma_{21}^2 \geq \sigma_{22}^2 \geq ...$ are the eigenvalues of V_2. To estimate $I_{m\ell}$ it remains now to apply **(7)** and to make some moment estimations.

The estimates of $\Delta_{nk}(a, r)$, $k = 1, 2, 3$ together prove the assertion of the theorem.

Estimate **(1)** has sense only if $E|X_1|^3 < \infty$. However one can construct similar meaningful estimates if only $E|X_1|^p < \infty$, $2 \leq p < 3$. E.g. (see [6]) in the notation of Theorem 1 for all p, $2 \leq p \leq 3$,

$$\Delta_n(a, r) \leq c(p) \left(\prod_1^6 \sigma_i^{-1} \right)^{p/3} \sigma^p E|X_1|^p (1 + |a|^p) n^{(2-p)/2}.$$

One can also construct estimates which sometimes give better bounds than **(1)**. Denote $d = ||a| - r|$, which is the distance from zero to the boundary of $B_r(a)$. Then in the notation of Theorem 1

$$\Delta_n(a, r) \leq c_1 \left[\sigma^{-1} \beta (1 + d^3)^{-1} + \left(\prod_{i=1}^6 \sigma_j^{-1} \right) \sigma^3 \beta (1 + \min(|a|^3, r^3)) \exp\{-c_2 d\} \right] n^{-\frac{1}{2}} \quad \textbf{(8)}$$

where α is a constant $\geq 1/5$ (see [7]). A similar estimate is claimed in [8]. Estimate (**8**) implies in particular that

$$\Delta_n(a, r) \leq c' \left(\prod_1^6 \sigma_i^{-1} \right) \sigma^3 \beta (1 + \min(|a|^3, r^3))(1 + d^3)^{-1} n^{-\frac{1}{2}}.$$

Consider now an application of Theorem 1 to the estimation of the rate of convergence of H–valued U–statistics. Let $T_1, T_2, ...$ be a sequence of i.i.d. random variables with values in a measurable space $(S, \mathcal{A})$, where $\mathcal{A}$ is a σ–algebra. Let $h(s, t)$ be a function on $S \times S$ with values in H which is symmetric, i.e. $h(s, t) = h(t, s)$ and measurable in the sense that $h^{-1}(E) \in \mathcal{A} \times \mathcal{A}$ for any Borel set $E \subset H$. Let U_n be the U–statistic of degree 2, viz.

$$U_n = \binom{n}{2}^{-1} \sum_{1 \leq i < j \leq n} h(T_i, T_j), \qquad n = 1, 2, \ldots .$$

Without loss of generality let $Eh(T_1, T_2) = 0$ and assume that $E|h(T_1, T_2)|^2 < \infty$. Let $Z_j = E(h(T_j, T_{j+1})|T_j)$, $j = 1, 2, ..., \tau^2 = E|Z_1|^2$, denote by W the covariance operator corresponding to Z_1 and let $\tau_1^2 \geq \tau_2^2 \geq ...$ be its eigenvalues. Finally let Y' be a $(0, W)$ Gaussian H–valued random variable. The following result was obtained in [9].

THEOREM 2. *For all* $a \in H$, $r \geq 0$, $n = 1, 2, ...$

$$\begin{aligned} D_n(a, r) &= \left| P((n^{\frac{1}{2}}/2) U_n \in B_r(a)) - P(Y' \in B_r(a)) \right| \qquad \textbf{(9)} \\ &\leq c \Bigg(\left(\tau_1^{-1} \tau_2^{-1} (\tau + |a|)^2 (E|h(T_1, T_2)|^2 - 2\tau^2) n^{-1} \right)^{1/3} \\ &\quad + \left(\prod_{j=1}^6 \tau_j^{-1} \right) \tau^3 E|Z_1|^3 (1 + |a|^3) n^{-\frac{1}{2}} \Bigg). \end{aligned}$$

If, in addition, $E|h(T_1, T_2)|^{2s} < \infty$ for a positive integer $s \geq 2$, then

$$D_n(a, r) = O\left(n^{-s/(2s+1)} \right). \qquad \textbf{(10)}$$

Theorem 2 implies obviously that if $|h(T_1, T_2)|$ has moments of all orders then $D_n(a,r) = O(n^{-1/2+\delta})$ for all $\delta > 0$.

Let us sketch the proof of Theorem 2. Denote

$$\hat{U}_n = \binom{n}{2} \sum_{1 \le i < j \le n} (Z_i + Z_j) = \frac{2}{n} \sum_{i=1}^{n} Z_i.$$

Letting $V_n = n^{\frac{1}{2}} U_n/2$, $\hat{V}_n = n^{\frac{1}{2}} \hat{U}_n/2$, we can write

$$\begin{aligned} D_n(a,r) &\le \left| P(V_n \in B_r(a)) - P(\hat{V}_n \in B_r(a)) \right| + \left| P(\hat{V}_n \in B_r(a)) - P(Y' \in B_r(a)) \right| \\ &= D_{n1}(a,r) + D_{n2}(a,r). \end{aligned}$$

Now for any $\varepsilon > 0$, $q > 0$ we have

$$\begin{aligned} D_{n1}(a,r) &= \left| P(V_n \in B_r(a)), \hat{V}_n \notin B_r(a)) - P(\hat{V}_n \in B_r(a), V_n \notin B_r(a)) \right| \\ &\le P(\hat{V}_n \in B_{r+\varepsilon}(a) \backslash B_r(a)) + P(\hat{V}_n \in B_r(a) \backslash B_{r-\varepsilon}(a)) + 2P(|V_n - \hat{V}_n| \ge \varepsilon) \\ &\le D_{n2}(a, r-\varepsilon) + D_{n2}(a, r+\varepsilon) \\ &\quad + P(Y \in B_{r+\varepsilon}(a) \backslash B_{r-\varepsilon}(a)) + 2\varepsilon^{-q} E|V_n - \hat{V}_n|^q \end{aligned}$$

where $B_{r-\varepsilon}(a) = \phi$ if $\varepsilon > r$. Thus

$$D_n(a,r) \le \sum_{i=1}^{3} D_{n2}(a, r_i) + P(Y \in B_{r+\varepsilon}(a) \backslash B_{r-\varepsilon}(a)) + 2\varepsilon^{-q} E|V_n - \hat{V}_n|^q,$$

where $r_1 = r - \varepsilon$, $r_2 = r$, $r_3 = r + \varepsilon$. To estimate D_{n2} we apply Theorem 1. Furthermore it is known that

$$P(Y' \in B_{r+\varepsilon}(a) \backslash B_{r-\varepsilon}(a)) \le c\tau_1^{-1}\tau_2^{-1}(\tau + |a|)\varepsilon$$

(see e.g. [10]), and a simple calculation gives

$$E|V_n - \hat{V}_n|^2 = 2^{-1}(n-1)^{-1}(E|h(T_1, T_2)|^2 - 2\tau^2).$$

Now choosing appropriately ε we obtain (9); (10) is proved similarly.

Acknowledgements.
Research of Madan L. Puri supported by the Office of Naval Research Contract N00014-91-J-1020.

REFERENCES:

1. Sazonov, V.V., Ulyanov, V.V. and Zalesskii, B.A. A precise estimate of the rate of convergence in the central limit theorem in Hilbert space. *Mat. Sbornik.*, **180**, 1587-1613 (1989); English transl. in *Math. USSR Sbornik,* **68** (1991).

2. Senatov, V.V. On the dependence of estimates of the convergence rate in the central limit theorem on the covariance operator of summands. *Teor. Verojatnost. i Primenen.,* **30**, 354-357 (1985); English transl. in *Theory Probab. Appl.,* **30**, (1985).

3. Senatov, V.V. Four examples of lower bounds in the multidimensional central limit theorem. *Teor. Verojatnost. i Primenen.,* **30**, 750-758 (1985); English transl. in *Theory Probab. Appl.,* **30** (1985).

4. Gotze, F. Asymptotic expansions for bivariate von Mises functional. *Z. Wahrscheinlichkeitstheor. verw. Gebiete* **50**, 333-355 (1979).

5. Yurinskii, V.V. On the accuracy of the normal approximation of the probability of hitting a ball. *Teor. Veroyatnost. i Primenen.* **27**, 270-278 (1982), English transl. *Theory Probab. Appl.* **27** (1982), 280-289.

6. Sazonov, V.V. and Ulyanov, V.V. Speed of convergence in the central limit theorem in Hilbert space under weakened moment conditions. *Probability Theory and Mathematical Statistics, Proceedings of the Fifth Vilnius Conference.* Editors: B. Grigelionis, Yu.V. Prohorov, V.V. Sazonov and V. Statulevicius. Mokslas, Vilnius, 394-410 (1991).

7. Sazonov, V.V. and Ulyanov, V.V. An improved estimate of the accuracy of Gaussian approximation in Hilbert space. *New Trends in Probability and Statistics,* Vol. **1**, Editors: V.V. Sazonov and T. Shervashidze, Mokslas, Vilnius, 123-136 (1991).

8. Senatov, V.V. Several remarks on the estimating the rate of convergence in the central limit theorem in Hilbert space. *Teor. Verojatnost. i Primenen.,* **36**, 382-385 (1991).

9. Puri, M.L. and Sazonov, V.V. On Hilbert space valued *U*–statistics. *Teor. Verojatnost. i Primenen.,* **36**, 604-605 (1991).

10. Paulauskas, V. and Račkauskas, A. *Approximation theory in the central limit theorem. Exact results in Banach spaces.* Mokslas, Vilnius (1987); English transl. Kluwer, Dordrecht, 1989.

Stat. Sci. & Data Anal., pp. 415-426
K. Matsusita *et al.* (Eds)

Asymptotics of the Perturbed Sample Quantile for a Sequence of m–dependent Stationary Random Process

MADAN L. PURI and SHAN SUN
Indiana University, Bloomington, U.S.A.

Abstract. Let $\{X_i, i \geq 1\}$ be a sequence of m–dependent stationary random variables having continuous cumulative distribution function F. The usual perturbed empirical distribution function is $\hat{F}_n(x) = n^{-1}\sum_{i=1}^{n} K_n(x-X_i)$, where K_n, $n \geq 1$ is a sequence of continuous distribution functions converging weakly to the distribution function of a unit mass at zero. In this paper, we study the perturbed sample quantile estimator $\hat{\xi}_{np} = \inf\{x \in I\!R, \hat{F}_n(x) \geq p\}$, $0<p<1$, based on a kernel k associated with K_n and a sequence of window–widths $a_n>0$ and we derive the necessary and sufficient conditions for the asymptotic normality of $\hat{\xi}_{np}$.

Key words. Perturbed empirical distribution functions; perturbed sample quantile; m–dependent stationary random variables; strong consistency; central limit theorem.

1. INTRODUCTION.

Let $\{X_i, i \geq 1\}$ be a sequence of m–dependent stationary random variables, that is, let the random vectors $(X_1, ..., X_i)$ and $(X_j, X_{j+1}, ...)$ be independent if $j-i > m$ where m is a non-negative integer. Further, let X_i and (X_i, X_{i+h}) have continuous cdfs (cumulative distribution functions) $F(x)$ and $F_{1,h}(x, y)$ respectively, $h = 1, 2, ..., m$ and $i = 1, 2, ...$. Denote the empirical cdf of $X_1, ..., X_n$ by $F_n(x)$. Thus $F_n(x) = \frac{1}{n}\sum_{i=1}^{n} u(x - X_i)$ where $u(t) = 1$ if $t \geq 0$, and $u(t) = 0$ otherwise. F_n is a natural estimator of F based on $X_1, ..., X_n$. However, if F is a smooth cdf (e.g. if F has a density), then it seems reasonable to consider smooth estimators which are better adapted to this situation. For the case $\{X_i, i \geq 1\}$ is a sequence of iid rvs (independent and identically distributed random variables), [1] initiated the study of estimators of the form

$$\hat{F}_n(x) = n^{-1}\sum_{i=1}^{n} K_n(x - X_i) \tag{1.1}$$

where $K_n(x) = \int_{-\infty}^{x} k_n(t)dt$, $k_n(t) = a_n^{-1}k(ta_n^{-1})$, $\{a_n, n \geq 1\}$ is a sequence of positive real numbers (window widths) such that $a_n \to 0$ as $n \to \infty$, and k is a probability density function.

Here, in this paper we consider an inverse problem: given $0 < p < 1$, define ξ_p (the pth quantile of F) by $F(\xi_p) = p$. That is, $\xi_p = \inf\{x \in I\!R, F(x) \geq p\}$. For estimating ξ_p when the distribution functions are smooth (at least locally), we consider the perturbed sample quantile $\hat{\xi}_{np}$ where

$$\hat{\xi}_{np} = \inf\left\{x \in I\!R, \hat{F}_n(x) \geq p\right\}. \tag{1.2}$$

In case the underlying random variables $\{X_i,\ i \geq 1\}$ are iid, the asymptotic normality of $\hat{\xi}_{np}$ was first obtained by [1]. For additional work in this direction, and some related work we refer to [2], [3], [4] and [5] among others. Recently [6] extended [1]'s result by weakening his conditions on a_n and the kernel k, and also providing the necessary conditions for the asymptotic normality of $\hat{\xi}_{np}$. [7] derived the asymptotic normality of the classical empirical sample quantile (i.e. $\inf\{x \in \mathbb{R},\ F_n(x) \geq p\}$) by using the results of [8] and [9] for the case of m–dependent random variables. Here in this paper, we study the asymptotic behavior of $\hat{\xi}_{np}$ for the case of m–dependent stationary sequences of random variables; in particular, we obtain (i) the strong consistency of $\hat{\xi}_{np}$, and (ii) the necessary and sufficient conditions for the asymptotic normality of $\hat{\xi}_{np}$. The results obtained are generalizations of [1] and [6], among others.

2. MAIN RESULTS.

Assumptions on the underlying cdf F, joint cdf $F_{1,h}$ ($h = 1, 2, ..., m$) and kernel k. We specify the following two sets of conditions:

A1 (i) f is bounded and $f(\xi_p) > 0$.
(ii) f is continuous in the neighborhood of ξ_p.
(iii) $\frac{\partial}{\partial x} F_{1,h}(x, y)$ and $\frac{\partial}{\partial y} F_{1,h}(x, y)$ are bounded ($h = 1, 2, ..., m$).

B1 $\int_{-\infty}^{\infty} |x| k(x) dx < \infty$ but $\int_{-\infty}^{\infty} x k(x) dx \neq 0$.

and

A2 (i) f is differentiable with bounded derivative f' on the support of F and $f(\xi_p) > 0$.
(ii) f' is continuous in a neighborhood of ξ_p and $f'(\xi_p) \neq 0$.
(iii) $\frac{\partial^2 F_{1,h}(x,y)}{\partial x^2}$, $\frac{\partial^2 F_{1,h}(x,y)}{\partial y^2}$ and $\frac{\partial^2 F_{1,h}(x,y)}{\partial x \partial y}$ are bounded ($h = 1, 2, ..., m$).

B2: $\int_{-\infty}^{\infty} x k(x) dx = 0$ and $\int_{-\infty}^{\infty} x^2 k(x) dx < \infty$.

THEOREM 2.1. *(Strong consistency of $\hat{\xi}_{n,p}$). Assume* **A1**(i), **A1**(ii) *and*

$$\int_{-\infty}^{\infty} |x| k_n(x) dx = O(n^{-\frac{1}{2}}) \quad as\ n \to \infty. \tag{2.1}$$

Then

$$P\left\{\lim_{n\to\infty} |\hat{\xi}_{np} - \xi_p| = 0\right\} = 1. \tag{2.2}$$

Define

$$\nu_m^2 = p(1-p) + 2\sum_{h=1}^{m} \left[F_{1,h}(\xi_p, \xi_p) - p^2\right] \tag{2.3}$$

and

$$\hat{Z}_n = n^{\frac{1}{2}} f(\xi_p)(\hat{\xi}_{np} - \xi_p). \tag{2.4}$$

The following central limit theorems are extensions of the corresponding results due to [6] for the case the underlying random variables $\{X_i,\ i \geq 1\}$ are independent and identically distributed.

THEOREM 2.2. *Under the assumptions* **A1** *and* **B1**, *if*

$$\nu_m^2 > 0, \tag{2.5}$$

then

$$\hat{Z}_n \xrightarrow{D} \mathcal{N}(0, \nu_m^2) \quad as \ n \to \infty \tag{2.6}$$

if and only if

$$\lim_{n\to\infty} n^{\frac{1}{2}} a_n = 0. \tag{2.7}$$

(By $\mathcal{N}(0,\sigma^2)$, we mean that a random variable having normal distribution with mean zero and variance σ^2).

THEOREM 2.3. *Assume assumptions* **A2**, **B2** *and* **(2.5)**. *Then* **(2.6)** *holds if and only if*

$$\lim_{n\to\infty} n^{\frac{1}{4}} a_n = 0. \tag{2.8}$$

3. PROOF OF THEOREM 2.1.

To prove Theorem 2.1 we use an inequality of [9] for the sum of m–dependent random variables.

Let $\varepsilon > 0$ be arbitrary, then

$$P\left\{|\hat{\xi}_{np} - \xi_p| > \varepsilon\right\} = P\left\{\hat{\xi}_{np} > \xi_p + \varepsilon\right\} + P\left\{\hat{\xi}_{np} < \xi_p - \varepsilon\right\} \tag{3.1}$$

and

$$\begin{aligned} P\left\{\hat{\xi}_{np} > \xi_p + \varepsilon\right\} &= P\left\{\hat{F}_n(\xi_p + \varepsilon) < p\right\} \\ &= P\left\{\sum_{i=1}^{n}[\mu_n - K_n(\xi_p + \varepsilon - X_i)] > n(\mu_n - p)\right\} = P\left\{\sum_{i=1}^{n} Y_{ni} > nt_n\right\} \end{aligned} \tag{3.2}$$

where

$$\mu_n = EK_n(\xi_p+\varepsilon-X_i), \quad Y_{ni} = \mu_n - K_n(\xi_p+\varepsilon-X_i),\ i = 1,2,...,n, \quad \text{and}\ t_n = \mu_n - p.$$

Here $\{Y_{ni},\ 1 \le i \le n,\ n \ge 1\}$ is a sequence of row-wise m–dependent, stationary double array of random variables.

It is clear that

$$\mu_n = \int_{-\infty}^{\infty} F(\xi_p + \varepsilon - x)\, k_n(x)dx \le 1. \tag{3.3}$$

Using **(3.3)** and $t_n > 0$ for large n together with an inequality of [9], we have

$$P\left\{\hat{\xi}_{np} > \xi_p + \varepsilon\right\} \le \exp\left\{-2\left[\frac{n}{m+1}\right] t_n^2\right\} \tag{3.4}$$

where $\left[\frac{n}{m+1}\right]$ denotes the largest integer not exceeding $\frac{n}{m+1}$. Note that t_n^2 can be decomposed as

$$t_n^2 = (u_n + v_n)^2 \ge v_n^2 - 2|u_n|v_n, \tag{3.5}$$

where

$$u_n = \mu_n - F(\xi_p + \varepsilon) \quad \text{and} \quad v_n = F(\xi_p + \varepsilon) - F(\xi_p)$$

Using **(3.3)** again, via **(2.1)** and **A1**(i), there exists a constant $T > 0$ and $c > 0$ such that

$$|u_n| = |\mu_n - F(\xi_p + \varepsilon)| \tag{3.6}$$
$$\leq \int_{-\infty}^{\infty} |F(\xi_p + \varepsilon - x) - F(\xi_p + \varepsilon)| k_n(x) dx$$
$$= \int_{-\infty}^{\infty} f(\theta_x)\,|x| k_n(x) dx \leq T \int_{-\infty}^{\infty} |x| k_n(x) dx \leq n^{-\frac{1}{2}} Tc$$

where $|\theta_x - \xi_p - \varepsilon| < |x|$.

$$\nu_n = F(\xi_p + \varepsilon) - F(\xi_p) = f(\theta_\varepsilon)\varepsilon \tag{3.7}$$

where $|\theta_\varepsilon - \xi_p| < \varepsilon$.

From **(3.5)** — **(3.7)** we obtain $t_n^2 \geq q_n$ where

$$q_n = [f(\theta_\varepsilon)\varepsilon]^2 - 2n^{-\frac{1}{2}} Tcf(\theta_\varepsilon)\varepsilon.$$

Therefore,

$$P\left\{\hat{\xi}_{np} > \xi_p + \varepsilon\right\} \leq \exp\left\{-2\left[\frac{n}{m+1}\right] q_n\right\}.$$

Using a similar argument, we are able to show that

$$P\left\{\hat{\xi}_{np} < \xi_p - \varepsilon\right\} \leq \exp\left\{-2\left[\frac{n}{m+1}\right] g_n\right\}.$$

where $g_n = [f(\theta_\varepsilon)\varepsilon]^2 + 2n^{-\frac{1}{2}} Tc\, f(\theta_\varepsilon)\varepsilon$. Using **A1**(i), we have

$$\lim_{n\to\infty} g_n = \lim_{n\to\infty} q_n = (f(\theta_\varepsilon)\varepsilon)^2 > 0, \quad \text{where} \quad |\theta_\varepsilon - \xi_p| < \varepsilon.$$

Thus,

$$P\left\{\sup_{n\geq k} |\hat{\xi}_{np} - \xi_p| > \varepsilon\right\}$$
$$\leq \sum_{n=k}^{\infty} P\left\{|\hat{\xi}_{np} - \xi_p| > \varepsilon\right\}$$
$$\leq \sum_{n=k}^{\infty} \left\{\exp\left(-2\left[\frac{n}{m+1}\right] q_n\right) + \exp\left(-2\left[\frac{n}{m+1}\right] g_n\right)\right\} \longrightarrow 0 \quad as\ k \to \infty.$$

Hence $\hat{\xi}_{np} \longrightarrow \xi_p$ with probability 1. The proof follows.

4. PROOFS OF THEOREMS 2.2 AND 2.3.

The proofs of Theorems 2.2 and 2.3 rest on the following lemmas.

The following lemma provides an Esséen's inequality.

LEMMA 4.1. *Let $\{\xi_i,\, i \geq 1\}$ be a stationary sequence of m-dependent random variables with $E\xi_i = 0$ $(i = 1, 2, ...)$ such that*

$$\tau_n^2 = E\left(\sum_{j=1}^{n} \xi_j\right)^2 \quad \text{and} \quad S_n = \frac{1}{\tau_n}\sum_{j=1}^{n} \xi_j.$$

Let

$$E|\xi_1|^3 < \infty.$$

If $0 < \tau^2 = \lim_{n\to\infty} \tau_n^2 < \infty$, then there exist constants T_1 and T_2 such that

$$\sup_t |P\{S_n \leq t\} - \Phi(t)| \leq T_1 \frac{b_m^2 E^{\frac{1}{3}}|\xi_1|^3}{\tau^3 \sqrt{n}} + T_2 \frac{m b_m E^{\frac{1}{3}}|\xi_1|^3 \log n}{\tau^2\, n} \tag{4.1}$$

where $b_m = \max\limits_{1\leq i\leq m+1} E^{\frac{1}{3}}|\sum\limits_{j=1}^{i} \xi_j|^3$ and $\Phi(t) = \int_{-\infty}^{t} \frac{1}{\sqrt{2\pi}}\, e^{-\frac{x^2}{2}} dx$.

PROOF. See [10].

Let us denote

$$\mu_n(x) = EK_n(x - X_1) = \int_{-\infty}^{\infty} F(x-t)k_n(t)dt \tag{4.2}$$

and

$$\sigma_n^2(x) = Var\, K_n(x - X_1) + \frac{2}{n}\sum_{h=1}^{m}(n-h)Cov(K_n(x - X_1), K_n(x - X_{1+h})). \tag{4.3}$$

LEMMA 4.2. *Under assumptions* **A1** *and* **B1**, *there exist positive constants c_1 and c_2 such that if $\|h\| = \sup\limits_x |h(x)|$, then*

$$\|\mu_n - F\| \leq c_1 a_n \tag{4.4}$$

and

$$\|\sigma_n^2 - F(1-F) - \frac{2}{n}\sum_{h=1}^{m}(n-h)(F_{1,h} - F^2)\| \leq c_2 a_n. \tag{4.5}$$

LEMMA 4.3. *Under conditions* **A2** *and* **B2**, *there exist positive constants d_1 and c_2 such that*

$$\|\mu_n - F\| \leq d_1 a_n^2 \tag{4.6}$$

and **(4.5)** *holds.*

Proof of Lemma 4.2:

$$\mu_n(x) = \int_{-\infty}^{\infty} F(x-t)k_n(t)dt,$$

where $k_n(t) = K_n'(t)$. We have

$$|\mu_n(x) - F(x)| \leq \int_{-\infty}^{\infty} f(\theta_x)|t|k_n(t)dt$$

where $|\theta_x - t| < |x|$.

Set $\gamma_n(x) = \mu_n(x) - F(x)$. By **A1**(i) and **B1** there exists a constant T, such that

$$\|\gamma_n\| \leq c_1 a_n \quad \text{where} \quad c_1 = T\int_{-\infty}^{\infty} |t|k(t)dt < \infty. \tag{4.7}$$

To prove **(4.5)**, we write

$$\sigma_n^2(x) = \sigma_{n1}^2(x) + \sigma_{n2}^2(x) \tag{4.8}$$

where

$$\sigma_{n1}^2(x) = Var\, K_n(x - X_1)$$

and

$$\sigma_{n2}^2(x) = \frac{2}{n}\sum_{h=1}^{m}(n-h)Cov\left(K_n(x - X_1),\, K_n(x - X_{1+h})\right).$$

Consider the stochastic process $\{W_{n,x}(s),\, -\infty < s < \infty\}$ defined by $W_{n,x}(s) = k_n(x)\, w(x - s - X_1)$, where $w(t) = 1$ for $t \geq 0$ and $w(t) = 0$ for $t < 0$.

Set $\alpha_n(x) = EK_n^2(x - X_1)$ and $\beta_n(x) = \alpha_n(x) - F(x)$, then clearly

$$\alpha_n(x) = \int_{-\infty}^{\infty}\int_{-\infty}^{\infty} F(x - \max(u,v))k_n(u)k_n(v)dudv.$$

Then proceeding essentially as in Lemma 3.1 in [6], we obtain

$$|\beta_n(x)| \leq c_3 a_n \tag{4.9}$$

where $c_3 = 2T\int_{-\infty}^{\infty} |t|k(t)dt < \infty$. Also, we have

$$\sigma_{n1}^2(x) = F(x)(1 - F(x)) + \beta_n(x) - \gamma_n^2(x) - 2\gamma_n(x)F(x). \tag{4.10}$$

Since

$$\begin{aligned} &Cov(K_n(x - X_1), K_n(x - X_{1+h})) \\ &\quad = \int_{-\infty}^{\infty}\int_{-\infty}^{\infty} \left[F_{1,h}(x-u, x-v) - F(x-u)F(x-v)\right] k_n(u)k_n(v)dudv, \end{aligned} \tag{4.11}$$

using **(4.9)** and the Taylor expansion up to first order terms, we have

$$\begin{aligned} \sigma_{n2}^2(x) = \frac{2}{n}\sum_{h=1}^{m}(n-h)\Bigg\{ & F_{1,h}(x,x) + \int_{-\infty}^{\infty}\int_{-\infty}^{\infty} \frac{\partial}{\partial x}F_{1,h}(\theta_{n1},\theta_{n2})(-u)k_n(u)k_n(v)dudv \\ & + \int_{-\infty}^{\infty}\int_{-\infty}^{\infty} \frac{\partial}{\partial y}F_{1,h}(\theta_{n1},\theta_{n2})(-v)k_n(u)k_n(v)dudv \\ & - F^2(x) + F(x)\int_{-\infty}^{\infty} vk_n(v)\left[f(\theta_{n1}) + f(\theta_{n2})\right] dv \\ & - \int_{-\infty}^{\infty}\int_{-\infty}^{\infty} uvf(\theta_{n1})f(\theta_{n2})k_n(u)k_n(v)dudv \Bigg\} \end{aligned} \tag{4.12}$$

where $|\theta_{n1} - u| < |x|$ and $|\theta_{n2} - v| < |x|$.

By **A1**(i) and **A1**(iii), there exist positive constants T, M_1 and M_2, such that

$$|f(x)| \leq T, \quad |\frac{\partial}{\partial x}F_{1,h}(x,y)| \leq M_1 \quad \text{and} \quad |\frac{\partial}{\partial y}F_{1,h}(x,y)| \leq M_2$$

for $h = 1, 2, ..., m$. Therefore, using **(4.8)**—**(4.10)** and **(4.12)**, since for sufficiently large n, $\gamma_n^2(x) \leq |\gamma_n(x)|$, we have

$$|\sigma_n^2(x) - F(x)(1 - F(x)) - \frac{2}{n}\sum_{h=1}^{m}(n-h)\left[F_{1,h}(x,x) - F^2(x)\right]| \tag{4.13}$$

$$\leq |\beta_n(x)| + 3|\gamma_n(x)| + \frac{2}{n}\sum_{h=1}^{m}(n-h)\Bigg\{(M_1 + M_2 + 2T)$$

$$\int_{-\infty}^{\infty} a_n|u|k(u)du + \left(a_n T\int|u|k(u)du\right)^2\Bigg\}.$$

This implies

$$\|\sigma_n^2 - F(1-F) - \frac{2}{n}\sum_{h=1}^{m}(n-h)\left[F_{1,h} - F^2\right]\| \leq c_2 a_n$$

where

$$c_2 = c_3 + 3c_1 + 2m\left[(M_1 + M_2 + 2T)\int_{-\infty}^{\infty}|u|k(u)du + T^2\left(\int_{-\infty}^{\infty}|u|k(u)du\right)^2\right].$$

The proof of Lemma 4.2 follows.

The proof of Lemma 4.3 is similar, and is therefore omitted.

We are now ready to prove our main theorems. We carry out in detail the proof of Theorem 2.2. The proof of Theorem 2.3 is similar except we shall use Taylor expansion on F and $F_{1,h}$ up to second order terms.

Denote

$$\Gamma_n(x) = P\left\{\hat{Z}_n/\nu_m \leq x\right\} = P\left\{B_n(x) \geq w_n(x)\right\} \tag{4.14}$$

where

$$B_n(x) = n^{-\frac{1}{2}}\sum_{i=1}^{n} Y_{ni}/\tau_n(x),$$

$$Y_{ni} = K_n(c_m x n^{-\frac{1}{2}} + \xi_p - X_i) - \mu_n(c_m x n^{-\frac{1}{2}} + \xi_p),$$

$$\mu_n(c_m x n^{-\frac{1}{2}} + \xi_p) = EK_n(c_m x n^{-\frac{1}{2}} + \xi_p - X_i),$$

$$\tau_n^2(x) = \sigma_n^2(c_m x n^{-\frac{1}{2}} + \xi_p) = Var\left\{n^{-\frac{1}{2}}\sum_{i=1}^{n} Y_{ni}\right\},$$

$$= \frac{1}{n}\left\{nEY_{n1}^2 + 2\sum_{h=1}^{m}(n-h)EY_{n1}Y_{n(1+h)}\right\},$$

$$w_n(x) = n^{\frac{1}{2}}(p - \mu_n(c_m x n^{-\frac{1}{2}} + \xi_p))/\tau_n(x) \text{ and } c_m = \frac{\nu_m}{f(\xi_p)}. \tag{4.15}$$

$$\begin{aligned}\Phi(x)-\Gamma_n(x) &= [P\{B_n(x)<w_n(x)\}-\Phi(w_n(x))]+[\Phi(x)-\Phi(-w_n(x))]\\ &= R_{n1}(x)+R_{n2}(x)\end{aligned}$$

To prove $\lim_{n\to\infty} R_{n2}(x)=0$, we have to show that $\lim_{n\to\infty} w_n(x)=-x$. This will follow if we prove

$$\lim_{n\to\infty}\tau_n^2(x)=\nu_m^2 \tag{4.16}$$

and

$$\lim_{n\to\infty} n^{\frac{1}{2}}[p-\mu_n(c_m x n^{-\frac{1}{2}}+\xi_p)]/\nu_m=-x. \tag{4.17}$$

We first prove **(4.16)**. To this end, we write

$$\begin{aligned}\tau_n^2(x)-\nu_m^2 = &\Bigg\{\sigma_n^2(c_m x n^{-\frac{1}{2}}+\xi_p)-F(c_m x n^{-\frac{1}{2}}+\xi_p)[1-F(c_m x n^{-\frac{1}{2}}+\xi_p)] \qquad (4.18)\\ &-\frac{2}{n}\sum_{h=1}^{m}(n-h)\left[F_{1,h}(c_m x n^{-\frac{1}{2}}+\xi_p,\, c_m x n^{-\frac{1}{2}}+\xi_p)-F^2(c_m x n^{-\frac{1}{2}}+\xi_p)\right]\Bigg\}\\ &+\Bigg\{F(c_m x n^{-\frac{1}{2}}+\xi_p)\left[1-F(c_m x n^{-\frac{1}{2}}+\xi_p)\right]\\ &+\frac{2}{n}\sum_{h=1}^{m}(n-h)\left[F_{1,h}(c_m x n^{-\frac{1}{2}}+\xi_p,\, c_m x n^{-\frac{1}{2}}+\xi_p)-F^2(c_m x n^{-\frac{1}{2}}+\xi_p)\right]\\ &-p(1-p)-2\sum_{h=1}^{m}\left[F_{1,h}(\xi_p,\xi_p)-p^2\right]\Bigg\}=H_{n1}+H_{n2},\quad \text{say.}\end{aligned}$$

By Lemma 4.2,

$$|H_{n1}|\le\|\sigma_n^2-F(1-F)-\frac{2}{n}\sum_{h=1}^{m}(n-h)\left[F_{1,h}-F^2\right]\|\le c_2 a_n\to 0 \quad \text{as} \quad n\to\infty. \tag{4.19}$$

We estimate H_{n2} by using a Taylor expansion:

$$\begin{aligned}
H_{n2} &= \left[p + c_m x n^{-\frac{1}{2}} f(\theta_n)\right]\left[(1-p) - c_m x n^{-\frac{1}{2}} f(\theta_n)\right] + \frac{2}{n}\sum_{h=1}^{m}(n-h)F_{1,h}(\xi_p,\xi_p) \\
&\quad + \frac{2}{n}\sum_{h=1}^{m}(n-h)c_m x n^{-\frac{1}{2}}\left[\frac{\partial}{\partial x}F_{1,h}(\theta_n,\theta_n) + \frac{\partial}{\partial y}F_{1,h}(\theta_n,\theta_n)\right] \\
&\quad - \frac{2}{n}\sum_{h=1}^{m}(n-h)p^2 - \frac{4}{n}\sum_{h=1}^{m}(n-h)p c_m x n^{-\frac{1}{2}} f(\theta_n) \\
&\quad - \frac{2}{n}\sum_{h=1}^{m}(n-h)c_m^2 x^2 n^{-1} f^2(\theta_n) - p(1-p) - 2\sum_{h=1}^{m}\left[F_{1,h}(\xi_p,\xi_p) - p^2\right] \\
&= \left\{(1-2p)c_m x n^{-\frac{1}{2}} f(\theta_n) - c_m^2 x^2 n^{-1} f^2(\theta_n)\right\} - \frac{2}{n}\sum_{h=1}^{m} h\, F_{1,h}(\xi_p,\xi_p) \\
&\quad + \frac{2}{n}\sum_{h=1}^{m}(n-h)c_m x n^{-\frac{1}{2}}\left[\frac{\partial}{\partial x}F_{1,h}(\theta_n,\theta_n) + \frac{\partial}{\partial y}F_{1,h}(\theta_n,\theta_n)\right] \\
&\quad + \frac{2}{n}\sum_{h=1}^{m} h p^2 - \frac{4}{n}\sum_{h=1}^{m}(n-h)p c_m x n^{-\frac{1}{2}} f(\theta_n) - \frac{2}{n}\sum_{h=1}^{m}(n-h)c_m^2 x^2 n^{-1} f^2(\theta_n) \\
&= \sum_{i=1}^{6} I_i
\end{aligned}$$

where $|\theta_n - \xi_p| < c_m |x| n^{-\frac{1}{2}}$.

We now estimate I_i $(i = 1, 2, ..., 6)$. Using **A1**(i), there exists a constant $T > 0$, such that

$$|I_1| \le (1-2p)\frac{\nu_m}{f(\xi_p)}|x| n^{-\frac{1}{2}} T + \frac{\nu_m^2 T^2}{f^2(\xi_p)} x^2 n^{-1} \longrightarrow 0 \quad \text{as} \quad n \to \infty.$$

It is clear $|I_2| = \frac{2}{n}\sum_{h=1}^{m} h\, F_{1,h}(\xi_p,\xi_p) \longrightarrow 0$ as $n \to \infty$.

Using **A1**(iii), there exist constants M_1 and $M_2 > 0$, such that

$$|I_3| \le \frac{2}{n}\sum_{h=1}^{m}(n-h)(M_1 + M_2)|x|\frac{\nu_m}{f(\xi_p)} n^{-\frac{1}{2}} \longrightarrow 0 \quad \text{as} \quad n \to \infty.$$

Next,

$$|I_4| = \frac{2}{n}\sum_{h=1}^{m} h\, p^2 \longrightarrow 0 \quad \text{as} \quad n \to \infty$$

and

$$|I_5| \le \frac{4}{n}\sum_{h=1}^{m}(n-h)p|x|\frac{\nu_m T}{f(\xi_p)} n^{-\frac{1}{2}} \longrightarrow 0 \quad \text{as} \quad n \to \infty$$

and finally,

$$|I_6| \le \frac{2}{n}\sum_{h=1}^{m}(n-h)\frac{\nu_m^2 T^2}{f^2(\xi_p)} x^2 n^{-1} \longrightarrow 0 \quad \text{as} \quad n \to \infty.$$

Thus

$$H_{n2} = \sum_{i=1}^{6} I_i \longrightarrow 0 \quad \text{as} \quad n \to \infty. \tag{4.20}$$

Consequently, from **(4.18)**—**(4.20)**, **(4.16)** follows.

We now prove **(4.17)**.

$$\begin{aligned} & n^{\frac{1}{2}}(p - \mu_n(c_m x n^{-\frac{1}{2}} + \xi_p)) \qquad (4.21)\\ &= n^{\frac{1}{2}}\left[F(\xi_p) - F(c_m x n^{-\frac{1}{2}} + \xi_p)\right] + n^{\frac{1}{2}}\left[F(c_m x n^{-\frac{1}{2}} + \xi_p) - \mu_n(c_m x n^{-\frac{1}{2}} + \xi_p)\right] \\ &= \Delta_{n1}(x) + \Delta_{n2}(x), \quad \text{say.} \end{aligned}$$

We have, using (2.7) and Lemma 4.2

$$|\Delta_{n2}(x)| \le n^{\frac{1}{2}}\|F - \mu_n\| \le c_1 a_n n^{\frac{1}{2}} \longrightarrow 0 \quad \text{as} \quad n \to \infty. \tag{4.22}$$

Also, by a Taylor expansion via **A1**(ii)

$$\Delta_{n1}(x) = n^{\frac{1}{2}}(-x)c_m n^{-\frac{1}{2}} f(\theta_n) = -x\frac{\nu_m}{f(\xi_p)} f(\theta_n) \longrightarrow -x\nu_m \quad \text{as} \quad n \to \infty. \tag{4.23}$$

where $|\theta_n - \xi_p| < c_m|x|n^{-\frac{1}{2}}$. **(4.23)** and **(4.16)** imply

$$\lim_{n\to\infty} \frac{\Delta_{n1}(x)}{\tau_n(x)} = \lim_{n\to\infty} \frac{\Delta_{n1}(x)}{\nu_m(x)} \cdot \frac{\nu_m}{\tau_n(x)} = -x. \tag{4.24}$$

From **(4.21)**, **(4.22)** and **(4.24)** we obtain **(4.17)**.

Finally, using **(4.16)** and **(4.17)**, we obtain

$$\lim_{n\to\infty} w_n(x) = \lim_{n\to\infty} \frac{n^{\frac{1}{2}}(p - \mu_n(c_m x n^{-\frac{1}{2}} + \xi_p))}{\nu_m} \cdot \frac{\nu_m}{\tau_n(x)} = -x.$$

Therefore,

$$\lim_{n\to\infty} R_{n2}(x) = \lim_{n\to\infty} [\Phi(x) - \Phi(-w_n(x))] = 0.$$

Next, we prove $\lim_{n\to\infty} R_{n1}(x) = 0$. To this end, we shall use Lemma 4.1.

Recall that $\{Y_{ni},\ 1 \le i \le n,\ n \ge 1\}$ is a sequence of row-wise m–dependent stationary random variables with $EY_{n1} = 0$ and $E|Y_{n1}|^3 \le 1$. From **(4.16)**,

$$\lim_{n\to\infty} \tau_n^2(x) = \nu_m^2 < \infty.$$

Now, using Lemma 4.1, we obtain

$$|R_{n1}(x)| \le \sup_t |P\{B_n \le t\} - \Phi(t)| \le T_1 \frac{b_m^2}{\nu_m^3 n^2} + T_2 \frac{m b_m \log n}{\nu_m^2 n^2}$$

where

$$b_m = \max_{1\leq i\leq m+1} E^{\frac{1}{3}} \left| \sum_{j=1}^{i} Y_{nj} \right|^3 < \infty.$$

Hence

$$\lim_{n\to\infty} R_{n1}(x) = 0. \qquad \textbf{(4.25)}$$

Therefore, $\lim_{n\to\infty}[\Phi(x) - \Gamma_n(x)] = \lim_{n\to\infty}[R_{n1}(x) + R_{n2}(x)] = 0$. The proof of the sufficiency part is completed.

Now to prove that the condition **(2.7)** is necessary, we proceed as follows:

We assume that $\hat{Z}_n \xrightarrow{D} \mathcal{N}(0, \nu_m^2)$ as $n \to \infty$. Then from **(4.15)** and **(4.25)**, we have

$$\lim_{n\to\infty} R_{n2}(x) = \lim_{n\to\infty} [\Phi(x) - \Phi(-w_n(x))] = 0. \qquad \textbf{(4.26)}$$

That is,

$$\lim_{n\to\infty} w_n(x) = -x. \qquad \textbf{(4.27)}$$

Now using **(4.21)**, it follows that

$$w_n(x)\tau_n(x) = n^{\frac{1}{2}} \left(p - \mu_n(xc_m n^{-\frac{1}{2}} + \xi_p)\right) = \Delta_{n1}(x) + \Delta_{n2}(x)$$

where $\Delta_{n1}(x)$ and $\Delta_{n2}(x)$ are given by **(4.21)**. Taking limit on both sides of the above equation via **(4.16)**, **(4.23)** and **(4.27)**, we obtain

$$-x\nu_m = -x\nu_m + \lim_{n\to\infty} \Delta_{n2}(x).$$

Thus

$$\lim_{n\to\infty} \Delta_{n2}(x) = 0.$$

But

$$\begin{aligned}
\Delta_{n2}(x) &= n^{\frac{1}{2}} \left[F(c_m x n^{-\frac{1}{2}} + \xi_p) - \mu_n(c_m x n^{-\frac{1}{2}} + \xi_p)\right] \\
&= n^{\frac{1}{2}} \int_{-\infty}^{\infty} \left[F(c_m x n^{-\frac{1}{2}} + \xi_p) - F(c_m x n^{-\frac{1}{2}} + \xi_p - s a_n)\right] k(s)ds \\
&= n^{\frac{1}{2}} a_n \int_{-\infty}^{\infty} f(\theta_n) s k(s) ds
\end{aligned}$$

where $|\theta_n - c_m x n^{-\frac{1}{2}} - \xi_p| < |s| a_n$.

By assumptions **A1**(i), **A1**(ii) and **B1** and applying Lebesgue's Dominated Convergence Theorem, we deduce

$$0 = \lim_{n\to\infty} \Delta_{n2}(x) = f(\xi_p) \int_{-\infty}^{\infty} s k(s) ds \lim_{n\to\infty} n^{\frac{1}{2}} a_n.$$

Hence

$$\lim_{n\to\infty} a_n n^{\frac{1}{2}} = 0.$$

The proof follows.

Acknowledgement.
The authors would like to thank the referee for pointing out typographical errors in the first draft of the manuscript. Research of Madan L. Puri supported by the Office of Naval Research Contract N00014-91-J-1020.

REFERENCES:

1. NADARAYA, E.A. Some new estimates for distribution functions. *Theor. Probability Appl.* (1964) **9**, 497-500.

2. CHANDA, K. Some comments on the asymptotic probabilities laws of sample quantiles. *Calcutta Stat. Assoc. Bulletin,* (1975) 123-126.

3. FALK, M. Asymptotic normality of the kernel quantile estimator. *Ann. Statist.* (1985) **13**, 428-433.

4. MACK, Y.P. Bahadur representation of sample quantiles based on smooth estimates of a distribution function. *Probab. and Math. Statist.* (1987) **8**, 183-189.

5. YAMATO, H. Uniform convergence of an estimator of a distribution function. *Bull. Math. Statist.* (1973) **15**, 69-78.

6. RALESCU, S. and SUN, S. Necessary and sufficient conditions for the asymptotic normality of perturbed sample quantile. *Jour. Stat. Plann. Inf.* (1992) (To appear).

7. SEN, P.K. Asymptotic normality of sample quantiles for m–dependent processes. *Ann. Math. Statist.* (1968) **39**, 1724-1730.

8. HOEFFDING, W. and ROBBINS, H. The central limit theorem for dependent random variables. *Duke Math. J.* (1948) **15**, 773-780.

9. HOEFFDING, W. Probability inequalities for sums of bounded random variables. *J. Amer. Statist. Assoc.* (1963). **58**, 13-30.

10. TIKHOMIROV, A.N. On the convergence rate in the central limit theorem for weakly dependent random variables. *Th. Prob. Applic.* (1980) **4**, 790-809.

Stat. Sci. & Data Anal., pp. 427-439
K. Matsusita *et al.* (Eds)

The L_1 Complete Convergence of Recursive Kernel Density Estimators Under Weak Dependence

LANH TAT TRAN
Indiana University
Bloomington, Indiana, U.S.A.

Abstract.
Let X_t, t=..., -1, 0, 1, ... be a strictly stationary sequence of random variables (r.v.'s) defined on a probability space (Ω, $\mathcal{F}$, P) and taking values in $\mathcal{R}^d$. Let $X_1, ..., X_n$ be n consecutive observations of X_t. Suppose that X_1 has a bounded density f. As an estimator of $f(x)$ we shall consider $\hat{f}_n(x) = n^{-1}\sum_{j=1}^n h_j^{-d}K((x - X_j)/h_j)$. Here K is a kernel function and h_n is a sequence of bandwidths tending to zero as $n \to \infty$. Under general conditions on the kernel functions and bandwidths, $\hat{f}_n$ is shown to converge completely to f . The process X_t is assumed to satisfy some weak dependence conditions. The results are applicable to a large class of time series models.

Key words: L_1 convergence, strong mixing, absolute regularity, density estimation, kernel, bandwidth.

1. INTRODUCTION.

Kernel-type density estimation for independent random variables (r.v.'s) has been studied extensively since the early works of Rosenblatt [1] and Parzen [2]. The purpose of this paper is to investigate the L_1 convergence of recursive kernel-type density estimators when the observations are dependent. The reader is referred to the book of Devroye and Györfi [3] for an in-depth treatment of the L_1 theory. Let X_t, t = ..., -1, 0, 1, ...be a strictly stationary sequence of random variables (r.v.'s) defined on a probability space (Ω, $\mathcal{F}$, P) and taking values in R^d . Let $X_1, ..., X_n$ be n consecutive observations of X_t. Assume that X_1 has a bounded density f . As an estimator of $f(x)$ we shall consider

$$\hat{f}_n(x) = n^{-1} \sum_{j=1}^{n} h_j^{-d} K((x - X_j)/h_j). \tag{1.1}$$

Here K is a kernel function and h_n is a sequence of bandwidths tending to zero as $n \to \infty$. Note that $\hat{f}_n(x)$ can be computed recursively by

$$\hat{f}_n(x) = \frac{n-1}{n} \hat{f}_{n-1}(x) + n h_n^{-d} K((x - X_n)/h_n) \ .$$

The estimator $\hat{f}_n$ does not achieve the smallest asymptotic variance within the class of recursive estimators. Deheuvels [4] introduced the following estimator

$$f_n^*(x) = (\sum_{i=1}^{n} h_i^d)^{-1} \sum_{j=1}^{n} K((x - X_j/h_j)) \ .$$

It can be shown that f_n^* has smaller asymptotic variance than $\hat{f}_n$ in the independent case. However, asymptotically $\hat{f}_n$ has smaller mean square error than f_n^* (see Wertz [5]).

The convergence of the L_1 distance considered here deals with the overall behavior of $\hat{f}_n$. We might want to compute $\hat{f}_n$ at an increasing number of x's as $n \to \infty$. This is only possible if the data is stored. Recursive estimates generally do not perform as well as their nonrecursive counterparts in this case. In fact, improvement in terms of the expected L_1 error can be almost always obtained by considering their nonrecursive counterparts (see Devroye and Györfi ([3], p. 283). However, recursive estimates are still often preferred since they can easily be updated with each additional observation.

The estimate $\hat{f}_n$ was introduced by Wolverton and Wagner [6,7], and Yamato [8]. In the independent case, $\hat{f}_n$ was studied by Davies [9], Deheuvels [4, 10, 11], Carroll [12], Ahmad and Lin [13], Devroye [14], Wegman and Davies [15]. The integral convergence of recursive estimators was obtained by Devroye [14] under weak assumptions.

In the dependent case, quadratic mean convergence and asymptotic normality of $\hat{f}_n$ were obtained by Masry [16] under various assumptions on the dependence of X_t. Strong pointwise consistency of $\hat{f}_n$ was proved in Györfi [17]. Takahata [18] and Masry and Györfi [19] obtained a.s. rates of $\hat{f}_n$ to f for the class of asymptotically uncorrelated processes, the definition of which can be found in Masry and Györfi [19]. Masry [20] established rates of almost sure convergence of $\hat{f}_n$ to f for vector valued stationary strong mixing processes under weak assumptions on the strong mixing condition. Recently, the L_1 and L_2 convergence of $\hat{f}_n$ have been studied by Györfi and Masry [21]. Throughout, X_t is assumed to satisfy either of the two dependence conditions defined below:

Definition 1.1. Let $\mathcal{F}_{-\infty}^0$ and $\mathcal{F}_n^\infty$ denote respectively the σ-fields generated by $X(t)$, $t \leq 0$ and by X_t, $t \geq n$. Then X_t is strong mixing if

$$\alpha(n) = \sup\{|P(A \cap B) - P(A)P(B)| : A \epsilon \mathcal{F}_{-\infty}^0 , B \epsilon \mathcal{F}_n^\infty\} \downarrow 0. \tag{1.2}$$

Definition 1.2. Let $\mathcal{F}_{-\infty}^0$ and $\mathcal{F}_n^\infty$ be the σ fields defined in Definition 1.1. Then $X(t)$ is absolutely regular if

$$\beta(n) = E\{\sup |P(A|M_{-\infty}^0) - P(A)| : A \epsilon \mathcal{F}_n^\infty\} \downarrow 0 \tag{1.3}$$

as $n \to \infty$.

For relevant literature on absolutely regular processes, the reader is referred to Yoshihara [22, 23]. Pham and Tran [24] and Pham [25] have shown that a large class of time series is absolutely regular. In particular, autoregressive moving average time series models and bilinear time series models are absolutely regular with $\beta(n)$ decaying to zero exponentially fast under weak conditions.

The strong mixing condition is weaker than the absolute regularity condition; however, some results for absolutely regular processes do not hold just under strong mixing. See for example, Berbee [26]. For more information on strong mixing processes, see Rosenblatt [1], or Roussas [27]. The letter C will be used to denote constants whose values are unimportant and may vary.

Recently, density estimation for weakly dependent processes, in particular strong mixing processes, has generated a considerable amount of interest. See for example, Robinson [28, 29], Ioannides and Roussas [30], Györfi, Hardle, Sarda and Vieu [31], Hart and Vieu [32], Masry [33, 16, 20], Masry and Györfi [19] and Roussas [27]. The reason is partly due to the fact that many stochastic processes, among them various useful time series models, satisfy the strong mixing property; and the strong mixing property is relatively easy to check. Gorodetskii [34], Withers [35], and Athreya and Pantula [36] have obtained conditions for linear processes and autoregressive processes to be strong mixing.

Let J_n be the L_1 distance defined by $J_n = \int |\hat{f}_n(x) - f(x)|dx$ where integration is over the entire space involved unless otherwise indicated. Our main purpose is to find conditions for J_n to converge completely to zero. In the independent case, $|\hat{f}_n(x) - f(x)|$ converges completely, almost all x, if $nh_n^d(\log n)^{-1} \to \infty$ and if K is a bounded density with integrable radial majorant (see Devroye and Györfi ([3], p.199)). In the independent case, Devroye and Györfi ([3], p.1) have pointed out that L_1 is a natural space for densities. This seems to be also the case under the strong mixing and absolute regularity conditions considered here. The L_1 distance converges to zero under simple conditions. Under various dependence conditions, Györfi and Masry [21] have shown that $\int |\hat{f}_n(x) - E\hat{f}_n(x)|dx$ converges a.s. to zero. The conditions imposed on the bandwidths in their paper are quite general; however, the kernel function there is assumed to be in the L_2 space . Note that in the present paper, K is assumed to be in the L_1 space and also that the complete convergence of J_n is obtained. Recently Tran [37] has studied the L_1 convergence of nonrecursive kernel type estimators under the strong mixing condition and the absolute regularity condition. The problem investigated here is much more involved since $\hat{f}_n(x)$ cannot be written as the sum of identically distributed r.v.'s as compared with the nonrecursive case. In Tran [37], simple approximations of absolutely regular r.v.'s and strong mixing r.v.'s by Bernoulli r.v.'s were employed. Such approximations do not work in the recursive case considered here. Note that the r.v.'s Δ_i's defined in Lemma 2.2 are not identically distributed. This is a difficulty which often arises in the nonrecursive setting. See, for example, Masry ([16], p. 262). Our method is based on approximations of weakly dependent r.v.'s by independent ones and is similar to those used by Yoshihara [22, 23] and Tran [37]. The paper is organized as follows: in Section 2, we provide some preliminaries and investigate recursive density estimators under the strong mixing condition. Weak conditions are obtained for the L_1 complete convergence of $\hat{f}_n(x)$ to $f(x)$. The absolute regularity condition is considered in Section 3.

2. COMPLETE CONVERGENCE OF $\hat{f}_n$ UNDER STRONG MIXING.

The following Lemma is from Bradley [38].

Lemma 2.1. Suppose X and Y are r.v.'s taking values on $\mathcal{S}$ and R , respectively, where $\mathcal{S}$ is a Borel space, and let U be a uniform-[0,1] r.v. independent of (X, Y); furthemore, suppose ξ and γ are positive numbers such that $\xi \leq \|Y\|_\gamma < \infty$. Then there exists a real-valued r.v. $Y^* = f(X, Y, U)$, where f is a measurable function defined on $\mathcal{S} \times R \times [0, 1]$, such that

i) Y^* is independent of X ,

ii) the probability distributions of Y and Y^* are identical, and

iii) $P(|Y^* - Y| \geq \xi) \leq 18(\|Y\|_\gamma/\xi)^{\gamma/(2\gamma+1)}[\alpha(\mathcal{B}(X), \mathcal{B}(Y))]^{2\gamma/2\gamma+1}$,
where $\mathcal{B}(X)$ and $\mathcal{B}(Y)$ are the σ-fields induced by X and Y, respectively, and $\|Y\|_\gamma = (E|Y|^\gamma)^{1/\gamma}$.

Consider the partition of R^d into sets B that are d-fold products of intervals of the form $[(i-1)h_n/N, ih_n/N)$, where i is an integer, and N is a positive number. Call the partition Ψ . Let A be a set in R^d and Z a r.v. on $(\Omega, \mathcal{F}, P)$ and taking values in R^d . Denote $v(A, Z) = I(A, Z) - E[I(A, Z)]$. Here $I(A, Z)$ is defined by $I(A, Z) = 1$ if Z takes values in A and $I(A, Z) = 0$ otherwise. Let μ be the probability measure associated with f.

Lemma 2.2. Suppose X_t satisfies the strong mixing condition with $\sum_{n=1}^{\infty}(\alpha(n))^{\delta/(2+\delta)} < \infty$ for some $\delta > 0$, and $h_n \downarrow 0$ slowly enough as $n \to \infty$ that,

$$nh_n^{2d(1+\delta)/(2+\delta)}(\log n)^{-1} \to \infty \; ; \qquad \textbf{(2.1)}$$

Let $0 < \tau < 1/2$ and let η be an arbitrary positive number. Denote

$$q(n,1) \equiv \{\alpha([nh_n^d/\eta \log n])\}^{2\tau}(\log n)h_n^{-d(1+\tau)} .$$

Let $\epsilon > 0$. Then for sufficiently large n,

$$P[|n^{-1}\sum_{i=1}^{n} h_i^{-d}v(B,X_i)| \geq \epsilon] \leq n^{-(1/16)\eta\epsilon} + C\epsilon^{-\tau}q(n,1), \tag{2.2}$$

where C is a positive constant.

Proof. Define $\Delta_i = n^{-1}h_i^{-d}v(B,X_i)$. Let $p = p(n) = [nh_n^d/(\eta \log n)]$. Without loss of generality assume that n is a multiple of 2p . Then $n = 2pq$ for some positive integer $q = q(n)$. The random variables $\Delta_i's$ can be successively grouped into 2q blocks of size p. Let $S_n = S_{n1} + S_{n2}$, where

$$S_{n1} = \sum_{j=1}^{q} V(n,2(j-1)),\ S_{n2} = \sum_{j=1}^{q} V(n,2j-1) \tag{2.3}$$

with

$$V(n,j) = \sum_{i=(j-1)p+1}^{jp} \Delta_i .$$

By assumption f is bounded and $h_n \downarrow 0$ as $n \to \infty$. Thus for $k \leq n$,

$$E\Delta_k^2 \leq n^{-2}h_k^{-2d}\mu(B) \leq n^{-2}h_n^{-2d}\mu(B) \leq Cn^{-2}h_n^{-d} . \tag{2.4}$$

By the Davydov inequality (see Deo [39]), we have for $k < \ell$,

$$|Cov\{\Delta_k,\Delta_\ell\}| \leq Cn^{-2}h_n^{-2d}h_n^{2d/(2+\delta)}\{\alpha(\ell-k)\}^{\delta/(2+\delta)}. \tag{2.5}$$

Denote $V(n,2(j-1))$ by V_j for simplicity. Observe that for all $1 \leq j \leq q$,

$$|V_j| \leq p(nh_n^d)^{-1} , \tag{2.6}$$

since $h_n \downarrow 0$ as n tends to infinity . Let $\gamma = \tau/(1-2\tau)$. By Lemma 2.1, there exists a sequence of independent r.v.'s W_i , $1 \leq i \leq q$ such that W_i has the same distribution as V_i and satisfies

$$P[|W_j - V_j| > \beta] \leq C\beta^{-\tau}\{\alpha(p)\}^{2\tau}p^\tau(nh_n^d)^{-\tau} , \tag{2.7}$$

where β is a positive number such that $\beta \leq \|V_j\|_\gamma < \infty$. Clearly,

$$P[|S_{n1}| > \epsilon/2] \leq P[\sum_{i=1}^{q} W_i > \epsilon/4] + P[-\sum_{i=1}^{q} W_i > \epsilon/4] + P[\sum_{i=1}^{q} |V_i - W_i| > \epsilon/4] . \tag{2.8}$$

Let $\lambda = \eta \log n/2$. Then $|\lambda W_i| \leq 1/2$. By Markov's inequality and the independence of the $W_i's$,

$$P[\sum_{i=1}^{q} W_i > \epsilon/4] \leq \exp[-(1/8)\eta\epsilon \log n]\exp[\lambda^2 \sum_{i=1}^{q} EW_i^2] . \tag{2.9}$$

By (2.4), (2.5) and the assumption that $\sum_{j=1}^{\infty}\{\alpha(j)\}^{\delta/(2+\delta)} < \infty$,

$$\sum_{i=1}^{q} EW_i^2 \le Cn^{-1}h_n^{-2d(1+\delta)/(2+\delta)} . \tag{2.10}$$

A simple computation using (2.1) and (2.10) shows that $\lambda^2 \sum_{i=1}^{q} EW_i^2 = o(\log n)$. Thus for sufficiently large n,

$$P[\sum_{i=1}^{q} W_i > \epsilon/4] \le n^{-(1/16)\eta\epsilon} . \tag{2.11}$$

The second term on the right hand side of (2.8) can be handled similarly. Clearly $q \le C\eta \log n/h_n^d$. From (2.7), for $\epsilon/(4q) \le \|V_i\|_\gamma$,

$$P[\sum_{i=1}^{q} |V_i - W_i| > \epsilon/4] \le \sum_{i=1}^{q} P[|V_i - W_i| > \epsilon/(4q)], \tag{2.12}$$

which is bounded $C\epsilon^{-\tau}q(n,1)$. If $\epsilon/(4q) > \|V_i\|_\gamma$. Then by Lemma 2.1,

$$P[\sum_{i=1}^{q} |V_i - W_i| > \epsilon/4] \le \sum_{i=1}^{q} P[|V_i - W_i| > \|V_i\|_\gamma], \tag{2.13}$$

which is again bounded by $C\epsilon^{-\tau}q(n,1)$. By (2.8), (2.11), (2.12) and (2.13), $P[|S_{n1}| > \epsilon/2]$ and similarly $P[|S_{n2}| > \epsilon/2]$ are bounded by the right hand side of (2.2). □

Lemma 2.3. Let $\epsilon > 0$, $\eta > 0$, $0 < \tau < 1/2$ and $E \subset R^d$ be a Borel set. Then for sufficiently large n

$$P[n^{-1}\sum_{i=1}^{n} h_i^{-d}v(E,X_i) > \epsilon h_n^{-d}] \le n^{-(1/16)\eta\epsilon} + C(\alpha([n/(\eta\log n)]))^{2\tau}\log n, \tag{2.14}$$

where C is a positive constant.

Proof. Define $\Delta_i = n^{-1}h_i^{-d}v(E,X_i)$ and let $p = [n/(\eta \log n)]$. The random variables Δ_i's can be successively grouped into 2q blocks of size p . Let S_n , S_{n1} , S_{n2} , V_j be as defined in Lemma 2.2. Clearly, for $1 \le k \le n$,

$$Var[\Delta_k] \le n^{-2}h_n^{-2d} ;$$

and for $1 \le k < \ell \le n$,

$$|Cov\{\Delta_k, \Delta_\ell\}| \le Cn^{-2}h_n^{-2d}\{\alpha(k-\ell)\}^{\delta/(2+\delta)} ,$$

Approximate the r.v.'s $V_1, ..., V_q$ by independent r.v.'s $W_1, ..., W_q$ as done in Lemma 2.2. Choose $\lambda = \eta(\log n)h_n^d/2$. Employing Markov inequality, the independence of the W_i's, the proof can be easily completed by an argument similar to that of Lemma 2.2. □

Assumption 1. The kernel function K is absolutely integrable with $\int K(z)dz = 1$ and $\int \|z\||K(z)|dz < \infty$.

Lemma 2.4. Define $\hat{g}_n(x) = E\hat{f}_n(x)$. Assume that Assumption 1 holds and $h_n \to 0$ as $n \to \infty$. Then

$$\int |\hat{g}_n(x) - f(x)|dx \to 0 \text{ as } n \to \infty . \tag{2.15}$$

Proof. Since $h_n \to 0$ as $n \to \infty$, we have $\hat{g}_n(x) \to f(x)$ at almost all x (see Devroye and Györfi [3], p.199). The lemma then follows from Scheffé Theorem (see Devroye and Györfi [3], p.10). □

Theorem 2.1. i) Suppose X_t satisfies the strong mixing condition as in Lemma 2.2 and that Assumption 1 holds. Suppose (2.1) holds and in addition

$$q(n,1)h_n^{-d} \to 0 \tag{2.16}$$

for some $0 < \tau < 1/2$ and any $\eta > 0$. Then $J_n \to 0$ in probability.

ii) If (2.16) is strengthened to $\sum_{n=1}^{\infty} q(n,1)h_n^{-d} < \infty$, then J_n converges completely to zero.

Proof. i) By (2.15) of Lemma 2.4, it is sufficient to show that $\int |\hat{f}_n(x) - \hat{g}_n(x)|dx \to 0$ in probability. By a slight variation of the argument of Devroye and Györfi ([3], p.15), it is then sufficient to show that for all rectangles A of R^d

$$\int |n^{-1} \sum_{i=1}^{n} h_i^{-d} v\{A, ((x - X_i)/h_i))\}|dx \to 0 \tag{2.17}$$

as $n \to \infty$. Let $a_1, ..., a_d$ be positive numbers. Denote

$$A = \prod_{i=1}^{d} [x_i, x_i + a_i), \min_{1 \leq i \leq d} a_i \geq 2/N; \; A^* = \prod_{i=1}^{d} [x_i + (1/N), x_i + a_i + (1/N)) \tag{2.18}$$

For $1 \leq i \leq n$, define

$$C_{x,i} = x + h_i A - \bigcup_{\substack{B \in \Psi \\ B \subseteq \bigcup_{1 \leq j \leq n} \{x + h_j A\}}} , \qquad C^*_{x,i} = x + h_i(A - A^*). \tag{2.19}$$

Using elementary set theory and the assumption that $h_i \geq h_n$ for $1 \leq i \leq n$, it is easily seen that

$$C_{x,i} \subseteq C^*_{x,i} \text{ for } 1 \leq i \leq n . \tag{2.20}$$

Denote

$$Q(n,1) = \int x + hA - \sum_{\substack{B \in \Psi \\ B \subseteq \bigcup_{1 \leq j \leq n} \{x + h_j A\}}} |n^{-1} \sum_{i=1}^{n} h_i^{-d} v\{B, X_i\}|dx; \tag{2.21}$$

$$Q(n,2) = n^{-1} \sum_{i=1}^{n} h_i^{-d} \int I(C^*_{x,i}, X_i)dx \; ; \; Q(n,3) = n^{-1} \sum_{i=1}^{n} h_i^{-d} \int\int I(C^*_{x,i}, y)f(y)dydx.$$

By (2.20) and (2.21) .

$$\int |n^{-1}\sum_{i=1}^{n} h_i^{-d} v\{(A,((x-X_i)/h_i))\}|dx \leq Q(n,1)+Q(n,2)+Q(n,3). \qquad \mathbf{(2.22)}$$

Let λ denote d-dimensional Lebesgue measure. For $1 \leq i \leq n$, $y \epsilon R^d$,

$$\int I(C^*_{x,i}, y)dx = \int I(C^*_{x,i}, X_i)dx = h_i^d \lambda(A - A^*) . \qquad \mathbf{(2.23)}$$

Let $\epsilon > 0$ be an arbitrarily small positive number. Arguing as in Devroye and Györfi [3], by choosing N sufficiently large, we have for $1 \leq i \leq n$,

$$\lambda(A - A^*) \leq \epsilon . \qquad \mathbf{(2.24)}$$

By (2.23) and (2.24), both $Q(n,2)$ and $Q(n,3)$ are bounded by ϵ .

Let $a > 0$ and

$$S = \prod_{i=1}^{d}[-a,a], \text{and } T = \prod_{i=1}^{d}[-2a,2a].$$

Choose a large enough so that $\mu(S^c) < \epsilon/2$, where S^c is the complement of S . Let $\emptyset$ denote the empty set. Define

$$F = \{B : B\epsilon\Psi, B\cap S \neq \emptyset,\ \mu(B) < (\epsilon/2)\lambda(B)/\lambda(T)\}, E = S^c \cup (\cup_{B\in F} B). \qquad \mathbf{(2.25)}$$

Note that $\cup_{B\in F} B \subseteq T$ for $h_n < Na$. Thus $\mu(\cup_{B\in F} B) < \epsilon/2$ and $\mu(E) < \epsilon$. Define

$$G = \{B\epsilon\Psi, B\cap S \neq \emptyset, B \notin F\}. \qquad \mathbf{(2.26)}$$

Clearly,

$$G \sqcup F = \{B\epsilon\Psi, B\cap S \neq \emptyset\} \text{ and } \Psi - G \subseteq E . \qquad \mathbf{(2.27)}$$

Denote

$$\gamma_n = \int_{B\subseteq x+h_nA} dx\ ,\ Q(n,4) = \sum_{B\epsilon G} |n^{-1}\sum_{i=1}^{n} h_i^{-d} v(B,X_i)|\gamma_n; \qquad \mathbf{(2.28)}$$

$$Q(n,5) = n^{-1}\sum_{i=1}^{n} h_i^{-d} v(E,X_i)\gamma_n\ ;\ Q(n,6) = 2n^{-1}\sum_{i=1}^{n} h_i^{-d}\int I(E,y)f(y)dy\ \gamma_n .$$

Using (2.27), we then have

$$Q(n,1) \leq Q(n,4) + \sum_{B\epsilon\Psi - G} |n^{-1}\sum_{i=1}^{n} h_i^{-d} v(B,X_i)|\gamma_n \qquad \mathbf{(2.29)}$$

$$\leq Q(n,4) + Q(n,5) + Q(n,6) .$$

To complete the proof of i), it is sufficient to show that the last three terms of (2.29) converge to zero in probability. Clearly

$$\int_{B\subseteq x+h_nA} dx \leq Ch_n^d \text{ and } \sum_{B\epsilon G}\mu(B) \leq 1 . \qquad \mathbf{(2.30)}$$

The number of sets in G is bounded by Ch_n^{-d} . Since f is bounded, $\mu(B) \leq Ch_n^d$. Assume $C = 1$ without loss of generality. By (2.30),

$$P[Q(n,4) > \epsilon] \leq P\left(\sum_{B\epsilon G} |n^{-1}\sum_{i=1}^{n} h_i^{-d} v(B, X_i)| \geq \epsilon h_n^{-d}\right) \tag{2.31}$$

$$\leq Ch_n^{-d} \sup_{B\epsilon G} P[|n^{-1}\sum_{i=1}^{n} h_i^{-d} v(B, X_i)| \geq \epsilon\mu(B) h_n^{-d}]$$

which is bounded by $Cq(n,1)h_n^{-d}$ by Lemma 2.2 and by noting $\mu(B) > Ch_n^{-d}$ for $B\epsilon G$. By (2.16), we have $Q(n,4) \to 0$ in probability.

Next, we show that $Q(n,5) \to 0$ in probability. Since h_n tends to zero as n tends to infinity,

$$nh_n^d/(\eta \log n) < n/(\eta \log n) \tag{2.32}$$

for sufficiently large n . By (2.32), for any $\eta > 0$,

$$\alpha([nh_n^d/(\eta \log n)]) \geq \alpha([n/(\eta \log n)]) \tag{2.33}$$

if n is sufficiently large . Using (2.16), (2.33) and Lemma 2.3, it follows that $Q(n,5) \to 0$ as $n \to \infty$.

Turning now to $Q(n,6)$, we have

$$2n^{-1}\sum_{i=1}^{n} h_i^{-d}[\int I(E,y)f(y)dy]\gamma_n \leq 2\mu(E) \leq 2\epsilon,$$

since h_n is nonincreasing in n . The proof of i) is completed since ϵ can be arbitrarily small.
ii) Part ii) follows easily from the proof of Part i). □

Remark 2.1. If X_t is a sequence of *i.i.d.* random variables, then $q(n,1) = 0$. It is not hard to see that (2.1) becomes $nh_n^d(\log n)^{-1} \to \infty$, which is the same condition obtained by Devroye and Györfi ([3], p. 199) for the pointwise complete convergence of $|\hat{f}_n - f|$. In view of this, it appears that Condition (2.1) is hard to improve upon.

Corollary 2.1. i) Suppose X_t satisfies the strong mixing condition with $\alpha(n) = O(n^{-\nu})$ for some $\nu > 1$. Suppose Assumption 1 holds and $h_n \downarrow 0$ slowly enough as $n \to \infty$ that

$$q(n,2) \equiv n^{2\tau\nu}(\log n)^{-(2\nu\tau+1)} h_n^{d(2+\tau+2\tau\nu)} \to \infty \tag{2.34}$$

for some $0 < \tau < 1/2$. Then $J_n \to 0$ in probability.
ii) Suppose (2.34) is strengthened to

$$q(n,2)/\ell(n) \to \infty , \tag{2.35}$$

where $\ell(n) \equiv n \log n(\log\log n)^{1+\epsilon}$, with ϵ being an arbitrarily small positive number. Then J_n converges completely to zero.

Proof. i) By (2.34), $1/q(n,2) \to 0$. Let η be an arbitrary positive number. A simple computation shows that

$$(nh_n^d/\eta \log n)^{-2\nu\tau}(\log n)h_n^{-d(2+\tau)} \to 0 ,$$

which implies that (2.16) holds.

Next, we show that (2.1) is satisfied. Let $\delta > 2/(\nu-1)$. Then $\sum_{k=1}^{\infty}\{\alpha(k)\}^{\delta/(2+\delta)} < \infty$. Thus it is sufficient to show that (2.1) holds for some $\delta > 2/(\nu-1)$. Let

$$r(n) \equiv nh_n^{d(\nu+1)/\nu}(\log n)^{-1} ,$$

$$s(n) \equiv h_n^{\{2d(1+\delta)/(2+\delta)\}-\{d(\nu+1)/\nu\}} .$$

Note that

$$nh_n^{2d(1+\delta)/(2+\delta)}(\log n)^{-1} = r(n) \times s(n) . \tag{2.36}$$

An easy computation shows that $2d(1+\delta)/(2+\delta) = d(\nu+1)/\nu$ if $\delta = 2/(\nu-1)$. Now, using the fact that $(2+\tau+2\tau\nu)/(2\tau\nu) \geq ((5/2)+\nu)/\nu$ for $0<\tau<1/2$, and h_n tends to zero as $n\to\infty$, we have, for sufficiently large n,

$$\begin{aligned}
&[r(n)\times s(n)]/[q(n,2)]^{1/[2\tau\nu]} \qquad (2.37)\\
&\quad\geq s(n)\times(\log n)^{-1+\{(2\nu\tau+1)/(2\tau\nu)\}}h_n^{\{-d(2+\tau+2\tau\nu)/(2\tau\nu)\}+\{d(\nu+1)/\nu\}}\\
&\quad\geq h_n^{\{2d(1+\delta)/(2+\delta)\}-\{d(\nu+1)/\nu\}}(\log n)^{1/(2\tau\nu)}h_n^{-3d/(2\nu)} \geq (\log n)^{1/(2\tau\nu)},
\end{aligned}$$

by choosing δ close enough to $2/(\nu-1)$ such that

$$\{2d(1+\delta)/(2+\delta)\} - \{d(\nu+1)/\nu\} \leq 3d/(2\nu) .$$

Thus (2.36) and (2.37) imply that (2.1) holds.

ii) By (2.35), $\sum_{n=1}^{\infty} q(n,2) < \infty$ since $\sum_{n=1}^{\infty}\{\ell(n)\}^{-1} < \infty$.

The proof of ii) follows immediately from Theorem 2.1 ii). □

Example 2.1. Suppose X_t satisfies the conditions of Corollary 2.1. Consider the popular case $h_n = n^{-\rho}$ for some $\rho > 0$. Then (2.35) is satisfied if

$$\rho < (\nu-1)d^{-1}(\nu+(5/2))^{-1} . \tag{2.38}$$

This can be argued as follows: (2.35) is satisfied if

$$n^{2\tau\nu-1-\rho d(2\tau\nu+2+\tau)}(\log n)^{-2(\tau\nu+1)}(\log\log n)^{-(1+\epsilon)} \to \infty. \tag{2.39}$$

Note that

$$2\tau\nu - 1 - \rho d(2\tau\nu+2+\tau) = \nu - 1 - \rho d(\nu+(5/2)), \tag{2.40}$$

when $\tau = 1/2$. By (2.38), the right hand side of (2.40) is positive. The left hand side of (2.40) is continuous as a function of τ and hence is positive for τ sufficiently close to $1/2$. Thus (2.39) is satisfied for τ sufficiently close to 1/2.

In the important case that $\alpha(n) = O(e^{-sn})$ for some $s>0$, Condition (2.38) holds if $\rho < 1/d$ since $\alpha(n) = O(n^{-\nu})$ for any $0<\nu<1/d$.

3. L_1 CONVERGENCE OF $\hat{f}_n(x)$ UNDER THE ABSOLUTE REGULARITY CONDITION.

Lemma 3.1. Suppose X_t satisfies the absolute regularity condition and $\sum_{n=1}^{\infty}(\beta(n))^{\delta/(2+\delta)} < \infty$, and $h_n \downarrow 0$ slowly enough that (2.1) holds. Let ϵ and η be positive numbers. Then for sufficiently large n,

$$P[|n^{-1}\sum_{i=1}^{n} h_i^{-d}v(B,X_i)| \geq \epsilon] \leq n^{-(1/8)\eta\epsilon} + Ch_n^{-d}(\log n)\beta([nh_n^d/(\eta\log n)]), \tag{3.1}$$

where C is a positive constant.

Proof. The argument is similar to that of Lemma 2.2, except that the approximation by independent r.v.'s here is simpler. We will use the notation of Lemma 2.2. Assume $n = 2pq$ where $p = [nh_n^d/(\eta \log n)]$; then set the random variables $\Delta_i' s$ into blocks in the same way as done in Theorem 2.1. By Lemma 3.1 of Yoshihara [23] ,

$$P[|S_{n1}| > \epsilon/2] \leq P[|\sum_{j=1}^{q} W_j| > \epsilon/2] + 4q\beta(p), \tag{3.2}$$

where the $W_j' s$ are independent random variables such that W_j has the same distribution as $V(n, 2(j-1))$, where $V(n, 2(j-1))$ is as defined in (2.3) . Corresponding to (2.11), we have

$$P[|\sum_{j=1}^{q} W_j| > \epsilon/2] \leq n^{-(1/8)\eta\epsilon}, \tag{3.3}$$

for sufficiently large n . Thus

$$P[|S_{n1}| > \epsilon/2] \leq n^{-(1/8)\eta\epsilon} + 4q\beta(p), \tag{3.4}$$

Clearly, $P[|S_{n2}| > \epsilon/2]$ is also upper bounded by the right hand sideof (3.4). □

Lemma 3.2. Let $\epsilon > 0$, $\eta > 0$. Then for sufficiently large n

$$P[n^{-1} \sum_{i=1}^{n} h_i^{-d} v(E, X_i) > \epsilon h_n^{-d}] \leq n^{-(1/8)\eta\epsilon} + C \log n \ \beta([n/\eta \log n]), \tag{3.5}$$

where C is a positive constant.

Proof. The proof uses the argument of Lemmas 2.3 and 3.1. Let $p = [n/\eta \log n]$. Then

$$P[|S_{n1}| > \epsilon h_n^{-d}/2] \leq P[|\sum_{j=1}^{q} W_j| > \epsilon h_n^{-d}/2] + Cq\beta(p) \ , \tag{3.6}$$

where the $W_j' s$ are independent random variables with W_j having the same distribution as V_j . Here $q \leq C \log n$. Following the argument of Lemma 2.3,

$$P[|\sum_{j=1}^{q} W_j| > \epsilon h_n^{-d}/2] \leq n^{-(1/8)\eta\epsilon} \ .$$

The rest of the proof is easy. □

Theorem 3.1. Suppose X_t satisfies the absolute regularity condition and $\sum_{n=1}^{\infty} (\beta(n))^{\delta/(2+\delta)} < \infty$. Suppose Assumption 1 holds and $h_n \downarrow 0$ slowly enough that (2.1) holds and

$$q(n, 3) \equiv h_n^{-2d} \log n \ \beta([nh_n^d/(\eta \log n)]) \to 0 \tag{3.7}$$

for some $\eta > 0$. Then $J_n \to 0$ in probability.

ii) If (3.7) is strengthened to $\sum_{n=1}^{\infty} q(n, 3) < \infty$, then J_n converges completely to zero.

Proof. Let $\epsilon > 0$ be an arbitrary positive number. By Lemma 3.1,

$$P[Q(n,4) > \epsilon] \le Ch_n^{-d} n^{-C\eta\epsilon} + Ch_n^{-2d} \log n \ \beta([nh_n^d/(\eta \log n)]) \ .$$

By Lemma 3.2, for sufficiently large n,

$$P[Q(n,5) > \epsilon] \le n^{-(1/8)\eta\epsilon} + C \log n \ \beta([n/\eta \log n]) \ . \tag{3.8}$$

Recall that $Q(n,4)$ and $Q(n,5)$ were defined in (2.28). It is not hard to see that, for sufficiently large C,

$$\log n \beta([n/(\eta \log n)]) \le Ch_n^{-2d} \log n \ \beta([nh_n^d/(\eta \log n)]), \tag{3.9}$$

since $n/(\eta \log n) > nh_n^d/(\eta \log n)$ for large n, and β is a nonincreasing function. The proof of i) now easily follows.
ii) Part ii) follows by the argument in i) and since $\sum_{n=1}^{\infty} q(n,3) < \infty$. □

Remark 3.1. The absolute regularity condition is stronger than the strong mixing condition, but (3.7) is weaker than (2.16). In addition, Theorem 3.1 is much easier to employ than Theorem 2.1. This will be pointed out in Example 3.1 below. In applications, it is important to obtain from the data some information about the rate of decay of $\beta(n)$ as $n \to \infty$. Once this rate is available, the condition on h_n can be obtained explicitly from (3.7). When the rate of decay is polynomial, we have:

Corollary 3.1. i) Suppose X_t satisfies the absolute regularity condition with $\beta(n) = O(n^{-\nu})$ for some $\nu > 1$. Suppose Assumption 1 holds and $h_n \downarrow 0$ slowly enough as $n \to \infty$ that

$$q(n,4) \equiv n^{\nu} h_n^{d(2+\nu)} (\log n)^{-(1+\nu)} \to \infty \ . \tag{3.10}$$

Then $J_n \to 0$ in probability.
ii) Suppose (3.10) is strengthened to

$$q(n,4)/\ell(n) \to \infty \ , \tag{3.11}$$

where $\ell(n)$ is defined in ii) of Corollary 2.1. Then J_n converges completely to zero.

The proof of Corollary 3.1 can be obtained by Theorem 3.1 and the argument in Corollary 2.1.

Example 3.1. Corollary 3.1 is much easier to employ than Corollary 2.1. Here we do not need to deal with the parameter τ, which is somewhat of a nuisance. Take $h_n = n^{-\rho}$ for $\rho > 0$. Then (3.11) is satisfied if

$$\rho < (\nu - 1) d^{-1} (2 + \nu)^{-1} \ . \tag{3.12}$$

If $\beta(n) = O(e^{-sn})$ for some $s > 0$, then **(3.12)** is satisfied if $\rho < 1/d$. Condition **(3.12)** is less restrictive than **(2.38)**.

REFERENCES:

[1] Rosenblatt, M.(1956). A central limit theorem and a strong mixing condition. *Proc. Nat. Acad. Sci.*, U.S.A. **42**, 43-47.

[2] Parzen, E. (1962). On estimation of a probability density and mode. *Ann. Math. Statist.* **33**, 1065-1076.

[3] Devroye, L. and Györfi, L. (1985). *Nonparametric Density Estimation: The L_1 View.* Wiley, New York.

[4] Deheuvels, P. (1973a). Sur une famille d'estimateurs de la densité d'une variable aléatoire. *C.R. Acad. Sci. Paris* Serie A, **276**, 1013-1015.

[5] Wertz, W. (1985). Sequential and recursive estimators of the probability density. *Statistics* **16**, 277-295.

[6] Wolverton, C.T. and Wagner, T.J. (1969). Asymptotically optimal discriminant functions for pattern classification. *IEEE Trans. Inform. Theory* **IT-15**, 258-265.

[7] Wolverton, C.T. and Wagner, T.J. (1969). Recursive estimates of probability densities. *IEEE Trans. Sys. Sci. Cyber.* **5**, p. 307.

[8] Yamato, H. (1971). Sequential estimation of a continuous probability density function and the mode. *Bull. Math. Stat.* **14**, 1-12.

[9] Davies, H.I. (1973).Strong consistency of a sequential estimator of a probability denstity function. *Bull. Math. Stat.* **15**, 49-53.

[10] Deheuvels, P. (1973). Sur l'estimation séquentielle de la densité. *C.R. Acad. Sci. Paris* Serie A, **276**, 1119-1121.

[11] Deheuvels, P. (1973). Conditions nécessaires et suffisantes de convergence ponctuelle presque sure sur des estimateurs de la densité. *C.R. Acad. Sci. Paris* Serie A,**278**, 1217-1220.

[12] Carroll, R. (1976). On sequential density estimation. *Z. Wahr. Verw. Gebiete* **36**, 136-151.

[13] Ahmad, I.A. and Lin P. (1976). Nonparametric sequential estimation of a multiple regression function. *Bull. Math. Stat.* **17**, 63-75.

[14] Devroye, L. (1979). On the pointwise and the integral convergence of recursive kernel estimates of probability densities. *Utilitas Mathematica* **15**, 113-128.

[15] Wegman, E.J. and Davies, H.I. (1979). Remarks on some recursive estimators of a probability density function. *Ann. Statist.* **7**, 316-327.

[16] Masry, E. (1986). Recursive probability density estimation for weakly dependent stationary processes. *IEEE Trans. Inform. Theory,* Vol. **IT 18**, 254-267.

[17] Györfi, L. (1981). Strong consistent density estimate from ergodic sample. *J. Multivar. Anal.* tenrmb 11, 81-84.

[18] Takahata, H. (1980). Almost sure convergence of density estimators for weakly dependent stationary processes. *Bull. Tokyo Gakugei Uni. (IV)*, **32**, 11-32.

[19] Masry, E. and Györfi, L. (1987). Strong consistency and rates for recursive probability density estimators of stationary processes. *J. Multivariate Anal.* **22**, 79-93.

[20] Masry, E. (1987). Almost sure convergence of recursive density estimators for stationary mixing processes. *Statist. & Probab. Lett.* **5**, 249-254.

[21] Györfi, L. and Masry, E. (1988). The L_1 and L_2 strong consistency of recursive kernel density estimation from dependent samples. Report, TU Budapest.

[22] Yoshihara, K. (1978). Probability inequalities for sums of absolutely regular processes and their applications. *Z. Wahrsch. Verw. Gebiete* **43**, 319-330.

[23] Yoshihara, K. (1984). Density estimation for samples satisfying a certain absolute regularity condition. *J. Statist. Plann. Inference* **9**, 19-32.

[24] Pham, D.T. and Tran, L.T. (1985). Some mixing properties of time series models. *Stoch. Proc. Appl.* **19**, 297-303.

[25] Pham, D.T. (1986). The mixing property of bilinear and generalized random coefficient autoregressive models. *Stoch. Proc. Appl.* **23**, 291-300.

[26] Berbee, H.C.P. (1979). Random walks with stationary increments and renewal theory. (*Mathematical Center Tract* No. tenrmb 112, Amsterdam).

[27] Roussas, G.G. (1988). Nonparametric estimation in mixing sequences of random variables. *J. Statist. Plann. Inference* **18**, 135-149.

[28] Robinson, P.M. (1983). Nonparametric estimators for time series. *J. Time Series Anal.* **4**, 185-207.

[29] Robinson, P.M. (1987). Time series residuals with application to probability density estimation. *J. Time Series Anal.* **8**, 329-344.

[30] Ioannides, D. and Roussas, G.G. (1987). Note on the uniform convergence of density estimates for mixing random variables. *Statist. Probab. Lett.* **5**, 279-285.

[31] Györfi, L., Hardle, W., Sarda, P. and Vieu, P. (1990). Nonparametric curve estimation from time series. *Lecture Notes in Statistics,* **60**. Springer Verlag, New York.

[32] Hart, J.D. and Vieu, P. (1990). Data driven bandwidth choice for density estimation based on dependent data. *Ann. Statist.* **18**, 873-890.

[33] Masry, E. (1983). Probability density estimation from sampled data. *IEEE Trans. Inform. Theory,* Vol. **IT 29**, 696-709.

[34] Gorodetskii, V.V. (1977). On the strong mixing properties for linear sequences. *Theory Probab. Appl.* **22**, 411-413.

[35] Withers, C.S. (1981). Central limit theorems for dependent variables I, *Z. Wahrsch. Verw. Gebiete* **57**, 509-534.

[36] Athreya, K.B. and Pantula, S.G. (1986). Mixing properties of Harris chains and autoregressive processes. *J. Appl. Prob.* **23**, 880-892.

[37] Tran, L.T. (1989). The L_1 convergence of kernel density estimates under dependence. *Canadian J. Statist.* **17**, 197-208.

[38] Bradley, R.C.(1983). Approximation theorems for strongly mixing random variables. *Michigan Math J.* **30**, 69-81.

[39] Deo, C.M. (1973). A note on empirical processes of strong mixing sequences. *Ann. Probab.* **1**, 870-875.

Stat. Sci. & Data Anal., pp. 441-449
K. Matsusita *et al.* (Eds)

MULTIVARIATE L_1-NORM ESTIMATION AND THE VULNERABLE BOOTSTRAP

PRANAB KUMAR SEN

Departments of Biostatistics and Statistics, University of North Carolina, Chapel Hill, NC 27599-7400, USA

Abstract. Rotation-invariant L_1-norm estimation of location of a multivariate distribution is considered; the multivariate general linear model is also considered along with. Although this method rests on bounded score functions in a multivariate M-estimation setup, its robustness properties depend on the distribution of the reciprocal distances (from the location parameter) of the sample observations as well as on interdependence of the coordinate variates. In this context, the delta method, jackknife, bootstrap and some other resampling methods can be adapted to estimate the asymptotic dispersion matrix, but these estimates are generally biased. A comparative picture of jackknifing and bootstrapping is presented.

Keywords: Asymptotic linearity; bootstrap; bounded score function; convexity; delta method; dispersion matrix; jackknife; L_1-norm; M-statistic; resampling schemes.

1. L_1-NORM ESTIMATOR

Let $X_1, ..., X_n$ be n independent and identically distributed random vectors (i.i.d.r.v.) with a distribution function (d.f.) $F_{\theta}(x) = F(x - \theta)$, $x \in R^p$, $\theta \in \Theta \subseteq R^p$, where the form of F does not depend on the location parameter θ. For a nonnegative definite (nnd) matrix **Q**, consider the quadratic norm

$$\|d\|_Q^2 = d'Qd \quad \text{and let} \quad \|d\| = \|d\|_I . \tag{1.1}$$

For simplicity, we work with $Q = I$, and define

$$\rho_n(\theta) = \sum_{i=1}^n \|X_i - \theta\| , \quad \theta \in \Theta . \tag{1.2}$$

The L_1-norm estimator $\hat{\theta}_n$ of θ is defined by

$$\hat{\theta}_n = \text{Arg. min}\{\rho_n(\theta): \theta \in R^p\} . \tag{1.3}$$

For $p = 1$, minimization of (1.2) with respect to θ leads to the estimator

$$\hat{\theta}_n = \text{median}(X_1, ..., X_n) ,$$

although the usual differentiation technique does not apply here. For $p \geq 2$, $\rho_n(\theta)$ is differentiable with respect to θ, and hence, the minimization process yields the following estimating equation:

$$\sum_{i=1}^n \|X_i - \hat{\theta}_n\|^{-1} (X_i - \hat{\theta}_n) = 0 , \tag{1.4}$$

where conventionally, we let $\|0\|^{-1}\, 0 = 0$, as may be needed to eliminate possible indeterminacy in (1.4) if in case $\hat{\theta}_n$ coincides with one of the observations. Generally, an iterative procedure is needed to solve for $\hat{\theta}_n$ in (1.4). It may be noted that (1.4) represents an estimating equation closely related to that in multivariate M-estimation of location (viz., Singer and Sen [1]), although the associated score function involves weights which depend in a rather involved manner on the estimator $\hat{\theta}_n$ (identifiable as the center of gravity). If we consider the group of transformations: $\mathbf{X} \to \mathbf{X}^* = \mathbf{AX}$ and $\theta \to \theta^* = \mathbf{A}\theta$ where $\mathbf{A}$ is an orthogonal matrix (i.e., $\mathbf{A}'\mathbf{A} = \mathbf{I}_p$), then (1.4) yields rotation-equivariance of $\hat{\theta}_n$. However, $\hat{\theta}_n$ may not be generally equivariant under affine transformations: $\mathbf{X} \to \mathbf{a} + \mathbf{BX}$, $\theta \to \mathbf{a} + \mathbf{B}\theta$ where $\mathbf{B}$ is nnd. In passing, we may remark that instead of the simple location model, we could have considered a general multivariate linear model:

$$\mathbf{Y}_i = \beta \mathbf{t}_i + \mathbf{e}_i, \quad i = 1, \dots, n, \tag{1.5}$$

where the $\mathbf{e}_i$ are i.i.d.r.v.'s with a d.f. F (defined on R^p) whose form does not depend on the unknown parameter β ($p \times q$ matrix), and the $\mathbf{t}_i$ are given q-vectors ($q \geq 1$). Parallel to (1.2), we could have taken the L_1-norm

$$D_{n1}(\beta) = \sum_{i=1}^{n} \| \mathbf{Y}_i - \beta \mathbf{t}_i \|, \quad \beta \in R^{p \times q}, \tag{1.6}$$

and noting that $D_{n1}(\beta)$ is convex in β, minimization of (1.6) with respect to β would have led us to the L_1-norm estimator of β:

$$\hat{\beta}_n = \text{Arg. min}\{D_{n1}(\beta): \beta \in R^{p \times q}\}. \tag{1.7}$$

The estimating equations leading to (1.7) may be expressed as

$$\sum_{i=1}^{n} \| \mathbf{Y}_i - \hat{\beta}_n \mathbf{t}_i \|^{-1} (\mathbf{Y}_i - \hat{\beta}_n \mathbf{t}_i) \mathbf{t}_i' = \mathbf{0}, \tag{1.8}$$

where both the sides of (1.8) are $p \times q$ matrices, and to eliminate possible indeterminacy, as in after (1.4), we (conventionally) let $\| \mathbf{0} \|^{-1} \mathbf{0} = \mathbf{0}$. Here also, for an orthogonal $\mathbf{A}$ (of order $p \times p$) the transformation $\mathbf{Y}_i \to \mathbf{Y}_i^*$ and $\beta \to \beta^* = \mathbf{A}\beta$, (1.7) yields rotation-equivariance of $\hat{\beta}_n$. However, affine-equivariance may not hold. The solution in (1.8) generally requires an iterative procedure which is presumably more complex than the one for (1.4). For the sake of simplicity of presentation, in the sequel, we consider the location model, although the results to follow pertain as well to the general regression model and will be indicated only briefly. In this context, we shall restrict ourselves to $p \geq 2$, so that $\rho_n(\theta)$ in (1.2) is differentiable with respect to θ. For $p = 1$, the results pertaining to (1.3) are all extensively studied in the literature, and hence, we do not repeat them here.

We conclude this section with a remark on the computation of $\hat{\theta}_n$ in (1.4) based on a simple geometric interpretation. Consider a sphere $S(\mathbf{a})$ of unit radius and center at $\mathbf{a}$ ($\in R^p$). Join each point $\mathbf{X}_i$ to $\mathbf{a}$ and obtain the point of intersection (or projection) of the line joining $\mathbf{X}_i$ and $\mathbf{a}$ on the sphere $S(\mathbf{a})$. In this manner, the set of n points $\mathbf{X}_1$, ..., $\mathbf{X}_n$ are mapped into a set of n points on the spherical surface of $S(\mathbf{a})$. Then the estimate $\hat{\theta}_n$ relates to that particular choice of $\mathbf{a}$ for which the projections of these n points on the spherical surface of $S(\hat{\theta}_n)$ along any direction has mean zero. As such, for solving for $\hat{\theta}_n$, we need to attend to simultaneous variations along all possible directions. This feature makes it computationally more complex than the L_2-norm estimator. The following figure relates to a simple bivariate situation (with $n = 6$):

Figure 1 Geometrical intepretation of the L_1-norm estimator

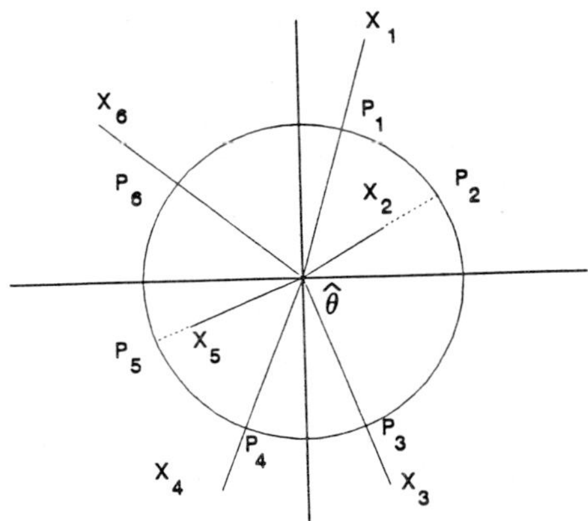

Each X_i continues to have a bounded influence in (1.4), although not solely on a coordinatewise basis. In this respect, the situation is different from the usual multivariate M-estimation of location, where coordinatewise influence functions dominate the scenario. Thus, the L_1-norm estimator may entail a lot of computational manipulations, albeit some iterative procedures work out well. The estimator is, however, highly non-linear. The situation is naturally more complex for the multivariate linear model in (1.5).

2. ASYMPTOTICS VIA LINEARITY APPROACH

Keeping in mind the estimating equations in (1.4), we define the following stochastic processes:

$$\mathbf{M}_n(\mathbf{t}) = n^{-1/2} \sum_{i=1}^{n} \| \mathbf{X}_i - \boldsymbol{\theta} - \mathbf{t} \|^{-1} (\mathbf{X}_i - \boldsymbol{\theta} - \mathbf{t}), \quad \mathbf{t} \in R^p . \tag{2.1}$$

Let then $\mathbf{Z}_n = \{\mathbf{Z}_n(\mathbf{u}),\ \mathbf{u} \in R^p\}$ be defined by

$$\mathbf{Z}_n(\mathbf{u}) = \{\mathbf{M}_n(n^{-1/2}\mathbf{u}) - \mathbf{M}_n(\mathbf{0}) + \mathbf{V}_n\mathbf{u}\}, \quad \mathbf{u} \in R^p , \tag{2.2}$$

where

$$\mathbf{V}_n = n^{-1} \sum_{i=1}^{n} \| \mathbf{X}_i - \boldsymbol{\theta} \|^{-1} \{\mathbf{I}_p - \| \mathbf{X}_i - \boldsymbol{\theta} \|^{-2} (\mathbf{X}_i - \boldsymbol{\theta})(\mathbf{X}_i - \boldsymbol{\theta})'\} . \tag{2.3}$$

Then, we have the following.

Theorem 2.1. Assume that there exists a $\delta > 0$, such that

$$\nu_F(\boldsymbol{\xi}) = E_F \| \mathbf{X} - \boldsymbol{\xi} \|^{-1} < \infty, \quad \text{uniformly in } \boldsymbol{\xi}: \| \boldsymbol{\xi} - \boldsymbol{\theta} \| < \delta . \tag{2.4}$$

Then, for every K: $0 < K < \infty$, as $n \to \infty$,

$$\sup\{ \| \mathbf{Z}_n(\mathbf{u}) \| : \| \mathbf{u} \| < K\} \to 0, \quad \text{in probability.} \tag{2.5}$$

Proof. Note that for every (fixed) $\mathbf{t} \in R^p$, $\mathbf{M}_n(\mathbf{t})$ involves independent summands (which are identically distributed too). Similarly, the $p \times p$ matrix $\mathbf{V}_n$ in (2.3) is an average of n i.i.d.r. matrices, each of whom has a finite first moment. Thus, the classical Khintchine strong law of large numbers can be invoked to show that $\mathbf{V}_n$ almost surely converges to $\boldsymbol{\Delta}_F$ ($= E\mathbf{V}_n$) as $n \to \infty$, where

$$\Delta_F = E_F\{ \| X-\theta \|^{-1}[I_p - \| X-\theta \|^{-2}(X-\theta)(X-\theta)']\} . \tag{2.6}$$

Hence, noting that for every (fixed) $u \in R^p$, $M_n(n^{-1/2}u) - M_n(0) + \Delta_F u$ has mean (vector) converging to 0 and dispersion matrix converging to a null matrix (as $n\to\infty$), we may conclude that $Z_n(u)\to 0$, in probability, as $n\to\infty$. Hence, it remains to show only that this convergence is uniform in u: $\| u \| < K$ (for every K: $0 < K < \infty$). This last assertion is a direct consequence of the convexity arguments of Pollard [2], Section 11, where the case of $p = 1$ has been treated in detail. Since this convexity argument remains in tact for every $p \geq 1$, the details are therefore omitted.

In passing, we may note that for $p = 1$, V_n is 0, so that the version in (2.2) and (2.5) is not true, but a more refined result (along the line of the classical Bahadur representation of sample quantiles) holds (viz., Jurečková and Sen [3, 4]). To make use of Theorem 2.1 in the context of the asymptotic distribution theory of $\hat{\theta}_n$, let us define

$$\Sigma_F = E_F\{ \| X-\theta \|^{-2} (X-\theta)(X-\theta)'\} \tag{2.7}$$

and

$$\Gamma_F = \Delta_F^{-1} \Sigma_F \Delta_F^{-1} , \tag{2.8}$$

where we assume that Δ_F and Σ_F are both nnd, so that Γ_F is also nnd. Note that by the Khintchine strong law of large numbers, whenever $E_F \| X-\theta \|^{-1} < \infty$, as $n\to\infty$,

$$V_n \to \Delta_F \text{ almost surely (a.s.)} \tag{2.9}$$

Also, by (2.1) and the multivariate central limit theorem, as $n\to\infty$,

$$M_n(0) \xrightarrow{\mathcal{D}} \mathcal{N}_p(0, \Sigma_F) . \tag{2.10}$$

Therefore, from (1.4), (2.2), (2.5) and (2.10), we obtain that as $n\to\infty$,

$$n^{1/2}(\hat{\theta}_n - \theta) \xrightarrow{\mathcal{D}} \mathcal{N}_p(0, \Gamma_F) , \tag{2.11}$$

which may be used to draw statistical conclusions on θ. For an alternative approach to the study of (2.11), we may refer to Bai et al. [5] where other pertinent references are also cited.

In the context of the multivariate linear model in (1.5), we may let

$$T_n = \sum_{i=1}^{n} t_i t_i' , \quad n \geq 1 , \tag{2.12}$$

and assume that the following conditions hold:

$$n^{-1}T_n \to T , \ T \text{ nnd}; \tag{2.13}$$

$$\max_{i \leq i \leq n} t_i' T_n^{-1} t_i \to 0 , \quad \text{as } n\to\infty ; \tag{2.14}$$

the later condition is known as the generalized Noether condition. Then, parallel to (2.11), we have

$$n^{1/2}(\hat{\beta}_n - \beta) \xrightarrow{\mathcal{D}} N_{pq}(0, \Gamma_F \otimes T^{-1}) , \tag{2.15}$$

where Γ_F is defined as in (2.8) and $\otimes$ stands for the Kronecker product. Since T_n is a specified nnd matrix, for both (2.11) and (2.15), the main concern is to estimate Γ_F consistently, which we consider in the next section.

3. STATISTICAL INFERENCE BASED ON L_1-NORM ESTIMATORS

In an asymptotic setup, in order to construct a confidence region for $\boldsymbol{\theta}$ (or $\boldsymbol{\beta}$) or to test for a null hypothesis on $\boldsymbol{\theta}$ (or $\boldsymbol{\beta}$), a valid use of (2.11) rests on a consistent estimator of Γ_F. Looking at (2.6), (2.7) and (2.8), we may propose the estimator

$$\hat{\Gamma}_n = \hat{\Delta}_n^{-1} \hat{\Sigma}_n \hat{\Delta}_n^{-1}, \tag{3.1}$$

where

$$\hat{\Delta}_n = \frac{1}{n} \sum_{i=1}^{n} \| X_i - \hat{\theta}_n \|^{-1} \{ I_p - \| X_i - \hat{\theta}_n \|^{-2} (X_i - \hat{\theta}_n)(X_i - \hat{\theta}_n)' \} \tag{3.2}$$

and

$$\hat{\Sigma}_n = \frac{1}{n} \sum_{i=1}^{n} \| X_i - \hat{\theta}_n \|^{-2} (X_i - \hat{\theta}_n)(X_i - \hat{\theta}_n)' . \tag{3.3}$$

It seems that to establish the consistency of $\hat{\Delta}_n$ and $\hat{\Sigma}_n$ (and hence, $\hat{\Gamma}_n$), we may need some extra regularity condition. Among other possibilities we assume that

$$E_{\theta} \Big\{ \sup \| \frac{1}{\| X - \theta - t \|} \{ I - \| X - \theta - t \|^{-2} (X - \theta - t)(X - \theta - t)' \}$$

$$- \{ I - \| X - \theta \|^{-2} ((X - \theta)(X - \theta)' \} \| : \ \| t \| < \delta \Big\} \to 0 \ \text{ as } \ \delta \downarrow 0 , \tag{3.4}$$

so that the consistency of $\hat{\theta}_n$ and (3.4) along with the Khintchine strong law of large numbers would imply the consistency of $\hat{\Gamma}_n$. Since the summands in (3.3) are points on the unit sphere $S(\hat{\theta}_n)$ (and the convention after (1.4) is adopted), for $\hat{\Sigma}_n$, the regularity conditions are simpler. However, for $\hat{\Delta}_n$, the scenario is dominated by $\hat{\nu}_n$ where

$$\hat{\nu}_n = n^{-1} \sum_{i=1}^{n} \| X_i - \hat{\theta}_n \|^{-1} \tag{3.5}$$

which has unbounded influence of the X_i "close to" $\hat{\theta}_n$ and thereby may have highly inflationary effect (i.e., upward bias) on the estimator $\hat{\nu}_n$. For this reason the classical "delta-method" of estimating Γ_F, as has been posed in (3.1) through (3.3), may not work out well, particularly, for small values of p. To make this point clear, we consider the case where F is a spherically symmetric d.f., so that by some routine analysis, we have

$$\Delta_F = \frac{p-1}{p} \nu_F I_p , \quad \Sigma_F = p^{-1} I_p \quad \text{and} \quad \Gamma_F = p(p-1)^{-2} \nu_F^{-2} I_p . \tag{3.6}$$

Thus, the problem reduces to that of estimating ν_F^{-2}. Recall that in (3.5), values of X_i "close to" $\hat{\theta}_n$ will have inflationary effect on $\hat{\nu}_n$, so that $\hat{\nu}_n^{-2}$ may have serious downward bias on an estimator of ν_F^{-2}. Let us consider an ϵ_n-neighborhood of $\hat{\theta}_n$:

$$J_n = \{ X_i : \ \| X_i - \hat{\theta}_n \| \le \epsilon_n , \ 1 \le i \le n \} , \tag{3.7}$$

where ϵ_n is positive and it converges to 0 at a suitable rate as $n \to \infty$. Let $\hat{\pi}_n$ be the sample measure of J_n (i.e., n^{-1} cardinality of the set J_n). Then the contribution of the r.v.'s in J_n towards $\hat{\nu}_n$ is bounded from below by $\hat{\pi}_n\epsilon_n^{-1}$, and depending on the choice of $\{\epsilon_n\}$ this may be made nonvanishing (stochastically) in the limit. Therefore the usual delta method of estimating ν_F^{-2} (or Γ_F) may result in considerable bias particularly for $p = 2$ or 3. As such, we examine the performance of the two other resampling methods in estimating Γ_F.

4. JACKKNIFING VS. BOOTSTRAPPING

The classical jackknife does not work out in the univariate case, although delete-d jackknifing, for d not too small, works out well. In the multivariate case, the differentiability picture is somewhat different, and hence, it is quite natural to inquire: How far jackknife can be prescribed for the estimation of Γ_F? We proceed as in Sen [6, 7].

Let $\hat{\theta}_n$ be defined as in (1.3), and based on $X_1, \cdots, X_{i-1}, X_{i+1}, \cdots, X_n$, let $\hat{\theta}_{n-1}^{(i)}$ be the corresponding L_1-norm estimator, for $i = 1, \ldots, n$. Then let the pseudovalues be defined by

$$\hat{\theta}_{n,i} = n\hat{\theta}_n - (n-1)\,\hat{\theta}_{n-1}^{(i)}, \quad i = 1, \ldots, n, \tag{4.1}$$

and let

$$\hat{\theta}_{nJ} = n^{-1} \sum_{i=1}^{n} \theta_{n,i}, \tag{4.2}$$

$$V_{nJ} = (n-1)^{-1} \sum_{i=1}^{n} (\hat{\theta}_{n,i} - \hat{\theta}_{nJ})(\hat{\theta}_{n,i} - \hat{\theta}_{nJ})' \tag{4.3}$$

be respectively the jackknifed version of $\hat{\theta}_n$ and the jackknifed variance matrix estimator. Further note that

$$\hat{\theta}_{nJ} = \hat{\theta}_n + (n-1)\{n^{-1} \sum_{i=1}^{n} (\hat{\theta}_n - \hat{\theta}_{n-1}^{(i)})\}, \tag{4.4}$$

where in a regular case (when the bias of $\hat{\theta}_n$ is $O(n^{-1})$), the second term on the right hand side of (4.4) adjusts it further, and reduces the bias of $\hat{\theta}_{nJ}$ to $o(n^{-1})$. In the current context, particularly, for small values of p, the bias of $\hat{\theta}_n$ may not be $O(n^{-1})$, and hence, the bias reduction role of jackknifing may not be that effective. Therefore, we mainly concentrate on the variance estimation role of jackknifing i.e., on the convergence properties of V_{nJ}.

We define $M_n(t)$ as in (2.1), and note that for every i: $1 \le i \le n$,

$$M_n(\hat{\theta}_{n-1}^{(i)} - \theta) = n^{-1/2} \| X_i - \hat{\theta}_{n-1}^{(i)} \|^{-1} (X_i - \hat{\theta}_{n-1}^{(i)}), \tag{4.5}$$

so that

$$\| M_n(\hat{\theta}_{n-1}^{(i)} - \theta) \| = n^{-1/2}, \quad \forall\, i = 1, \ldots, n, \tag{4.6}$$

and by (1.4), $M_n(\hat{\theta}_n - \theta) = 0$ (a.e.). Recalling that in (1.4), we have a bounded score function (relating to points on a spherical surface), and defining an "envelop" as in page 19 of Pollard [2], we obtain by a direct appearl to Theorem 8.3 of Pollard [2] that as $n \to \infty$,

$$\max_{1 \le i \le n} \| \hat{\theta}_{n-1}^{(i)} - \hat{\theta}_n \| \to 0 \text{ a.s.}, \quad \| \hat{\theta}_n - \theta \| \to 0 \text{ a.s.} \tag{4.7}$$

Next, in (3.2), we replace $\hat{\theta}_n$ by $\theta + t$, and define the ocrresponding matrix by $\Delta_n(t)$. Also, for every $k(=1, \ldots, n)$ and $t \in \mathbb{R}^p$, we define

$$V_{n-1}^{(k)}(t) = \frac{1}{n-1} \sum_{\substack{i=1 \\ (i \neq k)}}^{n} \| X_i - \theta - t \|^{-1} \{ I_p - \| X_i - \theta - t \|^{-2} (X_i - \theta - t)(X_i - \theta - t)' \} . \tag{4.8}$$

It may be recalled that by definition of $\Delta_n(t)$ and the $V_{n-1}^{(k)}(t)$, $n^{-1} \sum_{k=1}^{n} V_{n-1}^{(k)}(t) = \Delta_n(t)$, for every $t \in \mathbb{R}^p$ and $n \leq 2$. Let $\mathcal{C}_n$ be the sigma-field generated by the unordered collection $\{X_1, \ldots, X_n\}$ and by X_{n+j}, $j \geq 1$, for $n \geq 1$. If the X_i are real valued, then $\mathcal{C}_n = \mathcal{C}(X_{n:1}, \ldots, X_{n:n}; X_{n+j}, j \geq 1)$ where $X_{n:1} \leq \cdots \leq X_{n:n}$ are the ordered values corresponding to $X_1, \ldots, X_n$. Then $\mathcal{C}_n$ is nonincreasing in n (≥ 1). Also, given $\mathcal{C}_n$, the X_{n+j}, $j \geq 1$ are all fixed, while all possible permutations among the $X_1, \ldots, X_n$ are (conditionally) equally likely. As such, for every (fixed) $T \subset R^p$, it is easy to verify that

$$E\{[\Delta_n(t), t \in T] \mid \mathcal{C}_{n+1}\} = [\Delta_{n+1}(t), t \in T] . \tag{4.9}$$

so that by the convexity of the "sup norm", for every $T \subset R^p$,

$$\{\sup \{ \| \Delta_n(t) - \Delta_n(0) \| : t \in T \} , \mathcal{C}_n ; n \geq p\} \text{ is a nonnegative reversed sub-martingale.} \tag{4.10}$$

Allowing T to be a small neighborhood of $\mathbf{0}$ and making use of (4.7), (4.10) and (3.4), it follows that as $n \to \infty$

$$\max_{1 \leq k \leq n} \{ \sup_{t \in T} \| V_{n-1}^{(k)}(t) - \Delta_n(t) \| \} \xrightarrow{P} 0 , \tag{4.11}$$

$$\sup_{t \in T} \| \Delta_n(t) - \Delta_n(\mathbf{0}) \| \xrightarrow{P} 0 , \tag{4.12}$$

when $\text{diam}(T) \to 0$. Thus, if we make a first order Taylor's expansion of $M_n(\hat{\theta}_{n-1}^{(k)} - \theta)$ around $\hat{\theta}_n - \theta$, and make use of (4.7), (4.8), (4.11), (4.12) and (4.1), we obtain that as $n \to \infty$,

$$\max_{1 \leq k \leq n} \| \hat{\Delta}_n (\hat{\theta}_{n,k} - \hat{\theta}_n) - \| X_k - \hat{\theta}_n \|^{-1} (X_k - \hat{\theta}_n) \| \xrightarrow{P} 0 , \tag{4.13}$$

so that

$$\| V_{nJ} - \hat{\Delta}_n^{-1} \hat{\Sigma}_n \hat{\Delta}_n^{-1} \| \xrightarrow{P} 0 , \text{ as } n \to \infty . \tag{4.14}$$

Therefore, V_{nJ} converges in probability to Γ_F, as $n \to \infty$. In passing, we may remark that intuitively the reasons that jackknifing works in the current context are the following:

(i) Under the assumed regularity conditions, $X_1, \ldots, X_n$ are all distinct, with probability 1, and

(ii) $P\{\hat{\theta}_n = X_k \text{ for some k: } 1 \leq k \leq n\} = 0$.

These conditions, in turn, imply that $P\{\hat{\theta}_{n,k} = X_i$ for some $i = 1, \ldots, n$ and $k = 1, \ldots, n\} = 0$, and this controls the inflationary effect to a larger extent. On the other hand, faced with the remark on the order of bias of $\hat{\theta}_n$ made after (4.4), it seems more desirable to use delete-d jackknifing, for some $d \geq 1$.

For every $\mathbf{i} = (i_1, \ldots, i_d)$ of distinct indices [out of $(1, \ldots, n)$], we define

$$\hat{\theta}^{(d)}_{n,i} = d^{-1}\{n\hat{\theta}_n - (n-d)\,\hat{\theta}^{(i)}_{n-d}\}\ ,\quad i \in I\ , \tag{4.15}$$

where I is the set of all $\binom{n}{d}$ realizations of $\mathbf{i}$ and $\hat{\theta}^{(i)}_{n-d}$ is the L_1-norm estimator of θ based on the X_j , $j \in \{1, ..., n\} \setminus \mathbf{i}$. Let then

$$\hat{\theta}^{(d)}_{n,J} = \binom{n}{d}^{-1} \Sigma_{i\epsilon I}\, \hat{\theta}^{(d)}_{n,\,i} \qquad \text{(delete-d Jackknifed version)} \tag{4.16}$$

and

$$V^*_{n,d} = \binom{n}{d}^{-1} \Sigma_{i\epsilon I}\, (\hat{\theta}^{(d)}_{n,i} - \hat{\theta}^{(d)}_{n,J})\, (\hat{\theta}^{(d)}_{n,i} - \hat{\theta}^{(d)}_{n,J})'\ . \tag{4.17}$$

Then the delete-d jackknifed variance (matrix) estimator is

$$V^{(d)}_{n,J} = (n-d)^{-1}\,(n-1)\, dV^*_{n,d}\ . \tag{4.18}$$

By similar manipulations it can be shown that $V^{(d)}_{n,J}$ converges in probability to Γ_F as $n \to \infty$, and for d not too small, but d/n small, it has better robustness properties then $V_{n,J}$.

Let us examine the bootstrap method for the estimation of Γ_F . Note that one has to draw a sample of n units with replacement (SRSWR) from $X_1, ..., X_n$. If we denote these observations by $X^*_1, ..., X^*_n$, respectively, then we solve for (1.4) with the X_i replaced by X^*_i, $i \le n$. The corresponding estimator is denoted by $\hat{\theta}^*_n$. Suppose now that we draw a large number (say, M) of such (conditionally) independent samples of size n each from $X_1, ..., X_n$, and denote the resulting L_1-norm estimators by $\hat{\theta}^*_{n,1}, ..., \hat{\theta}^*_{n,M}$, respectively. Let then

$$V^*_{n,B} = nM^{-1} \sum_{i=1}^{M} (\hat{\theta}_{n,i} - \hat{\theta}_n)\,(\hat{\theta}^*_{n,i} - \hat{\theta}_n)' \tag{4.19}$$

be the bootstrap estimator of Γ_F . We are naturally tempted to follow the same type of manipulations as in (4.5) through (4.14) with the X_i and θ replaced by X^*_i and $\hat{\theta}_n$, respectively. In this context, parallel to (4.7), we may need to show that

$$\max_{1 \le k \le M} \| \hat{\theta}^*_{n,k} - \hat{\theta}_n \| \to 0\ , \text{ in probability, as } n \to \infty\ . \tag{4.20}$$

Since the $\hat{\theta}^*_{n,k}$ are conditionally (given $X_1, ..., X_n$) i.i.d.r.v.'s, and usually M is taken as O(n), a set of sufficient conditions for (4.20) to hold is that

$$n\, E\{ \| \hat{\theta}^*_n - \hat{\theta}_n \|^2 \mid X_1, ..., X_n\} = O(1) \text{ a.e.}; \tag{4.21}$$

$$n^{-1} \sum_{i=1}^{n} \| X^*_i - \hat{\theta}_n \|^{-1} - \hat{\nu}_n \xrightarrow{P} 0\ , \text{ as } n \to \infty\ , \tag{4.22}$$

where $\hat{\nu}_n$ is defined by (3.5). Actually, (4.22) stems from a conditional version of (3.4) [given $X_1, ..., X_n$] wherein θ and X are replaced by $\hat{\theta}_n$ and X^*, respectively. But $\hat{\nu}^*_n = n^{-1} \sum_{i=1}^{n} \| X^*_i - \hat{\theta}^*_n \|^{-1}$ is very vulnerable to the resampling process inherent in the classical bootstrapping. To make this point clear, note that for every k: $1 \le k \le n$,

$$P\{X^*_1 = \cdots = X^*_n = X_k \mid X_1, ..., X_n\} = n^{-n}, \tag{4.23}$$

and for $X^*_1 = \cdots = X^*_n = X_k$, $\hat{\theta}^*_n = X_k$, so that $\hat{\nu}^*_n = +\infty$. Thus, given $X_1, ..., X_n$, with a conditional probability $> n^{-(n-1)}$, $\hat{\nu}^*_n = +\infty$. In the other extreme case where $X^*_1, ..., X^*_n$ are all distinct (i.e., a permutation of $X_1, \cdots, X_n$), $\hat{\theta}^*_n = \hat{\theta}_n$ and $\hat{\nu}^*_n = \hat{\nu}_n$; but this has a probability $(n!)/n^n = n^{-(n-1)}(n-1)! \to 0$, as $n \to \infty$. Also, over the set of realizations of $X^*_1, ..., X^*_n$ for which $\hat{\theta}^*_n$ coincides with some X^*_r (which, in turn, is some X_k , $1 \le k \le n$), $\hat{\nu}^*_n = +\infty$. Even if $\hat{\theta}^*_n$ is not exactly equal to some X^*_r , but is arbitrarily close to some X^*_r , $\hat{\nu}^*_r$ can be arbitrarily large, and this possibility can not be ruled out, even for large n, as the conditional distribution of $X_1^*, \ldots, X_n^*$, given $X_1, \ldots, X_n$, is generated by the n^n (conditionally) equally likely realizations $\{X_{i_1}, \ldots, X_{i_n}\}$, where each i_j can take on the values 1, 2, ..., n with the common conditional probability n^{-1}. Thus, the vulnerability of the bootstrap method in this L_1-norm estimation problem is mainly due to the discreteness of the conditional distribution of X^*, given $X_1, \ldots, X_n$, which allows $\hat{\nu}_n^*$ to be $+\infty$ with a positive probability (however small it may be for large n), and this, in turn, makes $V_{n,B}^*$ in (4.19) generally underestimating Γ_F. This type of bias of bootstrap variance estimators is quite common for the univariate case too; see for example, Huang [8] for sample median in the discrete case, and Stangenhaus [9] for L_1-regression. Stangenhaus [9] proposed a smoothed bootstrap to overcome the problem of underestimating the asymptotic variance. Her procedure may be extended to the multivariate case to make similar adjustments. Other modifications of the classical bootstrap method may also work out better, and these are under active study.

ACKNOWLEDGEMENTS

Thanks are due to the referee for his (her) critical reading of the manuscript and most helpful comments.

REFERENCES

1. Singer, J. M. and Sen, P. K. (1985). M-methods in multivariate linear models. Journal of Multivariate Analysis 17, 168-184.

2. Pollard, D. (1990). Empirical Processes: Theory and Applications NSF-CBMS Regional Conference Series in Probability and Statistics, Volume 2. IMS-ASA, USA.

3. Jurečková, J. and Sen, P. K. (1987). A second order asymptotic distributional representation of M-estimators with discontinuous score functions. Annals of Probability 15, 814-823.

4. Jurečková, J. and Sen, P. K. (1990). Effect of the initial estimator on the asymptotic behavior of one-step M-estimators. Annals of the Institute of Statistical Mathematics 42, 345-347.

5. Bai, Z. D., Chen, X. R., Miao, B. Q. and Rao, C. R. (1990). Asymptotic theory of least distance estimates in multivariate analysis. Statistics 21, 503-519.

6. Sen, P. K. (1988a). Functional jackknifing: rationality and general asymptotics. Annals of Statistics 16, 450-469.

7. Sen, P. K. (1988b). Functional approaches in resampling plans: A review of some recent developments. Sankhya, Series A 50, 394-435.

8. Huang, J. S. (1991). Estimating the variance of the sample median, discrete case. Statistics & Probability Letters 11, 291-298.

9. Stangenhaus, G. (1987). Bootstrap and inference procedure for L_1-regression. In Statistical Data Analysis Based on the L_1-norm and Related Methods, Y. Dodge (editor), North Holland, pp. 323-332.

Stat. Sci. & Data Anal., pp. 451-465
K. Matsusita *et al.* (Eds)

Convergent Rates of M-Estimators for a Partly Linear Model

GUOYING LI and PEIDE SHI
(*Institute of Systems Science, Academia Sinica, Beijing 100080*)

Abstract. A partly linear model $Y = X^{\tau}\beta_0 + g_0(T) + u$ is considered, where T ranges over a nondegenerate compact interval, $X \in R^d$, Y is real-valued, u is an unobserved disturbance, β_0 is a d-vector of parameters to be estimated, and g_0 is an unknown smooth function whose mth derivative function satisfies the Hölder condition with exponent $\gamma \in (0,1]$. A piecewise polynomial is used to approximate g_0. The M - estimators of β and g_0 are defined, and their convergence rates are discussed. Under some mild conditions, it is shown that the underlying estimators of g_0 achieve the convergence rate as that of the optimal global rate of convergence for nonparametric regression according to Stone [6] and [7].

Key words and Phrases: Partly linear model, M-estimator, optimal global rate of convergence, piecewise polynomial.

1. INTRODUCTION

Consider the partly linear model

$$Y = X^{\tau}\beta_0 + g_0(T) + u, \tag{1.1}$$

where u is a random error, X is a d-vector of explanatory variables, β_0 is a vector of unknown parameters, T is another explanatory variable ranging over a nondegenerate compact interval, say [0, 1], and $g_0(\cdot)$ is an unknown smooth function. Since it contains a nonparametric component, this model is more flexible and attractive than the usual linear model. It belongs to the class of partial spline modes, which was proposed and applied to study the effect of weather on electricity demand by Engle et al. [1].

To estimate the unknown smooth function g_0, the smoothing spline method was proposed in the literature. Wahba [2] minimized the penalty function

$$\frac{1}{n}\sum_{i=1}^{n}(Y_i - X_i^{\tau}\beta - g(T_i))^2 + \lambda\int_0^1 g^{(m)}(t)^2\,dt$$

over $(\beta \in) R^d$ and $(g \in)$ a class of smooth functions to obtain the estimator $\widehat{\beta}$ of β_0 and the partial smoothing spline estimator $\widehat{g}$ of g_0. Heckman [3] proved that $\widehat{\beta}$ could achieve the $n^{-1/2}$ convergence rate if T is a non-random variable. Rice [4] studied the bias and the convergence rate of $\widehat{\beta}$ when X_i and T_i were correlated and shown that it was not generally possible to attain $n^{-1/2}$ convergence rate for the estimators of the parametric components.

Suppose that $g_0(\cdot)$ is $m(\geq 0)$ times continuously differentiable and its derivative satisfies the Hölder condition with exponent $\gamma \in (0,1]$. If $(X_1, T_1, Y_1), \cdots, (X_n, T_n, Y_n)$ are i.i.d. random observations, Chen [5] used a piecewise polynomial to approximate g_0 and obtained the estimators $\widehat{\beta}$ and $\widehat{g}_n$, respectively, for β_0 and g_0 via least square (LS) principle. He concluded that, under some mild conditions, $\sqrt{n}(\widehat{\beta} - \beta_0)$ is asymptotically normal and

$$\widehat{g}_n(t) - g_0(t) = O_P\left(n^{-\frac{m+\gamma}{2(m+\gamma)+1}}\right), \quad \text{for any given} \quad t \in [0,1].$$

Note that all the estimators above are constructed by using the LS method (or a criterion closely related to the LS method). The merit of LS estimators is that they are easy to compute and possess several desirable properties in the case that the random errors are independent and normally distributed. But in many practical applications, there may be outliers in the observations and the distribution of the random errors may be long-tailed. In these cases the LS estimators behave poorly.

Therefore a more "robust" procedure is deserved to be studied. A natural question raised here is that: if we adopt more robust estimators, what are their convergence rates?

In this paper, we use piecewise polynomials to approximate g_0 and obtain the M - estimators of β_0 and $g_0(\cdot)$. It will be proved that, under some mild conditions, the underlying estimators achieve the convergence rate as that of the optimal global rate of convergence of LS estimators for nonparametric regression according Stone [6] and [7]. The formal statements of the results are given in Section 2. Section 3 gives the theoretical proof of the main results.

2. STATEMENTS OF THE MAIN RESULTS

Consider model (1.1). Suppose that (T_i, X_i, Y_i), $1 \leq i \leq n$, are i.i.d. observations of (T, X, Y) with

$$Y_i = X_i^\tau \beta_0 + g_0(T_i) + u_i, \quad 1 \leq i \leq n, \tag{2.1}$$

where $u_1, \cdots, u_n$ are random errors and independent of $(T_1, X_1), \cdots, (T_n, X_n)$.

First, we need some notations. Let m be a nonnegative integer. For given positive integer M_n, denote $\delta_n = 1/(2M_n), p = (m+1)M_n$. Set

$$\begin{aligned} I_{nv} &= [2(v-1)\delta_n, 2v\delta_n), \quad 1 \leq v < M_n, \\ I_{nM_n} &= [1 - 2\delta_n, 1], \\ t_v &= (2v-1)\delta_n, \quad 1 \leq v \leq M_n, \\ I_{nv}(t) &= 1 \text{ if } t \in I_{nv}; \quad 0 \text{ if } t \notin I_{nv}, \\ \varphi_v(t) &= I_{nv}(t)(1, (t-t_v), \cdots, (t-t_v)^m)^\tau, \\ \varphi(t) &= (\varphi_1(t)^\tau, \cdots, \varphi_{M_n}(t)^\tau)^\tau. \end{aligned} \tag{2.2}$$

Obviously, $\varphi_v(t)^\tau \alpha_v (\alpha_v \in R^{m+1})$ is a polynomial of degree m on I_{nv}, and $\varphi(t)^\tau \alpha (\alpha \in R^p)$ is a piecewise polynomial of degree m on [0, 1].

The M-estimators of β_0 and $g_0(t)$ are respectively $\widehat{\beta}$ and $\widehat{g}_n(t) = \varphi(t)^\tau \widehat{\alpha}$, where $\widehat{\beta} \in R^d$ and $\widehat{\alpha} \in R^d$ are chosen to satisfy

$$\sum_{i=1}^{n} \rho(Y_i - X_i^\tau \widehat{\beta} - \varphi(T_i)^\tau \widehat{\alpha}) = \min_{\alpha,\beta} \sum_{i=1}^{n} \rho(Y_i - X_i^\tau \beta - \varphi(T_i)^\tau \alpha) \tag{2.3}$$

or

$$\begin{cases} \sum_{i=1}^{n} \Psi(Y_i - X_i^\tau \widehat{\beta}_n - \varphi(T_i)^\tau \widehat{\alpha}) X_i = 0, \\ \sum_{i=1}^{n} \Psi(Y_i - X_i^\tau \widehat{\beta}_n - \varphi(T_i)^\tau \widehat{\alpha}) \varphi(T_i) = 0. \end{cases} \tag{2.4}$$

$\widehat{g}_n(t)$ is called the piecewise polynomial M - estimator of $g_0(t)$.

It is well known that if function $\rho(\cdot)$ is convex and differentiable with derivative $\Psi(\cdot)$, then (2.3) and (2.4) are equivalent. However, as pointed out by Bai et al [8], if $\rho(\cdot)$ is not differentiable at least at one point, then (2.3) and (2.4) are not equivalent. In present paper, it is assumed that $\rho(\cdot)$ is convex and differentiable everywhere and $\Psi = \rho'$.

Let $\gamma \in (0, 1]$ be a number such that $m + \gamma > 1/2$, and let $M_0 \in (o, \infty)$. Let $\mathcal{H}$ stands for the collection of functions on [0, 1] such that, for every $h \in \mathcal{H}$, the mth derivative of h, denoted by $h^{(m)}$, exists and satisfies the Hölder condition with exponent γ:

$$|h^{(m)}(t) - h^{(m)}(s)| \leq M_0 |t - s|^\gamma, \quad \text{for} \quad 0 \leq t, s \leq 1.$$

We make the following assumptions:

A1. T is distributed according to a density function f, which satisfies

$$0 < b \leq \inf_{0 \leq t \leq 1} f(t) \leq \sup_{0 \leq t \leq 1} f(t) \leq B < \infty$$

for some constants b and B;

A2. $g_0 \in \mathcal{H}$;

A3. $EX = 0$, EXX^τ is a finite and positive definite matrix, and X and T are independent;

A4. Ψ is nondecreasing on R^1, $E\Psi(u) = 0$, and $E\Psi^2(u) < \infty$, where u is the random error in (1.1);

A5. There exist five positive constants c_1, c_2, c_3, d_1, and d_2 such that

$$\begin{aligned} & q \triangleq \mathcal{P}(|u| \leq c_1) > 0, \\ & D(s,t) \geq d_1, \quad \text{when} \quad |s| \leq c_1 \quad \text{and} \quad |t| \leq c_2, \\ & D(s,t) \leq d_2, \quad \text{when} \quad |t| \leq c_3, \end{aligned}$$

where

$$D(s,t) = \begin{cases} (\Psi(s+t) - \Psi(s))/t & t \neq 0 \\ c & t = 0. \end{cases}$$

A6. $\max_{1\leq i\leq n} |X_i| = o_P(n^\alpha)$ with $\alpha = (m+\gamma)/[2(m+\gamma)+1]$.

Let $\|h\|_{\mathcal{L}^2}$ denote the $\mathcal{L}^2$ norm, defined by $\|h\|^2_{\mathcal{L}^2} \triangleq \int_0^1 h^2(t)f(t)\,dt$. Let $|\cdot|$ denote either the Euclidean norm of a vector or the absolute value of a real number according to the context. Given positive numbers a_n and $b_n, n \geq 1$, let $a_n \sim b_n$ denote that a_n/b_n is bounded away from zero and infinite.

Then we have the following theorem.

Theorem 2.1. Let $(\widehat{\alpha}, \widehat{\beta})$ satisfy (2.4), and let $\widehat{\beta}$ and $\widehat{g}_n(t) = \varphi(t)^\tau \widehat{\alpha}$ be respectively the M-estimators of β_0 and $g_0(t)$. If A1-A6 hold and $M_n \sim n^{1/[2(m+\gamma)+1]}$, then

$$|\widehat{\beta} - \beta_0| = O_P(n^{-(m+\gamma)/[2(m+\gamma)+1]}),$$

$$\|\widehat{g}_n - g_0\|_{\mathcal{L}^2} = O_P(n^{-(m+\gamma)/[2(m+\gamma)+1]}).$$

Remark 2.1. According to Stone [6] and [7], the convergence rate of $\widehat{g}_n$ in Theorem 2.1 is the optimal global rate of convergence for nonparametric regression; while the convergence rate of $\widehat{\beta}$ here is only a result obtained in passing. In another paper, we show that $\widehat{\beta}$ achieves the usual parametric rates $n^{-1/2}$ (cf. Shi and Li [9]).

Remark 2.2. If $E|X_1|^{1/\delta} < \infty$ with $\delta = (m+\gamma)/[2(m+\gamma)+1]$. then from Theorem 2.3 of Chapter III in Rohatgi [10] (p.98), Condition A6 is true.

3. PROOF OF THEOREM 2.1

Before the proof of Theorem 2.1, some notations and results are needed. Let P denote the probability measure corresponding to the distribution of (T, X) and P_n denote the empirical probability measure associated with $(T_1, X_1), \cdots, (T_n, X_n)$. The notation Ph stands for the expectation of $h(T, X)$, i.e. $Ph \triangleq \int h\,dP$, and the analogues will be used. Put

$$\mathcal{F}_n = \{\varphi(t)^\tau \alpha : \quad \alpha \in R^p\}.$$

Lemma 3.1. If, for some number $c_0 > 0$, $\lim_{n\to\infty} n^{c_0-1} M_n^2 = 0$, then there is a positive constant M_1 depending upon only b, B, and m such that, except on an event whose probability tends to zero with n, for all $h \in \mathcal{H}$ and $g \in \mathcal{F}_n$,

$$\| g - h \|^2_{\mathcal{L}^2} \leq M_1(M_n^{-2(m+\gamma)} + P_n(g-h)^2).$$

Proof. See Shi and Li [11], and also Stone [12].

Let $A_1, \cdots, A_k$ respectively be $l_1 \times l_1, \cdots, l_k \times l_k$ matrices and let $\mathrm{DIAG}(A_1, \cdots, A_k)$ denote the matrix with submatrices $A_i (i = 1, \cdots, k)$ as the diagonal blocks and zero submatrices elsewhere. Set

$$\begin{aligned} D_0 &= \mathrm{diag}(1, \delta_n, \cdots, \delta_n^m)_{(m+1)\times(m+1)}, \\ D &= \mathrm{DIAG}(D_0, \cdots, D_0)_{p\times p}. \end{aligned}$$

Define

$$\left\{\begin{array}{ll}\pi_v(t) &= D_0^{-1}\varphi_v(t) \\ &= \left(I_{nv}(t)(1, \frac{t-t_v}{\delta_n}, \cdots, (\frac{t-t_v}{\delta_n})^m)\right)^\tau, \qquad 1 \le v \le M_n, \\ \pi(t) &= D^{-1}\varphi(t) = (\pi_1^\tau(t), \cdots, \pi_{M_n}^\tau(t))^\tau, \\ Z_n^\tau &= (\pi(T_1), \cdots, \pi(T_n))_{p\times n}, \\ H_{1n}^2 &= \sum_{i=1}^n X_i X_i^\tau, \quad H_{2n}^2 = 2M_n Z_n^\tau Z_n, \\ H_n &= \mathrm{DIAG}(H_{1n}, H_{2n}), \quad z_i = H_n^+ \begin{pmatrix} X_i \\ \pi(T_i)(2M_n)^{1/2} \end{pmatrix}, \end{array}\right\} \tag{3.1}$$

where H_n^+ is the Moore inverse of H_n and the analogues will be used. Denote respectively the smallest eigenvalues of matrices $\frac{1}{n}H_{1n}^2$ and $\frac{1}{n}H_{2n}^2$ by λ_{1n} and λ_{2n}. Set $\Lambda_n = \min(\lambda_{1n}, \lambda_{2n})$.

Proof of Theorem 2.1. Put

$$\begin{aligned}\alpha_{i0}^\tau &= (g_0(t_i), g_0^{(1)}(t_i)/1!, \cdots, g_0^{(m)}(t_i)/m!), \quad 1 \le i \le M_n, \\ \alpha_0 &= (\alpha_{10}^\tau, \cdots, \alpha_{M_n0}^\tau)^\tau.\end{aligned}$$

By Condition A2 and Taylor's Theorem,

$$g_0(t) = \varphi(t)^\tau \alpha_0 - R_{nt}, \qquad \sup_{0\le t\le 1} |R_{nt}|^2 \le M_0^2 \delta_n^{2(m+\gamma)}, \tag{3.2}$$

where

$$R_{nt} = \sum_{v=1}^{M_n} I_{nv}(t)\frac{(t-t_v)^m}{m!}(g_0^{(m)}(t_v) - g_0^{(m)}(t_v + \zeta_v(t-t_v)))$$

with some $\zeta_v \in (0,1), v = 1, \cdots, M_n$. Thus,

$$\begin{aligned}P_n(\widehat{g}_n(t) - g_0(t))^2 &= P_n(\varphi(t)^\tau\widehat{\alpha} - \varphi(t)^\tau\alpha_0 + R_{nv})^2 \\ &\le 2P_n[\varphi(t)^\tau((\widehat{\alpha} - \alpha_0)^2] + 2M_0^2\delta_n^{2(m+\gamma)}.\end{aligned} \tag{3.3}$$

It is clear that as a function of t on [0, 1], $\widehat{g}_n(t) \in \mathcal{F}_n$. By A2 and Lemma 3.1 there exists a positive constant M_1 depending upon only b, B, and m such that, except on an event whose probability tends to zero with n,

$$\|\widehat{g}_n - g_0\|_{\mathcal{L}^2}^2 \le M_1(M_n^{-2(m+\gamma)} + P_n(\widehat{g}_n(t) - g_0(t))^2). \tag{3.4}$$

From A3 and the law of large numbers, we can easily conclude

$$\liminf_{n\to\infty} \lambda_{1n} \ge \lambda_1/2, \quad \text{a.s.} \tag{3.5}$$

where λ_1 is the smallest eigenvalue of EXX^T. Set

$$\widehat{\theta} = H_n\begin{pmatrix}\widehat{\beta} - \beta_0 \\ D(\widehat{\alpha} - \alpha_0)\delta_n^{1/2}\end{pmatrix}.$$

Observe that

$$\begin{aligned}\widehat{\theta}^\tau\widehat{\theta} &= (\widehat{\beta}-\beta_0)^\tau H_{1n}^2(\widehat{\beta}-\beta_0) + (\widehat{\alpha}-\alpha_0)^\tau D Z_n^\tau Z_n D(\widehat{\alpha}-\alpha_0)\\ &= (\widehat{\beta}-\beta_0)^\tau H_{1n}^2(\widehat{\beta}-\beta_0) + nP_n[\varphi(t)^\tau(\widehat{\alpha}-\alpha_0)]^2\end{aligned}$$

Since $2\delta_n = M_n^{-1} \sim n^{-1/[2(m+\gamma)+1]}$ and $m+\gamma > 1/2$, from (3.3)–(3.5), to prove Theorem 2.1, we need only to check that

$$|\widehat{\theta}| = Op(M_n^{1/2}). \tag{3.6}$$

This will be given below in Proposition 3.1

Proposition 3.1. Under the conditions of Theorem 2.1, (3.6) holds.

The proof of this proposition will be given later, Now, we introduce some more preliminary lemmas.

Lemma 3.2. If, for some number $c_0 > 0$, $\lim_{n\to\infty} n^{c_0-1}M_n^2 = 0$ and Condition A1 holds, then

$$\lim_{n\to\infty} \lambda_{2n} \geq \lambda_2, \quad \text{a.s.}$$

where $\lambda_2 = b \inf_{|\beta|=1} \int_{-1}^{1} (\sum_{i=0}^{m} \beta_i z^i)^2 \, dz > 0.$

Proof. See Shi and Li [11].

Let $\lambda = \min(\lambda_1, \lambda_2)/2$. From (3.5) and Lemma 3.2, we deduce that when n is large enough with probability one

$$\Lambda_n \geq \lambda > 0. \tag{3.7}$$

For any $M^* > 0$, put

$$\mathcal{F}_{1n} = \left\{ \beta^\tau x I\{|x| \leq M^*\}\pi(t)^\tau \alpha M_n^{1/2} : \begin{array}{c} \alpha \in R^p, \beta \in R^d \\ |\alpha| \leq 1, |\beta| \leq 1 \end{array} \right\}.$$

Lemma 3.3. If $M_n \sim n^{1/[2(m+\gamma)+1]}$ with $m+\gamma > 1/2$, then for any given positive number M^*,

$$\sup_{g\in\mathcal{F}_{1n}} |(P_n - P)g| \to 0 \quad \text{a.s.} \quad n \to \infty$$

Proof. Set $c_0 = M^*((m+1)M_n)^{1/2}$. Note that $|\pi(t)|^2 \leq (m+1)$, then,

$$\sup_{x\in R^d,\, 0\leq t\leq 1} \sup_{g\in\mathcal{F}_{1n}} |g(x,t)| \leq c_0.$$

For given $\varepsilon > 0$, by a symmetrization argument similar to that in Pollard (1984), pp.14–15, we have that when $n > 8\varepsilon^{-2}c_0^2$

$$\mathcal{P}\{\sup_{g\in\mathcal{F}_{1n}} |P_n g - Pg| > \varepsilon\} \leq 4\mathcal{P}\{\sup_{\mathcal{F}_{1n}} |n^{-1}\sum_{i=1}^{n} \sigma_i h(X_i, T_i)| > \varepsilon/4\}, \tag{3.8}$$

where the sign random variables $\sigma_1, \cdots, \sigma_n$ are independent of $(X_1^\tau, T_1), \cdots,$ (X_n^τ, T_n) and are identically distributed, $\mathcal{P}\{\sigma_i = +1\} = \frac{1}{2}$ and $\mathcal{P}\{\sigma_i = -1\} = \frac{1}{2}$.

Define $S_1 = \{\beta : \beta \in R^d, |\beta| \le 1\}$ and $S_2 = \{\alpha : \alpha \in R^p, |\alpha| \le 1\}$. Write the cube $[-1, 1]^p$ as a union of K_n^p disjoint cubes Γ_{nv} of length $2K_n^{-1}$ and the cube $[-1, 1]^d$ as a union of K_n^{*d} disjoint cubes Γ_{nk}^* of length $2K_n^{*-1}$, where $K_n = [\sqrt{p}/q_0]_I + 1, K_n^* = [\sqrt{d}/q_0]_I + 1$, where $[a]_I$ denotes the largest integer part of real number a, and $q_0 = \varepsilon/(32c_0)$. Set

$$L_n = K_n^p \quad \text{and} \quad L_n^* = K_n^{*d}. \tag{3.9}$$

Choose arbitrarily $\alpha_i \in \Gamma_{ni}, \beta_j \in \Gamma_{nj}^*, i = 1, \cdots, L_n, j = 1, \cdots, L_n^*$. Let

$$\mathcal{H}_{1n} = \left\{\beta_j^\tau x\, I\{|x| \le M^*\}\pi(t)^\tau \alpha_i M_n^{1/2},\ 1 \le i \le L_n, 1 \le j \le L_n^*\right\}.$$

Note that for any $\beta, \beta_0 \in S_1, \alpha, \alpha_0 \in S_2$

$$\begin{aligned}
&|\beta_0^\tau x I\{|x| \le M^*\}\pi(t)^\tau \alpha_0 M_n^{1/2} - \beta^\tau x I\{|x| \le M^*\}\pi(t)^\tau \alpha M_n^{1/2}| \\
&\le |\beta - \beta_0| M^*(m+1)^{1/2} M_n^{1/2} + |\alpha - \alpha_0| M^*(m+1)^{1/2} M_n^{1/2} \\
&= c_0(|\beta - \beta_0| + |\alpha - \alpha_0|),
\end{aligned}$$

and that for all $\beta \in S_1$ and $\alpha \in S_2$

$$\begin{aligned}
&\min_{1 \le j \le L_n^*, 1 \le i \le L_n} c_0(|\beta - \beta_j| + |\alpha - \alpha_i|) \\
&= c_0 \left(\sqrt{4d\left(\left[\frac{\sqrt{d}}{q_0}\right]_I + 1\right)^{-2}} + \sqrt{4(m+1)M_n \left(\left[\frac{\sqrt{(m+1)M_n}}{q_0}\right]_I + 1\right)^{-2}} \right) \\
&\le c_0(2q_0 + 2q_0) = 4c_0 q_0 = \varepsilon/8.
\end{aligned}$$

Hence for any given sample $U^* \triangleq ((X_1, T_1), \cdots, (X_n, T_n))$

$$\min_{g \in \mathcal{H}_{1n}} P_n |g - h| \le \varepsilon/8 \quad \forall\, h \in \mathcal{F}_{1n}.$$

Observe, from this inequality, that for fixed U^*

$$\begin{aligned}
&\left| n^{-1} \sum_{i=1}^n \sigma_i h(X_i, T_i) \right| \\
&\le \min_{g \in \mathcal{H}_{1n}} \left(|n^{-1} \sum_{i=1}^n \sigma_i g(X_i, T_i)| + |n^{-1} \sum_{i=1}^n \sigma_i (g(X_i, T_i) - h(X_i, T_i))| \right) \\
&= \max_{g \in \mathcal{H}_{1n}} |n^{-1} \sum_{i=1}^n \sigma_i g(X_i, T_i)| + \varepsilon/8 \qquad \forall h \in \mathcal{F}_{1n}.
\end{aligned}$$

Therefore

$$\begin{aligned}
&\mathcal{P}\left\{\sup_{\mathcal{F}_{1n}} |n^{-1} \sum_{i=1}^n \sigma_i h(X_i, T_i)| \ge \varepsilon/4 \,\middle|\, U^*\right\} \\
&\le \#(\mathcal{H}_{1n}) \max_{g \in \mathcal{H}_{1n}} \mathcal{P}\left\{ |n^{-1} \sum_{i=1}^n \sigma_i g(X_i, T_i)| \ge \varepsilon/8 \,\middle|\, U^* \right\}.
\end{aligned} \tag{3.10}$$

Hoeffding's Inequality (see Pollard [13], Appendix B) gives that

$$\begin{aligned}
&\mathcal{P}\Big\{|n^{-1}\sum_{i=1}^{n}\sigma_i g(X_i,T_i)| \geq \varepsilon/8 \,\Big|\, U^*\Big\} \\
&\leq 2\exp\Big\{-2(n\varepsilon/8)^2 \Big/ \sum_{i=1}^{n}(2g(X_i,T_i))^2\Big\} \wedge 1 \\
&\leq 2\exp\Big\{-n\varepsilon^2 \Big/ 128c_0^2\Big\} \wedge 1 \quad \text{for all} \quad g \in \mathcal{H}_{1n}.
\end{aligned} \tag{3.11}$$

Combining (3.8) – (3.11), we obtain that when $p = (m+1)M_n \geq d$

$$\begin{aligned}
\sum_{n=1}^{\infty}\mathcal{P}\Big\{\sup_{\mathcal{F}_{1n}}|P_n h - Ph| \geq \varepsilon\Big\} &\leq 8\sum_{n=1}^{\infty}\exp\Big\{-\frac{\varepsilon^2}{128(m+1)M^{*2}}\frac{n}{M_n} \\
&\times\Big[1 - \frac{256M^{*2}(m+1)^2M_n^2}{n\varepsilon^2}\ln(32M^*(m+1)M_n/\varepsilon+1)\Big]\Big\} \wedge 1.
\end{aligned} \tag{3.12}$$

Since $M_n \sim n^{1/[2(m+\gamma)+1]}$ and $m+\gamma > 1/2$, from (3.10), we conclude that

$$\sum_{n=1}^{\infty}\mathcal{P}\Big\{\sup_{\mathcal{F}_{1n}}|P_n h - Ph| \geq \varepsilon\Big\} < \infty.$$

Lemma 3.3 follows from this inequality and Borel - Contelli Lemma.

Lemma 3.4. If A1 is satisfied and $M_n \sim n^{1/[2(m+\gamma)+1]}$ with $m+\gamma > 1/2$, then as $n \to \infty$

$$\sup_{|\alpha|\leq 1}|P_n(\pi(t)^\tau\alpha)^2 M_n - P(\pi(t)^\tau\alpha)^2 M_n| \longrightarrow 0, \quad \text{a.s.}$$

Proof. Set

$$\mathcal{F}_{2n} = \{(\pi(t)^\tau\alpha)^2 : \quad |\alpha| \leq 1\}.$$

By the argument used in the proof of Lemma 3.3, for any given $\varepsilon > 0$, we can find $\alpha_1, \cdots, \alpha_{L_n} \in S_2 \triangleq \{\alpha : \quad \alpha \in R^p, |\alpha| \leq 1\}$ with $L_n \leq K_n^p$ and $K_n \leq 16((m+1)M_n)^{3/2}/\varepsilon + 1$ such that

$$\min_{1\leq k\leq L_n}|\alpha - \alpha_k| < \varepsilon/[8(m+1)M_n] \quad \forall \alpha \in S_2.$$

Thus,

$$\begin{aligned}
&\min_{1\leq k\leq L_n}\Big|n^{-1}\sum_{i=1}^{n}\Big[((\pi(T_i)^\tau\alpha)^2 - P(\pi(t)^\tau\alpha)^2) - ((\pi(T_i)^\tau\alpha_k)^2 - P(\pi(t)^\tau\alpha_k)^2)\Big]\Big| \\
&\leq \min_{1\leq k\leq L_n} 4(m+1)|\alpha - \alpha_k| \leq \frac{\varepsilon}{2M_n} \qquad \forall \alpha \in S_2.
\end{aligned} \tag{3.13}$$

Put $\mathcal{H}_{2n} = \{(\pi(t)^\tau\alpha_k)^2 : \quad k = 1, \cdots, L_n\}$. Since, from (3.13), for any $g \in \mathcal{F}_{2n}$

$$\begin{aligned}
|P_n g - P g| &\leq \min_{h\in\mathcal{H}_{2n}}(|P_n h - Ph| + |(P_n - P)(g-h)|) \\
&\leq \min_{h\in\mathcal{H}_{2n}}\Big(\max_{h'\in\mathcal{H}_{2n}}|P_n h' - Ph'| + |(P_n - P)(g-h)|\Big) \\
&\leq \max_{h\in\mathcal{H}_{2n}}|P_n h - Ph| + \varepsilon/(2M_n),
\end{aligned}$$

we get

$$\mathcal{P}\ \left\{\sup_{\mathcal{F}_{2n}} |(P_n - P)g| \geq \varepsilon/M_n\right\} \leq \mathcal{P}\left\{\sup_{\mathcal{H}_{2n}} |(P_n - P)g| \geq \varepsilon/2M_n\right\} \\ \leq L_n \max_{g\in\mathcal{H}_{2n}} \mathcal{P}\left\{|(P_n - P)g| \geq \varepsilon/2M_n\right\}. \tag{3.14}$$

From A1 and the definition of $\pi(\cdot)$

$$\mathrm{Var}((\pi(T_i)^\tau\alpha)^2) \leq \mathcal{P}(\pi(T_i)^\tau\alpha)^4 = \sum_{k=1}^{M_n}\int_{I_{nk}}\left(\left(1, \frac{t-t_k}{\delta_n}, \cdots, \left(\frac{t-t_k}{\delta_n}\right)^m\right)\eta_k\right)^4 f(t)\,dt \\ \leq \frac{B(m+1)}{2M_n}\sum_{k=1}^{M_n}\int_{-1}^{1}((1, z, \cdots, z^m)\eta_k)^2\,dz,$$

where $\alpha = (\eta_1^\tau, \cdots, \eta_{M_n}^\tau)^\tau$. It is easily seen that $\int_{-1}^{1}((1,z,\cdots,z^m)\eta_k)^2\,dz \leq 2(m+1)|\eta_k|^2$. Hence $\mathrm{Var}((\pi(T_i)^\tau\alpha)^2) \leq B(m+1)^2/M_n$. Bernstein's Inequality (cf. Pollard [13], p.193) yields that uniformly in $g \in \mathcal{H}_{2n}$

$$\mathcal{P}\ \{|(P_n - P)g| \geq \varepsilon/2M_n\} \\ \leq 2\exp\left\{-n\left(\frac{\varepsilon}{2M_n}\right)^2\Big/[2(B(m+1)^2/M_n + 2(m+1)\varepsilon/M_n)]\right\} \\ \leq 2\exp\left\{-\frac{\varepsilon^2}{16B(m+1)^2}\frac{n}{M_n}\right\}$$

for $0 < \varepsilon \leq B(m+1)/2$. Therefore, from this inequality and (3.14)

$$\mathcal{P}\ \{\sup_{\mathcal{F}_{2n}} |(P_n - P)g| \geq \varepsilon/M_n\} \leq 2\exp\left\{-\frac{\varepsilon^2}{16B(m+1)^2}\frac{n}{M_n}\left(1 \right.\right. \\ \left.\left. -\frac{16B((m+1)^2M_n)^2}{n\varepsilon^2}\ln\left(\frac{16((m+1)M_n)^{3/2}}{\varepsilon}+1\right)\right)\right\}. \tag{3.15}$$

Noticing that $M_n \sim n^{1/[2(m+\gamma)+1]}$, $m+\gamma > 1/2$, from (3.15) we obtain

$$\sum_{n=1}^{\infty}\mathcal{P}\left\{\sup_{\mathcal{F}_{2n}}|(P_n - P)g| \geq \varepsilon/M_n\right\} < \infty.$$

Lemma 3.4 follows from this inequality and Borel - Cantelli Lemma.

Lemma 3.5. Suppose that A1 and A3 are satisfied. If $M_n \sim n^{1/[2(m+\gamma)+1]}$, $m+\gamma > 1/2$, then as $n \to \infty$

$$\sup_{|\alpha|\leq 1,\,|\beta|\leq 1} |n^{-1}\sum_{i=1}^{n}\beta^\tau X_i\pi(T_i)^\tau\alpha M_n^{1/2}|\Big/(\lambda_{2n}\,\lambda_{1n})^{1/2} \longrightarrow 0 \quad \text{a.s.}$$

Proof. From (3.7), we need only to show that as $n \to \infty$

$$\sup_{|\alpha|\leq 1,\,|\beta|\leq 1} |n^{-1}\sum_{i=1}^{n}\beta^\tau X_i\pi(T_i)^\tau\alpha M_n^{1/2}| \longrightarrow 0 \quad \text{a.s.}$$

Since X and T are independent and $E(X) = 0$, we have $P\,\beta^\tau x\pi(t)^\tau\alpha = 0$. Thus,

$$
\begin{aligned}
&\sup_{|\alpha|\le 1, |\beta|\le 1} |P_n\beta^\tau x\pi(t)^\tau\alpha M_n^{1/2}| \\
&= \sup_{|\alpha|\le 1, |\beta|\le 1} |P_n\beta^\tau x\pi(t)^\tau\alpha M_n^{1/2} - P\,\beta^\tau x\pi(t)^\tau\alpha M_n^{1/2}| \\
&\le \sup_{|\alpha|\le 1, |\beta|\le 1} |P_n\beta^\tau xI\{|x|\le M^*\}\pi(t)^\tau\alpha M_n^{1/2} - P\,\beta^\tau xI\{|x|\le M^*\}\pi(t)^\tau\alpha M_n^{1/2}| \\
&\quad +(P_n|x|^2I\{|x|>M^*\})^{1/2}(\sup_{|\alpha|\le 1}(P_n(\pi(t)^\tau\alpha)^2M_n)^{1/2} \\
&\quad +(P\,|x|^2I\{|x|>M^*\})^{1/2}\sup_{|\alpha|\le 1}(P(\pi(t)^\tau\alpha)^2M_n)^{1/2} \\
&\triangleq J_{1n} + J_{2n} + J_{3n}.
\end{aligned} \tag{3.16}
$$

By the argument similar to that used in the proof of Lemma 3.4, we can easily conclude

$$\sup_{|\alpha|\le 1} P\,(\pi(t)^\tau\alpha)^2M_n \le B(m+1), \tag{3.17}$$

$$\sup_{|\alpha|\le 1} P\,(\pi(t)^\tau\alpha)^2M_n \ge \frac{b}{2}\alpha^\tau DIAG(A,\cdots,A)\alpha, \tag{3.18}$$

where $A = \int_{-1}^1(1,z,\cdots,z^m)^\tau(1,z,\cdots,z^m)\,dz$. From a compactness argument, we can easily obtain

$$\inf_{|\xi^*|=1} \xi^{*\tau}A\xi^* > 0. \tag{3.19}$$

Since $P\,x^\tau x = \mathrm{trace}(P\,xx^\tau) < \infty$, for any given $\varepsilon > 0$, we can take $M^* > 0$ (fixed) such that

$$P\,|x|^2I\{|x|>M^*\}B(m+1) < \varepsilon.$$

This and (3.17) yields

$$J_{3n}^2 \le \varepsilon \qquad \text{for all } n. \tag{3.20}$$

From (3.18), (3.19), and Lemma 3.4, as $n \to \infty$

$$
\begin{aligned}
&\sup_{|\alpha|\le 1} |(P_n\,(\pi(t)^\tau\alpha)^2M_n)^{1/2} - (P(\pi(t)^\tau\alpha)^2M_n)^{1/2}| \\
&\le 2\sup_{|\alpha|\le 1} |P_n\,(\pi(t)^\tau\alpha)^2M_n - P(\pi(t)^\tau\alpha)^2M_n| \Big/ (b\inf_{|\xi^*|=1} \xi^{*\tau}A\xi^*)^{1/2} \\
&\longrightarrow 0. \quad \text{a.s.}
\end{aligned} \tag{3.21}
$$

Then, a simple calculation and the law of large numbers give that for any M^* fixed

$$|J_{2n}^2 - J_{3n}^2| \longrightarrow 0 \qquad a.s. \qquad (n\to\infty). \tag{3.22}$$

Thus, Lemma 3.5 follows from (3.7), (3.16), (3.20), (3.22) and Lemma 3.3.

Proof of Proposition 3.1. Set $B_n = \{\Lambda_n \ge \lambda\}$. From (3.7), $\mathcal{P}(B_n^c) \to 0$, as $n \to \infty$. Therefore, to prove Proposition 3.1, we need only to show that

$$\lim_{L\to\infty}\overline{\lim_{n\to\infty}}\,\mathcal{P}\{|\hat{\theta}| \ge LM_n^{1/2},\ B_n\} = 0. \tag{3.23}$$

By definitions of z_i and $\widehat{\theta}$, on B_n

$$\sum_{i=1}^{n} \Psi(u_i - z_i^{\tau}\widehat{\theta} - R_{ni})z_i = 0, \tag{3.24}$$

where $R_{ni} = R_{nT_i}$. For $l \in R$ and $\theta \in R^p$, define $U(\theta, l) = \sum_{i=1}^{n} \Psi(u_i - z_i^{\tau}\theta l - R_{ni})z_i^{\tau}\theta$. It is easily seen, from A4, that for $l_1 \geq l_2$ and $\forall\theta$, $U(\theta, l_1) \leq U(\theta, l_2)$. Observe, from this fact and (3.24), that when $|\widehat{\theta}| \geq M_n^{1/2}L$

$$\begin{aligned} 0 &= U\left(\frac{\widehat{\theta}}{|\widehat{\theta}|}, |\widehat{\theta}|\right) \leq U\left(\frac{\widehat{\theta}}{|\widehat{\theta}|}, M_n^{1/2}L\right) \\ &\leq \sup_{|\theta|=1} U(\theta, M_n^{1/2}L). \end{aligned}$$

Consequently

$$\mathcal{P}\{|\widehat{\theta}| \geq LM_n^{1/2},\ B_n\} \leq \mathcal{P}\{\sup_{|\theta|=1} U(\theta, M_n^{1/2}L) \geq 0,\ B_n\}. \tag{3.25}$$

Note that

$$|z_i|^2 \leq \max_{1\leq i\leq n} |X_i|^2 \Big/ n\lambda_{1n} + 2(m+1)M_n \Big/ n\lambda_{2n}. \tag{3.26}$$

By A4 and A6, (3.7), and inequality (3.26), for any given $\delta > 0$, we can first find $L_0 > 0$, then find $n_1 = n_1(\delta, L_0) > 0$ such that when $n \geq n_1$

$$\left\{\begin{array}{l} \dfrac{d_2}{2}\left(\dfrac{5}{4} + \dfrac{nM_0^2}{M_n^{2(m+\gamma)+1}}\right) \leq \dfrac{d_1 q L_0}{4}, \\ \dfrac{16\, E\,\Psi^2(u)\,((m+1)M_n + d)}{(d_1 q L_0)^2 M_n} < \delta/5, \\ \mathcal{P}\{\max\limits_{1\leq i\leq n} |z_i|^2((m+1)M_n + d) \geq \dfrac{\delta q}{80(1-q)}\} < \delta/5, \\ \mathcal{P}\Big\{L_0 M_n^{1/2} \max\limits_{1\leq i\leq n} |z_i| + M_0 \Big/ M_n^{(m+\gamma)} \geq \min\{c_2, c_3\}\Big\} < \delta/5. \end{array}\right\} \tag{3.27}$$

By (3.7) and Lemma 3.5, there exists $n_2 > 0$ such that when $n \geq n_2$

$$\begin{aligned} &\mathcal{P}\{\lambda_{1n} \leq \lambda\} + \mathcal{P}\{\lambda_{2n} \leq \lambda\} \\ &+\mathcal{P}\Big\{\sup_{\substack{|\alpha|\leq 1 \\ |\beta|\leq 1}} |n^{-1}\sum_{i=1}^{n} \beta^{\tau} X_i \pi(T_i)^{\tau}\alpha| M_n^{1/2} \Big/ (\lambda_{1n}\lambda_{2n})^{1/2} \geq 1/8\Big\} \\ &< \delta/5. \end{aligned} \tag{3.28}$$

Put

$$\begin{aligned} \Omega_n &= B_n \cap \{\max_{1\leq i\leq n} |z_i|^2((m+1)M_n + d) < \frac{\delta q}{80(1-q)}\} \\ &\cap\Big\{\sup_{\substack{|\alpha|\leq 1 \\ |\beta|\leq 1}} |n^{-1}\sum_{i=1}^{n} \beta^{\tau} X_i \pi(T_i)^{\tau}\alpha| M_n^{1/2} \Big/ (\lambda_{1n}\lambda_{2n})^{1/2} < 1/8\Big\} \\ &\cap\{L_0 M_n^{1/2} \max_{1\leq i\leq n} |z_i| + M_0 \Big/ M_n^{(m+\gamma)} < \min\{c_2, c_3\}\}. \end{aligned}$$

Thus, from (3.27) and (3.28) when $n \geq \max\{n_1, n_2\}$

$$\begin{aligned}&\mathcal{P}\left\{\sup_{|\theta|=1} U(\theta, L_0 M_n^{1/2}) \geq 0,\ B_n\right\}\\ &\leq \mathcal{P}\left\{\sup_{|\theta|=1} U(\theta, L_0 M_n^{1/2}) \geq 0,\ \Omega_n\right\} + 3\delta/5.\end{aligned} \tag{3.29}$$

From Lemma 3.6 below, when $n \geq \max\{n_1,\ n_2\}$

$$\mathcal{P}\left\{\sup_{|\theta|=1} U(\theta, L_0 M_n^{1/2}) \geq 0, \Omega_n\right\} < 2\delta/5.$$

Combining this inequality, (3.25), and (3.29), we complete the proof of (3.23).

Lemma 3.6. Under the notations above and the conditions of Theorem 2.1, when $n \geq \max\{n_1,\ n_2\} \triangleq n_0$

$$\mathcal{P}\left\{\sup_{|\theta|=1} U(\theta, L_0 M_n^{1/2}) \geq 0, \Omega_n\right\} < 2\delta/5. \tag{3.30}$$

Proof. Observe, from the definitions of $U(\theta, l)$ and $D(s,t)$ that

$$\begin{aligned} U(\theta, L_0 M_n^{1/2}) M_n^{-1/2} &= \sum_{i=1}^n \Psi(u_i) z_i^T \theta M_n^{-1/2}\\ &- L_0 \sum_{i=1}^n D(u_i, -L_0 M_n^{1/2} z_i^T\theta - R_{ni})(z_i^T\theta)^2\\ &- \sum_{i=1}^n D(u_i, -L_0 M_n^{1/2} z_i^T\theta - R_{ni}) z_i^T\theta R_{ni} M_n^{-1/2}.\end{aligned} \tag{3.31}$$

Put

$$V_n = d_1 \sum_{i=1}^n z_i z_i^T I\{|u_i| \leq c_1\}, \quad V_0 = d_1 q \sum_{i=1}^n z_i z_i^T.$$

We divide the proof of Lemma 3.6 into three steps.

Step 1 When $n \geq n_0$

(i) $\inf_{|\theta|=1} \sum_{i=1}^n D(u_i, -L_0 M_n^{1/2} z_i^T\theta - R_{ni})(z_i^T\theta)^2 I(\Omega_n) \geq \frac{3d_1 q}{4} I(\Omega_n) - \| V_n - V_0 \| I(\Omega_n)$,

(ii) $\mathcal{P}\{\| V_n - V_0 \| \geq \frac{d_1 q}{4}, \Omega_n \mid U^*\} < \delta/5$,

where $\| \cdot \|$ stands for the norm of matrices defined by $\|A\|^2 = \mathrm{trace}(AA^T)$ and $U^* = ((X_1, T_1), \cdots, (X_n, T_n))$.

Proof. Note, from the definition of Ω_n, that $\max_{1\leq i\leq n}\{L_0 M^{1/2}|z_i| + |R_{ni}|\} I(\Omega_n) \leq c_2$. By

A4, A5 and Lemma 3.5

$$\inf_{|\theta|=1} \sum_{i=1}^{n} D(u_i, -L_0 M_n^{1/2} z_i^T\theta - R_{ni})(z_i^T\theta)^2 I(\Omega_n)$$
$$\geq \inf_{|\theta|=1} \sum_{i=1}^{n} D(u_i, -L_0 M_n^{1/2} z_i^T\theta - R_{ni})(z_i^T\theta)^2 I\{|u_i| \leq c_1\} I(\Omega_n)$$
$$\geq d_1 \inf_{|\theta|=1} \sum_{i=1}^{n} (z_i^T\theta)^2 I\{|u_i| \leq c_1\} I(\Omega_n)$$
$$\geq \frac{3d_1 q}{4} I(\Omega_n) - \| V_n - V_0 \| I(\Omega_n).$$

This completes the proof of (i).

For (ii), by Tchebychev Inequality, conditioning on U^*, and the definition of Ω_n, we deduce

$$\mathcal{P}\{ \| V_n - V_0 \| \geq d_1 q/4, \Omega_n | U^*\} \leq \frac{16}{(d_1 q)^2} \mathcal{P}\{\| V_n - V_0 \|^2 I(\Omega_n) \Big| U^*\}$$
$$= \frac{16}{q^2} \mathrm{trace}\left(\mathcal{P}\left[\sum_{i=1}^{n}\sum_{j=1}^{n} z_i z_i^T z_j z_j^T (I\{|u_i| \leq c_1\} - q)(I\{|u_j| \leq c_1\} - q) \Big| U^* \right]\right) I(\Omega_n)$$
$$= \frac{16(1-q)}{q} \sum_{i=1}^{n} |z_i|^4 I(\Omega_n)$$
$$\leq \frac{16(1-q)((m+1)M_n + d)}{q} \max_{1\leq i \leq n} |z_i|^2 I(\Omega_n) < \delta/5.$$

Step 2. When $n \geq n_0$,

$$\sup_{|\theta|=1} \left| \sum_{i=1}^{n} D(u_i, -L_0 M_n^{1/2} z_i^T\theta - R_{ni}) z_i^T\theta R_{ni} M_n^{-1/2} \right| I(\Omega_n) < \frac{d_1 q L_0}{4}.$$

Proof. By definition of Ω_n, when $n \geq n_0$

$$\max_{1\leq i\leq n} \{L_0 M_n^{1/2} |z_i| + |R_{ni}|\} I(\Omega_n) \leq c_3.$$

Then using A5 and (3.27) we get

$$\sup_{|\theta|=1} \left| \sum_{i=1}^{n} D(u_i, -L_0 M_n^{1/2} z_i^T\theta - R_{ni}) z_i^T\theta R_{ni} M_n^{-1/2} \right| I(\Omega_n)$$
$$\leq d_2 \sup_{|\theta|=1} \sum_{i=1}^{n} |z_i^T\theta R_{ni}| M_n^{-1/2} I(\Omega_n) \leq \frac{d_2}{2} \sup_{|\theta|=1} \left(\sum_{i=1}^{n} (z_i^T\theta)^2 + \frac{n M_0^2}{M_n^{2(m+\gamma)+1}} \right) I(\Omega_n)$$
$$< \frac{d_1 q L_0}{4}.$$

Step 3. Inequality (3.30) is true.

Proof. From (3.31) and Steps 1 and 2, when $n \geq n_0$

$$\begin{aligned}
&\mathcal{P}\{\sup_{|\theta|=1} U(\theta, L_0 M_n^{1/2}) \geq 0, \Omega_n \Big| U^*\} \\
&\leq \mathcal{P}\{\sup_{|\theta|=1} \sum_{i=1}^{n} \Psi(u_i) z_i^T \theta M_n^{-1/2} - L_0 \frac{d_1 q}{2} + L_0 \| V_n - V_0 \| \geq 0, \Omega_n \Big| U^*\} \\
&\leq \mathcal{P}\{\sup_{|\theta|=1} \sum_{i=1}^{n} \Psi(u_i) z_i^T \theta M_n^{-1/2} \geq \frac{d_1 q L_0}{4}, \Omega_n \Big| U^*\} + \mathcal{P}\{\| V_n - V_0 \| \geq \frac{d_1 q}{4}, \Omega_n \Big| U^*\} \\
&\leq \mathcal{P}\{\sup_{|\theta|=1} |\sum_{i=1}^{n} \Psi(u_i) z_i^T \theta M_n^{-1/2}| \geq \frac{d_1 q L_0}{4}, \Omega_n \Big| U^*\} + \delta/5.
\end{aligned} \tag{3.32}$$

Condition A4, Tchebychev Inequality and (3.27) yield that when $n \geq n_0$

$$\begin{aligned}
&\mathcal{P}\{\sup_{|\theta|=1} |\sum_{i=1}^{n} \Psi(u_i) z_i^T \theta M_n^{-1/2}| \geq \frac{d_1 q L_0}{4}, \Omega_n \Big| U^*\} \\
&\leq \frac{16}{(d_1 q)^2 L_0^2 M_n} \mathcal{P}\{|\sum_{i=1}^{n} \Psi(u_i) z_i|^2 I(\Omega_n) \Big| U^*\} \\
&= \frac{16}{(d_1 q)^2 L_0^2 M_n} \mathrm{trace}\Big(\mathcal{P}\{\sum_{i=1}^{n} \Psi(u_i) z_i \sum_{j=1}^{n} \Psi(u_j) z_j^T I(\Omega_n) \Big| U^*\}\Big) \\
&= \frac{16((m+1)M_n + d)\mathcal{P}(\Psi^2(u))}{(d_1 q)^2 L_0^2 M_n} < \delta/5.
\end{aligned}$$

From this inequality and (3.32), we conclude (3.30).

Acknowledgment: This work was supported by the National Natural Science Foundation of China.

REFERENCES:

[1] R. F. Engle, C. W. J. Granger, J. Rice, and A. Weiss, Semiparametric estimates of the relation between weather and electricity sales, *J. Amer. Statist. Assoc.*, **81**, 310-320 (1986).

[2] G. Wahba, Cross validated spline methods for the estimation of multivariate functions from data on functionals, In *Statistics: an Appraisal, Proceedings* 50*th Anniversary Conference Iowa State Statistical Laboratory* (H. A. David and H. T. David, eds.) 205-235. Iowa State Univ. Press, Ames, Ia. (1984).

[3] N. E. Heckman, Spline smoothing in a partly linear model, *J. Roy. Statist. Soc. Ser. B*, **48**, 244-248 (1986).

[4] J. Rice, Convergence rates for partially spline models, *Statist. Probab. Lett.*, **4**, 203-208 (1986).

[5] H. Chen, Convergence rates for parametric components in a partly linear model, *Ann. Statist.*, **16**, 136-146 (1988).

[6] C. Stone, Optimal rates of convergence for nonparametric estimators, *Ann. Statist.*, **8**, 1348-1360 (1980).

[7] C. Stone, Optimal global rates of convergence for nonparametric regression, *Ann. Statist.*, **10**, 1040-1053 (1982).

[8] Z. D. Bai, Y. H. Wu, X. R. Chen, and B. Q. Miao, On solvability of an equation arising in the theory of M - estimates, *Commun. Statist. Theory Meth.*, **19**, 363-380 (1990).

[9] P. Shi and G. Li, Asymptotic normality of the M-estimates for parametric components in a partly linear model, submitted.

[10] V. K. Rohatgi, *An Introduction to Probability and Mathematical Statistics*, John Wiley & Sons, Inc. (1976).

[11] P. Shi and G. Li, Optimal global rates of convergence of M-estimates for nonparametric regression, Submitted.

[12] C. Stone, Additive regression and other nonparametric models, *Ann. Statist.*, **13**, 689-705 (1985).

[13] D. Pollard, *Convergence of Stochastic Processes*, Springer-Verlag, New York (1984).

Stat. Sci. & Data Anal., pp. 467-474
K. Matsusita *et al.* (Eds)

Discrete Distributions Related to Succession Events in a Two-State Markov Chain

SIGEO AKI and KATUOMI HIRANO

Department of Mathematical Science, Faculty of Engineering Science, Osaka University, Toyonaka 560, Japan

The Institute of Statistical Mathematics, 4-6-7 Minami-Azabu, Minato-ku, Tokyo 106, Japan

Abstract. Let $X_1, X_2, \cdots$ be a time-homogeneous { 0,1 } -valued Markov chain. Let E_0 be the event that a run of "0" of length r occurs and let E_1 be the event that a run of "1" of length k occurs in the sequence $X_1, X_2, \cdots$. Discrete distributions related to the events E_0 and E_1 are studied. The probability generating functions of the distributions of the waiting times of the sooner and later occurring events are given. The probability generating function of the distribution of the number of occurrences of E_1 in $X_1, X_2, \cdots, X_n$ is also obtained.

Key words: Sooner and later problems, generalized probability generating function, discrete distributions, Markov chain, geometric distribution of order k, binomial distribution of order k.

1. INTRODUCTION

In recent years exact discrete distribution theory related to succession events has been developed. Typical distributions are called discrete distributions of order k. For example, the geometric distribution of order k is the distribution of the number of trials until the first occurrence of the k-th consecutive success in a sequence of independent trials with success probability p. The binomial distribution of order k is the distribution of the number of occurrences of consecutive k successes until the n-th trial (cf. e.g., Philippou, Georghiou and Philippou[1] and Hirano[2]). The geometric distribution of order k is one of the simplest waiting time distributions. Several waiting time problems have been studied by many authors in more general situations (cf. e.g., Li[3], Gerber and Li[4], Aki[5], Ebneshahrashoob and Sobel[6], Móri[7] and Aki[8]).

The system called a consecutive-k-out-of-n:F system is an interesting example of succession events in a sequence of { 0,1 } -valued random variables (cf. e.g., Aki[5], Hirano[2] and Philippou[9]). In taking it into account, we are very interested in dependent sequences of random variables, since the i.i.d. case has been already studied. Hence, a time-homogeneous two-state Markov chain is a useful model because it contains independent trials and dependent stationary sequences as special cases. We think that investigating discrete distributions related to succession events in a two-state Markov chain is important not only for theoretical interest but also for the above applications.

Let $X_0, X_1, X_2, \cdots$ be a time-homogeneous { 0,1 } -valued Markov chain defined by

$$\begin{aligned} P(X_0 = 0) &= p_0, \qquad 0 < p_0 < 1, \\ P(X_0 = 1) &= p_1 = 1 - p_0, \end{aligned}$$

$$P(X_{i+1}=0|X_i=0)=p_{00}$$
$$P(X_{i+1}=1|X_i=0)=p_{01}$$
$$P(X_{i+1}=0|X_i=1)=p_{10}$$
$$P(X_{i+1}=1|X_i=1)=p_{11} \quad \text{for } i=0,1,2,\cdots,$$

where $p_{00}+p_{01}=1$ and $p_{10}+p_{11}=1$.

In Section 2, we consider the sooner and the later waiting time problems for the event that a run of "0" of length r occurs and the event that a run of "1" of length k occurs in the Markov chain. The probablity generating functions of the distributions of the sooner and the later waiting times are given. As a special case, the probability generating function of the distribution of the waiting time for a run of "1" of length k is also obtained. In Section 3, the distribution of the number of runs of "1" of length k until the n-th trial is studied. A recurrence relation of the probabilities of the distribution and the probability generating function of the distribution are given.

2. SOONER AND LATER WAITING TIME PROBLEMS

Let E_0 be the event that a run of "0" of length r occurs and let E_1 be the event that a run of "1" of length k occurs in the sequence $X_1, X_2, \cdots$. We are interested in these two events.

First, we solve the sooner waiting time problem by using the method of generalized probability generating function (gpgf) (cf. Ebneshahrashoob and Sobel[6]).

Definition 1. Let W be a nonnegative integer valued random variable defined on a probability space $(\Omega,\mathcal{F},P)$. Suppose that we are given a partition of Ω, i.e. Ω can be written as a disjoint union of finite or countably many subsets $\{\Omega_i\}$ of Ω. Suppose also that there exists one to one correspondence between $\{\Omega_i\}$ and a set of letters $\{x_i\}$. By the generalized probability generating function of W with markers $\{x_i\}$ we mean the formal expression

$$\phi(t)=\sum_{u=0}^{\infty}\{\sum_i P(\{W=u\}\cap\Omega_i)x_i\}t^u.$$

Let x be the marker which represents that the event E_1 comes sooner and let y be the marker for the other case. To be precise, we let $\Omega_1=\{E_1 \text{ comes sooner }\}$ and $\Omega_2=\{E_0 \text{ comes sooner }\}$ and we let x and y be the corresponding markers to Ω_1 and Ω_2 respectively. We let $\phi(t)$ denote the gpgf of the distribution of the waiting time for the occurrence of the sooner event. Let $\xi(t)$ be the gpgf of the conditional distribution of the waiting time given that $X_0=0$ and let $\psi(t)$ be the gpgf of the conditional distribution of the waiting time given that $X_0=1$. Let $\phi_i(t)$ be the gpgf of the conditional distribution of the waiting time given that we start with a run of "1" of length i and let $\phi^{(j)}(t)$ be the gpgf of the conditional distribution of the waiting time given that we start with a run of "0" of length j. Then, $\phi(t)$, $\xi(t)$, $\psi(t)$, $\phi_i(t); i=1,2,\cdots,k-1$, and $\phi^{(j)}(t); j=1,2,\cdots,r-1$ satisfy the following system of equations:

$$\phi(t)=p_0\xi(t)+p_1\psi(t) \tag{1}$$

$$\xi(t)=p_{01}t\phi_1(t)+p_{00}t\phi^{(1)}(t) \tag{2}$$

$$\psi(t)=p_{11}t\phi_1(t)+p_{10}t\phi^{(1)}(t) \tag{3}$$

$$\left\{\begin{array}{rcl}\phi_1(t) &=& p_{11}t\phi_2(t)+p_{10}t\phi^{(1)}(t)\\ &\vdots& \\ \phi_{k-2}(t) &=& p_{11}t\phi_{k-1}(t)+p_{10}t\phi^{(1)}(t)\\ \phi_{k-1}(t) &=& p_{11}tx+p_{10}t\phi^{(1)}(t)\end{array}\right. \tag{4}$$

$$\left\{\begin{array}{rcl}\phi^{(1)}(t) &=& p_{01}t\phi_1(t)+p_{00}t\phi^{(2)}(t)\\ &\vdots& \\ \phi^{(r-2)}(t) &=& p_{01}t\phi_1(t)+p_{00}t\phi^{(r-1)}(t)\\ \phi^{(r-1)}(t) &=& p_{01}t\phi_1(t)+p_{00}ty\end{array}\right. \tag{5}$$

By solving (4) and (5), we have

$$\phi_1(t)=(p_{11}t)^{k-1}x+B(t)p_{10}t\phi^{(1)}(t), \tag{6}$$

$$\phi^{(1)}(t)=(p_{00}t)^{r-1}y+A(t)p_{01}t\phi_1(t), \tag{7}$$

where

$$A(t)=\frac{1-(p_{00}t)^{r-1}}{1-p_{00}t} \text{ and } B(t)=\frac{1-(p_{11}t)^{k-1}}{1-p_{11}t}.$$

From (6) and (7), we obtain

Theorem 1. *The generalized probability generating functions $\phi(t)$, $\xi(t)$ and $\psi(t)$ are given by*

$$\begin{array}{rcl}\phi(t) &=& p_0\xi(t)+p_1\psi(t),\\ \xi(t) &=& p_{01}t\phi_1(t)+p_{00}t\phi^{(1)}(t),\\ \psi(t) &=& p_{11}t\phi_1(t)+p_{10}t\phi^{(1)}(t),\end{array}$$

where

$$\phi_1(t)=\frac{(p_{11}t)^{k-1}x+B(t)(p_{10}t)(p_{00}t)^{r-1}y}{1-A(t)B(t)(p_{01}t)(p_{10}t)}$$

and

$$\phi^{(1)}(t)=\frac{A(t)(p_{11}t)^{k-1}(p_{01}t)x+(p_{00}t)^{r-1}y}{1-A(t)B(t)(p_{01}t)(p_{10}t)}.$$

Remark 1. In Theorem 1, setting $x=y=1$, we have the (ordinary) probability genarating function. Further, by letting $r\to\infty$, we get the pgf of the waiting time for a run of "1" of length k (see the following corollary). This distribution is an extension of the geometric distribution of order k. Rajarshi[10] studied the distribution under the special initial condition by using the renewal theory (cf. Feller[11]).

Corollary. Let X be a random variable having the distribution of the waiting time for a run of "1" of length k in the sequence of the Markov chain. Then, the pgf and the probability function of X are written respectively as

$$\frac{(p_0p_{01}+p_1p_{11})p_{11}{}^{k-1}t^k+p_1(p_{01}p_{10}-p_{00}p_{11})p_{11}{}^{k-1}t^{k+1}}{1-(p_{00}t+p_{01}p_{10}t^2+p_{01}p_{10}p_{11}t^3+\cdots+p_{01}p_{10}p_{11}{}^{k-2}t^k)}$$

and

$$P(X=x)=\begin{cases} p_0p_{01}{p_{11}}^{k-1}+p_1{p_{11}}^k & \text{if } x=k,\\ p_0\cdot\sum_{x_1+2x_2+\cdots+kx_k=x-k}\binom{x_1+\cdots+x_k}{x_1,\cdots,x_k}{p_{00}}^{x_1}{p_{01}}^{x_2+\cdots+x_k+1} \\ \times {p_{10}}^{x_2+\cdots+x_k}{p_{11}}^{x_3+2x_4+\cdots+(k-2)x_k+k-1} \\ +p_1\cdot\sum_{j=1}^{k}{p_{11}}^{j-1}p_{10}\cdot\sum_{x_1+2x_2+\cdots+kx_k=x-k-j}\binom{x_1+\cdots+x_k}{x_1,\cdots,x_k} \\ \times {p_{00}}^{x_1}{p_{01}}^{x_2+\cdots+x_k+1}{p_{10}}^{x_2+\cdots+x_k} \\ \times {p_{11}}^{x_3+2x_4+\cdots+(k-2)x_k+k-1} & \text{if } x\ge k+1. \end{cases}$$

Next, we give a solution of the later waiting time problem, that is, we get the gpgf of the distribution of the waiting time for the later occurring event between E_0 and E_1. We denote by y the marker which means that the event E_0 occurs later and by x the marker which means the event E_1 occurs later. Let $\phi(t)$ be the gpgf of the distribution of the waiting time for the later event. Let $\xi(t)$ and $\psi(t)$ be the gpgf of the conditional distribution of the waiting time given that $X_0=0$ and that $X_0=1$, respectively. For $i=0,1,\cdots,k-1$, let $\phi_i(t)$ be the gpgf of the conditional distribution of the waiting time given that we start with a run of "1" of length i. For $j=0,1,\cdots,r-1$, let $\phi^{(j)}(t)$be the gpgf of the conditional distribution of the waiting time given that we start with a run of "0" of length j. For $i=0,1,\cdots,k-1$, let $\psi_i(t)$ be the gpgf of the conditional distribution of the waiting time given that the sooner event is E_1 and E_1 has already occurred and we are currently in a run of "1" of length i. For $j=0,1,\cdots,r-1$, let $\psi^{(j)}(t)$ be the gpgf of the conditional distribution given that the sooner event is E_1 and E_1 has already occurred and we are currently in a run of "0" of length j. For $i=0,1\cdots,k-1$, let $\xi_i(t)$ be the gpgf of the conditional distribution given that the sooner event is E_0 and E_0 has already occurred and we are currently in a run of "1" of length i. For $j=0,1,\cdots,r-1$, let $\xi^{(j)}(t)$ be the gpgf of the conditional distribution given that the sooner event is E_0 and E_0 has already occurred and we are currently in a run of "0" of length j. Note that $\phi_0(t)=\phi^{(0)}(t)$, $\psi_0(t)=\psi^{(0)}(t)$ and $\xi_0(t)=\xi^{(0)}(t)$. Then $\phi_0,\cdots,\phi_{k-1}$, $\phi^{(0)},\cdots,\phi^{(r-1)}$, $\psi_0,\cdots,\psi_{k-1}$, $\psi^{(0)},\cdots,\psi^{(r-1)}$, $\xi_0,\cdots,\xi_{k-1}$ and $\xi^{(0)},\cdots,\xi^{(r-1)}$ satisfy the following system of equations:

$$\phi(t)=p_0\xi(t)+p_1\psi(t) \tag{8}$$

$$\xi(t)=p_{01}t\phi_1(t)+p_{00}t\phi^{(1)}(t) \tag{9}$$

$$\psi(t)=p_{11}t\phi_1(t)+p_{10}t\phi^{(1)}(t) \tag{10}$$

$$\begin{cases} \phi_1(t) &= p_{11}t\phi_2(t)+p_{10}t\phi^{(1)}(t)\\ &\vdots\\ \phi_{k-2}(t) &= p_{11}t\phi_{k-1}(t)+p_{10}t\phi^{(1)}(t)\\ \phi_{k-1}(t) &= p_{11}t\psi_0(t)+p_{10}t\phi^{(1)}(t) \end{cases} \tag{11}$$

$$\begin{cases} \phi^{(1)}(t) &= p_{01}t\phi_1(t)+p_{00}t\phi^{(2)}(t)\\ &\vdots\\ \phi^{(r-2)}(t) &= p_{01}t\phi_1(t)+p_{00}t\phi^{(r-1)}(t)\\ \phi^{(r-1)}(t) &= p_{01}t\phi_1(t)+p_{00}t\xi_0(t) \end{cases} \tag{12}$$

$$\begin{cases} \psi_0(t) &= p_{11}t\psi_1(t) + p_{10}t\psi^{(1)}(t) \\ \psi_1(t) &= p_{11}t\psi_2(t) + p_{10}t\psi^{(1)}(t) \\ &\vdots \\ \psi_{k-2}(t) &= p_{11}t\psi_{k-1}(t) + p_{10}t\psi^{(1)}(t) \\ \psi_{k-1}(t) &= p_{11}t\psi_0(t) + p_{10}t\psi^{(1)}(t) \end{cases} \tag{13}$$

$$\begin{cases} \psi^{(1)}(t) &= p_{01}t\psi_1(t) + p_{00}t\psi^{(2)}(t) \\ &\vdots \\ \psi^{(r-2)}(t) &= p_{01}t\psi_1(t) + p_{00}t\psi^{(r-1)}(t) \\ \psi^{(r-1)}(t) &= p_{01}t\psi_1(t) + p_{00}ty \end{cases} \tag{14}$$

$$\begin{cases} \xi_0(t) &= p_{01}t\xi_1(t) + p_{00}t\xi^{(1)}(t) \\ \xi_1(t) &= p_{11}t\xi_2(t) + p_{10}t\xi^{(1)}(t) \\ &\vdots \\ \xi_{k-2}(t) &= p_{11}t\xi_{k-1}(t) + p_{10}t\xi^{(1)}(t) \\ \xi_{k-1}(t) &= p_{11}tx + p_{10}t\xi^{(1)}(t) \end{cases} \tag{15}$$

$$\begin{cases} \xi^{(1)}(t) &= p_{01}t\xi_1(t) + p_{00}t\xi^{(2)}(t) \\ &\vdots \\ \xi^{(r-2)}(t) &= p_{01}t\xi_1(t) + p_{00}t\xi^{(r-1)}(t) \\ \xi^{(r-1)}(t) &= p_{01}t\xi_1(t) + p_{00}t\xi^{(0)}(t) \end{cases} \tag{16}$$

The system of equations (8)-(12) has the same form as that of equations (1)-(5) if x and y are replaced by $\psi_0(t)$ and $\xi_0(t)$, respectively. If we get $\psi_0(t)$ and $\xi_0(t)$ from (13)-(16), we can obtain the gpgf by substituting them for x and y in the result of Theorem 1. By solving (13) and (14), we have

$$\psi_0(t) = \frac{\{B(t)(p_{10}t)(p_{00}t)^{r-1}(p_{11}t) + (p_{10}t)(p_{00}t)^{r-1}\}y}{1 - A(t)B(t)(p_{01}t)(p_{10}t) - (p_{11}t)^k - A(t)(p_{11}t)^{k-1}(p_{01}t)(p_{10}t)}. \tag{17}$$

Similarly, by solving (15) and (16), we get

$$\xi_0(t) = \frac{\{A(t)(p_{00}t)(p_{01}t)(p_{11}t)^{k-1} + (p_{01}t)(p_{11}t)^{k-1}\}x}{1 - A(t)B(t)(p_{01}t)(p_{10}t) - B(t)(p_{01}t)(p_{10}t)(p_{00}t)^{r-1} - (p_{00}t)^r}. \tag{18}$$

Hence, we obtain

Theorem 2. *The gpgf of the distribution of the later waiting time is written as*

$$\phi(t) = (p_0p_{01}t + p_1p_{11}t)C(t) + (p_0p_{00}t + p_1p_{10}t)D(t),$$

where

$$C(t) = \frac{(p_{11}t)^{k-1}\psi_0(t) + B(t)(p_{10}t)(p_{00}t)^{r-1}\xi_0(t)}{1 - A(t)B(t)(p_{01}t)(p_{10}t)},$$

$$D(t) = \frac{(p_{00}t)^{r-1}\xi_0(t) + A(t)(p_{11}t)^{k-1}(p_{01}t)\psi_0(t)}{1 - A(t)B(t)(p_{01}t)(p_{10}t)}$$

and $\psi_0(t)$ and $\xi_0(t)$ are given in (17) and (18).

3. NUMBER OF OCCURRENCES OF E_1 IN $X_1, X_2, \cdots, X_n$

In this section, we study the distribution of number of occurrences of the event E_1 in the sequence $X_1, X_2, \cdots, X_n$, where n is a given positive integer. When $X_1, X_2, \cdots, X_n$ are independent, the corresponding distribution is called the binomial distribution of order k.

Let $B_k^0(n, x)$ be the conditional probability that E_1 occurs x times until the n-th trial given that $X_0 = 0$. Let $B_k^1(n, x)$ be the conditional probability that E_1 occurs x times until the n-th trial given that $X_0 = 1$. Consider $(k+1)$ events $F_1, F_2, \cdots, F_k$ and G_k. For $j = 1, 2, \cdots, k$, F_j is the event that the first "0" occurs at the j-th trial and G_k is the event that "0" does not occur in the first k trials. Noting that $F_1, F_2, \cdots, F_k$ and G_k are mutually exclusive, we obtain

Proposition 1. The above conditional probabilities satisfy the following recurrence relation:

$$\left\{\begin{array}{lcl} B_k^0(n,0) & = & 1 \quad \text{if } 0 \le n < k \\ B_k^1(n,0) & = & 1 \quad \text{if } 0 \le n < k \\ B_k^0(n,0) & = & p_{00}B_k^0(n-1,0) + p_{01}p_{10}B_k^0(n-2,0) \\ & & + \sum_{m=1}^{k-2} p_{01}{p_{11}}^m p_{10} B_k^0(n-m-2,0) \qquad \text{if } n \ge k \\ B_k^1(n,0) & = & p_{10}B_k^0(n-1,0) + \sum_{m=0}^{k-2} {p_{11}}^{m+1} p_{10} B_k^0(n-m-2,0) \qquad \text{if } n \ge k \\ B_k^0(n,x) & = & p_{00}B_k^0(n-1,x) + \sum_{m=0}^{k-2} p_{01}p_{10}{p_{11}}^m B_k^0(n-m-2,x) \\ & & + p_{01}{p_{11}}^{k-1} B_k^1(n-k, x-1) \\ & & \quad \text{if } n \ge k \text{ and } x = 1, 2, \cdots, [\frac{n}{k}] \\ B_k^1(n,x) & = & p_{10}B_k^0(n-1,x) + \sum_{m=0}^{k-2} p_{10}{p_{11}}^{m+1} B_k^0(n-m-2,x) \\ & & + {p_{11}}^k B_k^1(n-k, x-1) \quad \text{if } n \ge k \text{ and } x = 1, 2, \cdots, [\frac{n}{k}]. \end{array}\right. \tag{19}$$

We set

$$\psi_n(t) = \sum_{x=0}^{[\frac{n}{k}]} B_k^0(n,x) t^x$$

and

$$\xi_n(t) = \sum_{x=0}^{[\frac{n}{k}]} B_k^1(n,x) t^x.$$

Then, (19) implies

Proposition 2. The conditional pgf's $\psi_n(t)$ and $\xi_n(t)$ satisfy the recurrence relation :

$$\left\{\begin{array}{lcl} \psi_n(t) & = & 1 \quad \text{if } 0 \le n < k, \\ \xi_n(t) & = & 1 \quad \text{if } 0 \le n < k, \\ \psi_n(t) & = & p_{00} + \sum_{m=0}^{k-2} p_{01}{p_{11}}^m p_{10} + p_{01}{p_{11}}^{k-1} t \qquad \text{if } n = k, \\ \xi_n(t) & = & p_{10} + \sum_{m=0}^{k-2} {p_{11}}^{m+1} p_{10} + {p_{11}}^k t \qquad \text{if } n = k, \\ \psi_n(t) & = & p_{00}\psi_{n-1}(t) + \sum_{m=0}^{k-2} p_{01}p_{10}{p_{11}}^m \psi_{n-m-2}(t) \\ & & + p_{01}{p_{11}}^{k-1} t \xi_{n-k}(t) \qquad \text{if } n > k, \\ \xi_n(t) & = & p_{10}\psi_{n-1}(t) + \sum_{m=0}^{k-2} p_{10}{p_{11}}^{m+1} \psi_{n-m-2}(t) \\ & & + {p_{11}}^k t \xi_{n-k}(t) \qquad \text{if } n > k. \end{array}\right. \tag{20}$$

We can solve (20) explicitly.

Theorem 3. *The conditional pgf's $\psi_n(t)$ and $\xi_n(t)$ are written explicitly as*

$$
\begin{aligned}
\psi_n(t) \;=\; & \textstyle\sum_{m=1}^{k-1}\sum_{n_1+2n_2+\cdots+(k+1)n_{k+1}=n-m}\binom{n_1+n_2+\cdots+n_{k+1}}{n_1,n_2,\cdots,n_{k+1}} \\
& \times p_{01}{p_{11}}^{m-1}{p_{00}}^{n_1}(p_{01}p_{10})^{n_2+\cdots+n_{k-1}}{p_{11}}^{\sum_{m=2}^{k+1}(m-2)n_m} \\
& \times(p_{01}p_{10}+{p_{11}}^2t)^{n_k}t^{n_{k+1}}(p_{01}p_{10}-p_{00}p_{11})^{n_{k+1}} \\
+ & \textstyle\sum_{n_1+2n_2+\cdots+(k+1)n_{k+1}=n}\binom{n_1+n_2+\cdots+n_{k+1}}{n_1,n_2,\cdots,n_{k+1}} \\
& \times{p_{00}}^{n_1}(p_{01}p_{10})^{n_2+\cdots+n_{k-1}}{p_{11}}^{\sum_{m=2}^{k+1}(m-2)n_m} \\
& \times(p_{01}p_{10}+{p_{11}}^2t)^{n_k}t^{n_{k+1}}(p_{01}p_{10}-p_{00}p_{11})^{n_{k+1}} \\
+ & \textstyle\sum_{n_1+2n_2+\cdots+(k+1)n_{k+1}=n-k}\binom{n_1+n_2+\cdots+n_{k+1}}{n_1,n_2,\cdots,n_{k+1}} \\
& \times(p_{01}-p_{11}){p_{11}}^{k-1}t{p_{00}}^{n_1}(p_{01}p_{10})^{n_2+\cdots+n_{k-1}}{p_{11}}^{\sum_{m=2}^{k+1}(m-2)n_m} \\
& \times(p_{01}p_{10}+{p_{11}}^2t)^{n_k}t^{n_{k+1}}(p_{01}p_{10}-p_{00}p_{11})^{n_{k+1}},
\end{aligned}
$$

and

$$
\begin{aligned}
\xi_n(t) \;=\; & \textstyle\sum_{m=1}^{k-1}\sum_{n_1+2n_2+\cdots+(k+1)n_{k+1}=n-m}\binom{n_1+n_2+\cdots+n_{k+1}}{n_1,n_2,\cdots,n_{k+1}} \\
& \times p_{01}{p_{11}}^{m-1}{p_{00}}^{n_1}(p_{01}p_{10})^{n_2+\cdots+n_{k-1}}{p_{11}}^{\sum_{m=2}^{k+1}(m-2)n_m} \\
& \times(p_{01}p_{10}+{p_{11}}^2t)^{n_k}t^{n_{k+1}}(p_{01}p_{10}-p_{00}p_{11})^{n_{k+1}} \\
+ & \textstyle\sum_{n_1+2n_2+\cdots+(k+1)n_{k+1}=n}\binom{n_1+n_2+\cdots+n_{k+1}}{n_1,n_2,\cdots,n_{k+1}} \\
& \times{p_{00}}^{n_1}(p_{01}p_{10})^{n_2+\cdots+n_{k-1}}{p_{11}}^{\sum_{m=2}^{k+1}(m-2)n_m} \\
& \times(p_{01}p_{10}+{p_{11}}^2t)^{n_k}t^{n_{k+1}}(p_{01}p_{10}-p_{00}p_{11})^{n_{k+1}} \\
+ & \textstyle\sum_{n_1+2n_2+\cdots+(k+1)n_{k+1}=n-k}\binom{n_1+n_2+\cdots+n_{k+1}}{n_1,n_2,\cdots,n_{k+1}} \\
& \times(p_{10}-p_{00}){p_{11}}^{k-1}{p_{00}}^{n_1}(p_{01}p_{10})^{n_2+\cdots+n_{k-1}}{p_{11}}^{\sum_{m=2}^{k+1}(m-2)n_m} \\
& \times(p_{01}p_{10}+{p_{11}}^2t)^{n_k}t^{n_{k+1}}(p_{01}p_{10}-p_{00}p_{11})^{n_{k+1}}.
\end{aligned}
$$

Proof. Define

$$\Psi(t,z)\equiv\sum_{n=0}^{\infty}\psi_n(t)z^n$$

and

$$\Xi(t,z)\equiv\sum_{n=0}^{\infty}\xi_n(t)z^n.$$

Then, from (20) we have

$$
\begin{cases}
(1-p_{00}z-\sum_{m=0}^{k-2}p_{01}p_{10}{p_{11}}^mz^{m+2})\Psi(t,z)-p_{01}{p_{11}}^{k-1}tz^k\Xi(t,z)=C_1 \\
-(p_{10}z+\sum_{m=0}^{k-2}p_{10}{p_{11}}^{m+1}z^{m+2})\Psi(t,z)+(1-{p_{11}}^ktz^k)\Xi(t,z)=C_2
\end{cases}
$$

where

$$C_1=\sum_{n=0}^{k-1}z^n+(p_{00}+\sum_{m=0}^{k-2}p_{01}{p_{11}}^mp_{10})z^k-p_{00}z\sum_{n=0}^{k-1}z^n-\sum_{m=0}^{k-2}p_{01}p_{10}{p_{11}}^mz^{m+2}\sum_{n=0}^{k-m-2}z^n$$

and

$$C_2=\sum_{n=0}^{k-1}z^n+(p_{10}+\sum_{m=0}^{k-2}{p_{11}}^{m+1}p_{10})z^k-p_{10}z\sum_{n=0}^{k-1}z^n-\sum_{m=0}^{k-2}p_{10}{p_{11}}^{m+1}z^{m+2}\sum_{n=0}^{k-m-2}z^n.$$

Noting that

$$\sum_{m=0}^{k-2} p_{01}p_{10}{p_{11}}^m z^{m+2} \sum_{n=0}^{k-m-2} z^n = \sum_{n=2}^{k} (p_{01}p_{10} \sum_{m=0}^{n-2} {p_{11}}^m) z^n,$$

we can rewrite

$$C_1 = 1 + \sum_{n=1}^{k-1} p_{01}{p_{11}}^{n-1} z^n \quad \text{and} \quad C_2 = 1 + \sum_{n=1}^{k-1} {p_{11}}^n z^n.$$

Then, from the above equations, we have

$$\Psi(t, z) = \frac{1 + \sum_{n=1}^{k-1} p_{01}{p_{11}}^{n-1} z^n + (p_{01} - p_{11}){p_{11}}^{k-1} t z^k}{1 - C(t, z)}$$

and

$$\Xi(t, z) = \frac{1 + \sum_{n=1}^{k-1} p_{01}{p_{11}}^{n-1} z^n + (p_{10} - p_{00}){p_{11}}^{k-1} t z^k}{1 - C(t, z)},$$

where

$$\begin{aligned} C(t, z) &= p_{00} z + \textstyle\sum_{m=0}^{k-3} p_{01}p_{10}{p_{11}}^m z^{m+2} \\ &\quad + {p_{11}}^{k-2}(p_{01}p_{10} + {p_{11}}^2 t) z^k - {p_{11}}^{k-1} t (p_{00}p_{11} - p_{01}p_{10}) z^{k+1}. \end{aligned}$$

After expanding $\Psi(t, z)$ and $\Xi(t, z)$ w.r.t. z, we have the desired result by taking the coefficient of z^n. This completes the proof.

Remark 2. In Theorem 3, setting $p_{00} = p_{10} = q$ and $p_{01} = p_{11} = p$, it is easy to check that $\psi_n(t)$ is equal to (2.7) in Aki and Hirano[12].

Acknowledgements
This research was partially supported by the ISM Cooperative Research Program (91-ISM-CRP-5) of the Institute of Statistical Mathematics.

REFERENCES

1. A.N. Philippou, C. Georghiou and G.N. Philippou, *Statist. Probab. Lett.*, **1**, 171-175 (1983).
2. K. Hirano, in: *Fibonacci Numbers and Their Applications*, A.N. Philippou, G.E. Gergum and A.F. Horadam(Eds.), pp. 43-53. Reidel, Dordrecht (1986).
3. S.-Y.R. Li, *Ann. Probab.*, **8**, 1171-1176 (1980).
4. H.U. Gerber and S.-Y.R. Li, *Stochastic Process. Appl.*, **11**, 101-108 (1981).
5. S. Aki, *Ann. Inst. Statist. Math.*, **37**, **A**, 205-224 (1985).
6. M. Ebneshahrashoob and M. Sobel, *Statist. Probab. Lett.*, **9**, 5-11 (1990).
7. T.F. Móri, *Probab. Th. Rel. Fields*, **87**, 313-323 (1991).
8. S. Aki, *Ann. Inst. Statist. Math.*, **44**, 363-378 (1992).
9. A.N. Philippou, in: *Fibonacci Numbers and Their Applications*, A.N. Philippou, G.E. Gergum and A.F. Horadam(Eds.), pp. 203-227. Reidel, Dordrecht (1986).
10. M.B. Rajarshi, *J. Appl. Probab.*, **11**, 190-192 (1974).
11. W. Feller, *An Introduction to Probability Theory and Its Applications*, vol.I, John Wiley, New York, 3rd ed. (1968).
12. S. Aki and K. Hirano, in: *Statistical Theory and Data Analysis II*, K. Matusita(Ed.), pp.211-222. North-Holland, Amsterdam (1988).

Stat. Sci. & Data Anal., pp. 475-488
K. Matsusita *et al.* (Eds)

How Large the Class of Waiting Distribution Can Be?

P.D. CHEN

Institute of Applied Mathematics, Academia Sinica Box. 2734, Beijing 100080, China

Abstract Gerber, H. U. and Li S-Y, R[1] obtained the probability generating function of the waiting time $\tau_{\mathbb{b}}$ of a given word $\mathbb{b}$ by means of periodic set $\mathbb{J}(\mathbb{b})$. Zbáganu, G. proposed an interesting conjecture: the structure of periodic sets doesn't depend on the alphabet set, and the problems "How large the class of W.D. can be?" and "How many words correspond to a given periodic set?" In this paper, we prove Zbáganu's conjecture and give an answer to his problems by recurrence procedures.

Key words: waiting distribution, periodic set.

1. INTRODUCTION

Let $S = \{a_1, a_2, \cdots, a_m\}$ be a given set of alphabet, each a_i occurs with probability $p_i > 0, i = 1, 2, \cdots, m$, and $\sum_{i=1}^m p_i = 1$. Consider a sequence of i.i.d. r.v.s. $\{X_k\}_{k \in \mathbb{N}}$ with their values in S. For a given word $\mathbb{b} = \{b_1, \cdots, b_n\} \in S^n$, the waiting time of $\mathbb{b}$, denoted by $\tau_{\mathbb{b}}$, is defined by

$$\tau_{\mathbb{b}} = \inf\{k : X_{k-n+1} = b_1, X_{k-n+2} = b_2, \cdots, X_k = b_n\}.$$

Its probability distribution $\mathbb{Q} = \mathbb{P} \circ \tau_{\mathbb{b}}^{-1}$ is what we mean by "Waiting Distribution". Gerber, H.U. and Li S-Y, R. [1] obtained the probability generating function of $\tau_{\mathbb{b}}$:

$$\varphi_{\mathbb{b}}(x) = \mathbb{E}x^{\tau_{\mathbb{b}}} = \sum_{k=n}^{\infty} x^k \mathbb{P}\{\tau_{\mathbb{b}} = k\} = \frac{x^n}{x^n + (1-x)\mathbb{b} \star \mathbb{b}(x)},$$

where $\mathbb{b} \star \mathbb{b}(x) = \sum_{j \in J(\mathbb{b})} \frac{x^j}{\prod_{i=1}^{n-j} \mathbb{P}(b_i)}$ is a polynomial with order smaller than n, $\mathbb{P}(b_i) = p_r$ for all $b_i = a_r, r = 1, 2, \cdots, m$; $J(\mathbb{b})$ is the periodic set of $\mathbb{b}$: an integer $j \in J(\mathbb{b})$ iff $0 \le j \le n-1$ and $b_{i+j} = b_i$ for all integers i such that $0 \le i \le n-j$. So everything is clear except the structure of periodic set $J(\mathbb{b})$.

Zbăganu, G. proposed the problem of "How large the class of W.D. can be?" to us during his visit to Beijing in the spring of 1991. He showed us his calculation on $C_n, 1 \le n \le 12$, the total number of Waiting Distribution for $S = \{0, 1\}$ and the length of words n being fixed, they are 1, 2, 3, 4, 6, 8, 10, 13, 17, 21, 27, 30 in the natural order. Of course, this is the result of classical case: $\mathbb{P}\{X = 0\} = \mathbb{P}\{X = 1\} = \frac{1}{2}$, otherwise such C_n is only the number of possible periodic sets, but this is essential. He conjectured an interesting result about C_n as follow:

"In classical case, that is, when $p_1 = p_2 = \cdots = p_m = \frac{1}{m}$, no matter what the alphabet set S is, only if $\#S = m \geq 2$, the corresponding C_n are always the same."

In this paper, we prove "Zbăganu's Conjecture" and give an answer to his problem by an inductive procedure to calculate C_n. We also discuss the word-classification, another problem also proposed by Zbăganu: how many words $\mathbb{b} \in S^n$ are there with a given J as their common periodic set?

2. STRUCTURE OF PERIODIC SETS

Let's put all of possible periodic sets of words with length n together and denoted by $\mathbb{J}_n$ (no matter the words come from which S). Given any $J \in \mathbb{J}_n$, list its elements in natural order,

$$J = \{j_0 = 0 < j_1 < \cdots < j_k\} \subset \{0, 1, 2, , \cdots, n-1\}.$$

0 must be a period, and it's the smallest one, so denote it by j_0. j_1 denotes the smallest positive period.

Lemma 1. Given integers r, s, t such that $0 \leq r < s \leq t \leq k-1$, set $d = j_s - j_r$, then $j_t + d, j_t + 2d, \cdots, j_t + [\frac{n-j_t-1}{d}]d \in J$.

Proof. It's sufficient to show that $j_t + d \in J$ whenever $j_t + d \leq n-1$, which follows from

$$b_{i+j_t+d} = b_{i+j_t+j_s-j_r} = b_{i+j_t-j_r} = b_{i+j_t} = b_i$$

for each i such that $1 \leq i \leq n - j_t - d$.

Lemma 2. Given integers r, s, t, u such that $0 \leq r < s \leq t < u \leq k$, set $d = j_s - j_r, \bar{d} = j_u - j_t$. If $\bar{d}|d$ and $j_t + d \leq n$, then $j_r + \bar{d} \in J$.

Proof. Denote $d = p\bar{d}, p$ is a positive integer. In case of $p = 1$, nothing need to be proved. Assume $p > 1$, given integer i such that $1 \leq i \leq n - j_r - \bar{d}$, we'll show that $b_{i+j_r+\bar{d}} = b_i$. To do this, let $i' = i - [\frac{i-1}{d}]d$, and $i'' = i' - [\frac{i'-1}{\bar{d}}]\bar{d}$, then $1 \leq i' \leq d = p\bar{d}, [\frac{i'-1}{\bar{d}}] < p$ and $1 \leq i'' \leq \bar{d}$. We have

$$b_{i+j_r+\bar{d}} = b_{i'+j_r+[\frac{i-1}{d}]d+\bar{d}} = b_{i'+\bar{d}} = b_{i''+[\frac{i'-1}{\bar{d}}]\bar{d}+\bar{d}},$$

either $[\frac{i'-1}{\bar{d}}] = p-1$,

$$b_{i''+[\frac{i'-1}{\bar{d}}]\bar{d}+\bar{d}} = b_{i''+p\bar{d}} = b_{i''+d} = b_{i''+j_s-j_r} = b_{i''},$$

or $[\frac{i'-1}{\bar{d}}] < p-1$, $i'' + [\frac{i'-1}{\bar{d}}]\bar{d} + \bar{d} \leq p\bar{d} = d$,

$$b_{i''+[\frac{i'-1}{\bar{d}}]\bar{d}+\bar{d}} = b_{i''+j_u-j_t+[\frac{i'-1}{\bar{d}}]\bar{d}} = b_{i''+j_u+[\frac{i'-1}{\bar{d}}]\bar{d}} = b_{i''}.$$

In both case,

$$b_{i+j_r+\bar{d}} = b_{i''} = b_{i''+j_t+[\frac{i'-1}{\bar{d}}]\bar{d}} = b_{i''+[\frac{i'-1}{\bar{d}}]\bar{d}} = b_{i'} = b_{i'+j_r+[\frac{i-1}{d}]d} = b_{i'+[\frac{i-1}{d}]d} = b_i.$$

We may use periodic increments $\{d_r = j_r - j_{r-1} : 1 \leq r \leq k\}$ to describe J, then properties of J given by Lemma 1 and Lemma 2 show that (i) $\{d_r\}$ is non-increasing. (ii) If $\{d_r\}$ includes some repeats such as $d_{r-1} > d_r = d_{r+1} = \cdots = d_{r+p-1}$, then the former increments $d_q, q < r$ must be larger than pd_r.

In the next lemma, assume $i, j \in J \in \mathbb{J}_n$ be two positive periods of some word with length $n, 0 < i < j, (i,j)$ denotes the Great Common Divisor. At the moment, the positions of i, j in the natural order of J are not important.

Lemma 3. If $i + j \leq n + (i,j)$, then $(i,j) \in J$.
Proof. Denote $j = q_1 i + r_1, i = q_2 r_1 + r_2, \cdots, r_{u-2} = q_u r_{u-1} + r_u, \cdots, r_{k-2} = q_k r_{k-1} + r_k, r_{k-1} = q_{k+1} r_k$, where

$$i = r_0 > r_1 > r_2 > \cdots > r_k = (i,j) \geq 1, \quad q_u \geq 1 \quad \text{for} \quad 1 \leq u \leq k$$

but $q_{k+1} \geq 2$, they are all positive integers. Since

$$n \geq i + j - (i,j) = i + j - r_k,$$

by Lemma 1, $i, 2i, \cdots, q_1 i$; $j = q_1 i + r_1, q_1 + 2r_1, \cdots, q_1 i + q_2 r_1$;

$$(q_1 + 1)i = q_1 i + q_2 r_1 + r_2, \qquad q_1 i + q_2 r_1 + 2r_2, \cdots$$

generally

$$\textstyle\sum_{l=1}^{u} q_l r_{l-1} + s r_u, \quad 1 \leq s \leq q_{u+1}, \quad 0 \leq u \leq k$$

are all periods (the only exception is $u = k$ and $s = q_{u+1}$, then $\sum_{l=1}^{k} q_l r_{l-1} + q_{k+1} r_k = i + j - (i,j)$ may equal to n). Use Lemma 2 repeatedly, we may get $\sum_{l=1}^{u} q_l r_{l-1} + r_k, u = k, k-1, \cdots, 2, 1, 0$, being periods successively.

It's well known that for a given infinite sequence $\mathbb{b} = \{b_1, b_2, \cdots, b_n, \cdots\}$, if i, j are its positive periods, so is (i,j). In case of finite sequence $\mathbb{b} = \{b_1, b_2, \cdots, b_n\}, i, j \in J(\mathbb{b})$ needn't imply $(i,j) \in J(\mathbb{b})$. $\mathbb{b} = \{1,1,0,1,1\}$ gives one simple example where $j(\mathbb{b}) = \{0,3,4\}$, but $(3,4) = 1 \notin J(\mathbb{b})$. In fact, for any given positive integers $i, j, (i,j) < i \wedge j$, and $n \leq i + j - (i,j) - 1$, we may always construct a word $\mathbb{b} \in \{0,1\}^n$ with $i \wedge j$ as its smallest positive period. In this sense, Lemma 3 couldn't be improved.

Now we turn to the construction of $\mathbb{J}_n$, when n is small, it's not difficult to construct them directly: $\mathbb{J}_1$ is a singleton, $\mathbb{J}_2 = \{\{0\}, \{0,1\}\}$. In order to give an inductive method to construct all $\mathbb{J}_n$, we need three criterions to prolong periodic sets.

Proposition 1 (criterion of non-complete prolongation). Assume that $J = \{0, j_1, \cdots, j_k\} \in \mathbb{J}_n$, then for $1 \leq u \leq k$, in order to $\overline{J} = \{0, j_u + J\} = \{0, j_u, j_u + j_1, \cdots, j_u + j_k\} \in \mathbb{J}_{n+j_u}$ iff $u = 1$ or $u > 1$ and $j_1 \nmid j_u$.
Proof. If $u > 1$ and $j_1 | j_u$, then by Lemma 2, j_1 is also a period, because $2j_u < n + j_u$ so j_u couldn't be the smallest positive period, that is to say $\overline{J} \notin \mathbb{J}_{n+j_u}$ which gives the necessity.

For the sufficiency, assume $\mathbb{b} = \{b_1, b_2, \cdots, b_n\} \in S^n$ be a word with J as its periodic set, let

$$\bar{\mathbb{b}} = \{b_1, \cdots, b_{j_u}; b_1, \cdots, b_{j_u}, \cdots, b_n\},$$

then each element of $\overline{J}$ is obviously a period of $\bar{\mathbb{b}}$. To show that $\overline{J}$ is just the periodic set of $\bar{\mathbb{b}}$, first notice that j_u is the smallest positive period of $\bar{\mathbb{b}}$. Otherwise if $j < j_u$ is the smallest positive period of $\bar{\mathbb{b}}$, then $j | j_u$ ($(j, j_u) < j$ and $j + j_u < 2j_u < n + j_u$ imply $(j, j_u) \in J(\bar{\mathbb{b}})$ contradicting to the smallest of j). On the other hand, $j_u \in \bar{\mathbb{b}} \Rightarrow$

$$\bar{\mathbb{b}} = \{b_1, \cdots, b_{j_u}, b_{j_u+1}, \cdots, b_n, b_{n-j_u+1}, \cdots, b_n\}$$

so $j \in J(\mathbb{b})$, and $j = j_r$ for some integer $r, 1 < r < u$. If $(j_1, j_r) = (j_1, j) = j_1$, then $j_1 | j_r \Rightarrow j_1 | j_u$ impossible, while $(j_1, j_r) < j_1, j_1 + j_r < 2j_r \leq n$ implies $(j_1, j_r) \in J(\mathbb{b})$ another contradiction. Having j_u as the smallest positive period of $\bar{\mathbb{b}}$, other period j of $\bar{\mathbb{b}}$ makes $j - j_u$ being period of $\mathbb{b}$ trivially, so $\overline{J} = J(\bar{\mathbb{b}})$.

Proposition 2. (criterion of complete prolongation) Assume that $J = \{0, j_1, \cdots, j_k\} \in \mathbb{J}_n$, then in order to $\overline{J} = \{0, n+J\} = \{0, n, n+j_1, \cdots, n+j_k\} \in \mathbb{J}_{2n}$, iff $j_1 \nmid n$.

Proof. $j_1|n$, and $n, n+j_1 \in \overline{J} \in \mathbb{J}_{2n} \Rightarrow j_1 \in \overline{J}$, so $j_1 \nmid n$ is necessary for $\overline{J}$ having the form $\{0, n+J\}$.

Turn to the sufficiency, suppose $\mathbb{b} = \{b_1, b_2, \cdots, b_n\} \in S^n$ be a word with J as its periodic set, let

$$\bar{\mathbb{b}} = \{b_1, b_2, \cdots, b_n; b_1, b_2, \cdots, b_n\}$$

then $\overline{J} \subset J(\bar{\mathbb{b}})$ trivially holds. If j is the smallest positive period of $\bar{\mathbb{b}}$ and $j < n$, then $j+n < 2n, (j,n) \in J(\bar{\mathbb{b}})$, so $j|n$ due to the smallestness of j. Similar to the argument in the proof of prop. 1, this implies $j \in J(\mathbb{b}) = J$ and $j_1|n$, a contradiction. Therefore the smallest positive period of $\bar{\mathbb{b}}$,must be n, and other period j of $\bar{\mathbb{b}}$ makes $j-n$ being period of $\mathbb{b}$ obviously, thus $\overline{J} = J(\bar{\mathbb{b}}) \in \overline{J}_{2n}$.

The reason of calling the above propositions "criterions" is that we can get any periodic set with the smallest positive period being not more than the half of the word from a shorter periodic set. In fact, assume $J = \{0, j_1, j_2, \cdots, j_k\} \in \mathbb{J}_n$, and $j_1 \leq \frac{n}{2}, \mathbb{b} = \{b_1, b_2, \cdots, b_n\}$ is a word such that $J = J(\mathbb{b})$, then $\mathbb{b}' = \{b_{j_1+1}, b_{j_1+2}, \cdots, b_n\}$ is a word with $J' = J(\mathbb{b}') = \{0, j_2 - j_1, \cdots, j_k - j_1\} \in \mathbb{J}_{n-j_1}$ and $(j_2 - j_1) \nmid j_1$. In other words, J is the prolongation of J' by means of prop. 1 (when $j_1 < \frac{n}{2}$) or Prop. 2 (when $j_1 = \frac{n}{2}$). Such prolongation can be done within the same alphabet set S. What happens when we encounter $J = \{0, j_1, j_2, \cdots, j_k\} \in \mathbb{J}_n$ with $j_1 > \frac{n}{2}$? If $\mathbb{b} = \{b_1, b_2, \cdot \cdots, b_{j_1}, b_{j_1+1}, \cdots, b_n\}$ is a word with $J = J(\mathbb{b})$, the problem comes from that $\mathbb{b}' = \{b_{j_1+1}, \cdots, b_n\}$ isn't long enough to recover the whole $\mathbb{b}$, it's possible that some letters of $\mathbb{b}$ are not included in $\mathbb{b}'$. If it's allowed to enlarge the alphabet set it's very easy to prolong a periodic set $J = \{0,, j_1, \cdots, j_k\} \in \mathbb{J}_n$ to get $\overline{J} = \{0, N + j_1, N + j_2, \cdots, N + j_k\} \in \mathbb{J}_{n+N}$ with $N > n$, choose $\mathbb{b} = \{b_1, \cdots, b_n\} \in S^n$ with $J = J(\mathbb{b})$, put a new letter Δ in S to get a larger alphabet set $\overline{S} = S + \{\Delta\}$, and let $\bar{\mathbb{b}} = \{b_1, \cdots, b_n, \Delta, \cdots, \Delta, b_1, \cdots, b_n\}$ where the number of Δ is $N - n$, then $J(\bar{\mathbb{b}}) = \overline{J}$. But the main purpose of our discussion is to prove "Zbăganu's conjecture", the key idea is to construct all periodic sets by using the words only come from the simplest alphabet set $S = \{0, 1\}$. So the next proposition may be regarded as the main result of this section.

Proposition 3 (criterion of over-complete prolongation). Assume that $J \doteq \{0, j_1, \cdots, j_k\} \in \mathbb{J}_n$ is a periodic set of some word $\mathbb{b} \in \{0,1\}^n$, then for any integer $N > n$.

$$\overline{J} = \{0, N + J\} = \{0, N, N + j_1, \cdots, N + j_k\} \in \mathbb{J}_{N+n}$$

is a periodic set of some word $\bar{\mathbb{b}} \in \{0,1\}^{N+n}$.

Proof. Without losing generality, we may assume $\mathbb{b}$ starting with 1, and discuss the only two possible cases where $\mathbb{b}$ is ended with 0 or 1 respectively.

Case 1. $\mathbb{b} = \underbrace{1\cdots1}_{\alpha_1}\underbrace{0\cdots0}_{\beta_1}\underbrace{1\cdots1}_{\alpha_2}\underbrace{0\cdots0}_{\beta_2}\cdots\underbrace{1\cdots1}_{\alpha_l}\underbrace{0\cdots0}_{\beta_l}.$

When $\mathbb{b}$ is ended with 0, it consists of l 1-runs and l 0-runs, the lengths of the i-th

1-run and the i-th 0-runs are denoted by α_i and $\beta_i, i = 1, 2, \cdots, l$. Let

$$\bar{b} = b\underbrace{0\cdots0}_{N-n}b = \underbrace{1\cdots1}_{\alpha_1}\underbrace{0\cdots0}_{\beta_1}\cdots\underbrace{1\cdots1}_{\alpha_l}\underbrace{0\cdots0}_{\beta_l}\underbrace{0\cdots0}_{N-n}\underbrace{1\cdots1}_{\alpha_1}\underbrace{0\cdots0}_{\beta_1}\cdots\underbrace{1\cdots1}_{\alpha_l}\underbrace{0\cdots0}_{\beta_l}$$

$$\tilde{b} = b\underbrace{1\cdots1}_{N-n}b = \underbrace{1\cdots1}_{\alpha_1}\underbrace{0\cdots0}_{\beta_1}\cdots\underbrace{1\cdots1}_{\alpha_l}\underbrace{0\cdots0}_{\beta_l}\overbrace{\underbrace{1\cdots1}_{N-n}\underbrace{1\cdots1}_{\alpha_1}}^{\tilde{\gamma}=N-n+\alpha_1}\underbrace{0\cdots0}_{\beta_1}\cdots\underbrace{1\cdots1}_{\alpha_l}\underbrace{0\cdots0}_{\beta_l}$$

We'll show that at least one of the following equalities holds

$$J(\bar{b}) = \bar{J} = \{0, N+J\}, \quad J(\tilde{b}) = \bar{J} = \{0, N+J\};$$

or equivalently, at least one of $\bar{b}, \tilde{b}$ possesses N as its smallest positive period.

If the smallest positive period j of $\bar{b}$ is smaller than N, since $\bar{b}_{j+1} = \bar{b}_1 = b_1 = 1$, we have $\sum_{r=1}^{i}(\alpha_r + \beta_r) \le j < \sum_{r=1}^{i}(\alpha_r + \beta_r) + \alpha_{i+1}$ for some $i, 1 \le i \le l-1$, and

$$\alpha_{i+1} + \sum_{r=1}^{i}(\alpha_r + \beta_r) - j = \alpha_1,$$

$$\alpha_{i+2} = \alpha_2, \cdots, \alpha_l = \alpha_{l-i}, \quad \alpha_1 = \alpha_{l-i+1}, \cdots, \alpha_{i+1} = \alpha_1, \cdots$$

which show $j = \sum_{r=1}^{i}(\alpha_r + \beta_r)$ for some $i, 1 \le i \le l-1$. Therefore, $\{\alpha_1, \cdots, \alpha_l, \alpha_1, \cdots, \alpha_l\}$ possesses i, l as its periods, as $i + l < 2l, d = (i, l)$ is also its period. $d \le i < l$ and $d|l$ imply $d \le \frac{l}{2}$. Notice that d is also a period of $\{\alpha_2, \cdots \alpha_l\}$.

Similarly, if the smallest period $\tilde{j}$ of $\tilde{b}$ is smaller than N, then $\sum_{r=1}^{\tilde{i}}(\alpha_r + \beta_r) \le \tilde{j} < \sum_{r=1}^{\tilde{i}}(\alpha_r + \beta_r) + \alpha_{\tilde{i}+1}$ for some $\tilde{i}, 1 \le \tilde{i} \le l-1$ and $\alpha_{\tilde{i}+1} + \sum_{r=1}^{\tilde{i}}(\alpha_r + \beta_r) - \tilde{j} = \alpha_1, \alpha_{\tilde{i}+2} = \alpha_2, \cdots, \alpha_l = \alpha_{l-\tilde{i}}, \tilde{\gamma} = \alpha_{l-\tilde{i}+1}, \alpha_2 = \alpha_{l-\tilde{i}+2}, \cdots, \alpha_{\tilde{i}+1} = \tilde{\gamma}, \cdots$ which show that $\tilde{i}, l$ are periods of $\{\alpha_2, \alpha_l, \tilde{\gamma}, \alpha_2, \cdots, \alpha_l\}$ and so is $\tilde{d} = (\tilde{i}, l)$ since $\tilde{i} + l \le 2l - 1$. Besides $\tilde{d}$ is also a period of $\{\alpha_2, \cdots, \alpha_l\}$, and $\tilde{d} \le \frac{l}{2}$.

Now if the smallest positive periods of both $\bar{b}$ and $\tilde{b}$ are all smaller than N, then both $d, \tilde{d}$ are periods of $\{\alpha_2, \cdots, \alpha_l\}$ and so is $p = (d, \tilde{d})$ because $d + \tilde{d} \le k \le k - 1 + (d, \tilde{d})$. Thus $\{\alpha_2, \cdots, \alpha_l\} = \{a_1, \cdots, a_p, a_1, \cdots, a_p, \cdots, a_1, \cdots, a_{p-1}\}$ for some natural number $a_1, \cdots, a_p$, and

$$\{\alpha_1, \cdots, \alpha_l, \alpha_1, \cdots, \alpha_l\} = \{\alpha_1, a_1, \cdots, a_p, a_1, \cdots, a_p, \cdots,$$
$$a_1, \cdots, a_{p-1}, \alpha_1, a_1, \cdots, a_p, \cdots\},$$
$$\{\alpha_2, \cdots, \alpha_l, \tilde{\gamma}, \alpha_2, \cdots, \alpha_l\} = \{a_1, \cdots, a_p, a_1, \cdots, a_p, \cdots,$$
$$a_1, \cdots, a_{p-1}, \tilde{\gamma}, a_1, \cdots, a_p, \cdots\},$$

hence $\alpha_1 = a_p = \tilde{\gamma} = N - n + \alpha_1 > \alpha_1$ contradiction.

Case 2. $b = \underbrace{1\cdots1}_{\alpha_1}\underbrace{0\cdots0}_{\beta_1}\cdots\underbrace{1\cdots1}_{\alpha_l}\underbrace{0\cdots0}_{\beta_l}\underbrace{1\cdots1}_{\alpha_{l+1}}.$

Let

$$\bar{b} = b\underbrace{0\cdots0}_{N-n}b = \underbrace{1\cdots1}_{\alpha_1}\underbrace{0\cdots0}_{\beta_1}\cdots\underbrace{1\cdots1}_{\alpha_{l+1}}\underbrace{0\cdots0}_{N-n}\underbrace{1\cdots1}_{\alpha_1}\underbrace{0\cdots0}_{\beta_1}\cdots\underbrace{1\cdots1}_{\alpha_{l+1}},$$

$$\tilde{b} = b\underbrace{1\cdots1}_{N-n}b = \underbrace{1\cdots1}_{\alpha_1}\underbrace{0\cdots0}_{\beta_1}\cdots\overbrace{\underbrace{1\cdots1}_{\alpha_{l+1}}\underbrace{1\cdots1}_{N-n}\underbrace{1\cdots1}_{\alpha_1}}^{\tilde{\gamma}=N-n+\alpha_1+\alpha_{l+1}}\underbrace{0\cdots0}_{\beta_1}\cdots\underbrace{1\cdots1}_{\alpha_{l+1}}.$$

The smallest positive period of $\bar{b}, j < N$, implies

$$\sum_{r=1}^{i}(\alpha_r+\beta_r) \le j < \sum_{r=1}^{i}(\alpha_r+\beta_r)+\alpha_{i+1} \quad \text{for some } 1 \le i \le l.$$

The similar argument as in Case 1 shown that both i and $l+1$ are periods of $\{\alpha_1,\alpha_2,\cdots,\alpha_{l+1},\alpha_1,\alpha_2,\cdots,\alpha_{l+1}\}$ and so is $d=(i,,l+1)$. Besides $d|l+1, d \le i < l+1, d \le [\frac{l+1}{2}]$, and d is also a period of $\{\alpha_2,\cdots,\alpha_l\}$.

If the smallest positive period of $\tilde{b}, \tilde{j} < N$, then

$$\sum_{r=1}^{\tilde{i}}(\alpha_r+\beta_r) \le \tilde{j} < \sum_{r=1}^{\tilde{i}}(\alpha_r+\beta_r)+\alpha_{\tilde{i}+1} \quad \text{for some } \tilde{i}, \quad 1 \le \tilde{i} \le l-1,$$

hence both $\tilde{i}$ and l are periods of $\{\alpha_2,\cdots,\alpha_l,\tilde{\gamma},\alpha_2,\cdots,\alpha_l\}$, so is $\tilde{d}=(\tilde{i},l)$ and $\tilde{d}$ is also a period of $\{\alpha_2,\cdots,\alpha_l\}$.

Thus if the smallest positive periods of both $\bar{b}$ and $\tilde{b}$ are all smaller than N, then both d and $\tilde{d}$ are periods of $\{\alpha_2,\cdots,\alpha_l\}$, and so is $(d,\tilde{d}) = 1\,(d|l+1, \tilde{d}|l \Rightarrow (d,\tilde{d})|(l,l+1)=1$, and $d+\tilde{d} \le [\frac{l+1}{2}]+[\frac{l}{2}] = l = l-1+(d,\tilde{d}))$. Hence $\alpha_2=\alpha_3=\cdots=\alpha_l$. Come back to the periodicity of $\{\alpha_1,\alpha_2,\cdots,\alpha_{l+1},\alpha_1,\alpha_2,\cdots,\alpha_{l+1},\}$ and $\{\alpha_2,\cdots,\alpha_l,\tilde{\gamma},\alpha_2,\cdots,\alpha_l\}$, we should have $\alpha_1=\alpha_2=\cdots=\alpha_{l+1}$ and $\alpha_2=\cdots=\alpha_l=\tilde{\gamma}=N-n+\alpha_1+\alpha_{l+1}$ but which is impossible. Therefore at least one of the $\bar{b}$ and $\tilde{b}$ must possess N as its smallest positive period.

Theorem 1. Given any $J \in \mathbb{J}_n$, there exists a word $b \in \{0,1\}^n$ with $J = J(b)$.

Proof. Assume $J=\{0,j_1,\cdots,j_k\}$, we'll prove it by induction on k. When $k=0$, take $b=\{1,\underbrace{0,\cdots,0}_{n-1}\}$. Suppose that the statement has been proved for $k \le K-1$ and any $n \in \mathbb{N}$, consider the case of $k=K$. Denote $J=\{0,j_1+J'\}$ where $J'=\{0,j_1'=j_2-j_1,\cdots,j_{k-1}'=j_k-j_1\} \in \mathbb{J}_{n-j_1}$. includes only $K-1$ positive periods, so there exists a word $b' \in \{0,1\}^{n-j_1}$ with $J'=J(b')$.

If $j_1 > \frac{n}{2}$, then $n-j_1 < \frac{n}{2}$, by Prof. 3, one of the

$$\bar{b} = b'\underbrace{0\cdots0}_{2j_1-n}b', \quad \tilde{b} = b'\underbrace{1\cdots1}_{2j_1-n}b'$$

satisfies $J=J(\bar{b})$ or $J=J(\tilde{b})$, and can be chosen as b.

If $j_1 \le \frac{n}{2}, b=\{b_1',\cdots,b_{j_1}',b'\}$ will do the job by Prop. 1 or Prop. 2. (It's easily seen that $(j_2-j_1)\not| j_1$)

Corollary 1. Zbăganu's conjecture is true.

Corollary 2. Suppose that i,j,n are positive integer with $0<i<j<n=i+j-(i,j)-1, j=q_1 i+r_1,\cdots,r_{u-2}=q_u r_{u-1}+r_u, u=2,\cdots,k-1, r_0=i, r_{k-1}=q_{k+1}r_k, r_k=(i,j)<r_{k-1}<\cdots<r_0$ all of them are integers, then

$$i,j \in J = \{\sum_{l=1}^{u} q_l r_{l-1}+vr_u < n : v=0,1,\cdots,q_{u+1}, u=0,1,\cdots,k\} \in \mathbb{J}_n$$

which means Lemma 3 couldn't be improved in some sense.

To sum up the main results of this section, it's only the following equality

$$\mathbb{J}_n \stackrel{\triangle}{=} \{J(b) : b \in S^n \text{ for some } S\} = \{J(b) : b \in \{0,1\}^n\}.$$

And from now on, we'll restrict ourselves to the simplest case of $S=\{0,1\}$, but it's easily seen that the method we are going to use is suitable for the general case.

3. CLASSIFICATION OF PERIODIC SETS AND WORDS

In this section, we are going to give a recurrence procedure to construct $\mathbb{J}_n$ and to calculate the number of words $W_n(J)$ with $J(\mathbb{b}) = J$: for a given $J \in \mathbb{J}_n$,

$$W_n(J) = \#\{\mathbb{b} \in \{0,1\}^n : J(\mathbb{b}) = J \in \mathbb{J}_n\}.$$

It's natural to classify $\mathbb{J}_n$ according to the smallest positive periods:

$$\mathbb{J}_n = \sum_{i=0}^{n-1} \mathbb{J}_n^f(i),$$

where $\mathbb{J}_n^f(i) = \{J = \{0, j_1, \cdots, j_k\} \in \mathbb{J}_n : j_1 = i\}$ for $1 \le i \le n-1$, while $\mathbb{J}_n^f(0) = \{\{0\} \in \mathbb{J}_n\}$ is the singleton consisting of the only periodic set without any positive period. If for $n \le n_0 - 1$, $\mathbb{J}_n$ have already been constructed, we may construct $\mathbb{J}_{n_0}^f(i)$ to finish the recurrence procedure as follow.

Lemma 4. For $i > \frac{n_0}{2}$, $\mathbb{J}_{n_0}^f(i) = \{J = \{0, i + J'\} : J' \in \mathbb{J}_{n_0 - i}\}$.

This is an immediate consequence of Lemma 3.

To construct $\mathbb{J}_{n_0}^f(i)$ for $i \le \frac{i_0}{2}$, we need some other subclasses of $\mathbb{J}_n$. First let

$$\mathbb{J}_n^p = \textstyle\sum_{i \not| n} \mathbb{J}_n^f(i) = \{J = \{0, j_1, \cdots, j_k\} \in \mathbb{J}_n : j_1 \not| n\},$$

then we have

Lemma 5. For $i | n_0$, denote $n_0 = qi$, $q \ge 2$,

$$J_{n_0}^f(i) = \{J = \{0, i, 2i, \cdots, (q-2)i, (q-1)i + J'\} : J' \in \mathbb{J}_i^p\}.$$

When $q = 2$, it's a direct consequence of Lemma 2, for general q, one may argue inductively by combining Lemma 2 and Lemma 1.

Since the recurrence procedure includes the construction of $\mathbb{J}_n^f(i)$, so both $\mathbb{J}_n^p$ and $\mathbb{J}_n$ are ready for smaller n. For the construction of the rest $\mathbb{J}_{n_0}^f(i) : (i, n_0) < i < \frac{n_0}{2}$, we need some more other subclasses of $\mathbb{J}_n$:

$$\begin{aligned}\mathbb{J}_n^q(i) =& \{J = \{0, j_1, \cdots, j_k\} \in \mathbb{J}_n : i \in J, j_1 \not| i \text{ whenever } i > j_1\} \\ =& \mathbb{J}_n^f(i) + \sum_{\substack{j=n-i+2 \\ j \not| i}}^{i-1} \mathbb{J}_n^f(j, i),\end{aligned}$$

where $\mathbb{J}_n^f(j, i) = \{J = \{0, j, j_2, \cdots, j_k\} \in \mathbb{J}_n : j < i \in J,\ j \not| i\}$. As $\mathbb{J}_n^f(j, i) \subset \mathbb{J}_n^f(j)$ are subsets of known subclasses $\mathbb{J}_n^f(j)$ for smaller n, so we may regard them being known, and $\mathbb{J}_n^q(i)$ too. By Lemma 1, we have

Lemma 6. For i such that $(i, n_0) < i < \frac{n_0}{2}$, denote $n_0 = bi + r, b \ge 2, 1 \le i < r$, then

$$\mathbb{J}_{n_0}^f(i) = \{J = \{0, i, 2i, \cdots, (b-2)i, (b-1)i + J'\} : \quad J' \in \mathbb{J}_{i+r}^q(i)\}.$$

Now we turn to the inductive calculation of $\{W_n(J)\}_{J \in \mathbb{J}_n}$. The easiest case is when $J \in \mathbb{J}_n^f(1), W_n(J) = 2$ for any $n \in \mathbb{N}$, because the only words are $\{0, \cdots, 0\}$ and $\{1, \cdots, 1\}$ with period 1 (for general $S = \{a_1, a_2, \cdots, a_m\}, W_n^{(m)}(J) = m$).

Next if $J \in \mathbb{J}_n^f(2), W_n(J) = 2$ for all $n \in \mathbb{N}$ either, the corresponding words are $\{1,0,1,0,\cdots\}$ and $\{0,1,0,1,\cdots\}$ (if $\#S = m, W_n^{(m)}(J) = m^2 - m$).

Of course, the calculation is not always so easy, but it's always possible if we calculate them step by step in the correct order. Since $\mathbb{J}_n^f(i)$ depends on two parameter n and i, we may calculate them in the natural order of n, as n being fixed, we should start from $i = 1,2,3,\cdots$, until $i = n-1$, finally calculate $i = 0$, that is $W_n(\{0\})$. For example, let's consider the case of $n = 4,5$ (when n is smaller, it becomes trivial).

$$\mathbb{J}_4 = \{\{0\},\{0,1,2,3\},\{0,2\},\{0,3\}\}.$$

$W_4(\{0,1,2,3\}) = W_4(\{0,2\}) = 2$ are already known, so we start from $\{0,3\} \in \mathbb{J}_4^f(3)$. Words with period 3 are decided by their first three letters, there are $2^3 = 8$ of them altogether, but some of them possess smaller positive period 1, so $W_4(\{0,3\}) = 2^3 - W_4(\{0,1,2,3\}) = 6$. Finally $W_4(\{0\}) = 2^4 - \sum_{J \neq \{0\}} W_4(J) = 6$. (If $\#S = m$, then $W_4^{(m)}(\{0,3\}) = m^3 - m, W_4^{(m)}(\{0\}) = m^4 - m^3 - m^2 + m$).

$$\mathbb{J}_5 = \{\{0\},\{0,1,2,3,4\},\{0,2,4\},\{0,3\},\{0,3,4\},\{0,4\}\}.$$

Now there are two member in $\mathbb{J}_5^f(3)$, $\{0,3\}$ and $\{0,3,4\}$, we may regard them as the over-complete prolongation of $\mathbb{J}_2 = \{\{0\},\{0,1\}\}$. Since $W_2(\{0\}) = W_2(\{0,1\}) = 2$, one half of the 2^3 words with period 3 possess period 4, but two of them possess smaller positive period 1. So

$$W_5(\{0,3,4\}) = 4 - 2 = 2, \quad W_5(\{0,3\}) = 4.$$

Similar argument as in the discussion of $\mathbb{J}_4$ shows that $W_5(\{0,4\}) = 10$ and $W_5(\{0\}) = 12$. (For $\#S = m, W_5^{(m)}(\{0,3,4\}) = m^2 - m, W_5^{(m)}(\{0,3\}) = m^3 - m^2, W_5^{(m)}(\{0,4\}) = m^4 - 2m^2 + m, W_5^{(m)}(\{0\}) = m^5 - m^4 - m^3 + m^2$).

Now we may state the general recurrence procedure as follow. Suppose that $W_k(J), J \in \mathbb{J}_k, k = 1,2,\cdots,n-1$, have been worked out, we want to calculate $W_n(J), J \in \mathbb{J}_n^f(i), i = 1,2,\cdots,n-1,0$.

Lemma 7. For $(i,n) < i < \frac{n}{2}$, denote $n = bi + r$, $b \geq 2, 1 \leq r \leq i-1$. $J \in \mathbb{J}_n^f(i)$ must have the form

$$J = \{0, i, 2i, \cdots, (b-2)i, (b-1)i + J'\},$$

where $J' \in \mathbb{J}_{i+r}^q(i)$, and $W_n(J) = W_{i+r}(J')$.

Lemma 8. For $i|n$, denote $n = bi, b \geq 2$. $J \in \mathbb{J}_n^f(i)$ must have the form

$$J = \{0, i, 2i, \cdots, (b-2)i, (b-1)i + J'\}$$

with $J' \in \mathbb{J}_i^p$, and $W_n(J) = W_i(J')$.

These are immediately consequence of Lemma 6 and Lemma 5, as $\mathbb{J}_{i+r}^p(i) \subset \mathbb{J}_{i+r}, \mathbb{J}_i^p \subset \mathbb{J}_i$ and $i + r < n, i < n$, so $W_{i+r}(J')$ and $W_i(J')$ are already know. Notice that for $1 \leq i \leq \frac{n}{2}$, we needn't calculate them one after the other in the natural order of i, the "second inductive procedure" of i is necessary only for $\frac{n}{2} < i \leq n-1$ as stated in the following lemma.

Lemma 9. For $\frac{n}{2} < i \leq n-1, J = \{0, i + J'\} \subset \mathbb{J}_n^f(i)$ with $J' \in \mathbb{J}_{n-i}$,

$$W_n(J) = 2^{2i-n} W_{n-i}(J') - \textstyle\sum_{\tilde{j} \in \mathbb{J}_n(J')} W_n(\tilde{j}),$$

where $\mathbb{J}_n(J') = \{J \in \mathbb{J}_n : \tilde{j} = \{0, j_1, \cdots, i + J'\}, j_1 < i\} \subset \sum_{j<i} \mathbb{J}_n^f(j)$.

The intuitive meaning of Lemma 9 is clear, when we prolong $J' \in \mathbb{J}_{n-i}$ to $J \in \mathbb{J}_n$, there are $2i - n$ free letters, they give us 2^{2i-n} possible choices, but some of them possess smaller positive periods, fortunately their numbers have been already calculated according our recurrence procedure.

Finally, when every $W_n(J), J \in \mathbb{J}_n^f(i), 1 \le i \le n-1$, have been worked out, then

$$W_n(\{0\}) = 2^n - \sum_{J \neq \{0\}} W_n(J).$$

The procedure is available for general $S = \{a_1, a_2, \cdots, a_m\}$ if we use m^{2i-n} instead of 2^{2i-n} in Lemma 9 and finally $W_n^{(m)}(\{0\}) = m^n - \sum_{J \neq \{0\}} W_n^{(m)}(J)$.

4. HOW LARGE THE CLASS OF WAITING DISTRIBUTION CAN BE?

If we only want to know how many members are there in $\mathbb{J}_n$, it's not necessary to list all the members in $\mathbb{J}_n$ as what we did last section, we discuss a recurrence method of calculating $C_n = \#\mathbb{J}_n$ this section.

It's natural to use the classification $\mathbb{J}_n = \sum_{i=0}^{n-1} \mathbb{J}_n^f(i)$, denote $C_n = \#\mathbb{J}_n$, $f_n(i) = \#\mathbb{J}_n^f(i)$, we get the first formula

$$C_n = \sum_{i=0}^{n-1} f_n(i) = 1 + \sum_{i=1}^{n-1} f_n(i).$$

When $i > \frac{n}{2}$, $J \in \mathbb{J}_n^f(i)$ has the form $J = \{0, i + J'\}$ with $J' \in \mathbb{J}_{n-i}$, and the correspondence between $\mathbb{J}_n^f(i)$ and $\mathbb{J}_{n-i}$ forms a one-to-one map, so

$$f_n(i) = \#\mathbb{J}_n^f(i) = \#\mathbb{J}_{n-i} = C_{n-i},$$

$$\sum_{\frac{n}{2} < i \le n-1} f_n(i) = \sum_{\frac{n}{2} < i \le n-1} C_{n-i} = \sum_{i=1}^{[\frac{n-1}{2}]} C_k,$$

thus

$$C_n = 1 + \sum_{i=1}^{[\frac{n-1}{2}]} C_k + \sum_{i=1}^{[\frac{n}{2}]} f_n(i),$$

which implies the inequality

$$C_n \ge 1 + \sum_{k=1}^{[\frac{n-1}{2}]} C_k$$

used by Zbăganu to show that C_n grow faster than any polynomial of n when $n \uparrow\uparrow \infty$. But he told me, he couldn't even prove that $\{C_n\}_{n \in \mathbb{N}}$ is an increasing sequence, and I couldn't neither.

Similarly, there are one-to-one correspondences between $\mathbb{J}_n^f(i)$ and $\mathbb{J}_i^p$ for $i|n$, and between $\mathbb{J}_n^f(i)$ and $\mathbb{J}_{i+r_i}^q(i)$ for $(i,n) < i < \frac{n}{2}$, $n = bi + r_i$, $b \ge 2$, $1 \le r_i \le i-1$. So if we denote $C_n^p = \#\mathbb{J}_n^p$, $C_n^q(i) = \#\mathbb{J}_n^q(i)$, then

$$C_n = 1 + \sum_{k=1}^{[\frac{n-1}{2}]} C_k + \sum_{i|n} C_i^p + \sum_{(i,n)<i<\frac{n}{2}} C_{i+r_i}^q(i).$$

The subscripts in the right hand side $k, i, i+r_i$ are all smaller than n (evenmore smaller that $\frac{2n}{3}$), so it looks like a recurrence formula. But it requires not only C_n, but also C_n^p and $C_n^q(i)$ for $i > \frac{n}{2}$ ($r_i \leq i-1 \Rightarrow i > \frac{i+r_i}{2}$). C_n^p is easy, since $\mathbb{J}_n^p = \sum_{i \nmid n} \mathbb{J}_n^f(i)$, so

$$C_n^p = 1 + \sum_{i \nmid n} f_n(i) = 1 + \sum_{k=1}^{[\frac{n-1}{2}]} C_k + \sum_{(i,n)<i<\frac{n}{2}} C_{i+r_i}^q(i),$$

similar to and simpler than C_n. As for $C_n^q(i)$, since

$$\mathbb{J}_n^q(i) = \sum_{\substack{j=n-i+2 \\ j \nmid i}}^{i-1} \mathbb{J}_n^f(j,i) + \mathbb{J}_n^f(i), \qquad \text{for } i > \frac{n}{2}.$$

Let $f_n(j,i) = \#\mathbb{J}_n^f(j,i)$, then

$$C_n^q(i) = f_n(i) + \sum_{\substack{j=n-i+2 \\ j \nmid i}}^{i-1} f_n(j,i) = C_{n-i} + \sum_{\substack{j=n-i+2 \\ j \nmid i}}^{i-1} f_n(j,i),$$

$f_n(j,i)$ similar to $f_n(i)$ but smaller than $f_n(i)$, and their calculation requires some smaller quantities to be calculated beforehand. To form a closed recurrence formulas, we have to divide $\mathbb{J}_n$ smaller and smaller. Let

$$\mathbb{J}_n(i_1, \cdots, i_l) = \{J \in \mathbb{J}_n,\ i_r \in J \text{ for } r = 1, \cdots, l \text{ and } i_1 < i_2 \cdots < i_l\},$$

$$\mathbb{J}_n^f(i, i_1, \cdots, i_l) = \{J = \{0, j_1, \cdots, j_k\} \in \mathbb{J}_n(i_1, \cdots, i_l) \text{ and } j_1 = i \nmid i_1\},$$

$$\mathbb{J}_n^q(i, i_1, \cdots, i_l) = \{J = \{0, j_1, \cdots, j_k\} \in \mathbb{J}_n(i_1, \cdots, i_l) \text{ and } i \in J,\ j_1 \nmid i \text{ if } j_1 \neq i\},$$

$$\mathbb{J}_n^p(i_1, \cdots, i_l) = \{J = \{0, j_1, \cdots, j_k\} \in \mathbb{J}_n(i_1, \cdots, i_l) \text{ and } j_1 \nmid n\},$$

$$C_n(i_1, \cdots, i_l) = \#\mathbb{J}_n(i_1, \cdots, i_l), \qquad f_n(i, i_1, \cdots, i_l) = \#\mathbb{J}_n^f(i, i_1, \cdots, i_l),$$

$$C_n^q(i, i_1, \cdots, i_l) = \#\mathbb{J}_n^q(i, i_1, \cdots, i_l), \qquad C_n^p(i_1, \cdots, i_l) = \#\mathbb{J}_n^p(i_1, \cdots, i_l),$$

then we may use f_n to calculate C_n, C_n^q, C_n^p, and in turn we may use C, C^q, C^p with smaller subscripts to calculate f_n. We state them into the following

Theorem 2. The following two groups of formulas form a closed recurrence procedure to calculate C_n and relative quantities f_n, C_n^p, C_n^q

$$\text{(I)} \begin{cases} C_n(i_1, \cdots, i_l) = f_n(i_1, \cdots, i_l) + \sum_{i=1}^{i_1-1} f_n(i, i_1, \cdots, i_l), \\ C_n^p(i_1, \cdots, i_l) = f_n(i_1, \cdots, i_l) + \sum_{\substack{i=2 \\ i \nmid i_1}}^{i_1-1} f_n(i, i_1, \cdots, i_l), \\ C_n^q(i, i_1, \cdots, i_l) = \widetilde{f}_n(i, i_1, \cdots, i_l) + \sum_{\substack{i_0=2 \\ i_0 \nmid i}}^{i \wedge i_1 - 1} \widetilde{f}_n(i_0, i, i_1, \cdots, i_l). \end{cases}$$

where $\widetilde{f}_n$ are principally f_n only the variables $i, i_1, \cdots, i_l$ or $i_0, i, i_1, \cdots, i_l$ are not necessary in natural orders, in more detail, for $i_1 < i_2 < \cdots < i_r$, if $\{j_1, j_2, \cdots, j_r\}$ is a permutation of $\{1, \cdots, r\}$, then

$$\widetilde{f}_n(i_{j_1}, i_{j_2}, \cdots, i_{j_r}) = f_n(i_1, i_2, \cdots, i_r).$$

$$
(\mathrm{II})\begin{cases}
f_n(i,i_1,\cdots,i_l) = C_{n-i}(i_1-i,\cdots,i_l-i) & \text{if } i>\frac{n}{2},\\
f_n(i,i_1,\cdots,i_l) = C_i^p(i_1+i-n,\cdots,i_l+i-n) & \text{if } i\mid n,\\
f_n(i,i_1,\cdots,i_l) = C_{i+r_i}^q(i,i_1+i+r_i-n,\cdots,i_l+i+r_i-n) & \\
\quad \text{for } (i,n)<i<\frac{n}{2},\ n=bi+r_i,\ b\geq 2,\ 1\leq r_i\leq i-1. &
\end{cases}
$$

For $l=0$ we have already shown the corresponding (I) and (II), and the reason is the same for general $l \in \mathbb{N}$, so we omit the detail of the demonstration. Put the procedure into practice, it's not so complicated as it looks like. What we need for l are in fact much smaller than n, for example, to calculate $C_n, n \leq 50, l=2$ is already enough. In the calculation of C_{50},

$$
C_{50} = 1+\sum_{k=1}^{24} C_k + \sum_{i\mid 50} C_i^p + \sum_{(i,50)<i<25} C_{i+r_i}^q(i),
$$

where $1+\sum_{k=1}^{24} C_k + \sum_{i\mid 50} C_i^p = 1625$ and $\sum_{(i,50)<i\leq 17} C_{i+r_i}^q(i) = 220$ is ready, the rest $C_{i+r_i}^q(i), 18 \leq i \leq 24$, can be calculated by using formulas with $l=2$.

To give some numerical impression, we list here the values of C_n, $13 \leq n \leq 50$, in the natural order,

37, 47, 57, 62, 75, 87, 102, 116, 135, 155, 180, 194, 220, 254, 289, 312, 359, 392, 438, 479, 538, 595, 664, 701, 772, 863, 956, 1005, 1115, 1205, 1317, 1414, 1552, 1677, 1836, 1920, 2074, 2249, $\cdots$ so evident an increasing sequence, what a pity fail to prove that theoretically!

Since the quantities decrease very quickly when l increases, so we may regard all quantities with larger l as 0 to get approximate formulas much simpler.

5. NON-PERIODIC WORDS AND HIGH PERIODIC WORDS

From last section, we know that C_n grow very quickly. On the other hand, in Section 2, we have seen that $W_n(J)$ are changeful for different $J \in \mathbb{J}_n$, some periodic sets are much more important than others (in the sense $W_n(J)$ much larger). In this section, we'll pay more attention to those periodic sets J with larger $W_n(J)$, especially $J=\{0\}$ — periodic set of Non-periodic words, and $J=\{0,n-1\} \in \mathbb{J}_n$ — periodic set of High periodic words.

In the example of $n=13$, we saw $W_{13}(\{0\}) = 2232$, $W_{13}(\{0,12\}) = 2462$: two periodic sets from 37 occupy 4694 words of 8192, with so high a percentage (nearly 57.3%). Is it a general fact for all $n \in \mathbb{N}$? The answer is affirmative, we'll show this by giving two recurrence formulas of $W_n(\{0\})$ and $W_n(\{0,n-1\})$.

If you need $W_n(J)$ for each $J \in \mathbb{J}_n$, the recurrence procedure given in Section 2 is all right. But if you are only interested in effective periodic sets such as $\{0\}$ and $\{0,n-1\}$, then the procedure is not so "effective", $W_n(\{0\})$ and $W_n(\{0,n-1\})$ are just the most difficult ones to calculate, we have to look for some new method.

For any subset $\mathbb{J}$ of $\mathbb{J}_n$, let $W_n(\mathbb{J}) = \sum_{J\in\mathbb{J}} W_n(J)$, then trivially,

$$
W_n(\mathbb{J}_n) = 2^n, \quad W_n(\mathbb{J}_n(i)) = 2^i, \qquad 1 \leq i \leq n-1.
$$

It's more interesting to know $W_n(\mathbb{J}_n^f(i)) =$? Fix i, consider sequence $\{W_n(\mathbb{J}_n^f(i))\}_{n\in\mathbb{N}}$, it's "tail definite", in fact, when $n \geq 2i-2$, $W_n(\mathbb{J}_n^f(i))$ doesn't depend on n any longer, denote the common value by $W(\mathbb{J}^f(i))$ or simply a_i. the sequence $\{a_i = W(\mathbb{J}^f(i))\}_{i\in\mathbb{N}}$ satisfies

$$\sum_{k|i} a_k = 2^i, \qquad \text{so} \qquad a_i = 2^i - \sum_{\substack{k|i\\k<i}} a_k$$

give a recurrence formula for $\{a_i\}_{i\in\mathbb{N}}$, thus

$$W(\mathbb{J}^f(1)) = 2, \qquad W(\mathbb{J}^f(2)) = 2, \qquad W(\mathbb{J}^f(3)) = 6, \qquad W(\mathbb{J}^f(4)) = 12$$

$$W(\mathbb{J}^f(5)) = 30, \qquad W(\mathbb{J}^f(6)) = 54, \qquad W(\mathbb{J}^f(7)) = 126, \qquad W(\mathbb{J}^f(8)) = 240, \cdots.$$

But still, when $n \le 2i-3$, it's not so easy to calculate $W_n(\mathbb{J}_n^f(i))$, we are going to discuss another extreme case $W_n(\mathbb{J}_n^f(n-k))$, especially $W_n(\mathbb{J}_n^f(n-1)) = W_n(\{0, n-1\})$.

Let's start from $W_n(\mathbb{J}_n^f(0)) = W_n(\{0\})$, to get a recurrence formula of $W_n(\{0\})$, $n \in \mathbb{N}$, it's convenient to introduce $\mathbb{J}_n^l(i) = \{J = \{0, j_1, \cdots, j_k\} \in \mathbb{J}_n : j_k = i\}$, the family of periodic sets with i as their largest positive period.

Lemma 10. For $i \ge [\frac{n+1}{2}]$

$$W_n(\mathbb{J}_n^l(i)) = 2^{2i-n} W_{n-i}(\{0\}).$$

Proof. $J(\mathbb{b}) = \{0\} \in \mathbb{J}_{n-i} \iff \bar{\mathbb{b}} = \mathbb{b}\mathbb{b}'\mathbb{b} \in \mathbb{J}_n^l(i)$, where $\mathbb{b}' \in \{0,1\}^{2i-n}$ can be freely taken.

Notice that $W_n(\{0\}) = 2^n - \sum_{i=[\frac{n+1}{2}]}^{n-1} W_n(\mathbb{J}_n^l(i))$, we have

Corollary. $W_n(\{0\}) = 2^n - \sum_{i=[\frac{n+1}{2}]}^{n-1} 2^{2i-n} W_{n-i}(\{0\}) = 2^n - \sum_{k=1}^{[\frac{n}{2}]} 2^{n-2k} W_k(\{0\})$

If let $p_n = \frac{1}{2^n} W_n(\{0\})$ denote the percentage of non-periodic words with length n, then

$$p_n = 1 - \sum_{k=1}^{[\frac{n}{2}]} \frac{1}{2^k} p_k, \qquad p_{2n} = p_{2n+1} = p_{2n-1} - \frac{1}{2^n} p_n, \; p_n \downarrow p_\infty.$$

$$p_1 = 1, \quad p_2 = p_3 = \frac{1}{2}, \quad p_4 = p_5 = \frac{3}{2^3}, \quad p_6 = p_7 = \frac{5}{2^4}, \quad p_8 = p_9 = \frac{23}{2^7},$$

$$p_{10} = p_{11} = \frac{71}{2^8}, \quad p_{12} = p_{13} = \frac{279}{2^{10}}, \quad p_{14} = p_{15} = \frac{553}{2^{11}}, \quad p_{16} = p_{17} = \frac{8811}{2^{16}},$$

$$p_{18} = p_{19} = \frac{17585}{2^{17}}, \quad p_{20} = p_{21} = \frac{70269}{2^{18}} = 0.26805\cdots,$$

$$p_{20} - p_\infty = \sum_{k=11}^{\infty} \frac{1}{2^k} p_k \le \frac{1}{2^{10}} p_{11} < 288 \cdot 10^{-6} \Rightarrow p_\infty > 0.267762$$

in turn

$$p_{20} - p_\infty \ge \frac{1}{2^{10}} \cdot 0.267762 \ge 0.000261 \;\Rightarrow\; p_\infty \le p_{20} - 0.000261 < 0.2678$$

thus $0.26776 < p_\infty < 0.2678$, take $\hat{p}_\infty = 0.26778$, it's an estimation with error smaller than $2 \cdot 10^{-5} = 1/50000$.

Now turn to the sequence $\{W_n(\{0, n-1\})\}_{n\in\mathbb{N}}$. Let

$$J_n^l(i, n-1) = \{J = \{0, j_1, \cdots, j_{k-1}, j_k\} \in \mathbb{J}_n : j_{k-1} = i, j_k = n-1\}$$

be the family of periodic sets with $i, n-1$ as their last two periods. The same reason as in Lemma 10 gives

$$W_n(\mathcal{J}_n^l(i,n-1)) = 2^{2i-n}W_{n-i}(\{0,n-i-1\}),$$

while $n = 2m+1, i = m$

$$W_n(\mathcal{J}_n^l(m,2m)) = W_{m+1}(\{0,m\}).$$

Corollary. $W_{2n}(\{0,2n-1\}) = 2^{2n-1} - \sum_{l=2}^{n} 2^{2n-2l}W_l(\{0,l-1\})$,

$$W_{2n+1}(\{0,2n\}) = 2^{2n} - \sum_{l=2}^{n} 2^{2n+1-2l}W_l(\{0,l-1\}) - W_{n+1}(\{0,n\})$$

Denote $q_n = \frac{1}{2^n}W_n(\{0,n-1\})$, then

$$q_{2n} = \frac{1}{2} - \sum_{l=2}^{n} \frac{1}{2^l}q_l, \qquad q_{2n+1} = \frac{1}{2} - \sum_{l=2}^{n} \frac{1}{2^l}q_l - \frac{1}{2^n}q_{n+1}.$$

Obviously $\{q_{2n}\}_{n\in\mathbb{N}}$ is decreasing, but $\{q_{2n+1}\}_{n\in\mathbb{N}}$ isn't. It's better to divide $\{q_{2n+1}\}_{n\in\mathbb{N}}$ into two sub-sequences $\{q_{4n+1}\}_{n\in\mathbb{N}}$ and $\{q_{4n+3}\}_{n\in\mathbb{N}}$, then

$$\begin{aligned} q_{4n+1} &= \frac{1}{2} - \sum_{l=2}^{2n} \frac{1}{2^l}q_l - \frac{1}{2^{2n}}q_{2n+1} \\ &= q_{4n-3} - \frac{1}{2^{2n-1}}q_{2n-1} - \frac{1}{2^{2n}}q_{2n} + \frac{1}{2^{2n-2}}q_{2n-1} - \frac{1}{2^{2n}}q_{2n+1} \\ &= q_{4n-3} - \frac{q_{2n}+q_{2n+1}-2q_{2n-1}}{2^{2n}} = q_{4n-3} - \frac{1}{2^{3n}}(2q_n - q_{n+1}) < q_{4n-3} \end{aligned}$$

show that $\{q_{4n+1}\}_{n\in\mathbb{N}}$, is also decreasing. But

$$\begin{aligned} q_{4n+3} &= \frac{1}{2} - \sum_{l=2}^{2n+1} \frac{1}{2^l}q_l - \frac{1}{2^{2n+1}}q_{2n+2} \\ &= q_{4n-1} - \frac{1}{2^{2n}}q_{2n} - \frac{1}{2^{2n+1}}q_{2n+1} + \frac{1}{2^{2n-1}}q_{2n} - \frac{1}{2^{2n+1}}q_{2n+2} \\ &= q_{4n-1} + \frac{2q_{2n} - q_{2n+1} - q_{2n+2}}{2^{2n+1}} > q_{4n-1}, \end{aligned}$$

that is, $\{q_{4n+3}\}_{n\in\mathbb{N}}$ is increasing. It's easily seen that $\{q_n\}_{n\in\mathbb{N}}$ is a Cauchy sequence, so the limits of three monotone subsequeces are the same, denoted by q_∞. Notice

$q_{2n} = q_{2n-2} - \frac{1}{2^n}q_n$, $\quad q_{2n+1} = q_{2n} - \frac{1}{2^n}q_{n+1}$,

$q_2 = \frac{1}{2}$, $\quad q_4 = \frac{3}{8}$, $\quad q_6 = \frac{11}{32}$, $\quad q_8 = \frac{41}{2^7}$, $\quad q_{10} = \frac{159}{2^9}$, $q_{12} = \frac{625}{2^{11}}$, $q_{14} = \frac{2481}{2^{13}}$ $q_{16} = \frac{9883}{2^{15}}$

$q_3 = \frac{1}{4}$ $\quad q_7 = \frac{19}{64}$ $\quad q_{11} = \frac{307}{2^{10}}$ $\quad q_{15} = \frac{4921}{2^{14}}$

$q_5 = \frac{5}{16}$ $\quad q_9 = \frac{77}{2^8}$ $\quad q_{13} = \frac{1231}{2^{12}}$ $\quad q_{17} = \frac{19689}{2^{16}}$

$$0.300369\cdots < \frac{4921}{2^{14}} = q_{15} < q_\infty < q_{17} = \frac{19689}{2^{16}} < 0.30043\cdots$$

take $\hat{q}_\infty = 0.3004$, we get an estimation of q_∞ with error smaller than $4 \cdot 10^{-5} = \frac{1}{25000}$

Thus, for any n, the percentage of Non-periodic words and High-periodic words (with periodic sets $\{0\}$ or $\{0, n-1\}$) is always not less than 56.8%.

The method can be used to discuss $\{W_n(\mathbb{J}_n^f(n-i))\}_{n\in\mathbb{N}}$ for i fixed to get the following recurrence formula

$$W_n(\mathbb{J}_n^f(n-i)) = 2^{n-i} - \sum_{l=i+1}^{[\frac{n-i}{2}]} (2^{n-2l} \vee 1) W_l(\mathbb{J}_l^f(l-i)).$$

Besides, general alphabet set $S = \{a_1, \cdots, a_m\}$ causes no trouble, $p_n^{(m)} = \frac{1}{m^n} W_n^{(m)}(\{0\})$ satisfy

$$p_n^{(m)} = 1 - \sum_{k=1}^{[\frac{n}{2}]} \frac{1}{m^k} p_k^{(m)} \downarrow p_\infty^{(m)},$$

$$p_\infty^{(m)} \geq 1 - \frac{1}{m} - \left(\sum_{k=2}^{\infty} \frac{1}{m^k} \right) p_2^{(m)} = 1 - \frac{1}{m} - \frac{m-1}{m} \cdot \frac{1}{m^2} \cdot \frac{1}{1-\frac{1}{m}} = \frac{m^2 - m - 1}{m^2},$$

even $m = 3$, $p_\infty^{(3)} \geq \frac{5}{9}$, most of the words are non-periodic ones. "English words", more than 96% of them are non-periodic words, " Russian" about 97%, " Japanese" about 98%. So we may get the CONCLUSION:

The most important (easier to encounter) WAITING DISTRIBUTIONS are the simplest WAITING DISTRIBUTION with generating function

$$\varphi_n(x) = \frac{x^n}{x^n + c(1-x)},$$

where $c = \mathbb{b} * \mathbb{b}(x) = \frac{1}{\prod_{i=1}^n P(b_i)}$, and when $S = \{0, 1\}$, also the Waiting distribution corresponding to $J = \{0, n-1\}$, with generating function

$$\varphi_n(x) = \frac{x^n}{x^n + (1-x)(c + dx^{n-1})},$$

where $d = \frac{1}{P(b_1)}$.

Acknowledgements

This research was supported by the National Natural Science Foundation of China.

REFERENCES

1. H.U. Gerber, S-Y. R. Li, *Stoch. Proc. Appl.*, **11** (1981), 101–108.
2. S-Y. R. Li, *Ann. Probab.*, **8** (1980), 1171–1176.

Stat. Sci. & Data Anal., pp. 489-499
K. Matsusita *et al.* (Eds)

The use of the inverse Gaussian model for analysing the lognormal data

EIJI YAMAMOTO and TAKEMI YANAGIMOTO
Dept. of Applied Math., Okayama Univ. of Science, 1-1 Ridai-cho, Okayama 700, Japan
Institute of Statistical Mathematics, 4-6-7 Minami-Azabu, Minato-ku, Tokyo 106, Japan

Abstract. By applying the inverse Gaussian model to the lognormal data, we can get practical procedures for inference of the mean and the squared coefficient of variation in the lognormal population. Performance of estimators of both the parameters and a test for the mean are examined analytically and numerically. Some examples are taken to show advantages of the procedures.

Key words: Estimation; test; mean; coefficient of variation.

1. INTRODUCTION

The lognormal distribution is known as one of typical distributions for a positively valued population such as the gamma, the inverse Gaussian and the Weibull distributions [1]. The probability density function of the lognormal distribution, $LN(\delta, \sigma^2)$, is defined by

$$f_{ln}(x;\delta,\sigma^2) = \frac{1}{\sqrt{2\pi}\sigma x} exp\{-\frac{(\log(x)-\delta)^2}{2\sigma^2}\}, \qquad 0 < x < \infty; -\infty < \delta < \infty; 0 < \sigma^2.$$

By definition, a log-transformed random variable, $log(x)$, follows a normal distribution, $N(\delta, \sigma^2)$. Therefore we can make statistical inference of (δ, σ^2) by using the ordinal statistical inference procedures for the normal population.

In some situations, we are interested in the distribution of the raw data, x, from the population under examination. Important parameters of a positively valued distribution are the mean, μ, and the squared coefficient of variation, ψ^2. They indicate scale and shape of the distribution. In the case of $LN(\delta, \sigma^2)$, they are given by

$$\mu = exp(\delta + \frac{\sigma^2}{2}),$$
$$\psi^2 = e^{\sigma^2} - 1.$$

For inference of μ, we are usually lazy to use the UMVUE and the UMP unbiased test because of their complicated forms. There exist the conventional inference procedures, the likelihood inference procedure and the moment method. Unfortunately, the MLE of μ and the LRT for μ do not have good properties. On the contrary, the sample mean is unbiased and robust. It is expected to have high efficiency for a small sample size and is not expected to lose an acceptable level in asymptotic efficiency when ψ is moderate, less than 1. For the estimation of ψ^2, the UMVUE has also a complicated function form. The MLE and the moment estimator of ψ^2 are not unbiased and do not keep the acceptable level in asymptotic efficiency. The test for ψ^2 is translated into the test for σ^2 in the normal distribution.

In this paper, we propose statistical inference procedures of (μ, ψ^2) by applying the inverse Gaussian model to the lognormal data. This claim is motivated from the fact that the inverse Gaussian distribution is close to the lognormal distribution [2]. In Section 2, we propose the inference procedures of (μ, ψ^2). The fact of closeness between the inverse Gaussian and the lognormal distributions is discussed in Section 3. Properties of the procedures are examined analytically and numerically in Sections 4 and 5. Examples are taken to show advantages of the procedures in Section 6. The final section is prepared for general discussions.

2. INFERENCE PROCEDURES

Let $x_1, x_2, \cdots, x_n$ be a sample of size n from the population with $LN(\delta, \sigma^2)$. We consider the inference problems of the estimation of μ, the test for μ and the estimation of ψ^2. From the fact of closeness between the inverse Gaussian and the lognormal distributions, which is discussed in the next section, we will recommend to apply the inverse Gaussian model even to the lognormal sample. This treatment enables us to enjoy the excellent statistical inference procedures for the inverse Gaussian distribution [3]. The probability density function of the inverse Gaussian distribution, $IG(\mu, \theta)$, is defined by

$$f_{ig}(x; \mu, \theta) = \frac{1}{\sqrt{2\pi\theta} \cdot x^{3/2}} exp\{-\frac{(x-\mu)^2}{2\theta\mu^2 x}\} \quad 0 < x < \infty; 0 < \mu; 0 < \theta.$$

The mean and the squared coefficient of variation are given by μ and $\theta\mu$, respectively. The recent development in conditional inference for $IG(\mu, \theta)$ [4] can lead us to an estimator of μ, $\hat{\mu} = \bar{x}$ and a test statistic for $H_0 : \mu = \mu_0$,

$$T = \frac{\left(\frac{\bar{x}}{\mu_0} + \frac{\mu_0}{\bar{x}} - 2\right)/\mu_0}{\hat{\theta}} \qquad where \quad \hat{\theta} = \sum \left(\frac{1}{x_i} - \frac{1}{\bar{x}}\right)/(n-1).$$

The test statistic T is compared to the F distribution with $(1, n-1)$ d.f.. The estimator of ψ^2 is given by

$$\hat{\psi}^2 = \hat{\mu}\hat{\theta} = \bar{x} \sum \left(\frac{1}{x_i} - \frac{1}{\bar{x}}\right)/(n-1).$$

We will propose the estimators $(\hat{\mu}, \hat{\psi}^2)$ and the test statistic T for μ in a unified way for the lognormal population. An advantage of using the inverse Gaussian model is the simple forms of the derived statistics.

3. CLOSE RELATIONS BETWEEN THE TWO DISTRIBUTIONS

It was mentioned that the proposed statistical inference procedure of (μ, ψ^2) in the lognormal population comes from the fact that the inverse Gaussian distribution is close to the lognormal distribution. In this section, we will examine this fact in details. Five evidences supporting this claim are listed up as follows.

(1) The probability density functions of the two distributions behave similarly. Both the values of the *pdf*s at the point $x = 0$ are 0. Both the shapes of the *pdf*s are unimodal. These two properties of the *pdf*s are not satisfied by the gamma and the Weibull distributions. In fact they have monotone decreasing *pdf*s when the squared coefficient of variation is greater than 1.

(2) The hazard functions of the two distributions behave similarly. Both the values of the *hf*s at the point $x = 0$ are 0. Both the shapes of the *hf*s are unimodal. A difference is found in the tail behaviour of the *hf*s. The limiting value of the *hf* of the lognormal distribution when x tends to become infinity is zero; while the one of the inverse Gaussian distribution is the finite value $1/2\mu\psi^2$. The tail behaviour of the *hf*s are quite different from the ones of the gamma and the Weibull distributions.

(3) There is a strong relation among the main three central moments μ_1, μ_2, μ_{-1} of the two distributions. If two of the three moments are common in the two distributions, then the remainder becomes common automatically. This means that the three moments expressed by functions of (μ, ψ^2) are common in the two distributions. We can easily prove this proposition since the reciprocal of a *r.v.* X from each of the two distributions takes the same expectation of $\frac{1}{\mu}(1 + \psi^2)$.

(4) A normalizing transformation of the inverse Gaussian distribution is given by $log(X)$. This means that $log(X)$ in the lognormal distribution follows exactly the normal distribution, while $log(X)$ in the inverse Gaussian distribution follows approximately the normal distribution [5][6]. Here recall that a normalizing transformation of the gamma distribution is the well known Wilson and Hilferty transformation $X^{1/3}$.

(5) The Kullback-Leibler information of the lognormal distribution to the inverse Gaussian distribution is small. Here the KL information is defined by

$$
\begin{aligned}
2KL(LN; IG) &= 2E\left(log \frac{f_{ln}(X; \delta, \sigma^2)}{f_{ig}(X; \mu, \theta)} | f_{ln}\right) \\
&= log\left(\frac{\psi^2}{log(1 + \psi^2)\sqrt{1 + \psi^2}}\right).
\end{aligned}
$$

It depends only on ψ^2 and converges to 0 as $\psi^2 \to 0$. The quantity of this information is much smaller than those to the other distributions, the gamma, the Weibull and the normal distributions. Table 1 and Figure 1 show the KL informations of the lognormal distribution to the four distributions for various values of ψ^2. When ψ^2 becomes zero, the KL informations to the inverse Gaussian, the gamma, the normal distributions converge to zero. On the contrary, the KL information to the Weibull distribution converges to the finite value $exp\{\psi(1) + \psi'(1)/2\} - \psi(1) - (log(2\pi\psi'(1)) + 1)/2$ where $\psi(z)$ is the di-gamma function. Note that the limiting distribution of the Weibull distribution when ψ^2 becomes zero is not the normal distribution but the double exponential distribution.

The above evidences suggest us to apply the inverse Gaussian model to the lognormal data for conducting a conventional statistical inference analysis. In the next section, we will show that the proposed inference procedures perform well as expected.

Table 1.
Kullback-Leibler Informations of the lognormal distribution to the four distributions

ψ^2	2KL(LN,IG)	2KL(LN,GA)	2KL(LN,WB)	2KL(LN,N)
0.02	0.00002	0.00338	0.24561	0.02972
0.04	0.00006	0.00686	0.21427	0.05890
0.06	0.00014	0.01042	0.19654	0.08754
0.08	0.00025	0.01407	0.18513	0.11569
0.10	0.00038	0.01780	0.17737	0.14334
0.20	0.00138	0.03744	0.16279	0.27487
0.40	0.00471	0.08046	0.16955	0.50942
0.60	0.00919	0.12643	0.18622	0.71419
0.80	0.01435	0.17384	0.20424	0.89603
1.00	0.01994	0.22178	0.22178	1.05966
1.50	0.03474	0.34066	0.26137	1.40918
2.00	0.04979	0.45526	0.29513	1.69771

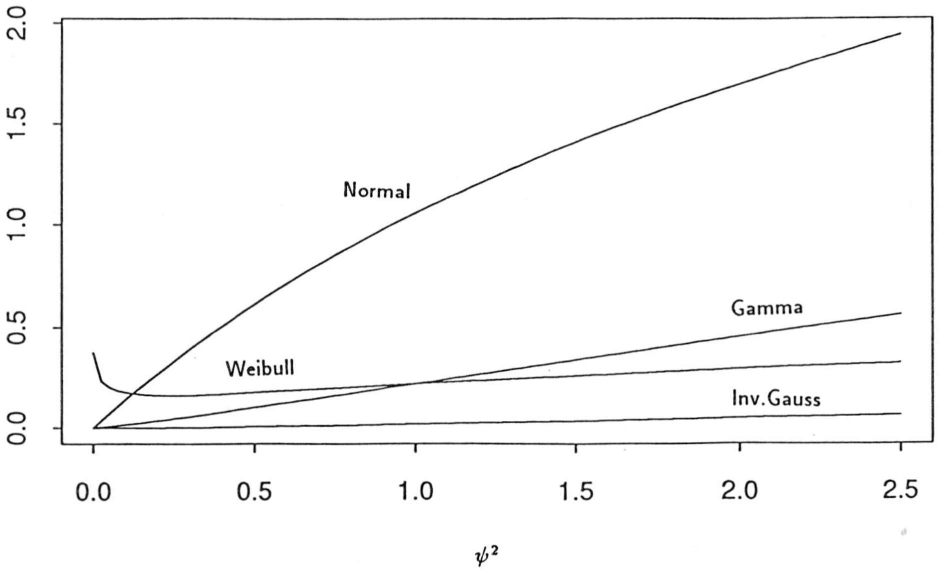

Figure 1. Comparison of the KL informations of the lognormal distribution to the four distributions.

4. ESTIMATORS OF (μ, ψ^2)

First we discuss the estimation of μ. The UMVUE is given by

$$\hat{\mu}_{mv} = e^{\tilde{x}} \cdot {}_0F_1\left(\frac{n-1}{2}; \frac{(n-1)^2}{4n}\tilde{s}^2\right)$$

where $\tilde{x} = \sum log(x_i)/n$, $\tilde{s}^2 = \sum(log(x_i)-\tilde{x})^2/(n-1)$ and ${}_0F_1(\alpha; z)$ is the hypergeometric function defined by

$${}_0F_1(\alpha; z) = \sum_{j=0}^{\infty} \frac{z^j}{(\alpha)_j j!}$$

with $(\alpha)_j = \alpha(\alpha+1)\cdots(\alpha+j-1) \quad j \geq 1, \ = 1 \quad j = 0$ [7]. The UMVUE of μ has the variance $\mu^2\{e^{\sigma^2/n} \cdot {}_0F_1(\frac{n-1}{2}; \frac{(n-1)^2}{4n^2}\sigma^4) - 1\}$ and the asymptotic one $\mu^2\{\frac{\sigma^2}{n}(1+\frac{\sigma^2}{2})\}$. The form looks discouragingly complicated for an actual use.

The MLE is commonly used for an estimator of μ, which is denoted by

$$\hat{\mu}_{ml} = e^{\tilde{x}+\tilde{s}^2(n-1)/2n}.$$

The modified version of the MLE

$$\hat{\mu}_{md} = e^{\tilde{x}+\tilde{s}^2/2}$$

is also familiar because of the simple form and the unbiasedness of $\tilde{s}^2$ to σ^2. A sample wise order among these three estimators holds

$$\hat{\mu}_{md} > \hat{\mu}_{ml} > \hat{\mu}_{mv}.$$

A potentially serious defect of the two ML estimators is found in the fact that the expectations of these estimators do not exist for a large σ^2. The expectation $E(\hat{\mu}_{ml})$ takes the value

$$E(\hat{\mu}_{ml}) = e^{\delta+\frac{\sigma^2}{2n}} \cdot \left(1 - \frac{\sigma^2}{n}\right)^{-\frac{n-1}{2}}.$$

Thus it does not exist when $\sigma^2 > n$. The expectation $E(\hat{\mu}_{md})$ takes the value

$$E(\hat{\mu}_{md}) = e^{\delta+\frac{\sigma^2}{2n}} \cdot \left(1 - \frac{\sigma^2}{n-1}\right)^{-\frac{n-1}{2}}.$$

It also does not exist when $\sigma^2 > n-1$.

We recommended $\hat{\mu} = \bar{x}$ in Section 2, which is unbiased and has the asymptotic efficiency

$$asym.\ eff(\hat{\mu}) = \frac{asymV(\hat{\mu}_{mv})}{V(\bar{X})} = \frac{ln(1+\psi^2) + \frac{1}{2}(ln(1+\psi^2))^2}{\psi^2}.$$

Table 2 shows the $asym.\ eff(\hat{\mu})$ for various values of ψ^2. The asymptotic efficiency looks high when ψ^2 is less than 1. The relative efficiency of $\hat{\mu}$ to $\hat{\mu}_{ml}$ of the case $n = 10$ is also given in Table 2 for various values of ψ^2. Note that when the sample size is small, the relative efficiency of $\hat{\mu}$ to $\hat{\mu}_{ml}$ is greater than 1. More specifically, we observe that it is increasing in ψ^2, though the asymptotic one is decreasing.

Table 2.
Asymptotic efficiency of $\hat{\mu}$ and $\hat{\psi}^2$ and Relative efficiency of $\hat{\mu}$ to $\hat{\mu}_{ml}$ when $n = 10$

ψ^2	Asym.eff. ($\hat{\mu}$)	Rel.eff. ($\hat{\mu}$)	Asym.eff. ($\hat{\psi}^2$)
0.2	0.995	1.001	0.997
0.4	0.983	1.002	0.991
0.6	0.967	1.005	0.982
0.8	0.951	1.009	0.972
1.0	0.933	1.013	0.961
1.2	0.916	1.018	0.950
1.4	0.899	1.024	0.939
1.6	0.883	1.030	0.927
1.8	0.866	1.037	0.916
2.0	0.851	1.043	0.905

Next we consider the estimation of ψ^2. The UMVUE of ψ^2 is given by

$$\hat{\psi}^2_{mv} = {}_0F_1\left(\frac{n-1}{2}; \frac{n-1}{2}\tilde{s}^2\right) - 1,$$

which has the variance $e^{\sigma^2}\{{}_0F_1(\frac{n-1}{2};\sigma^4) - 1\}$ and the asymptotic one $e^{2\sigma^2}2\sigma^4/n$. The form of the hypergeometric function looks again complicated for an actual use.

The MLE $\hat{\psi}^2_{ml}$ and its modification $\hat{\psi}^2_{md}$ are denoted by

$$\hat{\psi}^2_{ml} = e^{\tilde{s}^2(n-1)/n} - 1,$$
$$\hat{\psi}^2_{md} = e^{\tilde{s}^2} - 1.$$

We can easily prove that $\hat{\psi}^2_{md}$ is larger than $\hat{\psi}^2_{ml}$ and $\hat{\psi}^2_{mv}$. A serious defect also exists in the two MLEs of ψ^2 such that their expectations do not exist for a large σ^2. The expectations, if they exist, are as follows.

$$E(\hat{\psi}^2_{ml}) = \left(1 - \frac{2\sigma^2}{n-1}\right)^{-\frac{n-1}{2}} - 1 \quad for\ \sigma^2 < \frac{n-1}{2},$$
$$E(\hat{\psi}^2_{md}) = \left(1 - \frac{2\sigma^2}{n}\right)^{-\frac{n-1}{2}} - 1 \quad for\ \sigma^2 < \frac{n}{2}.$$

We proposed the estimator of ψ^2;

$$\hat{\psi}^2 = \bar{x}\sum\left(\frac{1}{x_i} - \frac{1}{\bar{x}}\right)/(n-1)$$

by applying the inverse Gaussian model in Section 2. This estimator is unbiased and has the variance

$$V(\hat{\psi}^2) = \frac{1}{n(n-1)}(e^{\sigma^2} - 1)^2(e^{2\sigma^2} + 2(n-1)e^{\sigma^2} + 1).$$

Then the asymptotic efficiency of $\hat{\psi}^2$ is given by

$$asym.\ eff(\hat{\psi}^2) = \frac{(\psi^2 + 1)(log(1 + \psi^2))^2}{\psi^4}.$$

The right column in Table 2 shows values of the asymptotic efficiency of $\hat{\psi}^2$ for various values of ψ^2. We observe that the asymptotic efficiency of $\hat{\psi}^2$ is acceptable for an actual use.

5. TEST FOR μ

In this section, we consider the test for the null hypothesis $H_0 : \mu = \mu_0$. We find a conditional UMP unbiased test statistic in a text book [1], which is given by

$$T_{mp} = \frac{(\tilde{x} - log(\mu_0))}{\sqrt{\tilde{s}^2(n-1)/n}}$$

under the condition that W is fixed to w where $W = ((n-1)^2\tilde{s}^2 + n^2(\tilde{x} - log(\mu_0))^2)^{1/2}$. The critical points are given by the roots of a complicated integral equation. It looks difficult for an actual use such as the UMVUEs of μ and ψ^2.

Table 3.
Empirical significance levels of the proposed test for $H_0 : \mu = \mu_0$ on various values of sample sizes, ψ^2 and various nominal levels

n	ψ^2	nominal levels		
		0.1	0.05	0.01
5	0.2	0.0972	0.0487	0.00995
	0.3	0.0994	0.0484	0.00926
	0.5	0.0983	0.0479	0.00907
	0.8	0.0938	0.0461	0.00920
	1.0	0.0926	0.0460	0.00883
	1.5	0.0920	0.0444	0.00833
	2.0	0.0865	0.0420	0.00782
20	0.2	0.0994	0.0498	0.00954
	0.3	0.1008	0.0504	0.00948
	0.5	0.0982	0.0481	0.00924
	0.8	0.0972	0.0474	0.00965
	1.0	0.0955	0.0471	0.00944
	1.5	0.0920	0.0459	0.00879
	2.0	0.0889	0.0432	0.00929
40	0.2	0.0999	0.0493	0.00927
	0.3	0.0998	0.0501	0.00973
	0.5	0.0990	0.0495	0.00977
	0.8	0.0995	0.0491	0.00994
	1.0	0.0956	0.0470	0.00934
	1.5	0.0940	0.0466	0.00921
	2.0	0.0916	0.0445	0.00980

The conventional test procedure, the LRT for $H_0 : \mu = \mu_0$ is given by

$$T_{lr} = -2ln\frac{l(\mu_0, \hat{\psi}^2_{H_0})}{l(\hat{\mu}, \hat{\psi}^2_{ml})} = -2ln\frac{l(\mu_0, \hat{\sigma}^2_{H_0})}{l(\hat{\delta}, \hat{\sigma}^2)}.$$

Unfortunately χ^2 approximation is not accurate enough to be used in a finite sample size.

Our proposed test procedure given in Section 2 is simply denoted by

$$T = \frac{(\frac{\bar{x}}{\mu_0} + \frac{\mu_0}{\bar{x}} - 2)/\mu_0}{\sum(\frac{1}{x_i} - \frac{1}{\bar{x}})/(n-1)},$$

which is compared to the F distribution with $(1, n-1)$ d.f.. For checking the accuracy of the empirical significance level of the test T, we conducted a Monte Carlo study with 100,000 replications. Table 3 shows empirical significance levels for various nominal significance levels on various sample sizes and various values of ψ^2. The accuracy of the empirical significance levels looks acceptable.

6. EXAMPLES

The data presented in Table 4, cited from Table 6.1 (p.95) in the text book [3], shows the results of experiments of high-speed turbine bearings of the type 1 to the type 5. In each experiment 10 bearings were tested and the failure times in units of millions of cycles were recorded. The proposed and the usual estimates of μ and ψ^2 are given in Tables 5 and 6. We observe that the three estimates $\hat{\mu}_{md}, \hat{\mu}_{ml}, \hat{\mu}_{mv}$ have the linear order and $\hat{\mu}$ has different behaviour from the others. For the estimation of ψ^2, $\hat{\psi}^2$ and $\hat{\psi}^2_{mv}$ are close, $\hat{\psi}^2_{ml}$ are smaller and $\hat{\psi}^2_{md}$ are larger than the former two estimates. The moment estimates $\hat{\psi}^2_{mo}$ are unstable.

The maximum log-likelihood functions of the five distributions for each type of data are summarized in Table 7. We can find the closeness between the lognormal and the inverse Gaussian distributions, even when the fitness of the two distributions to the data are worse. Figure 2 draws the probability density functions of the five distributions fitted to the type 3 data. The graphs clearly show the closeness of the two distributions and the difference from the other distributions.

Table 4.
Examples cited from Table 6.1 (P.95) of Inverse Gaussian Distribution (Chhikara and Folks [3])

Type	Data Set									
1	3.03	5.53	5.60	9.30	9.92	12.51	12.95	15.21	16.04	16.84
2	3.19	4.26	4.47	4.53	4.67	4.69	5.78	6.79	9.37	12.75
3	3.46	5.22	5.69	6.54	9.16	9.40	10.19	10.71	12.58	13.41
4	5.88	6.74	6.90	6.98	7.21	8.14	8.59	9.80	12.28	25.46
5	6.43	9.97	10.39	13.55	14.45	14.72	16.81	18.39	20.84	21.51

Table 5.
Estimates of μ

Data set	$\hat{\mu}$	$\hat{\mu}_{mv}$	$\hat{\mu}_{ml}$	$\hat{\mu}_{md}$
1	10.693	10.907	10.927	11.103
2	6.050	6.000	6.003	6.054
3	8.636	8.690	8.704	8.786
4	9.798	9.603	9.608	9.697
5	14.706	14.784	14.789	14.893

Table 6.
Estimates of ψ^2

Data set	$\hat{\psi}^2$	$\hat{\psi}^2_{mv}$	$\hat{\psi}^2_{ml}$	$\hat{\psi}^2_{md}$	$\hat{\psi}^2_{mo}$
1	0.365	0.364	0.333	0.376	0.207
2	0.183	0.182	0.165	0.185	0.238
3	0.204	0.204	0.185	0.208	0.147
4	0.203	0.197	0.179	0.200	0.364
5	0.149	0.149	0.135	0.151	0.111

Table 7.
Maximum log likelihoods of the five distributions

Data set	Lognormal	Inv. Gauss	Gamma	Weibull	Normal
1	-30.432	-30.487	-29.816	-29.242	-29.389
2	-21.964	-21.913	-22.588	-23.612	-24.361
3	-26.120	-26.110	-25.767	-25.379	-25.573
4	-26.966	-26.993	-28.199	-29.643	-31.252
5	-30.155	-30.172	-29.783	-29.329	-29.480

7. DISCUSSIONS

In this paper, we assumed the lognormal distribution as the population distribution under examination. However it is our understanding that there is no rigid evidence supporting the lognormal distribution in comparison with the inverse Gaussian distribution in most applications. This is because the two distributions have similar properties. The following is the summary of our recommendation to use the proposed estimators of $(\hat{\mu}, \hat{\psi}^2)$ and the test statistic T for μ.

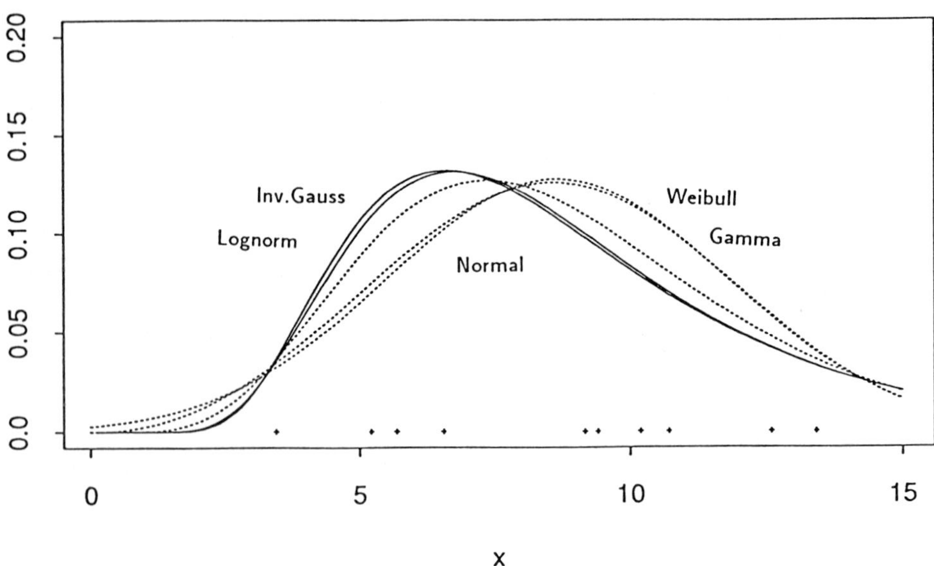

Figure 2. Probability density functions of the five distributions fitted to the data of the type 3 in Table 4.

Suppose that a lognormal distribution is reasonable to the population distribution. Then the proposed estimators and the test are regarded acceptably efficient, and robust. When an inverse Gaussian distribution is assumed to the population distribution, the proposed inference procedures are of course proper. In addition it is robust without reducing a large amount of efficiency on deviation to a lognormal distribution.

The lognormal distribution usually forces us to the log-transformed data, simply to apply standard inference based on the normal distribution. Such transformation makes meanings of the parameters less clear; for example they lose the dimension of data. When we are interested in the distribution of the raw data, the inverse Gaussian distribution looks more suitable for a population distribution. In this model, we can expect to proceed an advanced statistical analysis such as the regression analysis with the non-Gaussian errors [8].

Acknowledgements
The authors thank a referee for his valuable comments.

REFERENCES

1. E.L. Crow and K. Shimizu, *Lognormal Distributions.* Marcel Dekker, Inc., New York and Basel (1988).

2. R.S. Chhikara and J.L. Folks, *Tecnometrics*, **19**, 461-468 (1977).

3. R.S. Chhikara and J.L. Folks, *The Inverse Gaussian Distribution.* Marcel Dekker, Inc., New York and Basel (1989).

4. T. Yanagimoto and E. Yamamoto, *RM of ISM,* **368** (1990).

5. G.A. Whitemore and M. Yalovsky, *Technometrics,* **20**, 207-208 (1978).

6. R. Nishii, *TR of SRG/Hiroshima,* **279** (1990).

7. K. Shimizu and K. Iwase, *Commun. Statist.-Theor. Meth.,* **A10(11)**, 1127-1147 (1981).

8. G.A. Whitemore, *Appl. Statist.,* **35**, 8-15 (1986).

Stat. Sci. & Data Anal., pp. 501-511
K. Matsusita *et al.* (Eds)

AN ANALISYS OF DEGRADATION DATA OF A CARBON FILM AND THE PROPERTIES OF THE ESTIMATORS

KAZUYUKI SUZUKI, KOUJI MAKI and SINJI YOKOGAWA

Department of Communications and Systems Engineering,
The University of Electro-Communications, Chofu, Tokyo 182, Japan

Abstract : Reliability tests must be conducted with severe time and sample size constraints. Frequently no failures occur during the tests. Even if there is no failure, degradation measures contain information about product reliability. Comparing the precisions of estimators based on degradation data and time-to-failure data, this article shows the usefulness of degradation data. A nonlinear random effect model is applied for modeling a degradation data of a carbon film. The effects of censoring time, inspection intervals for measurement, variance ratio of a random error to a variation between items and their amounts on the precisions of estimators of $MTTF$ are examined.

Key words : Degradation data, Time-to-failure data, Nonlinear random effect model, Censored data.

1. INTRODUCTION

In developing new systems, it is necessary to assess its reliability with short development times and limited number of test pieces. That is, reliability tests must be conducted with severe *time* and *sample size* constraints. Frequently no failures occur during the tests. Therefore, it is difficult to assess reliability with traditional time-to-failure data. For some items, degradation measures, taken over time, contain information about product reliability. The failure time is related to an amount of degradation. Theoretically, several authors (e.g., Lu and Meeker [1], Nelson [2]) investigated the properties and analytical methods of degradation data. Shiomi and Yanagisawa [3] analyzed degradation data of a carbon film resistor. But as far as we know, no one has investigated an acceleration testing design of degradation data from the view point of longitudinal studies. Using the data of a carbon film resistor in [3], this article examines an usefulness of degradation data.

2. DEGRADATION DATA OF RESISTOR AND ITS MODELING

Resistance change x of a carbon film resistor under a thermal accelerated life test is shown in [3]. x is defined as $(r - r_0)/r_0$, where r_0 is an initial resistance and r is a current resistance value. Thermal levels(T) are 83 ℃ (356K=T_1), 133 ℃ (406K=T_2), and 173 ℃ (446K=T_3). For these levels, the following numbers of test pieces are tested ; 9 at T_1, 10 at T_2, and 10 at T_3. In the following, we define n_i to be the number of test pieces at the thermal level T_i. The sampling times (t) of measurement are 452(hours ; t_1), 1030(t_2),

4341(t_3), and 8084(t_4). For each resistor, four repeated measurements are done. For the fixed thermal levels, x vs. t plot and x vs. $\log t$ plot are done. Applying Bartlett's test, hypotheses of homoscedasty are rejected at the 5 percent level of significance. Throughout this article, the significant level is taken as 5 percent. Fig.1 shows $\log x$ vs. t plot. In the figure, the connected line shows the longitudinal data for one resistor. The variations become almost the same. Also, there seems a linear relationship between $\log x$ and t expect the data of time t_1. Actually, test of homoscedasty is not rejected by neither Bartlett's, Hartley's nor Cochran's test. Test of linearity (Rao [4]) for the data set except time t_1 is not rejected for each thermal level. Fig.2 shows $\log x$ vs. $\log t$ plot. Because of implementing an extrapolation, linearity must be held for the higher values of time and resistance change. But, the higher value of $\log x$ has a change point between times t_3 and t_4. Actually, test of linearity on the data set except time t_1 is rejected.

From the physical points of view, Arrhenius dependence on temperature is known (e.g. Nelson [2]), that is the reaction speed of degradation is related to $1/T$ (the reciprocal of an absolute temperature) through some function. Therefore, $\overline{\log x}$ (sample mean of $\log x$) vs. $1/T$ plot for the fixed t, and $\log(\overline{\log \kappa' x})$ (logarithm of the sample mean of $\log \kappa' x$, where κ' is a constant) vs. $1/T$ plot are down. Fig.3 shows them. Here we set up $\kappa' = 20$, temporarily. The lower one achieves a good linearity compared with the upper one. Therefore, the following models are considered;

$$E[\ \log x \ \mid t\ for\ fixed\ T\ ] = \alpha_1 t + \beta_1, \tag{1}$$

$$\log E[\ \log \kappa' x \ \mid T\ for\ fixed\ t\ ] = -\gamma_1/T + \beta_2, \tag{2}$$

$$i.e.\ \ E[\ \log x \ \mid T\ for\ \ fixed\ t\ ] = \beta' e^{-\gamma/T} - \log \kappa'. \tag{2'}$$

One possible model which satisfies (1) and (2') is,

$$E[\ \log x\ ] = (\alpha t + \beta)e^{-\gamma/T} + \kappa, \tag{3}$$

where $\kappa = -\log \kappa'$.

For the covariance structure of longitudinal data, the following models are investigated in the literature,

i) general multivariate models (e.g. Kleinbaum [5]),
ii) random effect models (e.g. Rao [6], Laird and Ware [7]) and
iii) autoregressive models (e.g. Box and Jenkins [8], Beach and Mackinnon [9]).

Since the design of the data set in Shiomi and Yanagisawa [3] is balanced on time, also homoscedasty is held after transformation, we apply random effect models on the data. The possible models for the observations are ;

Model **1**	$\log x = (\alpha t + \beta)e^{-\gamma/T} + \kappa + \eta + \varepsilon$	(4)
Model **2**	$\log x = (\alpha t + \beta + \eta)e^{-\gamma/T} + \kappa + \varepsilon$	(5)
Model **3**	$\log x = \{(\alpha + \eta)t + \beta\}e^{-\gamma/T} + \kappa + \varepsilon$	(6)
Model **1-2**	$\log x = (\alpha t + \beta + \eta_1)e^{-\gamma/T} + \kappa + \eta_2 + \varepsilon$	(7)
Model **1-3**	$\log x = \{(\alpha + \eta_1)t + \beta\}e^{-\gamma/T} + \kappa + \eta_2 + \varepsilon$	(8)
Model **2-3**	$\log x = \{(\alpha + \eta_1)t + \beta + \eta_2\}e^{-\gamma/T} + \kappa + \varepsilon$	(9)

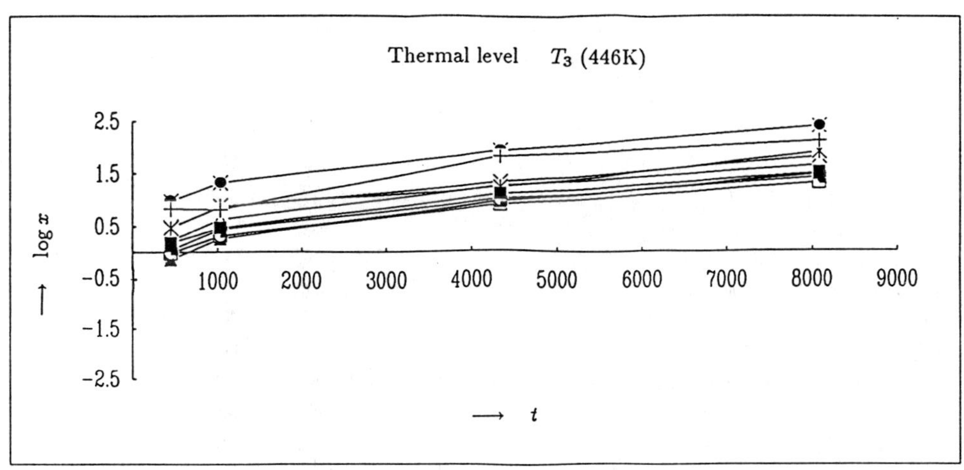

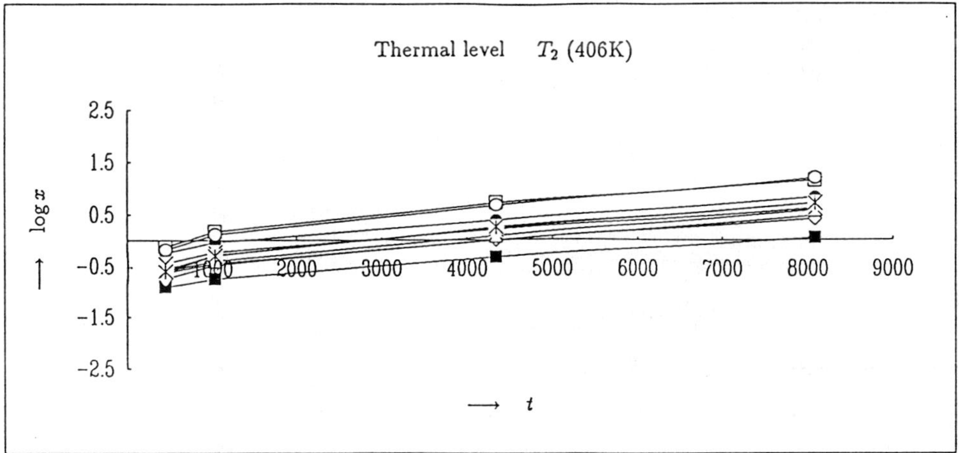

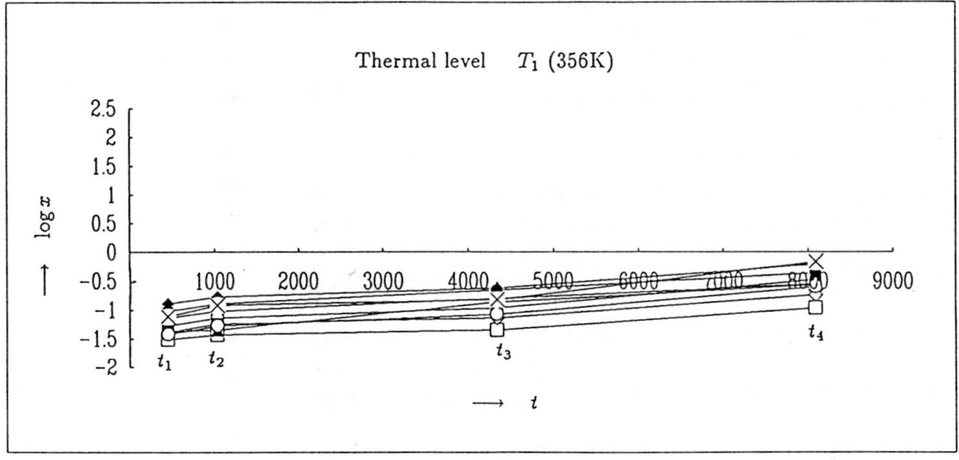

Fig. 1 $\log x$ vs. t (time) plots

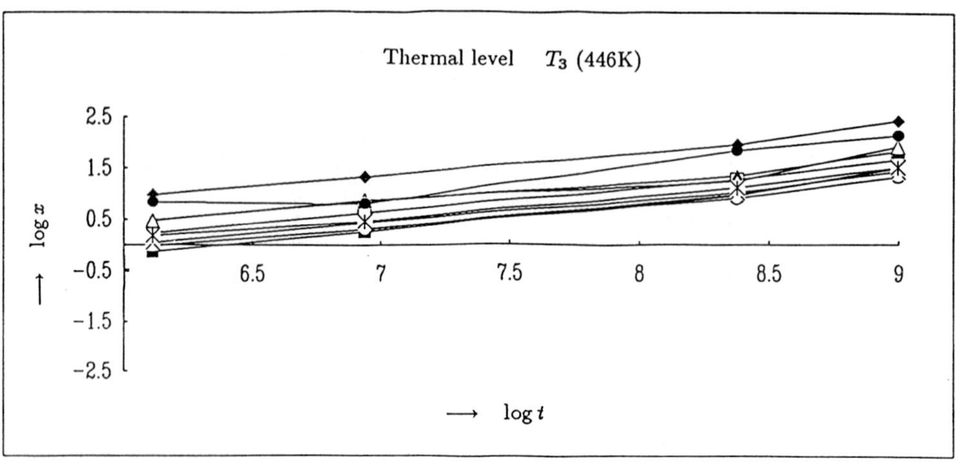

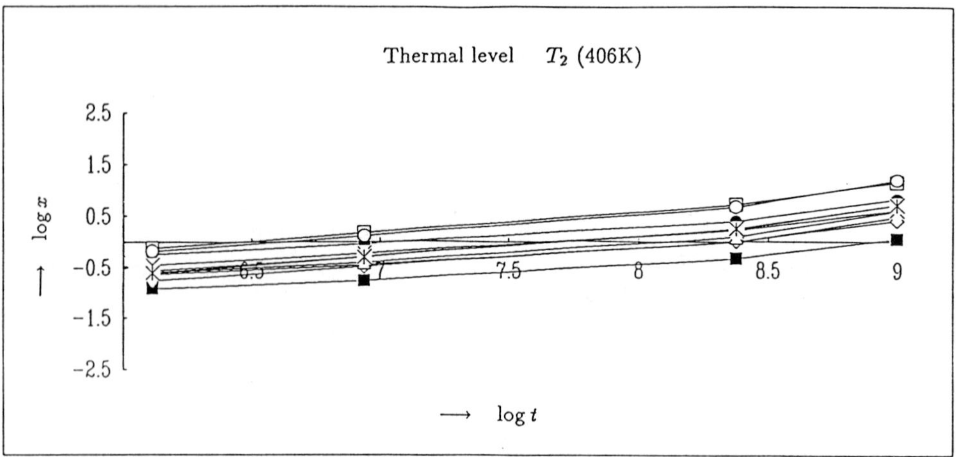

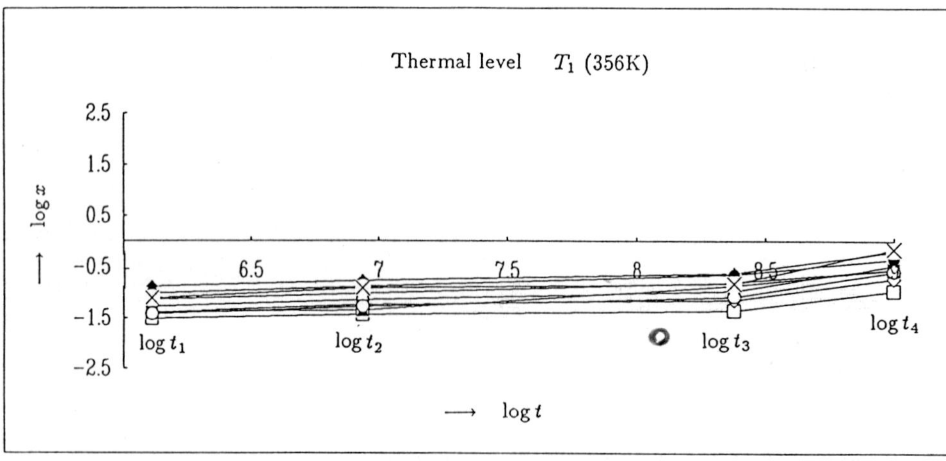

Fig. 2 log x vs. log t plots

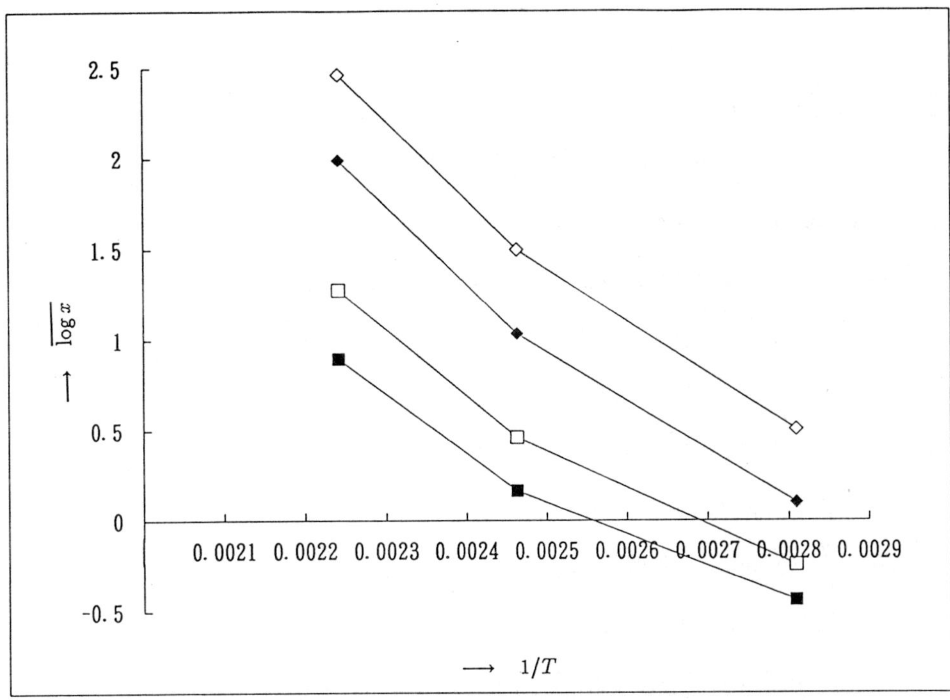

Fig. 3-(1) $\overline{\log x}$ vs. $1/T$ plots

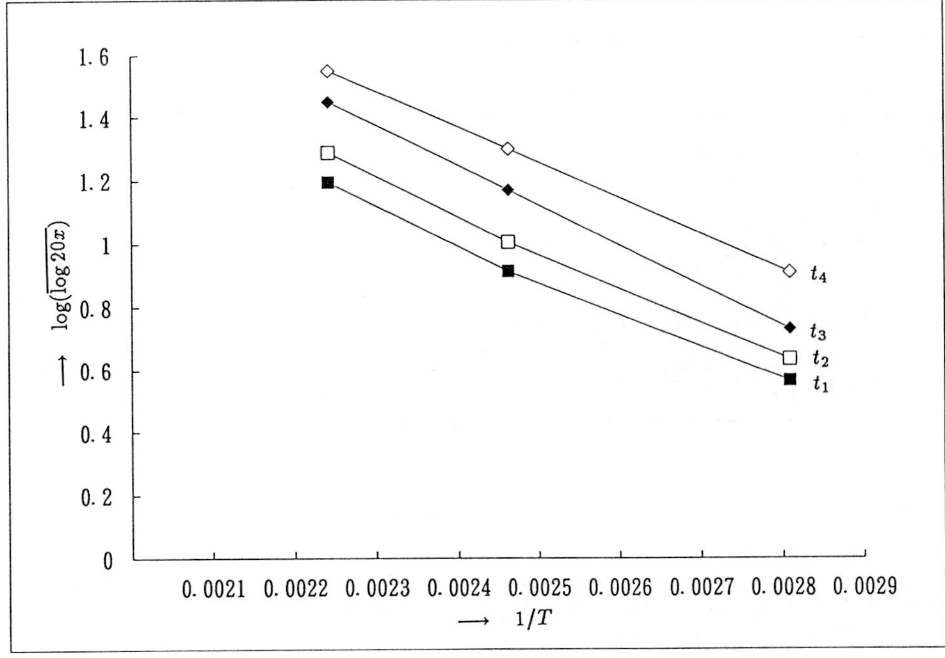

Fig. 3-(2) $\log(\overline{\log 20x})$ vs. $1/T$ plots

where η is a variable which shows a variation between items, and ε is a random error. In the case of Model **1**, the actual observation is written in the following,

$$\log x_{i(k)j} = (\alpha t_j + \beta)e^{-\gamma/T_k} + \kappa + \eta_{i(k)} + \varepsilon_{i(k)j}, \tag{10}$$

where $i(k)$: i-th item at thermal level k $(i=1,2,\cdots,n_k)$
j : suffix of sampling time for measurement $(j=1,2,\cdots,J)$
k : suffix of setting thermal level. $(k=1,2,\cdots,K)$

Probability plots of $\log x$ at fixed t and T show a good fit for a normal distribution. Also the variations of $\log x$ are almost the same for different t and T, therefore we set up the assumptions; $\eta_{i(k)} \rightsquigarrow i.i.d.N(0,\omega^2)$, $\varepsilon_{i(k)j} \rightsquigarrow i.i.d.N(0,\sigma^2)$.

Applying random effect models, the following multivariate normal distribution for $\log x_{i(k)j} \equiv y_{i(k)j}$ is set up,

$$\boldsymbol{y}_{i(k)} \equiv [\, y_{i(k)1}, \cdots, y_{i(k)J}\,]' \rightsquigarrow N_J(\boldsymbol{\mu}_k, \Sigma),$$

where

$$\boldsymbol{\mu}_k = \begin{pmatrix} (\alpha t_1 + \beta)e^{-\gamma/T_k} + \kappa \\ \vdots \\ \vdots \\ (\alpha t_J + \beta)e^{-\gamma/T_k} + \kappa, \end{pmatrix}, \quad \Sigma = \begin{pmatrix} \omega^2+\sigma^2 & \omega^2 & \cdots & \omega^2 \\ \omega^2 & \omega^2+\sigma^2 & \cdots & \omega^2 \\ \vdots & \vdots & \ddots & \vdots \\ \omega^2 & \omega^2 & \cdots & \omega^2+\sigma^2 \end{pmatrix}. \tag{11}$$

Σ is derived from the above assumptions. The log-likelihood is given by

$$\log L = const. - \frac{1}{2}\sum_{k=1}^{K}\sum_{i=1}^{n_k}[\, \log \mid \Sigma \mid + (\boldsymbol{y}_{i(k)} - \boldsymbol{\mu}_k)'\Sigma^{-1}(\boldsymbol{y}_{i(k)} - \boldsymbol{\mu}_k)\,]. \tag{12}$$

AIC of Model **1** for the data set [3] is listed in Table 1. Similarly, the results of other models are given in Table 1. Model **2** attains the minimum value of AIC. If we neglect the data at time t_1=452, AICs change into the values in the brackets. But the difference of AICs between Model **1** and Model **2** is less than 2.0. Also, Model **1** is easier to be handled with than Model **2**. From these reasons, Model **1** is applied for comparing the precisions of degradation data and time-to-failure data.

Table 1. AIC for six Models

	Model **1**	Model **2**	Model **3**	Model **1-2**	Model **1-3**	Model **2-3**
AIC {without t_1}	-276.7 {-217.7}	-278.5 {-219.1}	-151.8 {-124.8}	-276.7 {-217.2}	-274.7 {-215.7}	-276.5 {-217.1}

3. ESTIMATION OF *MTTF* USING DEGRADATION DATA

It is assumed the failure time is determined by a specified level of degradation in the case of resistance change of a carbon film[3]. Let X_M denote its critical level. Based on Model **1**, the failure time with a critical level X_M at a thermal level T is given by,

$$t^*_{|T,X_M} = \frac{1}{\alpha}(\log X_M - \kappa)e^{\gamma/T} - \frac{\beta}{\alpha} - (\eta+\varepsilon)\frac{e^{\gamma/T}}{\alpha}. \tag{13}$$

Therefore, its distribution is

$$t^*_{|T,X_M} \rightsquigarrow N\left(\frac{1}{\alpha}(\log X_M - \kappa)e^{\gamma/T} - \frac{\beta}{\alpha} , \frac{(\omega^2 + \sigma^2)}{\alpha^2} e^{2\gamma/T} \right). \tag{14}$$

Let $MTTF_{|T,X_M}$ represent $E[t^*_{|T,X_M}]$. Its estimate and asymptotic variance are obtained using the MLE in Table 2 and the δ-method as follows

$$\widehat{MTTF}_{|T,X_M} = \frac{1}{\hat{\alpha}}(\log X_M - \hat{\kappa})e^{\hat{\gamma}/T} - \frac{\hat{\beta}}{\hat{\alpha}},$$

$$\mathrm{Var}[\widehat{MTTF}_{|T,X_M}] \simeq [\frac{\partial}{\partial \boldsymbol{\theta}} MTTF_{|T,X_M}]' \boldsymbol{I}^{-1} [\frac{\partial}{\partial \boldsymbol{\theta}} MTTF_{|T,X_M}], \tag{15}$$

where

$$\boldsymbol{\theta} \equiv [\alpha , \beta , \gamma , \kappa , \omega^2 , \sigma^2]',$$

$$\boldsymbol{I} = E[-\frac{\partial^2}{\partial \boldsymbol{\theta} \partial \boldsymbol{\theta}'} \log L]. \tag{16}$$

4. ESTIMATION OF $MTTF$ USING TIME-TO-FAILURE DATA

Putting $a = -\frac{\beta}{\alpha}$, $b = \frac{1}{\alpha}(\log X_M - \kappa)$, $c = \gamma$, and $\xi = -(\eta + \varepsilon)\frac{e^{\gamma/T}}{\alpha}$ in (13),

$$t^*_{|T,X_M} = a + be^{c/T} + \xi, \tag{17}$$

is obtained. Therefore, failure time of an item i at a thermal level T_k is given by,

$$t^*_{i|T_k,X_M} = a + be^{c/T_k} + \xi_{i(k)}. \tag{18}$$

Here,

$$\xi_{i(k)} \rightsquigarrow N(0, v^2 e^{2c/T_k}), \tag{19}$$

where $v^2 = \frac{\omega^2 + \sigma^2}{\alpha^2}$. Let $t_{(i)kM}$ represent the order statistic of $t^*_{i|T_k,X_M}$, $i = 1, \cdots, n_k$.

The natural logarithm of the likelihood function of sample sizes n_k at thermal levels T_k $(k = 1, 2, \cdots, K)$ with the items having been censored at times t_C, is given by

$$\log L^* = \sum_{k=1}^{K} \left[-r_k \{ \frac{1}{2} \log 2\pi + \log v \} + \left\{ -\frac{1}{2} \sum_{i=1}^{r_k} Z^2_{(i)kM} \right\} + (n_k - r_k) \log\{1 - F(Z_{C,k})\} \right], \tag{20}$$

where r_k is a number of failure items at T_k, $F(\cdot)$ is a cumulative distribution function of a standard normal distribution, $Z_{(i)kM} = \frac{t_{(i)kM} - (a + be^{c/T_k})}{ve^{c/T_k}}$, and $Z_{C,k} = \frac{t_C - (a + be^{c/T_k})}{ve^{c/T_k}}$.

Let $\widetilde{MTTF}_{|T,X_M}$ be the MLE of $MTTF_{|T,X_M}$ based on time-to-failure data. It is given by,

$$\widetilde{MTTF}_{|T,X_M} = \tilde{a} + \tilde{b}e^{\tilde{c}/T},$$

where $\tilde{a}$, $\tilde{b}$ and $\tilde{c}$ are the MLEs of a, b and c. The asymptotic variance of $\widetilde{MTTF}_{|T,X_M}$, $\mathrm{Var}[\widetilde{MTTF}_{|T,X_M}]$, is given by like as eq.(15) using $\boldsymbol{\theta}^* = [a, b, c, v]'$ and $\log L^*$ instead of $\boldsymbol{\theta} = [\alpha, \beta, \gamma, \kappa, \omega^2, \sigma^2]'$ and $\log L$.

5. COMPARISON OF THE PRECISIONS BASED ON DEGRADATION DATA AND TIME-TO-FAILURE DATA

Using the data set of [3], the MLEs of the parameters of Model **1** are obtained as shown in Tables 2. The comparisons of the precisions of the estimators based on degradation data and time-to-failure data are examined under Model **1** and assuming the values of parameters in Table 2. We changed censored time t_C, variance ratio σ^2/ω^2, inspection interval t_I of degradation data and the number of sample size n_k at each thermal level T_k $(k = 1, \cdots, K)$.

Table 3 shows $\sqrt{\text{Var}[\widehat{MTTF}_{|T_0,X_M}]}$ based on degradation data and $\sqrt{\text{Var}[\widetilde{MTTF}_{|T_0,X_M}]}$ based on time-to-failure data. $q_k \equiv 1-F(Z_{C,k})$, $(k = 1, 2, 3)$ mean percentages of censored items which depend on the censored times t_C. Here we set up the number of thermal levels K=3 $(T_1 = 356K, T_2 = 406K, T_3 = 446K)$, and $T_0 = 335K$(62 ℃), where T_0 is a normal usage thermal level at which $MTTF$ of lifetimes is extrapolated [3]. From (14) and (19),

$$\text{Var}[\ t^*_{|T,X_M}] = \frac{(\omega^2 + \sigma^2)e^{2\gamma/T}}{\alpha^2} = v^2 e^{2c/T}.$$

Therefore, we changed the ratio of σ^2/ω^2 as shown in Table 3 subject to $\dfrac{\sigma^2 + \omega^2}{\alpha^2} = v^2 = 1.58 \times 10^4$ (constant) and $\alpha^2 = 5.66 \times 10^{-6}$ (constant). Inspection intervals, t_I, are 2000 in Table 3(a) and 500 in Table 3(b). That is, degradation data are obtained every 2000 or 500 hrs. Sample sizes are $n_1 = n_2 = n_3 = 10$ in both cases.

The precision of the estimators based on time-to-failure data is affected by censoring time more strongly than that of degradation data. If q_i $(i = 1, 2, 3)$ are small, the differences are not so large, but if q_1 and q_2 are large, a superiority of degradation data becomes clear. Inspection interval for degradation data affects the precision more strongly if q_1 and q_2 are large than if they are small, because in this case the number of available data becomes small and its number depends on an inspection interval. In the case of small values of σ^2/ω^2, that is, variation of random error σ^2 is smaller than variation between item ω^2, the information on the relationship of thermal levels and amounts of degradation is available even if q_1 and q_2 are large. But if q_1, q_2 and q_3 become small, the precision of $\widehat{MTTF}$ becomes better in the case of large values of σ^2/ω^2 because of increasing number of data.

Table 4 represents the effect of the numbers of items n_1, n_2 and n_3 at the thermal levels T_1, T_2 and T_3. Total number of items is 30 in every case. We set up $t_C = 8000$ $((q_1, q_2, q_3) = (1.00, .49, .00))$, $t_I = 2000$, $\dfrac{\sigma^2 + \omega^2}{\alpha^2} = v^2 = 1.58 \times 10^4$ and $\alpha^2 = 5.66 \times 10^{-6}$. The smallest variance is attained at $(n_1, n_2, n_3) = (20, 0, 10)$ in degradation data. This means the importance of the thermal slope β in the model. For time-to-failure data, variances become so large when $n_2 = 0$. As $(q_1, q_2, q_3) = (1.00, .49, .00)$, failure data is only observable at T_3. Therefore, the information on the thermal slope β becomes poor. This causes the large variances.

Table 2. MLE and S.E. based on Model **1**

	$\hat{\alpha}$	$\hat{\beta}$	$\hat{\gamma}$	$\hat{\kappa}$	$\hat{\omega}^2$	$\hat{\sigma}^2$
MLE	0.00238	44.7	1150.0	-3.05	0.0755	0.0140
S.E.	0.00074	9.26	128.9	0.390	0.0147	0.0004

Table 3(a). Comparisons of precisions based on time-to-failure data and degradation data ($t_I = 2000$)

t_C Censored Time	q_1 , q_2 , q_3	$\sqrt{\text{Var}[\widetilde{MTTF}]}$	$\sqrt{\text{Var}[\widehat{MTTF}]}$ $\sigma^2/\omega^2 = 1/5$	$\sigma^2/\omega^2 = 1/1$	$\sigma^2/\omega^2 = 5/1$
24000	.17 , .00 , .00	2.32	1.34	1.13	0.86
22000	.38 , .00 , .00	2.67	1.35	1.17	0.95
20000	.62 , .00 , .00	3.51	1.37	1.24	1.08
18000	.83 , .00 , .00	5.34	1.40	1.34	1.25
16000	.94 , .00 , .00	8.54	1.45	1.50	1.50
14000	.99 , .00 , .00	11.82	1.54	1.74	1.85
12000	1.00 , .03 , .00	13.00	1.69	2.13	2.36
10000	1.00 , .17 , .00	13.97	1.98	2.76	3.12
8000	1.00 , .49 , .00	17.96	2.56	3.83	4.30
6000	1.00 , .82 , .01	32.95	3.80	5.76	6.32
4000	1.00 , .97 , .11	92.31	6.93	9.69	10.50

($\times 10^3$)

Table 3(b). ($t_I = 500$)

t_C Censored Time	q_1 , q_2 , q_3	$\sqrt{\text{Var}[\widetilde{MTTF}]}$	$\sqrt{\text{Var}[\widehat{MTTF}]}$ $\sigma^2/\omega^2 = 1/5$	$\sigma^2/\omega^2 = 1/1$	$\sigma^2/\omega^2 = 5/1$
24000	.17 , .00 , .00	2.32	1.32	1.04	0.66
22000	.38 , .00 , .00	2.67	1.32	1.06	0.70
20000	.62 , .00 , .00	3.51	1.33	1.08	0.75
18000	.83 , .00 , .00	5.34	1.34	1.11	0.82
16000	.94 , .00 , .00	8.54	1.35	1.17	0.94
14000	.99 , .00 , .00	11.82	1.38	1.26	1.10
12000	1.00 , .03 , .00	13.00	1.42	1.41	1.37
10000	1.00 , .17 , .00	13.97	1.52	1.70	1.79
8000	1.00 , .49 , .00	17.96	1.73	2.22	2.47
6000	1.00 , .82 , .01	32.95	2.24	3.28	3.65
4000	1.00 , .97 , .11	92.31	3.89	5.65	5.88

($\times 10^3$)

$\widetilde{MTTF}$: MLE based on time-to-failure data
$\widehat{MTTF}$: MLE based on degradation data

Table 4. Effect of design of sample size on the precisions ($t_C = 8000$, $t_I = 2000$)

T_1 n_1	T_2 n_2	T_3 n_3	$\sqrt{\text{Var}[\widetilde{MTTF}]}$	$\sqrt{\text{Var}[\widehat{MTTF}]}$ $\sigma^2/\omega^2 = 1/5$	$\sigma^2/\omega^2 = 1/1$	$\sigma^2/\omega^2 = 5/1$
10	10	10	17.96	2.56	3.83	4.30
15	10	5	20.04	2.28	3.46	3.99
15	5	10	23.64	2.28	3.56	4.27
10	15	5	17.94	2.60	3.81	4.20
10	5	15	23.14	2.69	4.14	4.83
5	15	10	15.57	2.99	4.23	4.50
5	10	15	17.23	3.13	4.54	4.92
20	0	10	1130.93	2.07	3.37	4.29
15	0	15	1306.14	2.40	3.88	4.94
10	0	20	1599.69	2.93	4.75	6.04

($\times 10^3$)

For the degradation data, if σ^2 and ω^2 become $\lambda\sigma^2$ and $\lambda\omega^2$ $(0 < \lambda < 1)$, $\text{Var}[\widehat{MTTF}_{|T,X_M}]$ becomes $\lambda \cdot \text{Var}[\widehat{MTTF}_{|T,X_M}]$. This comes from the following fact. The (j,l)th element of Σ^{-1} in (12) is,

$$a_{jl} = \begin{cases} \dfrac{-\omega^2}{(J\omega^2+\sigma^2)\sigma^2} & if \;\; j \neq l, \\ \dfrac{(J-1)\omega^2+\sigma^2}{(J\omega^2+\sigma^2)\sigma^2} & if \;\; j = l. \end{cases}$$

Therefore a_{jl} becomes $\frac{1}{\lambda}a_{jl}$ if σ^2 and ω^2 become $\lambda\sigma^2$ and $\lambda\omega^2$. As $MTTF$ is a function of α, β, γ and κ, $\log|\Sigma|$ in (10) does not affect the right hand side of (15). Therefore, from (15) and (16), the proof is done. Decrease of σ^2 and ω^2 is done by quality improvement activities on production process of items, measurement methods and so on.

6. AN EXAMPLE OF A RESISTOR

In the data set of Shiomi and Yanagisawa [3], sampling times of measurement ($t_1 = 452, t_2 = 1030, t_3 = 4341, t_4 = 8084$) are different from those of §5. Also, the number of sample n_1 at T_1 is 9 instead of 10 (refer to §2). The MLEs of the unknown parameters are given in Table 2. Also, $\sqrt{\text{Var}[\widehat{MTTF}_{|T_0,X_M}]}=2.30 \times 10^3$ for degradation data and $\sqrt{\text{Var}[\widehat{MTTF}_{|T_0,X_M}]}=17.66 \times 10^3$ for time-to-failure data. These values are close to the values of Table 3(a) where $t_C = 8000$ and $\sigma^2/\omega^2 = 1/5$.

7. CONCLUSION

Applying the nonlinear random effect models to degradation data, comparisons of the precisions of the estimators based degradation data and time-to-failure data are done. The precision based on time-to-failure data is affected by censoring time more strongly than that of degradation data. If the percentage of censored items is high, the superiority of degradation data to time-to-failure data becomes clear. In this case, the smaller the inspection intervals for measurements are used, the better the precision is obtained. Also, in the situation of high percentage of censoring, the smaller the variation of random error becomes, the better the precision is obtained.

Acknowledgement

The authors wish to thank a referee for his valuable comments on this article.

REFERENCES

1. C.J. Lu and W.Q. Meeker, "Using Degradation Measures to Estimate a Time-to-Failure Distribution," Technical Report, Dept.of Stat., Iowa State Univ. (1990).
2. W. Nelson, "Analysis of Performance Degradation Data From Accelerated Tests," *IEEE Transactions on Reliability*, **R-30**, 149-155 (1981).
3. H. Shiomi and T. Yanagisawa, "On Distribution Parameter during Accelerated Life Test for a Carbon Film Resister," *Bulletin of the Electrotechnical Laboratory*, **43**, 330-345 (1979).
4. C.R. Rao, "Some Problems Involving Linear Hypotheses in Multivariate Analysis," *Biometrics*, **46**, 49-58 (1959).
5. D.G. Kleinbaum, "A Generalization of the Growth Curve Model which Allows Missing Data," *Journal of Multivariate Analysis*, **3**, 117-124 (1973).
6. C.R. Rao, "The Theory of Least Squares When the Parameters Are Stochastic and its Application to the Analysis of Growth Curves," *Biometrika*, **52**, 447-458 (1965).
7. N.M. Laird, and J.H. Ware, "Random Effects Models for Longitudinal Data," *Biometrics* **38**, 963-974 (1982).
8. G.E.P. Box and G.M. Jenkins, *Time Series Forecasting and Control*, San Francisco: Holden-Day (1970).
9. C.M. Beach and J.G. Mackinnon, "A Maximum likelihood procedure for regression with autocorrelated errors," *Econometrika*, **46**, 51-58 (1978).

Stat. Sci. & Data Anal., pp. 513-520
K. Matsusita *et al.* (Eds)

Bayesian Analysis for Exponential Zero-failure Data

DEJUN TANG and SHISONG MAO

Department of Mathematical Statistics

East China Normal University, Shanghai 200062, CHINA

abstract. On the basis of selecting a suitable prior distribution, this paper provides some choices to determine the parameter of the prior distribution in different cases, discusses the statistical inference for zero-failure data of the exponential distribution, and proves the optimality of estimators given in the paper.

keywords: zero-failure data, exponential distribution, Bayesian analysis, prior distribution, super-parameter.

INTRODUCTION

In order to shorten the test time and reduce the test expenses in life test, the time censored life test or the failure censored life test are usually adopted. With the appearances of high reliability units, the time censored test is more widely used. When test is time censored, it may occur, however, that none of the test units fail before the predetermined test time is reached. We call the phenomenon non-failure or zero-failure. Almost all existing methods of statistical inference are not of much use for zero-failure data and it is necessary to discuss some new method. Martz and Waller [1], [2] presented the Bayesian zero-failure reliability demonstration testing procedure which depended on some special prior information. This paper attempts to use Bayesian analysis for zero-failure data from an exponential population. In section 2, how to select the prior distribution will be described and some ways to determine the superparameter will be provided. In the third section, the Bayesian point estimators and interval estimators of the failure rate λ will be calculated. In the last section, it will be shown that the estimators given in the paper are strongly consistent and admissible. Although the discussion in the paper mainly focuses on the zero-failure data, the results are available for censored data with failure units.

LIKELIHOOD AND PRIOR

Suppose that n units are independently put on time censored life test with a predetermined time T_0, each unit has the cdf

$$F(x \mid \lambda) = 1 - \exp(-\lambda x), \quad x > 0, \ \lambda > 0$$

and the pdf

$$f(x \mid \lambda) = \lambda \exp(-\lambda x), \quad x > 0, \ \lambda > 0.$$

Let the censored failure data be $X_1 \leq X_2 \leq \cdots \leq X_{r_n}$, $0 \leq r_n \leq n$, where the number of failures

$$r_n = \begin{cases} \max\{i: X_i \leq T_0\}, & X_1 \leq T_0; \\ 0, & X_1 > T_0. \end{cases}$$

Then the likelihood function is

$$\begin{aligned} L(x_1, x_2, \cdots, x_{r_n} \mid \lambda) &= \frac{n!}{(n-r_n)!} \prod_{i=1}^{r_n} f(x_i \mid \lambda)[1 - F(T_0 \mid \lambda)]^{n-r_n} \\ &= \frac{n!}{(n-r_n)!} \lambda^{r_n} \exp(-\lambda W), \end{aligned} \tag{1}$$

where $W = \sum_{i=1}^{r_n} X_i + (n-r_n)T_0$ is the total time of the life test, and we defind $\sum_{i=1}^{0} X_i = 0$.

Now let us discuss how to select the prior distribution. Following a general process, the MLE of λ is easily obtained from $L(x_1, x_2, \cdots, x_{r_n} \mid \lambda)$ to be $\widehat{\lambda}_1 = r_n/W$. When $r_n = 0$, $\widehat{\lambda}_1 = 0$. That is to say the estimate of the average life is infinite. This estimate is obviously overestimated. Bartholomew [3] assumed that one unit just failed at T_0 when $r_n = 0$ and got another estimate $\widehat{\lambda}_2 = 1/W$, namely the estimate of the average life is W. Unforturnately, it is known to us to be underestimated because the true life of this unit is longer or even much longer than T_0 and the average life is generally more than W. We think that the true value of λ should be between $\widehat{\lambda}_1$ and $\widehat{\lambda}_2$.

On the other hand, if we deal with the problem in Bayesian's view, with respect to $g(\lambda)$, the prior density of λ, then the posterior density of λ is

$$g(\lambda \mid x_1, x_2, \cdots, x_{r_n}) = \frac{\lambda^{r_n} \exp(-\lambda W) g(\lambda)}{\int_0^\infty \lambda^{r_n} \exp(-\lambda W) g(\lambda) d\lambda}. \tag{2}$$

Let us consider the noninformation prior $g_1(\lambda) \propto 1/\lambda$ first, and then the uniform density on R^1 prior $g_2(\lambda) \propto 1$. With respect to these two priors, the Bayesian estimates of λ under squared-error loss are $\widehat{\lambda}_{B_1} = r_n/W = \widehat{\lambda}_1$ or $\widehat{\lambda}_{B_2} = (r_n+1)/W = \widehat{\lambda}_2$, respectively. Neither the former nor the latter is a good estimate according to the above discussion, and a suitable estimate would be between them. Thus we consider $g(\lambda) \propto \lambda^{-\alpha}$, where α is a superparameter whose value belongs to the interval $(0,1)$ having noticed that $1/\lambda = \lambda^{-1}$ and $1 = \lambda^0$. Obviously, $g(\lambda)$ is more reasonable than both $g_1(\lambda)$ and $g_2(\lambda)$. Several ways to determine α are available.

1. By historical data

The Bayesian estimator of λ, with respect to $g(\lambda) = \lambda^{-\alpha}$ $(0 < \alpha < 1)$, can be calculated to be $\widehat{\lambda}_B = (r_n + 1 - \alpha)/W$. If, in adition, the following historical data

$$(W_1, r_{n_1}), (W_2, r_{n_2}), \cdots, (W_k, r_{n_k}), \quad k \geq 2$$

were available, from the above relation somewhat transposed, $r_n + 1 = \alpha + W\widehat{\lambda}_B$ where α and $\widehat{\lambda}_B$ are considered as two parameters, we can determine the value of α by the least square principle

$$\alpha = 1 - \frac{\sum_{i=1}^{k}(W_i \cdot r_{n_i}) \cdot \sum_{i=1}^{k} W_i - \sum_{i=1}^{k} W_i^2 \cdot \sum_{i=1}^{k} r_{n_i}}{k \sum_{i=1}^{k} W_i^2 - (\sum_{i=1}^{k} W_i)^2}. \quad (3)$$

Making use of historical data will help us to obtain the value of α more objectively.

2. By experts' opinions

In most practical cases, engineers and other experts usually have rich knowledge about their products. Thus they, or together with statisticians, are able to determine a suitable value of α according to their special information. An acceptable method is if experts can determine the values of T and γ $(0 < \gamma < 1)$, which satisfy $P(X > T) \geq \gamma$, then $\gamma \leq \exp(-\lambda T)$. When one unit test is zero-failure, the estimate of λ is $\widehat{\lambda} = (1-\alpha)/T_0$, which should satisfy the condition given by experts, so that $\gamma \leq \exp(-\widehat{\lambda}T)$, $\alpha \geq 1 - (T_0/T)\log(1/\gamma)$. We can choose $\alpha = 1 - (T_0/T)\log(1/\gamma)$.

3. By a multistage prior

In case the prior information is not adequate to determine α, we suggest using of a multistage prior. Namely let α have a hyperprior distribution $h(\alpha)$. It is obvious that α has a very small probability in taking a value near 0 or 1 so that $h(\alpha)$ is chosen to be a Beta distribution. Let $h(\alpha) \propto [\alpha(1-\alpha)]^a$, $0 < \alpha < 1$, $a > 0$ being a constant. Thus

$$g(\lambda) \propto \int_0^1 \lambda^{-\alpha} h(\alpha) d\alpha \propto \int_0^1 \lambda^{-\alpha}[\alpha(1-\alpha)]^a d\alpha. \quad (4)$$

It is easily verified that the posterior $g(\lambda \mid x_1, x_2, \cdots, x_n)$ with respect to $g(\lambda)$ here is a density function for each r_n between 0 and n when $a > 0$.

The multistage prior can reduce, to a large extent, the sensitivity of α to the estimate of λ when $r_n = 0$. It is much better than taking the value of α without any reason in the absence of prior information.

BAYESIAN INFERENCE OF THE FAILURE RATE

In this section we discuss only the inference of the failure rate λ because other reliability quantities, such as $MTTF$, the reliability function $R(t)$ and the reliability lifetime t_R, are strictly monotonous functions of λ. We consider two situations according to the different prior distributions.

1. When α is a determined constant

Under the assumptions in section 2, $\lambda \mid x_1, x_2, \cdots, x_{r_n} \sim \Gamma(r_n+1-\alpha, W)$, the Bayesian estimate of λ for squared-error loss is

$$\widehat{\lambda}_B = E[\lambda \mid x_1, x_2, \cdots, x_{r_n}] = \frac{r_n + 1 - \alpha}{W}. \tag{5}$$

When $r_n = 0$, the result for zero-failure data is

$$\widehat{\lambda}_N = \frac{1-\alpha}{nT_0}. \tag{6}$$

Owing to the speciality of zero-failure data, the interval estimate is more important. Because $2W\lambda \mid x_1, x_2, \cdots, x_n \sim \chi^2(2r_n + 2 - 2\alpha)$, for given γ $(0 < \gamma < 1)$, and $1-\gamma$ Bayesian interval estimate of λ is

$$\left(\frac{1}{2W}\chi^2_{\frac{\gamma}{2}}(2r_n + 2 - 2\alpha)\ ,\ \frac{1}{2W}\chi^2_{1-\frac{\gamma}{2}}(2r_n + 2 - 2\alpha)\right), \tag{7}$$

where $\chi^2_\gamma(m)$ satisfies $P(\chi^2(m) \geq \chi^2_\gamma(m)) = \gamma$, in which m need not be an integer and its value can be looked up in the table of incomplete gamma functions or calculated by some approximation. When $r_n = 0$, the interval estimate is

$$\left(\frac{1}{2nT_0}\chi^2_{\frac{\gamma}{2}}(2-2\alpha)\ ,\ \frac{1}{2nT_0}\chi^2_{1-\frac{\gamma}{2}}(2-2\alpha)\right). \tag{8}$$

The $1-\gamma$ Bayesian upper limit or lower limit of λ will be obtained if $\gamma/2$ is replaced by γ.

It may be necessary to point out that the estimate, $\widehat{\lambda}_3 = 1/(3W)$ given by IEC [4] for exponential zero-failure data according to the engineering experience, is the special case of selecting $\alpha = 2/3$, that means to choose $\lambda^{-2/3}$ as the prior, but obviously it is rarely used.

2. When α is determined by a multistage prior

In this situation, we can obtain the point estimate of λ for squared-error loss to be

$$\begin{aligned}\widetilde{\lambda}_B &= \frac{\int_0^\infty \lambda^{r_n+1}\exp(-\lambda W)\{\int_0^1 \lambda^{-\alpha}[\alpha(1-\alpha)]^a d\alpha\}d\lambda}{\int_0^\infty \lambda^{r_n}\exp(-\lambda W)\{\int_0^1 \lambda^{-\alpha}[\alpha(1-\alpha)]^a d\alpha\}d\lambda} \\ &= \frac{\int_0^1[\int_0^\infty \lambda^{r_n+1-\alpha}\exp(-\lambda W)d\lambda][\alpha(1-\alpha)]^a d\alpha}{\int_0^1[\int_0^\infty \lambda^{r_n-\alpha}\exp(-\lambda W)d\lambda][\alpha(1-\alpha)]^a d\alpha}.\end{aligned} \tag{9}$$

If we denote $\Gamma(m) = \int_0^\infty t^{m-1}e^{-t}dt$, $C(r_n, W) = \int_0^1 W^\alpha[\alpha(1-\alpha)]^a\Gamma(r_n+1-\alpha)d\alpha$ and $K(r_n, W) = C(r_n+1, W)/C(r_n, W)$, then

$$\widetilde{\lambda}_B = \frac{C(r_n+1, W)}{C(r_n, W)} \cdot \frac{1}{W} = \frac{K(r_n, W)}{W}. \tag{10}$$

When $r_n = 0, W = nT_0$,

$$\widetilde{\lambda}_N = \frac{C(1,W)}{C(0,W)} \cdot \frac{1}{nT_0} = \frac{K(0,W)}{nT_0}. \quad \mathbf{(11)}$$

The values of $C(r_n, W)$ can be calculated by numerical integration. For instance, if $r_n = 0, W = 1000$ and $a = 1$, then $C(0,W) = 117.92$, $C(1,W) = 13.77$ and $\widetilde{\lambda}_N = 1.17 \times 10^{-4}$. In fact, $K(r_n, W)$ has the following properties.

Property 1. $r_n \le K(r_n, W) \le r_n + 1$.

Proof. According to $\Gamma(r_n + 2 - \alpha) = (r_n + 1 - \alpha)\Gamma(r_n + 1 - \alpha)$, we have

$$K(r_n, W) = r_n + 1 - \frac{\int_0^1 \alpha W^\alpha [\alpha(1-\alpha)]^a \Gamma(r_n + 1 - \alpha) d\alpha}{\int_0^1 W^\alpha [\alpha(1-\alpha)]^a \Gamma(r_n + 1 - \alpha) d\alpha}. \quad \mathbf{(12)}$$

Because $0 < \alpha < 1$, then

$$0 \le \frac{\int_0^1 \alpha W^\alpha [\alpha(1-\alpha)]^a \Gamma(r_n + 1 - \alpha) d\alpha}{\int_0^1 W^\alpha [\alpha(1-\alpha)]^a \Gamma(r_n + 1 - \alpha) d\alpha} \le 1. \quad \mathbf{(13)}$$

Taking (**12**) and (**13**) together, the property is proved. □

Property 2. *For fixed $r_n = r$, $K(r, W)$ is a strictly decreasing function of W.*

Proof. From (**12**),

$$\begin{aligned}
&\frac{dK(r,W)}{dW} \\
&= -\frac{\int_0^1 \alpha^2 W^{\alpha-1}[\alpha(1-\alpha)]^a \Gamma(r+1-\alpha)d\alpha \cdot \int_0^1 W^\alpha [\alpha(1-\alpha)]^a \Gamma(r+1-\alpha) d\alpha}{\{\int_0^1 W^\alpha [\alpha(1-\alpha)]^a \Gamma(r+1-\alpha) d\alpha\}^2} \\
&\quad + \frac{\int_0^1 \alpha W^\alpha [\alpha(1-\alpha]^a \Gamma(r+1-\alpha) d\alpha \cdot \int_0^1 \alpha W^{\alpha-1} [\alpha(1-\alpha)]^a \Gamma(r+1-\alpha) d\alpha}{\{\int_0^1 W^\alpha [\alpha(1-\alpha)]^a \Gamma(r+1-\alpha) d\alpha\}^2}.
\end{aligned}$$

Denoting $I(m) = \int_0^1 W^\alpha \alpha^{a+m-1}(1-\alpha)^a \Gamma(r+1-\alpha) d\alpha$,

$$\frac{dK(r,W)}{dW} = \frac{I^2(2) - I(3) \cdot I(1)}{W I^2(1)}. \quad \mathbf{(14)}$$

When $W > 0, \alpha \in (0,1)$ and $\alpha \ne 1/t$,

$$\begin{aligned}
0 &< [\alpha(1-\alpha)]^a W^\alpha \Gamma(r+1-\alpha)(\alpha t - 1)^2 \\
&= \alpha^{a+2}(1-\alpha)^a W^\alpha \Gamma(r+1-\alpha) \cdot t^2 - 2\alpha^{a+1}(1-\alpha)^a W^\alpha \Gamma(r+1-\alpha) \cdot t \\
&\quad + \alpha^a (1-\alpha)^a W^\alpha \Gamma(r+1-\alpha) \qquad \mathbf{(15)}
\end{aligned}$$

holds consistently for α. Let us integrate it from 0 to 1, then $I(3)t^2 - 2I(2)t + I(1) > 0$, $t \in R$, so that the discriminant of the quadratic equation $I^2(2) - I(3) \cdot I(1) < 0$. Therefore, $K(r, W)$ is a strictly decreasing function of W. □

These properties show that $\widetilde{\lambda}_B$ is between $(r_n + 1)/W$ and r_n/W, and it tends toward r_n/W with the increase of W.

Now we have the upper limit as a representative to discuss the interval estimator. For given γ $(0 < \gamma < 1)$, $\widetilde{\lambda}_U$, the $1 - \gamma$ Bayesian upper limit of λ, is the solution of the following equation.

$$\frac{\int_0^{\widetilde{\lambda}_U} \lambda^{r_n} \exp(-\lambda W)\{\int_0^1 \lambda^{-\alpha}[\alpha(1-\alpha)]^a d\alpha\} d\lambda}{\int_0^1 \lambda^{r_n} \exp(-\lambda W)\{\int_0^1 \lambda^{-\alpha}[\alpha(1-\alpha)]^a d\alpha\} d\lambda} = 1 - \gamma.$$

Denoting the incomplete gamma function $\Gamma(m; x) = \int_0^x t^{m-1} e^{-t} dt$ and $C(r_n, W; x) = \int_0^1 W^\alpha [\alpha(1-\alpha)]^a \Gamma(r_n + 1 - \alpha; x) d\alpha$, then

$$\frac{C(r_n, W; \widetilde{\lambda}_U)}{C(r_n, W)} = 1 - \gamma. \tag{16}$$

When $r_n = 0$, the upper limit $\widetilde{\lambda}_{NU}$ satisfies

$$\frac{C(0, nT_0; \widetilde{\lambda}_{NU})}{C(0, nT_0)} = 1 - \gamma. \tag{17}$$

Because $\Gamma(a; x)$ is the strictly increasing function of x, the solution of each of the above equations is unique.

OPTIMALITY OF ESTIMATORS

In this section we show the optimality of estimators presented above. At first we cite the conclution which is the direct corollory of Theorem 2.3 in Chen Jiading [5].

Lemma. *Suppose that $g(x)$ is a Borel measureable function and $\int_0^{T_0} | g(x) | F(x) < +\infty$. Then*

$$\lim_{n \to \infty} \frac{1}{n} \sum_{i=1}^{r_n} g(X_i) = \int_0^{T_0} g(x) dF(x) \ \ (a.s.). \tag{18}$$

Theorem 1. *$\widehat{\lambda}_B = (r_n + 1 - \alpha)/W$ converges almost surely to λ.*

Proof. Let us set $g(x) = x$ in Lemma first, and then $g(x) = 1$, we will, respectively, obtain

$$\begin{aligned}\lim_{n\to\infty}\frac{1}{n}\sum_{i=1}^{r_n}X_i &= \int_0^{T_0} x\,dF(x)\\ &= \int_0^{T_0} x\cdot\lambda\exp(-\lambda x)dx\\ &= \frac{1}{\lambda}F(T_0) - T_0\exp(-\lambda T_0)\quad (a.s.),\end{aligned} \tag{19}$$

and

$$\lim_{n\to\infty}\frac{r_n}{n} = \int_0^{T_0} dF(x) = F(T_0)\quad (a.s.). \tag{20}$$

$$\begin{aligned}\lim_{n\to\infty}\widehat{\lambda}_B &= \lim_{n\to\infty}\frac{r_n+1-\alpha}{W}\\ &= \frac{\lim_{n\to\infty}[(r_n+1-\alpha)/n]}{\lim_{n\to\infty}[W/n]}\\ &= \frac{F(T_0)}{\lim_{n\to\infty}[\frac{1}{n}\sum_{i=1}^{r_n}X_i + (1-\frac{r_n}{n})T_0]}\\ &= \frac{F(T_0)}{\frac{1}{\lambda}F(T_0) - T_0\exp(-\lambda T_0) + [1-F(T_0)]T_0}\\ &= \lambda\quad (a.s.).\end{aligned} \tag{21}$$

□

It is known to us that generalized Bayes rules may not be admissible. But the Bayesian estimators given above have the following property.

Theorem 2. *If there exists a positive constant d such that $W \geq d$ holds, then $\widehat{\lambda}_B$ is admissible for squared-error loss.*

Proof. The posterior risk of $\widehat{\lambda}_B$, if denoting $\underset{\sim}{x} = (x_1, x_2, \cdots, x_{r_n})$, is

$$R(\widehat{\lambda}_B \mid \underset{\sim}{x}) = E_{g(\lambda\mid\underset{\sim}{x})}[(\widehat{\lambda}_B - \lambda)^2] = \frac{r_n+1-\alpha}{W^2}, \tag{22}$$

and the marginal density of $\underset{\sim}{x}$ is

$$\begin{aligned}Q(\underset{\sim}{x}) &= \int_0^{\infty} L_n(\underset{\sim}{x}\mid\lambda)g(\lambda)d\lambda\\ &= \int_0^{\infty}\frac{n!}{(n-r_n)!}\lambda^{r_n}\exp(-\lambda W)\cdot\lambda^{-\alpha}d\lambda\\ &= \frac{n!}{(n-r_n)!}\cdot\frac{\Gamma(r_n+1-\alpha)}{W^{r_n+1-\alpha}}.\end{aligned} \tag{23}$$

Thus the Bayes risk of $\widehat{\lambda}_B$ can be calculated to be

$$\begin{aligned} R_H(\widehat{\lambda}_B) &= \int_\Omega R(\widehat{\lambda}_B \mid \underset{\sim}{x}) Q(\underset{\sim}{x}) d\underset{\sim}{x} \\ &= \int_\Omega \frac{n!\Gamma(r_n+1-\alpha)}{(n-r_n)! W^{r_n+3-\alpha}} d\underset{\sim}{x} \\ &\leq \int_\Omega \frac{n!\Gamma(r_n+1-\alpha)}{d^{r_n+3-\alpha}} d\underset{\sim}{x}. \end{aligned} \tag{24}$$

Because the integrand is a bounded function for fixed n, the Bayes risk of $\widehat{\lambda}_B$ is finite. It means that a rule minimizing the Bayes risk is equivelant to that minimizing the posterior risk. Now $\widehat{\lambda}_B$ is the unique rule minimizing the posterior risk under the squared-loss so that the Bayes rule we are discussing is unique. Then $\widehat{\lambda}_B$ is admissible according to the Theorem 8 in Section 4.8 of Berger [6]. □

Through the similar process, the optimality of $\widetilde{\lambda}_B$ can be proved.

Theorem 3. *$\widetilde{\lambda}_B$ defined by (10) converges almost surely to λ, and it is admissible for squared-error loss when $W \geq d > 0$.*

Acknowledgements
This research was supported by National Natural Science Foundation of China.

REFERENCES:

1. H.F. Martz and R.A. Waller, *A Bayesian zero-failure (BAZE) reliability demonstration testing procedure*, Journal of Quality Technology **11**, 128-138 (1979).

2. ———, *Bayesian Reliability Analysis*, John Wiley & Sons, New York (1982).

3. D.J. Bartholomew, *A problem in life testing*, J.A.S.A. **52**, 350-355 (1957).

4. IEC 605.4 (1982).

5. Jiading Chan, *Maximum likelihood estimation under censoring sample*, ACTA Mathematical Applicatal SINICA **3**, 306-321 (1980).

6. J.O. Berger, *Statistical Decision Theory and Bayesian Analysis*, 2nd Edition, Springer (1985).

Stat. Sci. & Data Anal., pp. 521-528
K. Matsusita *et al.* (Eds)

The Estimation of Distribution under a Particular Random Censoring

WENLIANG LU

Department of Statistics and Operational Research, Fudan University, Shanghai 200433, P.R.China

Abstract. This paper deals with the estimation problem of distribution function under a particular random censoring model (PRCM) which is useful in the study of the survival function of an industrial product in which the real operating time is different from its actual calender time. The PRCM has been discussed by Suzuki[1] in the parametric case and also in the case in which the interested variable takes only a finite number of values. In this article we discuss the PRCM with no assumption on the distribution function of the interested random variable.

Keywords. Particular random censoring model (PRCM), Modified Kaplan-Meier estimator, $D[0,\theta]$ space, Gaussion process.

1. INTRODUCTION

The ordinary censoring model is applicable to many areas, such as reliability analysis, survival analysis in medicine, etc., The ordinary random censoring model is describled as followes: Let (X_i, Y_i), $i = 1, 2, \cdots, N$, be independent, identically distributed non-negative random vectors, where X_i is the variable of interest, and Y_i is some independent censoring variable. The observed quantity is the pairs (Z_i, δ_i), where

$$Z_i = \min(X_i, Y_i), \qquad \delta_i = I(X_i \le Y_i), \qquad i = 1, 2, \cdots, N$$

($I(\cdot)$ is the indicator function of set $(\cdot)$). From these pairs, the distribution function of X_i is estimated. The very important estimator is the Kaplan-Meier[2] estimator.

In some practical cases, not all of the $(Z_i, \delta_i), i = 1, \cdots, N$ are observed. Suzuki[1] proposed a particular case in which all of the δ_i, $i = 1, 2, \cdots, N$, are observed but some of the $Z_i, i = 1, \cdots, N$, are not observed. Z_i is observed when $\delta_i = 1$ or $\delta_i = 0$ with $D_i = 1$, and Z_i is not observed when $\delta_i = 0$ with $D_i = 0$, where $D_i = (1$ or $0), i = 1, \cdots, N$, are known constants, not random variables. We call it the particular random censoring model (PRCM).

Example. During the 1980s, quality-assurance systems are becoming more popular throughout the world. For example, in the United States some automobile manufacturers offer warranties on every new car. In this cases the random variable of interest, X, is the mileage at the time of the first failure of the automobile. The censoring variable, Y, is the mileage that the automobile will have run during the warranty period. If the automobile failed during the warranty period, its owner will ask the manufacturer to repair it under contract. In that way the manufacturer can get a record of failures $(Z_i, \delta_i = 1)$. On the other hand, if there is no failure, the owner will not report the mileage that the car has run in the warranty period. Then the manufacturer will only have the record ($\delta_i = 0$) and Z_i will not be observed. Suppose now that the manufacturer follows up a fraction of the cars he sells. For these follow-up cars, we define $D_i = 1$, otherwise $D_i = 0$. With these cars, even if there has been no failures during the warranty period, the manufacturer will obtain the information about the mileage in the warranty period. If he surveys car owners whose cars have not failed(e.g. by postal reply cards), he will have some nonfailure data. Thus some of the unobserved data now become $(Z_i, \delta_i = 0)$. Therefore, the all obtained data are $(Z_i, \delta_i = 1)$, $(Z_i, \delta_i = 0)$ with $D_i = 1$ and $(\delta_i = 0)$ with $D_i = 0$.

Suzuki[1] gave the estimator of the distribution of X_i and discussed the properties of the estimator under the assumption that X_i is the discrete random variable taking a finite number of values. This paper gives the estimator of the distribution of X_i and the large sample properties with no assumption on the distribution of X_i.

2. ESTIMATION OF THE DISTRIBUTION FUNCTION

Denote the distribution functions of X_i and Y_i by $F(t)$, $G(t)$ respectively. The estimator of $F(t)$ is given by

$$1 - \widehat{F}(t) = \prod_{Z_{(i)} \leq t} (1 - \frac{m_i}{M_i})^{\delta_{(i)}}$$

where $Z_{(1)} \leq Z_{(2)} \leq \cdots \leq Z_{(n)}$ are the ordered statistics of the observed $Z_i, i = 1, \cdots, N$, $\delta_i, D_i, i = 1, \cdots, n$, are the corresponding δ_i, D_i of $Z_{(i)}, i = 1, 2, \cdots, n$, and

$$m_i = \begin{cases} 1, & \delta_{(i)} = 1 \\ 1 + \dfrac{n_{c1}}{n_{c2}}, & \delta_{(i)} = 0 \quad \text{and} \quad D_{(i)} = 1 \\ 0, & \text{otherwise} \end{cases}$$

$$M_i = \sum_{j=i}^{n} m_j, \qquad n_{c1} = \sum_{i=1}^{N} (1 - D_i)(1 - \delta_i), \qquad n_{c2} = \sum_{i=1}^{N} D_i(1 - \delta_i)$$

The estimator $\widehat{F}(t)$ is analogous with that of Suzuki[1]. It is a modified Kaplan-Meier estimator because it coincides with Kaplan-Meier estimator when $n_{c1} = 0$. It is also the generalized MLE of $F(t)$.

Assume that $(X_i, Y_i), i = 1, 2, \cdots, N$, are independent and identically distributed non-negative random vectors and that $Y_i, i = 1, 2, \cdots, N$, are also independent of $X_i, i = 1, 2, \cdots, N$. We also assume that $\frac{1}{N}\sum_{i=1}^{n} D_i$ converges to a constant p^* as N goes to infinite. Then we have the following properties of $\widehat{F}(t)$.

Theorem 1. *Let θ satisfy $H(\theta) < 1$, where $H(t) = P(Z_i \leq t)$, then*

$$\sup_{0 \leq t \leq \theta} |\widehat{F}(t) - F(t)| \longrightarrow 0 \quad (a.s.) \quad (N \to \infty)$$

We treat $\widehat{F}(t)$ as the random function in $D[0, \theta]$, the well-known space of functions on $[0, \theta]$ having jump discontinuities, using the Skorohod topology (Billingsley[3]).

Theorem 2. *If $F(t)$, $G(t)$ are continuous functions and θ satisfies $H(\theta) < 1$, then $\sqrt{N}(\widehat{F}(t) - F(t)) \in D[0,\theta]$ converges weakly to a Gaussion process $W_1(t)$ which has mean 0 and the covariance structure given by*

$$\mathrm{Cov}(W_1(s), W_1(t)) = [1 - F(s)][1 - F(t)]\Big\{ \int_0^{s\wedge t} \frac{d\Lambda(u)}{1 - H(u)} + \frac{1-p^*}{p^*}\frac{1}{\overline{p}} \int_0^s \int_0^t \frac{H_2(u \wedge v)\overline{H}_2(u \vee v)}{[1 - H(u)][1 - H(v)]} d\Lambda(u)\Lambda(v)\Big\}$$

where

$$H_2(t) = P(Z_i \le t, \delta_i = 0), \qquad \overline{p} = H_2(\infty), \qquad \overline{H}_2(t) = \overline{p} - H_2(t)$$

$$s \wedge t = \min(s,t), \quad s \vee t = \max(s,t), \quad \Lambda(t) = \int_0^t \frac{dF(u)}{1 - F(u-)}$$

Theorem 3. *Let θ be any positive constant and $H(\theta) < 1$, then $\sqrt{N}(\widehat{F}(t) - F(t))$ on $D[0,\theta]$ converges weakly to a Gaussion process $W_2(t)$ which has mean 0 and the covariance structure given by*

$$\mathrm{Cov}(W_2(s),W_2(t)) = [1 - F(s)][1 - F(t)]\Big\{ \int_0^{s\wedge t} \frac{d\Lambda(u)}{[1 - H(u-)][1 - \Delta\Lambda(u)]} + \frac{1-p^*}{p^*}\frac{1}{\overline{p}} \int_0^s \int_0^t \frac{H_2(u \wedge v-)\overline{H}_2(u \vee v-)d\Lambda(u)d\Lambda(v)}{[1 - H(u-)][1 - H(v-)][1 - \Delta\Lambda(u)][1 - \Delta\Lambda(v)]}\Big\}$$

where $\Delta\Lambda(t) = \Lambda(t) - \Lambda(t-)$, $\Lambda(t-)$ is the left limit of $\Lambda(t)$.

3. PROOF OF THE THEOREMS

Denote

$$\widehat{H}(t) = \frac{1}{N}\sum_{j=1}^{N}(I(Z_j \le t, \delta_j = 1) + (1 + \frac{n_{c1}}{n_{c2}})D_j I(Z_j \le t, \delta_j = 0))$$

$$\widehat{H}_1(t) = \frac{1}{N}\sum_{j=1}^{N} I(Z_j \le t, \delta_j = 1), \widehat{H}_2(t) = \frac{1}{p^* N}\sum_{j=1}^{N} D_j I(Z_j \le t, \delta_j = 0)$$

$$\widehat{M}(t) = \sqrt{N}\widehat{H}_1(t) - \int_0^t \sqrt{N}(1 - \widehat{H}(u-))d\Lambda(u), \quad \widehat{\Lambda}(t) = \int_0^t \frac{d\widehat{H}_1(u)}{1 - \widehat{H}(u-)}$$

The idea of proof of the preceding theorems is to express $\widehat{F}(t)$ as a smooth function of the empirical distribution and subdistribution functions $\widehat{H}(t)$, $\widehat{H}_1(t)$, $\widehat{H}_2(t)$, of which the asympototic properties have been known. The relation between $\widehat{F}(t)$ and $\widehat{H}(t)$, $\widehat{H}_1(t)$, $\widehat{H}_2$ is shown in Lemma 1.

Lemma 1. *For any $t < Z_{(n)}$, we have*

(1) $\widehat{H}(t) = \widehat{H}_1(t) + p^*(1 + \frac{n_{c1}}{n_{c2}})\widehat{H}_2(t)$

(2) $\sqrt{N}(\widehat{\Lambda}(t) - \Lambda(t)) = \int_0^t \frac{d\widehat{M}(u)}{1 - \widehat{H}(u-)}$

(3) $1 - \widehat{F}(t) = \prod_{s \leq t}(1 - \Delta\widehat{\Lambda}(s))$

(4) $\frac{\sqrt{N}(\widehat{F}(t) - F(t))}{1 - F(t)} = \int_0^t \frac{1 - \widehat{F}(s-)}{1 - F(s)} d\widehat{B}(s), \quad \widehat{B}(s) = \sqrt{N}(\widehat{\Lambda}(s) - \Lambda(s))$

The above identities are the same in appearance as those of Shorack and Wellner[4] and the proof is similar. In order to prove Theorem 1, the following lemmas are needed.

Lemma 2. *Let $p = 1 - \overline{p}$, $\overline{n}_{c1} = \frac{1}{N(1 - p^*)} n_{c1}$, $\overline{n}_{c2} = \frac{1}{Np^*} n_{c2}$, and $H_1(t) = P(Z_i \leq t, \delta_i = 1)$, then*

$$\overline{n}_{c1} \longrightarrow \overline{p}, \quad \overline{n}_{c2} \longrightarrow \overline{p} \quad (a.s.) \tag{3.1}$$

$$\sup_{0 \leq t < \infty} |\widehat{H}_1(t) - H_1(t)| \longrightarrow 0, \qquad \sup_{0 \leq t < \infty} |\widehat{H}_2(t) - H_2(t)| \longrightarrow 0 \quad (a.s.) \tag{3.2}$$

$$\sup_{0 \leq t < \infty} |\widehat{H}(t) - H(t)| \longrightarrow 0 \quad (a.s.) \quad (N \to \infty) \tag{3.3}$$

Proof. We note that $\overline{n}_{c1}, \overline{n}_{c2}$ are the sums of independent and identical random variables. Applying the strong law of large numbers, we obtain (3.1). The proof of (3.2) is similar to that of empirical distributions. (3.3) is obtained from (3.1), (3.2).

Utilizing Lemma 1 and Lemma 2, we obtain Lemma 3.

Lemma 3. *With the above assumptions, let θ satisfy $H(\theta) < 1$, then*

$$\sup_{0 \leq t \leq \theta} |\widehat{\Lambda}(t) - \Lambda(t)| \longrightarrow 0 \quad (a.s.), \qquad (N \to \infty)$$

Utilizing Lemma 3 and identity (4) of Lemma 1, we can obtain Theorem 1. The proof of Lemma 3 and Theorem 1 is similar to that of Shorack and Wellner[4] (Chapter 7).

The proof of Theorem 2 is also based on the identities of Lemma 1.

Let

$$\widehat{E}_1(t) = \sqrt{N}(\widehat{H}_1(t) - H_1(t)), \quad \widehat{E}_2(t) = \sqrt{N}(\widehat{H}_2(t) - H_2(t))$$

$$\widehat{n}_{c1} = \frac{1}{\sqrt{N}} \sum_{j=1}^{N} (1 - D_j)[I(\delta_j = 0) - \overline{p}], \quad \widehat{n}_{c2} = \frac{1}{\sqrt{N}} \sum_{j=1}^{N} D_j[I(\delta_j = 0) - \overline{p}]$$

We treat $\widehat{E}_1(t)$, $\widehat{E}_2(t)$, $\widehat{n}_{c1}$, $\widehat{n}_{c2}$ as the random functions in $D[0, \theta]$, where $\widehat{n}_{c1}, \widehat{n}_{c2}$ are not dependent on $t (t \in [0, \theta])$, and the combination $(\widehat{E}_1(t), \widehat{E}_2(t), \widehat{n}_{c1}, \widehat{n}_{c2})$ as the random element in the four-fold product space $D^4[0, \theta]$.

Lemma 4. *Assume that $F(t)$, $G(t)$ are continuous distribution functions, then $(\widehat{E}_1, \widehat{E}_2, \widehat{n}_{c1}, \widehat{n}_{c2})$ on $D^4[0, \theta]$ (for any $\theta > 0$) converges weakly to a multivariate Gaussian process*

$(E_1(t), E_2(t), \xi, \eta)$ which has mean 0 and the covariance structure given for $s \le t$ by

$$\mathrm{Cov}(E_1(s), E_1(t)) = H_1(s)(1 - H_1(t)), \mathrm{Cov}(E_2(s), E_2(t)) = \frac{1}{p^*} H_2(s)(1 - H_2(t))$$

$$\mathrm{Cov}(\xi, \xi) = (1 - p^*)p\bar{p}, \qquad \mathrm{Cov}(\eta, \eta) = p^* p\bar{p}$$

$$\mathrm{Cov}(E_1(s), E_2(t)) = -H_1(s)H_2(t), \qquad \mathrm{Cov}(E_1(t), E_2(s)) = -H_1(t)H_2(s)$$

$$\mathrm{Cov}(E_1(s), \xi) = -(1 - p^*)\bar{p}H_1(s), \qquad \mathrm{Cov}(E_1(s), \eta) = -p^*\bar{p}H_1(s)$$

$$\mathrm{Cov}(E_2(s), \xi) = 0, \qquad \mathrm{Cov}(E_2(s), \eta) = pH_2(s), \qquad \mathrm{Cov}(\xi, \eta) = 0$$

where $E_1(t)$, $E_2(t)$ are almost surely continuous.

The proof of this lemma is similar to that of Theorem 3 of Breslow and Crowley[5]. Like Breslow and Crowly[5], we replace $(\widehat{E}_1, \widehat{E}_2, \widehat{n}_{c1}, \widehat{n}_{c2})$ and (E_1, E_2, ξ, η) with a sequence of random functions having the same distribution for each N, but which satisfy also

$$\rho((\widehat{E}_1, \widehat{E}_2, \widehat{n}_{c1}, \widehat{n}_{c2}), (E_1, E_2, \xi, \eta)) \longrightarrow 0 \quad (a.s.), \qquad (N \to \infty)$$

where ρ is the Skorohod metric on $D^4[0, \theta]$. As $E_1(t)$, $E_2(t)$ are almost surely continuous, ρ is equivalent to the supremum metric on $[0, \theta]$ and this means that

$$\sup_{0 \le t \le \theta} |\widehat{E}_i(t) - E_i(t)| \longrightarrow 0 \quad (a.s.), \quad i = 1, 2$$

$$|\widehat{n}_{c1} - \xi| \longrightarrow 0, \qquad |\widehat{n}_{c2} - \eta| \longrightarrow 0 \quad (a.s.) \quad (N \to \infty)$$

Define $E(t) = E_1(t) + E_2(t) + H_2(t)\dfrac{p^*\xi + (1 - p^*)\eta}{p^*\bar{p}}$, $\widehat{E}(t) = \sqrt{N}(\widehat{H}(t) - H(t))$. For the new sequence, we have the following lemmas.

Lemma 5. *If $F(t)$, $G(t)$ are continuous distribution functions, then*

$$\sup_{0 \le t \le \theta} |\widehat{E}(t) - E(t)| \longrightarrow 0 \quad (a.s.), \qquad (N \to \infty), \quad \text{for any} \quad \theta > 0$$

The proof is immediate from Lemma 4 and the definition of $\widehat{E}(t)$ and $E(t)$.

Lemma 6. *Define $M(t) = E_1(t) + \int_0^t E(u)d\Lambda(u)$. Assume that F, G are continuous and θ satisfy $H(\theta) < 1$ and $\theta > 0$, then*

$$\sup_{0 \le t \le \theta} |\widehat{M}(t) - M(t)| \longrightarrow 0 \quad (a.s.), \qquad (N \to \infty)$$

Lemma 7. *Define $B(t) = \dfrac{M(t)}{1 - H(t)} - \displaystyle\int_0^t M(u)d\Big(\frac{1}{1 - H(u)}\Big)$. Under the assumptions of Lemma 6, we have*

$$\sup_{0 \le t \le \theta} |\widehat{B}(t) - B(t)| \longrightarrow 0 \quad (a.s.), \qquad (N \to \infty)$$

Lemma 8. *Under the assumptions of Lemma 7, we have*

$$\sup_{0\le t\le\theta} |\sqrt{N}(\widehat{F}(t) - F(t)) - W_1(t)| \longrightarrow 0 \quad (a.s.). \quad (N \to \infty)$$

where $W_1(t) = (1 - F(t))B(t)$

The Proof of Lemma 6,7,8 is very similar to that of Shorack and Wellner[4] (Chapter 7). From Lemma 8, we can see that $\sqrt{N}(\widehat{F}(t) - F(t))$ converges weakly to $W_1(t)$. Theorem 2 is valid because conclusions regarding the limiting distributions of the new sequence apply equally well to the same functions of the original sequence.

Suppose $F(t)$, $G(t)$ have discontinuities. We can enumerate all the discontinuities in a single sequence $t_1, t_2, \cdots$, say. The idea of proof of Theorem 3 is to spread the jump that $F(t)$ or $G(t)$ makes at t_j over a time interval which is inserted at this point by making a time transformation

$$\phi : [0,\infty) \longrightarrow [0,\infty)$$

$$\phi(t) = t + \sum_{m:t_m\le t} \alpha_m$$

where $\alpha_m > 0, m = 1,2,\cdots$, such that $\sum_{m=1}^{\infty} \alpha_m < \infty$.

Define distribution functions $F^*(t)$, $G^*(t)$ as follows

$$F^*(t^*) = \begin{cases} F(t), & \text{if } t^* = \phi(t) \text{ and } t \ne t_j,\ j = 1,2,\cdots \\ F(t_j-) + \Delta F(t_j)\frac{t^*-\phi(t_j-)}{\alpha_{j1}}, & \phi(t_j-) \le t^* \le \phi(t_j-) + \alpha_{j1} \\ F(t_j), & \phi(t_j-) + \alpha_{j1} < t^* \le \phi(t_j) \end{cases}$$

$$G^*(t^*) = \begin{cases} G(t), & \text{if } \ t^* = \phi(t) \ \text{ and } \ t \ne t_j, \ j = 1,2,\cdots \\ G(t_j-), & \phi(t_j-) \le t^* \le \phi(t_j-) + \alpha_{j1} \\ G(t_j-) + \Delta G(t_j)\frac{t^*-\phi(t_j-)-\alpha_{j1}}{\alpha_{j2}}, & \phi(t_j-) + \alpha_{j1} < t^* \le \phi(t_j) \end{cases}$$

where $\alpha_{j1}+\alpha_{j2} = \alpha_j$ and if t_j is a discontinuity of both F and G, then $\alpha_{j1} > 0$, $\alpha_{j2} > 0$; if t_j is a discontinuity of only F, then $\alpha_{j1} > 0$, $\alpha_{j2} = 0$; otherwise $\alpha_{j1} = 0$, $\alpha_{j2} > 0$.

From the definition of F^*, G^*, we note that F^*, G^* have the following properties:

(1). F^*, G^* are continuous on $[0,\infty)$

(2). $F^*(\phi(t)) = F(t)$, $G^*(\phi(t)) = G(t)$, for any $t \in [0,\infty)$

Let $U_{11}, U_{12}, \cdots, U_{1N}, U_{21}, U_{22}, \cdots, U_{2N}$ be independent and identically uniformly distributed random variables and also independent of (X_i, Y_i), $i = 1,2,\cdots,N$, and

$$X_i^* = I(X_i \ne t_j, \Delta F(t_j) > 0, j = 1,2,\cdots)\phi(X_i) + \sum_{j=1}^{\infty} I(X_i = t_j, \Delta F(t_j) > 0)[\phi(t_j-) + U_{1i}\alpha_{j1}]$$

$$Y_i^* = I(Y_i \ne t_j, \Delta G(t_j) > 0, j = 1,2,\cdots)\phi(Y_i) + \sum_{j=1}^{\infty} I(Y_i = t_j, \Delta G(t_j) > 0)[\phi(t_j-) + \alpha_{j1} + U_{2i}\alpha_{j2}]$$

$$\delta_i^* = I(X_i^* \le Y_i^*), \quad Z_i^* = \min(X_i^*, Y_i^*), \quad D_i^* = D_i, \quad i = 1, 2, \cdots, N$$

then $\delta_i^* = \delta_i$, and the distribution functions of X_i^*, Y_i^* are F^*, G^* respectively.

As the construction of $\widehat{F}(t)$ from observed $(Z_i, \delta_i), i = 1, 2, \cdots, N$, we can also construct the modified Kaplan-Meier estimator, $\widehat{F}^*(t)$, of $F^*(t)$ from the corresponding $(Z_i^*, \delta_i^*), i = 1, 2, \cdots, N$.

We note that $Z_i^* \le \phi(t)$ if and only if $Z_i < t$ and that $Z_i^* \le Z_j^*$ if and only if $Z_i \le Z_j$. By the definition of $\widehat{F}^*$, we can see that $\widehat{F}^*(\phi(t)) = \widehat{F}(t)$. Applying Theorem 2 to $\widehat{F}^*(t)$, we can see that $\sqrt{N}(\widehat{F}^*(t^*) - F^*(t^*))$ converges weakly to a Gaussion process $W_1^*(t^*)$. Replacing t^* with $\phi(t)$ and letting $W_2(t) = W_1^*(\phi(t))$, we can obtain Theorem 3.

The evaluation of the covariance structrue of the limiting process is given in the Appendix.

4. APPENDIX

A1. $\mathrm{Cov}(E_1(s), E(t)) = H_1(s)[1 - H(t)] \qquad (s \le t)$

$\mathrm{Cov}(E_1(t), E(s)) = H_1(s) - H(s)H_1(t) \qquad (s \le t)$

$\mathrm{Cov}(E(s), E(t)) = H(s)[1 - H(t)] + \dfrac{1-p^*}{p^*}\dfrac{1}{\overline{p}} H_2(s)\overline{H}_2(t) \qquad (s \le t)$

By the definition of $E(t)$ and the covariance structrue of (E_1, E_2, ξ, η) we can obtain the above results.

A2.
$$\mathrm{Cov}(M(s), M(t)) = \int_0^{s \wedge t} (1 - H(v))d\Lambda(v) + \frac{1-p^*}{p^*}\frac{1}{\overline{p}} \int_0^s \int_0^t H_2(u \wedge v)\overline{H}_2(u \vee v)d\Lambda(u)d\Lambda(v)$$

Proof. For $s \le t$

$$\mathrm{Cov}(M(s), M(t)) = I_1 + I_2 + I_3 + I_4$$

$$I_1 = \mathrm{Cov}(E_1(s), E_1(t)) = H_1(s)[1 - H_1(t)]$$

$$I_2 = \mathrm{Cov}(E_1(s), \int_0^t E(v)d\Lambda(v)) = \int_0^t \mathrm{Cov}(E_1(s), E(v))d\Lambda(v)$$
$$= \int_0^t [H_1(v) - H(v)H_1(s)]d\Lambda(v) + H_1(s)[H_1(t) - H_1(s)]$$

$$I_3 = \mathrm{Cov}(\int_0^s E(u)d\Lambda(u), E_1(t)) = \int_0^s \mathrm{Cov}(E(u), E_1(t))d\Lambda(u)$$
$$= \int_0^s [H_1(u) - H(u)H_1(t)]d\Lambda(u)$$

$$I_4 = \mathrm{Cov}(\int_0^s E(u)d\Lambda(u), \int_0^t E(v)d\Lambda(v)) = \int_0^s \int_0^t \mathrm{Cov}(E(u), E(v))d\Lambda(u)d\Lambda(v)$$
$$= \int_0^s H(v)[H_1(s) - H_1(v)]d\Lambda(v) + \int_0^s H(u)[H_1(t) - H_1(u)]d\Lambda(u)$$
$$+ \frac{1-p^*}{p^*}\frac{1}{\overline{p}} \int_0^s \int_0^t H_2(u \wedge v)\overline{H}_2(u \vee v)d\Lambda(u)d\Lambda(v)$$

$$I_1 + I_2 + I_3 + I_4 = \int_0^{s\wedge t} (1 - H(v))d\Lambda(v) + \frac{1-p^*}{p^*}\frac{1}{\overline{p}}\int_0^s \int_0^t H_2(u \wedge v)\overline{H}_2(u \vee v)d\Lambda(u)d\Lambda(v)$$

A3. $$\mathrm{Cov}(B(s), B(t)) = \int_0^{s\wedge t} \frac{d\Lambda(u)}{1 - H(u)} + \frac{1-p^*}{p^*}\frac{1}{\overline{p}}\int_0^s \int_0^t \frac{H_2(u \wedge v)\overline{H}_2(u \vee v)}{[1 - H(u)][1 - H(v)]}d\Lambda(u)d\Lambda(v)$$

The proof is similar to that of A3.

A4. $\mathrm{Cov}(W_1(s), W_1(t)) = [1 - F(s)][1 - F(t)]\mathrm{Cov}(B(s), B(t))$
The proof is immediate.

A5. $$\mathrm{Cov}(W_2(s), W_2(t)) = [1 - F(s)][1 - F(t)]\Big\{ \int_0^{s\wedge t} \frac{d\Lambda(u)}{[1 - H(u-)][1 - \Delta\Lambda(u)]} + \frac{1-p^*}{p^*}\frac{1}{\overline{p}}\int_0^s \int_0^t \frac{H_2(u \wedge v-)\overline{H}_2(u \vee v-)d\Lambda(u)d\Lambda(v)}{[1 - H(u-)][1 - H(v-)][1 - \Delta\Lambda(u)][1 - \Delta\Lambda(v)]}\Big\}$$

Proof.

$$\begin{aligned}\mathrm{Cov}(W_2(s), W_2(t)) &= \mathrm{Cov}(W_1^*(\phi(s)), W_1^*(\phi(t))) \\ &= [1 - F^*(\phi(s))][1 - F^*(\phi(t))]\Big\{ \int_0^{\phi(s)\wedge\phi(t)} \frac{d\Lambda^*(u)}{1 - H^*(u)} \\ &\quad + \frac{1-p^*}{p^*}\frac{1}{\overline{p}}\int_0^{\phi(s)} \int_0^{\phi(t)} \frac{H_2^*(u \wedge v)\overline{H}_2^*(u \vee v)}{[1 - H^*(u)][1 - H^*(v)]}d\Lambda^*(u)d\Lambda^*(v)\Big\}\end{aligned}$$

where H_1^*, H_2^*, $\overline{H}_2^*$, Λ^* are the corresponding H_1, H_2, $\overline{H}_2$, Λ of F^*, G^*.

By the definition of F^*, G^*, H_1^*, H_2^*, $\overline{H}_2^*$, Λ^*, we can obtain the covariance structrue of $W_2(t)$.

REFERENCES

1. K. Suzuki, *JASA*, **80**, 68-72(1985).
2. E. L. Kalpan and P. Meier, *JASA*, **53**, 457-481(1958).
3. P. Billingsley, *Convergence of Probability Measures*, John Wiley, New York, (1968).
4. G .R. Shorack and J. A. Wellner, *Empirical Processes with applications to statistics*, John Wiley, New York, (1986).
5. N. Breslow and J. Crowley, *Ann. Statist.*, **2**, 437-451(1974).

Stat. Sci. & Data Anal., pp. 529-542
K. Matsusita *et al.* (Eds)

New Main Effect Plus One Plans For 2^7 Factorial Experiments And Their Robustness Property Against Deletion Of Runs

SUBIR GHOSH
Department of Statistics, University of California, Riverside, CA 92521-0138, USA

Abstract. The paper presents new main effect plans with an additional property of searching and estimation of one nonnegligible interaction from two factor and higher order interactions for 2^7 factorial experiments. The plans are highly structured in the sense that they have both orthogonal arrays and balanced arrays as components. A robustness property against deletion of runs is proved. Using the robustness property, other plans with smaller number of runs are obtained.

Key Words and Phrases: Factorial designs, Linear models, Main effect plans, Robustness, Search designs.

1 INTRODUCTION

Main effect plans in fractional factorial experiments are very popular to experimenters (See Taguchi and Wu [1], Box, Hunter and Hunter [2]) in drawing inference on the general mean and main effects under the assumption that two-factor and higher order interactions are all zero. Such an assumption may or may not be valid in practice. Search designs introduced in Srivastava [3] permit the search for nonnegligible interactions with small number of runs from two factor and higher order interactions in addition to drawing inference on the general mean and main effects. Search designs with such an additional property are an improvement over the popular plans.

The construction of search design is a difficult and challenging task. The checking of search conditions is so time consuming and tedious that in many published research work including mine the proofs are not presented meticulously. This paper is unique in the sense that it is presenting every step in checking vividly. The paper is in the spirit of research in Ghosh [4] and Shirakura [5]. However, the paper not only presents new search designs for 2^7 factorial experiments but also presents proofs very clearly by developing new pertinent tools with the hope of its future generalizations for m factors.

Consider the following orthogonal main effect plan T_1 with 8 runs for 2^7 factorial experiments.

$$T_1 = \begin{bmatrix} 1 & 1 & 1 & 1 & 1 & 1 & 1 \\ 0 & 1 & 0 & 1 & 0 & 1 & 0 \\ 1 & 0 & 0 & 1 & 1 & 0 & 0 \\ 1 & 1 & 1 & 0 & 0 & 0 & 0 \\ 0 & 0 & 1 & 1 & 0 & 0 & 1 \\ 0 & 1 & 0 & 0 & 1 & 0 & 1 \\ 1 & 0 & 0 & 0 & 0 & 1 & 1 \\ 0 & 0 & 1 & 0 & 1 & 1 & 0 \end{bmatrix}. \tag{1}$$

Two more sets, T_2 of 8 runs and T_3 of 7 runs are given below.

$$T_2 = \begin{bmatrix} 1 & 1 & 0 & 1 & 1 & 0 & 0 \\ 0 & 1 & 1 & 1 & 0 & 1 & 0 \\ 1 & 0 & 0 & 1 & 0 & 1 & 1 \\ 1 & 1 & 1 & 0 & 0 & 0 & 1 \\ 0 & 0 & 1 & 1 & 1 & 0 & 1 \\ 0 & 1 & 0 & 0 & 1 & 1 & 1 \\ 1 & 0 & 1 & 0 & 1 & 1 & 0 \\ 0 & 0 & 0 & 0 & 0 & 0 & 0 \end{bmatrix}. \tag{2}$$

$$T_3 = \begin{bmatrix} 1 & 0 & 0 & 0 & 0 & 0 & 0 \\ 0 & 1 & 0 & 0 & 0 & 0 & 0 \\ 0 & 0 & 1 & 0 & 0 & 0 & 0 \\ 0 & 0 & 0 & 1 & 0 & 0 & 0 \\ 0 & 0 & 0 & 0 & 1 & 0 & 0 \\ 0 & 0 & 0 & 0 & 0 & 1 & 0 \\ 0 & 0 & 0 & 0 & 0 & 0 & 1 \end{bmatrix}. \tag{3}$$

Notice that both T_1 and T_2 are orthogonal arrays of strength 2 and T_3 is a balanced array of full strength (i.e., of strength 7).

Let D be a design with 23 runs in T_1, T_2 and T_3. The runs $1,\ldots,8$ in T_1, $9,\ldots,16$ in T_2 and $17,\ldots,23$ in T_3. The design D is in fact a balanced array of strength 2 and is highly structured. It is shown that the design D is an $MEP.1$ plan in the sense that the search and estimation of one nonnegligible interaction is possible from two factor and higher order interactions in addition to the estimation of the general mean and main effects. The designs $D^{(i)}$, $i = 9,\ldots,16$ with 22 runs, obtained from D by deleting the ith run (i.e., a run in T_2) are shown to be $MEP.1$ plans. The designs $D^{(i,16)}$, $i = 9,\ldots,15$ with 21 runs, obtained from $D^{(16)}$ by deleting the ith run are also shown to be $MEP.1$ plans. The designs $D^{(i)}$ and $D^{(i,16)}$, $i = 9,\ldots,15$, are not exactly balanced arrays but they are near balanced arrays. The design $D^{(16)}$ is a balanced array of strength 2. All designs are therefore structured. This is good particularly in view of optimum statistical properties.

The search linear model is given by

$$E(\underline{y}) = X_1\underline{\beta}_1 + X_2\underline{\beta}_2,\; V(\underline{y}) = \sigma^2 I, \tag{4}$$

where $\underline{y}(N \times 1)$ is a vector of observations, $X_1(N \times p_1)$ and $X_2(N \times p_2)$ are known matrices with Rank $X_1 = p_1, \sigma^2$ is an unknown nonnegative constant, $\underline{\beta}_1(p_1 \times 1)$ is a vector of unknown fixed parameters and $\underline{\beta}_2(p_2 \times 1)$ is a vector of partially known fixed parameters. It is known that at most K^* elements of $\underline{\beta}_2$ are nonzero but the identity of these nonzero elements are unknown. The problem is to search the nonzero elements of $\underline{\beta}_2$ and draw inference on them in addition to elements of $\underline{\beta}_1$. Throughout the paper, it is assumed that the elements of $\underline{\beta}_2$ are two factor and higher order interactions. For $MEP.1$ plans in 2^7 factorial experiments, $p_1 = 8$ and $p_2 = 2^7 - 8 = 120$. The values of N are 23 for D, 22 for $D^{(i)}$ and 21 for $D^{(i,16)}$.

2 *MEP*.1 PLANS

For the design T_u, $u = 1,2$, the $E(\underline{y})$ in (4) is written as

$$E(\underline{y}_u) = X_{u1}\underline{\beta}_1 + X_{u2}\underline{\beta}_2, \tag{5}$$

where $\underline{y}_u(8 \times 1)$, $X_{u1}(8 \times 8)$ and $X_{u2}(8 \times 120)$. Similarly for the design T_3,

$$E(\underline{y}_3) = X_{31}\underline{\beta}_1 + X_{32}\underline{\beta}_2, \tag{6}$$

where $\underline{y}_3(7 \times 1)$, $X_{31}(7 \times 8)$ and $X_{32}(7 \times 120)$. The matrices X_{11}, X_{21} and X_{31} are denoted as

$$\begin{aligned} X_{11} &= [\underline{d}_0, \underline{d}_1, \ldots, \underline{d}_7],\; X_{21} = [\underline{c}_0, \underline{c}_1, \ldots, \underline{c}_7] \\ \text{and } X_{31} &= [\underline{e}_0, \underline{e}_1, \ldots, \underline{e}_7], \end{aligned} \tag{7}$$

where $\underline{d}_0 = \underline{c}_0 = \underline{e}_0 = \underline{j}_8, \underline{j}_8$ is an (8×1) vector with all elements unity. Let $\underline{a}_1, \underline{a}_2$ and $\underline{a}_3$ be (8×1) vectors with elements +1 or -1 so that the (8×3) matrix $[\underline{a}_1, \underline{a}_2, \underline{a}_3]$ has the $2^3(= 8)$ vectors $(\pm 1, \pm 1, \pm 1)$ as rows. Suppose the $\underline{d}_1 = \underline{a}_1$, $\underline{d}_2 = \underline{a}_2$, $\underline{d}_3 = \underline{a}_1 * \underline{a}_2$, $\underline{d}_4 = \underline{a}_3, \underline{d}_5 = \underline{a}_1 * \underline{a}_3, \underline{d}_6 = \underline{a}_2 * \underline{a}_3$ and $\underline{d}_7 = \underline{a}_1 * \underline{a}_2 * \underline{a}_3$, where $*$ denotes the Hadamard product. Moreover, $\underline{c}_1 = \underline{d}_1$, $\underline{c}_2 = \underline{d}_2$, $\underline{c}_3 = -\underline{d}_5$, $\underline{c}_4 = \underline{d}_4$, $\underline{c}_5 = \underline{d}_7$, $\underline{c}_6 = -\underline{d}_3$ and $\underline{c}_7 = -\underline{d}_6$. The columns of

X_{u2}, $u = 1,2,3$, are determined by the elements of $\underline{\beta}_2$. For example, the columns of X_{12}, X_{22} and X_{32} are $\underline{d}_{i_1} * \underline{d}_{i_2} * \ldots * \underline{d}_{i_q}$, $\underline{c}_{i_1} * \underline{c}_{i_2} * \ldots * \underline{c}_{i_q}$ and $\underline{e}_{i_1} * \underline{e}_{i_2} * \ldots * \underline{e}_{i_q}$ for the q factor ($q = 2,3,\ldots,m$) interaction of factors $i_1, i_2, \ldots,$ and i_q. It can be seen that $X'_{11}X_{11} = X'_{21}X_{21} = 8I_8$.

Denote

$$Q_2 = X_{22} - X_{21}X_{11}^{-1}X_{12}, Q_3 = X_{32} - X_{31}X_{11}^{-1}X_{12} \text{and } Q' = [Q'_2 : Q'_3], \tag{8}$$

where Q_2, Q_3 and Q are (8×120), (7×120) and (15×120) matrices. It can be seen from Srivastava and Gupta [6] that D is an $MEP.1$ plan if and only if the matrix Q has the property P_2 i.e., every set of two columns of Q are linearly independent. The matrix Q_2 is highly structured. Table 1 presents the columns of X_{12} which are equal to $\underline{d}_i$, $i = 0,1,\ldots,7$. In Table 1, the column $i_1, i_2, \ldots, i_q$ of X means $\underline{d}_{i_1} * \underline{d}_{i_2} * \ldots * \underline{d}_{i_q}$. Thus the entry 123 in column 3 and Row 1 means $\underline{d}_1 * \underline{d}_2 * \underline{d}_3$. Notice that if the column $i_1, i_2, \ldots, i_q$ $(2 \leq q \leq 5)$ of X_{12} is identical to $\underline{d}_i$, then the column represented by the complement of $i_1, \ldots, i_q$ in $12, \ldots, 7$ is also identical to $\underline{d}_i$. For example, the column 123 of X_{12} is identical to $\underline{d}_0$. The complement of 123 in $12, \ldots, 7$ is 4567 and the column 4567 of X_{12} is also identical to $\underline{d}_0$. It can be seen from (8) and Table 1 that the columns of Q may be identified and grouped with respect to the elements of $\underline{\beta}_2$ and the columns of X_{u1}, $u = 2$ and 3. To illustrate this, the column of Q for the element of $\underline{\beta}_2$ as the three factor interaction of factors 1,2,3 has two components one in Q_2 and the other in Q_3. The component in Q_2 is $\underline{c}_1 * \underline{c}_2 * \underline{c}_3 - \underline{c}_0$ and the component in Q_3 is $\underline{e}_1 * \underline{e}_2 * \underline{e}_3 - \underline{e}_0$. The same is true for other columns of Q. Table 2 displays columns of Q_2 and shows that they are identical to $\underline{c}_u - \underline{c}_{u'}$ and $\underline{c}_u + \underline{c}_{u'}, u \neq u'$, $u, u' = 0,1,\ldots,7$. In Table 2, the column $(i_1 i_2 \ldots i_q, i_s)$ of Q_2 means the column $\underline{c}_{i_1} * \underline{c}_{i_2} * \ldots * \underline{c}_{i_q} - \underline{c}_{i_s}$. For example, the column (123,0) means the column $\underline{c}_1 * \underline{c}_2 * \underline{c}_3 - \underline{c}_0$. The column - (1237,7) means - $(\underline{c}_1 * \underline{c}_2 * \underline{c}_3 * \underline{c}_7 - \underline{c}_7)$.

Table 1

The Columns of X_{12}

i	$\underline{d}_i$	The columns of X_{12} are identical to $\underline{d}_i$
0	$\underline{d}_0$	123, 145, 167, 246, 257, 347, 356, 4567, 2367, 2345, 1357, 1346, 1256, 1247, 1234567
1	$\underline{d}_1$	23, 45, 67, 1246, 1257, 1347, 1356, 14567, 12367, 12345, 357, 346, 256, 247, 234567
2	$\underline{d}_2$	13, 1245, 1267, 46, 57, 2347, 2356, 24567, 367, 345, 12357, 12346, 156, 147, 134567
3	$\underline{d}_3$	12, 1345, 1367, 2346, 2357, 47, 56, 34567, 267, 245, 157, 146, 12356, 12347, 124567
4	$\underline{d}_4$	1234, 15, 1467, 26, 2457, 37, 3456, 567, 23467, 235, 13457, 136, 12456, 127, 123567
5	$\underline{d}_5$	1235, 14, 1567, 2456, 27, 3457, 36, 467, 23567, 234, 137, 13456, 126, 12457, 123467
6	$\underline{d}_6$	1236, 1456, 17, 24, 2567, 3467, 35, 457, 237, 23456, 13567, 134, 125, 12467, 123457
7	$\underline{d}_7$	1237, 1457, 16, 2467, 25, 34, 3567, 456, 236,, 23457, 135, 13467, 12567, 124, 123456

Table 2

The columns of Q_2 that are identical to $\underset{\sim}{c}_u - \underset{\sim}{c}_{u'}$ and $\underset{\sim}{c}_u + \underset{\sim}{c}_{u'}$, $u \neq u', u, u' = 0, 1, \ldots, 7$.

	Columns of Q_2	Columns of Q_2 identical to
1.	(123,0), -(1237,7)	$\underset{\sim}{c}_7 - \underset{\sim}{c}_0$
2.	(145,0), -(1245,2)	$\underset{\sim}{c}_2 - \underset{\sim}{c}_0$
3.	(167,0), -(1467,4)	$\underset{\sim}{c}_4 - \underset{\sim}{c}_0$
4.	(246,0), -(2346,3)	$\underset{\sim}{c}_3 - \underset{\sim}{c}_0$
5.	(257,0), -(2567,6)	$\underset{\sim}{c}_6 - \underset{\sim}{c}_0$
6.	(347,0), -(3457,5)	$\underset{\sim}{c}_5 - \underset{\sim}{c}_0$
7.	(356,0), -(1356,1)	$\underset{\sim}{c}_1 - \underset{\sim}{c}_0$
8.	-(4567,0), -(456,7)	$\underset{\sim}{c}_7 + \underset{\sim}{c}_0$
9.	-(2367,0), -(367,2)	$\underset{\sim}{c}_2 + \underset{\sim}{c}_0$
10.	-(2345,0), -(235,4)	$\underset{\sim}{c}_4 + \underset{\sim}{c}_0$
11.	-(1357,0), -(157,3)	$\underset{\sim}{c}_3 + \underset{\sim}{c}_0$
12.	-(1346,0), -(134,6)	$\underset{\sim}{c}_6 + \underset{\sim}{c}_0$
13.	-(1256,0), -(126,5)	$\underset{\sim}{c}_5 + \underset{\sim}{c}_0$
14.	-(1247,0), -(247,1)	$\underset{\sim}{c}_1 + \underset{\sim}{c}_0$
15.	-(23,1), -(1235,5)	$\underset{\sim}{c}_5 + \underset{\sim}{c}_1$
16.	-(45,1), -(1456,6)	$\underset{\sim}{c}_6 + \underset{\sim}{c}_1$
17.	-(67,1), -(1367,3)	$\underset{\sim}{c}_3 + \underset{\sim}{c}_1$
18.	-(1246,1), -(26,4)	$\underset{\sim}{c}_4 + \underset{\sim}{c}_1$
19.	-(1257,1), -(57,2)	$\underset{\sim}{c}_2 + \underset{\sim}{c}_1$

Table 2 - Continued

	Columns of Q_2	Columns of Q_2 identical to
20.	-(1347,1), -(34,7)	$\underline{c}_7 + \underline{c}_1$
21.	(14567,1), -(467,5)	$\underline{c}_5 - \underline{c}_1$
22.	(12367,1), -(237,6)	$\underline{c}_6 - \underline{c}_1$
23.	(12345,1), -(245,3)	$\underline{c}_3 - \underline{c}_1$
24.	(357,1), -(13457,4)	$\underline{c}_4 - \underline{c}_1$
25.	(346,1), -(12346,2)	$\underline{c}_2 - \underline{c}_1$
26.	(256,1), -(12567,7)	$\underline{c}_7 - \underline{c}_1$
27.	-(13,2), -(1234,4)	$\underline{c}_4 + \underline{c}_2$
28.	-(1267,2), -(16,7)	$\underline{c}_7 + \underline{c}_2$
29.	-(46,2), -(2456,5)	$\underline{c}_5 + \underline{c}_2$
30.	-(2347,2), -(47,3)	$\underline{c}_3 + \underline{c}_2$
31.	-(2356,2), -(35,6)	$\underline{c}_6 + \underline{c}_2$
32.	(24567,2), -(567,4)	$\underline{c}_4 - \underline{c}_2$
33.	(345,2), -(23457,7)	$\underline{c}_7 - \underline{c}_2$
34.	(12357,2), -(137,5)	$\underline{c}_5 - \underline{c}_2$
35.	(156,2), -(12356,3)	$\underline{c}_3 - \underline{c}_2$
36.	(147,2), -(12467,6)	$\underline{c}_6 - \underline{c}_2$
37.	-(12,3), -(1236,6)	$\underline{c}_6 + \underline{c}_3$

Table 2 - Continued

	Columns of Q_2	Columns of Q_2 identical to
38.	-(1345,3), -(14,5)	$\underset{\sim}{c}_5 + \underset{\sim}{c}_3$
39.	-(2357,3), -(25,7)	$\underset{\sim}{c}_7 + \underset{\sim}{c}_3$
40.	-(56,3), -(3456,4)	$\underset{\sim}{c}_4 + \underset{\sim}{c}_3$
41.	(34567,3), -(457,6)	$\underset{\sim}{c}_6 - \underset{\sim}{c}_3$
42.	(267,3), -(23567,5)	$\underset{\sim}{c}_5 - \underset{\sim}{c}_3$
43.	(146,3), -(13467,7)	$\underset{\sim}{c}_7 - \underset{\sim}{c}_3$
44.	(12347,3), -(127,4)	$\underset{\sim}{c}_4 - \underset{\sim}{c}_3$
45.	-(15,4), -(1457,7)	$\underset{\sim}{c}_7 + \underset{\sim}{c}_4$
46.	-(2457,4), -(27,5)	$\underset{\sim}{c}_5 + \underset{\sim}{c}_4$
47.	-(37,4), -(3467,6)	$\underset{\sim}{c}_6 + \underset{\sim}{c}_4$
48.	(23467,4), -(236,7)	$\underset{\sim}{c}_7 - \underset{\sim}{c}_4$
49.	(136,4), -(13456,5)	$\underset{\sim}{c}_5 - \underset{\sim}{c}_4$
50.	(12456,4), -(125,6)	$\underset{\sim}{c}_6 - \underset{\sim}{c}_4$
51.	-(1567,5), -(17,6)	$\underset{\sim}{c}_6 + \underset{\sim}{c}_5$
52.	-(36,5), -(3567,7)	$\underset{\sim}{c}_7 + \underset{\sim}{c}_5$
53.	(234,5), -(23456,6)	$\underset{\sim}{c}_6 - \underset{\sim}{c}_5$
54.	(12457,5), -(124,7)	$\underset{\sim}{c}_7 - \underset{\sim}{c}_5$
55.	-(24,6), -(2467,7)	$\underset{\sim}{c}_7 + \underset{\sim}{c}_6$
56.	(13567,6), -(135,7)	$\underset{\sim}{c}_7 - \underset{\sim}{c}_6$

The Lemmas given below are very useful for the main results of this paper.

Lemma 1. The following results are true for $u, u' = 0, 1, \ldots, 7$,

i. $\underline{c}_u * \underline{c}_{u'} * (\underline{c}_u + \underline{c}_{u'}) = \underline{c}_u + \underline{c}_{u'}$,

ii. $\underline{c}_u * \underline{c}_{u'} * (\underline{c}_u - \underline{c}_{u'}) = -(\underline{c}_u - \underline{c}_{u'})$.

The proof of Lemma 1 is easy. Lemma 1 explains the entries in Table 2. For example, in the row 50 of Table 2, the column (12456,4) i.e., $(\underline{c}_1 * \underline{c}_2 * \underline{c}_4 * \underline{c}_5 * \underline{c}_6 - \underline{c}_4)$ of Q_2 is identical to $\underline{c}_6 - \underline{c}_4$ and therefore the column $\underline{c}_6 * \underline{c}_4 * (\underline{c}_1 * \underline{c}_2 * \underline{c}_4 * \underline{c}_5 * \underline{c}_6 - \underline{c}_4)$, i.e., $(\underline{c}_1 * \underline{c}_2 * \underline{c}_5 - \underline{c}_6)$ of Q_2 is identical to - $(\underline{c}_6 - \underline{c}_4)$. Thus the column - (125,6) of Q_2 is in fact $\underline{c}_6 - \underline{c}_4$.

Lemma 2. Suppose that for a given u, $u = 0, 1, \ldots, 7$, $\underline{d}_u = \underline{d}_1^{\alpha_1} * \underline{d}_2^{\alpha_2} * \underline{d}_3^{\alpha_3} * \underline{d}_4^{\alpha_4} * \underline{d}_5^{\alpha_5} * \underline{d}_6^{\alpha_6} * \underline{d}_7^{\alpha_7}$, where $\alpha_i = 0, 1$, $\underline{d}_i^0 = \underline{d}_0$ and $\underline{d}_i^1 = \underline{d}_i$, $i = 1, \ldots, 7$, $\underline{\alpha}' = (\alpha_1, \ldots, \alpha_7)$, $\underline{j}' = (1, \ldots, 1)$ and $\underline{j}'\underline{\alpha} \geq 2$. There exists $\underline{\alpha}'$'s satisfying $\alpha_u = 0$, $\underline{j}'\underline{\alpha} = 2$ or 3 and $\underline{c}_1^{\alpha_1} * \underline{c}_2^{\alpha_2} * \underline{c}_3^{\alpha_3} * \underline{c}_4^{\alpha_4} * \underline{c}_5^{\alpha_5} * \underline{c}_6^{\alpha_6} * \underline{c}_7^{\alpha_7} = +\underline{c}_{u'}$ or - $\underline{c}_{u'}$ with $u \neq u'$, $u' = 0, 1, \ldots, 7$.

The proof of Lemma 2 follows from Table 1 and particularly from the first row for $i = 0$ in Table 1.

Lemma 3. There are two columns of Q_2 that are identical to any of $\underline{c}_u \pm \underline{c}_{u'}$, $u \neq u', u, u' = 0, 1, \ldots, 7$. Two columns except for signs are of the following forms: $[(ijk, 0)]$, $(ijku, u)]$, $[(ijk, u), (ijku, 0)]$, $[(ijk, u), (ijkuu', u')]$ and $[(ij, u), (ijuu', u')]$, where $i \neq j \neq k \neq u \neq u'$, $i, j, k, u, u' \epsilon \{1, \ldots, 7\}$.

The proof of Lemma 3 follows from Table 2.

Lemma 4. The distinct columns of Q_2 have the property P_2.

Proof. There are 56 distinct columns of Q_2. The columns are of the form $\underline{c}_u \pm \underline{c}_{u'}$, $u \neq u', u, u' \epsilon \{0, 1, \ldots, 7\}$. Any two of them can not be dependent since the columns $\underline{c}_u, u = 0, 1, \ldots, 7$ are mutually orthogonal.

Lemma 5. Consider the pairs of columns of $((\underline{e}_i * \underline{e}_j * \underline{e}_k - \underline{e}_o), (\underline{e}_i * \underline{e}_j * \underline{e}_k * \underline{e}_u - \underline{e}_u))$, $((\underline{e}_i * \underline{e}_j * \underline{e}_k - \underline{e}_u), (\underline{e}_i * \underline{e}_j * \underline{e}_k * \underline{e}_u * \underline{e}_{u'} - \underline{e}_{u'}))$, and $((\underline{e}_i * \underline{e}_j - \underline{e}_u), (\underline{e}_i * \underline{e}_j * \underline{e}_u * \underline{e}_{u'} - \underline{e}_{u'}))$, for $i \neq j \neq k \neq u \neq u', i, j, k, u, u' \epsilon \{1, \ldots, 7\}$. The following results are true.

i. $(\underline{e}_i * \underline{e}_j * \underline{e}_k - \underline{e}_o) \neq \pm(\underline{e}_i * \underline{e}_j * \underline{e}_k * \underline{e}_u - \underline{e}_u)$,

ii. $(\underline{e}_i * \underline{e}_j * \underline{e}_k - \underline{e}_u) \neq \pm(\underline{e}_i * \underline{e}_j * \underline{e}_k * \underline{e}_u * \underline{e}_{u'} - \underline{e}_{u'})$,

iii. $(\underline{e}_i * \underline{e}_j - \underline{e}_u) \neq \pm(\underline{e}_i * \underline{e}_j * \underline{e}_u * \underline{e}_{u'} - \underline{e}_{u'})$.

Moreover, $(\underline{e}_i * \underline{e}_j * \underline{e}_k - \underline{e}_o)$ is (-1) times $(\underline{e}_i * \underline{e}_j * \underline{e}_k * \underline{e}_u - \underline{e}_u)$ except for the uth row, $(\underline{e}_i * \underline{e}_j * \underline{e}_k - \underline{e}_u)$ is identical to $(\underline{e}_i * \underline{e}_j * \underline{e}_k * \underline{e}_u * \underline{e}_{u'} - \underline{e}_{u'})$ except for the uth row and $(\underline{e}_i * \underline{e}_j - \underline{e}_u)$ is also identical to $(\underline{e}_i * \underline{e}_j * \underline{e}_u * \underline{e}_{u'} - \underline{e}_{u'})$ except for the uth row.

Proof. Notice that $\underline{e}_u * (\underline{e}_i * \underline{e}_j * \underline{e}_k - \underline{e}_o) = (\underline{e}_i * \underline{e}_j * \underline{e}_k * \underline{e}_u - \underline{e}_u), \underline{e}_u * \underline{e}_{u'} * (\underline{e}_i * \underline{e}_j * \underline{e}_k - \underline{e}_u) =$

$(\underset{\sim}{e}_i * \underset{\sim}{e}_j * \underset{\sim}{e}_k * \underset{\sim}{e}_u * \underset{\sim}{e}_{u'} - \underset{\sim}{e}_{u'})$ and $\underset{\sim}{e}_u * \underset{\sim}{e}_{u'} * (\underset{\sim}{e}_i * \underset{\sim}{e}_j - \underset{\sim}{e}_u) = (\underset{\sim}{e}_i * \underset{\sim}{e}_j * \underset{\sim}{e}_u * \underset{\sim}{e}_{u'} - \underset{\sim}{e}_{u'})$. The rest is clear.

Theorem 1. The design D with 23 runs in T_1, T_2 and T_3 is an $MEP.1$ plan.

Proof. The proof of Theorem 1 follows from Lemmas 1-5.

The design D has 23 runs and the attempt is now made to reduce further the number of runs using the robustness property. It is to be noted that the design D is highly structured and the designs obtained by using the robustness property are close to being highly structured (i.e., near balanced). The robustness property of designs now stated for $MEP.1$ plans was introduced in Ghosh [7].

Definition: An $MEP.1$ plan is said to be robust against deletion of a set of s runs if the resulting plan after deletion of s runs remain an $MEP.1$ plan.

Theorem 2. The design D with 23 runs in T_1, T_2 and T_3 is not robust against deletion of any run in T_3.

Proof. The proof follows from Lemma 5 by observing that the deletion of any run in T_3 results in a Q matrix that does not have the property P_2.

Lemma 6. The following results are true.

i. In both $\underset{\sim}{c}_u + \underset{\sim}{c}_{u'}$ and $\underset{\sim}{c}_u - \underset{\sim}{c}_{u'}$, $u \neq u', u, u' \epsilon \{0, 1, \ldots, 7\}$, there are 4 nonzero elements and the nonzero elements are +2 or -2.

ii. In $\underset{\sim}{c}_0 \pm \underset{\sim}{c}_{u'}, u' \epsilon \{1, \ldots, 7\}$ there are 4 nonzero elements and they are +2. In $\underset{\sim}{c}_0 + \underset{\sim}{c}_{u'}$, the last element is 0 and in $\underset{\sim}{c}_0 - \underset{\sim}{c}_{u'}$, the last element is +2.

iii. In $\underset{\sim}{c}_u \pm \underset{\sim}{c}_{u'}, u \neq u', u, u' \epsilon \{1, \ldots, 7\}$, there are 4 nonzero elements, two are +2 and two are -2. In $\underset{\sim}{c}_u + \underset{\sim}{c}_{u'}$, the last element is -2 and in $\underset{\sim}{c}_u - \underset{\sim}{c}_{u'}$, the last element is 0.

Proof. The proof follows from the fact that T_2 is an orthogonal array of strength 2 and the last row of T_2 is $(0 \ldots 0)$.

Lemma 7. Consider one of $\underset{\sim}{c}_{u_1} \pm \underset{\sim}{c}_{u_2}$ and another of $\underset{\sim}{c}_{u_3} + \underset{\sim}{c}_{u_4}$. The following results are true.

i. If the nonzero elements of both vectors are at the same positions, then at two positions the elements of vectors are identical and at other two positions the elements of vectors are opposite in signs.

ii. If $t (0 \leq t \leq 3)$ non zero elements of both vectors occur at the same positions, then there are at least two out of eight positions so that (a) in one position the first vector has 0 and the second vector has either +2 or -2 and (b) in another position the first vector has either +2 or -2 and the second vector has 0.

iii. The last row in the (8×2) submatrix of Q_2 for the chosen two columns appears also as a row in the corresponding submatrix of Q_3.

Proof. The proof follows from the fact that T_2 is an orthogonal array of strength 2, $(\underline{c}_{u_1} \pm \underline{c}_{u_2}) \neq \pm(\underline{c}_{u_3} \pm \underline{c}_{u_4})$ and the structure of T_3.

There are columns of Q_3 that are identical. For example, the three columns $\underline{e}_i * \underline{e}_j - \underline{e}_k$, $\underline{e}_i * \underline{e}_k - \underline{e}_j$ and $\underline{e}_j * \underline{e}_k - \underline{e}_i$ for $i \neq j \neq k$, $i, j, k \epsilon \{1, \ldots, 7\}$. The columns of Q_2 corresponding to those three identical columns of Q_3 for values of i, j, k are given in Table 3. For given i, j, k, each of the pairs $(\underline{e}_i * \underline{e}_j - \underline{e}_k,\ \underline{e}_i * \underline{e}_k - \underline{e}_j)$, $(\underline{e}_i * \underline{e}_j - \underline{e}_k,\ \underline{e}_j * \underline{e}_k - \underline{e}_i)$ and $(\underline{e}_i * \underline{e}_k - \underline{e}_j,\ \underline{e}_j * \underline{e}_k - \underline{e}_i)$ have two rows whose deletion results in identical members of pairs. Table 4 presents run numbers for two rows in each pair.

Table 3

Columns of Q_2 corresponding to identical columns of Q_3

			Columns of Q_2 corresponding to identical columns of Q_3 as		
i	j	k	$\underline{e}_i * \underline{e}_j - \underline{e}_k$	$\underline{e}_i * \underline{e}_k - \underline{e}_j$	$\underline{e}_j * \underline{e}_k - \underline{e}_i$
1	2	3	$-\underline{c}_6 - \underline{c}_3$	$-\underline{c}_4 - \underline{c}_2$	$-\underline{c}_5 - \underline{c}_1$
1	4	5	$-\underline{c}_3 - \underline{c}_5$	$-\underline{c}_7 - \underline{c}_4$	$-\underline{c}_6 - \underline{c}_1$
1	6	7	$-\underline{c}_2 - \underline{c}_7$	$-\underline{c}_5 - \underline{c}_6$	$-\underline{c}_3 - \underline{c}_1$
2	4	6	$-\underline{c}_7 - \underline{c}_6$	$-\underline{c}_1 - \underline{c}_4$	$-\underline{c}_5 - \underline{c}_2$
2	5	7	$-\underline{c}_3 - \underline{c}_7$	$-\underline{c}_4 - \underline{c}_5$	$-\underline{c}_1 - \underline{c}_2$
3	4	7	$-\underline{c}_1 - \underline{c}_7$	$-\underline{c}_6 - \underline{c}_4$	$-\underline{c}_2 - \underline{c}_3$
3	5	6	$-\underline{c}_2 - \underline{c}_6$	$-\underline{c}_7 - \underline{c}_5$	$-\underline{c}_4 - \underline{c}_3$

Table 4
Runs whose deletions results in identical columns of Q_2 and hence of Q

From Table 3 i	j	k	Columns of Q_2	Runs whose deletion results in identical columns of Q_2
1	2	3	$-\underline{c}_6 - \underline{c}_3, -\underline{c}_4 - \underline{c}_2$	9, 15
			$-\underline{c}_6 - \underline{c}_3, -\underline{c}_5 - \underline{c}_1$	9, 10
			$-\underline{c}_4 - \underline{c}_2, -\underline{c}_5 - \underline{c}_1$	10, 15
1	4	5	$-\underline{c}_3 - \underline{c}_5, -\underline{c}_7 - \underline{c}_4$	11, 15
			$-\underline{c}_3 - \underline{c}_5, -\underline{c}_6 - \underline{c}_1$	11, 13
			$-\underline{c}_7 - \underline{c}_4, -\underline{c}_6 - \underline{c}_1$	13, 15
1	6	7	$-\underline{c}_2 - \underline{c}_7, -\underline{c}_5 - \underline{c}_6$	12, 15
			$-\underline{c}_2 - \underline{c}_7, -\underline{c}_3 - \underline{c}_1$	14, 15
			$-\underline{c}_5 - \underline{c}_6, -\underline{c}_3 - \underline{c}_1$	12, 14
2	4	6	$-\underline{c}_7 - \underline{c}_6, -\underline{c}_1 - \underline{c}_4$	9, 14
			$-\underline{c}_7 - \underline{c}_6, -\underline{c}_5 - \underline{c}_2$	9, 11
			$-\underline{c}_1 - \underline{c}_4, -\underline{c}_5 - \underline{c}_2$	11, 14
2	5	7	$-\underline{c}_3 - \underline{c}_7, -\underline{c}_4 - \underline{c}_5$	9, 12
			$-\underline{c}_3 - \underline{c}_7, -\underline{c}_1 - \underline{c}_2$	9, 13
			$-\underline{c}_4 - \underline{c}_5, -\underline{c}_1 - \underline{c}_2$	12, 13
3	4	7	$-\underline{c}_1 - \underline{c}_7, -\underline{c}_6 - \underline{c}_4$	10, 12
			$-\underline{c}_1 - \underline{c}_7, -\underline{c}_2 - \underline{c}_3$	10, 11
			$-\underline{c}_6 - \underline{c}_4, -\underline{c}_2 - \underline{c}_3$	11, 12
2	4	5	$-\underline{c}_2 - \underline{c}_6, -\underline{c}_7 - \underline{c}_5$	10, 13
			$-\underline{c}_2 - \underline{c}_6, -\underline{c}_4 - \underline{c}_3$	13, 14
			$-\underline{c}_7 - \underline{c}_5, -\underline{c}_4 - \underline{c}_3$	10, 14

Theorem 3. The designs $D^{(i)}, i = 9, \ldots, 16$, with 22 runs, obtained from D by deleting the ith run (in T_2) are $MEP.1$ plans.

Proof. The proof follows from Lemmas 6 and 7.

Theorem 4. The following results are true.

i. The designs $D^{(i,16)}, i = 9, \ldots, 15$, with 21 runs, obtained from D by deleting the ith run are $MEP.1$ plans.

ii. The designs $D^{(i,j)}, i, j = 9, \ldots, 15,\ i \neq j$, with 21 runs obtained from D by deleting the runs i and j are not $MEP.1$ plans.

Proof. The proof follows from Lemmas 6 and 7 and Tables 3 and 4.

3 DISCUSSIONS

It is to be emphasized that the designs presented here are highly structured and therefore very good in view of optimum statistical properties. Calling T_2 as T_1 and T_1 as T_2, it can be seen that $D^{(i)}, i = 1, \ldots, 8$ with 22 runs, obtained from D by deleting the ith run are also $MEP.1$ plans. Gupta [8] presented an $MEP.1$ plan with 22 runs.

The design is not structured and is highly unbalanced. Moreover, the proof is not given meticulously. However, it is to be noted without proof that there is one positive aspect of the design that is not given in the paper of Gupta [6] namely that the deletion of runs 12, 16 and 22 results in an $MEP.1$ plan with 19 runs. Again, the resulting plan is highly unstructured.

Acknowledgement
The author would like to thank Professor T. Hayakawa and the Pacific Statistical Institute for their kindness and help for making his visit to Tokyo possible. Thanks also go to the referee for comments. The work of the author is sponsored by the Air Force Office for Scientific Research under Grant AFOSR-91-0115.

REFERENCES

1. G. Taguchi and Y. Wu, *Introduction to off-line quality control.* Central Japan Quality Control Association. Tokyo (1985).

2. G.E.P Box, W.G. Hunter, J.S. and J. S. Hunter, *Statistics for experimenters, an introduction to design, data analysis and model building.* John Wiley & Sons, New York (1978).

3. J. N. Srivastava, in: *A survey of statistical design and linear models*, J. N. Srivastava (Ed.), pp. 507-519, North-Holland Publishing Co. (1975).

4. S. Ghosh, *Ann. Statist.* 8, 922-930 (1980).

5. T. Shirakura, *J. Statist. Plann. Inference*, 27, 65-74 (1991).

6. J. N. Srivastava, and B. C. Gupta, *J. Statist. Plann. Inference*, 3, 259-265 (1979).

7. S. Ghosh, *Sankhyā*, 40, Ser. B, Pts. 3 and 4, 204-208 (1979).

8. B. C. Gupta, *Commun.Statist. - Theory Meth.*, 20(9), 2955-2963 (1991).

Stat. Sci. & Data Anal., pp. 543-557
K. Matsusita *et al.* (Eds)

PROFILES OF 2^m FACTORIAL DESIGNS

Sumiyasu YAMAMOTO[1], Yoshio FUJII[1,2], Yoshifumi HYODO[1,2], and Hiromu YUMIBA[2]

International Institute for Natural Sciences[1], Kurashiki 710, JAPAN
Okayama University of Science[2], Okayama 700, JAPAN

Abstract The concepts of the frontage and the profile of a fractional 2^m factorial design are introduced. The frontage is a part of the information matrix related directly to the parameters to be estimated assuming that the remaining parameters are negligible. The profile is the remaining part of the information matrix related to all parameters including those not of interest. As a characteristic of the profile of a design which is invariant with respect to the permutation of factors as well as levels, the spot matrix of the design is defined. Several related invariants are also proposed.

Key words: Factorial design; Information matrix; Frontage; Profile; Spot matrix.

0. INTRODUCTION

In a factorial experiment with m factors each at two levels, a two-symbol orthogonal array T of even strength $t(=2p)$, size n, m constraints and index λ (denoted here as 2-OA($t = 2p, m, \lambda$)) has been recommended to use as a fractional factorial design composed of n assemblies. Under the assumption that $p + 1$ or more higher order interactions can be neglected, the best linear unbiased estimates (BLUE) of those factorial effects up to p-factor interactions can be obtained uncorrelatedly. Those estimates, however, cannot be unbiased if such an assumption is violated. In such a case, attention should be paid on the alias structure of the design (see, e.g., [3] and [4]).

According to our recent work on the construction, exhaustive enumeration and classification of two-symbol orthogonal arrays, all of those 1932 different 2-OA($t = 2, m = 5, \lambda = 4$)'s are classified into 11 isomorphic classes with respect to the permutation of symbols (levels) within columns (factors) and that of columns (see [5]). Representatives of those 11 isomorphic classes are given in Table 1.

Table 1. Representatives of 11 isomorphic classes of 2-OA($t = 2, m = 5, \lambda = 4$)

[1]	[2]	[3]	[4]	[5]	[6]	[7]	[8]	[9]	[10]	[11]
00101	00101	00100	00011	00011	00011	00011	00011	00011	00010	00001
00101	00101	00101	00101	00101	00101	00100	00100	00100	00011	00010
00110	00110	00110	00110	00110	00110	00110	00110	00101	00100	00100
00110	00110	00111	00111	00110	00110	00111	00111	00110	00101	00111
01001	01000	01000	01000	01001	01000	01000	01000	01000	01000	01000
01001	01001	01001	01001	01001	01001	01001	01001	01001	01001	01011
01010	01010	01010	01010	01010	01011	01011	01010	01010	01110	01101
01010	01011	01011	01100	01100	01100	01100	01101	01111	01111	01110
10000	10000	10000	10000	10000	10000	10000	10000	10000	10000	10000
10000	10001	10001	10001	10000	10001	10001	10001	10001	10001	10011
10011	10010	10010	10010	10011	10010	10010	10010	10010	10110	10101
10011	10011	10011	10100	10101	10101	10101	10101	10111	10111	10110
11100	11100	11100	11011	11010	11010	11010	11011	11011	11010	11001
11100	11100	11101	11101	11100	11100	11101	11100	11100	11011	11010
11111	11111	11110	11110	11111	11111	11110	11110	11101	11100	11100
11111	11111	11111	11111	11111	11111	11111	11111	11110	11101	11111

Attention, therefore, must be paid on two problems, i.e., (a) What is the statistical meaning of such an isomorphism of two arrays? and (b) Which design is the best preferable among those 11 representatives?

The former provides us a problem such that in what sense two isomorphic arrays are statistically the same? The latter provides us a problem such that, although every representative array yields the same orthogonal design of resolution III (see [1] and [2]), is there any difference among those 11 arrays?

1. THE FRONTAGE AND THE PROFILE OF A DESIGN

Consider a design T composed of n assemblies with respect to m factors and let $\boldsymbol{y}(T)$ be the observation vector subject to a linear model

$$\boldsymbol{y}(T) = E(T)\boldsymbol{\theta} + \boldsymbol{e},$$

where E is a known design matrix, $\boldsymbol{\theta}$ is a parameter vector and $\boldsymbol{e}$ is an error vector. Usually, $\boldsymbol{\theta}$ is divided into three parts $\boldsymbol{\theta}_1$, $\boldsymbol{\theta}_2$, and $\boldsymbol{\theta}_3$, where $\boldsymbol{\theta}_1$ is a vector of parameters to be estimated, $\boldsymbol{\theta}_2$ is that of unknown parameters not of interest and not assumed to be zero, and $\boldsymbol{\theta}_3$ is that of known parameters (which without loss of generality can be taken to be zero).

The observation vector can be expressed, in this case, as:

$$\boldsymbol{y}(T) = E_1(T)\boldsymbol{\theta}_1 + E_2(T)\boldsymbol{\theta}_2 + \boldsymbol{e},$$

and the information matrix $M(T)$ can be expressed in the partitioned form as:

$$M(T) = \begin{bmatrix} M_{11} & M_{12} \\ M_{21} & M_{22} \end{bmatrix},$$

where $M_{ij} = E_i(T)^t E_j(T)$ for $i, j = 1$ or 2.

Definition 1.1. The submatrix M_{11} of the information matrix $M(T)$ related directly to the estimation of $\boldsymbol{\theta}_1$ is said to be the *frontage* of the design T.

Definition 1.2. The remaining part $\begin{bmatrix} 0 & M_{12} \\ M_{21} & M_{22} \end{bmatrix}$ of the information matrix $M(T)$ is said to be the *profile* of the design T.

Suppose that the frontage M_{11} is nonsingular and that one can ignore the presence of $\boldsymbol{\theta}_2$ or the profile, then the BLUE of $\boldsymbol{\theta}_1$ is given by $\hat{\boldsymbol{\theta}}_1 = M_{11}^{-1} E_1(T)^t \boldsymbol{y}(T)$. If one cannot ignore the presence of $\boldsymbol{\theta}_2$, then, since $\text{Exp}[\hat{\boldsymbol{\theta}}_1] = \boldsymbol{\theta}_1 + M_{11}^{-1} M_{12} \boldsymbol{\theta}_2$, the estimate is no longer unbiased. The coefficient matrix $A = M_{11}^{-1} M_{12}$ of $\boldsymbol{\theta}_2$ is called the alias matrix of the design T relative to $\boldsymbol{\theta}_1$ and $\boldsymbol{\theta}_2$.

Customarily, serious attention has been focused on the construction of the frontage of a design. Namely, on the construction of orthogonal designs derived from orthogonal arrays, balanced designs derived from balanced arrays, and so on.

Although the frontage M_{11} of a design T constitutes an important part of $M(T)$ in estimating $\boldsymbol{\theta}_1$, the structure of its profile cannot be neglected in as much as the presence of $\boldsymbol{\theta}_2$ is not ignorable.

Those results given in Table 1 provide us 11 nonisomorphic orthogonal designs of resolution 3 with respect to five factors each at two levels. Frontages and profiles of those 11 designs can be seen in the *Appendix*. Since they are equivalent in their frontages in estimating the general mean and the main effects of five factors, our attention should be paid on the structure of their profiles in order to determine in what situation which design is the best preferable.

2. 2^m FACTORIAL DESIGNS

Consider a 2^m factorial experiment with m factors, $F(1), F(2), \ldots,$ and $F(m)$, each at two levels 0 and 1. Let $\theta\{\phi\}$; $\theta\{t_1\}$; and, in general, $\theta\{t_1, t_2, \ldots, t_k\}$, be various factorial effects called the general mean; the main effect of the factor $F(t_1)$; and the k-factor interaction ($k \geq 2$) of the factors $F(t_1), F(t_2), \ldots,$ and $F(t_k)$, for $2 \leq k \leq m$; respectively.

Suppose T and $\boldsymbol{y}(T)$ be a 2^m factorial design composed of n assemblies and the corresponding vector of observations, respectively, i.e.,

$$T = \begin{bmatrix} j_1^{(1)}, j_2^{(1)}, \ldots, j_m^{(1)} \\ \vdots \\ j_1^{(\alpha)}, j_2^{(\alpha)}, \ldots, j_m^{(\alpha)} \\ \vdots \\ j_1^{(n)}, j_2^{(n)}, \ldots, j_m^{(n)} \end{bmatrix} \text{ and } \boldsymbol{y}(T) = \begin{bmatrix} y(j_1^{(1)}, j_2^{(1)}, \ldots, j_m^{(1)}) \\ \vdots \\ y(j_1^{(\alpha)}, j_2^{(\alpha)}, \ldots, j_m^{(\alpha)}) \\ \vdots \\ y(j_1^{(n)}, j_2^{(n)}, \ldots, j_m^{(n)}) \end{bmatrix}.$$

Under an a priori assumption that $p+1$ $(1 \leq p \leq m)$ or more higher order interactions can be assumed to be zero, the observation vector can be expressed as

$$\boldsymbol{y}(T) = E(p, T)\boldsymbol{\theta}(p) + \boldsymbol{e}$$

in terms of the design matrix $E(p, T)$, the vector $\boldsymbol{\theta}(p)$ of various effects up to p-factor interactions, i.e.,

$$\begin{aligned} \boldsymbol{\theta}(p)^t \;=\; & \big(\theta\{\phi\}; \theta\{1\}, \ldots, \theta\{m\}; \theta\{1,2\}, \ldots, \theta\{m-1, m\}; \\ & \ldots; \theta\{1, 2, \ldots, p\}, \ldots, \theta\{m-p+1, m-p+2, \ldots, m\}\big), \end{aligned}$$

and the error vector $\boldsymbol{e}$. The αth row of $E(p,T)$ is expressed as:

$$\big(1; d(j_1^{(\alpha)}), \ldots, d(j_m^{(\alpha)}); d(j_1^{(\alpha)})d(j_2^{(\alpha)}), \ldots, d(j_{m-1}^{(\alpha)})d(j_m^{(\alpha)}); \\ \ldots ; d(j_1^{(\alpha)})d(j_2^{(\alpha)}) \cdots d(j_p^{(\alpha)}), \ldots, d(j_{m-p+1}^{(\alpha)})d(j_{m-p+2}^{(\alpha)}) \cdots d(j_m^{(\alpha)})\big),$$

where $d(j) = -1$ or 1 according as $j = 0$ or 1.

The normal equation for estimating $\boldsymbol{\theta}(\mathrm{p})$ is given by

$$M(p,T)\hat{\boldsymbol{\theta}}(p) = E(p,T)^t \boldsymbol{y}(T),$$

where $M(p,T) = E(p,T)^t E(p,T)$ is the information matrix of the design T relative to $\boldsymbol{\theta}(p)$.

Let $\varepsilon(U,V)$ be the $\theta\{U\}$-row, $\theta\{V\}$-column element of the information matrix $M(p,T)$ corresponding to the row and column indexed by $\theta\{U\}$ and $\theta\{V\}$, respectively, where U and V are the subsets of $\Omega = \{1, 2, \ldots, m\}$. Then,

$$\varepsilon(U,V) = \sum_{\alpha=1}^{n} \prod_{t \in U} d(j_t^{(\alpha)}) \prod_{t' \in V} d(j_{t'}^{(\alpha)}).$$

Since $d(j) = \pm 1$, this element of $M(p,T)$ can be reduced to a function $\gamma\{K\} = \sum_{\alpha=1}^{n} \prod_{i \in K} d(j_i^{(\alpha)})$ depending on T through those columns belonging to the symmetric difference $K = U \triangle V$ of two subsets U and V.

3. CHARACTERISTIC VECTOR OF THE INFORMATION MATRIX

Let $\boldsymbol{\gamma}(T)$ be a vector composed of the first column elements of the information matrix $M(m,T)$, i.e.,

$$\boldsymbol{\gamma}(T) = \begin{bmatrix} \boldsymbol{\gamma}_\phi(T) \\ \boldsymbol{\gamma}_1(T) \\ \vdots \\ \boldsymbol{\gamma}_k(T) \\ \vdots \\ \boldsymbol{\gamma}_m(T) \end{bmatrix}, \text{ where } \boldsymbol{\gamma}_k(T) = \begin{bmatrix} \gamma\{1, 2, \ldots, k\} \\ \vdots \\ \gamma\{i_1, i_2, \ldots, i_k\} \\ \vdots \\ \gamma\{m-k+1, m-k+2, \ldots, m\} \end{bmatrix}$$

is the $(k+1)$st $\binom{m}{k}$ dimensional component vector of $\boldsymbol{\gamma}(T)$. Hereafter $\binom{n}{r}$ denotes a binomial coefficient with the usual convention.

Such a 2^m dimensional vector $\boldsymbol{\gamma}(T)$ can be called the *characteristic vector* of an information matrix $M(m,T)$ since it determines $M(m,T)$ completely.

For every subset K of Ω with cardinality k, the frequency of the appearance of $\gamma\{K\}$ in $M(m,T)$, in (u,v) block of $M(m,T)$ corresponding to u-factor and v-factor interactions, and in $M(p,T)$ for $p \le m$, are given in the following:

Lemma 3.1. (a) *The frequency of the appearance of $\gamma\{K\}$ in $M(m,T)$ is 2^m irrespective of the argument subset K of Ω.*

(b) *The frequency of the appearance of $\gamma\{K\}$ in the (u,v) block of $M(m,T)$ is given by,*

$$\psi(k : u, v) = \begin{cases} \binom{m-k}{\frac{u+v-k}{2}} \binom{k}{\frac{k-|v-u|}{2}} & \text{if } u+v-k \text{ is even}, \\ 0 & \text{if } u+v-k \text{ is odd}, \end{cases}$$

where k is the cardinality of the subset K.

This implies the frequency is a function of u, v and k only. It is dependent on the argument set K only through its cardinality and it satisfies

$$\psi(k : m-u, m-v) = \psi(k : u, v).$$

(c) *The frequency of the appearance of $\gamma\{K\}$ in $M(p,T)$ for $1 \leq p \leq m$ is, therefore, given by*

$$\sum_{u=0}^{p}\sum_{v=0}^{p} \psi(k : u, v) \text{ for } 0 \leq k \leq \min(2p, m),$$

and zero otherwise.

Proof. (a) For every pair of subsets U and K of Ω, there exists a unique subset V of Ω such that the symmetric difference $U \Delta V$ of U and V is equal to K. Thus, for any subset K, there exists one and only one $\gamma\{K\}$ in every row of $M(m,T)$. The frequency of the appearance of $\gamma\{K\}$ in $M(m,T)$ is, therefore, equal to 2^m irrespective of the argument set K.

(b) The cardinality of the symmetric difference K of two subsets U and V of Ω with cardinality u and v is a nonnegative integer k if and only if $u+v-k$ is even. In such case the frequency of the appearance of $\gamma\{K\}$ in (u,v) block of $M(m,T)$ is, therefore, given by $\binom{m-k}{\frac{u+v-k}{2}}\binom{k}{\frac{k-|v-u|}{2}}$.

(c) Obvious.

Let $\boldsymbol{\gamma}^*(T)$ be a vector of $\gamma\{K\}$'s obtained by rearranging 2^m components of the characteristic vector $\boldsymbol{\gamma}(T)$ of the information matrix $M(m,T)$ of the design T in the binary order of argument sets and let $\boldsymbol{\nu}(T)$ be the *modular vector* of the array (design) T obtained by arranging 2^m components $\nu(j_1, j_2, \ldots, j_m)$'s representing respectively the frequencies of assemblies $(j_1, j_2, \ldots, j_m)$'s in T in the binary order of its argument binary vectors, i.e.,

$$\boldsymbol{\gamma}^*(T) = \begin{bmatrix} \gamma\{\phi\} \\ \gamma\{m\} \\ \vdots \\ \gamma\{i_1, i_2, \ldots, i_k\} \\ \vdots \\ \gamma\{1, 2, \ldots, m\} \end{bmatrix}, \text{ and } \boldsymbol{\nu}(T) = \begin{bmatrix} \nu(0,0,\ldots,0) \\ \nu(0,0,\ldots,1) \\ \vdots \\ \nu(j_1, j_2, \ldots, j_m) \\ \vdots \\ \nu(1,1,\ldots,1) \end{bmatrix}.$$

The modular vector $\boldsymbol{\nu}(T)$ is, of course, a faithful representation of the array (design) T.

Lemma 3.2. *Two vectors $\boldsymbol{\gamma}^*(T)$ and $\boldsymbol{\nu}(T)$ are linked with each other by the following formulas, i.e.,*

$$\boldsymbol{\gamma}^*(T) = D_{(m)}\boldsymbol{\nu}(T) \text{ and } \boldsymbol{\nu}(T) = \frac{1}{2^m} D_{(m)}^t \boldsymbol{\gamma}^*(T),$$

where $D_{(m)}$ is an m times Kronecker product, $D \otimes D \otimes \cdots \otimes D$, of the matrix $D = \begin{bmatrix} 1 & 1 \\ -1 & 1 \end{bmatrix}$ (see Theorem 3.2 of [6]).

4. SYMBOL OR LEVEL PERMUTATION WITHIN COLUMNS

Let $\boldsymbol{\sigma}$ be the group of symbol permutation within m factors (columns) of the design T and let $\boldsymbol{\sigma}_i$ be the symbol permutation applied on an ith column. Then the modular vector of the array $\sigma_i T$ obtained by applying σ_i is given by

$$\boldsymbol{\nu}(\sigma_i T) = (I_2 \otimes I_2 \otimes \cdots \otimes \begin{bmatrix} 0 & 1 \\ 1 & 0 \end{bmatrix} \otimes \cdots \otimes I_2)\boldsymbol{\nu}(T),$$

and $\boldsymbol{\gamma}^*(\sigma_i T)$ of the transformed information matrix is given by

$$\begin{aligned}\boldsymbol{\gamma}^*(\sigma_i T) &= (D \otimes D \otimes \cdots \otimes D)(I_2 \otimes I_2 \otimes \cdots \otimes \begin{bmatrix} 0 & 1 \\ 1 & 0 \end{bmatrix} \otimes \cdots \otimes I_2)\boldsymbol{\nu}(T) \\ &= (I_2 \otimes I_2 \otimes \cdots \otimes \begin{bmatrix} 1 & 0 \\ 0 & -1 \end{bmatrix} \otimes \cdots \otimes I_2)(D \otimes D \otimes \cdots \otimes D)\boldsymbol{\nu}(T) \\ &= (I_2 \otimes I_2 \otimes \cdots \otimes \begin{bmatrix} 1 & 0 \\ 0 & -1 \end{bmatrix} \otimes \cdots \otimes I_2)\boldsymbol{\gamma}^*(T).\end{aligned}$$

This implies a complementation occurs on every $\gamma\{K\}$ of K including $\{i\}$ by the symbol permutations σ_i.

Let $|\boldsymbol{\gamma}(T)|$ be a vector of $|\gamma\{K\}|$'s, $K \in 2^{\Omega}$. Then,

Lemma 4.1. *The vector $|\boldsymbol{\gamma}(T)|$ is invariant with respect to the symbol (level) permutation within columns (factors) of a design T.*

Lemma 4.2. *The set of vectors related to the characteristic vector $\boldsymbol{\gamma}(T)$ of information matrix $M(m,T)$ of a design T defined by*

$$[\boldsymbol{\gamma}(T)] = \{\boldsymbol{\gamma}(\sigma T)|^{\forall}\sigma \in \boldsymbol{\sigma}\}$$

is a maximal invariant function with respect to the group of symbol (or level) permutation σ within factors.

Proof. Obviously, $[\boldsymbol{\gamma}(T)]$ is invariant with respect to the group of symbol permutation $\boldsymbol{\sigma}$.

Suppose $[\boldsymbol{\gamma}(T_1)] = \{\boldsymbol{\gamma}(\sigma T_1)|^{\forall}\sigma \in \boldsymbol{\sigma}\} = [\boldsymbol{\gamma}(T_2)] = \{\boldsymbol{\gamma}(\sigma T_2)|^{\forall}\sigma \in \boldsymbol{\sigma}\}$, then there necessarily exists a permutation $\sigma_{12} \in \boldsymbol{\sigma}$ which satisfy $\boldsymbol{\gamma}(\sigma_{12}T_1) = \boldsymbol{\gamma}(T_2)$. This implies $\sigma_{12}T_1 = T_2$. The set of vectors $[\boldsymbol{\gamma}(T)]$ is, therefore, maximal invariant.

5. COLUMN PERMUTATION OR PERMUTATION OF FACTORS

Let

$$\boldsymbol{\tau} \ni \tau = \begin{pmatrix} 1 & 2 & \cdots & m \\ \tau(1) & \tau(2) & \cdots & \tau(m) \end{pmatrix}$$

be the group of column permutation among m factors (columns) of a design T.

A permutation τ transforms the characteristic vector $\boldsymbol{\gamma}(T)$ of the information matrix $M(m,T)$ of a design T to $\boldsymbol{\gamma}(\tau T)$ and its every component vector $\boldsymbol{\gamma}_k(T)$ to $\boldsymbol{\gamma}_k(\tau T) = P_k(\tau)\boldsymbol{\gamma}_k(T)$ for every k, where $P_k(\tau)$ is an $\binom{m}{k} \times \binom{m}{k}$ permutation matrix induced by τ over Ω_k , the family of all subsets of Ω having the same cardinality k. The following two lemmas are immediate:

Lemma 5.1. *The frequency distribution of $\gamma\{K\}$'s, $K \in \Omega_k$, over the (u, v) block of information matrix $M(m, T)$ is invariant with respect to the column permutation $\boldsymbol{\tau}$ of the design T for every u, v and k.*

In particular;

Lemma 5.2. *The frequency of nonzero $\gamma\{K\}$'s, $K \in \Omega_k$, over the (u, v) block of information matrix $M(m, T)$ of a design T is invariant with respect to the column permutation $\boldsymbol{\tau}$ for every u, v and k.*

Lemma 5.3. *The set of vectors related to the characteristic vector $\boldsymbol{\gamma}(T)$ of the information matrix $M(m, T)$ of a design T defined by*

$$[\boldsymbol{\gamma}(T)] = \{\boldsymbol{\gamma}(\tau T) | ^\forall \tau \in \boldsymbol{\tau}\}$$

is a maximal invariant with respect to the group of column permutation $\boldsymbol{\tau}$.

Proof. Obviously, $[\boldsymbol{\gamma}(T)]$ is invariant with respect to the group of column permutation $\boldsymbol{\tau}$.
Suppose $[\boldsymbol{\gamma}(T_1)] = \{\boldsymbol{\gamma}(\tau T_1)|^\forall \tau \in \boldsymbol{\tau}\} = [\boldsymbol{\gamma}(T_2)] = \{\boldsymbol{\gamma}(\tau T_2)|^\forall \tau \in \boldsymbol{\tau}\}$, then there exists a permutation $\tau_{12} \in \boldsymbol{\tau}$ satisfying $\boldsymbol{\gamma}(\tau_{12} T_1) = \boldsymbol{\gamma}(T_2)$. This implies $\tau_{12} T_1 = T_2$. The invariant set of vectors $[\boldsymbol{\gamma}(T)]$ is, therefore, maximal.

6. INVARIANTS WITH RESPECT TO THE SC-PERMUTATION

The symbol and column permutation (SC-permutation) group $\boldsymbol{\sigma} \times \boldsymbol{\tau}$ applied on a fractional 2^m factorial design T induces a transformation group on its information matrix $M(m, T)$ or on its characteristic vector $\boldsymbol{\gamma}(T)$.

Summarizing those have been investigated in the preceding sections, we have the following theorems:

Theorem 6.1. *The frequency distribution of $|\gamma\{K\}|$'s, $K \in \Omega_k$, over the (u, v) block of information matrix $M(m, T)$ of a design T is invariant with respect to the SC-permutation $\boldsymbol{\sigma} \times \boldsymbol{\tau}$ for every u, v and k.*

Theorem 6.2. *The frequency of nonzero $\gamma\{K\}$'s, $K \in \Omega_k$, over the (u, v) block of information matrix $M(m, T)$ of a design T is invariant with respect to the SC-permutation $\boldsymbol{\sigma} \times \boldsymbol{\tau}$ for every u, v and k.*

Theorem 6.3. *The orbit of $\boldsymbol{\gamma}(T)$ with respect to the SC-permutation of a design T defined by:*

$$[[\boldsymbol{\gamma}(T)]] = \{\boldsymbol{\gamma}(\sigma \tau T)|^\forall \sigma \in \boldsymbol{\sigma}, ^\forall \tau \in \boldsymbol{\tau}\}$$

is a maximal invariant function with respect to the group of SC-permutation.

7. INVARIANTS OF THE PROFILE OF 2^m FACTORIAL DESIGNS

Let z_{uv} be the frequency of nonzero $\boldsymbol{\gamma}\{K\}$'s sprinkled over the (u,v) block of the information matrix $M(p,T)$ of a 2^m factorial design T and consider a $(p+1)\times(p+1)$ matrix defined by

$$Z(p,T) = ||z_{uv}||,\ \ 0 \le u, v \le p(\le m).$$

Then, from Theorem 5.2, $Z(p,T)$ is invariant with respect to the group of SC-permutation of the design T.

Since every diagonal element of $M(p,T)$ is equal to $\gamma\{\phi\} = n \neq 0$ irrespective of the design T, the *spot matrix* $S(p,T)$ defined by the number of nonzero off diagonal elements or that of, say, spots, i.e.,

$$S(p,T) = ||s_{uv}||,\ \ s_{uv} = z_{uv} - \delta_{uv}\binom{m}{u},\ \ 0 \le u, v \le p\ (1 \le p \le m),$$

is also invariant with respect to the SC-permutation. Moreover, this spot matrix $S(p,T)$ can be considered to represent a kind of departure of a design T from an *ideal* orthogonal design for which the spot matrix might be the null matrix.

Table 2. $S(m,T)$ of 11 isomorphic classes of 2-OA$(t=2, m=5, \lambda=4)$

[1]						[2]						[3]						[4]					
0	0	0	2	1	0	0	0	0	3	2	0	0	0	0	1	0	0	0	0	0	3	0	1
0	0	6	4	4	1	0	0	9	8	6	2	0	0	3	0	2	0	0	0	9	0	11	0
0	6	6	12	4	2	0	9	12	18	8	3	0	3	0	6	0	1	0	9	0	28	0	3
2	4	12	6	6	0	3	8	18	12	9	0	1	0	6	0	3	0	3	0	28	0	9	0
1	4	4	6	0	0	2	6	8	9	0	0	0	2	0	3	0	0	0	11	0	9	0	0
0	1	2	0	0	0	0	2	3	0	0	0	0	0	1	0	0	0	1	0	3	0	0	0

[5]						[6]						[7]						[8]					
0	0	0	4	1	0	0	0	0	4	2	0	0	0	0	4	0	0	0	0	0	2	2	0
0	0	12	4	8	1	0	0	12	8	8	2	0	0	12	0	8	0	0	0	6	8	4	2
0	12	6	24	4	4	0	12	12	24	8	4	0	12	0	24	0	4	0	6	12	12	8	2
4	4	24	6	12	0	4	8	24	12	12	0	4	0	24	0	12	0	2	8	12	12	6	0
1	8	4	12	0	0	2	8	8	12	0	0	0	8	0	12	0	0	2	4	8	6	0	0
0	1	4	0	0	0	0	2	4	0	0	0	0	0	4	0	0	0	0	2	2	0	0	0

[9]						[10]						[11]					
0	0	0	1	2	1	0	0	0	0	1	0	0	0	0	0	0	1
0	0	3	8	7	2	0	0	0	4	0	1	0	0	0	0	5	0
0	3	12	16	8	1	0	0	6	0	4	0	0	0	0	10	0	0
1	8	16	12	3	0	0	4	0	6	0	0	0	0	10	0	0	0
2	7	8	3	0	0	1	0	4	0	0	0	0	5	0	0	0	0
1	2	1	0	0	0	0	1	0	0	0	0	1	0	0	0	0	0

Those $S(m,T)$ including, of course, all $S(p,T)$ of the 11 representatives of 2-OA$(t = 2, m = 5, \lambda = 4)$ can be seen in Table 2. It can be verified from Table 2 that the above invariant is maximal in that there is no pair of representatives where both $S(p,T)$ are coincide each other for any p satisfying $2 \le p \le 5$. Those invariants, therefore, characterize the class of orthogonal arrays of strength 2 isomorphic to each other with respect to the SC-permutation.

Since all of those designs considered here belong to the class of all 2-OA($t = 2, m = 5, \lambda = 4$), every $S(1,T)$ is equal to zero matrix. Thus, every $S(p,T)$ for $p \geq 2$ is a function of the profile of the design T. In order to compare those 11 classes through this maximal invariant $S(p,T)$, several characteristics or norms such as:

$$||S(p,T)||_1 = \sum_{u=0}^{p}\sum_{v=0}^{p} |s_{uv}|, \quad ||S(p,T)||_2 = \left\{\sum_{u=0}^{p}\sum_{v=0}^{p} s_{uv}^2\right\}^{\frac{1}{2}},$$

etc., may be useful. The smaller the norm may be the better.

Table 3 shows the computational results of those norms with respect to the 11 isomorphic classes.

Table 3. Norms of 11 isomorphic classes of 2-OA($t = 2, m = 5, \lambda = 4$)

$||S(p,T)||_1$

p	[1]	[2]	[3]	[4]	[5]	[6]	[7]	[8]	[9]	[10]	[11]
2	18	30	6	18	30	36	24	24	18	6	**0**
3	60	100	**20**	80	100	120	80	80	80	**20**	**20**
4	90	150	**30**	120	150	180	120	120	120	**30**	**30**
5	96	160	**32**	128	160	192	128	128	128	**32**	**32**

$||S(p,T)||_2$

p	[1]	[2]	[3]	[4]	[5]	[6]	[7]	[8]	[9]	[10]	[11]
2	10.4	17.5	4.2	12.7	18.0	20.8	17.0	14.7	12.7	6.0	**0.0**
3	21.7	35.3	**9.6**	41.8	39.7	43.5	38.4	28.0	30.8	10.2	14.1
4	24.7	40.2	**10.9**	46.4	45.0	49.4	43.5	32.0	34.6	11.8	15.8
5	24.9	40.5	**11.0**	46.6	45.4	49.8	43.8	32.3	34.8	11.8	15.9

Note: Boldfaced indicates the smallest norm in each case.

References

[1] G.E.P. Box and J.S. Hunter, The 2^{k-p} fractional factorial designs. I. *Technometrics*, **3**, 311-351 (1961a).

[2] G.E.P. Box and J.S. Hunter, The 2^{k-p} fractional factorial designs. II. *Technometrics* , **3**, 449-458 (1961b).

[3] G.E.P. Box and K.B. Wilson, On the experimental attainment of optimum conditions. *J. Roy. Statist. Soc., B*, **13**, 1-45 (1951).

[4] A. Hedayat, B.L. Raktoe and W.T. Federer, On a measure of aliasing due to fitting an incomplete model. *Ann. Statist.*, **12**, 650-660 (1974).

[5] T. Namikawa, Y. Fujii and S. Yamamoto, Computational study on the classification of two-symbol orthogonal arrays of strength t, $m = t + e$ constraints for $e \leq 3$. *SUT J. Math.*, **25**, 179-195 (1989).

[6] S. Yamamoto, T. Shirakura and M. Kuwada, Balanced arrays of strength 2ℓ and balanced fractional 2^m factorial designs. *Ann. Inst. Statist. Math.* **27**, 143-157 (1975).

APPENDIX. Information matrices of 11 representative designs

[1]

θ{U} \ θ{V}	φ	1	2	3	4	5	12	13	14	15	23	24	25	34	35	45	123	124	125	134	135	145	234	235	245	345	1234	1235	1245	1345	2345	12345
φ	16	·	·	·	·	·	·	·	·	·	·	·	·	·	·	·	16	·	·	·	·	16	·	·	·	·	·	·	·	·	16	·
1	·	16	·	·	·	·	·	·	·	·	16	·	·	·	·	16	·	·	·	·	·	·	·	·	·	·	·	·	·	·	·	16
2	·	·	16	·	·	·	·	16	·	·	·	·	·	·	·	·	·	·	·	·	·	·	·	·	·	16	·	·	16	·	·	·
3	·	·	·	16	·	·	16	·	·	·	·	·	·	·	·	·	·	·	·	·	·	·	·	·	16	·	·	·	·	16	·	·
4	·	·	·	·	16	·	·	·	·	16	·	·	·	·	·	·	·	·	·	·	·	·	·	16	·	·	16	·	·	·	·	·
5	·	·	·	·	·	16	·	·	16	·	·	·	·	·	·	·	·	·	·	·	·	·	16	·	·	·	·	16	·	·	·	·
12	·	·	·	16	·	·	16	·	·	·	·	·	·	·	·	·	·	·	·	·	·	·	·	·	16	·	·	·	·	16	·	·
13	·	·	16	·	·	·	·	16	·	·	·	·	·	·	·	·	·	·	·	·	·	·	·	·	·	16	·	·	16	·	·	·
14	·	·	·	·	·	16	·	·	16	·	·	·	·	·	·	·	·	·	·	·	·	·	16	·	·	·	·	16	·	·	·	·
15	·	·	·	·	16	·	·	·	·	16	·	·	·	·	·	·	·	·	·	·	·	·	·	16	·	·	16	·	·	·	·	·
23	·	16	·	·	·	·	·	·	·	·	16	·	·	·	·	16	·	·	·	·	·	·	·	·	·	·	·	·	·	·	·	16
24	·	·	·	·	·	·	·	·	·	·	·	16	·	·	16	·	·	·	16	16	·	·	·	·	·	·	·	·	·	·	·	·
25	·	·	·	·	·	·	·	·	·	·	·	·	16	16	·	·	·	16	·	·	16	·	·	·	·	·	·	·	·	·	·	·
34	·	·	·	·	·	·	·	·	·	·	·	·	16	16	·	·	·	16	·	·	16	·	·	·	·	·	·	·	·	·	·	·
35	·	·	·	·	·	·	·	·	·	·	·	16	·	·	16	·	·	·	16	16	·	·	·	·	·	·	·	·	·	·	·	·
45	·	16	·	·	·	·	·	·	·	·	16	·	·	·	·	16	·	·	·	·	·	·	·	·	·	·	·	·	·	·	·	16
123	16	·	·	·	·	·	·	·	·	·	·	·	·	·	·	·	16	·	·	·	·	16	·	·	·	·	·	·	·	·	16	·
124	·	·	·	·	·	·	·	·	·	·	·	·	16	16	·	·	·	16	·	·	16	·	·	·	·	·	·	·	·	·	·	·
125	·	·	·	·	·	·	·	·	·	·	·	16	·	·	16	·	·	·	16	16	·	·	·	·	·	·	·	·	·	·	·	·
134	·	·	·	·	·	·	·	·	·	·	·	16	·	·	16	·	·	·	16	16	·	·	·	·	·	·	·	·	·	·	·	·
135	·	·	·	·	·	·	·	·	·	·	·	·	16	16	·	·	·	16	·	·	16	·	·	·	·	·	·	·	·	·	·	·
145	16	·	·	·	·	·	·	·	·	·	·	·	·	·	·	·	16	·	·	·	·	16	·	·	·	·	·	·	·	·	16	·
234	·	·	·	·	·	16	·	·	16	·	·	·	·	·	·	·	·	·	·	·	·	·	16	·	·	·	·	16	·	·	·	·
235	·	·	·	·	16	·	·	·	·	16	·	·	·	·	·	·	·	·	·	·	·	·	·	16	·	·	16	·	·	·	·	·
245	·	·	·	16	·	·	16	·	·	·	·	·	·	·	·	·	·	·	·	·	·	·	·	·	16	·	·	·	·	16	·	·
345	·	·	16	·	·	·	·	16	·	·	·	·	·	·	·	·	·	·	·	·	·	·	·	·	·	16	·	·	16	·	·	·
1234	·	·	·	·	16	·	·	·	·	16	·	·	·	·	·	·	·	·	·	·	·	·	·	16	·	·	16	·	·	·	·	·
1235	·	·	·	·	·	16	·	·	16	·	·	·	·	·	·	·	·	·	·	·	·	·	16	·	·	·	·	16	·	·	·	·
1245	·	·	16	·	·	·	·	16	·	·	·	·	·	·	·	·	·	·	·	·	·	·	·	·	·	16	·	·	16	·	·	·
1345	·	·	·	16	·	·	16	·	·	·	·	·	·	·	·	·	·	·	·	·	·	·	·	·	16	·	·	·	·	16	·	·
2345	16	·	·	·	·	·	·	·	·	·	·	·	·	·	·	·	16	·	·	·	·	16	·	·	·	·	·	·	·	·	16	·
12345	·	16	·	·	·	·	·	·	·	·	16	·	·	·	·	16	·	·	·	·	·	·	·	·	·	·	·	·	·	·	·	16

[2]

θ{U} \ θ{V}	φ	1	2	3	4	5	12	13	14	15	23	24	25	34	35	45	123	124	125	134	135	145	234	235	245	345	1234	1235	1245	1345	2345	12345
φ	16	·	·	·	·	·	·	·	·	·	·	·	·	·	·	·	16	·	·	·	·	8	·	·	8	·	·	·	·	8	8	·
1	·	16	·	·	·	·	·	·	·	·	16	·	·	·	·	8	·	·	·	·	·	·	·	·	·	8	·	·	8	·	·	8
2	·	·	16	·	·	·	·	16	·	·	·	·	·	·	·	8	·	·	·	·	·	·	·	·	·	8	·	·	8	·	·	8
3	·	·	·	16	·	·	16	·	·	·	·	·	·	·	·	·	·	·	·	·	·	8	·	·	8	·	·	·	·	8	8	·
4	·	·	·	·	16	·	·	·	·	8	·	·	8	·	·	·	·	·	·	·	8	·	·	8	·	·	16	·	·	·	·	·
5	·	·	·	·	·	16	·	·	8	·	·	8	·	·	·	·	·	·	·	8	·	·	8	·	·	·	·	16	·	·	·	·
12	·	·	·	16	·	·	16	·	·	·	·	·	·	·	·	·	·	·	·	·	·	8	·	·	8	·	·	·	·	8	8	·
13	·	·	16	·	·	·	·	16	·	·	·	·	·	·	·	8	·	·	·	·	·	·	·	·	·	8	·	·	8	·	·	8
14	·	·	·	·	·	8	·	·	16	·	·	·	·	·	8	·	·	·	8	·	·	·	16	·	·	·	·	8	·	·	·	·
15	·	·	·	·	8	·	·	·	·	16	·	·	·	8	·	·	·	8	·	·	·	·	·	16	·	·	8	·	·	·	·	·
23	·	16	·	·	·	·	·	·	·	·	16	·	·	·	·	8	·	·	·	·	·	·	·	·	·	8	·	·	8	·	·	8
24	·	·	·	·	·	8	·	·	·	·	·	16	·	·	8	·	·	·	8	16	·	·	·	·	·	·	·	8	·	·	·	·
25	·	·	·	·	8	·	·	·	·	·	·	·	16	8	·	·	·	8	·	·	16	·	·	·	·	·	8	·	·	·	·	·
34	·	·	·	·	·	·	·	·	·	8	·	·	8	16	·	·	·	16	·	·	8	·	·	8	·	·	·	·	·	·	·	·
35	·	·	·	·	·	·	·	·	8	·	·	8	·	·	16	·	·	·	16	8	·	·	8	·	·	·	·	·	·	·	·	·
45	·	8	8	·	·	·	·	8	·	·	8	·	·	·	·	16	·	·	·	·	·	·	·	·	·	·	·	·	·	·	·	16
123	16	·	·	·	·	·	·	·	·	·	·	·	·	·	·	·	16	·	·	·	·	8	·	·	8	·	·	·	·	8	8	·
124	·	·	·	·	·	·	·	·	·	8	·	·	8	16	·	·	·	16	·	·	8	·	·	8	·	·	·	·	·	·	·	·
125	·	·	·	·	·	·	·	·	8	·	·	8	·	·	16	·	·	·	16	8	·	·	8	·	·	·	·	·	·	·	·	·
134	·	·	·	·	·	8	·	·	·	·	·	16	·	·	8	·	·	·	8	16	·	·	·	·	·	·	·	8	·	·	·	·
135	·	·	·	·	8	·	·	·	·	·	·	·	16	8	·	·	·	8	·	·	16	·	·	·	·	·	8	·	·	·	·	·
145	8	·	·	8	·	·	8	·	·	·	·	·	·	·	·	·	8	·	·	·	·	16	·	·	·	·	·	·	·	·	16	·
234	·	·	·	·	·	8	·	·	16	·	·	·	·	·	8	·	·	·	8	·	·	·	16	·	·	·	·	8	·	·	·	·
235	·	·	·	·	8	·	·	·	·	16	·	·	·	8	·	·	·	8	·	·	·	·	·	16	·	·	8	·	·	·	·	·
245	8	·	·	8	·	·	8	·	·	·	·	·	·	·	·	·	8	·	·	·	·	·	·	·	16	·	·	·	·	16	·	·
345	·	8	8	·	·	·	·	8	·	·	8	·	·	·	·	·	·	·	·	·	·	·	·	·	·	16	·	·	16	·	·	·
1234	·	·	·	·	16	·	·	·	·	8	·	·	8	·	·	·	·	·	·	·	8	·	·	8	·	·	16	·	·	·	·	·
1235	·	·	·	·	·	16	·	·	8	·	·	8	·	·	·	·	·	·	·	8	·	·	8	·	·	·	·	16	·	·	·	·
1245	·	8	8	·	·	·	·	8	·	·	8	·	·	·	·	·	·	·	·	·	·	·	·	·	·	16	·	·	16	·	·	·
1345	8	·	·	8	·	·	8	·	·	·	·	·	·	·	·	·	8	·	·	·	·	·	·	·	16	·	·	·	·	16	·	·
2345	8	·	·	8	·	·	8	·	·	·	·	·	·	·	·	·	8	·	·	·	·	16	·	·	·	·	·	·	·	·	16	·
12345	·	8	8	·	·	·	·	8	·	·	8	·	·	·	·	16	·	·	·	·	·	·	·	·	·	·	·	·	·	·	·	16

APPENDIX. Information matrices of 11 representative designs - continued

[3]

θ{U} \ θ{V}	φ	1	2	3	4	5	12	13	14	15	23	24	25	34	35	45	123	124	125	134	135	145	234	235	245	345	1234	1235	1245	1345	2345	12345
φ	16	·	·	·	·	·	·	·	·	·	·	·	·	·	·	·	16	·	·	·	·	·	·	·	·	·	·	·	·	·	·	·
1	·	16	·	·	·	·	·	·	·	·	16	·	·	·	·	·	·	·	·	·	·	·	·	·	·	·	·	·	·	·	·	·
2	·	·	16	·	·	·	·	16	·	·	·	·	·	·	·	·	·	·	·	·	·	·	·	·	·	·	·	·	·	·	·	·
3	·	·	·	16	·	·	16	·	·	·	·	·	·	·	·	·	·	·	·	·	·	·	·	·	·	·	·	·	·	·	·	·
4	·	·	·	·	16	·	·	·	·	·	·	·	·	·	·	·	·	·	·	·	·	·	·	·	·	·	16	·	·	·	·	·
5	·	·	·	·	·	16	·	·	·	·	·	·	·	·	·	·	·	·	·	·	·	·	·	·	·	·	·	16	·	·	·	·
12	·	·	·	16	·	·	16	·	·	·	·	·	·	·	·	·	·	·	·	·	·	·	·	·	·	·	·	·	·	·	·	·
13	·	·	16	·	·	·	·	16	·	·	·	·	·	·	·	·	·	·	·	·	·	·	·	·	·	·	·	·	·	·	·	·
14	·	·	·	·	·	·	·	·	16	·	·	·	·	·	·	·	·	·	·	·	·	·	16	·	·	·	·	·	·	·	·	·
15	·	·	·	·	·	·	·	·	·	16	·	·	·	·	·	·	·	·	·	·	·	·	·	16	·	·	·	·	·	·	·	·
23	·	16	·	·	·	·	·	·	·	·	16	·	·	·	·	·	·	·	·	·	·	·	·	·	·	·	·	·	·	·	·	·
24	·	·	·	·	·	·	·	·	·	·	·	16	·	·	·	·	·	·	·	16	·	·	·	·	·	·	·	·	·	·	·	·
25	·	·	·	·	·	·	·	·	·	·	·	·	16	·	·	·	·	·	·	·	16	·	·	·	·	·	·	·	·	·	·	·
34	·	·	·	·	·	·	·	·	·	·	·	·	·	16	·	·	·	16	·	·	·	·	·	·	·	·	·	·	·	·	·	·
35	·	·	·	·	·	·	·	·	·	·	·	·	·	·	16	·	·	·	16	·	·	·	·	·	·	·	·	·	·	·	·	·
45	·	·	·	·	·	·	·	·	·	·	·	·	·	·	·	16	·	·	·	·	·	·	·	·	·	·	·	·	·	·	·	16
123	16	·	·	·	·	·	·	·	·	·	·	·	·	·	·	·	16	·	·	·	·	·	·	·	·	·	·	·	·	·	·	·
124	·	·	·	·	·	·	·	·	·	·	·	·	·	16	·	·	·	16	·	·	·	·	·	·	·	·	·	·	·	·	·	·
125	·	·	·	·	·	·	·	·	·	·	·	·	·	·	16	·	·	·	16	·	·	·	·	·	·	·	·	·	·	·	·	·
134	·	·	·	·	·	·	·	·	·	·	·	16	·	·	·	·	·	·	·	16	·	·	·	·	·	·	·	·	·	·	·	·
135	·	·	·	·	·	·	·	·	·	·	·	·	16	·	·	·	·	·	·	·	16	·	·	·	·	·	·	·	·	·	·	·
145	·	·	·	·	·	·	·	·	·	·	·	·	·	·	·	·	·	·	·	·	·	16	·	·	·	·	·	·	·	·	16	·
234	·	·	·	·	·	·	·	·	16	·	·	·	·	·	·	·	·	·	·	·	·	·	16	·	·	·	·	·	·	·	·	·
235	·	·	·	·	·	·	·	·	·	16	·	·	·	·	·	·	·	·	·	·	·	·	·	16	·	·	·	·	·	·	·	·
245	·	·	·	·	·	·	·	·	·	·	·	·	·	·	·	·	·	·	·	·	·	·	·	·	16	·	·	·	·	16	·	·
345	·	·	·	·	·	·	·	·	·	·	·	·	·	·	·	·	·	·	·	·	·	·	·	·	·	16	·	·	16	·	·	·
1234	·	·	·	·	16	·	·	·	·	·	·	·	·	·	·	·	·	·	·	·	·	·	·	·	·	·	16	·	·	·	·	·
1235	·	·	·	·	·	16	·	·	·	·	·	·	·	·	·	·	·	·	·	·	·	·	·	·	·	·	·	16	·	·	·	·
1245	·	·	·	·	·	·	·	·	·	·	·	·	·	·	·	·	·	·	·	·	·	·	·	·	·	16	·	·	16	·	·	·
1345	·	·	·	·	·	·	·	·	·	·	·	·	·	·	·	·	·	·	·	·	·	·	·	·	16	·	·	·	·	16	·	·
2345	·	·	·	·	·	·	·	·	·	·	·	·	·	·	·	·	·	·	·	·	·	16	·	·	·	·	·	·	·	·	16	·
12345	·	·	·	·	·	·	·	·	·	·	·	·	·	·	·	16	·	·	·	·	·	·	·	·	·	·	·	·	·	·	·	16

[4]

θ{U} \ θ{V}	φ	1	2	3	4	5	12	13	14	15	23	24	25	34	35	45	123	124	125	134	135	145	234	235	245	345	1234	1235	1245	1345	2345	12345
φ	16	·	·	·	·	·	·	·	·	·	·	·	·	·	·	·	8	8	8	·	·	·	·	·	·	·	·	·	·	·	·	-8
1	·	16	·	·	·	·	·	·	·	·	8	8	8	·	·	·	·	·	·	·	·	·	·	·	·	·	·	·	·	·	-8	·
2	·	·	16	·	·	·	·	8	8	8	·	·	·	·	·	·	·	·	·	·	·	·	·	·	·	·	·	·	·	-8	·	·
3	·	·	·	16	·	·	8	·	·	·	·	·	·	·	·	·	·	·	·	·	·	·	·	·	·	·	8	8	-8	·	·	·
4	·	·	·	·	16	·	8	·	·	·	·	·	·	·	·	·	·	·	·	·	·	·	·	·	·	·	8	-8	8	·	·	·
5	·	·	·	·	·	16	8	·	·	·	·	·	·	·	·	·	·	·	·	·	·	·	·	·	·	·	-8	8	8	·	·	·
12	·	·	·	8	8	8	16	·	·	·	·	·	·	·	·	·	·	·	·	·	·	·	·	·	·	-8	·	·	·	·	·	·
13	·	·	8	·	·	·	·	16	·	·	·	·	·	·	·	·	·	·	·	·	·	·	8	8	-8	·	·	·	·	·	·	·
14	·	·	8	·	·	·	·	·	16	·	·	·	·	·	·	·	·	·	·	·	·	·	8	-8	8	·	·	·	·	·	·	·
15	·	·	8	·	·	·	·	·	·	16	·	·	·	·	·	·	·	·	·	·	·	·	-8	8	8	·	·	·	·	·	·	·
23	·	8	·	·	·	·	·	·	·	·	16	·	·	·	·	·	·	·	·	8	8	-8	·	·	·	·	·	·	·	·	·	·
24	·	8	·	·	·	·	·	·	·	·	·	16	·	·	·	·	·	·	·	8	-8	8	·	·	·	·	·	·	·	·	·	·
25	·	8	·	·	·	·	·	·	·	·	·	·	16	·	·	·	·	·	·	-8	8	8	·	·	·	·	·	·	·	·	·	·
34	·	·	·	·	·	·	·	·	·	·	·	·	·	16	·	·	8	8	-8	·	·	·	·	·	·	·	·	·	·	·	·	8
35	·	·	·	·	·	·	·	·	·	·	·	·	·	·	16	·	8	-8	8	·	·	·	·	·	·	·	·	·	·	·	·	8
45	·	·	·	·	·	·	·	·	·	·	·	·	·	·	·	16	-8	8	8	·	·	·	·	·	·	·	·	·	·	·	·	8
123	8	·	·	·	·	·	·	·	·	·	·	·	·	8	8	-8	16	·	·	·	·	·	·	·	·	·	·	·	·	·	·	·
124	8	·	·	·	·	·	·	·	·	·	·	·	·	8	-8	8	·	16	·	·	·	·	·	·	·	·	·	·	·	·	·	·
125	8	·	·	·	·	·	·	·	·	·	·	·	·	-8	8	8	·	·	16	·	·	·	·	·	·	·	·	·	·	·	·	·
134	·	·	·	·	·	·	·	·	·	·	8	8	-8	·	·	·	·	·	·	16	·	·	·	·	·	·	·	·	·	·	8	·
135	·	·	·	·	·	·	·	·	·	·	8	-8	8	·	·	·	·	·	·	·	16	·	·	·	·	·	·	·	·	·	8	·
145	·	·	·	·	·	·	·	·	·	·	-8	8	8	·	·	·	·	·	·	·	·	16	·	·	·	·	·	·	·	·	8	·
234	·	·	·	·	·	·	·	8	8	-8	·	·	·	·	·	·	·	·	·	·	·	·	16	·	·	·	·	·	·	8	·	·
235	·	·	·	·	·	·	·	8	-8	8	·	·	·	·	·	·	·	·	·	·	·	·	·	16	·	·	·	·	·	8	·	·
245	·	·	·	·	·	·	·	-8	8	8	·	·	·	·	·	·	·	·	·	·	·	·	·	·	16	·	·	·	·	8	·	·
345	·	·	·	·	·	·	-8	·	·	·	·	·	·	·	·	·	·	·	·	·	·	·	·	·	·	16	8	8	8	·	·	·
1234	·	·	·	8	8	-8	·	·	·	·	·	·	·	·	·	·	·	·	·	·	·	·	·	·	·	8	16	·	·	·	·	·
1235	·	·	·	8	-8	8	·	·	·	·	·	·	·	·	·	·	·	·	·	·	·	·	·	·	·	8	·	16	·	·	·	·
1245	·	·	·	-8	8	8	·	·	·	·	·	·	·	·	·	·	·	·	·	·	·	·	·	·	·	8	·	·	16	·	·	·
1345	·	·	-8	·	·	·	·	·	·	·	·	·	·	·	·	·	·	·	·	·	·	·	8	8	8	·	·	·	·	16	·	·
2345	·	-8	·	·	·	·	·	·	·	·	·	·	·	·	·	·	·	·	·	8	8	8	·	·	·	·	·	·	·	·	16	·
12345	-8	·	·	·	·	·	·	·	·	·	·	·	·	8	8	8	·	·	·	·	·	·	·	·	·	·	·	·	·	·	·	16

APPENDIX. Information matrices of 11 representative designs - continued

[5]

θ{U} \ θ{V}	ϕ	1	2	3	4	5	12	13	14	15	23	24	25	34	35	45	123	124	125	134	135	145	234	235	245	345	1234	1235	1245	1345	2345	12345
ϕ	16	-	-	-	-	-	-	-	-	-	-	-	-	-	-	-	8	8	-	-	8	8	-	-	-	-	-	-	-	-	16	-
1	-	16	-	-	-	-	-	-	-	-	8	8	-	-	8	8	-	-	-	-	-	-	-	-	-	-	-	-	-	-	-	16
2	-	-	16	-	-	-	-	8	8	-	-	-	-	-	-	-	-	-	-	-	-	-	-	-	-	16	-	8	8	-	-	-
3	-	-	-	16	-	-	8	-	-	8	-	-	-	-	-	-	-	-	-	-	-	-	-	-	16	-	8	-	-	8	-	-
4	-	-	-	-	16	-	8	-	-	8	-	-	-	-	-	-	-	-	-	-	-	-	-	16	-	-	8	-	-	8	-	-
5	-	-	-	-	-	16	-	8	8	-	-	-	-	-	-	-	-	-	-	-	-	-	16	-	-	-	-	8	8	-	-	-
12	-	-	-	8	8	-	16	-	-	-	-	-	-	-	-	-	-	-	-	-	-	-	-	8	8	-	-	-	-	16	-	-
13	-	-	8	-	-	8	-	16	-	-	-	-	-	-	-	-	-	-	-	-	-	-	8	-	-	8	-	-	16	-	-	-
14	-	-	8	-	-	8	-	-	16	-	-	-	-	-	-	-	-	-	-	-	-	-	8	-	-	8	-	16	-	-	-	-
15	-	-	-	8	8	-	-	-	-	16	-	-	-	-	-	-	-	-	-	-	-	-	-	8	8	-	16	-	-	-	-	-
23	-	8	-	-	-	-	-	-	-	-	16	-	-	-	-	16	-	-	8	8	-	-	-	-	-	-	-	-	-	-	-	8
24	-	8	-	-	-	-	-	-	-	-	-	16	-	-	16	-	-	-	8	8	-	-	-	-	-	-	-	-	-	-	-	8
25	-	-	-	-	-	-	-	-	-	-	-	-	16	16	-	-	8	8	-	-	8	8	-	-	-	-	-	-	-	-	-	-
34	-	-	-	-	-	-	-	-	-	-	-	-	16	16	-	-	8	8	-	-	8	8	-	-	-	-	-	-	-	-	-	-
35	-	8	-	-	-	-	-	-	-	-	-	16	-	-	16	-	-	-	8	8	-	-	-	-	-	-	-	-	-	-	-	8
45	-	8	-	-	-	-	-	-	-	-	16	-	-	-	-	16	-	-	8	8	-	-	-	-	-	-	-	-	-	-	-	8
123	8	-	-	-	-	-	-	-	-	-	-	-	8	8	-	-	16	-	-	-	-	16	-	-	-	-	-	-	-	-	8	-
124	8	-	-	-	-	-	-	-	-	-	-	-	8	8	-	-	-	16	-	-	16	-	-	-	-	-	-	-	-	-	8	-
125	-	-	-	-	-	-	-	-	-	-	8	8	-	-	8	8	-	-	16	16	-	-	-	-	-	-	-	-	-	-	-	-
134	-	-	-	-	-	-	-	-	-	-	8	8	-	-	8	8	-	-	16	16	-	-	-	-	-	-	-	-	-	-	-	-
135	8	-	-	-	-	-	-	-	-	-	-	-	8	8	-	-	-	16	-	-	16	-	-	-	-	-	-	-	-	-	8	-
145	8	-	-	-	-	-	-	-	-	-	-	-	8	8	-	-	16	-	-	-	-	16	-	-	-	-	-	-	-	-	8	-
234	-	-	-	-	-	16	-	8	8	-	-	-	-	-	-	-	-	-	-	-	-	-	16	-	-	-	-	8	8	-	-	-
235	-	-	-	-	16	-	8	-	-	8	-	-	-	-	-	-	-	-	-	-	-	-	-	16	-	-	8	-	-	8	-	-
245	-	-	-	16	-	-	8	-	-	8	-	-	-	-	-	-	-	-	-	-	-	-	-	-	16	-	8	-	-	8	-	-
345	-	-	16	-	-	-	-	8	8	-	-	-	-	-	-	-	-	-	-	-	-	-	-	-	-	16	-	8	8	-	-	-
1234	-	-	-	8	8	-	-	-	-	16	-	-	-	-	-	-	-	-	-	-	-	-	-	8	8	-	16	-	-	-	-	-
1235	-	-	8	-	-	8	-	-	16	-	-	-	-	-	-	-	-	-	-	-	-	-	8	-	-	8	-	16	-	-	-	-
1245	-	-	8	-	-	8	-	16	-	-	-	-	-	-	-	-	-	-	-	-	-	-	8	-	-	8	-	-	16	-	-	-
1345	-	-	-	8	8	-	16	-	-	-	-	-	-	-	-	-	-	-	-	-	-	-	-	8	8	-	-	-	-	16	-	-
2345	16	-	-	-	-	-	-	-	-	-	-	-	-	-	-	-	8	8	-	-	8	8	-	-	-	-	-	-	-	-	16	-
12345	-	16	-	-	-	-	-	-	-	-	8	8	-	-	8	8	-	-	-	-	-	-	-	-	-	-	-	-	-	-	-	16

[6]

θ{U} \ θ{V}	ϕ	1	2	3	4	5	12	13	14	15	23	24	25	34	35	45	123	124	125	134	135	145	234	235	245	345	1234	1235	1245	1345	2345	12345
ϕ	16	-	-	-	-	-	-	-	-	-	-	-	-	-	-	-	8	8	-	-	8	-	-	-	8	-	-	-	-	8	8	-
1	-	16	-	-	-	-	-	-	-	-	8	8	-	-	8	-	-	-	-	-	-	-	-	-	-	8	-	-	8	-	-	8
2	-	-	16	-	-	-	-	8	8	-	-	-	-	-	-	8	-	-	-	-	-	-	-	-	-	8	-	8	-	-	-	8
3	-	-	-	16	-	-	8	-	-	8	-	-	-	-	-	-	-	-	-	-	-	8	-	-	8	-	8	-	-	-	8	-
4	-	-	-	-	16	-	8	-	-	-	-	-	-	8	-	-	-	-	-	-	8	-	-	8	-	-	8	-	-	8	-	-
5	-	-	-	-	-	16	-	8	-	-	-	8	-	-	-	-	-	-	-	8	-	-	8	-	-	-	-	8	8	-	-	-
12	-	-	-	8	8	-	16	-	-	-	-	-	-	-	-	-	-	-	-	-	-	8	-	8	-	-	-	-	-	8	8	-
13	-	-	8	-	-	8	-	16	-	-	-	-	-	-	-	8	-	-	-	-	-	-	8	-	-	-	-	-	8	-	-	8
14	-	-	8	-	-	-	-	-	16	-	-	-	-	-	8	-	-	-	8	-	-	-	8	-	-	8	-	8	-	-	-	-
15	-	-	-	8	-	-	-	-	-	16	-	-	-	8	-	-	-	8	-	-	-	-	-	8	8	-	8	-	-	-	-	-
23	-	8	-	-	-	-	-	-	-	-	16	-	-	-	-	8	-	-	8	8	-	-	-	-	-	8	-	-	8	-	-	-
24	-	8	-	-	-	8	-	-	-	-	-	16	-	-	8	-	-	-	-	8	-	-	-	-	-	-	-	8	-	-	-	8
25	-	-	-	-	8	-	-	-	-	-	-	-	16	8	-	-	8	-	-	-	8	8	-	-	-	-	8	-	-	-	-	-
34	-	-	-	-	-	-	-	-	-	8	-	-	8	16	-	-	8	8	-	-	-	8	-	8	-	-	-	-	-	-	-	-
35	-	8	-	-	-	-	-	-	8	-	-	8	-	-	16	-	-	-	8	-	-	-	8	-	-	-	-	-	-	-	-	8
45	-	-	8	-	-	-	-	8	-	-	8	-	-	-	-	16	-	-	8	8	-	-	-	-	-	-	-	-	-	-	-	8
123	8	-	-	-	-	-	-	-	-	-	-	-	8	8	-	-	16	-	-	-	-	8	-	-	8	-	-	-	-	8	-	-
124	8	-	-	-	-	-	-	-	-	8	-	-	-	8	-	-	-	16	-	-	8	-	-	8	-	-	-	-	-	-	8	-
125	-	-	-	-	-	-	-	-	8	-	8	-	-	-	8	8	-	-	16	8	-	-	8	-	-	-	-	-	-	-	-	-
134	-	-	-	-	-	8	-	-	-	-	8	8	-	-	-	8	-	-	8	16	-	-	-	-	-	-	-	8	-	-	-	-
135	8	-	-	-	8	-	-	-	-	-	-	-	8	-	-	-	-	8	-	-	16	-	-	-	-	-	8	-	-	-	8	-
145	-	-	-	8	-	-	8	-	-	-	-	-	8	8	-	-	8	-	-	-	-	16	-	-	-	-	-	-	-	-	8	-
234	-	-	-	-	-	8	-	8	8	-	-	-	-	-	8	-	-	-	8	-	-	-	16	-	-	-	-	-	8	-	-	-
235	-	-	-	-	8	-	8	-	-	8	-	-	-	8	-	-	-	8	-	-	-	-	-	16	-	-	-	-	-	8	-	-
245	8	-	-	8	-	-	-	-	-	8	-	-	-	-	-	-	8	-	-	-	-	-	-	-	16	-	8	-	-	8	-	-
345	-	8	8	-	-	-	-	-	8	-	8	-	-	-	-	-	-	-	-	-	-	-	-	-	-	16	-	8	8	-	-	-
1234	-	-	-	8	8	-	-	-	-	8	-	-	8	-	-	-	-	-	-	-	8	-	-	-	8	-	16	-	-	-	-	-
1235	-	-	8	-	-	8	-	-	8	-	-	8	-	-	-	-	-	-	-	8	-	-	-	-	-	8	-	16	-	-	-	-
1245	-	8	-	-	-	8	-	8	-	-	8	-	-	-	-	-	-	-	-	-	-	-	8	-	-	8	-	-	16	-	-	-
1345	8	-	-	-	8	-	8	-	-	-	-	-	-	-	-	-	8	-	-	-	-	-	-	8	8	-	-	-	-	16	-	-
2345	8	-	-	8	-	-	8	-	-	-	-	-	-	-	-	-	-	8	-	-	8	8	-	-	-	-	-	-	-	-	16	-
12345	-	8	8	-	-	-	-	8	-	-	-	8	-	-	8	8	-	-	-	-	-	-	-	-	-	-	-	-	-	-	-	16

APPENDIX. Information matrices of 11 representative designs - continued

[7]

θ{U} \ θ{V}	Φ	1	2	3	4	5	12	13	14	15	23	24	25	34	35	45	123	124	125	134	135	145	234	235	245	345	1234	1235	1245	1345	2345	12345
Φ	16	·	·	·	·	·	·	·	·	·	·	·	·	·	·	·	8	8	·	·	8	-8	·	·	·	·	·	·	·	·	·	·
1	·	16	·	·	·	·	·	·	·	·	8	8	·	·	8	-8	·	·	·	·	·	·	·	·	·	·	·	·	·	·	·	·
2	·	·	16	·	·	·	·	8	8	·	·	·	·	·	·	·	·	·	·	·	·	·	·	·	·	·	·	8	-8	·	·	·
3	·	·	·	16	·	·	8	·	·	8	·	·	·	·	·	·	·	·	·	·	·	·	·	·	·	·	8	·	·	-8	·	·
4	·	·	·	·	16	·	8	·	·	-8	·	·	·	·	·	·	·	·	·	·	·	·	·	·	·	·	8	·	·	8	·	·
5	·	·	·	·	·	16	·	8	-8	·	·	·	·	·	·	·	·	·	·	·	·	·	·	·	·	·	·	8	8	·	·	·
12	·	·	·	8	8	·	16	·	·	·	·	·	·	·	·	·	·	·	·	·	·	·	·	8	-8	·	·	·	·	·	·	·
13	·	·	8	·	·	8	·	16	·	·	·	·	·	·	·	·	·	·	·	·	·	·	8	·	·	-8	·	·	·	·	·	·
14	·	·	8	·	·	-8	·	·	16	·	·	·	·	·	·	·	·	·	·	·	·	·	8	·	·	8	·	·	·	·	·	·
15	·	·	·	8	-8	·	·	·	·	16	·	·	·	·	·	·	·	·	·	·	·	·	·	8	8	·	·	·	·	·	·	·
23	·	8	·	·	·	·	·	·	·	·	16	·	·	·	·	·	·	·	8	8	·	·	·	·	·	·	·	·	·	·	·	-8
24	·	8	·	·	·	·	·	·	·	·	·	16	·	·	·	·	·	·	-8	8	·	·	·	·	·	·	·	·	·	·	·	8
25	·	·	·	·	·	·	·	·	·	·	·	·	16	·	·	·	8	-8	·	·	8	8	·	·	·	·	·	·	·	·	·	·
34	·	·	·	·	·	·	·	·	·	·	·	·	·	16	·	·	8	8	·	·	-8	8	·	·	·	·	·	·	·	·	·	·
35	·	8	·	·	·	·	·	·	·	·	·	·	·	·	16	·	·	·	8	-8	·	·	·	·	·	·	·	·	·	·	·	8
45	·	-8	·	·	·	·	·	·	·	·	·	·	·	·	·	16	·	·	8	8	·	·	·	·	·	·	·	·	·	·	·	8
123	8	·	·	·	·	·	·	·	·	·	·	·	8	8	·	·	16	·	·	·	·	·	·	·	·	·	·	·	·	·	-8	·
124	8	·	·	·	·	·	·	·	·	·	·	·	-8	8	·	·	·	16	·	·	·	·	·	·	·	·	·	·	·	·	8	·
125	·	·	·	·	·	·	·	·	·	·	8	-8	·	·	8	8	·	·	16	·	·	·	·	·	·	·	·	·	·	·	·	·
134	·	·	·	·	·	·	·	·	·	·	8	8	·	·	-8	8	·	·	·	16	·	·	·	·	·	·	·	·	·	·	·	·
135	8	·	·	·	·	·	·	·	·	·	·	·	8	-8	·	·	·	·	·	·	16	·	·	·	·	·	·	·	·	·	8	·
145	-8	·	·	·	·	·	·	·	·	·	·	·	8	8	·	·	·	·	·	·	·	16	·	·	·	·	·	·	·	·	8	·
234	·	·	·	·	·	·	·	8	8	·	·	·	·	·	·	·	·	·	·	·	·	·	16	·	·	·	·	-8	8	·	·	·
235	·	·	·	·	·	·	8	·	·	8	·	·	·	·	·	·	·	·	·	·	·	·	·	16	·	·	-8	·	·	8	·	·
245	·	·	·	·	·	·	-8	·	·	8	·	·	·	·	·	·	·	·	·	·	·	·	·	·	16	·	8	·	·	8	·	·
345	·	·	·	·	·	·	·	-8	8	·	·	·	·	·	·	·	·	·	·	·	·	·	·	·	·	16	·	8	8	·	·	·
1234	·	·	·	8	8	·	·	·	·	·	·	·	·	·	·	·	·	·	·	·	·	·	·	-8	8	·	16	·	·	·	·	·
1235	·	·	8	·	·	8	·	·	·	·	·	·	·	·	·	·	·	·	·	·	·	·	-8	·	·	8	·	16	·	·	·	·
1245	·	·	-8	·	·	8	·	·	·	·	·	·	·	·	·	·	·	·	·	·	·	·	8	·	·	8	·	·	16	·	·	·
1345	·	·	·	-8	8	·	·	·	·	·	·	·	·	·	·	·	·	·	·	·	·	·	·	8	8	·	·	·	·	16	·	·
2345	·	·	·	·	·	·	·	·	·	·	·	·	·	·	·	·	-8	8	·	·	8	8	·	·	·	·	·	·	·	·	16	·
12345	·	·	·	·	·	·	·	·	·	·	-8	8	·	·	8	8	·	·	·	·	·	·	·	·	·	·	·	·	·	·	·	16

[8]

θ{U} \ θ{V}	Φ	1	2	3	4	5	12	13	14	15	23	24	25	34	35	45	123	124	125	134	135	145	234	235	245	345	1234	1235	1245	1345	2345	12345
Φ	16	·	·	·	·	·	·	·	·	·	·	·	·	·	·	·	8	8	·	·	·	·	·	·	·	·	·	-8	8	·	·	·
1	·	16	·	·	·	·	·	·	·	·	8	8	·	·	·	·	·	·	·	·	·	·	·	-8	8	·	·	·	·	·	·	·
2	·	·	16	·	·	·	·	8	8	·	·	·	·	·	·	·	·	·	·	·	-8	8	·	·	·	·	·	·	·	·	·	·
3	·	·	·	16	·	·	8	·	·	·	·	·	·	·	·	·	·	·	-8	·	·	·	·	·	·	·	8	·	·	·	·	8
4	·	·	·	·	16	·	8	·	·	·	·	·	·	·	·	·	·	·	8	·	·	·	·	·	·	·	8	·	·	·	·	-8
5	·	·	·	·	·	16	·	·	·	·	·	·	·	·	·	·	-8	8	·	·	·	·	·	·	·	·	·	8	8	·	·	·
12	·	·	·	8	8	·	16	·	·	·	·	·	·	·	-8	8	·	·	·	·	·	·	·	·	·	·	·	·	·	·	·	·
13	·	·	8	·	·	·	·	16	·	·	·	·	-8	·	·	·	·	·	·	·	·	·	8	·	·	·	·	·	·	·	8	·
14	·	·	8	·	·	·	·	·	16	·	·	·	8	·	·	·	·	·	·	·	·	·	8	·	·	·	·	·	·	·	-8	·
15	·	·	·	·	·	·	·	·	·	16	-8	8	·	·	·	·	·	·	·	·	·	·	·	8	8	·	·	·	·	·	·	·
23	·	8	·	·	·	·	·	·	·	-8	16	·	·	·	·	·	·	·	·	8	·	·	·	·	·	·	·	·	·	8	·	·
24	·	8	·	·	·	·	·	·	·	8	·	16	·	·	·	·	·	·	·	8	·	·	·	·	·	·	·	·	·	-8	·	·
25	·	·	·	·	·	·	·	-8	8	·	·	·	16	·	·	·	·	·	·	·	8	8	·	·	·	·	·	·	·	·	·	·
34	·	·	·	·	·	·	·	·	·	·	·	·	·	16	·	·	8	8	·	·	·	·	·	·	·	·	·	8	-8	·	·	·
35	·	·	·	·	·	·	-8	·	·	·	·	·	·	·	16	·	·	·	8	·	·	·	·	·	·	·	8	·	·	·	·	8
45	·	·	·	·	·	·	8	·	·	·	·	·	·	·	·	16	·	·	8	·	·	·	·	·	·	·	-8	·	·	·	·	8
123	8	·	·	·	·	-8	·	·	·	·	·	·	·	8	·	·	16	·	·	·	·	·	·	·	·	8	·	·	·	·	·	·
124	8	·	·	·	·	8	·	·	·	·	·	·	·	8	·	·	·	16	·	·	·	·	·	·	·	-8	·	·	·	·	·	·
125	·	·	·	-8	8	·	·	·	·	·	·	·	·	·	8	8	·	·	16	·	·	·	·	·	·	·	·	·	·	·	·	·
134	·	·	·	·	·	·	·	·	·	·	8	8	·	·	·	·	·	·	·	16	·	·	·	8	-8	·	·	·	·	·	·	·
135	·	·	-8	·	·	·	·	·	·	·	·	·	8	·	·	·	·	·	·	·	16	·	8	·	·	·	·	·	·	·	8	·
145	·	·	8	·	·	·	·	·	·	·	·	·	8	·	·	·	·	·	·	·	·	16	-8	·	·	·	·	·	·	·	8	·
234	·	·	·	·	·	·	·	8	8	·	·	·	·	·	·	·	·	·	·	·	8	-8	16	·	·	·	·	·	·	·	·	·
235	·	-8	·	·	·	·	·	·	·	8	·	·	·	·	·	·	·	·	·	8	·	·	·	16	·	·	·	·	·	8	·	·
245	·	8	·	·	·	·	·	·	·	8	·	·	·	·	·	·	·	·	·	-8	·	·	·	·	16	·	·	·	·	8	·	·
345	·	·	·	·	·	·	·	·	·	·	·	·	·	·	·	·	8	-8	·	·	·	·	·	·	·	16	·	8	8	·	·	·
1234	·	·	·	8	8	·	·	·	·	·	·	·	·	·	8	-8	·	·	·	·	·	·	·	·	·	·	16	·	·	·	·	·
1235	-8	·	·	·	·	8	·	·	·	·	·	·	·	8	·	·	·	·	·	·	·	·	·	·	·	8	·	16	·	·	·	·
1245	8	·	·	·	·	8	·	·	·	·	·	·	·	-8	·	·	·	·	·	·	·	·	·	·	·	8	·	·	16	·	·	·
1345	·	·	·	·	·	·	·	·	·	·	8	-8	·	·	·	·	·	·	·	·	·	·	·	8	8	·	·	·	·	16	·	·
2345	·	·	·	·	·	·	·	8	-8	·	·	·	·	·	·	·	·	·	·	·	8	8	·	·	·	·	·	·	·	·	16	·
12345	·	·	·	8	-8	·	·	·	·	·	·	·	·	·	8	8	·	·	·	·	·	·	·	·	·	·	·	·	·	·	·	16

APPENDIX. Information matrices of 11 representative designs – continued

[9]

θ{U} \ θ{V}	φ	1	2	3	4	5	12	13	14	15	23	24	25	34	35	45	123	124	125	134	135	145	234	235	245	345	1234	1235	1245	1345	2345	12345
φ	16	·	·	·	·	·	·	·	·	·	·	·	·	·	·	·	8	·	·	·	·	·	·	·	·	·	-8	-8	·	·	·	-8
1	·	16	·	·	·	·	·	·	·	·	8	·	·	·	·	·	·	·	·	·	·	·	-8	-8	·	·	·	·	·	·	-8	·
2	·	·	16	·	·	·	·	8	·	·	·	·	·	·	·	·	·	·	·	-8	-8	·	·	·	·	·	·	·	·	-8	·	·
3	·	·	·	16	·	·	8	·	·	·	·	·	·	·	·	·	·	-8	-8	·	·	·	·	·	·	·	·	·	-8	·	·	·
4	·	·	·	·	16	·	·	·	·	·	·	·	·	·	·	·	-8	·	·	·	·	·	·	·	·	·	8	-8	·	·	·	-8
5	·	·	·	·	·	16	·	·	·	·	·	·	·	·	·	·	-8	·	·	·	·	·	·	·	·	·	-8	8	·	·	·	-8
12	·	·	·	8	·	·	16	·	·	·	·	·	·	-8	-8	·	·	·	·	·	·	·	·	·	·	-8	·	·	·	·	·	·
13	·	·	8	·	·	·	·	16	·	·	·	-8	-8	·	·	·	·	·	·	·	·	·	·	·	-8	·	·	·	·	·	·	·
14	·	·	·	·	·	·	·	·	16	·	-8	·	·	·	·	·	·	·	·	·	·	·	8	-8	·	·	·	·	·	·	-8	·
15	·	·	·	·	·	·	·	·	·	16	-8	·	·	·	·	·	·	·	·	·	·	·	-8	8	·	·	·	·	·	·	-8	·
23	·	8	·	·	·	·	·	·	-8	-8	16	·	·	·	·	·	·	·	·	·	·	-8	·	·	·	·	·	·	·	·	·	·
24	·	·	·	·	·	·	·	-8	·	·	·	16	·	·	·	·	·	·	·	8	-8	·	·	·	·	·	·	·	·	-8	·	·
25	·	·	·	·	·	·	·	-8	·	·	·	·	16	·	·	·	·	·	·	-8	8	·	·	·	·	·	·	·	·	-8	·	·
34	·	·	·	·	·	·	-8	·	·	·	·	·	·	16	·	·	·	8	-8	·	·	·	·	·	·	·	·	·	-8	·	·	·
35	·	·	·	·	·	·	-8	·	·	·	·	·	·	·	16	·	·	-8	8	·	·	·	·	·	·	·	·	·	-8	·	·	·
45	·	·	·	·	·	·	·	·	·	·	·	·	·	·	·	16	-8	·	·	·	·	·	·	·	·	·	-8	-8	·	·	·	8
123	8	·	·	·	-8	-8	·	·	·	·	·	·	·	·	·	-8	16	·	·	·	·	·	·	·	·	·	·	·	·	·	·	·
124	·	·	·	-8	·	·	·	·	·	·	·	·	·	8	-8	·	·	16	·	·	·	·	·	·	·	-8	·	·	·	·	·	·
125	·	·	·	-8	·	·	·	·	·	·	·	·	·	-8	8	·	·	·	16	·	·	·	·	·	·	-8	·	·	·	·	·	·
134	·	·	-8	·	·	·	·	·	·	·	·	8	-8	·	·	·	·	·	·	16	·	·	·	·	-8	·	·	·	·	·	·	·
135	·	·	-8	·	·	·	·	·	·	·	·	-8	8	·	·	·	·	·	·	·	16	·	·	·	-8	·	·	·	·	·	·	·
145	·	·	·	·	·	·	·	·	·	·	-8	·	·	·	·	·	·	·	·	·	·	16	-8	-8	·	·	·	·	·	·	8	·
234	·	-8	·	·	·	·	·	·	8	-8	·	·	·	·	·	·	·	·	·	·	·	-8	16	·	·	·	·	·	·	·	·	·
235	·	-8	·	·	·	·	·	·	-8	8	·	·	·	·	·	·	·	·	·	·	·	-8	·	16	·	·	·	·	·	·	·	·
245	·	·	·	·	·	·	·	-8	·	·	·	·	·	·	·	·	·	·	·	-8	-8	·	·	·	16	·	·	·	·	8	·	·
345	·	·	·	·	·	·	-8	·	·	·	·	·	·	·	·	·	·	-8	-8	·	·	·	·	·	·	16	·	·	8	·	·	·
1234	-8	·	·	·	8	-8	·	·	·	·	·	·	·	·	·	-8	·	·	·	·	·	·	·	·	·	·	16	·	·	·	·	·
1235	-8	·	·	·	-8	8	·	·	·	·	·	·	·	·	·	-8	·	·	·	·	·	·	·	·	·	·	·	16	·	·	·	·
1245	·	·	·	-8	·	·	·	·	·	·	·	·	·	-8	-8	·	·	·	·	·	·	·	·	·	·	8	·	·	16	·	·	·
1345	·	·	-8	·	·	·	·	·	·	·	·	-8	-8	·	·	·	·	·	·	·	·	·	·	·	8	·	·	·	·	16	·	·
2345	·	-8	·	·	·	·	·	·	-8	-8	·	·	·	·	·	·	·	·	·	·	·	8	·	·	·	·	·	·	·	·	16	·
12345	-8	·	·	·	-8	-8	·	·	·	·	·	·	·	·	·	8	·	·	·	·	·	·	·	·	·	·	·	·	·	·	·	16

[10]

θ{U} \ θ{V}	φ	1	2	3	4	5	12	13	14	15	23	24	25	34	35	45	123	124	125	134	135	145	234	235	245	345	1234	1235	1245	1345	2345	12345
φ	16	·	·	·	·	·	·	·	·	·	·	·	·	·	·	·	·	·	·	·	·	·	·	·	·	·	-16	·	·	·	·	·
1	·	16	·	·	·	·	·	·	·	·	·	·	·	·	·	·	·	·	·	·	·	·	-16	·	·	·	·	·	·	·	·	·
2	·	·	16	·	·	·	·	·	·	·	·	·	·	·	·	·	·	·	·	-16	·	·	·	·	·	·	·	·	·	·	·	·
3	·	·	·	16	·	·	·	·	·	·	·	·	·	·	·	·	·	-16	·	·	·	·	·	·	·	·	·	·	·	·	·	·
4	·	·	·	·	16	·	·	·	·	·	·	·	·	·	·	·	-16	·	·	·	·	·	·	·	·	·	·	·	·	·	·	·
5	·	·	·	·	·	16	·	·	·	·	·	·	·	·	·	·	·	·	·	·	·	·	·	·	·	·	·	·	·	·	·	-16
12	·	·	·	·	·	·	16	·	·	·	·	·	·	-16	·	·	·	·	·	·	·	·	·	·	·	·	·	·	·	·	·	·
13	·	·	·	·	·	·	·	16	·	·	·	-16	·	·	·	·	·	·	·	·	·	·	·	·	·	·	·	·	·	·	·	·
14	·	·	·	·	·	·	·	·	16	·	-16	·	·	·	·	·	·	·	·	·	·	·	·	·	·	·	·	·	·	·	·	·
15	·	·	·	·	·	·	·	·	·	16	·	·	·	·	·	·	·	·	·	·	·	·	·	·	·	·	·	·	·	·	-16	·
23	·	·	·	·	·	·	·	·	-16	·	16	·	·	·	·	·	·	·	·	·	·	·	·	·	·	·	·	·	·	·	·	·
24	·	·	·	·	·	·	·	-16	·	·	·	16	·	·	·	·	·	·	·	·	·	·	·	·	·	·	·	·	·	·	·	·
25	·	·	·	·	·	·	·	·	·	·	·	·	16	·	·	·	·	·	·	·	·	·	·	·	·	·	·	·	·	-16	·	·
34	·	·	·	·	·	·	-16	·	·	·	·	·	·	16	·	·	·	·	·	·	·	·	·	·	·	·	·	·	·	·	·	·
35	·	·	·	·	·	·	·	·	·	·	·	·	·	·	16	·	·	·	·	·	·	·	·	·	·	·	·	·	-16	·	·	·
45	·	·	·	·	·	·	·	·	·	·	·	·	·	·	·	16	·	·	·	·	·	·	·	·	·	·	·	-16	·	·	·	·
123	·	·	·	·	-16	·	·	·	·	·	·	·	·	·	·	·	16	·	·	·	·	·	·	·	·	·	·	·	·	·	·	·
124	·	·	·	-16	·	·	·	·	·	·	·	·	·	·	·	·	·	16	·	·	·	·	·	·	·	·	·	·	·	·	·	·
125	·	·	·	·	·	·	·	·	·	·	·	·	·	·	·	·	·	·	16	·	·	·	·	·	·	-16	·	·	·	·	·	·
134	·	·	-16	·	·	·	·	·	·	·	·	·	·	·	·	·	·	·	·	16	·	·	·	·	·	·	·	·	·	·	·	·
135	·	·	·	·	·	·	·	·	·	·	·	·	·	·	·	·	·	·	·	·	16	·	·	·	-16	·	·	·	·	·	·	·
145	·	·	·	·	·	·	·	·	·	·	·	·	·	·	·	·	·	·	·	·	·	16	·	-16	·	·	·	·	·	·	·	·
234	·	-16	·	·	·	·	·	·	·	·	·	·	·	·	·	·	·	·	·	·	·	·	16	·	·	·	·	·	·	·	·	·
235	·	·	·	·	·	·	·	·	·	·	·	·	·	·	·	·	·	·	·	·	·	-16	·	16	·	·	·	·	·	·	·	·
245	·	·	·	·	·	·	·	·	·	·	·	·	·	·	·	·	·	·	·	·	-16	·	·	·	16	·	·	·	·	·	·	·
345	·	·	·	·	·	·	·	·	·	·	·	·	·	·	·	·	·	·	-16	·	·	·	·	·	·	16	·	·	·	·	·	·
1234	-16	·	·	·	·	·	·	·	·	·	·	·	·	·	·	·	·	·	·	·	·	·	·	·	·	·	16	·	·	·	·	·
1235	·	·	·	·	·	·	·	·	·	·	·	·	·	·	·	-16	·	·	·	·	·	·	·	·	·	·	·	16	·	·	·	·
1245	·	·	·	·	·	·	·	·	·	·	·	·	·	·	-16	·	·	·	·	·	·	·	·	·	·	·	·	·	16	·	·	·
1345	·	·	·	·	·	·	·	·	·	·	·	·	-16	·	·	·	·	·	·	·	·	·	·	·	·	·	·	·	·	16	·	·
2345	·	·	·	·	·	·	·	·	·	-16	·	·	·	·	·	·	·	·	·	·	·	·	·	·	·	·	·	·	·	·	16	·
12345	·	·	·	·	·	-16	·	·	·	·	·	·	·	·	·	·	·	·	·	·	·	·	·	·	·	·	·	·	·	·	·	16

APPENDIX. Information matrices of 11 representative designs - continued

[11]

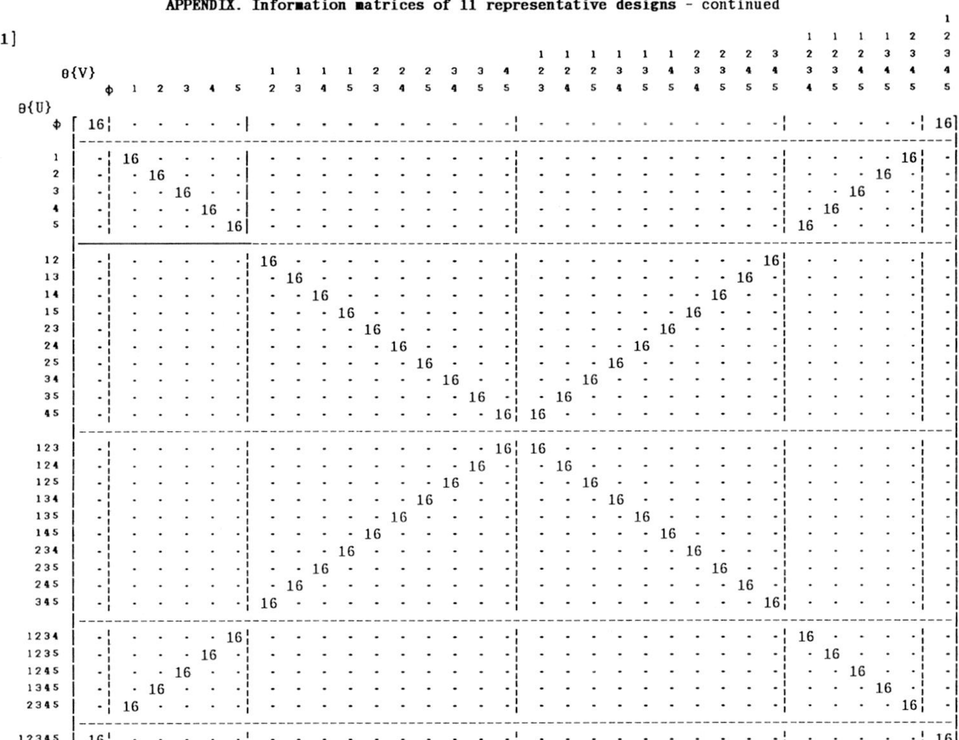

θ{U} \ θ{V}	φ	1	2	3	4	5	12	13	14	15	23	24	25	34	35	45	123	124	125	134	135	145	234	235	245	345	1234	1235	1245	1345	2345	12345
φ	16	·	·	·	·	·	·	·	·	·	·	·	·	·	·	·	·	·	·	·	·	·	·	·	·	·	·	·	·	·	·	16
1	·	16	·	·	·	·	·	·	·	·	·	·	·	·	·	·	·	·	·	·	·	·	·	·	·	·	·	·	·	·	16	·
2	·	·	16	·	·	·	·	·	·	·	·	·	·	·	·	·	·	·	·	·	·	·	·	·	·	·	·	·	·	16	·	·
3	·	·	·	16	·	·	·	·	·	·	·	·	·	·	·	·	·	·	·	·	·	·	·	·	·	·	·	·	16	·	·	·
4	·	·	·	·	16	·	·	·	·	·	·	·	·	·	·	·	·	·	·	·	·	·	·	·	·	·	·	16	·	·	·	·
5	·	·	·	·	·	16	·	·	·	·	·	·	·	·	·	·	·	·	·	·	·	·	·	·	·	·	16	·	·	·	·	·
12	·	·	·	·	·	·	16	·	·	·	·	·	·	·	·	·	·	·	·	·	·	·	·	·	·	16	·	·	·	·	·	·
13	·	·	·	·	·	·	·	16	·	·	·	·	·	·	·	·	·	·	·	·	·	·	·	·	16	·	·	·	·	·	·	·
14	·	·	·	·	·	·	·	·	16	·	·	·	·	·	·	·	·	·	·	·	·	·	·	16	·	·	·	·	·	·	·	·
15	·	·	·	·	·	·	·	·	·	16	·	·	·	·	·	·	·	·	·	·	·	·	16	·	·	·	·	·	·	·	·	·
23	·	·	·	·	·	·	·	·	·	·	16	·	·	·	·	·	·	·	·	·	·	16	·	·	·	·	·	·	·	·	·	·
24	·	·	·	·	·	·	·	·	·	·	·	16	·	·	·	·	·	·	·	·	16	·	·	·	·	·	·	·	·	·	·	·
25	·	·	·	·	·	·	·	·	·	·	·	·	16	·	·	·	·	·	·	16	·	·	·	·	·	·	·	·	·	·	·	·
34	·	·	·	·	·	·	·	·	·	·	·	·	·	16	·	·	·	·	16	·	·	·	·	·	·	·	·	·	·	·	·	·
35	·	·	·	·	·	·	·	·	·	·	·	·	·	·	16	·	·	16	·	·	·	·	·	·	·	·	·	·	·	·	·	·
45	·	·	·	·	·	·	·	·	·	·	·	·	·	·	·	16	16	·	·	·	·	·	·	·	·	·	·	·	·	·	·	·
123	·	·	·	·	·	·	·	·	·	·	·	·	·	·	·	16	16	·	·	·	·	·	·	·	·	·	·	·	·	·	·	·
124	·	·	·	·	·	·	·	·	·	·	·	·	·	·	16	·	·	16	·	·	·	·	·	·	·	·	·	·	·	·	·	·
125	·	·	·	·	·	·	·	·	·	·	·	·	·	16	·	·	·	·	16	·	·	·	·	·	·	·	·	·	·	·	·	·
134	·	·	·	·	·	·	·	·	·	·	·	·	16	·	·	·	·	·	·	16	·	·	·	·	·	·	·	·	·	·	·	·
135	·	·	·	·	·	·	·	·	·	·	·	16	·	·	·	·	·	·	·	·	16	·	·	·	·	·	·	·	·	·	·	·
145	·	·	·	·	·	·	·	·	·	·	16	·	·	·	·	·	·	·	·	·	·	16	·	·	·	·	·	·	·	·	·	·
234	·	·	·	·	·	·	·	·	·	16	·	·	·	·	·	·	·	·	·	·	·	·	16	·	·	·	·	·	·	·	·	·
235	·	·	·	·	·	·	·	·	16	·	·	·	·	·	·	·	·	·	·	·	·	·	·	16	·	·	·	·	·	·	·	·
245	·	·	·	·	·	·	·	16	·	·	·	·	·	·	·	·	·	·	·	·	·	·	·	·	16	·	·	·	·	·	·	·
345	·	·	·	·	·	·	16	·	·	·	·	·	·	·	·	·	·	·	·	·	·	·	·	·	·	16	·	·	·	·	·	·
1234	·	·	·	·	·	16	·	·	·	·	·	·	·	·	·	·	·	·	·	·	·	·	·	·	·	·	16	·	·	·	·	·
1235	·	·	·	·	16	·	·	·	·	·	·	·	·	·	·	·	·	·	·	·	·	·	·	·	·	·	·	16	·	·	·	·
1245	·	·	·	16	·	·	·	·	·	·	·	·	·	·	·	·	·	·	·	·	·	·	·	·	·	·	·	·	16	·	·	·
1345	·	·	16	·	·	·	·	·	·	·	·	·	·	·	·	·	·	·	·	·	·	·	·	·	·	·	·	·	·	16	·	·
2345	·	16	·	·	·	·	·	·	·	·	·	·	·	·	·	·	·	·	·	·	·	·	·	·	·	·	·	·	·	·	16	·
12345	16	·	·	·	·	·	·	·	·	·	·	·	·	·	·	·	·	·	·	·	·	·	·	·	·	·	·	·	·	·	·	16

Note: Dot '·' indicates '0' throughout this appendix

Stat. Sci. & Data Anal., pp. 559-570
K. Matsusita *et al.* (Eds)

Characterization and Optimality of Block Designs with Estimation of Parameters Under Mixed Models

SANPEI KAGEYAMA and ROMAN ZMYŚLONY

Department of Mathematics, Hiroshima University, Hiroshima 734, Japan
Institute of Mathematics, Polish Academy of Sciences, Wroclaw, Poland

Abstract. A problem of estimation of either treatment or block effects in mixed models regarding block designs is considered. Necessary and sufficient conditions for the existence of best linear unbiased estimators of some parametric functions are completely characterized in the concrete form of the incidence matrix of designs. The resulting form of designs also has the trace- or determinant-optimality in the usual sense. Moreover, best quadratic unbiased estimators of variance components in our models are obtained. Finally, it is shown that those estimators are best unbiased under the assumption of normality.

Key words: best linear unbiased estimator (BLUE); best quadratic unbiased estimator (BQUE); quadratic estimability; incidence matrix; connected; trace-optimal; determinant-optimal

1. INTRODUCTION

The problem of estimation for some parametric functions should be taken so that all the existence conditions for the estimators are described in a computational attractive form from a practical point of view. Under a general linear model, Gnot [1], Seely [2] and Zmyślony [3] and [4] have discussed the existence of best linear unbiased estimators (BLUE) and/or best quadratic unbiased estimators (BQUE) for estimating some effects and/or variance components from a similar standpoint.

One of our purpose is first to make a complete characterization of BLUE in mixed models based on block designs, as a further development to the results due to Zmyślony [4]. Necessary and sufficient conditions for the existence of BLUE for fixed parameters of block effects or treatment effects in the explicit form of incidence matrix of designs are given. Next in the class of designs satisfying such necessary and sufficient conditions, we characterize trace-(or determinant-)optimal designs, i.e. designs in which the minimum trace (or determinant) of covariance matrix of BLUE is attained. It is also shown that for some models of such designs, there exists BQUE for variance components of random block or treatment effects. Finally, we show the best unbiasedness for our estimators under normality.

The observations presented here include the results due to Gnot [1] and Seely [5] as special cases in some sense (see Remark 4.2 in Section 4).

2. CHARACTERIZATION OF DESIGNS RELATED TO BLUE

Zmyślony [4] made a purely mathematical argument for the general linear model, and to obtain a BLUE for every estimable function of either treatment or block effects in the model, he considered the usual block design model and partially characterized a necessary and sufficient condition for the existence of such BLUE in terms of the incidence matrix of a block design. We here characterize completely such conditions in the concrete form of the incidence matrix. This complete characterization is meaningful from a practical point of view.

Consider an experiment with v treatments arranged in b blocks according to the $v\times b$ incidence matrix N with the total number of experimental units n. Let $\mathbf{y} = (y_1,\ldots,y_n)'$ be an $n\times 1$ observation vector and follow the model

$$\mathbf{y} = [\Delta':D'](\underline{\gamma}',\underline{\beta}')' + \mathbf{e} \tag{2.1}$$

where Δ' is the $n\times v$ matrix with elements $\delta_{i\ell}$ being 1 or 0 according as y_i receives the ℓth treatment or not; D' is the $n\times b$ matrix with elements d_{ij} being 1 or 0 according as y_i is in the jth block or not; $\underline{\gamma}$ and $\underline{\beta}$ are the $v\times 1$ vector of treatment effects and $b\times 1$ vector of block effects, respectively. Note that $N = \Delta D' = \|n_{ij}\|$, where $n_{ij} = 1$ or 0 according as the ith treatment occurs in the jth block or not. The $n\times 1$ random error vector $\mathbf{e}$ is assumed to satisfy the conditions: $E(\mathbf{e}) = \mathbf{0}$, $E(\mathbf{ee}') = \sigma_e^2 I_n$, where σ_e^2 is unknown. As usual, let $\mathbf{r} = (r_1,\ldots,r_v)' = N\mathbf{1}_b$ and $\mathbf{k} = (k_1,\ldots,k_b)' = N'\mathbf{1}_v$ stand for vectors of treatment replications r_i and block sizes k_j, respectively, where $\mathbf{1}_c$ is a $c\times 1$ vector with all elements 1. Further let $R = \text{diag}\{r_1,\ldots,r_v\}$ and $K = \text{diag}\{k_1,\ldots, k_b\}$. Note that $\Delta\Delta' = R$, $DD' = K$, $D'\mathbf{1}_b = \mathbf{1}_n$, $D\mathbf{1}_n = \mathbf{k}$, $\Delta'\mathbf{1}_v = \mathbf{1}_n$ and $\Delta\mathbf{1}_n = \mathbf{r}$.

We shall deal with mixed models in the two cases following the classification due to Zmyślony [4].

(a) γ is a vector of fixed effects and β is a random vector satisfying $E(\beta) = \mathbf{0}$, $E(\beta\beta') = \sigma^2_\beta I_b$, $E(\beta\mathbf{e}') = 0$, where σ^2_β is unknown.

(b) β is a vector of fixed effects and γ is a random vector satisfying $E(\gamma) = \mathbf{0}$, $E(\gamma\gamma') = \sigma^2_\gamma I_v$, $E(\gamma\mathbf{e}') = 0$, where σ^2_γ is unknown.

As a necessary and sufficient condition for the existence of BLUE of every estimable function of fixed parameters in the model, Zmyślony [4] proved that in case (a) $U = NKN' - NN'K^{-1}NN'$ is a zero matrix, and in case (b) $U = N'RN - N'NK^{-1}N'N = 0$. This result will be used to derive the present results. However, regarding the case (b), we have $U = N'(R - NK^{-1}N')N = N'CN$ with $C = R - NK^{-1}N'$ being positive semi-definite which is well-known as a coefficient matrix of the usual adjusted normal equation in a block design. This observation shows that the case (b) can be characterized simply by dualizing the results obtained in case (a), i.e. replacing R with K, K with R, N with N', respectively. Hence the results for the case (b) will not be presented here. The detail can be found in Zmyślony and Kageyama [6].

We shall characterize the incidence matrix N of a block design satisfying $U = 0$ in case (a) only.

Theorem 2.1. In a block design with parameters v, b, r_i, k_j $(i=1,\ldots,v; j=1,\ldots,b)$ having the incidence matrix N, in case (a), $U = 0$ if and only if for some integer $g \geq 1$

$$N = \text{diag}\{\frac{1}{b_1}[r_1\mathbf{1}_{b_1}:\ldots:r_{v_1}\mathbf{1}_{b_1}]',\ldots,\frac{1}{b_g}[r_c\mathbf{1}_{b_g}:\ldots:r_v\mathbf{1}_{b_g}]'\}, \quad (2.2)$$

where $c = v_1+\ldots+v_{g-1}+1$ with some partitions as $v = v_1+\ldots+v_g$ and $b = b_1+\ldots+b_g$.

Proof. Now $U = N(K-N'R^{-1}N)N' = NHN'$ by letting $H = K-N'R^{-1}N$. Since H is a positive semi-definite matrix, it is obvious that $U = 0$ iff $NH = 0$. Furthermore, $H\mathbf{1}_b = \mathbf{0}$ and hence $\text{rank}(H) \leq b-1$. We then consider our problem by separating into two cases.

(i) rank(H) = b-1. In this case, 0 is a simple eigenvalue of H with the eigenvector $\mathbf{1}_b$. Then NH = 0 holds iff the incidence matrix is of the form

$$N = [\lambda_1 \mathbf{1}_b : \dots : \lambda_v \mathbf{1}_b]'$$

for some real $\lambda_1, \dots, \lambda_v$. Since $N\mathbf{1}_b = \mathbf{r}$, we get $\lambda_i = r_i/b$ for i = 1,...,v. Hence U = 0 iff $N = \frac{1}{b}[r_1 \mathbf{1}_b : \dots : r_v \mathbf{1}_b]' = \frac{1}{b}\mathbf{r}\mathbf{1}_b'$.

(ii) rank(H) = b-g for g ≥ 1. In this case, from a property of connectedness in a block design (cf. Kageyama [7]), we can express the incidence matrix N as

$$N = \text{diag}\{N_1, \dots, N_g\}$$

with $g \le \min\{v,b\}$, where N_i is a $v_i \times b_i$ incidence submatrix with $\text{rank}(H_i) = b_i - 1$ for $H_i = K_i - N_i' R_i^{-1} N_i$. Here $K = \text{diag}\{K_1, \dots, K_g\}$ and $R = \text{diag}\{R_1, \dots, R_g\}$ corresponding to the above-mentioned partition of N. (For the later use, put $\mathbf{k}_i = K_i \mathbf{1}_{b_i}$ and $\mathbf{r}_i = R_i \mathbf{1}_{v_i}$ for i=1,...,g.) Then it holds that $H = \text{diag}\{H_1, \dots, H_g\}$. Hence U = 0 iff $N_i H_i = 0$ for all i = 1,...,g. An argument similar to case (i) yields $N_i = \frac{1}{b_i}\mathbf{r}_i \mathbf{1}_{b_i}'$ for all i. Therefore, it follows that U = 0 iff (2.2) holds. This completes the proof. □

Remark 2.1. In the proof of Theorem 2.1 the case (i) is included in the case (ii), but the case (i) is specially given from a point of view of special attention for practical usefulness related to connectedness of a block design.

Remark 2.2. It can also be shown that if the condition (2.2) is satisfied, then the BLUE of γ has the form $\hat{\gamma} = (\Delta\Delta')^{-1}\Delta\mathbf{y}$ (see Baksalary and Kala [8], Kala [9], Zyskind [10]).

Corresponding to the usual definition of connectedness of a block design in terms of the matrix C (= $R - NK^{-1}N'$), we here call a block design connected if rank(H) = b-1. Otherwise, it is said to be disconnected. In this case, we have the following.

Corollary 2.1. In a connected block design with parameters v, b, r_i, k_j having the incidence matrix N, in case (a), U = 0 iff $N = \frac{1}{b}\mathbf{r}\mathbf{1}_b'$ with constant block size $(1/b)\sum_{i=1}^{v} r_i$.

Note that the resulting design in Corollary 2.1 is necessarily proper for size $k_1 = \dots = k_b = (1/b)\sum_{i=1}^{v} r_i$.

Corollary 2.2. In a connected block design with parameters v, b, r, k_j having the incidence matrix N, in case (a), U = 0 iff $N = (r/b)\mathbf{1}_v\mathbf{1}_b'$.

When $R = rI_v$ and $K = kI_b$ for positive integers r and k, Zmyślony [4] shows that in case (a), for each treatment parameter there exists a BLUE iff $\{1/(rk)\}NN'$ is idempotent. Corollary 2.2 can complete his characterization in the concrete form of the incidence matrix of connected block designs.

We thus have characterized $U = 0$ completely. In some sense $U = 0$ gives strong restriction to the structure of the incidence matrix of a block design. After an experimentation by use of a block design in case (a), if by chance we get

$$U = \begin{bmatrix} U_{11} & 0_{v_1 \times v_2} \\ 0_{v_2 \times v_1} & 0_{v_2 \times v_2} \end{bmatrix} \tag{2.3}$$

where $0_{s\times t} = \mathbf{0}_s\mathbf{0}_t'$ and $\mathbf{0}_\ell$ is a column zero vector of size ℓ, then from Theorem 1 of Zmyślony [4] it follows that for the second group of v_2 treatment parameters there exist BLUE, and that there does not exist any BLUE for the first group of v_1 treatment parameters. But it appears that this observation also has some statistically practical meaning. Hence we now give the characterization of block designs having U of the form (2.3).

Remark 2.3. If we are interested in a given set of treatment parameters in advance, then exchanging the order of rows and/or columns of the matrix Δ in the model will lead to the pattern of the matrix U as in (2.3) without loss of generality.

In case (a), let $N = [M_1' : M_2']'$, where M_i is a $v_i \times b$ incidence submatrix, i=1,2, with $v_1+v_2 = v$. In this case

$$U = NHN' = \begin{bmatrix} M_1HM_1' & M_1HM_2' \\ M_2HM_1' & M_2HM_2' \end{bmatrix}.$$

For (2.3), it is sufficient only to consider a case of $M_2HM_2' = 0_{v_2\times v_2}$. Because this relation necessarily implies $M_1HM_2' = 0$. It follows that $M_2HM_2' = 0$ iff $M_2H = 0_{v_2\times b}$. Then we can follow the proof of Theorem 2.1 in the same way to derive the following.

Theorem 2.2. In a connected block design with parameters v, b, r_i, k_j (i=1,...,v; j=1,...,b) having the incidence matrix $N = [M_1' : M_2']'$ in case (a),

$$U = \begin{bmatrix} U_{11} & 0_{v_1 \times v_2} \\ 0_{v_2 \times v_1} & 0_{v_2 \times v_2} \end{bmatrix}$$

iff

$$N = [M_1' : \frac{1}{b} \mathbf{1}_b(\mathbf{r}_2', \ldots, \mathbf{r}_g')]'$$

where M_1 is a $v_1 \times b$ incidence submatrix of N.

The disconnected case, i.e. rank(H) < b-1, can also be obtained similarly and hence it is omitted here.

3. TRACE- AND DETERMINANT-OPTIMAL DESIGNS

In this section, under the model (2.1) of the case (a), we shall investigate designs satisfying U = 0 under a trace or determinant optimum criterion for given v, b and $n = \sum_{i=1}^{v} r_i = \sum_{j=1}^{b} k_j$ (total number of experimental units). Denote by $\Omega_{v,b,n}$ a class of the above-mentioned block designs.

It is known that under (2.1) an unbiased estimator of the vector $\underset{\sim}{\tau}$ of treatment effects is given by $\hat{\underset{\sim}{\tau}} = (\Delta\Delta')^{-1}\Delta\mathbf{y} = R^{-1}\Delta\mathbf{y}$, and further Var $\mathbf{y} = D'D\sigma_\beta^2 + I_n\sigma_e^2$. Then it follows that

$$\begin{aligned} \text{Var } \hat{\underset{\sim}{\tau}} &= R^{-1}\Delta(D'D\sigma_\beta^2 + I_n\sigma_e^2)\Delta'R^{-1} \\ &= R^{-1}NN'R^{-1}\sigma_\beta^2 + R^{-1}\sigma_e^2. \end{aligned} \tag{3.1}$$

By Corollary 2.1, a connected block design with parameters v, b, r_i, k_j ($i=1,\ldots,v$; $j=1,\ldots,b$) satisfying U = 0 has the incidence matrix of the following form

$$N = \frac{1}{b}\mathbf{r}\mathbf{1}_b'$$

which, from (3.1), yields

$$\text{Var } \hat{\underset{\sim}{\tau}} = \frac{1}{b}\mathbf{1}_v\mathbf{1}_v'\sigma_\beta^2 + R^{-1}\sigma_e^2. \tag{3.2}$$

Now we search optimal designs by taking a trace-criterion, i.e. to obtain the minimum of $\text{tr}(\text{Var } \hat{\underset{\sim}{\tau}}) = (v/b)\sigma_\beta^2 + (\sum_{i=1}^{v} r_i^{-1})\sigma_e^2$. Since unknown parameters σ_β^2 and σ_e^2 are not dependent upon the design, it is obvious that the minimum of $\text{tr}(\text{Var } \hat{\underset{\sim}{\tau}})$ is attained only when $r_1 = \ldots = r_v$ (= r, say). We thus have established the following.

Theorem 3.1. An equireplicate connected block design (if it exists) is trace-optimal in the class $\Omega_{v,b,n}$.

Remark 3.1. Corollary 2.2 and Theorem 3.1 show that a trace-optimal block design is given by the incidence matrix of the form $(r/b)\mathbf{1}_v\mathbf{1}_b'$, i.e. orthogonal design.

A characterization of disconnceted designs satisfying $U = 0$ is given in Theorem 2.1 with $g \geq 2$ as in (2.2), a class of which is denoted by $\Omega_{v,b,n,g}$. Furthermore, values $v_1,\ldots,v_g$ and $b_1,\ldots,b_g$ are assumed to be fixed in partitions $v = v_1+\ldots+v_g$ and $b = b_1+\ldots+b_g$. In this case, similarly we can get

$$\text{Var}\ \hat{\gamma} = \text{diag}\{\frac{1}{b_1}\mathbf{1}_{v_1}\mathbf{1}_{v_1}',\ldots,\frac{1}{b_g}\mathbf{1}_{v_g}\mathbf{1}_{v_g}'\}\sigma_\beta^2 + R^{-1}\sigma_e^2$$

which yields

$$\text{tr}(\text{Var}\ \hat{\gamma}) = \textstyle\sum_{i=1}^{g}(v_i/b_i)\sigma_\beta^2 + \sum_{i=1}^{v} r_i^{-1}\sigma_e^2.$$

An argument similar to Theorem 3.1 leads to the following.

Theorem 3.2. An equireplicate disconnected block design is trace-optimal in the class $\Omega_{v,b,n,g}$ for $g \geq 2$.

Note that a design in Theorem 3.2 is given by

$$N = \text{diag}\{\ \frac{r}{b_1}\mathbf{1}_{v_1}\mathbf{1}_{b_1}',\ldots,\ \frac{r}{b_g}\mathbf{1}_{v_g}\mathbf{1}_{b_g}'\ \}$$

with $v = v_1+\ldots+v_g$ and $b = b_1+\ldots+b_g$.

We next search optimal designs by taking a determinant-criterion, i.e. to obtain the minimum of $\det(\text{Var}\ \hat{\gamma})$. In a connected block design, by (3.2)

$$\begin{aligned}\det(\text{Var}\ \hat{\gamma}) &= \det(\frac{\sigma_\beta^2}{b}\mathbf{1}_v\mathbf{1}_v' + R^{-1}\sigma_e^2)\\ &= \det(R^{-1}\sigma_e^2)\{1 + \frac{\sigma_\beta^2}{b\sigma_e^2}\ \mathbf{1}_v'R\mathbf{1}_v\}\\ &= (\sigma_e^2)^v(1 + \frac{\sigma_\beta^2 k}{\sigma_e^2})/\prod_{i=1}^{v} r_i\end{aligned}$$

since $n = \mathbf{1}_v'R\mathbf{1}_v = bk$ in Corollary 2.1. Thus, the minimum of $\det(\text{Var}\ \hat{\gamma})$ is attained when $\prod_{i=1}^{v} r_i$ is in maximum, i.e. when $r_1 = \ldots = r_v$ (= r, say). This leads to the following.

Theorem 3.3. If in the class $\Omega_{v,b,n}$ an equireplicate connected block design exists, then it is determinant-optimal.

Remark 3.2. A determinant-optimal connected block design is also given by the incidence matrix of the form $(r/b)\mathbf{1}_v\mathbf{1}_b'$.

Similarly, we can obtain the following.

Theorem 3.4. If in the class $\Omega_{v,b,n,g}$, $g \geq 2$, with fixed $v_1,\ldots,v_g$ and fixed $b_1,\ldots,b_g$ for $\sum_i v_i = v$ and $\sum_i b_i = b$, an equireplicate disconnected block design exists, then it is determinant-optimal.

Finally note that other discussion on optimality under mixed model (case(a)) can be found in Shah and Sinha [11].

4. BEST QUADRATIC UNBIASED ESTIMATORS OF VARIANCE COMPONENTS

Throughout this section we assume that the vector $\mathbf{y}$ in the model (2.1) is normally distributed. We here deal with the problems of estimation of variance components σ_β^2 and σ_e^2 in the case (a). In the class of designs, $\Omega_{v,b,n}$ and $\Omega_{v,b,n,g}$ defined in Section 3, we characterize block designs for which there exist BQUE for σ_β^2 and σ_e^2. They are used in the sense of a best unbiased estimator in the class of estimators of the following form $\mathbf{y}'A\mathbf{y} = tr(A\mathbf{y}\mathbf{y}')$, where A is a symmetric matrix. It is well known (Zmyślony [3]) that the condition U = 0 is equivalent to MD'D = D'DM = MD'DM, where

$$M = I - \Delta' R^{-1}\Delta \tag{4.1}$$

which in turn implies that BQUE's are invariant, in the sense that the following condition holds: $(\mathbf{y}+\Delta'\mathbf{b})'A(\mathbf{y}+\Delta'\mathbf{b}) = \mathbf{y}'A\mathbf{y}$ for all $\mathbf{b} \in R^v$ or equivalently MAM = A (see Rao [12]). Following a definition of quadratic estimability (which corresponds to A-estimability of Seely [5]), we introduce the following notations. Let W_1 = MD'D and W_2 = M, where M is as in (4.1). Further let a 2×2 matrix T = $\|tr(W_iW_j)\|$; i,j = 1,2. Then it can easily be shown that in the connected case, $W_1 = D'D - (1/b)\mathbf{1}_n\mathbf{1}_n'$ and

$$T = \begin{bmatrix} k^2(b-1) & k(b-1) \\ k(b-1) & n-v \end{bmatrix}. \tag{4.2}$$

Recall that k is the constant block size of the resulting block design as seen from Corollary 2.1.

Furthermore, in the disconnected case, Theorem 2.1 yields, after some calculation,

$$W_1 = \mathrm{diag}\{D_1'D_1 - \frac{1}{b_1}\mathbf{1}_{k_1^*b_1}\mathbf{1}_{k_1^*b_1}', \ldots, D_g'D_g - \frac{1}{b_g}\mathbf{1}_{k_g^*b_g}\mathbf{1}_{k_g^*b_g}'\}$$

where k_i^* is the constant block size of the ith component $N_i = \frac{1}{b_i}\mathbf{r}_i\mathbf{1}_{b_i}'$ in (2.2). Noting that

$$(D_i'D_i - \frac{1}{b_i}\mathbf{1}_{k_i^*b_i}\mathbf{1}_{k_i^*b_i}')^2 = k_i^*(D_i'D_i - \frac{1}{b_i}\mathbf{1}_{k_i^*b_i}\mathbf{1}_{k_i^*b_i}'),\ i = 1,\ldots,g,$$

it follows that

$$T = \begin{bmatrix} \sum_{j=1}^{g} k_j^{*2}(b_j-1) & n-k^* \\ n-k^* & n-v \end{bmatrix} \tag{4.3}$$

with $k^* = \sum_{j=1}^{g} k_j^*$. Note that $n-k^* = \sum_{j=1}^{g} k_j^*(b_j-1)$.

The following two propositions are given as the evidence of the existence of quadratic unbiased estimators.

Proposition 4.1. In a connected case σ_β^2 and σ_e^2 are quadratic estimable iff $b > 1$ and $n > v+b-1$.

Proof. Following Theorem 3 of Seely [2] for σ_β^2 and σ_e^2, there exists a BQUE iff $\det(T) > 0$. Thus by (4.2) we can get $\det(T) = k^2(b-1)(n-v-b+1)$ which completes the proof. □

Remark 4.1. For a connected design, it is known (cf. Kageyama [7]) that $n \geq v+b-1$. Furthermore, the usual block design has at least two blocks. In this sense, the necessary and sufficient condition in Proposition 4.1 has a very wide coverage and is reasonable.

Similarly, using (4.3) in a disconnected case we have the following.

Proposition 4.2. In a disconnected case σ_β^2 and σ_e^2 are quadratic estimable iff

$$(n-v)\sum_{j=1}^{g} k_j^{*2}(b_j-1) > (n-k^*)^2 \quad \text{with} \quad k^* = \sum_{j=1}^{g} k_j^*.$$

Under the model (2.1) we introduce the following notations: $y_o = \sum_{i=1}^{n} y_i$; y_{T_i} is the total yield of observations over the ith treatment for $i = 1,\ldots,v$; y_{B_j} is the total yield of observations over the jth block for $j = 1,\ldots,b$. We can obtain the following.

Theorem 4.1. For any model (2.1) corresponding to designs in $\Omega_{v,b,n}$ with $b > 1$ and $n-v-b+1 > 0$, there exist BQUE, for σ_e^2 and σ_β^2, given by

$$\hat{\sigma}_e^2 = \frac{1}{n-v-b+1}\left\{\sum_{i=1}^{n} y_i^2 - \sum_{i=1}^{v}\frac{y_{T_i}^2}{r_i} - \frac{1}{k}\left(\sum_{j=1}^{b} y_{B_j}^2 - \frac{1}{b}y_o^2\right)\right\} \tag{4.4}$$

and

$$\hat{\sigma}_\beta^2 = \frac{1}{k^2(b-1)}\left(\sum_{i=1}^{v} y_{T_i}^2 - \frac{1}{b}y_o^2\right) - \frac{1}{k}\hat{\sigma}_e^2. \tag{4.5}$$

Proof. Note that the existence of quadratic unbiased estimators follows from Proposition 4.1. To prove the first part (existence of BQUE), it is sufficient to prove that $\mathcal{B} = sp\{W_1, W_2\}$ is a quadratic subspace, in view of Theorem 3 of Seely [5].

Since $W_2W_1 = W_1$, $\mathscr{B}$ is a quadratic subspace iff $W_1^2 = cW_1$ for some real $c \neq 0$. In the present case, it is easy to check that $W_1^2 = kW_1$, where k is the common block size as in (4.2). This completes the proof of the first part. To derive (4.4) and (4.5), note that $\hat{\sigma}_\beta^2$ and $\hat{\sigma}_e^2$ are solutions of the equation $T\underline{\sigma} = \mathbf{z}$, where $\underline{\sigma} = (\sigma_\beta^2, \sigma_e^2)'$ and $\mathbf{z} = (z_1, z_2)'$ with $z_1 = \mathbf{y}'W_1\mathbf{y}$ and $z_2 = \mathbf{y}'W_2\mathbf{y}$. Since it follows that

$$z_1 = \sum_{j=1}^{b} y_{B_j}^2 - \frac{1}{b}y_o^2 \quad \text{and} \quad z_2 = \sum_{i=1}^{n} y_i^2 - \sum_{i=1}^{v} \frac{y_{T_i}^2}{r_i},$$

we get (4.4) and (4.5). □

Remark 4.2. Seely [5] has shown that in case a balanced incomplete block design is adopted as a block design in our mixed model, there exists a best invariant quadratic unbiased estimator (BIQUE). Gnot [1] has found a necessary and sufficient condition for the existence of BIQUE for σ_β^2 and σ_e^2. It is clear that our resulting designs characterized in Section 2 satisfy his condition. But our estimators are simply BQUE. Note that in both cases of Gnot [1] and Seely [5] a condition $U = 0$ for the existence of a BLUE of every estimable function does not hold in general.

In a disconnected case we shall need additional notations. Let $\mathbf{y} = (\mathbf{y}^{(1)}, \ldots, \mathbf{y}^{(p)}, \ldots, \mathbf{y}^{(g)})$ be a partition of $\mathbf{y}$ corresponding to the partition of D, i.e.

$$D' = \mathrm{diag}\{D_1, \ldots, D_g\}$$

where D_i are of size $b_ik_i^* \times b_i$, $i = 1, \ldots, g$. Further let $y_o^{(p)}$ denote the total yield over the observations $\mathbf{y}^{(p)}$, $p = 1, \ldots, g$. Thus in a disconnected case referring to (4.3) we have the following.

Theorem 4.2. For any model (2.1) corresponding to designs in $\Omega_{v,b,n,g}$, $g \geq 2$, with $(n-v)\sum_{j=1}^{g} k_j^{*2}(b_j - 1) > (n-k^*)^2$, there exist BQUE of σ_e^2 and σ_β^2 iff $k_1^* = \ldots = k_g^*$ (= k, say). The estimators are given by

$$\hat{\sigma}_e^2 = \frac{1}{n-v-b+g}\{\sum_{i=1}^{n} y_i^2 - \sum_{i=1}^{v}\frac{y_{T_i}^2}{r_i} - \frac{1}{k}(\sum_{j=1}^{b} y_{B_j}^2 - \frac{1}{k}\sum_{p=1}^{g}\frac{1}{b_p}y_o^{(p)2})\} \tag{4.6}$$

and

$$\hat{\sigma}_\beta^2 = \frac{1}{k^2(b-g)}\{\sum_{j=1}^{b} y_{B_j}^2 - \sum_{p=1}^{g}\frac{1}{b_p}y_o^{(p)2}\} - \frac{1}{k}\hat{\sigma}_e^2. \tag{4.7}$$

Proof. First note that in a disconnected case

$$W_1^2 = \mathrm{diag}\{k_1^*(D_1'D_1 - \frac{1}{b_1}\mathbf{1}_{k_1^*b_1}\mathbf{1}_{k_1^*b_1}'),\ldots,k_g^*(D_g'D_g - \frac{1}{b_g}\mathbf{1}_{k_g^*b_g}\mathbf{1}_{k_g^*b_g}')\}.$$

Then it follows that $W_1^2 = cW_1$ for some real $c \neq 0$ iff $k_1^* = \ldots = k_g^*$ (= k, say). In view of (4.3), we get

$$T = \begin{bmatrix} k^2(b-g) & k(b-g) \\ k(b-g) & n-v \end{bmatrix}.$$

Moreover, with $k_i^* = k$, $i = 1,\ldots,g$, we have

$$z_1 = y'W_1y = \sum_{j=1}^{b} y_{B_j}^2 - \sum_{p=1}^{g} \frac{1}{b_g} y_o^{(p)2}.$$

Thus the same argument as in the proof of Theorem 4.1 completes the proof. □

Remark 4.3. It may be worth-while to point out that the assumption of normality of $\mathbf{y}$ is used only to obtain the covariance of $\mathbf{yy}'$. In other words, the stated results, i.e. Theorems 4.1 and 4.2, hold provided moments up to 4-th order coincide with moments of $\mathbf{y}$ under normality.

Under the assumption of normality, using Theorem 4.1 of Seely [13], we can easily obtain the following.

Corollary 4.1. For any model (2.1) corresponding to designs in $\Omega_{v,b,n}$, estimators $\hat{\gamma} = R^{-1}\Delta y$, $\hat{\sigma}_e^2$ and $\hat{\sigma}_\beta^2$ given by (4.4) and (4.5), respectively, are best unbiased estimators.

In a disconnected case we have

Corollary 4.2. For any model (2.1) corresponding to designs in $\Omega_{v,b,n,g}$ satisfying $k_1^* = \ldots = k_g^*$, estimators $\hat{\gamma} = R^{-1}\Delta y$, $\hat{\sigma}_e^2$ and $\hat{\sigma}_\beta^2$ given by (4.6) and (4.7), respectively, are best unbiased estimators.

CONCLUDING REMARK. Not all orthogonal designs, but only proper designs have the following property that there exist BLUE of treatment effects and BQUE of variance components. In the class of such block designs equireplicate designs are trace-optimal as well as determinant-optimal, and the BLUE of treatment effects have equal variances and equal covariances.

Acknowledgements

The authors wish to thank Professor R. Kala for his comments and helpful suggestions. Thanks are also due to a referee for his constructive comments.

REFERENCES

1. S. Gnot, Quadratic estimation of variance components in mixed block designs. *J. Statist. Plann. Inference* **16**, 231-236 (1987).
2. J. Seely, Vector spaces and unbiased estimation - application to the mixed linear model. *Ann. Math. Statist.* **41**, 1735-1748 (1970).
3. R. Zmyślony, On estimation of parameters in linear models. *Applicationes Mathematicae* **XV**, **3**, 271-276 (1976).
4. R. Zmyślony, A characterization of best linear unbiased estimators in the general linear model. *Lecture Notes in Statistics* Vol. **2**, pp.365-373. Springer-Verlag, New York (1980).
5. J. Seely, Quadratic subspaces and completeness. *Ann. Math. Statist.* **42**, 710-721 (1971).
6. R. Zmyślony and S. Kageyama, Estimation of parameters in block designs under mixed models. Institute of Mathematics, Polish Academy of Sciences, *Preprint* **379** (1987).
7. S. Kageyama, Connected designs with the minimum number of experimental units. *Lecture Notes in Statistics* Vol. **35**, pp. 99-117. Springer-Verlag, New York (1985).
8. J. K. Baksalary and R. Kala, On equalities between BLUEs, WLSEs, and SLSEs. *Canad. J. Statist.* **11**, 119-123 (1983).
9. R. Kala, Projectors and linear estimation in general linear models. *Commun. Statist. -Theor. Meth.* **A 10**, 849-873 (1981).
10. G. Zyskind, On canonical forms, non-negative covariance matrices and best and simple least squares linear estimators in linear models. *Ann. Math. Statist.* **38**, 1092-1109 (1967).
11. K. R. Shah and B. K. Sinha, *Theory of Optimal Designs*. Springer-Verlag, New York (1989).
12. C. R. Rao, Estimation of variance and covariance components - MINQE theory. *J. Multivariate Analysis* **1**, 257-275 (1971).
13. J. Seely, Minimal sufficient statistics and completeness for multivariate normal families. *Sankhyā* **A 39**, 170-185 (1977).